T0191564

Advances in Intelligent Systems and Computing

Volume 454

Series editor

Janusz Kacprzyk, Polish Academy of Sciences, Warsaw, Poland
e-mail: kacprzyk@ibspan.waw.pl

About this Series

The series "Advances in Intelligent Systems and Computing" contains publications on theory, applications, and design methods of Intelligent Systems and Intelligent Computing. Virtually all disciplines such as engineering, natural sciences, computer and information science, ICT, economics, business, e-commerce, environment, healthcare, life science are covered. The list of topics spans all the areas of modern intelligent systems and computing.

The publications within "Advances in Intelligent Systems and Computing" are primarily textbooks and proceedings of important conferences, symposia and congresses. They cover significant recent developments in the field, both of a foundational and applicable character. An important characteristic feature of the series is the short publication time and world-wide distribution. This permits a rapid and broad dissemination of research results.

Advisory Board

Chairman

Nikhil R. Pal, Indian Statistical Institute, Kolkata, India
e-mail: nikhil@isical.ac.in

Members

Rafael Bello, Universidad Central "Marta Abreu" de Las Villas, Santa Clara, Cuba
e-mail: rbellop@uclv.edu.cu

Emilio S. Corchado, University of Salamanca, Salamanca, Spain
e-mail: escorchado@usal.es

Hani Hagras, University of Essex, Colchester, UK
e-mail: hani@essex.ac.uk

László T. Kóczy, Széchenyi István University, Győr, Hungary
e-mail: koczy@sze.hu

Vladik Kreinovich, University of Texas at El Paso, El Paso, USA
e-mail: vladik@utep.edu

Chin-Teng Lin, National Chiao Tung University, Hsinchu, Taiwan
e-mail: ctlin@mail.nctu.edu.tw

Jie Lu, University of Technology, Sydney, Australia
e-mail: Jie.Lu@uts.edu.au

Patricia Melin, Tijuana Institute of Technology, Tijuana, Mexico
e-mail: epmelin@hafsamx.org

Nadia Nedjah, State University of Rio de Janeiro, Rio de Janeiro, Brazil
e-mail: nadia@eng.uerj.br

Ngoc Thanh Nguyen, Wroclaw University of Technology, Wroclaw, Poland
e-mail: Ngoc-Thanh.Nguyen@pwr.edu.pl

Jun Wang, The Chinese University of Hong Kong, Shatin, Hong Kong
e-mail: jwang@mae.cuhk.edu.hk

More information about this series at http://www.springer.com/series/11156

Valentina Emilia Balas · Lakhmi C. Jain
Xiangmo Zhao
Editors

Information Technology and Intelligent Transportation Systems

Volume 1, Proceedings of the 2015 International Conference on Information Technology and Intelligent Transportation Systems ITITS 2015, held December 12–13, 2015, Xi'an China

Springer

Editors
Valentina Emilia Balas
Department of Automation and Applied Informatics, Faculty of Engineering
Aurel Vlaicu University of Arad
Arad
Romania

Lakhmi C. Jain
Bournemouth University
Poole
UK

Xiangmo Zhao
School of Information Engineering
Chang'an University
Xi'an
China

ISSN 2194-5357 ISSN 2194-5365 (electronic)
Advances in Intelligent Systems and Computing
ISBN 978-3-319-38787-1 ISBN 978-3-319-38789-5 (eBook)
DOI 10.1007/978-3-319-38789-5

Library of Congress Control Number: 2016945142

© Springer International Publishing Switzerland 2017
This work is subject to copyright. All rights are reserved by the Publisher, whether the whole or part of the material is concerned, specifically the rights of translation, reprinting, reuse of illustrations, recitation, broadcasting, reproduction on microfilms or in any other physical way, and transmission or information storage and retrieval, electronic adaptation, computer software, or by similar or dissimilar methodology now known or hereafter developed.
The use of general descriptive names, registered names, trademarks, service marks, etc. in this publication does not imply, even in the absence of a specific statement, that such names are exempt from the relevant protective laws and regulations and therefore free for general use.
The publisher, the authors and the editors are safe to assume that the advice and information in this book are believed to be true and accurate at the date of publication. Neither the publisher nor the authors or the editors give a warranty, express or implied, with respect to the material contained herein or for any errors or omissions that may have been made.

Printed on acid-free paper

This Springer imprint is published by Springer Nature
The registered company is Springer International Publishing AG Switzerland

Preface

These volumes constitute the Proceedings of the 2015 International Conference on Information Technology and Intelligent Transportation Systems (ITITS 2015) held in Xi'an, China during December 12–13, 2015. The Conference ITITS 2015 was sponsored by Shaanxi Computer Society and co-sponsored by Chang'an University, Xi'an University of Technology, Northwestern Poly-technical University, CAS, Shaanxi Sirui Industries Co., Ltd.

The book covers a broad spectrum of intelligent techniques, theoretical and practical applications employing knowledge and intelligence to find solutions for intelligent transportation systems and other applications.

The conference papers included in these proceedings, published post-conference, were grouped into the following parts:

Volume I—Part II: Theory Research in Intelligent Transportation Systems
Volume I—Part III: Application and Technologies in Intelligent Transportation
Volume II—Part I: Management Issues on Intelligent Transportation
Volume II—Part II: Information Technology, Electronic and Control System

At ITITS 2015 we had 12 eminent keynote speakers: Profs. Asad J. Khattak (USA), Robert L. Bertini (USA), Heng Wei (USA), Ping Yi (USA), Haizhong Wang (USA), Jonathan Corey (USA), Zhixia (Richard) Li (USA), Guohui Zhang (USA), Luke Liu (USA), Yu Zhang (USA), Valentina E. Balas (Romania), and Lakhmi C. Jain (UK). Their summary talks are included in this book.

Intelligent transport systems vary in technologies applied, from basic management systems to more application systems, and information technology also plays tightly with intelligent transportation systems including wireless communication, computational technologies, floating car data/floating cellular data, sensing technologies, and video vehicle detection; and technologies of intelligent transportation systems also include topics from theoretical and application topics, such as emergency vehicle notification systems, automatic road enforcement, collision avoidance systems, and some cooperative systems. The conference also fostered cooperation among organizations and researchers involved in the merging fields by inviting worldwide well-known professors to further explore these topics and discuss in depth the technical presentations with the presenters, including 12 invited speakers and over 200 participants. The conference received overwhelming response with 330 submissions from five countries and regions, and each paper was doubly peer reviewed by at least three reviewers, and finally more than 120 papers were accepted.

We would like to thank the authors of the submitted papers for keeping the quality of the ITITS 2015 Conference at high levels. The editors of this book would like to acknowledge all the authors for their contributions and also the reviewers. We have received invaluable help from the members of the International Program Committee and the Chairs responsible for different aspects of the Workshop.

Special thanks go to Janusz Kacprzyk (Editor in Chief, Springer, Advances in Intelligent Systems and Computing Series) for the opportunity to organize these guest-edited volumes.

We are grateful to Springer, especially to Dr. Thomas Ditzinger (Senior Editor, Applied Sciences & Engineering Springer-Verlag) for the excellent collaboration, patience, and help during the evolvement of this volume.

We hope that the volumes will provide useful information to professors, researchers, and graduate students in the area of intelligent transportation.

Arad, Romania — Valentina Emilia Balas
Poole, UK — Lakhmi C. Jain
Xi'an, China — Xiangmo Zhao

Contents

Part I
Invited Keynote Papers

Integrating Big Data in Metropolitan Regions to Understand Driving Volatility and Implications for Intelligent Transportation Systems

Asad J. Khattak

Abstract Higher driving volatility, e.g., hard accelerations or hard braking, can imply unsafe outcomes, more energy use, and higher emissions. This presentation will demonstrate how large-scale data, increasingly available from sensors, can be transformed into useful knowledge. This is done by creating a framework for combining data from multiple sources and comparing counties/regions in terms of driving volatility of resident drivers. The unique database was created from four sources that include large-scale travel surveys, historical traffic counts from California and Georgia Department of Transportation, socio-demographic information from Census, and geographic information from Google Earth. The database provides a rich resource to test hypothesis and model driving decisions at the micro-level, i.e., second-by-second. The database has 117,022 trips made by 4,560 drivers residing in 78 counties of 4 major US metropolitan areas across two states. They represent significant variations in land use types and populations; all trips were recorded by in-vehicle GPS devices giving 90,759,197 second-by-second speed records. The data integration helps explore links between driving behaviors and various factors structured in hierarchies, i.e., the data are structured at the levels of trips, drivers, counties, and regions. Appropriate hierarchical models are estimated to study correlates of driving performance and to compare traffic performance across regions. The implications of our analysis for intelligent transportation systems will be discussed.

A.J. Khattak (✉)
The University of Tennessee, Knoxville, TN 37996-2313, USA
e-mail: akhattak@utk.edu

© Springer International Publishing Switzerland 2017
V.E. Balas et al. (eds.), *Information Technology and Intelligent Transportation Systems*, Advances in Intelligent Systems and Computing 454,
DOI 10.1007/978-3-319-38789-5_1

Asad J. Khattak

1 Short Biography

Dr. Asad J. Khattak is Beaman Professor of Civil and Environmental Engineering, University of Tennessee, Knoxville and Transportation Program Coordinator in the Department. He is affiliated with the UT Center for Transportation Research, where he works on research and educational projects related to the Southeastern Transportation Center and NURail University Transportation Center. He has recently established the Initiative for Sustainable Mobility, a campus-wide organized research unit. Dr. Khattak's research focuses on various types of innovations related to (1) intelligent transportation systems, (2) transportation safety, and (3) sustainable transportation. Dr. Khattak received his Masters and Ph.D. degrees in Civil Engineering from Northwestern University in 1988 and 1991, respectively. Dr. Khattak is: (1) Editor of Science Citation Indexed Journal of Intelligent Transportation Systems, with a 5-year impact factor of 1.841 in 2013. (2) Associate Editor of SCI-indexed International Journal of Sustainable Transportation.

Connected Autonomous Vehicle Control Optimization at Intersections

Guohui Zhang

Abstract Connected and Autonomous Vehicle (CAV)-enabled traffic system has demonstrated great potential to mitigate congestion, reduce travel delay, and enhance safety performance. According to the U.S. Based on seamless Vehicle-To-Vehicle (V2V) and Vehicle-To-Infrastructure communication as well as autonomous driving technologies, traffic management and control will be revolutionized. The existing studies indicate that traffic lights will be eliminated and 75 % of vehicles will be autonomous vehicles by 2040. National Highway Traffic Safety Administration (NHTSA) plans to mandate inter-vehicle communication technologies on every single vehicle by 2016. However, one should note that the current research regarding CAV system management and control is still in its early stage. The presented study concentrates on the VISSIM-based simulation platform development to enable an innovative autonomous intersection control mechanism and optimize CAV operations at intersections without signal lights. Simulation-based investigation on traffic system operations provides a cost-effective, risk-free means of exploring optimal management strategies, identifying potential problems, and evaluating various alternatives. In the study, a VISSIM-based simulation platform is developed for simulating individual-CAV-conflict-based traffic control optimization at intersections. A novel external module will be developed via VISSIM Component Object Model (COM) interfaces. A new CAV-based control algorithm entitled a Discrete Forward-Rolling Optimal Control (DFROC) model, is developed and implemented through the VISSIM COM server. This external module can provide sufficient flexibility to satisfy any specific demands from particular researchers and practitioners for CAV control operations. Research efforts will be made to calibrate driving behavior parameters in the simulation model using drivers' characteristic data to further strengthen the simulation creditability. Furthermore, a method for statistically analyzing simulation outputs and examining simulation reliability is developed. The methodology developed is applicable for quantitatively evaluating the impacts of various CAV control strategies on urban arterials.

G. Zhang (✉)
University of New Mexico, Albuquerque, NM 87131, USA
e-mail: guohui@unm.edu

© Springer International Publishing Switzerland 2017

V.E. Balas et al. (eds.), *Information Technology and Intelligent Transportation Systems*, Advances in Intelligent Systems and Computing 454,
DOI 10.1007/978-3-319-38789-5_2

Guohui Zhang

1 Short Biography

Dr. Guohui Zhang is an Assistant Professor in the Department of Civil Engineering at the University of New Mexico (UNM). Dr. Zhang received his Ph.D. from the University of Washington in 2008. Dr. Zhang's research focuses on large-scale transportation systems modeling, customized traffic simulation, travel delay estimation, traffic safety and accident modeling, congestion pricing, traffic detection and sensor data analysis, and sustainable transportation infrastructure design and maintenance. Dr. Zhang has published nearly 50 peer-reviewed journal articles, conference papers, and technical reports and presented his research contributions numerous times at prestigious international and national conferences.

Modeling Potential Consequences of Connected and Automated Vehicle to Future Travel Behaviors and Patterns Changes: A Fuzzy Cognitive Map Approach

Haizhong Wang

Abstract The authors examined changes that are likely to affect transportation behaviors in the future, developed a "fuzzy cognitive map" (FCM) of the relationships, and used the FCM model to investigate the effects of those relationships. This new FCM method enables modeling the potential consequences of new technologies and services using a variant of the fuzzy cognitive map (FCM) approach, which enables problems involving imprecise and uncertain information to be modeled. Significant modifications to the standard FCM approach were made to address deficiencies found in applying the standard approach. The new approach retains some basic FCM characteristics, but it deviates substantially in a number of ways as well. It has been found that this produces well-behaved models that can be explained in common-sense terms, be easily configured, run many scenarios quickly, and used to analyze scenarios of disruptive change. The results of the study show that FCM models offer a promising method for transportation planners to enhance their ability to reason about system effects when quantitative information is limited and uncertain. More specifically, the results provide some initial guidance on the potential impacts of disruptive changes on future travel, which may help in targeting limited research funds on the most consequential potential changes.

Haizhong Wang

H. Wang (✉)
Oregon State University, Corvallis, OR 97330, USA
e-mail: Haizhong.Wang@oregonstate.edu

© Springer International Publishing Switzerland 2017

V.E. Balas et al. (eds.), *Information Technology and Intelligent Transportation Systems*, Advances in Intelligent Systems and Computing 454,
DOI 10.1007/978-3-319-38789-5_3

1 Short Biography

Dr. Haizhong Wang is an Assistant Professor of Transportation Engineering within the School of Civil and Construction Engineering at Oregon State University, Corvallis, OR. Dr. Wang received M.S. and Ph.D. degrees from University of Massachusetts, Amherst in Applied Mathematics and Civil Engineering (Transportation), and B.S. and M.S. degrees from Hebei University of Technology and Beijing University of Technology, China. Dr. Wang's research areas include (1) stochastic traffic flow models, traffic system planning and analysis in particular the impacts of emerging technologies such as connected and automated vehicles on traffic operations and future travel behavior; (2) an agent-based modeling and simulation (ABMS) to model behavioral heterogeneity (i.e., when, how, where to evacuate) for life safety and post-disaster mobility in multi-hazard emergency evacuation and disaster response; (3) a network of network (NON) approach to model interdependency for resilient lifeline infrastructure systems; (4) Complex adaptive system (CAS) for large-scale system modeling and simulation; (5) Mileage-based road user charge for alternative financing; and (6) Dada driven smart city and big data applications for urban mobility. Dr. Wang has published over 40 journal and major conference papers. He is a member for two TRB standing committee: ABJ70 Artificial Intelligence and Advanced Computing Applications and ABR 30: Emergency Evacuation and AHB45 (3) Subcommittee on Connected and Automated Vehicles through Traffic Flow Theory and Characteristics. He is the most recent receipt of the Outstanding Reviewer for ASCE Journal of Transportation Engineering for 2014.

Synthetic Approach for Scenario-Based Performance Estimation of Connected Vehicles Operating at Highway Facilities

Heng Wei

Abstract Connected Vehicle (CV) systems are envisioned to enhance a wide range of safety, mobility and environmental aspects to highway traffic. A critical research need lies in clarifying the cause-and-effect mechanism between the CV information and driver behaviors and subsequent adaptive resiliency of improvement in operation, safety, and emissions reductions. A novel approach is hence created via developing the simulation-based tool, Synthetic Adaptive V2X Effect (SAVE) Estimator, to explore the interactions between the CV system and transportation performance. This presentation will introduce preliminary results from the speaker's on-going research on identifying factors possibly affecting travel behavior and rationale of their aggregated impact on mobility, safety, and vehicle emission in support of simulated outcomes of synthesized scenarios at a freeway and ramp conjunction facility in the Cincinnati are, Ohio (USA). The framework and associated modeling methodology for the development of the SAVE Estimator will be also introduced alongside the discussion of the case study.

H. Wei (✉)
The University of Cincinnati, Cincinnati, OH 45221-0071, USA
e-mail: heng.wei@uc.edu

© Springer International Publishing Switzerland 2017

V.E. Balas et al. (eds.), *Information Technology and Intelligent Transportation Systems*, Advances in Intelligent Systems and Computing 454,
DOI 10.1007/978-3-319-38789-5_4

Heng Wei

1 Short Biography

Dr. Heng Wei is a Professor of Transportation Systems and Engineering, and Director of Advanced Research in Transportation Engineering & Systems (ART-EngineS) Laboratory at The University of Cincinnati (UC). He has a wide spectrum of research interests and expertise in intelligent transportation systems (ITS). Since his faculty appointment at UC, he has secured a great number of research projects from ODOT, FHWA, NSF, EPA, OTC, NEXTRANS Center, and UC URC/FDC. His research has resulted in 167 peer-reviewed papers and 9 professional books/chapters. He has been honored with UC College of Engineering and Applied Science Distinguished Researcher Award for Excellence in Research and Engineering Master Educator Award for Excellence in Teaching, as well as Honored Faculty/Staff Who Made a Real Difference in the Life of a UC Student. In addition, he is a member of numerous outstanding professional committees, such as TRB AND20, ABJ70 and ABE90 Committees, ASCE T & DI Committees on Advanced Technology Committee and Transportation Safety and on Sustainability and Environment. He is the Chair of IEEE ITSS Travel Information and Traffic Management Committee and Past President of Chinese Overseas Transportation Association (COTA). Dr. Wei has successfully organized and/or chaired 31 international conferences/sessions and symposiums.

Accessing and Integrating CV and AV Sensor Data into Traffic Engineering Practice

Jonathan Corey

Abstract Autonomous vehicles (AV) and connected vehicles (CV) are being designed with numerous sensors, including cameras, radar and Lidar, to enable features like adaptive cruise control, blind spot monitoring, collision avoidance and navigation. As AVs and CVs enter the vehicle fleet, practitioners are going to have the opportunity to monitor operations for freeways, intersections and urban environments to a degree that has not been possible or practical previously. But, the very availability of data will threaten practitioners with information overload. To properly use this newly abundant data, new algorithms and systems designs will be needed to automate data collection and processing into formats that practitioners can use.

Jonathan Corey

J. Corey (✉)
The University of Cincinnati, Cincinnati, OH 45221-0071, USA
e-mail: Jonathan.corey@uc.edu

© Springer International Publishing Switzerland 2017

V.E. Balas et al. (eds.), *Information Technology and Intelligent Transportation Systems*, Advances in Intelligent Systems and Computing 454,
DOI 10.1007/978-3-319-38789-5_5

1 Short Biography

Dr. Jonathan Corey is a civil engineering professor at the University of Cincinnati's Department of Civil and Architectural Engineering and Construction Management. He specializes in transportation engineering with focuses on sensors, data collection and data management. His current work focuses on vehicle sensors and how data collected from those sensors can be used by practitioners to improve traffic safety and operations.

Intelligent Techniques for Improving the Aviation Operations

Lakhmi C. Jain

Abstract Air travel in modern passenger aircraft has become extremely safe. This is largely due to the engine reliability, on-board computing system reliability and excellent flight crew training. Flight crews are highly trained to operate in the technical and human environments of the cockpit. Despite all these measures, accidents do happen. This talk presents the development of intelligent flight data monitoring system for improving the safety of aviation operations. The progress made in the development of an in-flight agent to monitor pilot situation awareness is also presented.

Lakhmi C. Jain

L.C. Jain (✉)
University of Canberra, Canberra, Australia
e-mail: jainlc2002@yahoo.co.uk

L.C. Jain
Bournemouth University, Poole, UK

© Springer International Publishing Switzerland 2017

V.E. Balas et al. (eds.), *Information Technology and Intelligent Transportation Systems*, Advances in Intelligent Systems and Computing 454,
DOI 10.1007/978-3-319-38789-5_6

1 Short Biography

Lakhmi C. Jain, PhD, ME, BE(Hons), Fellow (Engineers Australia) serves as a Visiting Professor in Bournemouth University, United Kingdom, and University of Canberra, Australia. Dr. Jain founded the KES International for providing a professional community the opportunities for publications, knowledge exchange, cooperation and teaming. Involving around 10,000 researchers drawn from universities and companies world-wide, KES facilitates international cooperation and generate synergy in teaching and research. KES regularly provides networking opportunities for professional community through one of the largest conferences of its kind in the area of KES. His interests focus on the artificial intelligence paradigms and their applications in complex systems, security, e-education, e-healthcare, unmanned air vehicles and intelligent agents.
http://www.kesinternational.org/organisation.php

Support for Connected Vehicle Testing in Urban Environment

Luke Liu

Abstract Connected Vehicle Safety Pilot Model Deployment presented a unique opportunity to demonstrate DSRC-based vehicle safety applications in real-world driving scenarios. A diverse team of industry, public agencies and academic institutions is involved in the planning and delivery of the project. The presentation illustrates the collaborative effort from the Safety Pilot and forthcoming deployment projects. The discussion also contributes to the exploration of future research opportunities to leverage the skills and experience from past projects and the infrastructure support in Ann Arbor.

Luke Liu

1 Short Biography

Luke brings professional experience and expertise in planning and implementation of ITS applications and coordination with active traffic operations. He graduated from Michigan State University in 2007 with a doctoral degree in civil engineering. Luke

L. Liu (✉)
City of Ann Arbor, Michigan, MI 48107, USA
e-mail: yliu@a2gov.org

© Springer International Publishing Switzerland 2017

V.E. Balas et al. (eds.), *Information Technology and Intelligent Transportation Systems*, Advances in Intelligent Systems and Computing 454,
DOI 10.1007/978-3-319-38789-5_7

has since held positions in the consulting industry and public sector in Michigan for the past nine years. His skills and experience range from signal control systems, ITS applications, traffic modeling and analysis, transportation management center operations and data management. His current responsibilities include the operations of the City's traffic signal network and the expansion of the SCOOT adaptive signal system. Luke is passionate about sharing knowledge and skills and has delivered senior and graduate level course at Western Michigan University and guest lecturer at the University of Michigan.

Impact Study of Vehicle Platooning and Gap Management on Traffic Operation Through Automated Vehicles

Ping Yi

Abstract Since the advent of automated vehicle technologies, the current trend of practice in this fast-evolving field has started to move from basic research and development in a lab environment to field trials and pilot testing. While a number of studies on V2V communications and vehicle control systems have been reported for the purpose of enhancing traffic safety, this research focuses on the efficiency benefit of the technologies in traffic operations when implemented even in a small number of vehicles in the traffic stream. Specifically, this presentation discusses the effects of automated vehicles in a traffic flow mixed with regular (human operated) vehicles on platoon formation and gap acceptance to increase roadway capacity and reduce delay. The theoretical basis for such improvements is reviewed first, followed by case studies involving intersection dilemma reduction, side street gap selection, and bottleneck management at a work zone. The resultant benefits are quantified under different rates of market penetration of the automated vehicles, which are distributed randomly in the traffic stream. Preliminary findings are summarized, including the pros and cons of the implementation.

Ping Yi

P. Yi (✉)
The University of Akron, Akron, OH 44325-3905, USA
e-mail: pyi@uakron.edu

© Springer International Publishing Switzerland 2017

V.E. Balas et al. (eds.), *Information Technology and Intelligent Transportation Systems*, Advances in Intelligent Systems and Computing 454,
DOI 10.1007/978-3-319-38789-5_8

1 Short Biography

Dr. Ping Yi is a professor in the Department of Civil Engineering of The University of Akron. His education experience in the US includes a Ph.D. from University of Minnesota and a M.S. from Washington State University. His main areas of research include traffic control and safety, sensor technology and data mining/fusion, and information systems and technology. Dr. Yi was a research scientist and principal in the Minnesota DOT's IVHS/ITS Office, where he managed several federally funded ITS operational test projects over sensors testing, adaptive signals, parking information systems, and incident and special event management. After joining the academia, Dr. Yi has published widely in refereed journals and completed many federally and state funded projects. He has served many professional societies and committees such as ASCE, TRB, AASHTO, NRC-IDEA, etc.

Toward Assessing State Department of Transportation Readiness for Connected Vehicle/Cooperative Systems Deployment Scenarios: An Oregon Case Study

Robert L. Bertini

Abstract As connected vehicle research moves into deployment, state, local and transit agencies, metropolitan planning organizations (MPOs) and the private sector will start experiencing the effects of vehicles, after-market devices, mobile devices, and infrastructure with dedicated, short-range wireless communications (DSRC) and other wireless connectivity at their cores. Along with other states and regions, the Oregon Department of Transportation (ODOT) can benefit from preliminary scoping, evaluation, and assessment of the impact of connected vehicles and infrastructure and a wide range of potential cooperative system applications. With this in mind, ODOT is aiming to determine whether or not to pursue the next phases of federal connected vehicle application funding. To assist ODOT in this assessment, a survey was distributed within the agency to gauge perception of connected and automated vehicle technology. However, many had concerns with cyber security and system failure having catastrophic consequences. Likewise, many voiced concerns about ODOT's preparedness for connected or automated vehicles. ODOT can use these findings to help prepare for a better future of connected and automated vehicles.

Robert L. Bertini

R.L. Bertini (✉)
California Polytechnic State University, San Luis Obispo, CA 93407-0353, USA
e-mail: rbertini@calpoly.edu

© Springer International Publishing Switzerland 2017

V.E. Balas et al. (eds.), *Information Technology and Intelligent Transportation Systems*, Advances in Intelligent Systems and Computing 454,
DOI 10.1007/978-3-319-38789-5_9

1 Short Biography

Robert L. Bertini, Ph.D., P.E. is a Professor of Civil and Environmental Engineering at the California Polytechnic State University, San Luis Obispo. Dr. Bertini's primary research interests are in sustainable transportation solutions, traffic flow theory informed by empirical and experimental measurements, intelligent transportation systems (ITS), multimodal transportation "big data" for improving performance measurement, planning and operations, and proactive traffic management and operations. Dr. Bertini recently completed a 6-year term as chair of the Transportation Research Board (TRB) Committee on Traffic Flow Theory and Characteristics (AHB45) and is currently the chair of the TRB Operations Section (AHB00), overseeing 13 committees. Dr. Bertini received the National Science Foundation CAREER Award in 2002, where he developed an online multimodal transportation data repository that is a platform for performance measurement, modeling and prediction. The recipient of many awards, he received the DeFazio Transportation Hall of Fame Award and was invited to deliver the Ogden Lecture at Monash University in Australia in 2014. He was the director of the Portland State University Intelligent Transportation Systems Laboratory and of the Oregon Transportation Research and Education Consortium (OTREC), which is a statewide, federally funded university transportation center. He also served in the Obama Administration as Deputy Administrator of the Research and Innovative Technology Administration (RITA) at the U.S. Department of Transportation where he also led the Intelligent Transportation Systems Joint Program Office and chaired the Department's Innovation Council. He received his B.S. in Civil Engineering from California Polytechnic State University San Luis Obispo, an M.S. in Civil Engineering from San Jose State University, and a Ph.D. in Civil Engineering from the University of California at Berkeley. Dr. Bertini is a licensed professional engineer in the states of California and Oregon.

Automatic Control of the Traffic Flow

Valentina E. Balas

Abstract Automate driving is enhancing the driving performance and reducing the crash risks. The presentation illustrates a new method for the management of the traffic flow on highways, based on the constant time to collision criterion. This criterion is addressing the car fallowing issue and it offers a speed adapted planner for the distance gap between cars. This method is able also to support a highway traffic flow management. The interface's decision block is implemented by a fuzzy interpolative controller that is estimating the collision risk, taking into account the traffic intensity.

Valentina E Balas

1 Short Biography

Valentina E. Balas is currently Full Professor in the Department of Automatics and Applied Software at the Faculty of Engineering, University "Aurel Vlaicu" Arad (Romania). She holds a Ph.D. in Applied Electronics and Telecommunications from

V.E. Balas (✉)
Aurel Vlaicu University of Arad, Arad, Romania
e-mail: balas@drbalas.ro

© Springer International Publishing Switzerland 2017

V.E. Balas et al. (eds.), *Information Technology and Intelligent Transportation Systems*, Advances in Intelligent Systems and Computing 454,
DOI 10.1007/978-3-319-38789-5_10

Polytechnic University of Timisoara. She is author of more than 180 research papers in refereed journals and International Conferences. Her research interests are in Intelligent Systems, Fuzzy Control, Soft Computing, Smart Sensors, Information Fusion, Modeling and Simulation. She is the Editor-in Chief to International Journal of Advanced Intelligence Paradigms (IJAIP) and to International Journal of Computational Systems Engineering (IJCSysE), member in Editorial Board member of several national and international journals and is evaluator expert for national and international projects. Dr. Balas participated in many international conferences as General Chair, Organizer, Session Chair and member in International Program Committee. She was a mentor for many student teams in Microsoft (Imagine Cup), Google and IEEE competitions in the last years. She is a member of EUSFLAT, ACM and a Senior Member IEEE, member in TC—Fuzzy Systems (IEEE CIS), member in TC—Emergent Technologies (IEEE CIS), member in TC—Soft Computing (IEEE SMCS).

Dr. Balas is Vice-president (Awards) of IFSA International Fuzzy Systems Association Council and Join Secretary of Joint Secretary of the Governing Council of Forum for Interdisciplinary Mathematics (FIM)—A Multidisciplinary Academic Body, India.

Impacts of Autonomous Vehicle to Airport Landside Terminal Planning and Design

Yu Zhang

Abstract Income from parking and rental car facilities, for most of commercial airports in the U.S., are significant components in their revenue. Airports design the parking capacity and calculate the parking fees according to passenger throughput and mode split forecast, as well as the leasing rate of the rental car facilities. Nevertheless, with the emerging AV transportation modes, the fundamentals could change. In the future, if financially more economical, passengers may send their AVs back to their house instead of parking at the airports. Rental car industry could follow completely new business model and may not need to lease space on airport property. In addition, the flexibility of driverless may encourage more car sharing and real-time ridesharing. Given the information of mode split and parking information of one hub airport in the U.S., this study applies statistical and simulation methods to estimate potential parking needs and provide insights for future airport landside terminal planning and design.

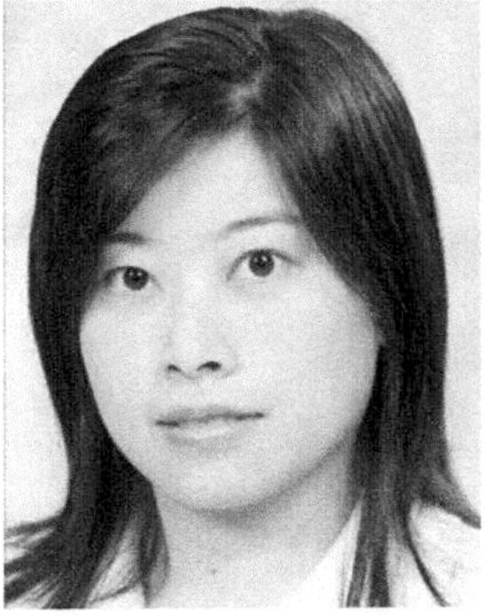

Yu Zhang

Y. Zhang (✉)
University of South Florida, Tampa, FL 33620-5350, USA
e-mail: yuzhang@usf.edu

© Springer International Publishing Switzerland 2017

V.E. Balas et al. (eds.), *Information Technology and Intelligent Transportation Systems*, Advances in Intelligent Systems and Computing 454,
DOI 10.1007/978-3-319-38789-5_11

1 Short Biography

Dr. Zhang's main research areas are: Transportation system modeling, analysis, and simulation; Resilient system design and operations; Air transportation and global airline industry; Multimodal transportation planning and sustainable transportation. Dr. Zhang applies mathematical programming and optimization techniques, simulation, econometric and statistical tools to solve the problems for resilient, efficient, and sustainable transportation systems. Her research projects are funded by government agencies, such as NSF, FAA, FHWA, FDOT and also local industry companies. Dr. Zhang is the recipient of the 2010 Fred Burggraf Award, for excellence in transportation research by researchers 35 years of age or younger, presented by TRB of the National Academies of Science. She has published papers in top transportation journals such as Transportation Research Part B, Part C, Part D, and Part E. Dr. Zhang is serving on the editorial board for Transportation Research Part C and is a reviewer for Transportation Science, Transportation Research Part A, Part B, Part C, Part D, Part E, Journal of Air Transport Management, Journal of Intelligent Transportation Systems, European Journal of Operation Research etc.

Dr. Zhang is actively involved in professional organizations. She is the committee member, research and paper review coordinators for Transportation Research Board (TRB) Airfield and Airspace Capacity and Delay (AV060) committee, and also the committee member of TRB Aviation System Planning (AV020) committee. Dr. Zhang is also serving as the Elected President for Chinese Overseas Transportation Association (COTA) (term 2016–2017). Dr. Zhang holds Ph.D. and M.S. from the University of California Berkeley in Civil and Environmental Engineering and Bachelors from Southeast University of China in Transportation Engineering.

Next-Generation Intersection Control Powered by Autonomous and Connected Vehicle Technologies

Zhixia Li

Abstract Urban intersections are one of the key bottlenecks that cause recurring congestions. Traditional signalized control is effective but capacity-restrained. Availability of autonomous and connected vehicle technologies provides the possibility to improve intersection capacity. Powered by autonomous and connected vehicle technologies, a next-generation intersection control strategy ACUTA was developed by employing a reservation-based centralized control strategy. ACUTA converts the conflicts between traffic movements into conflicts between individual vehicles, hence enhancing intersection capacity. Comparison between ACUTA and optimized signal control revealed that ACUTA increased the intersection capacity by 33 %, resulting substantially lower delays. Particularly, comparison of the v/c ratios indicated that ACUTA could process 163 more vehicles per hour per lane without being oversaturated when compared to optimized signal control. Sustainability-wise, as ACUTA minimizes vehicle stops at intersections, it reduces emission and energy consumption as well. Sustainability effects compared with signalized intersection control include: (1) ACUTA reduces CO and PM 2.5 emissions by about 5 % under low to moderate volume conditions and by about 3 % under high volume condition; and (2) energy consumption is reduced by about 4 % under low to moderate volume conditions and by about 12 % under high volume condition. All these enhancements validate the potential benefits of implementing the next-generation intersection control.

Z. Li (✉)
University of Louisville, Louisville, Ky 40292, USA
e-mail: richard.li@louisville.edu

© Springer International Publishing Switzerland 2017

V.E. Balas et al. (eds.), *Information Technology and Intelligent Transportation Systems*, Advances in Intelligent Systems and Computing 454,
DOI 10.1007/978-3-319-38789-5_12

Zhixia Li

1 Short Biography

Dr. Zhixia Li is an Assistant Professor in the Department of Civil and Environmental Engineering, University of Louisville. His research spans in the areas of traffic operations and control, Traffic safety, traffic simulation, GIS-Transportation, ITS, and sustainable transportation. Research grant proposals he wrote or contributed have successfully secured research fund from FHWA, NCHRP, TRB's NCHRP IDEA, NHTSA, and multiple state DOTs. So far, Dr. Li's research has produced more than 50 peer-reviewed journal and conference publications as well as a book chapter in ITE's Traffic Control Devices Handbook. Particularly, his research was highlighted in Washington Post, NBC, Yahoo, The Times of India, The Vancouver Sun, and The Ottawa Citizen. In addition, Dr. Li was recipient of six professional and student awards at international, national, and regional levels, including the International ITE's Danial Fambro Best Student Paper Award. Dr. Li serves for The Transportation Research Board by sitting in on two standing committees ABJ50 and ABE80, and as panelist of NCHRP project 03-113. He is member of Technical Committee on Travel Information and Traffic Management of IEEE ITS Society. Dr. Li obtained his Ph.D. degree in Civil Engineering from University of Cincinnati, and Bachelor's degree in Electrical Engineering from Sun Yat-sen University. He received his post-doctoral training from University of Wisconsin-Madison.

Part II
Proceedings Papers: Theory Research in Intelligent Transportation Systems

High Accuracy Solutions of the Modified Helmholtz Equation

Hu Li and Jin Huang

Abstract We study the numerical solutions for modified Helmholz equation. Based on the potential theory, the problem can be converted into a boundary integral equation. Mechanical quadrature method (MQM) is presented for solving the equation, which possesses high accuracy order $O(h_{max}^3)$ and low computing complexities. Moreover, the multivariate asymptotic error expansion of MQM accompanied with $O(h_i^3)$ for all mesh widths h_i is got. Hence, once discrete equations with coarse meshes are solved in parallel, the higher accuracy order of numerical approximations can be at least $O(h_{max}^5)$ by splitting extrapolation algorithm (SEA). The numerical examples support our theoretical analysis.

Keywords Mechanical quadrature method · Splitting extrapolation algorithm · Modified Helmholtz equation

1 Introduction

Time-harmonic acoustic wave scattering or radiation by a cylindrical obstacle is essentially a two-dimensional problem and is often described in acoustic media by modified Helmholtz equation with associated boundary condition. We consider modified Helmholtz equation with Dirichlet boundary condition:

$$\begin{cases} \Delta u(x) - \alpha^2 u(x) = 0, & x \in \Omega, \\ u_m(x) = g_m(x), & x \in \Gamma_m(\Gamma = \cup_{m=1}^{d}), \end{cases} \tag{1}$$

This work is Supported by the National Natural Science Foundation of China(11371079).

H. Li (✉) · J. Huang
School of Mathematical Sciences, University of Electronic Science and Technology of China, Chengdu 611731, Sichuan, People's Republic of China
e-mail: lihuxiwangzhixing@163.com

© Springer International Publishing Switzerland 2017

V.E. Balas et al. (eds.), *Information Technology and Intelligent Transportation Systems*, Advances in Intelligent Systems and Computing 454,
DOI 10.1007/978-3-319-38789-5_13

where $\Omega \subset \Re^2$ is a bounded, simply connected domain with a piecewise smooth boundary Γ, and $\Gamma = \cup_{m=1}^{d} \Gamma_m, d > 1$ is a closed curve, and the function $g_m(x) = g(x)|_{\Gamma_m}$ is known on Γ_m.

By the potential theory, the solutions of (1) can be represented as a single-layer potential

$$u(y) = \int_{\Gamma} K^*(y, x)v(x)ds_x, \quad y = (y_1, y_2) \in \Omega, \tag{2}$$

where $x = (x_1, x_2)$, $K^*(y, x)$ is the foundation solution of modified Helmholtz equation [1]

$$K^*(y, x) = -\frac{1}{2\pi} K_0(\alpha \mid x - y \mid), \tag{3}$$

where K_0 is a modified Bessel function

$$K_0(z) = -\ln z + \ln 2 - \gamma, \quad z \to 0, \tag{4}$$

where $\gamma = 0.57721\ldots$ is Euler constant. $v(x)$ is the solution of the following equation

$$g(y) = \int_{\Gamma} K^*(y, x)v(x)ds_x, \quad y = (y_1, y_2) \in \Gamma. \tag{5}$$

Equation (5) is weakly singular BIE system of the first kind, whose solution exists and is unique as long as $C_T \neq 1$ [2], where C_T is the logarithmic capacity. As soon as $v(x)$ is solved from (5), the function $u(y)(y \in \Omega)$ can be calculated by (2).

The kernels and solutions of (5) have singularities at both the points $x = y$ and the corner points of Γ, which degrade the rate of convergence in numerical methods. Several numerical methods have been proposed to overcome this difficulty, such as Galerkin methods and collocation methods [2–5]. However, the discrete matrix is full and each element has to calculate the weakly singular integral for collocation methods or the double weakly singular integral for Galerkin methods, which imply CPU-time expended by calculating discrete matrix is so more as to exceed to solve discrete equations. When the numerical methods are applied, the accuracy of numerical solutions is lower at singular points [6], and the corresponding numerical results becomes to be unreliable any more, because the condition numbers are very large.

In the paper, MQM is proposed to calculate weakly singular integrals by Sidi quadrature rules [7], which makes the calculation of the discrete matrix become very simple and straightforward without any singular integrals. MQM retains the optimal convergence order $O(h_{max}^3)$ and possesses the optimal condition number $O(h_{min}^{-1})$. Since MQM possesses the multivariate asymptotic expansion of errors, we can construct SEA to obtain the convergence order $O(h_{max}^5)$. Once discrete equations on some coarse meshes are solved in parallel, the accuracy of numerical solutions can be greatly improved by SEA.

This paper is organized as follows: in Sect. 2, the MQM is described. In Sect. 3, we can obtain multi-parameter asymptotic expansion of errors and SEA is described. In Sect. 4, numerical examples are provided to verify the theoretical results.

2 Mechanical Quadrature Method

Let Γ_m be describe by the parameter mapping: $x_m(s) = (x_{1m}(s), x_{2m}(s)) : [0, 1] \to \Gamma_m, 0 \leq s \leq 1$, with $| x_m'(s) | = \left[(x_{1m}'(s))^2 + (x_{2m}'(s))^2\right]^{1/2} > 0, m = 1, \ldots, d$. Then (5) can be written as:

$$g_q(t) = -\frac{1}{2\pi}\sum_{m=1}^{d}\int_0^1 K_0(\alpha \mid x_m(t) - x_q(s) \mid) \mid x_m'(s) \mid v_m(s)ds, \quad q = 1, \ldots, d, \tag{6}$$

where $v_m(s) = v(x_m(s)), g_q(t) = g_q(x_q(t))$. Using Sidi periodic transformation [8]:

$$\psi_p(\tau) = \vartheta_p(\tau)/\vartheta_p(1) : [0, 1] \to [0, 1], \quad p \in N, \tag{7}$$

where $\vartheta_p(\tau) = \int_0^\tau (\sin(\pi\rho))^p d\rho$. Define the integral operations on [0, 1],

$$(K_{qm}\omega_m)(t) = \int_0^1 k_{qm}(t, s)\omega_m(s)ds, \quad q, m = 1, \ldots, d, \tag{8}$$

where $k_{qm}(t, s) = -\frac{1}{2\pi}K_0(\alpha|x_q(\psi_p(t)) - x_m(\psi_p(s))|)$, $\omega_m(s) = v_m(\psi_p(s))|x_m'(\psi_p(s))|\psi_p'$. Then (6) can be converted into a matrix operator equation

$$K\omega = G, \tag{9}$$

where $K = [K_{qm}]_{q,m=1}^d$, $\omega = (\omega_1(s), \ldots, \omega_d(s))^T$, $G = (G_1(t), \ldots, G_d(t))^T$, and $G_m(t) = g_m(\psi_p(t))$.

Let $K_{mm} = A_{mm} + B_{mm}$. The kernel of A_{mm} is $a_{mm}(t, s) = -\frac{1}{2\pi}\ln|2e^{-\frac{1}{2}} \sin\pi(t - s)|$ and the kernel $b_{qm}(t, s)$ of B_{qm} satisfy

$$b_{qm}(t, s) = \begin{cases} -\frac{1}{2\pi}\ln\left|\frac{x_q(t)-x_m(s)}{2e^{-\frac{1}{2}}\sin\pi(t-s)}\right| - \ln(2\alpha) - \gamma & for \quad q = m, \\ -\frac{1}{2\pi}K_0(\alpha|x_q(t) - x_m(s)|) & for \quad q \neq m. \end{cases} \tag{10}$$

Thus, (9) can be split into a singularity and compact perturbation part

$$(A + B)\omega = G, \tag{11}$$

where $A = diag(A_{11}, \ldots, A_{dd})$. The operator $A_{mm}(m = 1, \ldots, d)$ is an isometry operator from $H^r[0, 1]$ to $H^{r+1}[0, 1]$ for any real number r and $\| A_{mm}\omega_m \|_{r+1} = \|$

$\omega_m \|_r$. A is also an isometry operator from $(H^r[0,1])^d$ to $(H^{r+1}[0,1])^d$. Hence, A is invertible, (11) is equivalent to

$$(E + A^{-1}B)\omega = A^{-1}G. \tag{12}$$

Let $h_m = 1/n_m (n_m \in N, m = 1, \ldots, d)$ be mesh widths, and $t_{mj} = (j - 1/2) h_m (j = 1, \ldots, n_m)$ be nodes.

(1)Since k_{qm} are the smooth function on [0, 1], by the trapezoidal or the midpoint rule [9], we can construct the Nyström approximate operator $K^h_{qm}(q \neq m)$ of K_{qm}, defined by

$$(K^h_{qm}\omega_m)(t) = h_m \sum_{j=1}^{n_m} k_{qm}(t, t_{mj})\omega_m(t_{mj}), \tag{13}$$

and the errors

$$(K^h_{qm}\omega_m)(t) - (K_{qm}\omega_m)(t) = O\left(h_m^{2l}\right), \quad l \in N. \tag{14}$$

(2)Since $K_{mm} = A_{mm} + B_{mm}$ have the singularities on [0, 1], by Sidi quadrature formula [7], we get the following approximations A^h_{mm} of A_{mm},

$$\begin{aligned}(A^h_{mm}\omega_m)(t) = &-\frac{1}{2\pi} h_m \sum_{j=1, t \neq t_{mj}}^{n_m} \ln \mid 2e^{-1/2} \sin \pi(t - t_{mj}) \mid \omega_m(t_{mj}) \\ &- \frac{h_m}{2\pi} \ln \mid 2\pi e^{-1/2} h_m/(2\pi) \mid \omega_m(t),\end{aligned} \tag{15}$$

and the errors

$$(A^h_{mm}\omega_m)(t) - (A_{mm}\omega_m)(t) = -\frac{2}{\pi} \sum_{\mu=1}^{2l-1} \frac{\zeta'(-2\mu)}{(2\mu)!} [\omega_m(t)]^{2\mu} h_m^{2\mu+1} + O\left(h_m^{2l}\right), \tag{16}$$

where $\zeta'(t)$ is the derivative of the Riemann zeta function, and

$$\begin{aligned}(B^h_{mm}\omega_m)(t) = &h_m \sum_{j=1, t \neq t_{mj}}^{n_m} b_{mm}(t, t_{mj})\omega_m(t_{mj}) \\ &+ \frac{h_m}{2\pi} (\ln \mid \frac{e^{1/2} h_m x'_m(t_{mj}) \psi'_p(t_{mj})}{2\pi} \mid + \epsilon_\alpha)\omega_m(t),\end{aligned} \tag{17}$$

where $\epsilon_\alpha = -\ln(2\alpha) - \gamma$. Hence, the approximates K^h_{mm} of K_{mm} are defined by

$$(K^h_{mm}\omega_m)(t) = h_m \sum_{j=1, t \neq t_{mj}}^{n_m} k_{mm}(t, t_{mj})\omega_m(t_{mj})$$

$$+\frac{h_m}{2\pi}\Big(\ln\mid\frac{e^{1/2}h_m x_m'(t_{mj})\psi_p'(t_{mj})}{2\pi}\mid$$
$$+\epsilon_\alpha - \ln\mid 2\pi e^{-1/2}h_m/(2\pi)\mid\Big)\omega_m(t).$$

Then (11) can be rewritten as

$$(A^h + B^h)\omega^h = G^h, \tag{18}$$

where $A^h = diag(A_{11}^h, \ldots, A_{dd}), G^h = [G_q^h]_{q=1}^d, G_q^h = (g_q(t_{q1}), \ldots, g_q(t_{qn_q}))$, and $B^h = [B_{qm}^h]_{q,m=1}^d$.

Obviously, (18) is a system of linear equations with $n(= \sum_{m=1}^d n_m)$ unknowns. Once ω^h is solved by (18), the solution $u(y)(y \in \Omega)$ can be computed by

$$u^h(y) = -\frac{1}{2\pi}\sum_{m=1}^{d} h_m \sum_{j=1}^{n_m} K_0\big[\alpha|y - x(t_{mj})|\big]\omega_m(t_{mj}). \tag{19}$$

A_{mm}^h are symmetric circular matrices and have the form of

$$A_{mm}^h = circulate\left(-\frac{h_m}{2\pi}\ln|2e^{-1/2}h_m/\pi|, \ldots, -\frac{h_m}{2\pi}\ln|2e^{-1/2}\sin(\pi(n_m-1)h_m)|\right). \tag{20}$$

Lemma 1 (see [11]) *(1) There exist a positive $c_1 > 0$ so that the eigenvalues $\lambda_\beta(\beta = 1, \ldots, n_m)$ of A_{mm}^h satisfy: $c_1 > \lambda_\beta > 1/(2\pi n_m)$. (2) The condition number of A_{mm}^h is $O(n_m)$. (3) A_{mm} is invertible, and $(A_{mm}^h)^{-1}$ is uniformly bounded with the spectral norm $\| (A_{mm}^h)^{-1} \| = O(n_m)$.*

Based on Lemma 1, we immediately get the following corollary.

Corollary 1 (see [11]) *(1) A^h is invertible, and $(A^h)^{-1}$ is uniformly bounded with the spectral norm $\| (A^h)^{-1} \| = O(n_0)$. (2) The condition number of A^h is $O(n_0)$, where $n_0 = \max_{m=1}^d n_m$*

From corollary 1, we know (18) is equivalent to

$$(E^h + (A^h)^{-1}B^h)\omega^h = (A^h)^{-1}G^h, \tag{21}$$

where E^h denotes the unit matrix.

For the stability of MQM, we have the following corollary.

Corollary 2 (see [11]) *Let $\Gamma = \cup_{m=1}^d \Gamma_m$ with $C_\Gamma \neq 1$, Γ_m $(m = 1, \ldots, d)$ be smooth curve, A^h and B^h be the discrete matrices defined by (15) and (17), respectively, and $\lambda_i(i = 1, \ldots, n)$ be the eigenvalues of discrete matrix $K^h{=}A^h{+}B^h$. Then there exists the bound of condition number*

$$\mathrm{Cond}(K^h) = \frac{\max_{1\le i\le n} |\lambda_i(K^h)|}{\min_{1\le i\le n} |\lambda_i(K^h)|} = O\left(h_{min}^{-1}\right). \tag{22}$$

where $h_{\min} = \min_{m=1}^d h_m$, *and* $h_m = 1/n_m$, *is the mesh step size of a curved edge* Γ_m.

3 Multi-parameter Asymptotic Expansion of Errors and SEA

In this section, we derive the multivariate asymptotic expansion of solution errors and describe SEA. We first provide the main result.

Theorem 1 (see [11]) *Let* $\Gamma = \cup_{m=1}^d \Gamma_m$ *with* $C_\Gamma \neq 1$. *There exists a vector function* $\Phi = (\phi_1, \ldots, \phi_d)^T$ *independent of* $h = (h_1, \ldots, h_d)$ *so that the following multi-parameter asymptotic expansion hold at nodes*

$$\omega^h - \omega = diag\left(h_1^3, \ldots, h_d^3\right)\Phi + o\left(h_0^3\right), \quad h_0 = max_{1\le m\le d} h_m, \tag{23}$$

The multi-parameter asymptotic expansion (23) means that SEA can be applied to solved (5), that is, a higher order accuracy $o(h_0^3)$ at coarse grid points can be obtained by solving some discrete equations in parallel. The process of SEA is as follows [12]:

Step 1. Take $h^{(0)} = (h_1^{(0)}, \ldots, h_d^{(0)})$ and $h^{(m)} = (h_1^{(0)}, \ldots, h_m^{(0)}/2, \ldots, h_d^{(0)})$, and solve (21) under mesh parameters $h^{(m)}$ in parallel to get the numerical solutions $\omega^{h^{(0)}}(t_{mj})$ and $\omega^{h^{(m)}}(t_{mj})$, $m = 1, \ldots, d$, $j = 1, \ldots, n_m$.

Step 2. Compute $u^{h^{(0)}}(y)$ and $u^{h^{(m)}}(y)(y \in \Omega)$, by (19), $\omega^{h^{(0)}}(t_{mj})$ and $\omega^{h^{(m)}}(t_{mj})$.

Step 3. Compute a extrapolation on the coarse grids as follows:

$$u^*(y) = \frac{8}{7}\left[\sum_{m=1}^d u^{h^{(m)}}(y) - \left(d - \frac{7}{8}\right) u^{h^{(0)}}(y)\right]. \tag{24}$$

4 Numerical Examples

In this section, we carry out some numerical examples for the modified Helmholtz equation by MQM and SEA, in order to verify the error and stability analysis in the previous sections. Let $error = |u^h - u|$, and SEA-error denotes the error after SEA once. $n_{1,2}$ denote the mesh nodes.

Example 1 Consider modified Helmholtz equation with $\alpha = \sqrt{2}$ on a plate domain Ω. We describe the boundary $\Gamma = \cup_{m=1}^2 \Gamma_m$ with $\Gamma_1 = \{(x_1, x_2) = (t, 0) : 0 \le t \le 1\}$, $\Gamma_2 = \{(x_1, x_2) = (0.5\cos \pi t + 0.5, 0.5 \sin \pi t) : 0 \le t \le 1\}$. The analytic solu-

Table 1 The simulation results for Example 1

(n_1, n_2)	(0.3, 0.2)	(0.4, 0.25)	(0.5, 0.2)	(0.6, 0.15)
(16, 16)	5.203e−5	3.072e−5	3.020e−5	8.087e−5
(32, 16)	2.210e−5	4.797e−5	5.401e−5	4.177e−5
(16, 32)	3.674e−5	1.295e−5	1.900e−5	4.377e−5
SEA-error	3.412e−7	5.304e−7	1.185e−6	6.212e−6
(32, 32)	6.753e−6	4.348e−6	4.869e−6	4.721e−6
(64, 32)	2.896e−6	6.652e−6	6.677e−6	5.180e−6
(32, 64)	4.738e−6	1.793e−6	1.227e−6	1.060e−7
SEA-error	4.217e−8	3.653e−8	3.249e−8	2.920e−8

Table 2 The condition number for Example 1

(n_1, n_2)	$(2^3, 2^3)$	$(2^4, 2^4)$	$(2^5, 2^5)$	$(2^6, 2^6)$
$\|\lambda_{min}\|$	3.070e−3	1.489e−3	7.389e−4	3.687e−4
$\|\lambda_{max}\|$	0.4510	0.4525	0.4528	0.4529
Cond	1.469e+002	3.038e+002	6.129e+002	1.228e+003

Table 3 The simulation results for Example 2

(n_1, n_2)	(0.5, 0.3)	(0.5, 0.5)	(0.4, 0.5)	(0.5, 0.4)
(8, 8)	5.054e−3	6.940e−3	5.711e−3	5.571e−3
(16, 8)	3.662e−3	7.008e−3	5.462e−3	5.252e−3
(8, 16)	2.007e−3	8.069e−4	1.060e−3	1.030e−3
SEA-error	1.849e−5	8.864e−6	1.111e−4	1.685e−5
(16, 16)	5.932e−4	8.494e−4	7.928e−4	6.863e−4
(32, 16)	4.579e−4	8.526e−4	7.591e−4	6.464e−4
(16, 32)	2.094e−4	1.028e−4	1.327e−4	1.256e−4
SEA-error	1.266e−7	2.331e−7	1.244e−7	1.239e−7

tions $e^{-x_1-x_2}$. We compute the numerical solution u^h by $\psi_3(t)$. The numerical results are listed in Tables 1 and 2.

Example 2 Consider modified Helmholtz equation with $\alpha = \sqrt{2}$ on a plate domain Ω. We describe the boundary $\Gamma = \cup_{m=1}^{2} \Gamma_m$ with $\Gamma_1 = \{(x_1, x_2) = (0.5\cos(\pi t + \pi) + 0.5, 0.5\sin(\pi t + \pi) + 0.5) : 0 \le t \le 1\}$, $\Gamma_2 = \{(x_1, x_2) = (0.5\cos \pi t + 0.5, \sin \pi t + 0.5) : 0 \le t \le 1\}$. The analytic solutions $e^{x_1+x_2}$. We compute the numerical solution u^h by $\psi_3(t)$. The numerical results are listed in Tables 3 and 4.

From Tables 1 and 3, we can known the convergence rates of u^h are $O(h_{max}^3)$ for MQM and the convergence rates of u^h are at least $O(h_{max}^5)$ for SEA. From

Table 4 The condition number for Example 2

(n_1, n_2)	$(2^3, 2^3)$	$(2^4, 2^4)$	$(2^5, 2^5)$	$(2^6, 2^6)$
$\|\lambda_{min}\|$	9.947e−3	4.928e−3	2.458e−3	1.228e−3
$\|\lambda_{max}\|$	0.3738	0.3779	0.3789	0.3791
Cond	3.758e+001	7.668e+001	1.541e+002	3.086e+002

Tables 2 and 4, we can see $Cond|_{(2^{k+1},2^{k+1})}/Cond|_{(2^k,2^k)} \approx 2, (k = 3, 4, 5, 6)$ to indicate corollary 2. It verifies the stability of convergent theory for MQM.

5 Concluding Remarks

To close this paper, let us make a few concluding remarks.

(1) Evaluation on entries of discrete matrices is very simple and straightforward, without any singular integrals by MQM.

(2) The numerical experiments show that MQM retains the optimal convergence order $O(h_{max}^3)$ and possesses the optimal condition number $O(h_{min}^{-1})$ which shows MQM own the excellent stability. The approximate solutions accuracy order is at least $O(h_{max}^5)$ after splitting extrapolation once, which is a greatly improvement in accuracy.

References

1. Atkinson KE, Sloan IH (1991) The numerical solution of first kind logarithmic kernel integral equation on smooth open arcs. Math Comp 56:119–139
2. Yi Y, Sloan I (1988) On integral equations of the first kind with logarithmic kernel. J Integral Equ Appl 1:549–579
3. Anselone PM (1971) Collectively compact operator approximation theory. Prentice-Hall, Englewood Cliffs
4. Sloan IH, Spence A (1988) The galerkin method for integral equations of the first kind with logarithmic kernel. IMA J Numer Anal V 8:105–122
5. Yi Y (1990) The collocation method for the first kind with boundary integral equation on polygonal region. Math Comp 54:139–154
6. Atkinson K (1977) The numerical solution of integral equations of the second kind. Cambridge University Press, Cambridge
7. Sidi A, Israrli M (1988) Quadrature methods for periodic singular fredholm integral equations. J Sci Comput 3:201–231
8. Sidi A (1993) A new variable transformation for numerical integration. In integration IV. Birkhaurer, Frankfurt, pp 359–374
9. Davis P (1984) Methods of numerical integration, 2nd edn. Academic Press, New York
10. Huang J, Lu T (2006) Splitting extrapolations for solving boundary integral equations of linear elasticity dirichlet problems on polygons by mechanical quadrature methods. J Comput Math, Sinica 24(1):9–18

11. Huang J, Wang Z (2009) Extraplotion algorithms for solving mixed boundary integral equations of the Helmholz equation by mechanical quadrature methods. SIAM J Sci Comput 31:4115–4129
12. Lin Q, Lu T (1983) Splitting extrapolation for multidimensional problems. J Comp Math 1:45–51
13. Lu T, Huang J (2000) Quadrature methods with high accuarcy and extrapolation for solving boundary integral equations of first-kind. J Math Num Sinica 1:59–72
14. Lu T, Lu J (2002) Splitting extrapolation for solving the second order elliptic system with curved boundary in rd by using quadratic isoparametric finite element. Appl Numer Math 40:467–481
15. Rui Zhu, Jin Huang, Tao Lu (2006) Mechanical quadrature methods and their splitting extrapolations for solving boundary integral equations of axisymmetric laplace mixed boundary value problems. Eng Anal Bound Elem 30(5):391–398
16. Stakgold I (1979) Greens functions and boundary value problems. Wiley, New York
17. Yi Y (1990) Cosine change of variable for symms integral equation on open arcs. IMA J Numer Anal 10:521–535
18. Pozrikidis C (1992) Boundary integral and singularity methods for linearized viscous flow. Cambridge University Press, Cambridge
19. Huang J, Lu T (2004) The mechanical quadrature methods and their extrapolations for solving BIEs of Steklov eigenvalue problems. J Comput Math 22:719–726

Cooperative Trajectory Planning for Multiple UAVs Using Distributed Receding Horizon Control and Inverse Dynamics Optimization Method

Yu Zhang, Chao Wang, Xueqiang Gu and Jing Chen

Abstract This paper studies the problem of generating obstacle avoidance trajectories through complex 3-D environments on-board for a group of non homonymic Unmanned Aerial Vehicles (UAVs). First, the collision-free multi-vehicle cooperative trajectory planning problem is mathematically formulated as a decentralized receding horizon optimal control problem (DRH-OCP). Next, a real-time trajectory planning framework based on a decentralized planning scheme which only uses local information, and an inverse dynamics direct method which has high computational efficiency and good convergence properties, is designed to solve the DRH-OCP. Finally, the simulation results demonstrate that the proposed planning strategies successfully generates the trajectories that satisfy the given mission objectives and requirements.

Keywords Cooperative trajectory planning · Inverse dynamics optimization · Distributed receding horizon control · Unmanned aerial vehicles

1 Introduction

The research and development activities on cooperative control and coordination of multiple unmanned aerial vehicles (UAVs), have experienced a significant increase in the last decade in both military and civilian areas [1]. The main motivation for

Y. Zhang (✉) · C. Wang · X. Gu · J. Chen
College of Mechatronic Engineering and Automation, National University of Defense Technology, Changsha 410073, People's Republic of China
e-mail: redarmy_zy@163.com

C. Wang
e-mail: zhongwangchao@163.com

X. Gu
e-mail: xqgu_nudt@163.com

J. Chen
e-mail: chenjing001@vip.sina.com

© Springer International Publishing Switzerland 2017

V.E. Balas et al. (eds.), *Information Technology and Intelligent Transportation Systems*, Advances in Intelligent Systems and Computing 454,
DOI 10.1007/978-3-319-38789-5_14

multi-UAV cooperation stems from the fact that, in many complicated applications, such as search and localization [2], and ground moving target tracking [3], the group performance is expected to exceed the sum of the performance of the individual UAVs.

For virtually any multi-vehicle system, there is an essential requirement: vehicles must avoid collisions with objects (e.g. obstacles and restricted areas) and other vehicles in the environment while executing their own tasks. In practice, vehicles are subject to actuation limits and constrained dynamics, such as flight envelope limits, which can restrict their maneuverability. These limitations are especially important for teams deployed in crowded environments (e.g. urban or dense air defense environments). Consequently, trajectory planners developed for such environments must explicitly account for these constraints in order to guarantee safe navigation.

One of the main challenges in multi-vehicle cooperative trajectory planning problems involving vehicles with kinodynamic constraints is the computational complexity. In large teams, the computational resources required to solve the associated centralized planning problem can become prohibitive. This necessitates the development of de-centralized (distributed) planners, where each vehicle makes local decisions. Much of the current research on decentralized planning assumes that each vehicle solves a local problem and communicates this intent information to its neighbors. Various approaches have been proposed, including prioritized planning, exchanging coordination variable, and coupling penalty functions, etc. [4–12]. Some other approaches are based on a receding horizon control approach (RHC). The main idea in decentralized RHC is to divide the centralized RHC planner into the local planning problems of small sizes. The computation time of these small size optimization problems is dramatically reduced.

Different decentralized RHC strategies have been proposed in the literature. In [13], a distributed RHC scheme is applied to formation stabilization problem with quadratic cost and no coupling constraints. By introducing a compatibility constraint, they ensure that no vehicle will diverge too far from the behavior that others expect. In [14], each decoupled subsystem optimizes locally for itself as well as for every neighbor at each update, resulting in an increase in the computing time and a decrease in the decentralization. In [8], the decentralized receding horizon planner is solved using mixed-integer linear programming (MILP). Some extensions of this approach are proposed in [5, 15], which can generate a local plan over a short horizon while guaranteeing the robust feasibility of the entire fleet under the action of external disturbances. The main disadvantage of these approaches is that during each iteration vehicles solve their planning problems one by one in a specified order, rather than in parallel. This will lead to a relatively long execution horizons and a single point of failure.

In addition, a inverse dynamics method has recently been used for real-time onboard calculation of near-optimal trajectories [16–18], which has many potential advantages. First, it can use any model and any performance index [19] (i.e., it is not subject to the curse of dimensionality and does not require differentiability of the performance index as many other direct methods do). Second, it can transform the OCP into a NLP problem of very low dimension (typically <20). Some numerical simulations [20] suggest that its computational speed could be more than an order of magnitude faster than the pseudospectral methods, with small loss of optimality, and its robustness and convergence properties are also better.

Motivated by the above advantages, this paper presents a novel real-time, decentralized-coordination trajectory planning framework, which effectively combines the benefits of inverse dynamics optimization method and receding horizon optimal control technique.

2 Problem Formulation

The multi-vehicle cooperative planning problem formulation usually consists of three parts: the system dynamics, the internal and external constraints, and the objective function. These parts are covered in the following subsections.

To begin, some basic notations in this paper are defined as follows. There are a total of N_{UAV} UAVs and N_{obs} obstacles. The index or super/subscript p, q denotes the UAV index. Unless otherwise noted, $\forall p$ denotes $\forall p = 1, \ldots, N_{UAV}$ and $\forall q$ denotes $\forall q = 1, \ldots, N_{UAV}$ but $q \neq p$. We assume that each UAV p knows its initial location $\mathbf{R}^p_{\text{init}} = (x^p_{\text{init}}, y^p_{\text{init}}, z^p_{\text{init}})$ and its goal location $\mathbf{R}^p_{\text{goal}} = (x^p_{\text{goal}}, y^p_{\text{goal}}, z^p_{\text{goal}})$. Here the goal location is chosen by some type of high-level commands, such as the output of a task allocation algorithm [21].

2.1 UAV Dynamical Model

In this paper, the essence of the problem is the flight-path control design, therefore, a detailed 3-DOF Euler-angle-based point-mass model is sufficiently accurate. The kinematic and dynamical equations of the UAV are given by [22]:

$$\begin{cases} \dot{x} = V\cos\gamma\cos\psi \\ \dot{y} = V\cos\gamma\sin\psi \\ \dot{z} = V\sin\gamma \end{cases} \quad \begin{cases} \dot{V} = g(n_X - \sin\gamma) \\ \dot{\gamma} = \frac{g}{V}(n_Z\cos\mu - \cos\gamma) \\ \dot{\psi} = \frac{g}{V\cos\gamma} n_Z \sin\mu \end{cases} \tag{1}$$

where (x, y, z) are the aircraft inertial coordinate, i.e., the latitude, longitude and altitude, V is the true airspeed, γ is the flight-path angle, ψ is the heading, μ is the roll angle, and g is the acceleration due to gravity. n_x and n_Z denote tangential and normal components of the load factor in the wind coordinate frame, respectively.

Note that, in this model, we assume that the UCAV can be controlled directly with the load factor and the roll angle. Thus, the state vector x is $[x, y, z, V, \gamma, \psi]^T$ and the control vector u is $[\mu, n_X, n_Z]^T$.

Furthermore, in order to ensure that the generated trajectories are dynamically feasible, the above UAV model is constrained to incorporate the platform performance limits, which can be described by the following state constraints and control input constraints which are modeled as lower and upper bounds:

$$\begin{cases} 0 \le z \le z_{\max} \\ V_{\min} \le V \le V_{\max} \end{cases} \begin{cases} |\mu| \le \mu_{\max} \\ n_{X\min} \le n_X \le n_{X\max} \\ n_{Z\min} \le n_Z \le n_{Z\max} \end{cases} \begin{cases} \dot{\mu}_{\min} \le \dot{\mu} \le \dot{\mu}_{\max} \\ \dot{n}_{X\min} \le \dot{n}_X \le \dot{n}_{X\max} \\ \dot{n}_{Z\min} \le \dot{n}_Z \le \dot{n}_{Z\max} \end{cases} \tag{2}$$

where $\dot{n}_X$ represents the thrust build-up/decay times, $\dot{\mu}$ and $\dot{n}_Z$ represent the characteristics of UAVs control system.

2.2 *Internal and External Constraints*

For the multi-UAV optimal control problem, there exist several essential internal and external constraints, including collision avoidance constraints and inter-vehicle constraints. Modeling an obstacles exact shape and size is very complex and highly unnecessary. In this work, the -norm [23] is used to mathematically model the shapes of the obstacles. Accordingly, the obstacles can be expressed by the following path constraints in the OCP formulation:

$$\begin{aligned} h_i(x, y, z) = \ln\left(\frac{1}{(k_4^i)^{\rho_i}}\left(\left(\frac{x - x_c^i(t)}{k_1^i + b_s}\right)^{\rho_i} + \left(\frac{y - y_c^i(t)}{k_2^i + b_s}\right)^{\rho_i} + \left(\frac{z - z_c^i(t)}{k_3^i + b_s}\right)^{\rho_i}\right)\right) \\ \ge 0, i = 1, 2, \ldots, N_{\text{obs}} \end{aligned} \tag{3}$$

where $h + i$ represents the ith obstacle, (x_c^i, y_c^i, z_c^i) indicate the location of the geometric center of the obstacle i, $h_i(x, y, z)$ is the distance between a point (x, y, z) and the boundary of the obstacle i, $k_1^i \sim k_4^i$ are the constant parameters chosen to define the size of the obstacle i, and b_s represents the width of a safe buffer which accounts for the size of the vehicle.

Compared to single UAV, there is an essential additional requirement for a multi-UAV system that the UAVs should keep a safe distance with each other to avoid collisions. The inter-vehicle constraints can be denoted as follows:

$$d\left(\mathbf{R}^p(t), \mathbf{R}^q(t)\right) = \left\|\left(x^p - x^q, y^p - y^q, z^p - z^q\right)\right\|_2^2 \ge r_{\text{safe}}^p + r_{\text{safe}}^q, \forall p, q \tag{4}$$

where $\mathbf{R}^p(t) = (x^p, y^p, z^p)$ represents the inertial coordinate of UAV p, and $r_{\text{safe}}^p \in \Re^+$ is the predetermined minimum safety radius of UAV p (based on the vehicle size, maneuver agility, and worst-case turbulence, etc.). $\mathbf{R}^p(t)$ and r_{safe}^q are similar.

In addition to the constraint on UAV position for collision avoidance, there is another crucial coupling constraint, i.e., the communication links between some UAVs must be maintained during the flight.

In general, the communication topology of UAVs can be modeled by a graph structure in the formulation. Let $\mathbf{G}(t) = \{\mathbf{\Theta}, \mathrm{E}\}$ denotes an undirected interconnection graph, where $\boldsymbol{\Theta} = \{\mathbf{1}, \ldots, \mathbf{N}_{\mathrm{UAV}}\}$ is a set of nodes representing the UAVs, and $\mathrm{E} \subseteq \boldsymbol{\Theta} \times \boldsymbol{\Theta}$ is an edge set of pairs of nodes that encodes the communication links. Also, let $\mathbf{Q}_p = \{q \in \boldsymbol{\Theta} \,|(p, q) \in \mathrm{E}\}$ defines the communication neighbor set of UAV_p.

For each edge $(p, q) \in \mathrm{E}$, there is one corresponding communication constraint between UAVs p and q. This constraint relates to the limited range of transmitters and receivers, and can be written as follows:

$$d\left(\mathbf{R}^p(t), \mathbf{R}^q(t)\right) \leq \min(r^p_{\mathrm{com}}, r^q_{\mathrm{com}}) \tag{5}$$

where $r^p_{\mathrm{com}}, r^q_{\mathrm{com}} \in \Re^+$ are the maximum broadcasting range of UAVs p and q respectively. Here each r^p_{com} is assumed to be strictly larger than $r^p_{\mathrm{safe}} + r^q_{\mathrm{safe}}(p \neq q, \forall q)$.

2.3 Cost Function

In this work, the desired behavior for the UAVs is to navigate safely through the air-space in which the missions are to be performed, and then reach their respective goals as quickly as possible. So, the following specific objective function is proposed for UAV p:

$$J_p = \int_{t_0}^{t_0+T_{PH}} \left[w_{sm} L^p_{SM} + w_{ru} L^p_{RU}\right] dt + w_{ctg} \Omega^p_{CTG} \tag{6}$$

where, t_0 is the start time of the current plan, and T_{PH} is the prediction horizon. The objective function is defined by the weighted sum of three separate cost terms with appropriate weighting factors $(w_{sm}, w_{ru}, w_{ctg})$.

The first element of the running cost, L^p_{SM}, penalizes the dramatic changes on attitude (yaw, pitch, and roll), which tend to smooth the flight trajectory and associated control inputs. The trajectory smoothing cost is defined as

$$L^p_{SM} = w_{sm,1}(\dot{\gamma}^p)^2 + w_{sm,2}(\dot{\psi}^p)^2 + w_{sm,3}(\dot{\mu}^p)^2 \tag{7}$$

where w_{sm1}, w_{sm2}, and w_{sm3} are the weighting factors.

The second component in running cost, L^P_{RU}, ensures robustness of UAV collision avoidance against model uncertainty and control interpolation errors, which could otherwise cause the vehicle to collide with an obstacle. The robustness cost is defined as [24]

$$L^p_{RU} = \sum_{i=1}^{N_{obs}} l_{ru,i}(e^{e^{-h_i(x,y,z)}} - 1) \tag{8}$$

where $l_{ru,i}$ are constant factors that defines the maneuver robustness level for the ith obstacle.

The third term, Ω^p_{CTG}, is the cost-to-go from the terminal state $\mathbf{x}^p(t+T_{PH})$ in the prediction horizon to the goal state $\mathbf{R}^p_{\text{goal}}$. The quality of that cost-to-go strongly influences the performance of the resulting control policy, either in terms of goal-reaching of the vehicle using RHC. Therefore, in this work a sophisticated method based on 3-D visibility graph [25] is used to generate a good cost-to-go estimation.

2.4 Decentralized Receding Horizon Optimal Control Problem

In the RHC strategy, a cost function is optimized over a finite time (denoted as the prediction horizon T_{PH}) and the first portion of the generated optimal input is applied to the vehicle during a period of time called the execution horizon, $T_{EH}(\leq T_{PH})$ repeating this procedure yields a closed loop solution.

Given UAV ps nonlinear decoupled dynamics, constraints and objective function, a decentralized receding horizon optimal control problem (DRH-OCP) can be formulated as follows:

Problem 1 ($DRH - PCP_p$):

$$\min_{\mathbf{x}^p(t),\mathbf{u}^p(t)} J_p(t) \tag{9}$$

for $\forall t \in [t_0, t_0 + T_{PH}]$, subject to:

$$\dot{\mathbf{x}}^p(t) = \mathbf{F}\left(\mathbf{x}^p(t), \mathbf{u}^p(t)\right) \tag{10}$$

$$\mathbf{x}^p(t_0) = \hat{\mathbf{x}}^p_{0+1}, \mathbf{u}^p(t_0) = \hat{\mathbf{u}}^p_{0+1} \tag{11}$$

$$\begin{aligned} &\mathbf{x}^p_{\min} \leq \mathbf{x}^p(t) \leq \mathbf{x}^p_{\max}, \mathbf{u}^p_{\min} \leq \mathbf{u}^p(t) \leq \mathbf{u}^p_{\max} \\ &\dot{\mathbf{u}}^p_{\min} \leq \dot{\mathbf{u}}^p(t) \leq \dot{\mathbf{u}}^p_{\max} \end{aligned} \tag{12}$$

$$h_i\left(x^p, y^p, z^p\right) \geq 0, i = 1, 2, \ldots, N_{\text{obs}} \tag{13}$$

$$d\left(\mathbf{R}^p(t), \mathbf{R}^q(t)\right) \geq r^p_{\text{safe}} + r^q_{\text{safe}}, \forall q \tag{14}$$

$$d\left(\mathbf{R}^p(t), \mathbf{R}^q(t)\right) \leq \min(r^p_{\text{com}}, r^q_{\text{com}}), \forall q \in \mathbf{Q}_p \tag{15}$$

where $J_p(t)$ is defined by Eq. (6), and $\mathbf{F}(.,.)$ represents the vehicle dynamics (see Eqs. (1) and (2) in [26]). Equation (11) defines the boundary conditions, and $\hat{\mathbf{x}}_{0+1}$ and are respectively the predicted states and controls of the vehicle at next horizon update step based on $\hat{\mathbf{u}}_{0+1}$ the current measurement. Equation (12) are the state and control constraints (see Eqs. (3) and (4) in [26]).

3 Decentralized Receding Horizon Trajectory Planning Scheme

As in the above decentralized receding horizon formulations the neighbor UAVs are coupled through the constraints (Eqs. (14) and (15)), each UAV needs frequently updated information of its neighbors at each sampling time (execution time) to solve the optimization problem $DRH - OCP_p$ and generate its trajectory otherwise the trajectories may not be feasible. Moreover, due to the mismatch between predicted and actual neighbor trajectories, this local optimization problem is not guaranteed to be always feasible. Therefore, a coordination strategy is needed to manage information exchange and maintain data consistency between UAVs.

This section first presents a decentralized coordination strategy based on two-stage iterative optimization. And then this strategy is incorporated in the decentralized RH trajectory planning framework.

3.1 Decentralized Coordination Strategy

At each update in a decentralized receding horizon planning process, every UAV p must predict some preferred trajectories for other UAVs($\forall q$) in order to plan its optimal conflict-free trajectory. The primary challenge is to remain consistent in predicted and actual neighbor trajectories.

Several approaches have been developed to address this issue. One of the most common approaches is prioritized planning [4–8]. It avoids plan conflicts through sequential planning, where the planning order is determined by the priority assigned to each vehicle [4]. At each time step, each vehicle accommodates the latest plans of those vehicles earlier in the sequence and predicted plans of those later in the sequence. The main disadvantage of these approaches is that sequential planning can limit system performance. For example, a busy node or a fault node may block the entire planning process. And, as the number of vehicles increases, the required execution horizon length will rapidly also increase, thus seriously reducing the over systems real-time performance.

This paper employs a two-stage iterative optimization strategy [7] to resolve this issue. At each receding horizon update (step i), every UAV does the following:

In the first stage, each UAV p first senses its own current state and senses or receives the current state of its neighbors, and computes its intention (also called predicted) trajectory and corresponding control input, denoted by $\left\{\hat{\mathbf{x}}_i^p, \hat{\mathbf{u}}_i^p\right\} = \left\{\hat{\mathbf{x}}_i^p(t), \hat{\mathbf{u}}_i^p(t) \,|t \in [t_i, t_i + T_{PH}]\right\}$. This predicted trajectory is obtained by solving the optimization problem $DRH - OCP_p$ without taking the inter-vehicle constraints into account. The simplified version of the problem is denoted by $DRH - OCP_p^{'}$. And then UAV p transmits this predicted plan to all of its neighbors and receives the predicted plans from each neighbor.

In the second stage, each UAV p uses $\{\hat{\mathbf{x}}_i^p, \hat{\mathbf{u}}_i^p\}$ as an initial guess, and resolves the complete problem $DRH-OCP_p$. To eliminate the potential effects of the mismatch between predicted and actual neighbor trajectories, a suitable-size safe interval (also referred to as safe corridor) σ is added around every predicted trajectory, which is then incorporated in the inter-vehicle coupling constraints (Eqs. (14) and (15)), which can be rewritten as

$$d\left(\mathbf{R}^p(t), \mathbf{R}^q(t)\right) \geq r_{\text{safe}}^p + (r_{\text{safe}}^q + \sigma_q), \forall q \tag{16}$$

$$d\left(\mathbf{R}^p(t), \mathbf{R}^q(t)\right) \leq \min(r_{\text{com}}^p, r_{\text{com}}^q) - \sigma_q, \forall q \in \mathbf{Q}_p \tag{17}$$

In addition, the replanned solution of this stage also requires to be restricted within the corresponding safe corridor. Accordingly, a additional constraint is added to the problem formulation, given by

$$d\left(\mathbf{R}^p(t), \hat{\mathbf{R}}^p(t)\right) \leq \sigma_p \tag{18}$$

where $\hat{\mathbf{R}}^p(t) = (\hat{x}^p, \hat{y}^p, \hat{z}^p) \in \hat{\mathbf{x}}_i^p$. This constraint enforces the consistency between the final resulting trajectory and the predicted trajectory which is the trajectory that other UAVs rely on.

Beyond these, a local coordination strategy is also considered in the process of decentralized coordination planning to further reduce the communication requirements and the computational complexity. The main motivation from the fact that each vehicle may not need to directly coordinate with each other in the fleet to effect formation cooperative behavior [9]. In this strategy, every UAV only requires to exchange information with other UAVs that may have direct conflicts (i.e., may produce a collision or may lose communication). For each UAV p, the collision conflict set $\mathbf{\Phi}_{\text{coll}}^p$ and communication conflict set $\mathbf{\Phi}_{\text{comm}}^p$ are defined as follows:

$$\begin{aligned}
\mathbf{\Phi}_{\text{coll},i}^p &= \left\{ q \left| \begin{array}{l} d\left(\mathbf{R}^p(t), \mathbf{R}^q(t)\right) \leq (r_{\text{safe}}^p + T_{PH} V_{\max}^p) + \\ (r_{\text{safe}}^q + T_{PH} V_{\max}^q) \\ \forall q, t \in [t_i, t_i + T_{PH}] \end{array} \right. \right\} \\
\mathbf{\Phi}_{\text{comm},i}^p &= \left\{ q \left| \begin{array}{l} d\left(\mathbf{R}^p(t), \mathbf{R}^q(t)\right) \geq \min(r_{\text{com}}^p, r_{\text{com}}^q) - \\ T_{PH}(V_{\max}^p + V_{\max}^q) \\ \forall q \in \mathbf{Q}_p, t \in [t_i, t_i + T_{PH}] \end{array} \right. \right\}
\end{aligned} \tag{19}$$

Accordingly, in Eqs. (16) and (17) the neighbor sets are replaced by $\mathbf{\Phi}_{\text{coll}}^p$ and $\mathbf{\Phi}_{\text{comm}}^p$, respectively. The new problem formulation is denoted by, and the final optimal conflict-free solution is denoted by $\{\mathbf{x}_i^p, \mathbf{u}_i^p\}$.

This coordination strategy is described by the flowchart given in Fig. 1. We can see that, the coordinated planning process is fully distributed and decentralized. This allows multiple vehicles to plan trajectories without a severe increase in computational requirements over single-vehicle trajectory planning.

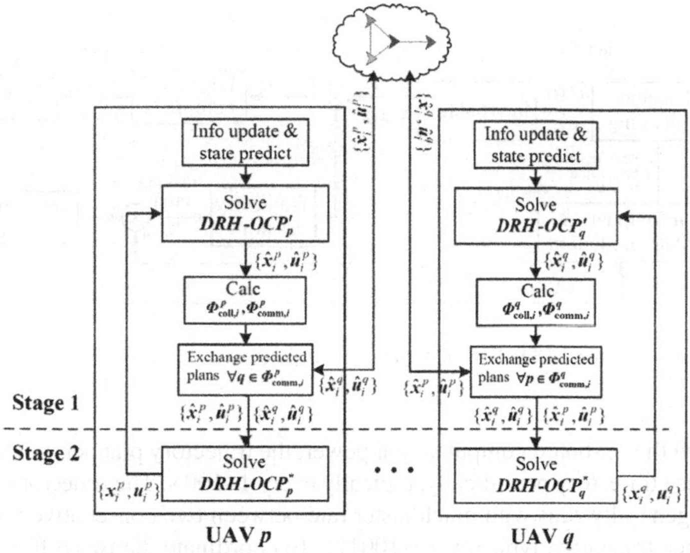

Fig. 1 Flowchart of the decentralized coordination strategy

3.2 Decentralized RH Trajectory Planning Framework

This subsection describes briefly the integration of the decentralized RH trajectory planner with the vehicles low-level control system.

In order to further provide robust performance in the face of these disturbances and uncertainties, a 2-degree-of-freedom (2-DOF) control scheme is used to integrate the trajectory planner with the flight control system (FCS) of vehicle. The similar control scheme has been applied to autonomous operation of a quadrotor UAV in [19]. Figure 2 shows the block diagram of this control system architecture, which provides two feedback control loops (i.e., 2 DOF) to compensate for the uncertainty. The inner loop (shown with pink shaded area in Fig. 2) is used for stabilizing around and tracking the reference trajectory generated by the trajectory planner in outer loop, which takes care of small errors due to external disturbances, parametric uncertainties, unmodeled dynamics, etc., and the outer loop (shown with green shaded area in Fig. 2) adopts the RHC strategy to provide periodical updates of this reference trajectory, which takes responsibility for handling the large discrepancy caused by mission change, pop-up threats, moving target, etc.

This control scheme has the advantage that the slow outer-loop trajectory planning can be separated from the fast inner-loop trajectory tracking and control. This allows the outer loop to operate at the rate of several orders less than that of the inner loop while still maintaining the closed-loop stability. Depending on the

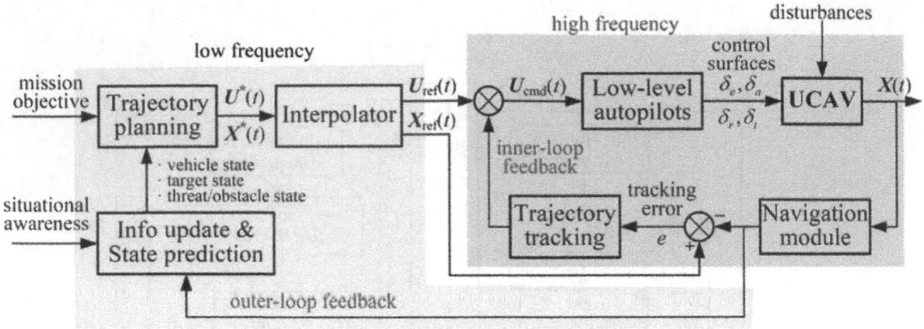

Fig. 2 Schematic of the 2-DOF RHC design

mission and the on-board computational power, the trajectory planner would update the reference trajectory periodically, typically every 1–100 s. The trajectory tracking controller generally runs with much faster rate between two consecutive updates of the reference trajectory, typically 10–100 Hz. To coordinate the two different rates, the interpolator produces samples of the reference trajectory at the desired (high frequency) rate. Furthermore, this 2-DOF control system architecture, by allocating part of the workload to the trajectory tracking, can leave more computation time for the trajectory planning process that takes complex constraints and costs into account.

This paper focuses on the outer-loop cooperative RH trajectory planning. The design and implementation of the inner-loop trajectory tracking is discussed in [27], and will not be covered in this paper.

Figure 3 provides a graphical illustration of the whole running process of the trajectory planning framework for UAV p. As can be seen, the outer-loop RH trajectory planner updates the current reference command (generated in previous cycle, $\mathbf{x}_{\text{ref}}^{i-1}$, $\mathbf{u}_{\text{ref}}^{i-1}$) to be executed by the inner-loop controller at a fixed rate of every T_{EH} seconds. During each cycle, the current environment information and vehicles state are first updated, and then the updated vehicles state x_i is propagated to the start of the next execute cycle, yielding $\hat{\mathbf{x}}_{i+1}$. Subsequently, using the propagated states $\hat{\mathbf{x}}_{i+1}$ as the initial condition, the optimization is performed over a fixed length of time, T_{PH} ($\geq 2T_{EH}$). The optimization computation time is restricted to $T_{EH} - \Delta t_{\text{pre}} - \Delta t_{\text{output}}$ seconds. If the solution fails to converge, the planner will check the feasibility of the part of the current reference command that is beyond t_{i+1} (shown in yellow in Fig. 3), according to the latest environment information. If this part is still feasible, it is chosen as the reference command for next cycle. Otherwise, the planner will generate an emergency safety trajectory (e.g. loiter maneuver [28]) to attempt to keep the vehicle in a safe state. The above process is repeated until the vehicle reaches the goal state (i.e., enters the guided bombs LAR).

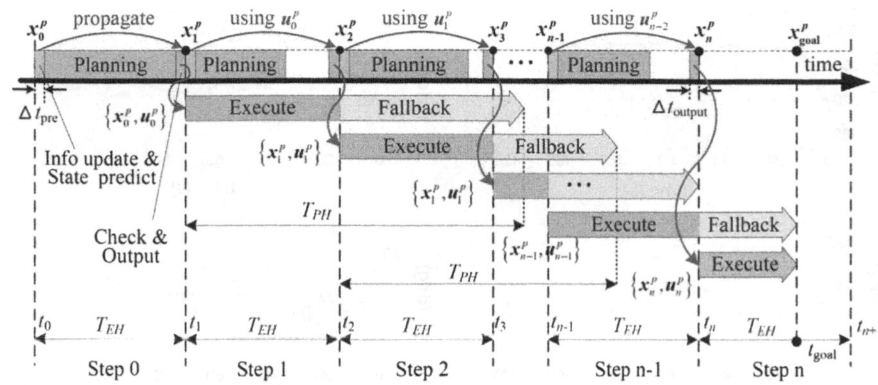

Fig. 3 Illustration of the decentralized RH trajectory planning process

4 Trajectory Planning Based on Inverse Dynamics Optimization Method

To solve the aforementioned DRH-OCP within the required time intervals, an efficient computational direct method, called inverse dynamic in virtual domain (IDVD), is introduced. It was first proposed by Yakimenko [16]. The method is based on a combination of differential flatness theory, polynomial interpolation, space and time parameterization decoupling, and nonlinear programming. Some numerical simulations [20] suggest that its computational speed is more than an order of magnitude faster than that of the pseudospectral methods, at small loss of optimality, and its robustness and convergence properties are also better.

The operational routine that convert the OCP to a NLP problem of low dimension are given below and more detail can be found in [16, 29].

5 Experiments and Results

In this section, we perform a typical numerical flight simulation to illustrate the feasibility and applicability of the proposed method. The experimental test environment is a square of area 40 × 40 km^2, which contains several no-fly zone (NFZ) and one moving obstacle, shown in Fig. 4. The specifications of the UAV are taken from data originally presented in [30]. For the sake of simplicity, we use the most conservative values for the given UAV as fixed constraints to approximate the original time-varying non-linear flight envelope constraints (see Table 1 in [26] for more). All the results presented below are generated using TOMLAB/SNOPT software toolbox on a 2.40GHz Core 2 Duo CPU computer running Windows 7 and MATLAB R2012a. Table 1 summarizes the parameters used in this algorithm.

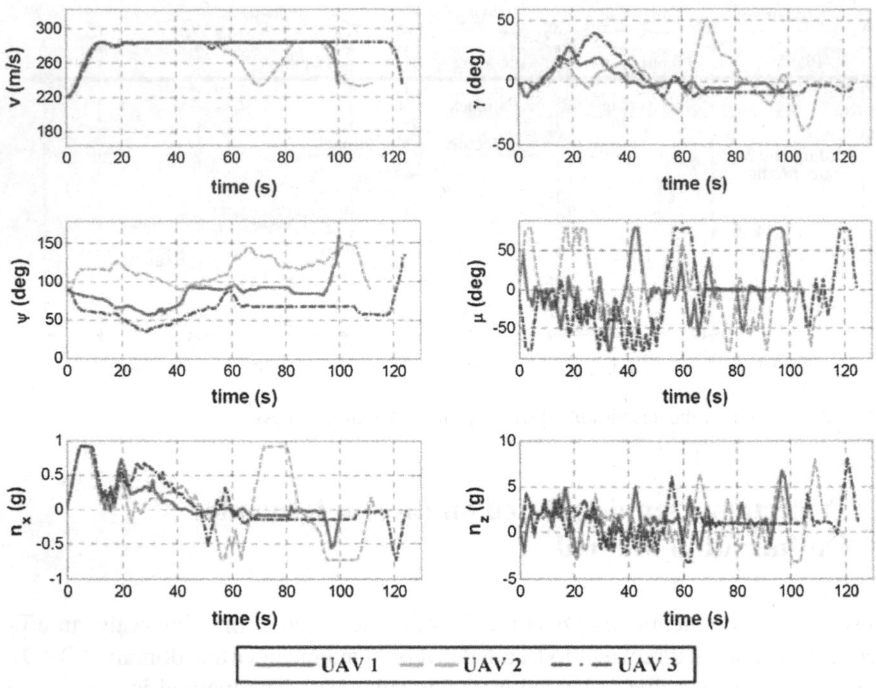

Fig. 4 Three collision-free UAV trajectories in 3D view

Table 1 Algorithm parameters

Parameter	Value	Parameter	Value
$T_{CH}(s)$	2	$(w_{sm}, w_{ru}, w_{ctg})$	(0.2, 0.5, 0.3)
$T_{PH}(s)$	20	$(w_{sm1}), w_{sm2}, w_{sm3}$	(2, 1, 1)
N_S	100	$r_{safe}(m)$	500
b_s	200		

The scenario is that three UAVs begins at initial point (IP), and attempt to reach specified goal locations respectively, while avoiding a series of static obstacles/threats en route. The overall collision-free trajectories of the UAVs and their arrival time, are shown in Fig. 4 And the time histories of the UAVs state variables and control inputs are shown in Fig. 5. Figure 6 shows the distance between each pair of UAVs, where the minimum distance is longer than the minimum safety radius. Obviously, the generated trajectories are feasible and satisfy the given mission objectives and requirements.

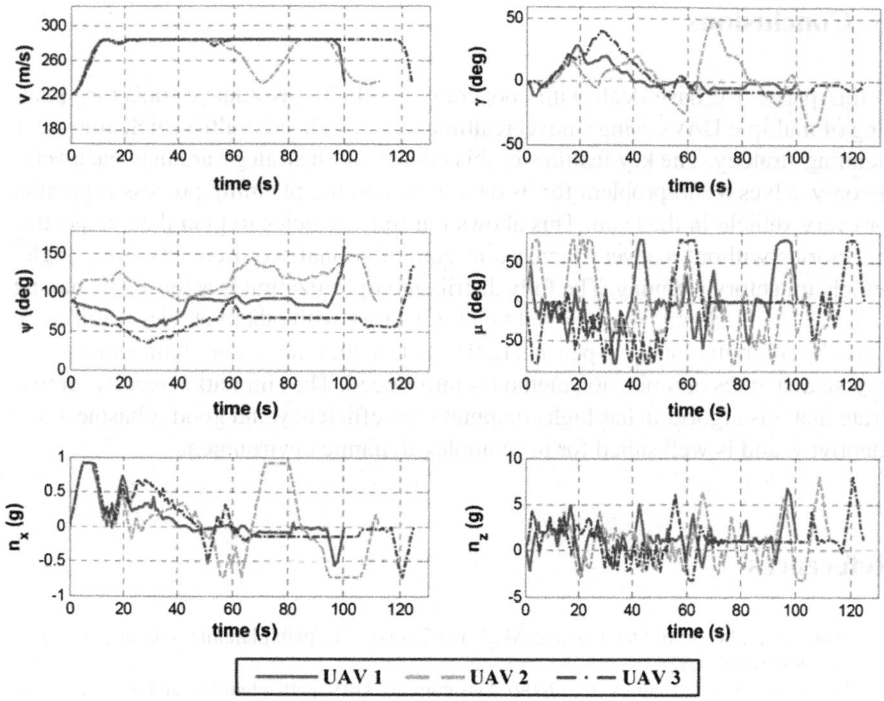

Fig. 5 State and control time histories of the three UAVs

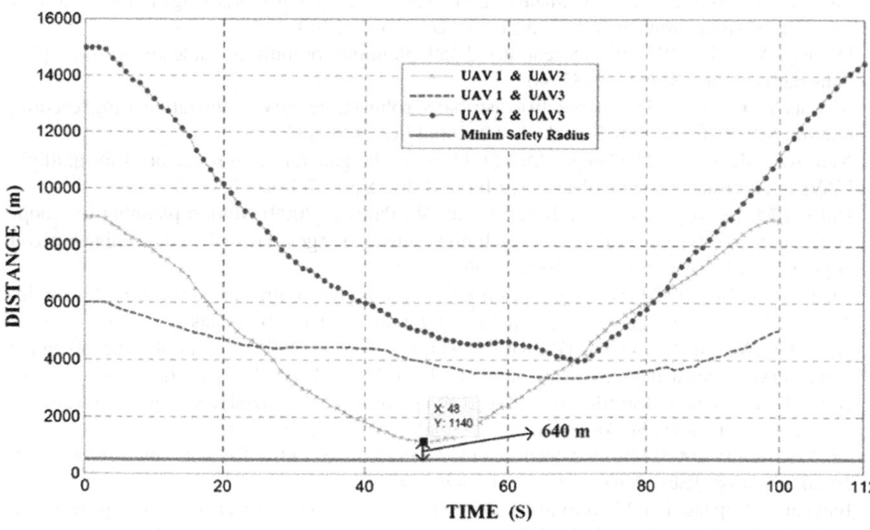

Fig. 6 Distance between each pair UAVs

6 Conclusions

In this paper, we have dealt with cooperative collision-avoidance trajectory planning of multiple UAVs using a novel realtime de-centralized multi-vehicle trajectory planning strategy. The key features of this coordination strategy are that, each vehicle only solves a sub-problem for its own plan, and the planning process is parallel for every vehicle in the team. This allows multiple vehicles to plan their respective trajectories without a severe increase in computational requirements over single-vehicle trajectory planning. The fully distributed optimization is achieved by having each vehicle ex-change its predicted flight trajectory with other vehicles. Besides, to solve the computationally expensive DRH-OCP within the required time intervals, a inverse dynamics optimization method is introduced. The simulation results demonstrate that this algorithm has high computational efficiency and good robustness and adaptivity, and is well suited for the complex dynamic environment.

References

1. Tsourdos A, White B, Shanmugavel M (2011) Cooperative path planning of unmanned aerial vehicles. Wiley, UK
2. Tisdale J, Kim Z, Hedrick JK (2009) Autonomous UAV path planning and estimation: an online path planning framework for cooperative search and localization. IEEE Robot Autom Mag 16:35–42
3. Shaferman V, Shima T (2008) Unmanned aerial vehicles cooperative tracking of moving ground target in urban environments. J Guid Control Dyn 31(5):1360–1371
4. Desaraju VR, How JP (2012) Decentralized path planning for multi-agent teams with complex constraints. Auton Robot 32:385–403
5. Kuwata Y, How JP (2011) Cooperative distributed robust trajectory optimization using receding horizon MILP. IEEE Trans Control Syst Technol 9(2):423–431
6. Neto AA, Macharet DG, Campos MFM (2009) On the generation of trajectories for multiple UAVs in environments with obstacles. J Intell Robot Syst 57(1–4):123–141
7. Defoort M, Kokosy A, Floquet T, Perruquetti W, Palos J (2009) Motion planning for cooperative unicycle-type mobile robots with limited sensing ranges: a distributed receding horizon approach. Robot Auton Syst 57:1094–1106
8. Richards A, How J (2004) Decentralized model predictive control of cooperating UAVs. In: 43rd IEEE conference on decision and control, Paradise Island, Bahamas
9. Beard RW, Mclain TW, Nelson DB, Kingston D, Johanson D (2006) Decentralized cooperative aerial surveillance using fixed-wing miniature UAVs. Proc IEEE 94(7):1306
10. Keviczky T, Vanek B, Borrelli F, Balas GJ (2006) Hybrid decentralized receding horizon control of vehicle formations. In: American control conference, Minnesota, USA
11. Jia D, Krogh B (2002) Min-max feedback model predictive control for distributed control with communication. IEEE Am Control Conf 6:4507–4545
12. Inalhan G, Stipanovic DM, Tomlin CJ (2002) Decentralized optimization, with application to multiple aircraft coordination. In: IEEE conference on decision and control, Las Vegas, NV
13. Dunbary WB, Murray RM (2006) Distributed receding horizon control for multi-vehicle formation stabilization. Automatica 42(4):549–558
14. Keviczky T, Borrelli F, Balas GJ (2006) Decentralized receding horizon control for large scale dynamically decoupled systems. Automatica 42:2105–2115

15. Kuwata Y, Richards A, Schouwenaars T, How JP (2007) Distributed robust receding horizon control for multivehicle guidance. IEEE Trans Control Syst Technol 15(4):627–641
16. Yakimenko OA (2000) Direct method for rapid prototyping of near-optimal aircraft trajectories. J Guid Control Dyn 23(5):865–875
17. Lai C-K, Lone M, Thomas P, Whidborne J, Cooke A (2011) On-board trajectory generation for collision avoidance in unmanned aerial vehicles. In: Proceedings of the ieee aerospace conference
18. Kaminer II, Yakimenko OA, Pascoal AM (2006) Coordinated control of multiple UAVs for time-critical applications. In: Proceedings of the 27th ieee aerospace conference, Big Sky, Montana
19. Cowling ID, Yakimenko OA, Whidborne JF, Cooke AK (2010) Direct method based control system for an autonomous quadrotor. J Intell Robot Syst 60(2):285–316
20. Yakimenko OA, Xu Y, Basset G (2008) Computing short-time aircraft maneuvers using direct methods. In: AIAA guidance, navigation and control conference and exhibit, Honolulu, Hawaii
21. Edison E, Shima T (2011) Integrated task assignment and path optimization for cooperating uninhabited aerial vehicles using genetic algorithms. Comput Oper Res 38(1):340–356
22. Etkin B (1972) Dynamics of atmospheric flight. Wiley, New York
23. Lewis L-PR (2006) Rapid motion planning and autonomous obstacle avoidance for unmanned vehicles. Naval Postgraduate School
24. Hurni MA, Sekhavat P, Karpenko M, Ross IM (2010) A pseudospectral optimal motion planner for autonomous unmanned vehicles. In: American control conference, Baltimore, MD, USA
25. Kuwata Y (2007) Trajectory planning for unmanned vehicles using robust receding horizon control. Massachusetts institute of technology
26. Yu Z, Jing C, Lincheng S (2013) Real-time trajectory planning for UCAV air-to-surface attack using inverse dynamics optimization method and receding horizon control. Chin J Aeronaut 26(4):1038–1056
27. Kaminer I, Yakimenko O, Dobrokhodov V, Pascoal A, Hovakimyan N, Cao C, Young A, Patel V (2007) Coordinated path following for time-critical missions of multiple UAVs via L1 adaptive output feedback controllers. In: AIAA guidance, navigation and control conference, Hilton Head, South Carolina
28. Narayan P, Campbell D, Walker R (2008) Multi-objective UAS flight management in time constrained low altitude local environments. In: 46th AIAA aerospace sciences meeting and exhibit, Reno
29. Drury RG (2010) Trajectory generation for autonomous unmanned aircraft using inverse dynamics. Cranfield University
30. Storm Shadow UCAV Performance the American Institute of Aeronautics and Astronautics, http://www.aerospaceweb.org/design/ucav/main.shtml

Research on the Measurement Method of the Detection Range of Vehicle Reversing Assisting System

Bowei Zou and Xiaochuan Cui

Abstract This paper introduces the measurement method on detection range of reversing assisting system, including reversing radar system and reversing vision system, which using laser distance measurementor and plannimeter. Fit the farthest positions that radar can detect smoothly in combination with the work principle of ultrasonic radar to realize visualization of detection range of reversing radar. A test bench was designed and established to simulate different installation size and location of the reversing assisting system of different types of vehicles. The deviation of test method was verified which can be controlled in a reasonable scope.

Keywords Reversing assisting system · Detection range · Ultrasonic sensor · Reversing vision system

1 Introduction

Since the 21st century, intelligent vehicle technology is maturing progressively; it is an important part of Intelligent Transportation Systems (ITS). The main purpose of intelligent vehicle research and development is to reduce accident rates [1]. According to the domestic and international statistical data, the vehicle in reverse accident occurred frequently. A certain Chinese area traffic control department have done a micro statistics that there was 564 reverse collision accidents occurred within 20 days, almost 30 cases in one day, which not including the accidents solved privately. NHTSA statistics show that there are nearly 292 deaths accidents and 18000 injured accidents caused by vehicle reverse collision in one year. Behind these data is a potential cause, including but not limited to: vehicle performance and the extent of the driver's perception of reversing environment [2].

B. Zou (✉) · X. Cui
China Automotive Technology and Research Center, Tianjin, China
e-mail: zoubowei@catarc.ac.cn

X. Cui
e-mail: cuixiaochuan@catarc.ac.cn

© Springer International Publishing Switzerland 2017
V.E. Balas et al. (eds.), *Information Technology and Intelligent Transportation Systems*, Advances in Intelligent Systems and Computing 454,
DOI 10.1007/978-3-319-38789-5_15

The NHTSA have extended requirements on the range of vehicle reversing radar and reversing image vision on the basis of FMVSS No.111, and accomplished large number of tests. ECE R46 and Japan and South Korea standards allow vehicles to install reverse assisting system, but is not in the scope of testing [3] Chinese standard, GB 15084-2013 motor vehicle indirect vision device performance and installation requirements, also doesn't put forward performance requirements for reversing assisting system. This paper has proposed a measurement method for reversing assisting system in order to make up the merge of relative Chinese standards.

2 Working Principle of Reversing Radar

The most commonly used sensors for reversing radar system are ultrasonic sensor. In addition, there are infrared and electromagnetic induction sensors. However, due to the weakness in anti-interference and limitation in usage, the latter ones are gradually eliminated by the market. The reversing assisting system is composed of ultrasonic sensors, control unit and warning unit. Ultrasonic sensors send signal and then receive the reflected signal when it encounter obstacles. The distance information is send to driver by warning unit after the calculation by control unit. The basic principle of ultrasonic ranging is called transit time method. Ultrasonic sensor signal strong electrical excitation, ultrasonic reflection waves are formed after the object being measured. Then the sensor calculates the distance to the objects detected according to the transmitting and receiving time. The transmitting time is t, and the transmitting velocity is c. Thus, the distance D, between sensor and object is:

$$D = \frac{1}{2} c \cdot t \tag{1}$$

The acoustic energy along the axis of the sensor is the strongest and gradually fade out on both sides along the direction. Ultrasound beam width is defined as the angle where maximum acoustic energy down to the half. The narrower is the ultrasound beam width, the smaller is the detection range [4] Thus, the directivity of ultrasound is more obvious, either is the accuracy of detection range. There is a direct relationship between resonance frequency f and ultrasound beam width, and radiation area R. The relationship is expressed below:

$$\theta = 2arcsin\frac{0.26c}{Rf} \tag{2}$$

The commonly used frequency of ultrasonic sensor in reversing assisting system are 40, 48 and 58 kHZ [5]. The sensor with different frequency can measure different distance. The lower the frequency, the longer of the longitudinal length and the greater the vertical height can sensor measure. The OEMs have to consider the shape and size of vehicles to choose the appropriate reversing radar system to avoid the interference

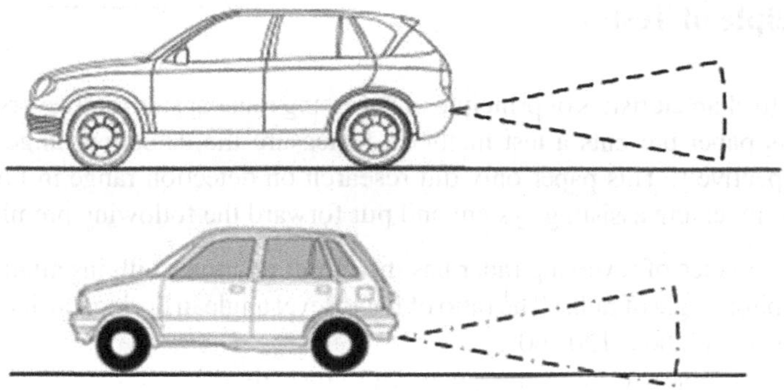

Fig. 1 Interference phenomenon

phenomenon. As it shows in Fig. 1, the SUV and limousine were equipped with the same system. However, the installation height of limousine is too low, which result in a false alarming. Considering that the installation position of passenger car is relatively low and that the detection range of vertical direction is bigger, this paper only focuses on the research on measurement method of detection range in the horizontal direction.

3 Working Principle of Reversing Vision

Comparing to the reversing radar system, reversing vision system provides more intuitive information. Most middle and high ranking passenger cars are equipped with both systems, which can improve the safety when reversing. Reversing vision system is composed of camera and display system. The camera is usually installed on the top of plate, pointing rearward. The display system is integrated to the central control panel. The information of vehicles tail environment is provided to driver directly. The working principle of reversing vision system can be explained by simple pinhole camera model. However, the model is an ideal linear model. Actually, reversing vision system uses exaggerated lens to gain wider field of vision, which brings in image distortion. Image distortion includes radial distortion and tangential distortion. There is no radial distortion in the center of the image plane, which is more and more serious when moving along the radial direction of image. Additionally, tangential distortion is due to the reasons for lens manufacturing flaws, causing that lens itself and the image plane are not parallel. In fact, in addition to the radial distortion and tangential distortion, there are many other distortion that can be neglected because of having no obvious influence [6].

4 Principle of Test

According to characteristics of principle of reversing radar system and reverse vision system, this paper presents a test method to measure the detection range of both system respectively. This paper only did research on detection range in horizontal direction of reversing assisting system and put forward the following premises:

- The performance of reversing radar has the closet relations with installation position and pitch angle of radar. The ratio of basic level angle in horizontal and vertical direction is 2:1, about 120°:60°.
- The detected object used in this paper can cover the vertical detection point of view. Thus, this paper did not carry out on vertical detection range measurement.
- This paper discusses the rear area of vehicle that cannot be observed from the rearview mirror, which is often located in the center of image, where with not serious distortion. Thus, the influence of lens distortion can be neglected.

4.1 Measurement Method on Detection Range of Reversing Radar System

Make a car equipped with four-radar reversing assisting system as an example. Measure the installation height and angle of four radars before the test. In accordance with the premises, measure the detection range of each reversing radar.

The specified test method as follow:

(a) Horizontal detection distance measurement in radial direction. Power on only one radar sensor. Place the calibration object outside the area where radar can detect objects, and make sure that calibration object is in right front of radar and the radar emitting surface is parallel with the calibration object plane. Move the calibration object towards radar slowly until the warning system alarm constant audio cues. Then, stop moving the calibration object, and mark the position where it stops as point A. The distance from sensor to point A is defined as the radial detection distance (Fig. 2).

(b) Horizontal lateral direction detection range measurement. Set sensor surface as center, choose one direction every (counter-clockwise in Fig. 3) five degree at a time, such as the direction 1, direction 2 and direction 3 direction X shown in Fig. 5. Use the same method mentioned above to measure detection distance in these directions and mark the positions (B, C and D in Fig. 3) respectively until the warning system no longer alarm. Similarly, the other side detection range can be measured with same method.

(c) Connect all the points marked in each direction. The area surrounded by points and sensor is detection range of single radar.

Repeat the measurement on other radars, then can acquire all radar detection ranges. Take the collection of all radar ranges, namely the entire vehicle detection range radar system.

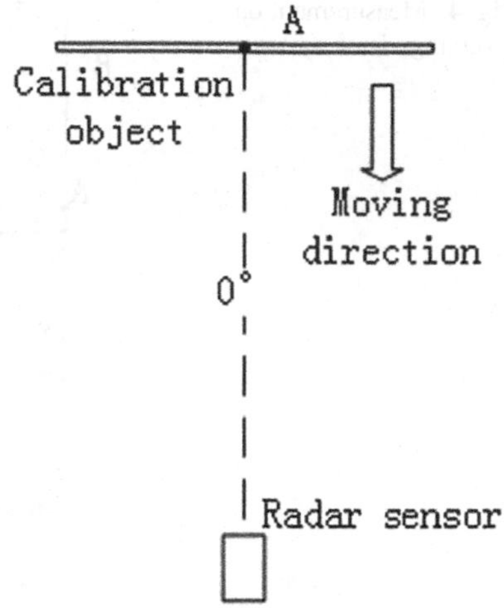

Fig. 2 Horizontal detection distance measurement in radial direction

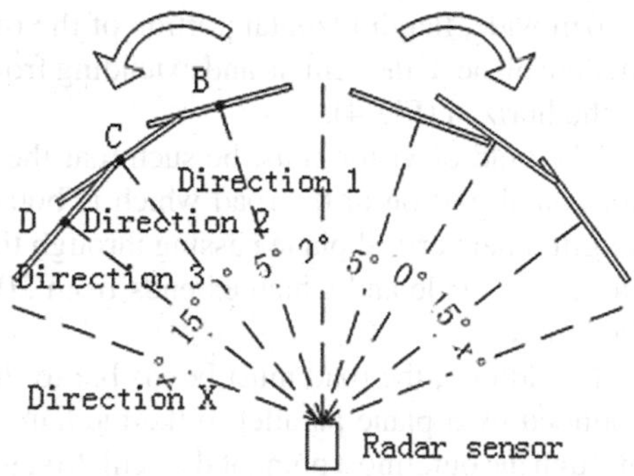

Fig. 3 Horizontal lateral direction detection range measurement

4.2 *Measurement Method on Detection Range of Reversing Vision System*

The area directly behind the vehicle is significantly considered when measuring the detection range of reversing vision system. The steps of measurement are explained as following:

(a) Determine the blind area directly behind vehicle according to the width of vehicle body, such as the shaded area shown in Fig. 8.

(b) In the reversing image, determine four points, A, B, C and D, which are boundary points of blind area that could be seen from reversing image.

(c) The area surrounded by four connected points is the reversing vision system can detected in the blind zone directly behind vehicle.

Chinese mandatory standard GB 15084-2013 Motor vehicles- Devices for indirect vision- Requirements of performance and installation puts forward regulations on interior view mirrors and main exterior view mirrors for passenger cars. The number of installation, mounting strength and field of view are specified in the standard.

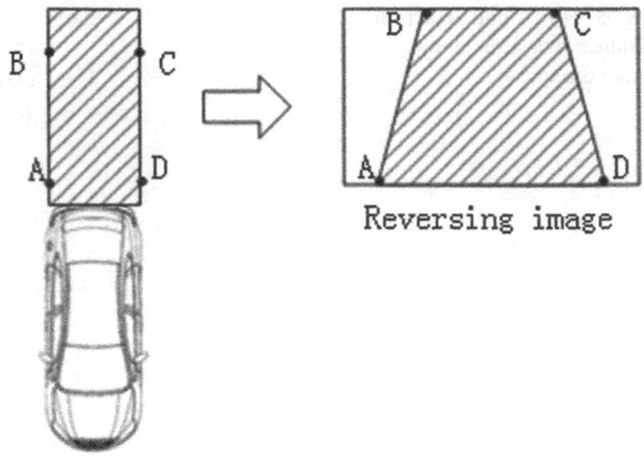

Fig. 4 Measurement on reversing vision system

In the standard, The field of vision must be such that the driver can see at least a 20 m wide, flat, horizontal portion of the road centred on the vertical longitudinal median plane of the vehicle and extending from 60 m behind the driver's ocular points to the horizon (Fig. 4).

The field of vision must be such that the driver can see at least a 4 m wide flat, horizontal portion of the road which is bounded by a plane parallel to the median longitudinal vertical plane passing through the outermost point of the vehicle on the passenger's side and which extends from 20 m behind the driver's ocular points to the horizon.

In addition, the road must be visible to the driver over a width of 1 m, which is bounded by a plane parallel to the median longitudinal vertical plane and passing through the outermost point of the vehicle starting from a point 4 m behind the vertical plane passing through the driver's ocular points [7].

As can be seen from this standard, the requirements for the field of vision behind the vehicle areas specified area only after the driver's eye point 60 000 mm extending to the ground, while the driver's eye point to 60000 mm area is not made demands. Define the shortest distance from vehicle tail to ground where driver can see from interior mirror as $S1$. D is the width of vehicle. The blind area directly behind vehicle that is not equipped with reversing assisting system S_0 is:

$$S_0 = S1 \times D \tag{3}$$

Define S_D is the area reversing assisting system can detect. The blind area directly behind vehicle that is equipped with reversing assisting system S

$$S = S_0 - S_D \tag{4}$$

The most vulnerable group in revers collision accident is children according to the accidents statistics. This paper choose three different height of calibration object, including 50, 70 and 100 cm, to represent different posture and height of children.

Measure the shortest distance from calibration object to vehicle tail when driver can see the object through interior mirror in the situation of different height of calibration object. The distance stands for the maximum length of blind zone in each situation. Finally, the proportion of detection range and blind zone can be measured and calculated.

5 Verification Test

5.1 Test Bench Design

In this paper, test method was validated based on the test bench (see in Fig. 5). The test bench is constructed by aluminum structure and can simulate any dimension of vehicle tails equipped with different kinds of reversing assisting system. The test bench is composed of bench skeleton, brackets of sensor and laser emission system. The most significant is that the bracket of sensor can adjust the pitch angle and yaw angle of sensor (radar or camera) accurately.

The main role of bench skeleton is to ensure the sensor can be adjusted in the direction of the three translational directions. For instance, in the vehicle coordinate system, the height from reversing sensors to ground and the distance between sensors in x and y directions can be altered. The function of bracket is to fix radar and camera, and to adjust the pitch angle and yaw angle of sensors (see in Fig. 6a). The horizontal and vertical compasses are used to indicate the installation pitch and yaw angle. The laser emission system has two functions, the first one is to display detection angle range of ultrasonic sensor, and the second one is to be the baseline which has different kinds of angle with normal line of sensor surface. The green lines in the Fig. 6b is the boundary of detection angle range of four ultrasonic radars.

Fig. 5 Test bench

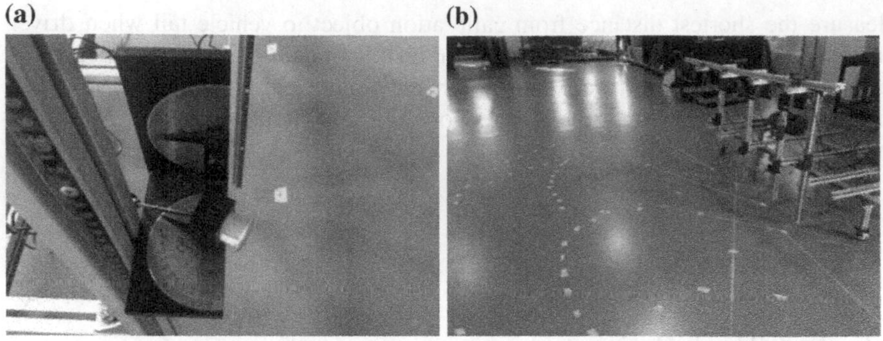

Fig. 6 Sensor bracket (a)and laser emission system (b)

5.2 Test on Reversing Radar System

This paper selected two passenger cars equipped with reversing radar system and reversing vision system as prototype. One is a midsize sedan, another is a SUV. Two different kinds of reversing assisting system products were selected for verification test. The two products both have reversing radar and reversing vision function. At the beginning of the test, the pressure of tire should be Tire pressure should be adjusted to the recommended cold tire pressure [8]. Then, accurately measure the installation dimension on prototype vehicle by using coordinate instrument. In accordance with the measurement data from coordinate instrument, adjust test bench to make the installation of sensors to be same with those on prototype vehicle.(see in Fig. 9a) The deviation can be controlled within 5 mm. Following the experimental method described earlier, the detection range points of each radar were marked on the ground (see in Fig. 6b). Connect the points of each radar on ground and add the boundary of vehicle width (see in Fig. 7). Then use planimeter to calculate the area of shadow part where radar system cannot detect. The results of tests are shown in Tables 1 and 2. This paper selected 3-D machine to detect the shortest distance that drive can observe from interior mirror. Using the principle of light reflection, in a position to determine whether the driver can observe the position of the driver by observing whether the light bulbs can be seen from interior mirror.(See in Fig. 9) The test results of vision minimum distance under the situation of three different calibration objects are indicated in Table 3. According to the minimum vision distance, Table 4 shows the proportion of blind area of different heights of calibration object (Fig. 8).

At last, the test method of the radar detection range is verified, calculating the deviation of fitted boundary. Place the calibration objection the fitting boundary, if the warning system alarms, move the object outwards until alarm stop, recording the distance to the reversing radar. The difference between the distance from boundary to radar and the real distance from alarming position to radar is the deviation of the method. This paper chose several points on the boundary randomly to verify the deviation and the results were indicated in the Fig. 9. There is indication that the deviation of measurement can be controlled in 2 %.

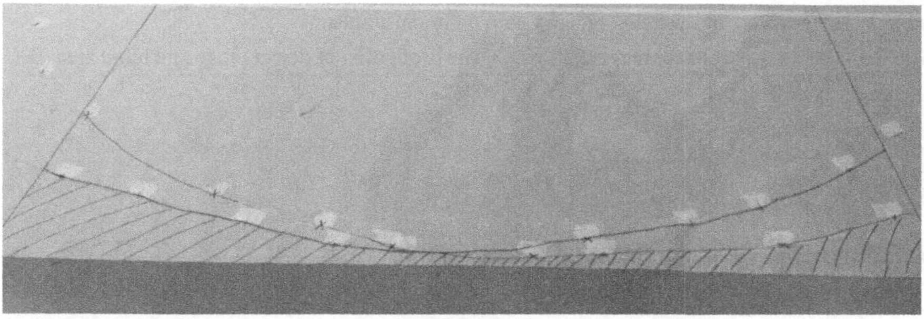

Fig. 7 The part of blind area radar cannot detect

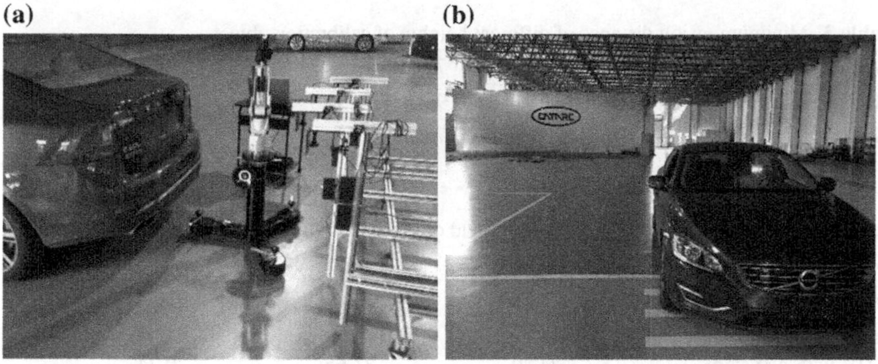

Fig. 8 Instrument used in test procedure. **a** Coordinate instrument. **b** D machine

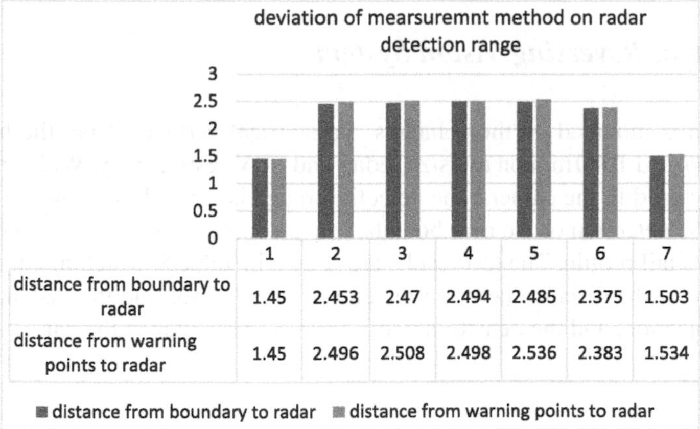

Fig. 9 Deviation of measurement method on radar detection range

Table 1 Performance of reversing radar system on midsize sedan

Midsize sedan	Detect range unit: m^2	The proportion of detect range and blind area (%)
Ultrasonic sensor A	4.46	11.33
Ultrasonic sensor B	4.43	11.26

Table 2 Performance of reversing radar system on SUV

SUV	Detect range unit: m^2	The proportion of detect range and blind area (%)
Ultrasonic sensor A	4.75	15.62
Ultrasonic sensor B	4.71	15.49

Table 3 Minimum vision distance of different heights of calibration objects

Vehicle type	50 cm	70 cm	100 cm
Midsize sedan	13.3 m	9.3 m	2.47 m
SUV	9.7 m	7.6 m	3.4 m

Table 4 Proportion of blind area and vision field of different heights of calibration

Vehicle type	50 cm	70 cm	100 cm
Midsize sedan with A system	18.28	26.20	98.67
Midsize sedan with B system	18.27	26.03	97.34
SUV with A system	24.79	31.65	70.74
SUV with B system	24.58	31.38	70.14

5.3 *Test on Reversing Vision System*

The camera is mounted on the vehicle's symmetrical vertical plane, the height of which is 947 and 1080 mm on midsize sedan and SUV respectively. Wide angle lens is generally used in the camera, the detection range is far and the lens distortion is large, but the detection of the near boundary is not coincident with the projection of the vehicle's tail profile. The test results are shown in Table 5, which mainly reflects the distance of the distance between the end of the detector and the distance to the surface projection, and the detection width nearest to the rear of the car.

6 Conclusion

In this paper, a method for measuring the range of the assisting system is proposed, which is based on the reverse radar and reverse image system. In order to verify

Table 5 The performance of reversing vision system on different vehicles

Vehicle type	Midsize sedan with A system	Midsize sedan with A system	SUV with A system	SUV with B system
Camera mounting height	947 mm	947 mm	1080 mm	1080 mm
Distance between rear of vehicle to nearest vision detection boundary	310 mm	110 mm	225 mm	0 mm
Vision width nearest to the rear of the car	1690 mm	2450 mm	2045 mm	3130 mm

the measurement method, the test carried out is based on the test bench, which can simulate installation size and location of the reversing assisting system of different types of vehicles. This paper used two types of vehicle, midsize sedan and SUV, for the tests. During the test, some conclusions can be concluded as follow:

- For the vehicle equipped with four-ultrasonic radar system, the detection range of the two sides of radar is less than those in the middle.
- For the vehicle equipped with reversing radar system, the blind area of vehicle is related to the distance of radar detection, but little to do with the detection angle range of radar.
- For the vehicle equipped with reversing vision system, the blind area of vehicle is related to the mounting height and mounting pitch angle.

In the future, a measurement method based on vehicle should be put forward. To improve test efficiency, a more simplified measurement method should be developed due to the long test cycle, although the current test method can meet high accuracy.

References

1. Hattori H (2000) Stereo for 2D visual navigation. Proc IEEE Intell Veh Symp, pp 31–38
2. Department of transportation, National Highway Traffic Safety Administration, Federal Motor Vehicle Safety Standard; Rear Visibility; Final Rule (2014) 79(66), 19180–19181
3. Department of transportation, National Highway Traffic Safety Administration. Low-speed Vehicles Phase-in Reporting Requirements, Federal Motor Vehicle Safety Standard (2010) pp 36–37
4. Zhang H, Gao Y (2011) The research of ultrasonic ranging technique, instrument technology, 2011–09, pp 58–60
5. Feng Y, Wang Y (2011) Working principle of reversing radar. Operation and Maintenance, pp 18–20
6. Ding H, Zou B (2013) Comparison of several lane marking line recognition methods. In: IEEE intelligent control and information processing, pp 53–58

7. GB 15084-2013 Motor vehicles-devices for indirect vision-requirements of performance and installation (2013)
8. GB 1589-2004 Limits of dimensions, axle load and masses for road vehicles (2004)

Studies on Optimized Algorithm for SINS Under High Dynamic

Zhao-Fei Zhang, Jian-Jun Luo and Bai-Chun Gong

Abstract As to improve the attitude solving accuracy of strapdown inertial navigation system (SINS) boarded on the spacecraft for high dynamic application, a novel optimized coning error compensation algorithm is developed. Based on the rotation vector concept, a general scheme of N measurement samples coning error compensation by utilizing the last P measurement samples is proposed. Then, an optimized compensation algorithm is given in fixed frequency environment. And the key coefficients of this algorithm can be acquired by a simple matrix calculus rather than a complicated derivation. Finally, the algorithm is verified and tested by physical experiment. The results show that the coning error compensation accuracy does can be improved.

Keywords SINS · Rotation vector · Coning error · Fixed frequency

1 Introduction

Attitude solution is playing a key role in SINS because the computed attitude is continuously used to transform spacecrafts acceleration measured by accelerometer from the body frame to the inertial frame. The noncommutativity error of finite rotations is one of the major error sources of numerical solutions of the attitude equations. It is also inevitable to update the attitude incorporated into digital data processing. In general, the noncommutativity error can be reduced by increasing the number of computation updates with an efficient algorithm. This requires a higher-speed airborne computer. However, it is not a good trade-off to use a much better computer and even it can not be achieved sometimes. Thus designing computational efficient algorithms has become an attractive research topic [1–4].

Z.-F. Zhang (✉) · J.-J. Luo · B.-C. Gong
National Key Laboratory of Aerospace Flight Dynamics, Northwestern Polytechnical University, Xian 710072, China
e-mail: 348830139@qq.com

© Springer International Publishing Switzerland 2017

V.E. Balas et al. (eds.), *Information Technology and Intelligent Transportation Systems*, Advances in Intelligent Systems and Computing 454,
DOI 10.1007/978-3-319-38789-5_16

Commonly used attitude updating algorithms for strapdown systems are the Euler method, the direction cosine method, the quaternion method and rotation vector method. Among them, the rotation vector method is quite popular due to its advantages of nonsingularity, simplicity, and computational efficiency. The first detailed optimization algorithms for the coning integral was proposed by Miller [1]. The 4 samples compensation algorithm was introduced by Lee [2], and the algorithms utilizing accumulated gyro increments between the last and the current update time was proposed. The N measurement samples compensation scheme utilizing M accumulated gyro increments was introduced by Z.F. Zhang [5]. This approach can improve the compensation accuracy without increasing calculated amount.

And the objective of this paper is to further the development of the scheme introduced by Z.F. Zhang, obtain the general formulation of the algorithm and verify the algorithm by physical experiments.

2 Coning Motion and Rotation Vector

When the rotating time of rigid body is finite, although the accumulated incremental angle is pretty small, it cannot be ignored. Because the angle position of rigid body is related with the order of rotation, the finite rotations will introduce noncommutativity error. It has been proven that the quaternion updating method with the rotation vector can effectively reduce the noncommutativity error. The rotation vector differential equation for small Φ can be approximated as [6–9].

$$\dot{\Phi} = \omega + \frac{1}{2}\Phi \times \omega + \frac{1}{12}\Phi \times (\Phi \times \omega) \tag{1}$$

Without loss of the generality, the classical coning motion can be described by the rotation vector:

$$\Phi = \begin{bmatrix} 0 \\ \phi\cos(\omega t) \\ \phi\sin(\omega t) \end{bmatrix} \tag{2}$$

$$\omega = \begin{bmatrix} -2\omega\sin^2(\frac{\phi}{2}) \\ -\omega\sin(\phi)\sin(\omega t) \\ \omega\sin(\phi)\cos(\omega t) \end{bmatrix} \tag{3}$$

where ϕ represents the coning half-apex angle and ω represents the coning frequency. Then the angular velocity ω describing the coning motion can be expressed by Eq. (3).

3 The N Samples Algorithm Utilizing the Last P Samples

Generally speaking, the noncommutativity error can be reduced by increasing the sample frequency of the gyroscope. However, there is an up limit of the sample frequency. Thus designing a computational efficient algorithm has become an attractive research topic. Based on the N samples algorithm by utilizing M last accumulated gyro increments [2–5], a modified algorithm is proposed: by utilizing the last samples of gyroscope, the estimated rotation vector can be achieved by increasing the number of gyroscope samples without increasing the sample frequency. Then the N samples algorithm by utilizing the last P samples is provided.

$$\Delta\hat{\Phi} = \Delta\theta + \sum_{i=1}^{N-1} K_i\left(\Delta\theta_i \times \Delta\theta_N\right) + \sum_{j=1-P}^{0}\sum_{k>j}^{N} K_{j,k}\left(\Delta\theta_j \times \Delta\theta_k\right) P >= 1 \qquad (4)$$

where $\Delta\hat{\Phi}$ represents the compensation amount of the rotation vector, $\Delta\theta_i$ and $\Delta\theta_k$ represents the ith and the kth gyroscope sample in the attitude update period. $\Delta\theta_N$ represents the Nth gyroscope sample in the attitude update period. As to N samples algorithm, $\Delta\theta = \sum_{i=1}^{N} \Delta\theta_i$ is the accumulated incremental gyroscope sample in attitude update period. In the Fig. 1, $\Delta\theta_j$ represents the last jth gyroscope measurement, while $j = 0$ denotes the last 1st of the gyroscope measurement, $j = -1, -2, \ldots$ stand for the 2nd, 3rd of the previous gyroscope sample, K_i and K_j is the weighting coefficient of the compensation. Then the problem reduces to determine the coefficients, K_i and K_j which are dependent on the classical coning motion and used to minimize the magnitude of estimation error contained in $\Delta\hat{\Phi}$.

$$\Delta\theta_j = \int_{t+(j-1)*h/N}^{t+j*h/N} \omega(\tau)d\tau \mathrm{j} = 1-P, 2-P, \ldots, -1, 0$$

$$= \begin{bmatrix} -\frac{2}{N}(\omega h)\sin^2\left(\frac{\phi}{2}\right) \\ -2\sin(\phi)\sin\left(\frac{\omega h}{2N}\right)\sin\left[\omega\left(t + \frac{2j-1}{2N}h\right)\right] \\ 2\sin(\phi)\sin\left(\frac{\omega h}{2N}\right)\cos\left[\omega\left(t + \frac{2j-1}{2N}h\right)\right] \end{bmatrix} \qquad (5)$$

where h represents the period of attitude update. $\Delta\theta_j$ and $\Delta\theta_i$ have the same format with Eq. (5). Therefore, Eq. (4) can be easily simplified as

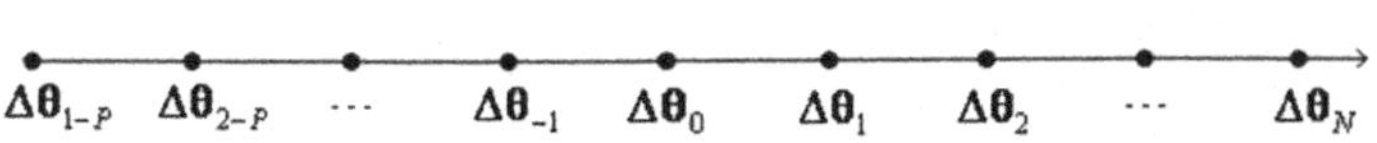

Fig. 1 The N samples and P samples from previous

$$\Delta\hat{\Phi} = \Delta\theta + \sum_{i=1}^{N-1} K_i\left(\Delta\theta_i \times \Delta\theta_N\right) + \sum_{j=1-p}^{0} K_j\left(\Delta\theta_j \times \Delta\theta_N\right)$$
$$= \Delta\theta + \sum_{i=1-p}^{N-1} K_i\left(\Delta\theta_i \times \Delta\theta_N\right) P >= 1 \tag{6}$$

The error criterion is naturally defined as the difference between the true updating rotation vector $\Delta\Phi$ and its estimation value $\Delta\hat{\Phi}$. However, as shown in Eq. (3), there is solely a non-zero average error along the coning axis. Thus, the criterion can be reduced to the non-zero average component of the difference between these two vectors [3].

$$\phi_\varepsilon = \Delta\phi_1 - \Delta\hat{\phi}_1 \tag{7}$$

Because ϕ is pretty small, we can have the approximation for the sine and cosine value of ϕ, $\sin(\phi) \approx \phi$, $\sin(\phi/2) \approx \phi/2$. $[\Delta\theta^m]_1$ and $\left[\sum_{i=1-P}^{N-1} K_i\left(\Delta\theta_i^m \times \Delta\theta_N^m\right)\right]_1$, that the only non periodic component, denote the first dimension of $\Delta\theta^m$ and $\sum_{i=1-P}^{N-1} K_i\left(\Delta\theta_i^m \times \Delta\theta_N^m\right)$. Let $\lambda = \omega h/N$, then

$$\begin{aligned}
\phi_\varepsilon &= \Delta\phi_1 - \Delta\hat{\phi}_1 \\
&= -\frac{\phi^2}{2}\sin(N\lambda) - [\Delta\theta]_1 - \left[\sum_{i=1-P}^{N-1} K_i\left(\Delta\theta_i \times \Delta\theta_N\right)\right]_1 \\
&= -\frac{\phi^2}{2}\sin(N\lambda) + \frac{\phi^2}{2}N\lambda - \frac{\phi^2}{2}\left[4K_{1-P}(1-\cos\lambda)\sin((N-1+P)\lambda)\right. \\
&\quad \left. + 4K_{2-P}(1-\cos\lambda)\sin((N-2+P)\lambda) + \cdots + 4K_{N-1}(1-\cos\lambda)\sin\lambda\right] \\
&= \frac{\phi^2}{2}\left[-\sin(N\lambda) + N\lambda\right. \\
&\quad - K_{1-P}(4\sin((N-1+P)\lambda) - 2\sin((N+P)\lambda) - 2\sin((N-2+P)\lambda)) \\
&\quad - K_{2-P}(4\sin((N-2+P)\lambda) - 2\sin((N-1+P)\lambda) \\
&\quad -2\sin((N-3+P)\lambda)) - \cdots \\
&\quad \left. - K_{N-1}(4\sin(\lambda) - 2\sin(2\lambda))\right]
\end{aligned} \tag{8}$$

By introducing the Taylor series expansion to the sine terms, Eq. (8) can be approximated as the following

$$\phi_\varepsilon = \frac{\phi^2}{2}\left[-\frac{(N\lambda)^3}{3!}\ \frac{(N\lambda)^5}{5!}\ -\frac{(N\lambda)^7}{7!}\cdots\right]\left(F\cdot\begin{bmatrix}K_{1-P}\\K_{2-P}\\\vdots\\K_0\\K_1\\K_2\\\vdots\\K_{N-1}\end{bmatrix}-\begin{bmatrix}1\\1\\\vdots\\1\\1\\1\\\vdots\\1\end{bmatrix}\right) \tag{9}$$

where

$$F=\begin{bmatrix}A_{1,1-P} & A_{1,2-P} & \cdots & A_{1,0} & A_{1,1} & A_{1,2} & \cdots & A_{1,N-1}\\ A_{2,1-P} & A_{2,2-P} & \cdots & A_{2,0} & A_{2,1} & A_{2,2} & \cdots & A_{2,N-1}\\ \vdots & \vdots & \cdots & \vdots & \vdots & \vdots & \cdots & \vdots\end{bmatrix}$$

$$A_{i,j}=\frac{-4(N-j)^{(2i+1)}+2(N-j+1)^{(2i+1)}+2(N-j-1)^{(2i+1)}}{(N)^{(2i+1)}}$$

As λ is much smaller than 1, then in order to get a smallest estimation error ϕ_ε, the small power terms should be 0 as close as possible. It is equivalent to let the right side of the previous $N+P-1$th of Eq. (9) equal to zeros, let F_{N+P-1} denote the previous $N+P-1$th of F. Then

$$F_{N+P-1}\cdot\left[K_{1-P}\ K_{2-P}\ \cdots\ K_0\ K_1\ \cdots\ K_{N-1}\right]^{\mathrm{T}}=\left[1\ 1\ 1\ \cdots\ 1\right]^{\mathrm{T}} \tag{10}$$

Once N and P are selected, the corresponding optimal coning compensation algorithm can be designed as shown in Eq. (11). The typical coefficients of the N samples algorithm utilizing the last P samples were shown in Table 1.

$$\left[K_{1-P}\ K_{2-P}\ \cdots\ K_0\ K_1\ \cdots\ K_{N-1}\right]^{\mathrm{T}}=F_{N+P-1}^{-1}\cdot\left[1\ 1\ 1\ \cdots\ 1\right]^{\mathrm{T}} \tag{11}$$

Table 1 The typical coefficients of the N samples algorithm utilizing the last P samples

N	P	K_{-1}	K_0	K_1	K_2	Algorithm error (ϕ^2)
	P = 0					$1/12*(\omega h^3$
N = 1	P = 1		1/12			$1/60*(\omega h)^5$
	P = 2	−1/60	7/60			$1/280*(\omega h)^7$
	P = 0			2/3		$1/960*(\omega h)^5$
N = 2	P = 1		−1/30	11/15		$1/17920*(\omega h)^7$
	P = 2	1/140	−13/210	323/420		$1/17920*(\omega h)^9$
	P = 0			9/20	27/20	$1/204120*(\omega h)^7$
N = 3	P = 1		3/280	57/140	393/280	$1/8266860*(\omega h)^9$
	P = 2	−1/420	1/40	157/420	1207/840	$1/327367970*(\omega h)^{11}$

4 The Optimized Algorithm in Fixed Frequency

The accuracy of classical coning algorithms monotonically decreases with the increment of coning frequency. In practical application, the motions of SINS are in a certain fixed frequency range, and these motions mainly generate the system coning errors. Based on the classical coning algorithms formula, the design criterion of classic algorithms is modified and two new algorithm optimization principles are introduced to match the fixed frequency motion. The proposed optimal coning algorithm can get the minimized point in the specified frequency band and be no longer monotonic. So the system attitude accuracy can be improved significantly. The new algorithm neither change the formation of classic algorithms, nor complicate the algorithms realization [10].

According to Eq. (8), two new algorithms optimization principles are put forward based on the classical coning algorithms formula.

$$\phi_\varepsilon(\lambda_0) = 0 \tag{12}$$

$$\phi_\varepsilon(\lambda_0) = 0 \tag{13}$$

$$\lambda_0 = \omega_0 h / N \tag{14}$$

where ϕ_ε represents coning error which has been compensated, ϕ'_ε represents the differential of ϕ_ε, and ω_0 stands for the fixed frequency. These two formulas can be expressed as the following.

$$\phi_\varepsilon(\lambda) = \frac{\phi^2}{2}\left[-\sin(N\lambda) + N\lambda - \sum_{i=1-P}^{N-1} 4K_i(1-\cos\lambda)\sin(N-i)\lambda\right] \tag{15}$$

$$\begin{aligned}\phi'_\varepsilon(\lambda) = \frac{\phi^2}{2}\{&-N\cos(N\lambda) + N \\ &- \sum_{i=1-P}^{N-1} 4K_i[\sin\lambda\sin(N-i)\lambda + (1-\cos\lambda)(N-i)\cos(N-i)\lambda]\}\end{aligned} \tag{16}$$

Equation (12) represents the coning error equal zeros when coning frequency $\omega = \omega_0$. Equation (13) represents that the coning error is minimum in coningfrequency ω_0, i.e., the coning error will as small as possible neighborhood of ω_0. Add Eqs. (12) and (13) to in regular sequence, then get rid of the last two equation of the N samples by utilizing P samples from the previous algorithm. The coefficients of the optimized algorithm in fixed frequency can be calculated as follows.

$$\left[K_{1-P}\ K_{2-P}\ \cdots\ K_0\ K_1\ \cdots\ K_{N-1}\right]^{\mathrm{T}} = F_{N+P-1}^{-1} \cdot \left[1\ 1\ 1\ \cdots\ 1\right]^{\mathrm{T}} \tag{17}$$

$$F_{N+P-1} = \begin{bmatrix} C_{1,1-P} & C_{1,2-P} & \cdots & C_{1,0} & C_{1,1} & C_{1,2} & \cdots & C_{1,N-1} \\ D_{1,1-P} & D_{1,2-P} & \cdots & D_{1,0} & D_{1,1} & D_{1,2} & \cdots & D_{1,N-1} \\ A_{1,1-P} & A_{1,2-P} & \cdots & A_{1,0} & A_{1,1} & A_{1,2} & \cdots & A_{1,N-1} \\ A_{2,2-P} & A_{2,2-P} & \cdots & A_{2,0} & A_{2,1} & A_{2,2} & \cdots & A_{2,N-1} \\ \vdots & \vdots & \cdots & \vdots & \vdots & \vdots & \cdots & \vdots \\ A_{N+P-3,1-P} & A_{N+P-3,2-P} & \cdots & A_{N+P-3,0} & A_{N+P-3,1} & A_{N+P-3,2} & \cdots & A_{N+P-3,N-1} \end{bmatrix}$$

$$C_{1,j} = \frac{4\left(1-\cos(\lambda)\right)\sin\left[(N-j)\lambda\right]}{N\lambda - \sin(N\lambda)}$$

$$D_{1,j} \frac{4\sin(\lambda)\sin\left[(N-j)\lambda\right] + \left(1-\cos(\lambda)\right)(N-i)\cos\left[(N-j)\lambda\right]}{N\lambda - \sin(N\lambda)}$$

$$A_{ij} = \frac{-4(N-j)^{(2i+1)} + 2(N-j-1)^{(2i+1)} + 2(N-j-1)^{(2i+1)}}{N^{(2i+1)}}$$

5 Experiment Analyze

The performance of the specific force transformation algorithm was tested by the classical coning motion. A certain type of laser SINS was fixed in double swing installation. X–Y–Z axes are in the direction of north-up-east. First LSINS was stillness 400 s, then X axis and Y axis was added sine wave 300 s, where $\phi = 6°$, $f = 1\,\text{Hz}$,

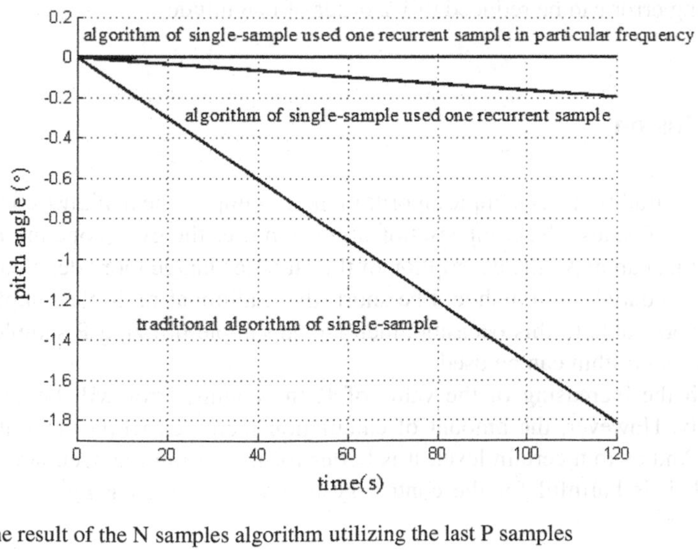

Fig. 2 The result of the N samples algorithm utilizing the last P samples

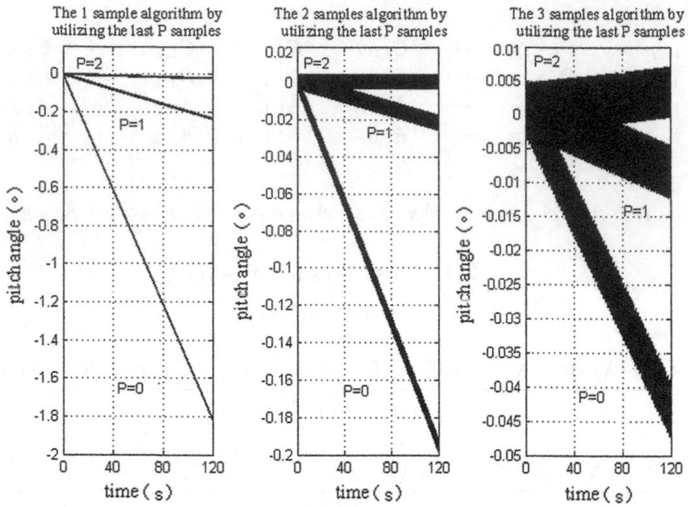

Fig. 3 Compensation results achieved by the three different algorithms

gyroscope sample time is $T_0 = 5\,\text{ms}$ and attitude update time $h = 30\,\text{ms}$. Lastly, double swing installation was taken to origin.

The result of the N samples algorithm utilizing the last P samples were shown in Fig. 2. The effect of the N samples algorithm utilizing the last P samples was better than the traditional algorithm. The traditional algorithm of single sample, the single sample algorithm utilizing the last one sample and the optimized algorithm in a fixed frequency were shown in Fig. 3 respectively. As shown in Fig. 3, the optimized algorithm in a fixed frequency is the best algorithm of all, by using which the magnitude of the coning error can be reduced to 1 2 order of magnitude.

6 Conclusion

(1) For the traditional N sample algorithm, more samples mean higher compensation accuracy. Because the compensation algorithm uses the gyroscope information increased, that can make the description of the attitude change more detailed. When the attitude updating period should be short, the traditional multiple sample algorithm is inapplicable to this occasion, then single sample utilizing P samples from the previous algorithm can be used.

(2) With the increasing of the value of P, the coning error will be decreased theoretically. However, the amount of calculation increases while the value of P increases. And as to a certain level, it is better for increasing the accuracy; On the other hand, it is harmful for the control system when P increasing, because the

delaying problem becomes more serious when P becomes too large. Therefore, the value of P should be determined according to the environment and control system.

(3) The coning error will decreases by using traditional coning error compensation algorithm when the frequency of coning motion is decreasing, and the error of the optimized N samples utilizing P samples from the previous algorithm is no longer monotonous. Optimization algorithm of a specific frequency does not change the structure of classic algorithms, and does not increase the difficulty of realization Error analysis. And experiment results show that the optimal algorithm can effectively improve the attitude accuracy of strapdown system at the fixed frequency. Thus, the proposed algorithm can be used to optimize the compensation algorithm, and for the strapdown system with a narrow range of motion frequency, the using of this algorithm can effectively improve the accuracy of attitude on the particular frequency.

References

1. Miller RB (1983) A new strapdown attitude algorithm. J Guid Control Dyn 6(4):287–291
2. Lee JG, Yoon YJ (1990) Extension of strapdown attitude for high-frequency base motion. J Guid Control Dyn 13(4):738–743
3. Jiang YF, Lin YP (1992) Improved strap-down coning algorithms. IEEE Trans Aerosp Electron Syst 19(2):424–429
4. Lin XY, Liu JY, Liu H (2003) An improve rotation vector attitude algorithm for laser strapdown inertial navigation system. Trans Nanjing Univ Aeronaut Astronaut 20(1):47–52
5. Zhang ZF, Yang MX (2012) An improved attitude algorithm for strapdown inertial navigation system. J Project Rockets Missiles Guid 6(3):6–10
6. Savage PG (2011) Explicit frequency-shaped coning algorithms for pseudo-coning environments. J Guid Control Dyn 34(3):774–782
7. Savage PG (2010) Coning algorithm design by explicit frequency shaping. J Guid Control Dyn 33(4):1123–1132
8. Qin YY (2006) Inertial navigation. Science Press, Beijing, pp 305–326
9. Savage PG (2010) Coning algorithm design by explicit frequence shaping. J Guid Control Dyn 33(4):1123–1132
10. Pan XF, Wu MP (2008) Optimization coning algorithms to the fixed frequency motion. J Chin Inert Technol 2(1):20–23

Algorithm on Scan Trajectory for Laser Rapid Prototyping Based on Square Lattice

Jingchi Zhang, Hongru Li, Weizhi Zhang and Qing Yuan

Abstract Laser Rapid Prototyping (LRP) can make the material accumulating by layer. And during the process, it has been known that shrinking stress may cause curl distortion. We have attempted to improve the processing speed and reduce the frequency of ON/OFF the laser. In this paper, we propose a new trajectory scan algorithm based on square lattice for Laser Rapid Prototyping (LRP). First, side-length of the square lattice is determined by the metal type and the material properties, and a set of regular square lattices is arranged based on the slices. Then, following the principles of not repeating and fewer breakpoints in solving Traveling Salesman Problem (TSP), the scanning sequence for each area is calculated. Therefore, a certain scan strategy can be determined, followed by the output of trajectory scan. We present the specific steps of the algorithm. The experiment proves the feasibility and superiority of the algorithm.

Keywords Laser rapid prototyping · Scan trajectory · Shrinking stress · Algorithm

Laser rapid prototyping (LRP) technology is a new manufacturing method for 3D products. Compared to conventional methods, this technology can be applied to the design of complex parts which could be very challenging for conventional methods. Also, it can greatly shorten the processing time [1]. With the development of rapid prototyping technologies, the algorithm on scan trajectory has become a hot issue. Appropriate scan trajectory has been of much significance to the quality of the parts and the fabrication efficiency, in terms of three areas, as follows:

J. Zhang (✉) · H. Li · W. Zhang · Q. Yuan
College of Information Science and Engineering, Northeastern University, Shenyang 110819, People's Republic of China
e-mail: 695275717@qq.com

H. Li
e-mail: lihongru@mail.neu.edu.cn

W. Zhang
e-mail: 1348221834@qq.com

Q. Yuan
e-mail: 1042270331@qq.com

© Springer International Publishing Switzerland 2017

V.E. Balas et al. (eds.), *Information Technology and Intelligent Transportation Systems*, Advances in Intelligent Systems and Computing 454,
DOI 10.1007/978-3-319-38789-5_17

(1) Improve the precision of the products
(2) Decrease the curl distortion due to shrinking stress
(3) Enhance the processing efficiency and prolong the laser lifetime by reducing the frequency of laser ON/OFF.

Currently, two scan modes have been commonly used. They are parallel scan mode and offset scan modes [2]. However, these modes require frequent laser ON/OFF, which can reduce the lifetime of the laser. Meanwhile, frequent jumping of the cavity would lead to a lot of unnecessary empty travel. And this can seriously affect the scanning time. More importantly, single point energy transportation is used for both parallel scan mode and offset scan mode. However, this transportation may have its limitations. For instance, the material forms at different times in different locations, the corresponding heating time, cooling time, cooling rate and shrinking stress will be different. In other words, different locations in the slice layer will have different volume contraction, resulting in warpage in section, greatly influencing material and mechanical properties [3].

To solve those problems above, [4] proposed a selective laser melting scanning strategy, which can reduce shrinking stress and warpage, also improve fabrication efficiency. But this method is based on powder-laying system, which may affect the density or precision of the parts. Meanwhile, it may cause the parts porous and fail to meet the requirements. Moreover, the application of this strategy may be constrained because of too many partitions, where the strength at the joints of thin parts may be weakened. [5] Proposed algorithm based on partition so the forming sequence synchronization can be reached in a section, effectively improving the precision and minishing the shrinking stress and warpage.

Based on [5, 6] proposed a new algorithm of regular polygon grid path. Using different scanning sequence of different polygons to raise the temperature balance of the sintering parts, and minish the shrinking stress and the warpage. However, the defect of the algorithm of regular polygon grid path is that scanning trajectory between polygons is discontinuous, resulting in the laser be turned on and off frequently, plus more empty travels. For a part contains many layers, empty travels in each layer would be accumulated so that a lot of time would be wasted. These can greatly influence scanning efficiency. Inspired by the algorithm of regular polygon grid path, a novel stable algorithm on scan trajectory is presented in this paper, based on square lattice.

1 Basic Ideas of Algorithm on Scan Trajectory for LRP Based on Square Lattice

Our basic ideas are: the algorithm scan trajectory is determined by square lattice, rather than the regular polygon. The side-length of the square is obtained according to the material and the requirement on the property. Also, we have to ensure that the scanning path is most suitable for scanning in order to reduce the shrinking stress. By optimizing the scan trajectory referring to TSP, following the principles of no repeating and fewer breakpoints, we can improve the continuity of scanning and ensure scanning lines by using perpendicular adjacent lattice in adjacent lattice be perpendicular., which will further reduce the shrinking stress and improve the fabrication efficiency. The innovations of the algorithm are described in the following two aspects:

(1) Algorithm of regular polygon grid path divides a complex graphic into several regular pieces. Moreover, the algorithm on scan trajectory based on square lattice adopts the regular polygon into square lattice, and the side-length of the square is also optimized. The shrinking stress, therefore, decreased in every scan path. Besides, shrinking stress can also be reduced effectively by optimizing the length of scanning path when parallel scanning mode is applied in each square lattice.

(2) Algorithm of regular polygon grid path scans each polygon independently to meet the requirement of the property and to ensure the scanning lines of perpendicular adjacent lattice, which greatly reduces the efficiency. After combining TSP with characteristic on laser rapid prototyping, the algorithm on scan trajectory based on square lattice presented in this paper designed the scanning trajectory in every lattice and linked more lattices, which break through the limitation of the algorithm of regular polygon grid path and improve the fabrication efficiency and reduce the frequency of ON/OFF the laser and prolong the service time of the laser. For the purpose of linking adjacent lattices and ensuring the scanning lines in perpendicular adjacent lattices, the side-length of a square lattice should be odd times of the scanning line width. Meanwhile, in order to minish the shrinking stress, the length of scanning lines should be close to the optimal length of scanning line [7]. So the side length of the square L should be the greatest odd times of the scanning line width but no longer than the optimal length and the purposes of continuity of the scan are accomplished, as is shown in Fig. 1.

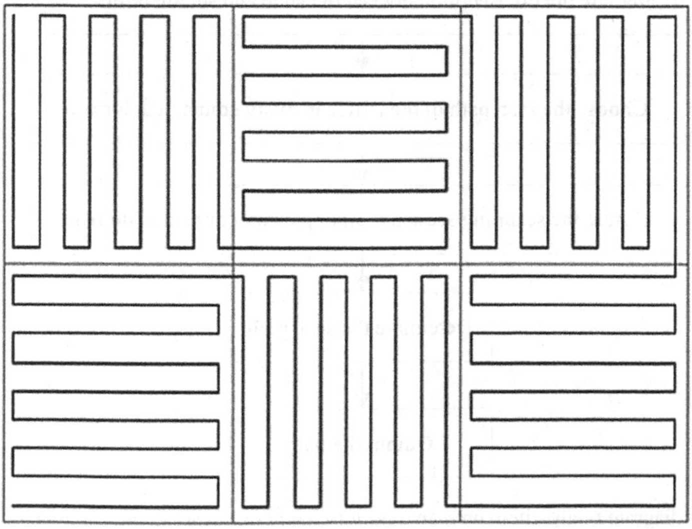

Fig. 1 Schematic diagram of square lattice

2 The Implementation of Algorithm on Scan Trajectory for LRP Based on Square Lattice

The proposed algorithm on scan trajectory is shown in Fig. 2.

In Fig. 3 for example, we introduce the main steps of the algorithm for a single slice.

Step 1: Arrange the original square lattice and obtain the bounding volume. Bounding volume is the smallest polygon, consisting of several square lattices, that surrounds the graphic, as is shown in Fig. 4. By traversing every coordinate data point

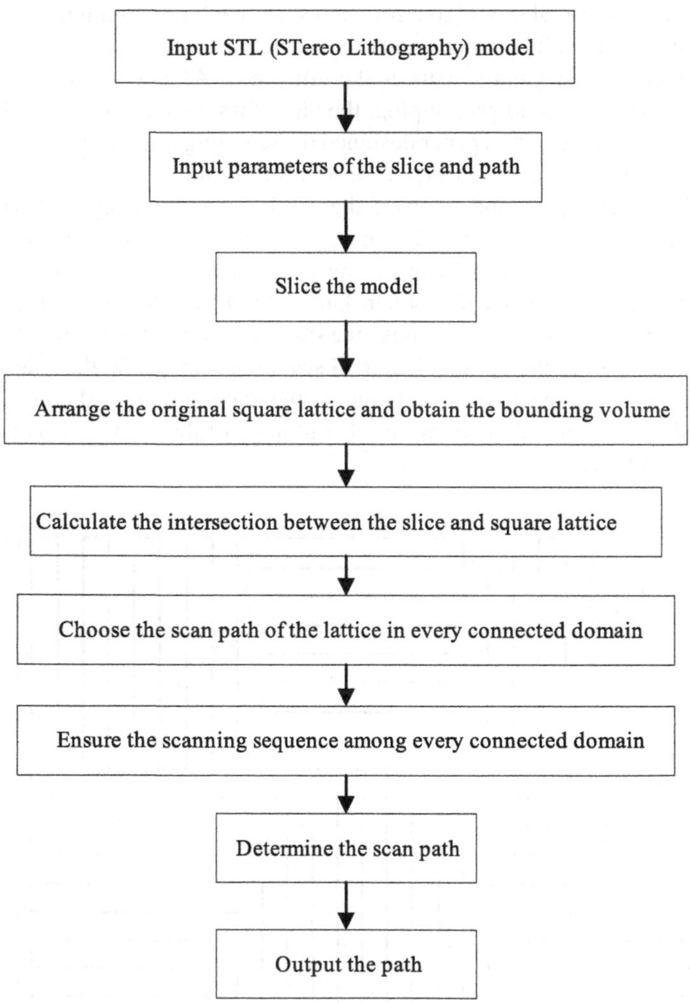

Fig. 2 Flow diagram of algorithm on scan trajectory based on square lattice

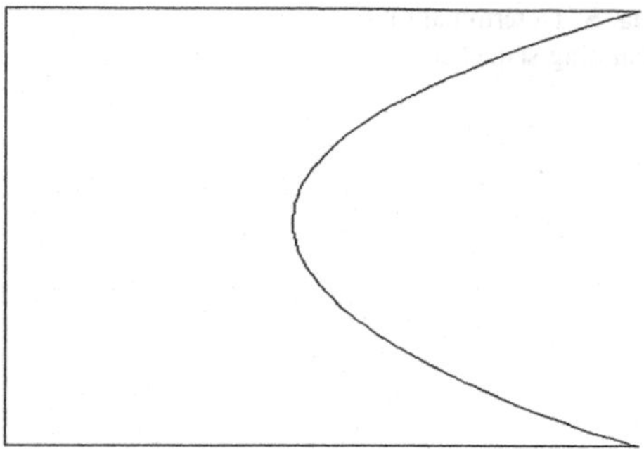

Fig. 3 Sliced graphics

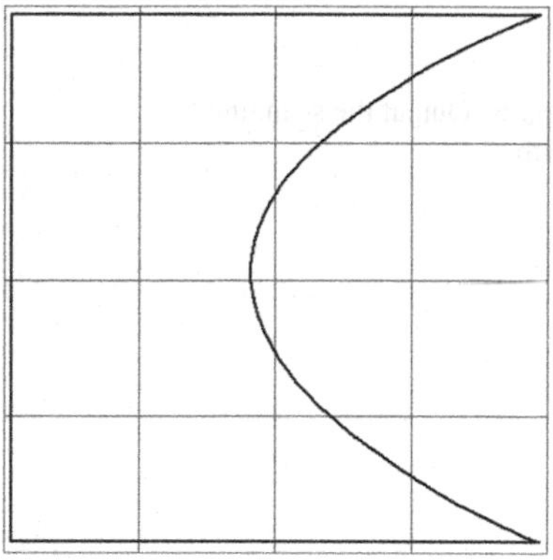

Fig. 4 Bounding box comprised by square lattices

X, Y on the graphic, we can find the minimum values in both directions. And then, the endpoint coordinates of the bounding box can be determined by the minimum values, so that the square lattices can be placed till the entire graphic is completely surrounded by the lattices. According to different requirements on the properties, the side length of the square should be L-the greatest odd times of the scanning line width but no longer than the optimal length of the scanning lines.

Step 2: Calculate the intersection through between the slice and square lattice. Calculate the intersection in terms of Boolean operation, between the graph of the slice and the preset square lattices, denoted by set S. This step is used to determine the direction of the scanning lines in every lattice and the scanning path in linking lattices.

Step 3: Choose the scan path of the lattice in every connected domain. As is shown in Fig. 5, the scanning sequences of square lattices are chosen. The purposes of optimization the scanning path of the section are to reduce the frequency of ON/OFF the laser, as well as the empty travel, so that the fabrication efficiency can be improved. If it comes to the boundary or the lattice is intact, the scanning path will not connect to next lattice in the ordinary way. Following the principles of not repeating and fewer breakpoints in solving TSP, the most appropriate adjacent lattice will be chosen, and the scanning algorithm will be repeated. If the scanning

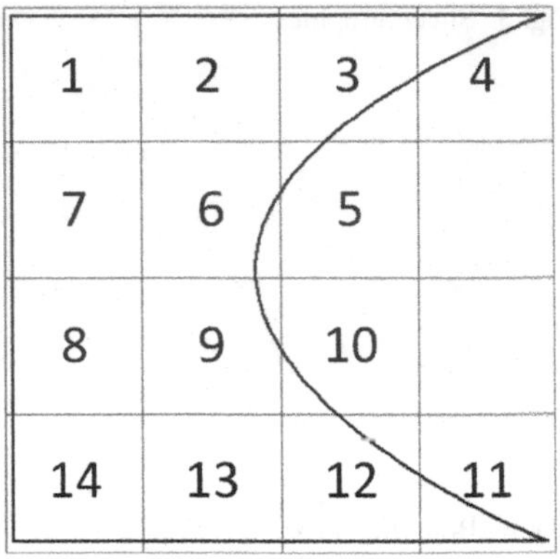

Fig. 5 Determination of scanning sequence

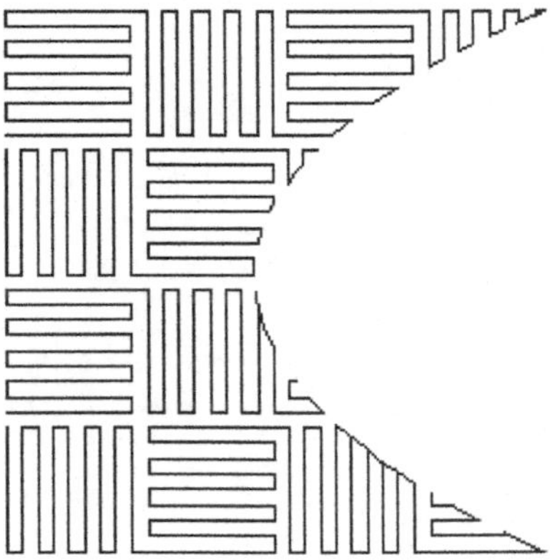

Fig. 6 Output the scanning path

path cannot be able to continuously go through the next lattice, the laser has to be turned off, and search for another starting point and restart scanning.

Step 4: Ensure the scanning sequence among every connected domain. When multiple connected domains appear in the slice, empty travel between connected domains should be optimized. So the sequence of scanning connected domains should be calculated because empty travel differs for different scanning sequence. Once the scanning path of the lattice in every connected domain is chosen, the starting point and end point of each connected domain will be determined. Referring to solutions on TSP, shortest distance between each connected domain can be obtained.

Step 5: Output the path. By following the steps above, the final scanning trajectory can be determined, as shown in Fig. 6. Empty travel is included in the information about final scanning trajectory, so the information about empty travel will also be output to machinery equipment.

3 Experiment and Analysis

Choose a test graphic as shown in Fig. 7, input the graphic into MATLAB and calculate the scan trajectory according to the steps of the algorithm above, the result is shown in Fig. 8.

As shown in Fig. 8, Horizontal and vertical lines intersect, most lengths of the scan lines are close to the optimal length of scan line, which would improve the precision effectively by minishing the shrinking stress and warpage. Meanwhile, scanning trajectory has a good continuity, with 4 breakpoints in Fig. 8, while there would be at least 14 if algorithm of regular polygon grid is used. It can be concluded that the empty travel is reduced and fabrication efficiency is improved. Since a part may contain many layers which needs to be scanned with the same scanning mode, a lot of filling path will be made by the scanning system, also the empty travel and the laser off time would be enormous after accumulation. The simulation demonstrates a good effect on optimization over time, and the fabrication efficiency has also been improved.

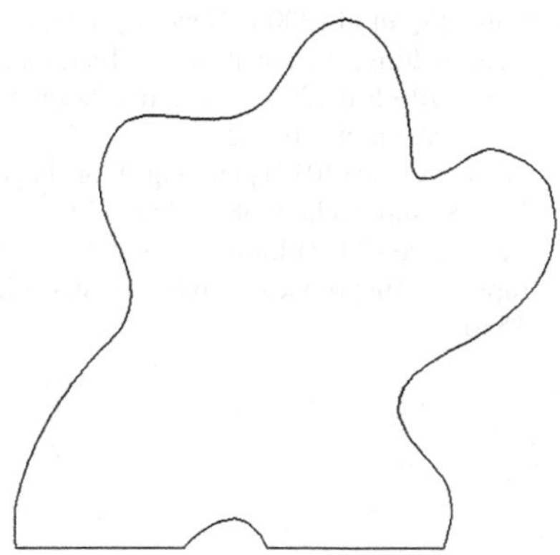

Fig. 7 Test graphic

Fig. 8 Outcome of algorithm application

4 Conclusion

Algorithm on scan trajectory based on square lattice has been further developed based on the algorithm of regular polygon grid. It can be more helpful to minish the shrinking stress and warpage, and improve scanning precision. At the same time, Experimental results show that this algorithm can reduce the frequency of ON/OFF the laser, and the empty travel as well. And therefore, the fabrication efficiency has been enhanced.

References

1. Xue Y, Gu P (1996) A review of rapid prototyping technologies and systems. Computer Aided Design 28(4):307–318
2. Shi Y et al (2002) Research and implement of a new kind of scanning mode for selective laser sintering. Jixie Gongcheng Xuebao/Chin J Mech Eng 38(2):35–39
3. Huang X et al (2008) Path planning for parallel raster scanning and filling. J Comput Aided Design Comput Graph 20(3):326–331
4. Yang YQ et al (2006) Newest progress of direct rapid prototyping of metal part by selective laser melting. Aeronaut Manuf Technol 2:73–76+97
5. Kruth JP et al (2004) Selective laser melting of iron-based powder, vol 149. Elsevier Ltd., Amsterdam, pp 616–622
6. Huang X et al (2013) planning algorithm of regular polygon grid path for laser rapid prototyping. Forg Stamp Technol 38(3):152–155
7. Lai Y et al (2013) Influencing factors of residual stress of Ti-6.5Al-1Mo-1V-2Zr alloy by laser rapid forming process. Xiyou Jinshu Cailiao Yu Gongcheng/Rare Metal Mater Eng 42(7):1527–1530

Increasing Robustness of Differential Evolution by Passive Opposition

Yiping Cheng

Abstract The differential evolution (DE) algorithm is known to be fairly robust among various global optimization algorithms. However, the application of this algorithm to an extensive set of test functions shows DE fails at least partially on 19 % of the test functions, and completely on 6.9 % of the test functions. The opposition-based DE in the literature, while aimed to improve the efficiency of DE, has a robustness even worse than that of the classic DE. In this paper we describe a new variant of DE called "passive oppositional differential evolution" (PODE) which utilizes opposition in such a way that a set of opposite vectors, while not part of the population, are used in mutation to gain diversity. Numerical experiments show that compared to DE and ODE, it has a much better robustness and a similar speed of convergence.

Keywords Differential evolution · Passive opposition · Global optimization · Evolutionary computing

1 Introduction

Differential evolution (DE) is a stochastic global optimization algorithm proposed by Price and Storn in 1995 [1–3]. According to [4], DE outperforms particle swarm optimization (PSO) and most evolutionary algorithms. It was found to be simple, fast converging, and robust in the sense that it finds the optimum in almost every run. DE has also gained reputation as a robust algorithm from engineers' experience in solving real-world optimization problems, for example, finding best design parameters, training artificial neural networks, etc.

Y. Cheng (✉)
School of Electronic and Information Engineering,
Beijing Jiaotong University, Beijing, China
e-mail: ypcheng@bjtu.edu.cn

© Springer International Publishing Switzerland 2017

V.E. Balas et al. (eds.), *Information Technology and Intelligent Transportation Systems*, Advances in Intelligent Systems and Computing 454,
DOI 10.1007/978-3-319-38789-5_18

However, this does not mean that the robustness of DE is nearly perfect and cannot be improved further. Reference [5] described the "opposition-based differential evolution" (ODE) algorithm for improving DE, and it contained the results of numerical experiments comparing the performances of DE and ODE on a big set of 58 benchmark functions. From the results, one can derive the following data relating to the robustness of DE and ODE: When the scheme is DE/rand/1/bin (the classic version), DE fails at least partially on 11 functions, and completely on 4 functions, and ODE fails at least partially on 15 functions, and completely on 4 functions. The situation is similar when the other three scheme are used. Although many of the test functions are designed intentionally to be hard ones, these data suggest that there is much room for further improvement of the algorithm's robustness.

In this paper, we shall present such an improvement, which is a new variant of DE called *passive oppositional differential evolution* (PODE). Actually, like ODE, PODE is also opposition-based, but the opposition is used in a different way. The essential idea here is that, unlike in ODE, the opposite vectors do not actively take part in the competition to be candidate solution vectors, that is, they do not belong to the population. However, they do take part in the mutation operation, to increase diversity of the population. Our original purpose of conceiving this approach was to reduce the number of function calls but ironically it served more to increase the robustness of the algorithm. We will later give results of PODE applied to the same 58 test functions and see the considerable reduction in the number of failed functions.

The rest of the paper is organized as follows: Sect. 2 gives a sketchy introduction of the basic DE algorithm. Section 3 introduces the concept of opposition-based learning and the ODE algorithm. In Sect. 4, we describe our PODE algorithm in full detail. Numerical results and discussions are provided in Sect. 5. Finally, Sect. 6 concludes the paper.

2 Basic Differential Evolution

DE works by carrying out iterations on a population of candidate solutions (called individuals) until a suitable convergence criterion is met. Let D denote the dimension of the solution space, N_p denote the size of the population, $\mathbf{x}_i^g$ denote the ith individual of the population of the gth generation (or iteration), and $x_{i,j}^g$ denote the jth entry of $\mathbf{x}_i^g$. In the beginning, the 1st population of individuals are generated by

$$x_{i,j}^1 = a_j + \text{rand}()(b_j - a_j) \text{ for } i = 1, \ldots, N_p \text{ and } j = 1, \ldots, D \tag{1}$$

where a_j are b_j are lower and upper limits of the jth dimension, respectively, rand() is random number generator with uniform distribution in [0, 1).

During each iteration (let it be the gth iteration), the following three operations are done successively to each individual $\mathbf{x}_i^g$. The details of the operations depend on the scheme used, however in this paper we only use the classic scheme: DE/rand/1/bin.

2.1 Mutation

A mutant vector is generated by

$$\mathbf{v}_i^g = \mathbf{x}_{r_1}^g + F(\mathbf{x}_{r_2}^g - \mathbf{x}_{r_3}^g) \tag{2}$$

where the indices r_1, r_2, r_3 are drawn randomly from $1, \ldots, D$ and they must be mutually different and also different from i, the real coefficient F is taken from $(0, 2)$. If any entry of $\mathbf{v}_i^g$ goes out of bound, then it is set to a random value within the boundary.

2.2 Crossover

A trial vector $\mathbf{u}_i^g$ is obtained by combining the target vector $\mathbf{x}_i^g$ and the mutant vector $\mathbf{v}_i^g$ by

$$u_{i,j}^g = \begin{cases} v_{i,j}^g, & \text{if } j = j^* \text{ or rand}() < C_r \\ x_{i,j}^g, & \text{otherwise} \end{cases} \tag{3}$$

where j^* is chosen randomly from $1, \ldots, D$ to ensure that $\mathbf{u}_i^g$ and $\mathbf{x}_i^g$ differ at least at one entry, C_r is a user-chosen crossover constant in $(0, 1]$.

2.3 Selection

A greedy selection strategy is used:

$$\mathbf{x}_i^{g+1} = \begin{cases} \mathbf{u}_i^g, & \text{if } f(\mathbf{u}_i^g) < f(\mathbf{x}_i^g) \\ \mathbf{x}_i^g, & \text{otherwise} \end{cases} \tag{4}$$

That is, the trial vector is accepted as new target vector if and only if it is fitter than the old target vector.

3 Opposition-Based Differential Evolution

The concept of opposition-based learning was first proposed in [6]. Further exposition of this concept was provided in [7, 8], and the ODE algorithm was proposed in [5, 7] which embeds opposition-based learning into differential evolution. Let us begin with the definition of *opposite number*.

Definition 1 Let x be a real number in an interval $[a, b]$, the opposite (with respect to the interval) of x, denoted by $\breve{x}$, is defined by $\breve{x} = a + b - x$. Let $\mathbf{x}$ be a point in the box $[a_1, b_1] \times \cdots \times [a_D, b_D]$, the opposite (with respect to the box) of $\mathbf{x}$, denoted by $\breve{\mathbf{x}}$, is defined by $\breve{\mathbf{x}} = (\breve{x}_1, \ldots, \breve{x}_D)$ with $\breve{x}_i = a_i + b_i - x_i$ for $i = 1, \ldots, D$.

The rationale for proposing opposite numbers/points, for the 1-D case, is as follows: Suppose x_s is the solution, and x is a random guess of x_s, then "the probability that the opposite point is closer to the solution is higher than a second random guess, assuming the original guess is not closer" [8]. However, it was discovered in [9] that a quasi-opposite point is more likely to be closer to the solution than the opposite point. A quasi-opposite point is a number evenly randomly chosen from the interval between $\frac{a+b}{2}$ and $\breve{x}$.

The ODE algorithm is the attempt made in [5, 7] to accelerate the convergence rate of differential evolution. In ODE, opposite numbers are used during population initialization and also for generating new populations during the evolutionary process. Due to space limitation, full description of ODE will not be given here. The interested reader can consult [5]. The main points that are new in ODE are listed below:

- Population initialization is quite different from the original DE: A population P with size N_p is randomly generated as usual. Then an opposite population OP with the same size N_p is calculated consisting of the opposite vectors of the vectors in P. And the N_p fittest individuals from the set $\mathrm{P} \cup \mathrm{OP}$ are selected as the initial population. This phase requires $2N_p$ function calls.
- At the end of each iteration, there is a probability J_r (jumping rate) that the population will jump to a new population. First, an opposite population is calculated in a similar way it was in population initialization, but this time the lower and upper limits used in the definition of opposite numbers are no longer the predefined interval boundary values, but are the minimum and maximum values of each coordinate in the current population, that is

$$\mathrm{OP}_{i,j} = \mathrm{MIN}_j + \mathrm{MAX}_j - \mathrm{P}_{i,j} \text{ for } i = 1, \ldots, N_p \text{ and } j = 1, \ldots, D. \quad (5)$$

  Then, the N_p fittest individuals from $\mathrm{P} \cup \mathrm{OP}$ are selected as the new population. This again requires $2N_p$ function calls, but this occurs only in $1/J_r$ of the iterations, and hopefully this occasional jumping will help to reduce the number of iterations.

4 Passive Oppositional Differential Evolution

It is our perception that ODE is not an ideal replacement for DE. Although ODE has increased speed of convergence for most of the functions that DE is already successful, its performance regarding robustness becomes even worse. According to [5], of the 58 test functions, the number of functions that DE fails at least partially is 11, and 15 for ODE; the number of functions that DE fails completely is 4, and

4 for ODE. The failure of ODE in improving robustness may be attributed to its definition of opposite vectors using dynamic lower and upper limits. This practice, albeit helpful in keeping the population within the reduced region it already goes in thus speeding up convergence, is detrimental in maintaining population diversity and makes it easy to converge prematurely around a local optimizer.

Actually, opposite vectors can also be used as a means to enhance population diversity, and this has been overlooked in the existing literature. Recently, there is growing literature about using Levy flight to gain diversity, see e.g. [10]. However, in this paper we shall only use opposition, and currently it is still open which of the diversity enhancement methods is the best.

Another motivation for our seeking a different approach is we want to avoid function calls as they are costly. Can we use the opposite vectors without function calls on them?

The outcome of these two thoughts is our passive oppositional differential evolution algorithm. Its key points are given as follows:

- Population initialization is the same as in classic DE, so there is no extra N_p function calls.
- At the beginning of each iteration, the opposite vectors of the individuals of the population are built. They do not belong to the population, so again there will not be N_p additional function calls. But the opposite vectors take part in mutation, i.e., they can be the $\mathbf{x}_{r_1}^g$, $\mathbf{x}_{r_2}^g$, $\mathbf{x}_{r_3}^g$ vectors in (2). There is no population jumping, no jumping rate to specify, because building opposite vectors is almost costless and hence can be done for every iteration.
- To strike a fair balance between global exploration and local exploitation. In the algorithm there are two kinds of opposite vectors: global/static opposite vectors and local/dynamic opposite vectors.

A detailed pseudocode of PODE is given in Algorithm 1. Note there that $\mathbf{x}_1, \ldots, \mathbf{x}_{N_p}$ are population vectors but $\mathbf{x}_{N_p+1}, \ldots, \mathbf{x}_{2N_p}$ are not part of the population though they can be stored in memory just succeeding the population. Note also that when calculating global opposite vectors we actually use the quasi-opposite vectors, as we found they perform much better; but quasi-opposite vectors are not employed in calculating local opposite vectors.

5 Numerical Results

In this section, PODE is applied to solve an extensive set of benchmark functions to test its performance. To allow fair comparison with DE and ODE, we choose the same set of test functions used in [5]. This set contains 58 functions, which are described in [5, 11]. The other settings and parameters are also the same as in [5] except for the crossover probability C_r.

Algorithm 1 Passive Oppositional Differential Evolution

```
function x* = PODE(f, [a_1, b_1] × ··· × [a_D, b_D])
  g ← 1
  Generate N_p vectors x_1^1, ..., x_{N_p}^1 according to (1)
  for i = 1, ..., N_p do
    Y_i ← f(x_i^1)
  end for
  while convergence criterion not met do
    for i = 1, ..., N_p do
      if rand() < 0.5 then
        for j = 1, ..., D do
          x^g_{N_p+i,j} ← (a_j+b_j)/2 + rand()((a_j+b_j)/2 − x^g_{i,j})
        end for
      else
        for j = 1, ..., D do
          x^g_{N_p+i,j} ← MIN_j + MAX_j − x^g_{i,j}
        end for
      end if
    end for
    for i = 1, ..., N_p do
      Randomly generate integers r_1 ≠ r_2 ≠ r_3 ≠ i from 1, ..., 2N_p
      Randomly generate integer j* from 1, ..., D
      for j = 1, ..., D do
        if j = j* or rand() < C_r then
          u^g_{i,j} ← x^g_{r_1,j} + F(x^g_{r_2,j} − x^g_{r_3,j})
          Bound u^g_{i,j} within [a_j, b_j]
        else
          u^g_{i,j} ← x^g_{i,j}
        end if
      end for
      y ← f(u_i^g)
      if y < Y_i then
        x_i^{g+1} ← u_i^g
        Y_i ← y
      else
        x_i^{g+1} ← x_i^g
      end if
    end for
    g ← g + 1
  end while
  x* ← the fittest vector of the vectors x_1^g, ..., x_{N_p}^g
end function
```

- Mutation constant $F = 0.5$ and crossover probability $C_r = 1$. We found the algorithm performs better with this setting than with the usual setting $F = 0.5$ and $C_r = 0.9$.
- Population size $N_p = 100$.
- Convergence criterion: The iteration will stop if the fittest individual reaches a function-specific *value to reach* (VTR) or the number of function calls (NFC)

reaches a predefined MAX_{NFC}. Here, VTR=exact minimum of the function$+1E-8$ and $MAX_{NFC}=10^6$. A successful run is one that reaches VTR within MAX_{NFC} function calls.

- Since different run yields different NFC, in the experiment we do 50 runs, and report the average NFC. The success rate (SR) is defined as the ratio of successful runs to total runs. The acceptance rate (AR) of PODE is defined to be $\frac{NFC_{DE}}{NFC_{PODE}}$.

Our experimental results are now reported in Tables 1 and 2, along with previous results of DE and ODE taken from [5] to facilitate comparison. The following observations can be drawn from the results:

- In the table the best NFC or SR is marked with a different color. Thus we can see that DE performs the best on 12 functions, ODE performs the best on 21 functions, and PODE performs the best on 25 functions. So PODE is the best in this respect.
- In the table the acceptance rate is calculated only when the success rates are equal. If for one particular function one algorithm has a higher/lower success rate than a compared algorithm, then this algorithm is deemed accepted/rejected over the compared algorithm with no regard of NFCs. This is because NFCs have been calculated without failed runs, and the actual NFC will be much larger if the failed

Table 1 Results for functions f_1–f_{20}

			DE		ODE			PODE		
F	Name	D	NFC	SR	NFC	SR	AR	NFC	SR	AR
f_1	Sphere	30	87748	1	47716	1	1.84	**14728**	1	5.96
f_2	Axis	30	96488	1	53304	1	1.81	**16458**	1	5.86
f_3	Schwefel 1.2	20	177880	1	168680	1	1.05	**36356**	1	4.89
f_4	Rosenbrock	30	**403112**	1	–	0	R	966807	0.28	R
f_5	Rastrigin	10	328844	1	70389	0.76	R	**44286**	1	7.43
f_6	Griewangk	30	113428	1	69342	0.96	R	**19092**	1	5.94
f_7	Sum of Power	30	25140	1	8328	1	3.02	**3664**	1	6.86
f_8	Ackley	30	169152	1	98296	1	1.72	**28864**	1	5.86
f_9	Beale	2	**4324**	1	4776	1	0.91	10748	1	0.40
f_{10}	Colville	4	**16600**	1	19144	1	0.87	41106	1	0.40
f_{11}	Easom	2	8016	1	**6608**	1	1.21	11200	1	0.72
f_{12}	Hartmann 1	3	**3376**	1	3580	1	0.94	12232	1	0.28
f_{13}	Hartmann 2	6	–	0	–	0	–	**35804**	1	A
f_{14}	Six Hump	2	5352	1	**4468**	1	1.20	12308	1	0.43
f_{15}	Levy	30	101460	1	**70408**	1	1.44	222118	0.98	R
f_{16}	Matyas	2	3608	1	3288	1	1.10	**2694**	1	1.34
f_{17}	Perm	4	549850	0.04	311800	0.12	A	**43324**	1	A
f_{18}	Michalewicz	10	**191340**	0.76	213330	0.56	R	–	0	R
f_{19}	Zakharov	30	385192	1	369104	1	1.04	**230342**	1	1.60
f_{20}	Branin	2	**4884**	1	5748	1	0.85	28752	1	0.17

Table 2 Results for functions f_{21}–f_{58}

F	Name	D	DE NFC	DE SR	ODE NFC	ODE SR	ODE AR	PODE NFC	PODE SR	PODE AR
f_{21}	Schwefel 2.22	30	187300	1	155636	1	1.20	**31328**	1	5.98
f_{22}	Schwefel 2.21	30	570290	0.28	72250	0.88	A	**33402**	1	17.1
f_{23}	Step	30	41588	1	23124	1	1.80	**6514**	1	6.38
f_{24}	Quartic	30	818425	0.15	**199810**	1	A	–	0	R
f_{25}	Kowalik	4	13925	0.80	15280	0.60	R	**32724**	1	A
f_{26}	Shekel 5	4	–	0	–	0	–	**23764**	1	A
f_{27}	Shekel 7	4	–	0	–	0	–	**23418**	1	A
f_{28}	Shekel 10	4	4576	1	**4500**	1	1.02	23518	1	0.19
f_{29}	Tripod	2	**10788**	1	11148	1	0.97	28394	1	0.38
f_{30}	De Jong 4	2	1016	1	996	1	1.02	**772**	1	1.32
f_{31}	Alpine	30	411164	1	337532	1	1.22	**31844**	1	12.9
f_{32}	Schaffer 6	2	7976	1	**5092**	1	1.57	16826	1	0.47
f_{33}	Pathological	5	2163	0.88	**2024**	1	A	647484	1	A
f_{34}	Inverted Cosine	5	38532	1	**16340**	1	2.36	35546	1	1.08
f_{35}	Aluffi-Pentini	2	2052	1	**1856**	1	1.11	7392	1	0.28
f_{36}	Becker-Lago 5	2	8412	1	**5772**	1	1.46	25362	1	0.33
f_{37}	Bohachevsky 1	2	5284	1	4728	1	1.12	**3828**	1	1.38
f_{38}	Bohachevsky 2	2	5280	1	4804	1	1.10	**3890**	1	1.36
f_{39}	Camel Back	2	3780	1	3396	1	1.11	**2700**	1	1.40
f_{40}	Dekkers-Aarts 4	2	2424	1	**2152**	1	1.13	23256	1	0.10
f_{41}	Exponential	10	19528	1	15704	1	1.24	**7702**	1	2.54
f_{42}	Goldstein-Price	2	4780	1	**4684**	1	1.02	10836	1	0.44
f_{43}	Gulf Research	3	**6852**	1	8484	1	0.81	22184	1	0.31
f_{44}	Helical Valley	3	7036	1	**6172**	1	1.14	14292	1	0.49
f_{45}	Hosaki	2	3256	1	**3120**	1	1.04	7104	1	0.46
f_{46}	Levy-Montalvo 1	3	6184	1	**5472**	1	1.13	12832	1	0.48
f_{47}	McCormick	2	2976	1	**2872**	1	1.04	7540	1	0.39
f_{48}	Miele-Cantrell 2	4	**1108**	1	1232	1	0.90	5402	1	0.21
f_{49}	Multi-Gaussian	2	5232	1	5956	0.92	R	**248**	1	21.1
f_{50}	Neumaier 2	4	**379900**	1	250260	0.20	R	665818	1	0.57
f_{51}	Odd Square	10	–	0	**16681**	0.84	A	703254	1	A
f_{52}	Paviani 6	10	**14968**	1	15104	1	0.99	70718	1	0.21
f_{53}	Periodic	2	7888	1	**4272**	1	1.85	8206	1	0.96
f_{54}	Powell Quadratic	4	8856	1	**7504**	1	1.18	7710	1	1.15
f_{55}	Price Transistor	9	78567	0.24	97536	**0**.56	A	761100	0.02	R
f_{56}	Salomon	10	37824	1	**24260**	1	1.56	747800	0.16	R
f_{57}	Schaffer 2	2	30704	1	55980	0.76	R	**14616**	1	2.10
f_{58}	Wood	4	**16600**	1	19144	1	0.87	41716	1	0.40

runs are included. Now both ODE and PODE are compared with DE. ODE has 6 SR acceptances, 8 SR rejections, and the geometric average AR is 1.21. This roughly means that ODE is worse in robustness but is 1.21 times more efficient for the easy functions. PODE has 7 SR acceptances, 6 SR rejections, and the geometric average AR is 1.11. This roughly means that PODE is better in robustness and is 1.11 times more efficient for the easy functions than basic DE.

- Finally, we must notice that the number of functions that PODE fails at least partially is 6, the number of functions that PODE fails completely is 2. Both numbers are significantly smaller than the corresponding numbers of DE and ODE, which are 11/4 and 15/4, respectively. These clearly show that PODE has significantly better robustness than both DE and ODE.

6 Conclusion

We have given a new algorithm PODE, which has a significantly better robustness and a similar speed of convergence for easy functions, as compared to basic DE and a state-of-the-art algorithm ODE. The key idea used is passive opposition. It thus shows that opposition can be a means to give the diversity that an optimization algorithm needs to find the optimum. However, how it relates to other diversity enhancement techniques, and which is the best, is still open and deserves further study. We also believe that further fine tuning in both the opposite vector generation and mutation parts will yield an even more robust and more efficient algorithm.

Acknowledgments This work was supported by the National Natural Science Foundation of China (NSFC) under grant 61227002.

References

1. Storn R, Price K (1995) Differential evolution - a simple and efficient adaptive scheme for global optimization over continuous spaces. Technical Report TR-95-012, University of California, Berkeley (1995)
2. Storn R, Price K (1997) Differential evolution - a simple and efficient heuristic for global optimization over continuous spaces. J Optim 11(4):341–359
3. Price K, Storn RM, Lampinen JA (2005) Practical approach to global optimization. Natural computing series. Springer, New York
4. Vesterstrom J, Thomsen, R (2004) A comparative study of differential evolution, particle swarm optimization, and evolutionary algorithms on numerical benchmark problems. In: Proceedings 2004 IEEE congress on evolutionary computation (CEC2004), vol 2. IEEE, pp 1980–1987
5. Rahnamayan S, Tizhoosh H, Salama M (2008) Opposition-based differential evolution. IEEE Trans Evolut Comput 12(1):64–79
6. Tizhoosh H (2005) Opposition-based learning: a new scheme for machine intelligence. In: Proceedings international conference on computational intelligence for modelling, control and automation, vol I, IEEE, pp 695–701

7. Rahnamayan S, Tizhoosh H, Salama M (2008) Opposition versus randomness in soft computing techniques. Appl Soft Comput 8(2):906–918
8. Ventresca M, Rahnamayan S, Tizhoosh H (2010) A note on "opposition versus randomness in soft computing techniques". Appl Soft Comput 10(3):956–957
9. Rahnamayan S, Tizhoosh H, Salama M (2007) Quasi-oppositional differential evolution. In: Proceedings IEEE congress on evolutionary computation, CEC 2007, IEEE, pp 2229–2237
10. Sharma H, Jadon S, Bansal J, Arya K (2013) Levy flight based local search in differential evolution. In: Swarm, evolutionary, and memetic computing, vol 8297. Lecture notes in computer science. Springer, Berlin, pp 248–259
11. Ali M, Khompatraporn C, Zabinsky Z (2005) A numerical evaluation of several stochastic algorithms on selected continuous global optimization test problems. J Glob Optim 31(4):635–672

Studies on Two Classes of Skew Cyclic Codes Over Rings

Yan Li and Xiuli Li

Abstract Two classes of skew cyclic codes over rings are studied in this paper. According to their features, we present proper automorphisms. Combining with the given automorphisms, we construct skew cyclic codes and discuss the properties of the codes.

Keywords Automorphisms · Skew cyclic codes

1 Introduction

Recently, the appearance of skew cyclic codes has led coding theory into a new direction. Many scholars have drawn certain useful research results with the help of algebra knowledge. So far, research results are mainly obtained on Gauss Ring, $Z_2 + \mu Z_2 + \mu^2 Z_2$, $F_4 + \nu F_4$, $F_p + \nu F_p (\nu^2 = 1)$ [1, 2]. Based on the former and original findings, we keep on looking for new conclusions to increase and enrich coding theory [5–8].

In this paper, we are interested in discussing two classes of skew cyclic codes based on rings with restrictions. One is the ring $R_1 = F_p + \nu F_p = \{a + b\nu | a, b \in F_p\}$ with $\nu^2 = \nu$, the other one is the ring $R_2 = F_{2^m} + \nu F_{2^m} = \{a + b\nu | a, b \in F_{2^m}\}$ with $\nu^2 = \nu$, where F_p and F_{2^m} are finite fields.

In this section, R is an ordinary ring, θ is an automorphism of R.

Research supported by reward fund for outstanding young and middle-aged scientists of ShanDong(BS2011DX011).

Y. Li · X. Li (✉)
School of Mathematics and Physics, Qingdao University of Science and Technology, Qingdao 266042, China
e-mail: lixiuli2004@tom.com

Y. Li
e-mail: 1273070631@qq.com

© Springer International Publishing Switzerland 2017

V.E. Balas et al. (eds.), *Information Technology and Intelligent Transportation Systems*, Advances in Intelligent Systems and Computing 454,
DOI 10.1007/978-3-319-38789-5_19

Definition 1 Let $R[x,\theta] = \{a_0 + a_1x + a_2x^2 + \cdots + a_nx^n | a_i \in R, 0 \le i \le n\}$. In $R[x,\theta]$, we definite two operations: addition and multiplication. The addition is defined to be the usual addition of polynomials and the multiplication is defined by the rule $(ax^i) * (bx^j) = a\theta^i(b)x^{i+j}$, and extended to all elements of $R[x,\theta]$ by associative law and distributing. It is easy to check that the set $R[x,\theta]$ contains unit. Then we call $R[x,\theta]$ a skew polynomial ring and each element in $R[x,\theta]$ a skew polynomial. Obviously, $R[x,\theta]$ is non-commutative.

For $f(x), g(x) \in R[x,\theta]$ which are non-zero polynomials and the leading coefficient of $g(x)$ is a unit, there exist unique polynomials $p(x), q(x) \in R[x,\theta]$, such that $f(x) = p(x) * g(x) + q(x)$ with $q(x) = 0$ or $deg(q(x)) < deg(g(x))$. If $q(x) = 0$, then $g(x)$ is a right divisor of $f(x)$, $g(x)$ divides $f(x)$ from right. Similarly, we also give the definition of left division.

Let $R_n = R[x,\theta]/\langle x^n - 1\rangle$. And note

$$R_n = \{r(x) | r(x) = r_0 + r_1x + \cdots + r_{n-1}x^{n-1}, r_i \in R, 0 \le i \le n-1\}.$$

Define the map ρ,

$$\begin{aligned} \rho : R[x,\theta] &\to R[x,\theta]/\langle x^n - 1\rangle, \\ f(x) &\to r(x), \end{aligned}$$

where $r(x) = f(x) + p(x)(x^n - 1)$, $p(x)$ is a polynomial of $R[x,\theta]$. Clearly,we have $f(x) \equiv r(x) mod (x^n - 1)$. That is, the image of $f(x)$ under map ρ is the remainder of $f(x)$ divided by $x^n - 1$ from right.

Definition 2 Let R^n be the direct product of R and $R^n = \{(r_0, r_1, \ldots, r_{n-1}) | r_i \in R, 0 \le i \le n-1\}$. Let C be a subset of R^n, then we call C a skew cyclic code of length n, if
(1) C is a R–submodule of R^n;
(2) If $(c_0, c_1, \ldots, c_{n-1}) \in C$, then $(\theta(c_{n-1}), \theta(c_0), \ldots, \theta(c_{n-2})) \in C$.

We call any vector element$(c_0, c_1, \ldots, c_{n-1})$ a codeword, n is the length of the codewords.

Definition 3 Let C be a skew cyclic code and θ be an automorphism of R. For all elements in R, if there exists an integer t, such that $\theta^t(a + b\nu) = a + b\nu$, then we call the minimal positive integer t as the order of θ and note $|\langle\theta\rangle|$.

For example, let $R = R_1$, we give two automorphisms θ_1, θ_2 over R_1,

$$\begin{aligned} &\theta_1 : \theta_1(a + b\nu) = a - b\nu; \\ &\theta_2 : \theta_2(a + b\nu) = a + b(1 - \nu). \end{aligned}$$

By the above definition, $|\langle\theta_1\rangle| = |\langle\theta_2\rangle| = 2$. Obviously, for all $a \in F_p, \theta(a) = a$.

Similarly, let $R = R_2$, the automorphism θ are defined as follows:

$$\theta : \theta(a + b\nu) = a^{2^q} + b^{2^q}\nu, \text{ (q being positive integer and $q \leq m$).} \quad (1)$$

According to the definition, $|\langle\theta\rangle| = \frac{m}{(m,q)}$. Suppose that $q = 1$, then $|\langle\theta\rangle| = m$.

2 Skew Cyclic Codes Over R_1

Let $R_1 = F_p + \nu F_p = \{a + b\nu | a, b \in F_p\}$ with $\nu^2 = \nu$, θ is an automorphism over it.

Lemma 1 *If $n|t$, then $\langle x^n - 1\rangle$ is a two-side ideal of the skew polynomial ring $R[x, \theta]$.*

Proof For any polynomial $f(x) \in R[x, \theta]$, let

$$f(x) = a_0 + a_1x + a_2x^2 + \cdots + a_mx^m,$$

where $a_i \in R$ and $i = 0, 1, \ldots, m-1$,then

$$\begin{aligned}(x^n - 1) * f(x) &= (x^n - 1) * (a_0 + a_1x + a_2x^2 + \cdots + a_mx^m)\\ &= x^n * a_0 + \cdots + x^n * (a_mx^m) - (a_0 + \cdots + a_mx^m)\\ &= \theta^n(a_0)x^n + \cdots + \theta^n(a_m)x^{n+m} - (a_0 + \cdots + a_mx^m).\end{aligned}$$

Since $\theta^n(a_i) = a_i$, we get

$$\begin{aligned}(x^n - 1) * f(x) &= a_0x^n + a_1x^{n+1} + \cdots + a_mx^{n+m} - (a_0 + \cdots + a_mx^m)\\ &= (a_0 + a_1x + \cdots + a_mx^m) * (x^n - 1)\\ &= f(x) * (x^n - 1).\end{aligned}$$

Hence, $x^n - 1$ commutes with each element of $R[x, \theta]$, so $x^n - 1$ is the center of $R[x, \theta]$. Thus $\langle x^n - 1\rangle$ is a two-side ideal.

Theorem 1 *A code C of length n over R is a skew cyclic code if and only if C is a left ideal of $R_n = R[x, \theta]/\langle x^n - 1\rangle$.*

Proof C is a skew cyclic code over R of length n. For any vector $(c_0, c_1, \ldots, c_{n-1}) \in C$, we have $(\theta(c_{n-1}), \theta(c_0), \ldots, \theta(c_{n-2})) \in C$. Accordingly, if $f(x) = c_0 + c_1x + \cdots + c_{n-1}x^{n-1} \in C$, then

$$\theta(c_{n-1}) + \theta(c_0)x + \cdots + \theta(c_{n-2})x^{n-1} \in C.$$

Clearly, C is a R−submodule of R^n; for any $f(x) \in C$,

$$\begin{aligned} x * f(x) &= x * (c_0 + c_1 x + \cdots + c_{n-1} x^{n-1}) \\ &= x * c_0 + x * (c_1 x) + \cdots + x * (c_{n-1} x^{n-1}) \\ &= \theta(c_{n-1}) + \theta(c_0) x + \cdots + \theta(c_{n-2}) x^{n-1} \in C, \end{aligned}$$

Then for any $r(x) \in R_n$,$r(x) * f(x) \in C$. In summary, C is a left ideal of R_n. Conversely, suppose that C is a left ideal of R_n. For any $f(x) \in C$,

$$\begin{aligned} x * f(x) &= x * (c_0 + c_1 x + \cdots + c_{n-1} x^{n-1}) \\ &= x * c_0 + x * (c_1 x) + \cdots + x * (c_{n-1} x^{n-1}) \\ &= \theta(c_{n-1}) + \theta(c_0) x + \cdots + \theta(c_{n-2}) x^{n-1} \in C, \end{aligned}$$

That is, for any $(c_0, c_1, \ldots, c_{n-1}) \in C$, we get $(\theta(c_{n-1}), \theta(c_0), \ldots, \theta(c_{n-2})) \in C$. This implies that C is a skew cyclic code over R.

Theorem 2 *Let C be a skew cyclic code of length n over R, and $g(x)$ be a monic polynomial with minimal degree in C, then $C = \langle g(x) \rangle_l$, $g(x)|_r(x^n - 1)$. (In this paper, $\langle\rangle_l$ stands for left ideal, $|_r$ represents right division).*

Proof For any polynomial $f(x) \in C$, $f(x) = p(x) * g(x) + q(x)$, then $q(x) = 0$. (If $q(x) \neq 0$, by the linear operation of code words, we have $q(x) = f(x) - p(x) * g(x) \in C$. Since $deg(q(x)) < deg(g(x))$, it produce a contradiction). Therefore, $f(x) = p(x) * g(x)$, $C = \langle g(x) \rangle_l$.

Let $x^n - 1 = p(x) * g(x) + q(x)$. Since $q(x) = 0$, $x^n - 1 = p(x) * g(x)$, we know that $g(x)$ is a right factor of $x^n - 1$. Hence $g(x)|_r(x^n - 1)$.

Theorem 3 *$F_p[x^t]$ is the center of $R_1[x, \theta]$.*

Proof For all x in R_1, we have $\theta^t(x) = x$. Let $f(x) = f_0 + f_1 x + \cdots + f_i x^i + \cdots + f_m x^m$, where $f(x) \in R_1[x, \theta]$,$f_i \in R_1$ and $0 \le i \le m$; let $g(x) = g_0 + g_1 x^t + \cdots + g_j x^{tj} + \cdots + g_n x^{tn}$, where $g(x) \in F_p[x^t]$,$g_j \in F_p$ and $0 \le j \le n$; then we get $(f_i x^i) * (g_j x^{tj}) = f_i \theta^i(g_j) x^{i+tj}$,$(g_j x^{tj}) * (f_i x^i) = g_j \theta^{tj}(f_i) x^{i+tj}$.

Since $\theta^{tj}(f_i) = f_i$,$\theta^i(g_j) = g_j$,it follows that $(f_i x^i) * (g_j x^{tj}) = (g_j x^{tj}) * (f_i x^i)$, $f(x) * g(x) = g(x) * f(x)$. Therefore $F_p[x^t]$ is the center of $R_1[x, \theta]$.

3 Skew Cyclic Codes Over R_2

Let $R_2 = F_{2^m} + \nu F_{2^m} = \{a + b\nu | a, b \in F_{2^m}\}$ with $\nu^2 = \nu$, θ is an automorphism defined as (1). Similar to the proof of Theorem 2 and 3, we have the following results.

Theorem 4 *$R^*[x^t, \theta]$ is the center of $R_2[x, \theta]$. (Here $R^* = F_2 + \nu F_2$)*

If the automorphism θ over R_2 satisfies $\theta(\nu) = \nu$, $\theta(\nu - 1) = \nu - 1$, we have the following conclusion.

For the ring $R_2 = F_{2^m} + \nu F_{2^m}(\nu^2 = \nu)$, it consists of two maximal ideals $I_1 = \langle \nu - 1 \rangle$, $I_2 = \langle \nu \rangle$ and $R_2/\langle \nu - 1 \rangle \cong F_{2^m}$, $R_2/\langle \nu \rangle \cong F_{2^m}$. According to the Chinese remainder theorem, $R_2 = \langle \nu - 1 \rangle \oplus \langle \nu \rangle$, then $R_2 = \{r | r = c(\nu - 1) + d\nu, c, d \in F_{2^m}\}$. Any non-zero element r of R_2 is reversible if and only if $c, d \in F_{2^m}^*$.

Let A, B be codes over R_2, and $A \oplus B = \{a + b | a \in A, b \in B\}$. Let C be the subset of R_2^n. If C is a R_2−submodule of R_2^n, then C a linear code of length n over R_2, we define:

$$C_\nu = \{c \in F_{2^m}^n | \exists d \in F_{2^m}^n, c(\nu - 1) + d\nu \in C\},$$
$$C_{\nu-1} = \{d \in F_{2^m}^n | \exists c \in F_{2^m}^n, c(\nu - 1) + d\nu \in C\},$$

Clearly, C_ν and $C_{\nu-1}$ are linear codes of R_2; by the definition we get

$$C = (\nu - 1)C_\nu \oplus \nu C_{\nu-1}.$$

Theorem 5 *Let $C = (\nu - 1)C_\nu \oplus \nu C_{\nu-1}$ be a linear code of length n over R_2, then C is a skew cyclic code of length n over R_2 if and only if C_ν and $C_{\nu-1}$ are all skew cyclic codes of length n over F_{2^m}.*

Proof Note that any element $c = (c_0, \ldots, c_i, \ldots, c_{n-1})$ and $c_i = (\nu - 1)a_i + \nu b_i$, where $a_i, b_i \in F_{2^m}$ and $0 \le i \le n - 1$. Meanwhile note $a = (a_0, \ldots, a_i, \ldots, a_{n-1}) \in C_\nu$, $b = (b_0, \ldots, b_i, \ldots, b_{n-1}) \in C_{\nu-1}$.

Since C is a skew cyclic code over R_2, we have $\vartheta(c) = (\theta(c_{n-1}), \theta(c_0), \ldots, \theta(c_{n-2})) \in C$. And

$$\begin{aligned}\vartheta(c) &= ((\nu - 1)\theta(a_{n-1}) + \nu\theta(b_{n-1}), (\nu - 1)\theta(a_0) + \nu\theta(b_0), \ldots, (\nu - 1)\theta(a_{n-2}) \\ &\quad + \nu\theta(b_{n-2})) \\ &= ((\nu - 1)\theta(a_{n-1}), (\nu - 1)\theta(a_0), \ldots, (\nu - 1)\theta(a_{n-2})) + (\nu\theta(b_{n-1}), \nu\theta(b_0), \\ &\quad \ldots, \nu\theta(b_{n-2})) \\ &= (\nu - 1)\vartheta(a) + \nu\vartheta(b) \in C.\end{aligned}$$

we have

$$\vartheta(a) = (\theta(a_{n-1}), \theta(a_0), \ldots, \theta(a_{n-2})) \in C_\nu,$$
$$\vartheta(b) = (\theta(b_{n-1}), \theta(b_0), \ldots, \theta(b_{n-2})) \in C_{\nu-1},$$

Hence C_ν and $C_{\nu-1}$ are all skew cyclic codes of length n over F_{2^m}.

Conversely, the same is as above.

Theorem 6 *Let $C = (\nu - 1)C_\nu \oplus \nu C_{\nu-1}$ be a skew cyclic code of length n over R_2, then $C = \langle (\nu - 1)g_1(x), \nu g_2(x) \rangle_l$ where $g_1(x)$ and $g_2(x)$ are the generators of C_ν and $C_{\nu-1}$ respectively.*

Proof From the above conclusion, we know that C_ν and $C_{\nu-1}$ are all skew cyclic codes of length n over F_{2^m}. Therefore $C_\nu = \langle g_1(x) \rangle_l$, $C_{\nu-1} = \langle g_2(x) \rangle_l$. And there exist $f_1(x) \in \langle g_1(x)\rangle_l$, $f_2(x) \in \langle g_2(x)\rangle_l$, such that $C = \{c(x)|c(x) = (\nu - 1)f_1(x) + \nu f_2(x)\} \subseteq \langle(\nu - 1)g_1(x), \nu g_2(x)\rangle_l$.

Conversely, we take two elements of $R_2[x, \theta]$, $k_1(x) = a_1(x) + b_1(x)\nu$, $k_2(x) = a_2(x) + b_2(x)\nu$.

Let $r_1(x) = a_1(x) \in F_{2^m}[x, \theta]$, $r_2(x) = a_2(x) + b_2(x) \in F_{2^m}[x, \theta]$, then

$$(\nu - 1)k_1(x) * g_1(x) = (\nu - 1)a_1(x) * g_1(x) = (\nu - 1)r_1(x) * g_1(x),$$
$$\nu k_2(x) * g_2(x) = \nu(a_2(x) + b_2(x)) * g_2(x) = \nu r_2(x) * g_2(x),$$

We have

$$(\nu - 1)k_1(x) * g_1(x) + \nu k_2(x) * g_2(x) \subseteq C,$$

Hence

$$C = \langle(\nu - 1)g_1(x), \nu g_2(x)\rangle_l.$$

Theorem 7 *Let $C = (\nu - 1)C_\nu \oplus \nu C_{\nu-1}$ is a skew cyclic code over R_2, then there exists unique $g(x)$, such that $C = \langle g(x)\rangle_l$, $g(x)|_r(x^n - 1)$.*

Proof From the above conclusion, we know that $C = \langle(\nu - 1)g_1(x), \nu g_2(x)\rangle_l$ where $g_1(x)$ and $g_2(x)$ are generators of C_ν and $C_{\nu-1}$ respectively, then $g_1(x)|_r(x^n - 1)$, $g_2(x)|_r(x^n - 1)$. Therefore there exist $r_1(x), r_2(x) \in F_{2^m}[x, \theta]$ such that

$$x^n - 1 = r_1(x) * g_1(x), x^n - 1 = r_2(x) * g_2(x)$$

Let $g(x) = (\nu - 1)g_1(x) + \nu g_2(x)$. Clearly, $\langle g(x)\rangle_l \subseteq C$; and

$$(\nu - 1)g(x) = (\nu - 1)^2 g_1(x) = (\nu - 1)g_1(x),$$
$$\nu g(x) = \nu^2 g_2(x) = \nu g_2(x),$$

hence $\langle g(x)\rangle_l \supseteq C$, we get $C \doteq \langle g(x)\rangle_l$.

Since $g_1(x)$ and $g_2(x)$ are unique, $g(x)$ is also unique, and

$$\begin{aligned} x^n - 1 &= [(\nu - 1)r_1(x) + \nu r_2(x)] * [(\nu - 1)g_1(x) + \nu g_2(x)] \\ &= (\nu - 1)^2 r_1(x) * g_1(x) + \nu^2 r_2(x) * g_2(x) \\ &= (\nu - 1)r_1(x) * g(x) + \nu r_2(x) * g(x) \\ &= [(\nu - 1)r_1(x) + \nu r_2(x)] * g(x) \end{aligned}$$

Hence $g(x)|_r(x^n - 1)$.

4 Applications

Example 1 Let $R = F_4 + \nu F_4$, we define an automorphism $\theta : \theta(a + bv) = a^2 + b^2\nu$, then $|\langle\theta\rangle| = 2$. Consider the generators of skew cyclic code of length 4. Since $x^4 + 1 = (x^2 + \omega x + \omega^2 + \nu) * (x^2 + \omega x + \omega + \nu)$ where $\omega^2 + \omega + 1 = 0$. Let $g(x) = x^2 + \omega x + \omega + \nu$ be a generator, then the generator matrix is

$$\mathbf{G} = \begin{pmatrix} \omega + \nu & \omega & 1 & 0 \\ 0 & \omega^2 + \nu & \omega^2 & 1 \end{pmatrix}$$

we get a [2–4] code.

5 Conclusion

In this paper, we mainly discuss two classes of skew cyclic codes by giving different automorphisms. We get the existence conditions of skew cyclic codes over $R_1 = F_p + \nu F_p = \{a + b\nu | a, b \in F_p\}$ with $\nu^2 = \nu$ and its generating polynomials. We also study the direct sum decomposition of skew cyclic codes over $R_2 = F_{2^m} + \nu F_{2^m} = \{a + b\nu | a, b \in F_{2^m}\}$ with $\nu^2 = \nu$. The automorphisms have key effect on the properties of skew cyclic codes. Therefore, looking for fine automorphisms is important. The following works will be related with finding new automorphisms over rings.

References

1. Li J, Zhu S (2011) Skew cyclic codes over $Z_2 + \mu Z_2 + \mu^2 Z_2$. J Hefei Univ Technol (Natural Science) 34:1745–1748
2. Lin J (2014) Skew cyclic codes over rings $F_p + \nu F_p(\nu^2 = 1)$. J Electron 31:227–231
3. Xu X, Zhu S (2011) Skew cyclic codes over $F_4 + \nu F_4$. J Hefei Univ Technol (Natural Science) 34:1429–1432
4. Gao J (2013) Skew cyclic codes over $F_p + \nu F_p$. Appl Math Inf 31:337–342
5. Pless V, Sole P, Qian Z (1997) Cyclic self-dual Z_4- codes. Finite Fields Appl 3:334–352
6. Wolfmann J (1999) Negacyclic and cyclic codes over Z_4 codes. IEEE Trans Inf Theory 45:2527–2532
7. Boucher D, Geiselmann W, Ulmer F (2007) Skew cyclic codes. Appl Algebr Eng Commun Comput 18(4):379–389
8. Boucher D, Ulmer F (2009) Coding with skew polynomial rings. J Symbol Comput 44:1644–1656

C-Logit Stochastic System Optimum Traffic Assignment

Gui Yong and Yan Xu

Abstract The C-logit model proposed by Cascetta et al. in 1996 aims at resolving the route overlapping problem in logit-based stochastic traffic assignment, being lack of economic meanings. In this paper, we formulate the C-logit system optimum as a mathematical program through maximizing the system's economic benefit. Numerical studies are conducted for illustrating C-logit system optimum and user equilibrium.

Keywords Route overlapping · C-logit model · Stochastic system optimum · Net economic benefit

1 Introduction

In traffic network modeling and analyses, it is usually assumed that the travelers have perfect information about the whole traffic, and then try to minimize their perceived travel costs by choosing favorable routes, departure times or transport modes. If the perceived travel costs are exactly the actual ones, this assumption leads to the deterministic user equilibrium (DUE) assignment. In reality, travelers traveling on the same route may perceive different travel times or costs due to their inherent perception diversities, external information availabilities or other unmeasured factors. If the perceived route costs are treated as Gumbel random variables and follow an independent and identical distribution (IID), all users' route choices for minimizing their own perceived travel costs result in multinomial logit-based stochastic user equilibrium (MNL-SUE) [1]. The MNL-SUE model has a simple closed-form of the route

G. Yong (✉)
School of Economics and Management, Beihang University, Beijing 100191, China
e-mail: baoyonggui@163.com

G. Yong · Y. Xu
School of Statistics and Mathematics, Inner Mongolia University of Finance and Economics, Huhhot 010070, China
e-mail: buaayanxu@163.com

© Springer International Publishing Switzerland 2017

V.E. Balas et al. (eds.), *Information Technology and Intelligent Transportation Systems*, Advances in Intelligent Systems and Computing 454,
DOI 10.1007/978-3-319-38789-5_20

choice probability expression, but has been criticized for giving unrealistic results because the route overlapping problem is not well resolved [2]. In the MNL-SUE model, two partly overlapped routes are independently handled, which is obviously unreasonable.

To overcome the above deficiency, some modifications to the MNL-SUE model have been conducted [3–9]. Cascetta et al. [10] proposed a so-called C-logit SUE (CL-SUE) model, aiming at resolving the overlapping problem by adding a commonality factor into the deterministic part of the utility formula while remaining the single-level tree structure of the logit model. Zhou et al. [11] developed an equivalent mathematical programming formula of the length-based CL-SUE model.

The stochastic user equilibrium (SUE) solution corresponds to the DUE solution with drivers choosing the route which minimizes their personal perceived travel cost. Maher et al. [12] formulates the stochastic social optimum (SSO) which relates to the SUE solution in the same way as the deterministic system optimum (DSO) solution relates to the DUE solution. The SSO solution is that flow pattern which minimizes the total of the travel costs perceived by drivers. Just as the DSO solution generally requires some drivers traveling on paths which are not the minimum cost paths for that OD pair, so the SSO solution generally requires some drivers to be assigned to paths that are not their minimum perceived cost path. Yang [13] has characterized the SSO solution of logit model as that which maximizes consumer surplus.

The paper is organized as follows. For completeness, Sect. 2 briefly reviews the CL-SUE model and derives the formulation of the CL-SSO. In Sect. 3, we give a simple example. Section 4 concludes the paper.

2 CL-SSO Model

Define a strongly connected transportation network (N, A), where N and A are the sets of nodes and links, respectively. Let W denote the set of all origin-destination (OD) pairs, R_w be the set of all routes connecting OD pair $w \in W$, R_w be the travel demand between OD pair $w \in W$, f_{rw} be the flow on route $r \in R_w$, $w \in W$, v_a be the flow on link $a \in A$. Set an indicator δ_{ar} which is 1 if route r uses link a and 0 otherwise. The following relationships and constraints hold:

$$d_w = \sum_{r \in R_w} f_{rw}, w \in W, \tag{1}$$

$$f_{rw} \geq 0, r \in R_w, w \in W, \tag{2}$$

$$v_a = \sum_{w \in W} \sum_{r \in R_w} f_{rw} \delta_{ar}, a \in A. \tag{3}$$

In this paper, we consider a separable and increasing link travel time function $t_a(v_a)$, $a \in A$. Let c_{rw} be the actual travel time of route $r \in R_w$, $w \in W$, which is the sum of travel time on all links that constitute the route. We then have

$$c_{rw} = \sum_{a \in A} t_a(v_a)\delta_{ar}, r \in R_w, w \in W. \tag{4}$$

To overcome the deficiency of the MNL-SUE model, Cascetta et al. [10] proposed a CL-SUE model by modifying the deterministic part of the utility. The CL-SUE model considers the similarity between overlapping routes by adding a commonality factor in the deterministic part of the utility function, while keeping the analytical closed-form expression of route choice probabilities. Route choice probability P_{rw} is given below

$$P_{rw} = \frac{\exp(-\theta(c_{rw} + cf_{rw}))}{\sum_{k \in R_w} \exp(-\theta(c_{kw} + cf_{kw}))}, r \in R_w, w \in W, \tag{5}$$

where cf_{rw} is a commonality factor of route $r \in R_w$. The commonality factor is computed by equation [10]

$$cf_{rw} = \beta \ln \sum_{k \in R_w} \left(\frac{L_{rk}}{\sqrt{L_r} \cdot \sqrt{L_k}} \right)^{\gamma}, r \in R_w, w \in W, \tag{6}$$

where L_{rk} is the 'length' of links commonly contained in routes r and k, L_r and L_k are the overall 'lengths' of routes r and k, respectively, in this the 'lengths' represent the free-flow time. The route flow assignment is given by

$$f_{rw} = d_w P_{rw}, r \in R_w, w \in W. \tag{7}$$

If the commonality factor is assessed based on the flow-independent attributes, e.g., physical length or free-flow travel time, the CL-SUE model with fixed OD demand can be formulated as an equivalent minimization problem [11]:

$$\min_{f \in \Omega_f} Z(f) = \sum_{a \in A} \int_0^{v_a} t_a(\omega) d\omega + \frac{1}{\theta} \sum_{w \in W} \sum_{r \in R_w} f_{rw} \ln f_{rw} + \sum_{w \in W} \sum_{r \in R_w} f_{rw} cf_{rw}. \tag{8}$$

It is known that for the logit-type model, Eq. (5), the expected indirect utility received by a randomly sampled individual can be expressed as [14, 15]

$$S_w = \mathrm{E}\{\max_{r \in R_w}[-(c_{rw} + cf_{rw}) + \xi_{rw}]\} = -\frac{1}{\theta} \sum_{r \in R_w} \exp[-\theta(c_{rw} + cf_{rw})]. \tag{9}$$

According to the representative consumer theory of the logit model [16], the behavior of consumers with different tastes can be described by the choices made by a single individual who has a preference for diversity. The direct utility function of traveler can be expressed as

$$U = -\frac{1}{\theta}\sum_{w\in W}\sum_{r\in R_w} f_{rw}\ln f_{rw} - \sum_{w\in W}\sum_{r\in R_w} f_{rw}cf_{rw} + T_0, \tag{10}$$

where T_0 is a constant representing the amount spent on non-travel items. The Eq. (10) consistent with the logit model is an entropy-type function that has also been used as a benefit measure in terms of interactivity in trip distribution. The total social cost is given by $TC = \sum_{a\in A} t_a(v_a)\cdot v_a$.Then, the net economic benefit could be measured as $(U - TC)$, minimizing negative value of the net economic benefit, we have the optimization program

$$\min \sum_{a\in A} t_a(v_a)\cdot v_a + \frac{1}{\theta}\sum_{w\in W}\sum_{r\in R_w} f_{rw}\ln f_{rw} + \sum_{w\in W}\sum_{r\in R_w} f_{rw}cf_{rw}. \tag{11}$$

Subject to

$$d_w = \sum_{r\in R_w} f_{rw},\ w\in W, \tag{12}$$

$$v_a = \sum_{w\in W}\sum_{r\in R_w} f_{rw}\delta_{ar},\ a\in A, \tag{13}$$

$$f_{rw}\ge 0,\ r\in R_w,\ w\in W. \tag{14}$$

The following Kuhn–Tucker conditions for any $f_{rw}>0$ are also sufficient to obtain the unique optimal solution

$$\sum_{a\in A}\bar{t}_a(v_a)\delta_{ar} + \frac{1}{\theta}(\ln f_{rw}+1) + cf_{rw} - \lambda_w = 0,\ r\in R_w,\ w\in W, \tag{15}$$

where λ_w is the Lagrange multiplier associated with constraint (12), and

$$\bar{t}_a(v_a) = t_a(v_a) + \hat{t}_a(v_a), \tag{16}$$

where

$$\hat{t}_a(v_a) = v_a\frac{\partial t_a(v_a)}{\partial v_a}. \tag{17}$$

Evidently, the first term of the right-hand side of Eq. (16) is the actual link travel time of traveler and the second term is the additional travel time that a traveler imposes on all other travelers in the link. Setting $\bar{c}_{rw} = \sum_{a\in A}\bar{t}_a(v_a)\delta_{ar}$ and $\hat{c}_{rw} = \sum_{a\in A}\hat{t}_a(v_a)\delta_{ar},\ r\in R_w,\ w\in W$, then $\bar{c}_{rw} = c_{rw} + \hat{c}_{rw}$. According Eq. (15), we have

$$\ln f_{rw} = \theta(\lambda_w - \bar{c}_{rw} - cf_{rw}) - 1. \tag{18}$$

Simplify Eq. (18) obtain

$$P_{rw} = \frac{\exp(-\theta(\overline{c}_{rw} + cf_{rw}))}{\sum_{k \in R_w} \exp(-\theta(\overline{c}_{kw} + cf_{kw}))}, r \in R_w, w \in W. \tag{19}$$

From the above derivation, we can easily observe that the formulation of CL-SSO be equivalent to the expression which be similar to route choice probability of CL-SUE.

3 Numerical Example

Figure 1 illustrates an example network for the CL-SSO. The network has one OD pair (A, B) connected by 3 routes. Route 2-3 overlaps with route 2-4. Setting travel time functions for all segments:$t_1 = 1 + 0.02x$, $t_2 = 0.6 + 0.01x$, $t_3 = 0.4 + 0.01x$ and $t_4 = 0.2 + 0.01x$.

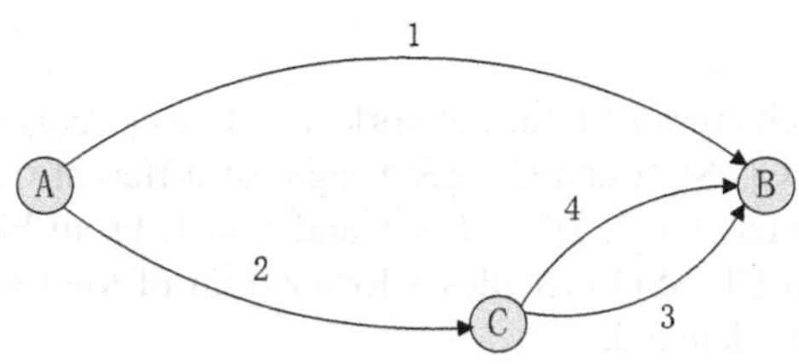

Fig. 1 An example network

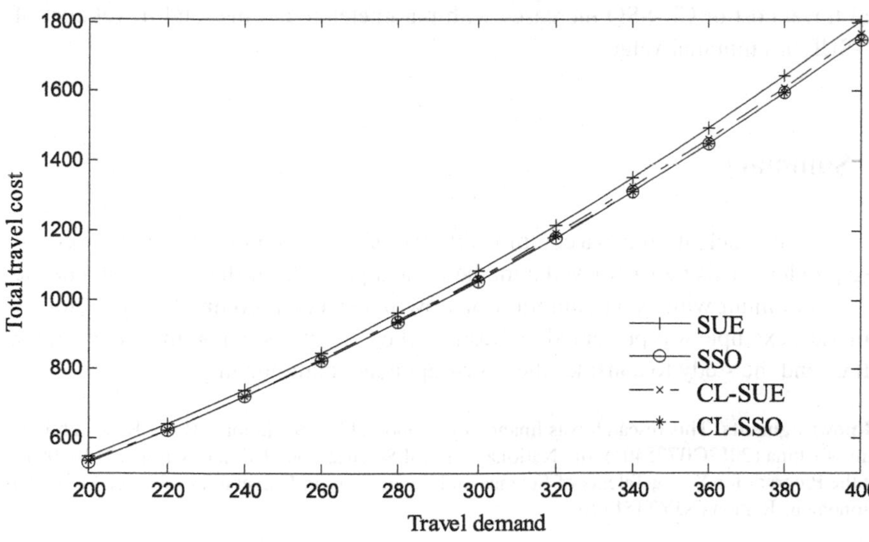

Fig. 2 Changes of total network travel cost against demand

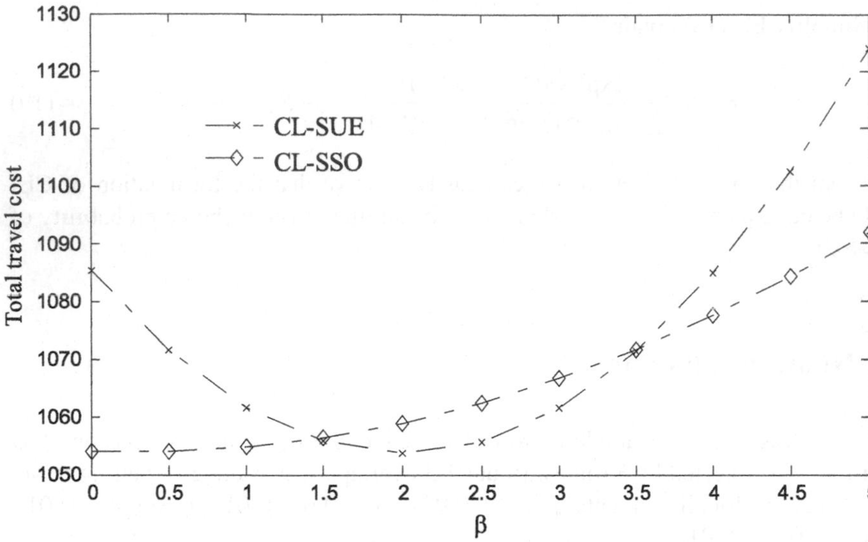

Fig. 3 Changes of total network travel cost against parameter β

Figure 2 plots the changes of total network travel costs (corresponding, respectively, to SUE, SSO, CL-SUE and CL-SSO) against different travel demand for a given value of parameter $\theta = 3.2064$, $\beta = 1$ and $\gamma = 1$. From Fig. 2 we know that the total travel cost of CL-SSO provides a lower limit of total network travel cost associated with a given demand.

Figure 3 shows the changes of total network travel cost against parameter β for a given demand D=300, $\theta = 3.2064$ and $\gamma = 1$. From this figure, we know that the total travel cost of CL-SSO increases with parameter β and the total travel cost of CL-SUE has minimal value.

4 Summary

The C-logit model introduces a commonality factor for overcoming the route overlapping problem. This paper derived a mathematical program of the C-logit stochastic system optimum with the assumption of flow-independent commonality factor. A numerical example was presented to illustrate the C-logit system optimum. Next, we will extend the study to consider the flow-dependent commonality factor.

Acknowledgments This research was financially supported by the National Basic Research Program of China (2012CB725401), the National Natural Science Foundation of China (71401083), and the Program for Young Talents of Science and Technology in Universities of Inner Mongolia Autonomous Region (NJYT15B06).

References

1. Sheffi Y (1985) Urban transportation networks. Prentice-Hall, Englewood Cliffs
2. Daganzo CF, Sheffi Y (1977) On stochastic models of traffic assignment. Transp Sci 11:253–274
3. Bekhor S, Prashker JN (2001) Stochastic user equilibrium formulation for generalized nested logit model. Transp Res Rec 1752:84–90
4. Huang HJ, Wang SY, Bell MGH (2001) A bi-level formulation and quasi-Newton algorithm for stochastic equilibrium network design problem with elastic demand. J Syst Sci Complex 14:34–50
5. Chen P, Kasikitwiwat Z, Ji Z (2003) Solving the overlapping problem in route choice with paired combinatorial logit model. Transp Res Rec 1857:65–73
6. Papola A (2004) Some developments on the cross-nested logit model. Transp Res B-Meth 38B:833–851
7. Huang HJ, Liu TL, Guo XL, Yang H (2011) Inefficiency of logit-based stochastic user equilibrium in a traffic network under ATIS. Netw Spat Econ 11:255–269
8. Pravinvongvuth S, Chen A (2005) Adaptation of the paired combinatorial logit model to the route choice problem. Transportmetrica 1:223–240
9. Hoogendoorn S, Bovy P (2007) Modeling overlap in multimodal route choice by including trip part-specific path size factors. Transp Res Rec 2003:78–83
10. Cascetta E, Nuzzolo A, Russo F, Vitetto A (1996) A modified logit route choice model overcoming path overlapping problems: specification and some calibration results for interurban networks. In: Proceedings of the 13th international symposium on transportation and traffic theory, Lyon, pp 697–711
11. Zhou Z, Chen A, Bekhor S (2012) C-logit stochastic user equilibrium model: formulations and solution algorithm. Transportmetrica. 8:17–41
12. Maher M, Stewart K, Rosa A (2005) Stochastic social optimum traffic assignment. Transp Res B-Meth 39B:753–767
13. Yang H (1999) System optimum, stochastic user equilibrium, and optimal link tolls. Transp Sci 33:354–360
14. Williams HCWL (1977) On the formation of travel demand models and economic evaluation measures of user benefits. Environ Plann A 9:285–344
15. Small KA, Rosen HS (1981) Applied welfare economics with discrete choice models. Econometrica. 49:105–130
16. Oppenheim N (1995) Urban travel demand modeling: from individual choices to general equilibrium. Wiley, New York

Intelligent Robot Finger Vein Identification Quality Assessment Algorithm Based on Support Vector Machine

Yu Chengbo, Yuan Yangyu and Yang Rumin

Abstract The identity recognition technology as an important aspect in the field of artificial intelligence. Especially with the development of intelligent robot has the function of identification, the application fields of further widening. This paper presents an identification algorithm of finger vein quality assessment based on the support vector machine. Through the analysis of some existing features, and analyses the three characteristic parameters of great influence on the finger vein image quality (image contrast, image gradient covariance feature values, and thc effective area). Vein image by establishing a model of support vector machine to the known training quality, and then, classify the test image random sampling. The experimental results show that this algorithm can well distinguish the vein image high and low quality for enhance the performance of the finger vein identification system.

Keywords Intelligent · Robot · SVM · Identification

1 Introduction

The identification on the robot, and it has the ability to identify the active research topic in the field of intelligent robot research. The existing recognition system such as: face recognition: the robot through the collection of face images are matched to determine the identity of the visitors. Also by the visitor's fingerprint, voice, iris recognition method, but this belongs to the biological external features are easily copied is not conducive to the protection of visitor information. However, advantages of finger vein recognition. In robot vision, the more common finger vein recognition

Y. Chengbo · Y. Yangyu (✉) · Y. Rumin
Chongqing University of Technology, Chongqing, China
e-mail: 743376143@qq.com

Y. Chengbo
e-mail: yuchengbo@cqut.edu.cn

Y. Rumin
e-mail: 47883105@163.com

© Springer International Publishing Switzerland 2017
V.E. Balas et al. (eds.), *Information Technology and Intelligent Transportation Systems*, Advances in Intelligent Systems and Computing 454,
DOI 10.1007/978-3-319-38789-5_21

system conclude of the following steps: (1) the finger vein image acquisition, and the transmission of infrared light is more common. (2) the size of the collected image is normalized. (3) image enhancement. In general by the histogram equalization, gray transformation, denoising, convolution and other ways to improve the contrast of the vein in the image. (4) feature extraction. Mainly divided into two categories of global feature extraction and local feature extraction. (5) pattern matching. Principal component analysis method based on feature extraction.

Because the difference of individual and the influence of the factors such as light, extrusion, position offset and so on. It will be reduce the quality of the finger vein image These low quality images will affect the performance of the whole finger vein recognition system, and low quality images will increase the processing of the enhanced part. In contrast, high quality image enhancement part is simple, and the parameters of feature extraction are less, so it is necessary to divide the vein image into two categories of low quality and high quality.

Existing finger vein quality assessment method is mainly extract the vein characteristics and it make all feature weighted fusion. We obtained a quality score, then the provisions of a quality score threshold to distinguish between high quality and low quality, in the weighted fusion, Get accurate weight, it is complex and not enough to reflect the extent of the impact of image quality. So the final assessment inaccurate results. Although these methods are widely used, but the results of the quality assessment of accuracy and reliability have certain limitations, resulting in this reason there are three points: (1) In the aspect of feature extraction, only the overall extraction, can not bring the image to extract local block, so that the extraction method not only has high complexity, and the effect is not good. (2) a lot of parameters are extracted by using weighted fusion. But it is not the more the better extraction parameters, because some quality parameters and image quality is not very good. (3) single and multiple parameters weighted linear combination can well reflect mutual relationship between the quality and the characteristic parameters, and the problem is no longer a linearly separable, so the quality assessment model is attributed to nonlinear classification.

Based on the above reasons, this paper proposes a method based on support vector machine to evaluate the quality of vein image. Through the analysis of experimental data, three parameters were found (image contrast, image gradient covariance feature values, and effective area). Among them, the covariance features of image contrast and image gradient values can reflect the changes of the image gray value.

2 Select Feature Quality Assessment

2.1 The Image of Gradient Covariance Eigenvalues

In the airspace, the gradient value of the image pixel is a characterization of the ridge—valley direction clarity characteristics. In particular, a given single image is split into several pieces on-overlapping area. For each area, referred to as H,

$g_s = (g_s^x, g_s^y)$ is as gradient covariance matrix, of which g_x is the horizontal gradient at the point, g_y the the vertical gradient value at that point. For this region, all 256 pixels, its covariance matrix can be expressed as:

$$\frac{1}{256} \sum_{x \in H} g_x{}^H g_y = \begin{bmatrix} a_{11} & a_{12} \\ a_{21} & a_{22} \end{bmatrix} \tag{1}$$

The covariance matrix is a semi-positive definite matrix, meaning that value is greater than or equal to zero. The determinant of the covariance matrix eigenvalues:

$$\lambda_1 = \frac{(a_{11} + a_{22}) + \sqrt{a_{11}{}^2 + a_{22}{}^2 + 2a_{11}a_{22} - 4\det(g_x{}^H g_y)}}{2} \tag{2}$$

$$\lambda_2 = \frac{(a_{11} + a_{22}) - \sqrt{a_{11}{}^2 + a_{22}{}^2 + 2a_{11}a_{22} - 4\det(g_x{}^H g_y)}}{2} \tag{3}$$

To reflect the ridge—valley direction, the above definition eigenvalues normalized as follows:

$$X = (x_{s_i}, x_{c_i}, x_{g_i}) \tag{4}$$

Experiments show that the closer k approaches one, the better a ridge—valley direction clarity, that is $\lambda_1 \geq \lambda_2$, the k more close to 0, the worse the ridge—valley direction clarity, that is $\lambda_1 \approx \lambda_2$, now the mass fraction $Q1$ is used to show the overall feature image quality conditions in this value:

$$Q_1 = \frac{1}{s} \sum_{i=1}^{s} k_i \tag{5}$$

s represents the share of the total number of blocks, and $k_i (i = 1, 2, 3, \ldots, s)$ shows normalized feature value for each region. In this study, 250 high quality images and low-quality image were tested, high-quality content $Q1$ is mainly distributed in [0.55, 0.67], and low-quality content $Q1$ mainly in [0.22, 0.41]. Figure 1 is mass fraction distribution of high and low image quality.

2.2 Image Contrast

Image contrast is a reaction of the statistical feature of a picture's grey level. If a image is divided into 16×16 regions, each piece of contrast is defined as follows:

$$C_j = \sqrt{\frac{\sum_{i=1}^{N} (x_i - \overline{x})^2}{N}} \tag{6}$$

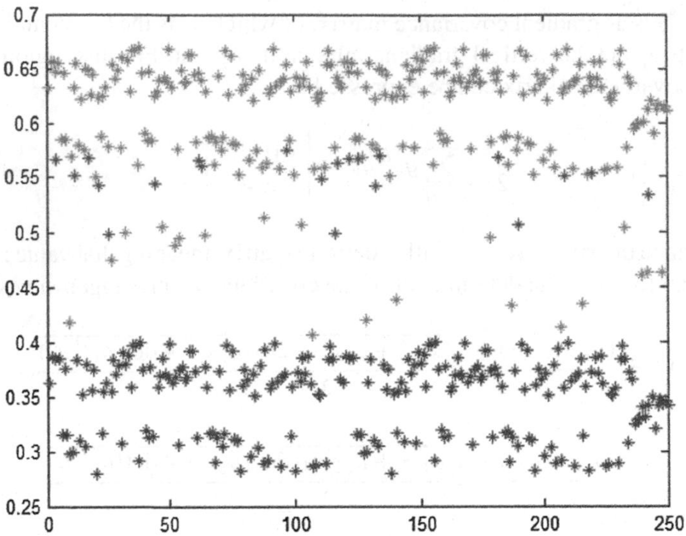

Fig. 1 Mass fraction distribution of high and low image quality

N represents the total number of pixels inside each piece, and x_i represents a pixel value, and represents the average gray value of each block area. The entire image contrast is the average value of each piece of contrast:

$$Q_2 = \frac{1}{S}\sum_{j=1}^{S} C_j \tag{7}$$

In the experiment, 250 high and low quality of the finger vein image are used to calculate the contrast of the image size, and finally get the mass fraction under this feature. For high-quality images, the mass fraction ranges from 14–16. For low-quality images, the mass fraction ranges from 11–14. Figure 2 is 250 high quality low quality scores vein image distribution.

2.3 *Effective Area*

Visibly, high-quality images is obvious in very obvious ridge- valley direction from high-quality image and low-quality images, that is clearly visible to the vein (Figs. 3 and 4), and in the low-quality image it is almost not clear (Figs. 5 and 6).

From the above analysis, it shows that the image of all ridge—Valley area accounting for the percentage of the entire image is treated as an effective area (Fig. 7):

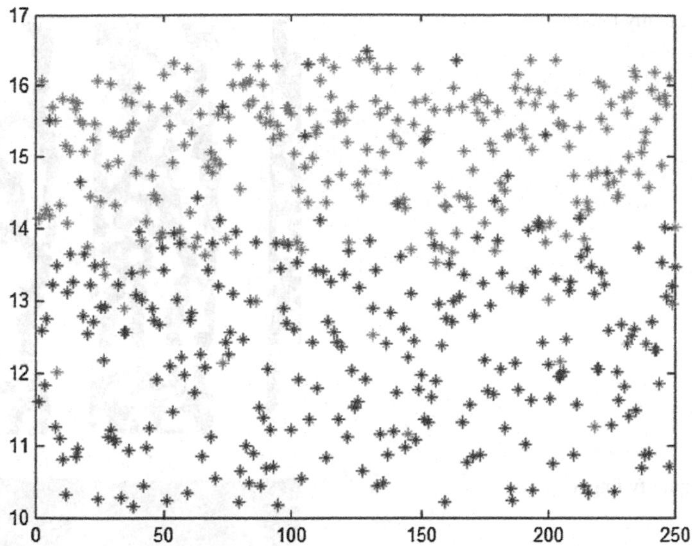

Fig. 2 250 Mass fraction distribution of high and low image quality

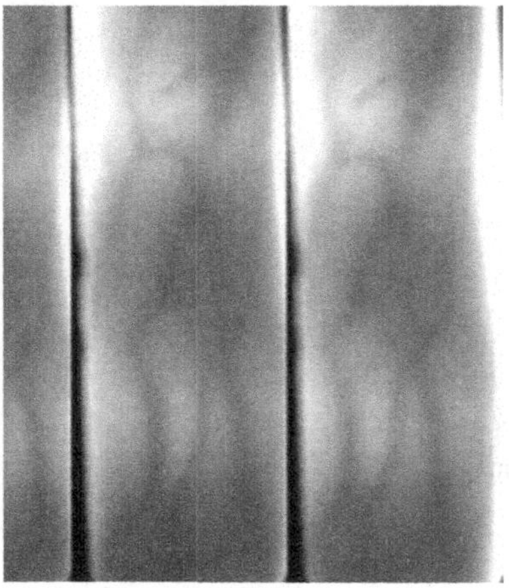

Fig. 3 High quality finger vein

$$S_{eff} = \frac{S'}{(high \times weigh)} \times 100\,\% \tag{8}$$

In formula (8), S_{eff} represents the effective area. S' represents the ridge—valley area, which refers to the number of pixels of the foreground image in the binarized finger vein image. represent the height and width of the image. In the experiment, 250 pieces' images of high and low quality are tested. The effective area of high-quality images ranges from 0.41 to 0.53,while that of low-quality images is less than 0.39 (Fig. 3).

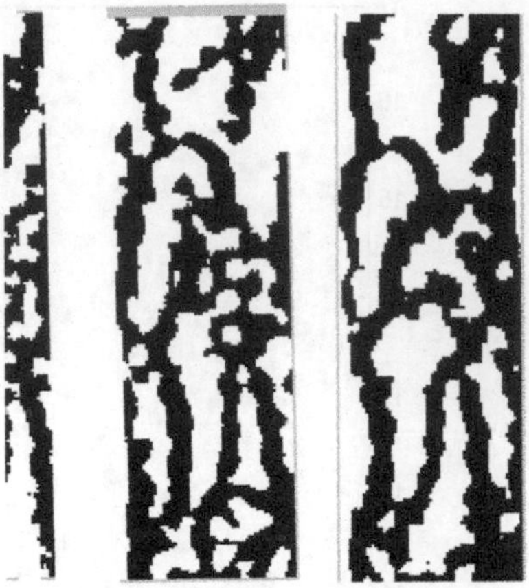

Fig. 4 High quality binary image

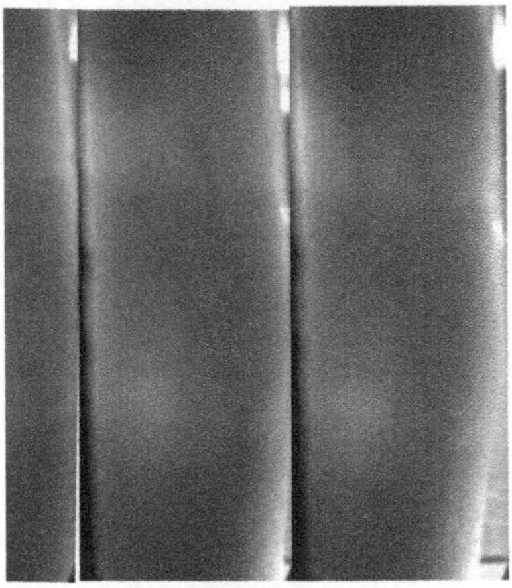

Fig. 5 Low quality finger vein

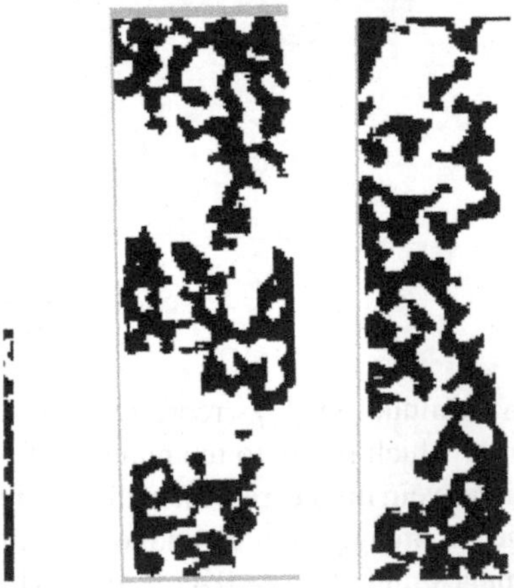

Fig. 6 High quality binary image

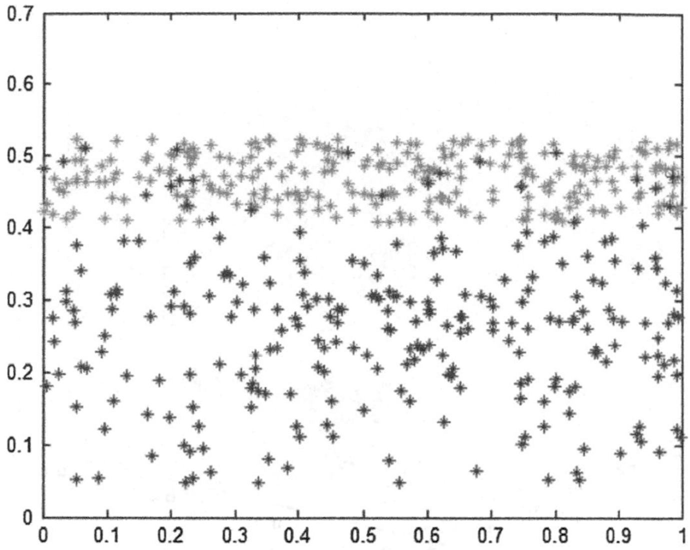

Fig. 7 250 high and low quality of the finger vein image effective area

3 SVM Theory and Analysis

Support vector machine theory was put forward firstly by the Cortes and Vapnik in 1995, used widely in pattern recognition, linear and nonlinear classifier field to find an optimal strategy to make the experience of risk and confidence risk minimization. In this paper, the problem is to conduct a convex quadratic programming finger vein set, meaning being divided into two categories with high quality and low quality. From SVM theory, it is a typical two kinds of classifiers, answering only for positive or negative type of problems, so this method can solve the problem of text classification in this paper.

To solve this problem, we will do some quantitative processing with output of quality assessment, referring to $g(x)$ as an output, recording threshold of high and low quality of images as $+1$, -1.

$$g(x) = \begin{cases} high - quality\, images\ g(x) > 1 \\ high - quality\, images\ g(x) \leq -1 \end{cases} \tag{9}$$

We will also make the following quantization processing for three characteristics of vein image the text referring, recording x_{si} to represent the effective area of the finger vein, x_{ci} showing the finger vein contrast and x_{gi} representing the eigenvalues finger vein gradient. If the three feature points mapped on the three-dimensional coordinates, they would mean a point of the three-dimensional coordinates $X = (x_{s_i}, x_{c_i}, x_{g_i})$, $(i = 1, 2, 3, \ldots, 500)$ its axis of x_{si}, x_{ci}, x_{gi} represent the effective area, contrast, gradient characteristic value, as shown in Fig. 8.

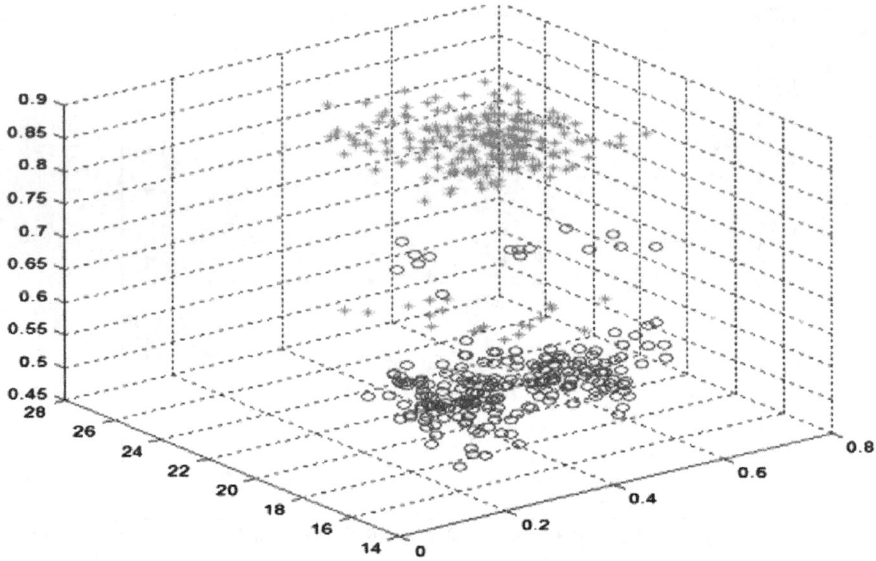

Fig. 8 Different characteristic parameters described vein image

Through the above analysis, we can see the total can be found in the three-dimensional space in an optimal classification surface, by which the finger vein image is divided into two types of high quality and low quality. The decision surface is expressed as follows:

$$g(x) = \omega X + b \tag{10}$$

In the formula $X = (x_{s_i}, x_{c_i}, x_{g_i})$, ω is a variable, b is a real number. To find the optimal classification surface,we must determine the two parameters. Because the sample point have been determined, the decision surface is also determined, and ω is determined by the sample point. ω express as:

$$\omega = AX \tag{11}$$

A represent lagrange multiplier formula, $A = (a_1, a_2, \ldots, a_n)$.

Observing positive samples and negative samples position of Fig. 4, we fix position of all points, but just one of the positive sample point change as a negative sample point (that is, the shape of a point from circular to square). Then three lines have to be moved because the three lines of this requirement is to point to the correct square and round apart, indicating that ω is not only concerned with the location of sample points, but also concerned with sample category. Therefore, the Eq. (10) is rewritten as:

$$\omega = AYX \tag{12}$$

$Y = (y_1, y_2, \ldots\ldots, y_n)$, $y_i \in \{1, -1\}$,then turn A into to get:

$$\omega = a_1 y_1 x_1 + a_2 y_2 x_2 + \cdots\cdots + a_n y_m x_n \tag{13}$$

y_i represents samples categories of the i, and about a cluster of Lagrange multipliers of the Eq. (13), only a small part is not equal to zero. This part Lagrange multiplier that is not equal to 0 multiply sample points, in fact, which fall on the H1 and H2, and it is this portion of the sample determine the classification function only. Of course, some of these samples can determine. This part of the sample point is really needed is the support vector.

The (10) of the decision surface $g(x)$ rewritten as:

$$g(x) = \sum_{i=1}^{n} a_1 y_i x_i \cdot X + b \tag{14}$$

Because x_i, X is only the vector in Eq. (14), the others parameters are real numbers, then the formula can be simplified to:

$$g(x) = \sum_{i=1}^{n} a_i y_i \langle x_i, x \rangle + b \tag{15}$$

By SVM, the finger vein quality assessment nonlinear programming problem is transformed into a linear regulation of the problem. But, inevitably, there will be the classification error because of manual classification. This kind of point can not be satisfied by the formula and the solution to the problem. To solve these fault tolerance problems of these error sample point, we introduce slack variable ζ and punishment factor c, which indicates the degree of fault tolerance. The final classification problem becomes:

$$\min \left\{ \mu(\omega, \zeta) = \frac{1}{2}\|\omega\|^2 + c \sum_{i=1}^{n} \zeta_i \right\} \tag{16}$$

Make this equation to get min:

$$\sum_{i=1}^{n} a_i y_i \langle x_i, x \rangle + b + \zeta \geq 1 \tag{17}$$

To make the results less bias,in experiment of this paper, the choice of the kernel function is radial basis function, with which this function calculate the inner product kernel. And then we combinate kernel function with "a" worked out by certain ways to gain decision surface $g(x)$.

4 Results and Analysis

4.1 First Experiment

As described above, using three features represent the quality of a finger vein, but choosing different characteristics will get different assessment results. In this experiment, a single feature, two characteristics, and three different features are used to get the comparison of experimental results. The experimental purposes is how to choose the most appropriate combination of features makes the assessment results close to the actual results, that is to say, the higher accuracy. The results are shown in Table 1.

From Table 1, to evaluate the accuracy of the intravenous quality by three kinds of features is the most suitable, especially the gradient covariance eigenvalues, the contrast, the effective area. but it can also be seen the accuracy by a single accuracy or two features are more than 60 %, which indicates that these features reflect the vein image quality level in the maximum extent.

4.2 Second Experiment

In this part of the test, using 250 high quality pictures and 250 low quality pictures as the database, from which we selected 200 high and 100 low quality pictures, 100 high and 200 low quality, and 200 high and 50 low quality, 50 high and 200 low-quality images to train the SVM for classification. In the training process, radial basis function, Sigmoid, polynomial function were selected as kernel function, for detecting impact of different kernel functions in training Effect degree. In addition, to other classification methods do the appropriate comparison, such as: Bayesian classifier, cluster analysis, decision tree learning, in Table 2.

Table 1 The accuracy of the different characteristics

Feature	Accuracy (%)
The gradient covariance eigenvalues	72.68
The contrast	80.25
The effective area	85.69
The gradient covariance eigenvalues, the effective area	87.52
The gradient covariance eigenvalues, the effective area	87.88
The contrast, the effective area	89.21
The gradient covariance eigenvalues, the contrast, the effective area	92.65
The gradient covariance eigenvalues, the contrast, information	91.44
The gradient covariance eigenvalues, the contrast, SNR	90.08

Table 2 The accuracy of the different classification algorithms, and the number of different vein

Classification	The accuracy of high quality image (%)	The accuracy of low quality image (%)	High-image	Low-image	Kernel function	The total accuracy (%)
Bayes classifier	92.02	88.21	200	100		90.11
Bayes classifier	88.63	91.56	100	200		90.09
Bayes classifier	90.25	86.87	150	50		88.56
Bayes classifier	88.02	90.44	50	150		89.23
Cluster analysis	77.15	75.65	200	100		76.40
Cluster analysis	76.89	78.45	100	200		77.67
Cluster analysis	78.20	76.58	150	50		77.39
Cluster analysis	77.66	78.67	50	150		78.17
SVM	98.63	86.25	200	100	RBF	92.44
SVM	90.12	92.69	100	200	RBF	91.41
SVM	100	82.23	150	50	RBF	91.12
SVM	92.56	93.69	50	150	RBF	93.13
SVM	81.32	71.32	200	100	Sigmoid	76.32
SVM	80.36	79.65	100	200	Sigmoid	80.01
SVM	83.66	71.69	150	50	Sigmoid	77.68
SVM	79.32	78.62	50	150	Sigmoid	78.97
SVM	76.12	63.57	200	100	Polynomial	69.85
SVM	74.69	69.21	100	200		71.95
SVM	77.42	70.58	150	50	Polynomial	74.00
SVM	75.10	68.66	50	150		71.88

The classification algorithm is divided into two categories, meaning learning experience and no learning experience. Through training to learn to get this type of algorithm, accuracy is better than that in another class. It can be drawn experienced learn is appropriate for solving nonlinear classification problems. Compared with other algorithms, SVM classification algorithm, can get the higher accuracy in all, but the accurate of low quality picture is low. The reason is that so many good pictures are as training samples, fewer low-quality image as training samples, which makes decisions plane close to minority category.

SVM classification algorithm accuracy is not only related to the training samples, but also to the choice of kernel, since the different kernel functions will make the original nonlinear models into linear models with varying degrees of good and bad. And ultimately the accurate rate is very low in the classification of the test sample time. In the data used for this experiment, the choice of radial basis function as the kernel function is the most stable and accurate rate is high. The other kernel function is not suitable for this experiment, but may be applied to other data. In a word radial

Table 3 The error of different databases

Database	Intravenous number	Error
Database1	1420	0.08021
Database2	812	0.04386

basis function is chosen, because it wont get a large deviation in test results, even though the different data is used.

4.3 Third Experiment

Quality assessment can effectively improve the performance of the finger vein recognition system. This experiment shows that the quality assessment is how to effective role in the identification system. And it use the database, people vein image of 71 age of 20–26 years old, and each person collected 20, a total of 1420 images. The database is recorded database1.

In short, choosing a image at random from the database, after SVM quality assessment algorithm classification, if they are classified as high-quality image, then retains the image. On the contrary, if it is classified as low-quality image, the image is deleted. Eventually establishment of a finger vein database contains only high-quality images. The database is recorded database2.

Database1 and database2 are as a test database used to identify system performance. In order to reflect the identification system, it uses FRR probability equal to probability of false rate, that is equal error rate as performance indicators. Table 3 is equal error rate in using the same recognition algorithm under different the database.

5 Conclusions

In this paper, we propose a method based on support vector machine to assess the amount of finger vein, and with three experiments to prove the validity of this method. The main results of the following four points: First of all: the influence of finger vein image extraction quality large three characteristic parameters: Contrast, covariance eigenvalues, effective area. Then, by extracting a low-quality picture experiments and high quality pictures of three characteristic parameters. And then establish SVM model of the finger vein image classification, It will gain high quality and low quality of the vein image, and finally the quality evaluation method applied to the recognition experiments show that can improve the recognition accuracy.

References

1. Bojic N, Pang KK (2000) Adaptive skin segmentation for head and shoulder video sequences. Vis Commun Image Process SPIE 4067:704–711
2. Cui JJ et al (2009) On the vein image capturing system based on near-infrared image quality assessment. J Northeast Univ Nat Sci 30(8):1099–1102
3. Guan FX et al (2011) Bi-direction weighted (2D) 2PCA with eigenvalue normalization one for finger vein recognition. Pattern Recognit Artif Intell 24(3):417–424
4. Hashimoto J (2006) Finger vein authentication technology and its future. In: Symposium on VLSI circuits digest of technical papers. Institute of Electrical and Electronics Engineers Inc., Piscataway, pp 5–8
5. Kejun W, Jingyu L, Hui M, Xuefeng L (2011) Finger vein image quality evaluation. J Intell Syst 6(4):324–327
6. Kejun W, Hui M, Guan FX, Xuefeng L (2012) Based on image acquisition quality evaluation of fingerprint and finger vein bimodal recognition decision level fusion method. Pattern Recognit Artif Intell 25(4):669–675
7. Lee EC, Jung H, Kim D (2011) New finger biometric method using near infrared imaging. Sensors 11(12):2319–2333
8. Liu Z et al (2010) Finger vein recognition with manifold learning. J Netw Comput Appl 33(3):275–282
9. Vezhnevets V, Sazonov V, Andreeva AA (2003) Survey on pixel-based skin color detection-techniques. Graphicon. Moscow, Russia, p 8592
10. Wang KJ et al (2011) Finger vein image quality assessment method. CAAI Trans Intell Syst 6(4):324–327
11. Wu JD, Liu CT (2011) Finger-vein pattern identification using principal component analysis and the neural network technique. Expert Syst Appl 38(5):5423–5427
12. Wu JD, Liu CT (2011) Finger-vein pattern identification using SVM and neural network technique. Expert Syst Appl 38(11):14284–14289
13. Wu JD, Ye SH (2009) Driver identification using finger-vein patterns with radon transform and neural network. Expert Syst Appl 36(3):5793–5799
14. Xiaofei J (2013) Voice information identification and instruction identification method. Comput Syst Appl 22(4)
15. Yang J, Lu W, Waibel A (1998) Skin color modeling and adaptation. In: Asian conference on computer vision, Hong Kong, China, pp 687–694
16. Zhao Y, Sheng MY (2011) Application and analysis on quantitative evaluation of hand vein image quality. In: Proceedings of the international conference on multimedia technology. IEEE Computer Society, Piscataway, pp 5749–5751

A Design of Floating Vehicle Data Acquisition System Based-on ZigBee

Zhuang Yan

Abstract Internet + will become normal. Every social or commercial stage has a normal and development trends. The normal stage before "Internet +" is introduced is in the big background that thousands of enterprises need to transform and upgrade. In the next period of time, Internet + will greatly develop in macro aspects. The widespread use of Internet + technology, Internet + service, and O2O has gradually become the mainstream of information society. As for the government, the implementation of "Internet +" will become the main battlefield of realizing the Thirteen five planning. Wireless sensor network based on ZigBee technology has won an extensive popularity among engineers for the characteristics of low power consumption, self organized network, large network capacity and high security. In particular, the CC2430/31 wireless sensor network solution provided by Chipcon Corporation brings more conveniences to developers. In this paper, well design a wireless sensor network system based-on ZigBee technology that can give timely and accurate fore-warnings of the road congestion situations based on the information and analysis of the current existing congestion situation. Well also introduce the technology of wireless sensor networks and ZigBee, and describe the specific contents including network topology, hardware design, node design and software design, etc.

Keywords Zigbee · Big data · Internet +

1 Introduction

1.1 Background and Significance of the Research

The traffic index is obtained by in-depth processing the dynamic vehicle location information (referred to as the floating car data) distributed in the city streets, particularly in Beijing, is through the city's more than 30,000 taxi GPS dynamic data that

Z. Yan (✉)
Baicheng Normal University, Baicheng, China
e-mail: 381954@qq.com

© Springer International Publishing Switzerland 2017

V.E. Balas et al. (eds.), *Information Technology and Intelligent Transportation Systems*, Advances in Intelligent Systems and Computing 454,
DOI 10.1007/978-3-319-38789-5_22

transmits to the data processing center. First of all, the vehicles moving speed on all different classes of roads though the position data, and then the weight of the road in the whole traffic net is calculated according to the roads function and traffic flow value. Finally, the 0–10's index value is given by the people's perception judgment as to the congestion situation. Currently popular navigation software is based on the dynamic data provided by the software company's collaborative units such as buses, taxi companies or private owners. Generally speaking, it is the judgment of the running speed of the vehicle to make a congestion index. While in different sections of roads people have different feeling for the speeds, so in order to distinguish these grades thousands of staff have to carry GPS instruments and travel all over the streets, and then gave the final traffic index of different road sections on account of both the on-site experience and traffic data.

In this paper what is designed is a kind of calculation method that is based on the floating vehicle data, that is, the number of vehicles in a unit area. The data provided by this method project the real situation and is actually an abstract map showing of the actual situation happens under the camera. Compared with the GPS positioning data by the cooperative units, you can cover all registered vehicles, so not only in the developed big cities but in the communities or in the small alleyway, it can still continue to play its advantages. Of course, this also requires all vehicles implanting a chip at the same time when registering the license plate. But the cost is not only to provide floating car data, but also as the identity card of the car and play an important role in the subsequent traffic enforcement function, which is also an imperative way to implement "Internet +". Compared with the original monitoring equipment it has the advantages of low cost, accurate data and needing less personnel equipped with.

1.2 Wireless Sensor Networks

Overview of Wireless Sensor Networks Sensor Networks Wireless (WSN) is a kind of wireless network, which integrates the sensor, embedded computing, network and information processing technology. It can perceive and collect information of various kinds of environment or monitoring objects in real time, and it can be transmitted in wireless mode, and it can be transmitted to the user terminal in the form of self organized multi hops to meet the growing complexity of human society.

Architecture of Wireless Sensor Networks In wireless sensor networks, a large number of nodes are deployed inside or surrounding the perception objects. These nodes form a wireless network perceiving and processing specific information in a cooperative way. The network, which is composed of self organized form, transmits the data to the SINK node through multi hop relay. Finally, the data is transmitted to the remote control center for processing through the SINK link. The architecture of a typical wireless sensor network consists of sensor nodes, SINK nodes, and the Internet, etc., as shown in Fig. 1.

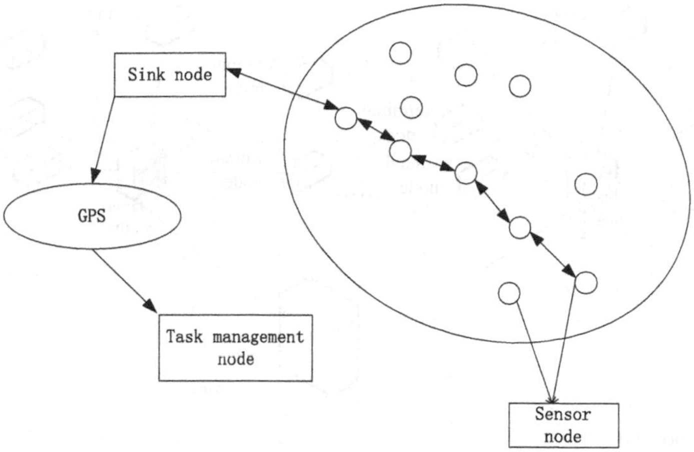

Fig. 1 Typical wireless sensor network architecture

1.3 Introduction of IEEE802.15.4/ZigBee

ZigBee technology is a group of communication technology based on IEEE802.15.4 wireless standard for networking, security and application software development. IEEE802.15.4 specification is an economical, efficient, low data rate (<250 kbps), wireless technology working in 2.4 GHz and 868/928 MHz. It is the basis of ZigBee application layer and network layer protocol. Mainly used for short-range wireless connection. It is based on the 802.15.4 standard, in the coordination of thousands of tiny sensors to achieve communication. These sensors only need a little energy in order to relay the data from a network node to another node through radio waves, so their communication efficiency is very high.

1.4 Basic Functions

System basic functions:

- Rea-time collection and return vehicle location information in the form of a chart or table;
- real-time storage and analysis and upload data;
- ensure the number of nodes and the reorganization ability after the failure of some node.

The main contents of this paper includes the design of ZigBee wireless network, the design of the network topology, the design of the positioning scheme and nodes, the design of the information processing function.

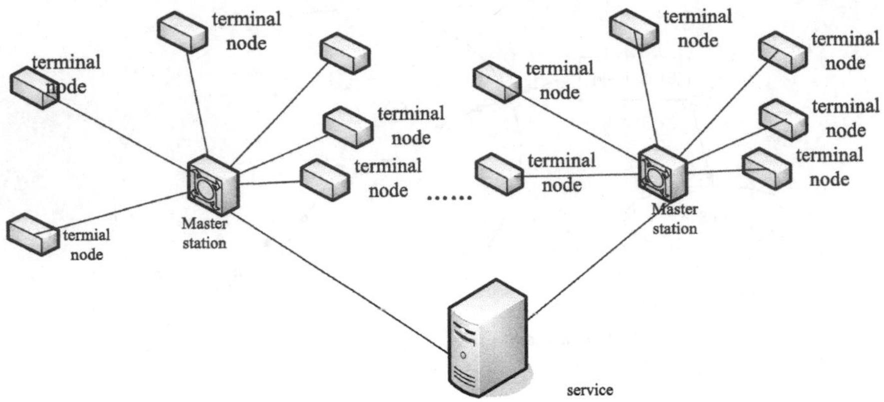

Fig. 2 Network architecture

2 Design Scheme

2.1 ZigBee Network Components and Topology Design

Figure 2 in the street lamp pole, FFD is installed as the master station equipment while RFD is installed as the terminal node device to save energy consumption. The main equipment is responsible for analyzing and processing data it has collected and then transmits it to the command center through GPS or WIFI network center. Each FFD device and a range of general terminal nodes it can reach form a network which is responsible for collecting data in the region, while the vehicle's RFD device automatically join the network of the nearest FFD device.

Each network uses a star network topology for the reason that each network coordinator, that is, the main network equipment, communicates with the terminal nodes and the terminal nodes do not need communicate with each other. The master station equipment can be supplied by the power system, and the terminal nodes due to the simple functions and the small amount of data can choose the cheap RFD equipment.

2.2 Positioning Scheme and Node Design

In this paper, a method based on the acceptance signal strength indicator (RSSI) is proposed. In this method, the transmitting signal intensity of the emitted nodes is calculated, and the receiving node calculates the transmission loss of the signal according to the intensity of received signal. The transmission loss is converted into a distance by using theoretical or experimental empirical model.

The position of the nodes is calculated by using the three edge measurement method or the triangulation method. Thus, the method requires multiple reference nodes. The reason for choosing this method is that location of each master station, also known as the sink node, can be precisely determined which is the emitted node mentioned above and is called reference node. Each terminal node can be determined and checked position by the surrounded vehicles. This kind of terminal node is called blind node. The blind node is moving node. The reference node must be configured correctly in the coordinates of the location area, which provides a package including X, Y coordinates that contains its own position and RSSI information to the blind node so that the blind node can calculate the coordinates of its own based-on the information it receives and then send the appropriate information to the gateway.

The positioning algorithm is based on the chip's hardware positioning engine. It requires at least 3 reference nodes and the distance between the actual wire poles is about 50 m. Usually a blind node is surrounded by 3 or 4 reference nodes. The positioning algorithm can be correct to the resolution of 0.25 m, which meets the need of this paper. Also the time consumes less than 50 uS, which can meets the need of positioning compared with the relative speed of the vehicle. The location can be completely covered within the 64 m × 64 m area. The algorithm uses distributed computing method to avoid the large number of network transmission and communication delays caused by centralized computing method.

In addition, a back-up FFD equipment is added to the master station node so that there are 2 FFD devices in a master station node. This design has 4 reasons. Firstly, a device failure can be performed by second devices. Secondly, it can be used by the vehicle to determine the direction of the vehicle. The data uploaded including the direction information can provide accurate travelling routes for users or navigation software. Thirdly, the positioning data will be more accurate and fast. Finally, if the system is applied to traffic law enforcement the time difference between these two nodes can provide speed information. In the specific network building the nodes near the traffic lights can be used as gateways and coordinators.

2.3 *Software Design*

Network software development is mainly responsible for the two aspects. One is the transmission of data by both the sink nodes and common nodes. The other is the management function mainly by the coordinator node, that is, the sink nodes. The management includes configuration addresses for newly added nodes, maintenance of routing table, response to binding requirement, defining communication channel, etc. Another function is to collect the data from other nodes and transmit to upper node.

The main function modules of the reference node contain:

An information processing function that processes the information the reference node receives according to the cluster ID.

A response function that responses to the requests from the blind node. A configuration function that completes the configuration of the reference node parameters.

A RSSI collection function that adds and initializes the RSSI value of the network address.

The main function modules of the blind node contain:

An information processing function that receives the data, processes data, do corresponding work as the cluster ID and replies with a RSSI value.

A configuration function that completes the configuration of the blind node parameters.

A configuration sending function that send responses to requests to configuration information.

A position computing parameter collection function that sends out multiple commands to collect RSSI values.

A position computing function that starts the positioning engine related registers and read out the position information of its own.

A coordinates sending function that sends its own coordinates information to the coordinator in some time cycle.

The main function of the gateway includes node data processing, collection of node network address, effective data length, cluster ID, and coordinate data, etc., check and analysis of the data packet and forwarding data the upper server through GPS or WIFI network.

3 Conclusions

There have been many research results on the data acquisition of floating car, but the action of the actual operation is not carried out. This paper attempts to discuss the practical application of wireless sensor network and CC2430 development kit and propose an energy saving, high speeding, low cost and feasible scheme which not only can solve the problem of data collection of floating vehicles, but also provides a wider range of expansion to the traffic law enforcement in the future.

References

1. Sohrabi K et al (2000) Protocols for Self-Organization of a Wireless Sensor Network. IEEE Personal Communication pp 16–27
2. Chen B, Kyle J, Balakrishnan et al (2012) Span. An energy efficient coordination algorithm for topology maintenance in ad hoc wireless networks. ACM Wirel Netw 8(5):481–494
3. Wei S, Canyang W (2009) Zigbee technology practice guide, Beihang University Press, Beihang
4. Chipcon. SmartRF CC2430[EB/OL]. www.chipcon.com 2005/2006
5. (TI).TLV320AIC1107 Data Sheet[EB/OL]. http://www.ti.com (2007)
6. Aamodt K, CC2431 Location EngineTexas Instruments, ApplicationNote AN042, SWRA095
7. Zhu S, Ding Z, Karina, M (2010) TOA based jointsynchronization and localization. In: Proceedings of the IEEE international conference on communications, ICC 2010, Article ID 5502036
8. http://www.ti.com

Design and Analysis of Efficient Algorithm for Counting and Enumerating Cycles in LDPC Codes

Kongzhe Yang, Bangning Zhang, Ye Zhan and Daoxing Guo

Abstract Since short cycles and trapping sets are culprits of performance and error floor of low-density parity-check (LDPC) codes, it is necessary to obtain information about the number and the distribution of all cycles in Tanner graphs. However, the established algorithms could not efficiently search all cycles on account of restrictions from both the structure and the girth of Tanner graphs. The proposed algorithm solve the above problems with message-passing schedule, counting and enumerating the cycles simultaneously. With information derived from the proposed algorithm, performance will be enhanced and error floor will be lower, which is meaningful for both adjustment and design of LDPC codes. Furthermore, the proposed algorithm can be applied on general bipartite graphs.

Keywords Low-density parity-check (ldpc) codes · Bipartite graphs · Cycle · Trapping sets · Enumeration

1 Introduction

After the booming development of iterative decoding algorithms, low-density parity-check (LDPC) codes were rediscovered and gradually accepted from industry to academia. And LDPC codes have been widely used in various communication systems, for their capacity-approaching error correction over many importance chan-

K. Yang · B. Zhang · Y. Zhan · D. Guo (✉)
College of Communication Engineering, PLA University of Science and Technology, Nanjing 210007, China
e-mail: rgsc2014@163.com; rgsc@126.com

K. Yang
e-mail: kongzheyang@gmail.com

B. Zhang
e-mail: bangn_zhang@163.com

Y. Zhan
e-mail: ye_zhan@163.com

© Springer International Publishing Switzerland 2017
V.E. Balas et al. (eds.), *Information Technology and Intelligent Transportation Systems*, Advances in Intelligent Systems and Computing 454,
DOI 10.1007/978-3-319-38789-5_23

nels. However, the performance of iterative decoding algorithm depends largely on the structure of Tanner graphs, which is an important tool for studying LDPC codes. Numerous publications have shown that the cycle structure of Tanner graphs is crucial to the performance of LDPC codes. The girth, the percentage of short cycles, and the distribution of cycles are all the factors which may influence the ability of error correction. Furthermore, trapping sets, whose subsets are mainly short cycles, are the vital factors influencing the error floor of the codes. Since the established algorithms either have difficulty in computation complexity and storage complexity, or be restricted by the girth of Tanner graphs, it is necessary to design efficient algorithm to count and enumerate all cycles in the graphs.

Numerical publications have made use of the knowledge of cycles in Tanner graphs to estimate the performance of LDPC codes or design good LDPC codes with better performance and lower error floor [1–5]. However, counting and enumerating cycles in a graph are both known to be NP-hard problems [6]. Plenty of work has been made to lower the complexity of solving these problems [7–11], especially counting and enumerating problems in bipartite graphs or in Tanner graphs [12–15].

The mentioned algorithms successively reduced their complexity, even made breakthroughs in Tanner graphs due to the sparsity of LDPC codes. Nevertheless, most of them cannot count or enumerate long cycles on account of restrictions from the girth of Tanner graphs. The proposed algorithm derived from the algorithm in [15]. And the improvement of the proposed algorithm is counting and recording the exact trails of cycles simultaneously, while deleting the unqualified walks which have same nodes at the same time.

2 Definitions and Notations

2.1 *Bipartite Graphs*

A bipartite graph $G = (V, E)$ is defined as two sets of nodes U and W, where $U \bigcup W = V, U \bigcap W = \emptyset$, and a set of edges E, where E is some subset of the pairs $\{(u, w) \mid u \in U, w \in W\}$. And a Tanner graphs is indeed a certain kind of bipartite graphs, whose U and W are referred to as variable nodes and check nodes in LDPC codes. Figure 1 shows an example.

2.2 *Simple Cycles*

In graph theory, a walk of length k in G is a sequence of successive nodes $\{v_1, v_2, \ldots, v_{k+1}\}$ in V such that $(v_i.v_{i+1}) \in E$ for all $i \in \{1, \ldots, k\}$ [15]. A cycle or a $closedwalk$ is a type of walks whose two end nodes are identical. It should be noted that all nodes except the two end ones are not necessary distinguish

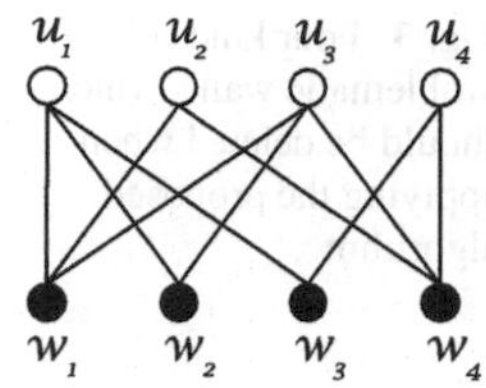

Fig. 1 An example of Tanner graphs

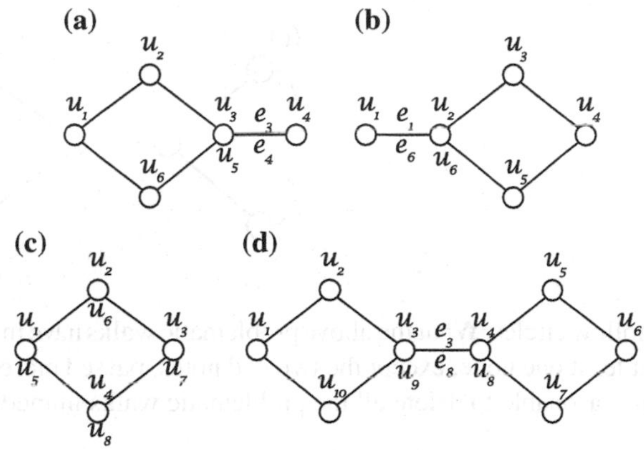

Fig. 2 Some example of non-simple cycles. **a** a non-backtrackless cycle, **b** a non-tailless cycle, **c** a non-primitive cycle, **d** a tbc walk is not always a simple cycle

with each other in this definition. In another aspect, a cycle C of length k can be described by corresponding sequence of k edges $\{e_1, e_2, \ldots, e_k\}$.

Some definitions of special cycles have been mentioned in [15], such as *backtrackless* cycles, *tailless* cycles, *primitive* cycles and *simple* cycles. Figure 2 shows some examples of the above special cycles. In this design, we mostly restrict our attention to *simple* cycles, and *tailless backtrackless closed* walk is abbreviated to *tbc* walk.

3 The Proposed Algorithm

3.1 Message Passing

A message-passing algorithm operates in a bipartite graph by computing messages at the nodes from *U(W)* and passing them along the edges to the adjacent nodes from *W(U)*. In this case, a complete cycle of message-passing from U to W and then from W to U is called one *iteration* [15].

Four kinds of problematic walks are summarized in Fig. 3, which should be deleted when applying the proposed algorithm. The hollow circles represent the nodes in bipartite graphs. It should be noted that the solid line in the figure means there is only an edge connects the adjacent two hollow circles, while the dotted line means there can be none or more than one nodes connected one by one between the adjacent two

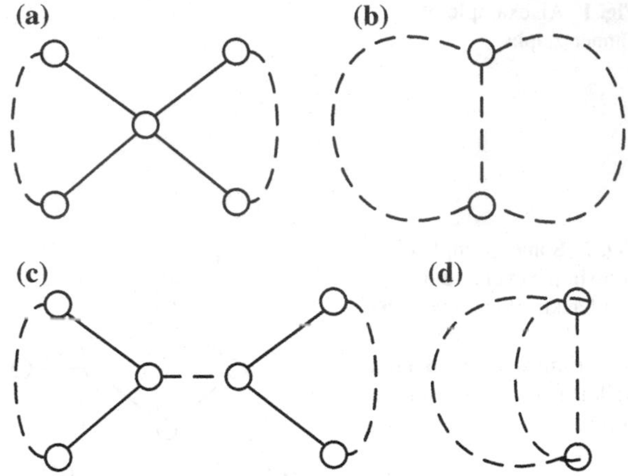

Fig. 3 Four kinds of problematic walks which should be deleted when applying the proposed algorithm

hollow circles. What the above problematic walks have in common is that they all have at least one node, except the two end notes, passed more than once. Consequently, it is reasonable to delete all the problematic walks immediately they are found.

3.2 Counting and Enumerating Cycles

According to the lemma proved in [15], the number of 2k-cycles in the design can be expressed as

$$N_{2k}^{v} = \frac{1}{2}\sum_{j=1}^{d_v}\sum_{\substack{i=1\\ i\neq j}}^{d_v} N_{2k}^{v;e_i,e_j} - \sum_{\substack{l=1\\ u_l\neq v}}^{n} N_{2k}^{v;u_l}, \qquad (1)$$

where

N_{2k}^{v} is the number of 2k-cycles whose starting node is v;

$N_{2k}^{v;e_i,e_j}$ is the number of 2k-cycles whose starting node is v, starting edge is e_i, ending edge is e_j;

$N_{2k}^{v,u}$ is the number of 2k-cycles whose starting node is v;

$u_l, l = \{1, 2, \ldots, n\}$ in $N_{2k}^{v,u}$ is one of the multi-passed nodes which has the smallest serial number l, so that the number of cycles will not be miscount;

n in Eq. (1) is the number of nodes in the graph.

3.3 A Simple Example

We illustrate the proposed algorithm by a simple example in Fig. 4. Compared with the example used in [15], we delete the problematic walks the moment we find them. That is the core reason why the algorithm proposed in [15] can only count 2k-cycles

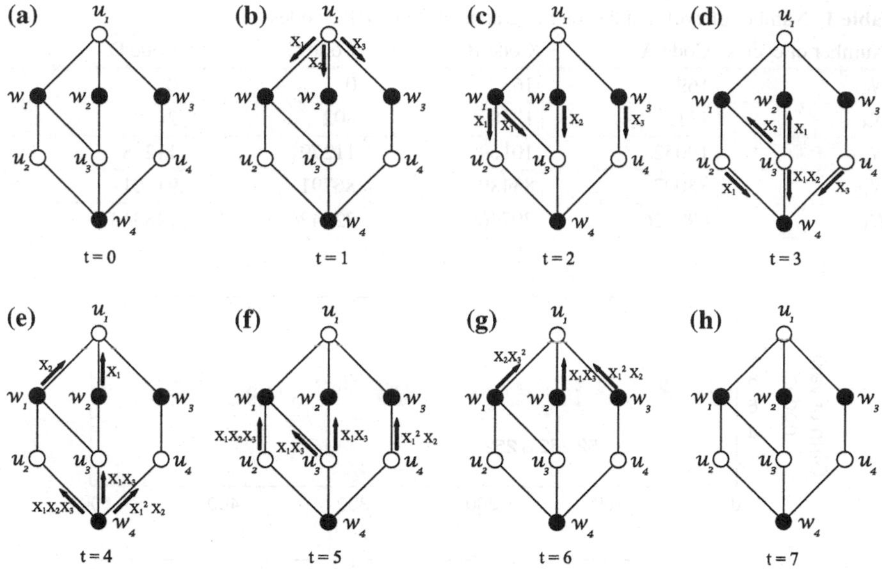

Fig. 4 Message passing of the proposed algorithm in the graph of Fig. 1

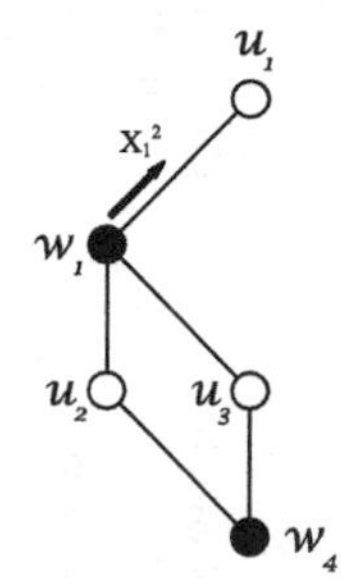

Fig. 5 A problematic walks is deleted at $t = 6$ because node w_1 is passed twice

where k is smaller than g. Additionally, the proposed algorithm efficiently enumerate all the cycles in the graph by deleting the problematic walks immediately, such as an example shown in Fig. 5.

4 Numerical Results

Numerical results were obtained by applying the proposed algorithm to the Tanner graphs of four LDPC codes from [16]. Codes A to D are listed in [16] as 252.252.3.252, 504.504.3.504, $PEGReg252x504$ and $PEGReg504x1008$, respectively.

For convenience, the number of cycles with length from 6 to 14 in the Tanner graphs of these codes was listed in Table 1. The algorithm was run on the machine with a 1.3 GHz CPU and 8 GB of RAM. The girth histogram of Codes A to D were shown in Fig. 6.

Table 1 Number of cycles in the Tanner graphs of four LDPC codes

Number of cycles	Code A	Code B	Code C	Code D
N_6	169	165	0	0
N_8	1312	1258	802	2
N_{10}	10052	10169	11279	11238
N_{12}	83007	83489	86791	91101
N_{14}	699526	707468	723426	748343

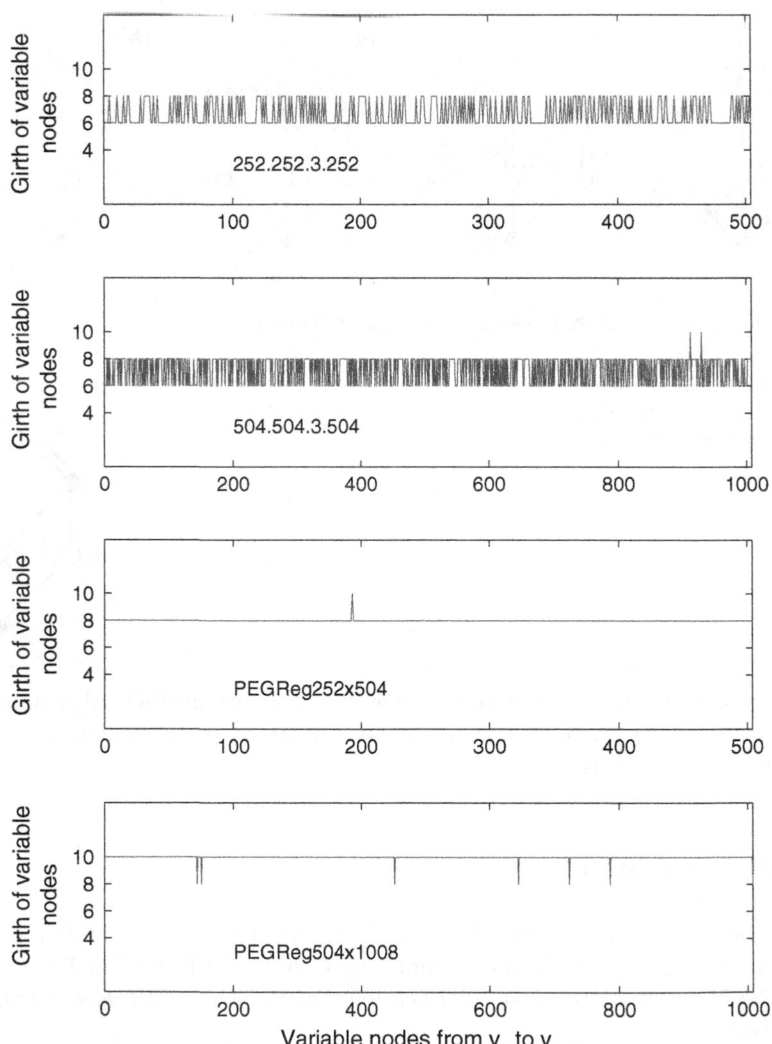

Fig. 6 Girth histograms of 252.252.3.252, 504.504.3.504, $PEGReg252x504$ and $PEGReg504x1008$, respectively

5 Conclusions

In this paper, we proposed an improved efficient algorithm derived from [15] to count and enumerate cycles in Tanner graphs. Compared with the original one, the proposed algorithm was not restricted by the girth of the graph, and can not only count, but also enumerate cycles of any length if exist. However, the complexity of the proposed algorithm increases because we enumerate and count the cycles in the meantime. Moreover, the proposed algorithm can be applied on general bipartite graphs.

Although our main concern was enumerating and counting simple cycles, there may be applications where one is interested in estimating and designing LDPC codes with better performance, or analyze the impact of dominant trapping sets on LDPC codes with the results of the algorithm.

References

1. Mao Y, Banihashemi AH (2001) A heuristic search for good low-density parity-check codes at short block lengths. Proc IEEE Int Conf Commun 1:41–44
2. Hu X-Y, Eleftheriou E, Arnold D-M (2001) Progressive edge-growth Tanner graphs. Proc IEEE Global Telecommun Conf 1:995–1001
3. Karimi M, Banihashemi AH (2012) Efficient algorithm for finding dominant trapping sets of LDPC codes. IEEE Trans Inf Theory 58(11):6942–6958
4. Xiao H, Banihashemi AH (2009) Error rate estimation of low-density parity-check codes on binary symmetric channels using cycle enumeration. IEEE Trans Commun 57(6):1550–1555
5. Asvadi R, Banihashemi AH, Ahmadian-Attari M (2011) Lowering the error floor of LDPC codes using cyclic liftings. IEEE Trans Inf Theory 57(4):2213–2224
6. Flum J, Grohe M (2002) The parameterized complexity of counting problems. Proc IEEE Symp Found Comput Sci 1:538–547
7. Tarjan R (1973) Enumeration of the elementary circuits of a directed graph. J SIAM 2:211–216
8. Johnson DB (1975) Find all the elementary circuits of a directed graph. J SIAM 4:77–84
9. Liu H, Wang J (2006) A new way to enumerate cycles in graph. In: Proceedings of the advanced international conference on telecommunications and international conference on internet and web applications and services (AICT/ICIW 2006)
10. Bax ET (1994) Algorithms to count paths and cycles. Inf Process Lett 52:249–252
11. Alon N, Yuster R, Zwick U (1997) Finding and counting given length cycles. Algorithmica 17(3):209–223
12. Fan J, Xiao Y (2006) A method of counting the number of cycles in LDPC codes. Proc Int Conf Signal Process 3:2183–2186
13. Chen R, Huang H, Xiao G (2007) Relation between parity-check matrices and cycles of associated Tanner graphs. IEEE Commun Lett 11(8):674–676
14. Halford TR, Chugg KM (2006) An algorithm for counting short cycles in bipartite graphs. IEEE Trans Inf Theory 52(1):287–292
15. Karimi M, Banihashemi AH (2013) Message-passing algorithms for counting short cycles in a graph. IEEE Trans Commun 61(2):485–495
16. D. MacKay's Gallager Code Resources [Online]. http://www.inference.phy.cam.ac.uk/mackay/codes/

Soft Fusion with Second-Order Uncertainty Based on Vague Set

Jianhong Wang and Tao Li

Abstract The vague set theory can overcome the shortcomings of fuzzy set by describing the membership from two sides of both TRUE and FALSE, rather than only by a single membership value, with the further generalization of fuzzy set theory. It is superior in mathematical analysis of system with uncertainty, since vague sets can provide more information than fuzzy sets. And it is more powerful in the describing and processing of uncertain, inaccurate, even conflicting information. In this paper, focused on the second-order uncertainty (SOU) issues in a weak knowledge environment which comes from generalized or ubiquitous sensors, a novel method of fuzzy decision fusion was presented based on vague set. The new method is more efficient and powerful to fulfill decision fusion with uncertain and inaccurate information.

Keywords Vague set · Second-order uncertainty · Fuzzy set · Decision making · Decision fusion · Soft fusion

1 Introduction

Data fusion is defined as processes or technologies which analyze, optimize, and integrate the data from the multiple sensors or multiple sources, focused on the problems of decision making. Therefor, we can obtain more accurate and more stable decision result than the result obtained by using the sources solely. Many effective theories and approaches have been developed since the data fusion is proposed, most of which are based on the theory of probability. However, the decision is often made under

J. Wang (✉)
Institute of Geospatial Information, Information Engineering University, Baoji 68210, Zhengzhou, People's Republic of China
e-mail: JianhongWang@163.com; jianhong72@163.com

T. Li
Institute of Geospatial Information, Information Engineering University, Dalian 61206, Zhengzhou, People's Republic of China
e-mail: TaoLi73@163.com

© Springer International Publishing Switzerland 2017

V.E. Balas et al. (eds.), *Information Technology and Intelligent Transportation Systems*, Advances in Intelligent Systems and Computing 454,
DOI 10.1007/978-3-319-38789-5_24

weak knowledge environment because of using soft sensors. Thus, these problems exhibit their fuzziness other than randomness, even uncertainty in the uncertainty (second-order uncertainty) which make it difficult to calculate the probability, it is hard to solve only by probability; even it maybe not have an accurate result in some cases [1]. Fortunately, the vague set (VS) theory [2] can remedy the shortage of fuzzy set [3] (FS) by describing the membership from two sides of both TRUE and FALSE, rather than only by a single membership value. Many researchers have been interested in the vague set theory in recent years, and got ahead in theory in many fields, such as vague sets description, vague set operators, translation between vague set and other sets, measures of similarity between vague sets, decision making and approximate reasoning based on vague sets [3–5]. On the other hand, vague set has been applied to target/pattern recognition, data fusion and so on successfully. With further development of the information technology, some new approaches based on the knowledge and cognition, such as data mining, machine learning and reasoning, and intelligent decision, will become the main development trend to uncertainty information processing, which will make more use of the advantage of vague sets in uncertain information processing.

2 Traditional Fusion Model

2.1 Problem Description

Reference [6] Assuming that O is the pending objects set, such as alternative schemes or recognition, U is the corresponding attributes or features set, S is a decision-maker (DM) set such as sensors or person of DMs, where $\mathbf{O} = \{O_1,\ O_2,\ \ldots,\ O_m\}$, $\mathbf{U} = \{U_1,\ U_2,\ \ldots,\ U_l\}$, $\mathbf{S} = \{S_1,\ S_2,\ \ldots,\ S_n\}$.

m, l, n is the number of objects, features and DMs respectively. The normalized weights set of attributes, targets and sensors is written as

$$\mathbf{W} = \{w_1, w_2, \ldots, w_m\}, \ \mathbf{V} = \{v_1, v_2, \ldots, v_l\}, \ \lambda = \{\lambda_1, \lambda_2, \ldots, \lambda_n\}$$

here the weighted values can be either real number or VS value. If they are real number, they will satisfy

$$w_1 + w_2 + \cdots + w_m = 1, \quad v_1 + v_2 + \cdots + v_l = 1, \quad \lambda_1 + \lambda_2 + \cdots + \lambda_n = 1$$

The date set likes the data table as follow

$$D = \begin{array}{cc} & \begin{array}{ccc} U_1 & \cdots & U_l \\ w_1 & \cdots & w_l \end{array} \\ \begin{array}{cc} S_1 & \lambda_1 \\ \vdots & \vdots \\ S_n & \lambda_n \end{array} & \begin{pmatrix} u_{11} & \ldots & u_{1l} \\ \vdots & \ddots & \vdots \\ u_{n1} & \cdots & u_{nl} \end{pmatrix} \end{array} \tag{1}$$

Now, the problem is how to making the fusion using the data set?

2.2 *Traditional Algorithms Steps*

Generally, a simply computational steps is given as follow:

Step 1. Select the fusion model and criterion.

Step 2. Input the original data O, U, S, W, V, λ.

Step 3. Translating the raw data derived from sensors (or other sources) and attributes in vague set style.

Step 4. Aggregating the attributes and constructing the synthetic attribute values, then converting to fuzzy style matrix.

Step 5. Calculating vague comprehensive evaluation vector can be calculated by using one of decision models.

Step 6. Output the decision making result according to criterion.

3 Modeling Soft Fusion Based on Vague Set

3.1 *Definitions*

References [7–10] **Definition 1. Ideal Target**. Assuming that sensor can identify target completely (100 %), the target can be defined as an ideal target, which can be written in Vague Set value [1, 1] as

$$\mathbf{O}^* = \{(S_1, [1, 1]), (S_2, [1, 1]), \ldots, (S_n, [1, 1])\} \tag{2}$$

Also as

$$\mathbf{O}^* = \{s_j, \bigvee_{i=1}^{m} [t_{ij}, 1 - f_{ij}]\} = \{s_j, \bigvee_{i=1}^{m} [t_{o^*j}, 1 - f_{o^*j}]\} \quad (j = 1, 2, \ldots, n) \tag{3}$$

Definition 2. Evaluation Function. Reference [7] proposed two methods to measure the degree of similarity between VSs—general evaluation function and weighted evaluation function based on score function.

Let $\mathbf{S} = \{S_1, S_2, \ldots, S_n\}$ is a sensors set, $\mathbf{W} = \{w_1, w_2, \ldots, w_n\}$ is a normalized weights set, $\mathbf{O} = \{O_1, O_2, \ldots, O_m\}$ is a targets set, and $\mathbf{O}^*$ is the ideal target, then the similarity measure function between the targets needing decision and the ideal target can be described as

$$T(\mathbf{O}, \mathbf{O}^*) = \sum_{i=1}^{n} \left(w_i * \left|(t_O(u_i) + f_O^*(u_i)) - (t_{O^*}(u_i) + f_{O^*}^*(u_i))\right|\right) \tag{4}$$

Assuming a Vague Set A is defined as

$$q_A^i = t_A(u_i) + \rho[f_A^*(u_i) - t_A(u_i)] \tag{5}$$

Then, the evaluation function is

$$T(\mathbf{O}, \mathbf{O}^*) = \sum_{i=1}^{n} \left(w_i * \left|q_O^i - q_{O^*}^i\right|\right) \tag{6}$$

Definition 3. Least Distance and Criterion. For the target needing decision set and the ideal target, if the least distance exists, it can be defined as

$$\mathbf{T}_{\min}\left(\mathbf{O}, \mathbf{O}^*\right) = \min\left\{T\left(O_1, \mathbf{O}^*\right), T\left(O_2, \mathbf{O}^*\right), \ldots, T\left(O_m, \mathbf{O}^*\right)\right\} = T\left(O_i, \mathbf{O}^*\right) \tag{7}$$

Criterion I: The target corresponding to the least distance is the decision result.

3.2 Soft Fusion Based on Vague Set

As above, objects set O, attributes or features set U, decision-maker set S, weights set W, is defined respectively as follow $\mathbf{O} = \{O_1, O_2, \ldots, O_m\}, \mathbf{U} = \{U_1, U_2, \ldots, U_l\}, \mathbf{S} = \{S_1, S_2, \ldots, S_n\}$

$$\mathbf{W} = \{w_1, w_2, \ldots, w_l\}, \quad \lambda = \{\lambda_1, \lambda_2, \ldots, \lambda_n\}$$

Then, the data set D is

$$\begin{array}{ccc} & & \begin{array}{ccc} U_1 & \cdots & U_l \\ [w_1^L,\ w_1^U] & \cdots & [w_l^L,\ w_l^U] \end{array} \\ \begin{array}{c} S_1 \\ \vdots \\ S_n \end{array} & \begin{array}{c} [\lambda_1^L,\ \lambda_1^U] \\ \vdots \\ [\lambda_n^L,\ \lambda_n^U] \end{array} & \begin{pmatrix} [u_{11}^L,\ u_{11}^U] & \ldots & [u_{1l}^L,\ u_{1l}^U] \\ \vdots & \ddots & \vdots \\ [u_{n1}^L,\ u_{n1}^U] & \cdots & [u_{nl}^L,\ u_{nl}^U] \end{pmatrix} \end{array} \tag{8}$$

Note that value are VS style value. In the kth batch data gathered by kth DM or sensors, u_{ij}^k stands for the score of DM S_j to feature U_i, which can be written in VS as

$$u_{ij}^k = [u_{ij}^L,\ u_{ij}^U]^k = (S_j,\ [t_{ij},\ 1 - f_{ij}])^k \tag{9}$$

The preference decision matrix given by S_j is written as [?]

$$A_j^k = \begin{matrix} & U_1 \cdots U_l \\ \begin{matrix} O_1 \\ \vdots \\ O_m \end{matrix} & \begin{pmatrix} [t_{11}, 1 - f_{11}] & \ldots & [t_{1l}, 1 - f_{1l}] \\ \vdots & \ddots & \vdots \\ [t_{m1}, 1 - f_{m1}] & \cdots & [t_{ml}, 1 - f_{ml}] \end{pmatrix} \end{matrix} \tag{10}$$

Aggregating the sensors and attributes, constructing the synthetic attribute values (sensor S_j measures to target O_i) as

$$r_{ij} = (S_j,\ [t_{ij},\ 1 - f_{ij}]) \tag{11}$$

Also, like as the follow.

$$O_i = \{(S_1,\ [t_{i1},\ 1 - f_{i1}]),\ (S_2,\ [t_{i2},\ 1 - f_{i2}]),\ \ldots,\ (S_n,\ [t_{in},\ 1 - f_{in}])\},\ \ (i = 1,\ 2,\ \ldots,\ m)$$

Then, the VS normalized matrix (synthetic decision matrix) of S to O is written as

$$R = \begin{matrix} & S_1 \ldots S_n \\ \begin{matrix} O_1 \\ \vdots \\ O_m \end{matrix} & \begin{pmatrix} [t_{11}, 1 - f_{11}] & \ldots & [t_{1n}, 1 - f_{1n}] \\ \vdots & \ddots & \vdots \\ [t_{m1}, 1 - f_{m1}] & \cdots & [t_{mn}, 1 - f_{mn}] \end{pmatrix} \end{matrix} \tag{12}$$

Here, t_{ij} and f_{ij} stand for S_j pros and cons to U_i and O_i respectively. $t_{ij} \in [0, 1]$, $f_{ij} \in [0, 1]$, $t_{ij} + f_{ij} \leq 1$, $(i = 1, 2, \ldots m,\ j = 1, 2, \ldots, n)$. The value of $\mathbf{W} = \{w_1,\ w_2,\ \ldots,\ w_l\}$, $\lambda = \{\lambda_1,\ \lambda_2,\ \ldots,\ \lambda_n\}$ is a real number in the interval [0, 1], even a VS data.

Finally, we can fusion calculation by using one of decision models, and output the decision making result according to criterion.

4 Example

In a mission, there are three different kinds of planes, such as fighter, bomber and transporter, which forms object set $\mathbf{O} = \{O_1,\ O_2,\ O_3\}$. PW, RF, NE, WW and WS form the attributes set $\mathbf{U}$u$_1$,u$_2$, …, u$_5$. The generalized sensors is $\mathbf{S} = \{S_1,\ S_2,\ S_3\}$, and weight set is $\{\lambda_1,\ \lambda_2,\ \lambda_3\}$.

Table 1 Input data set

	S		O_1		O_2		O_3	
Sj	wjL	wjU	t1j	1−f1j	t2j	1−f2j	t3j	1−f3j
S1	0.4	0.6	0.6	0.7	0.7	0.8	0.5	0.6
S2	0.2	0.4	0.5	0.6	0.6	0.9	0.3	0.4
S3	0.1	0.3	0.4	0.5	0.6	0.8	0.5	0.6

Step 1 Receive the data and organize in VS style as Table 1.

Step 2 Convert the input data into VS style in matrix form by formula (8) as follow,

$$R_V = \begin{matrix} \\ O_1 \\ O_2 \\ O_3 \end{matrix} \begin{matrix} \begin{matrix} S_1 & \quad S_2 & \quad S_3 \end{matrix} \\ \begin{pmatrix} [0.6, 0.7] & [0.5, 0.6] & [0.4, 0.5] \\ [0.7, 0.8] & [0.6, 0.9] & [0.6, 0.8] \\ [0.5, 0.6] & [0.3, 0.4] & [0.5, 0.6] \end{pmatrix} \end{matrix}$$

And the weights set of the sensors is

$$\lambda_V = \{[0.4,\ 0.6],\ [0.2,\ 0.4],\ [0.1,\ 0.3]\}$$

For demonstration purposes, let x_V and x_F VS value and FS value respectively. If VS data is $x_V = [t_x, 1 - f_x]$, we can convert it to FS data by the function as

$$x_F = (t_x + 1 - f_x)/2$$

Thus, the weight set can convert from VS to FS form $\lambda_F = \{0.5,\ 0.3,\ 0.2\}$.

Step 3 Refer to traditional fuzzy comprehensive evaluation method, calculating vague comprehensive evaluation vector $B = \lambda \circ R$. Ideal target value calculated by formula (2), evaluation function selected formula (7), and evaluation model selected as

$$M(\bullet, +) : b_j = \sum_{i=1}^{n} (\lambda_i \bullet r_{ij})$$

Then, calculating evaluation vector as $B = (0.84, 0.51, 1.02)$.

Step 4 Using maximum membership principle (MMP), we can determine the decision result. In this case, the second element is the minimum value, so its corresponding object is O_2.

5 Conclusion

Focused the second-order uncertainty issues under a weak knowledge environment which comes from generalized or ubiquitous sensors, the novel method based on vague set can obtain coherent decision result. It is more efficient and powerful to fulfill decision fusion with uncertain and inaccurate information. Universality, as vague set theory can describe the soft data, it is more powerful than fuzzy method to solve decision making and soft fusion problems.

References

1. Hall D, Llinas J (2008) A framework for dynamic hard/soft fusion. In: Proceedings of the 11th international conference on information fusion (FUSION 2008). Cologne, Germany
2. Gau WL, Buehrer DJ (1993) Vague sets. IEEE Trans Syst Man Cybern 23(2):610–614
3. Zadeh LA (1965) Fuzzy sets. Inf Control 8:338–353
4. Bellman R, Zadeh LA (1990) Decision making in a fuzzy environment. Manage Sci 17:41–164
5. Chen SM (1995) Measures of similarity between vague sets. Fuzzy Sets Syst 74:217–223
6. Hong DH, Choi CH (2000) Multicriteria fuzzy decision making problems based on vague set theory. Fuzzy Sets Syst 74:103–113
7. Wang JH, Li X (2004) Fuzzy data fusion based on vague set. J Huazhong Univ Sci Technol (Nat Sci Ed) 32(8):54–56
8. Wang JH et al (2005) Multi-attribute, multi-sensor and multi-target data fusion based on Vague Set. In: Proceedings of the 4th international conference on machine learning and cybernetics (ICMLC2005), Guangzhou, vol. 4, pp 2511–2517
9. Wang JH (2006) Target recognition based on rough set and data fusion in remote sensing image. In: Proceedings of the 6th World congress on intelligent control and automation (WCICA06), pp 10420–10424
10. Wang JH, Liu L (2012) Modeling fuzzy decision fusion based on Vague set. Appl Mech Mater Appl Sci Eng 197:7–12

Vehicle Positioning Method of Modern Tram Based on TETRA Communication System

Sun Yongmei, Wang Fuzhang and Zhong Yishun

Abstract Vehicle positioning system is an important component of traffic dispatch and intelligent control of modern tram. The selection of vehicle positioning scheme and wireless communication scheme decided the performance of the system. According to the characteristics and the requirements of modern tram vehicle positioning, the paper proposed the positioning method of modern tram vehicle by utilizing combined positioning method with satellite positioning and inertial navigation, and the communication scheme between various terminal is realized by Terrestrial Trunked Radio (TETRA) wireless communication system of rail transit, The proposed vehicle positioning scheme can save the cost of the whole project, ensure the precision of the system. Moreover, the vehicle positioning system have the merits of continuity, reliability and maintainability.

Keywords Modern tram · Vehicle positioning · Communication · TETRA

1 Introduction

As a new type of urban transport, Modern Tram has the advantages of green environmental protection, moderate traffic volume, comfortable ride, low cost and short construction period, etc. Modern Tram is getting more and more attention of many cities. At present, the lines which already on service in domestic including Shenyang Hunnan tram, Tianjin Binhai New District tram, Dalian tram, Shanghai Zhangjiang tram, etc. The cities which tram project is planned or under construction including

S. Yongmei (✉) · W. Fuzhang · Z. Yishun
Institute of Computer Technologies, China Academy of Railway Sciences, Beijing, China
e-mail: sunym1118@sina.com

W. Fuzhang
e-mail: wfzh2013@wo.cn

Z. Yishun
e-mail: zhongyishuncz@126.com

© Springer International Publishing Switzerland 2017

V.E. Balas et al. (eds.), *Information Technology and Intelligent Transportation Systems*, Advances in Intelligent Systems and Computing 454,
DOI 10.1007/978-3-319-38789-5_25

Beijing, Zhuhai, Chengdu, Shenzhen, Xuzhou, Weihai and Huaian etc. Compared with traditional tram, modern tram has great breakthrough in vehicle design and operation control. The body of modern tram is consist of numbers of modules, which is lightweight designed and have smaller turning radius and larger climbing ability. Considering of the power supply mode, modern tram can adopt ground third rail power supply, super capacitor and battery power supply besides traditional overhead contact network power supply. The application of new materials and new technologies can effectively reduce the vibration noise of vehicles [1–3].

The information of vehicle position and speed of modern tram vehicles in operation is very important for the dispatch system and the passenger information system. Accurate positioning of vehicles can effectively ensure the safety of vehicles, improve operating efficiency and shorten the distance between vehicles, moreover, it can also promote the management modernization, improve the comprehensive transport capacity and service quality. There are a lot of ways of vehicles positioning in the area of rail transit, the different positioning method suitable for different environment, and the cost and the positioning accuracy are different. Shenyang Hunnan New District modern tram dispatching assistant system using integrated positioning (satellite positioning, beacon, pulse mileage sensor) scheme, and the wireless transmission scheme using a set of wireless transmission system for signal specialty [3]. Shanghai Pudong Zhangjiang tram project adopt GPS as the main positioning mode, when entering the blind area, it can automatically switch to auxiliary positioning module. Utilizing GPRS as the communication platform, sent the real-time data of vehicle positioning to control center server [4]. This paper introduces the vehicle positioning method used in the current rail transit system, analyzes the special requirements of the modern tram to the vehicle positioning system, and puts forward the vehicle positioning scheme based on the TETRA communication network.

2 Rail Transit Vehicle Positioning Schemes

The selection of rail transit vehicle positioning scheme mainly includes two aspects, one is the positioning system scheme, the other is wireless communication scheme. Vehicle positioning method includes track circuit, transponder, odometer, electronic map, GPS, inertial navigation, dead reckoning [5–7], etc. The single vehicle location method cannot meet the requirement of vehicle positioning. At present, the vehicle positioning is mainly used in combination, the specific positioning program including GPS/ODO combination positioning [8], DR/GPS/MM combination positioning [9], GPS/GLONASS combination positioning [10], etc.

The information collected from multiple sensors is required to be transmitted to the control terminal for data analysis and data processing through wireless communication. There are several common wireless communication schemes in rail transit system as followed.

2.1 Vehicle Positioning Based on GSM-R Network

Reference [11] designs a fingerprint positioning system based on GSM-R net-work. Vehicle terminal equipment (GPS receiver, gyroscope, odometer, GSM-R receive antennas) acquire the location data and the intensity of the base station signal, through data processing, obtain the vehicle geographic coordinates, time, angular velocity, speed and signal strength information. The information is encapsulated in ARM embedded system, and transmitted to the GSM-R wireless communication network in the form of short message. GSM-R network establish a data channel according to the communication protocol between the vehicle terminal and the monitoring center, through this channel the information is sent to the monitoring center. Monitoring center analyze the short message to get the vehicle information, through fingerprint positioning technology and data fusion to get the estimated position, then the positioning information return to the vehicle terminal through the data channel, the terminal control the vehicle running based on the positioning information.

2.2 Vehicle Positioning Based on Wireless Spread Spectrum Communication

Wireless spread spectrum vehicle positioning system consist of wireless terminal, wireless station (including terrestrial wireless base station) and vehicular wireless apparatus. The position information is transmitted from the vehicular radio station to trackside radio station. The distance between the vehicular radio station and the ground radio station is calculated by using the DSP information processing technology, furthermore, the position coordinates of the vehicle is calculated through distance. Wireless spread spectrum vehicle positioning is based on advanced wireless spread spectrum communication, pseudo code ranging and computer information processing technology, which can realize real-time and accurate positioning of vehicle in complex environment [12].

2.3 Vehicle Positioning Based on Special Wireless Communication System

Special wireless communication system is to provide a reliable means of communication for rail transit fixed users (control center, vehicle dispatcher, station attendant) and mobile users (train drivers, disaster prevention personnel, maintenance personnel and other mobile personnel). Special wireless communication system provides call function, data function, assistant business function and network management function. The vehicular radio station receive the vehicle positioning signals of GPS satellite signal and other navigation signal, and send it to location server via special wireless communication network. The location server parse the received data, and obtain the location information of the vehicle through data fusion. This kind of wireless communication does not require extra investment and project construction, it can save the system cost effectively.

2.4 Renting Public Network or Laying Special Network

The wireless communication scheme select Global system for mobile communication (GSM), or CDMA, GPRS. Both the base station and the wireless coverage equipment use the existing telecom operators equipment. It should laying special line or renting channel for the data transfer between the telecom operator room and the modern tram control center, and doing extended development of wireless network software. The information of satellite positioning and related information of vehicle transmit through the public wireless network to the communication controller of telecom operator room, then pass through the special line or rented channel to the server of control center. It can also lay signal system and vehicle positioning data transmission special line, this scheme is stable and reliable, but it will.

3 Requirements of Modern Tram Positioning

Compared to high-speed rail, subway and light rail, modern tram vehicles mainly running on ground lines, the conditions are relatively complex. Most modern tram select mixed lane or semi closed lane, the right of way are open relatively. Therefore it cannot copy the existing rail transit train positioning scheme. In the design of modern tram vehicle positioning system, it needs to be combined with the characteristics of the railway line, and considered the feasibility, reliability and maintainability of the system.

3.1 Reduce the Trackside Equipment as Far as Possible

Modern tram has variety right of way, and it always parallel with public traffic network. Too much trackside equipment will influence the road traffic, and does not easy to maintenance. In the choice of positioning scheme, the vehicle equipment is the preferred selection.

3.2 *Ensure the Continuity of Positioning*

It including the continuous of space and time. As long as the vehicle running on the main line, the vehicle position information should be display in real-time any time any place. In the choice of sensors and communication network, it need to consider whether the sensor data output continuous, whether the sensor can work reliably when the vehicle passing through the special terrain, such as the tunnel, elevated, U-shaped slot, etc.

3.3 *Ensure the Accuracy of Positioning*

The location information of the vehicle is provided to vehicle dispatching system, passenger information system, signal system, and other related driving systems, which further realize the vehicle monitoring, arrival forecasting and other functions. In the selection of the positioning scheme, the accuracy of data acquisition and data fusion method of different sensors should be considered to meet the design requirements.

4 Vehicle Positioning Method Based on TETRA System

4.1 *TETRA Digital Trunked Communication System*

Terrestrial Trunked Radio (TETRA) digital trunked communication system is a multi-functional digital trunked radio standard developed by European Telecommunications Standards Institute (ESTI). It adopt TDMA technique, and provide trunked and non-trunked voice, circuit data, short message, packet data service and direct mode operation communication. The system has the advantages of good compatibility, good openness, high frequency spectrum utilization and strong security. TETRA digital trunked communication system has wide applications in urban rail transit, which provides reliable communication for exchange of voice and data information between the users of rail transit. The signal of TETRA covered throughout the vehicle running area. Under the TETRA trunked communication mode, the application of vehicle positioning can be realized by equipping GPS receiver and other positioning sensors on vehicular radio station, and utilizing the short data to transmit the positioning information.

4.2 Integrated Vehicle Positioning Scheme

The operating environment of modern tram is mainly on the urban ground road, some of the line has a certain distance of the underground tunnels and elevated sections. For vehicle positioning system, the best choice is GPS satellite positioning method. In case of GPS signal coverage, it has higher positioning accuracy. But considering the influence of high buildings, trees and tunnels on the operating line of modern tram, it needs to use other positioning method to complement the non-continuous output when the GPS signal is blocked, to ensure the continuity of position parameters. Inertial navigation measurement unit (IMU) is composed of gyroscope and accelerometer. It can provide effective attitude, angular velocity and acceleration, furthermore, the speed and position of the vehicle can be calculated by the above parameters. The inertial navigation system can work continuously and seldom appear hardware failure. But the error of the inertial navigation system will accumulate with time.

The performance of GPS and IMU is complementary. IMU can make up for the signal interruption of GPS, and GPS can restrain and correct the accumulated error of inertial navigation. Therefore, in order to realize the continuous, accurate and reliable output, this paper uses the GPS/IMU combined positioning method.

4.3 System Architecture of Vehicle Positioning Based on TETRA

The vehicle positioning system is composed of inertial navigation system, satellite signal receiver, location server (main/standby redundancy), core switch (main/standby redundancy), external interface, base station and transmission channel. System architecture is shown in Fig. 1.

4.4 Vehicle Positioning Equipment

Modern tram vehicular radio station consist of vehicular radio station host, control box, interface circuit and power module. The vehicular radio station can realize voice communication and data communication between the driver and the dispatch and the station staff. The satellite data received module and the inertial navigation module are installed in the vehicular radio station, and received the satellite data and the inertial navigation data in real-time. The ARM embedded system complete the acquisition, parsing and caching of various data. The data of sensors is transmitted to the location server via the TETRA wireless communication network. The schematic diagram of the vehicle positioning device is shown in Fig. 2.

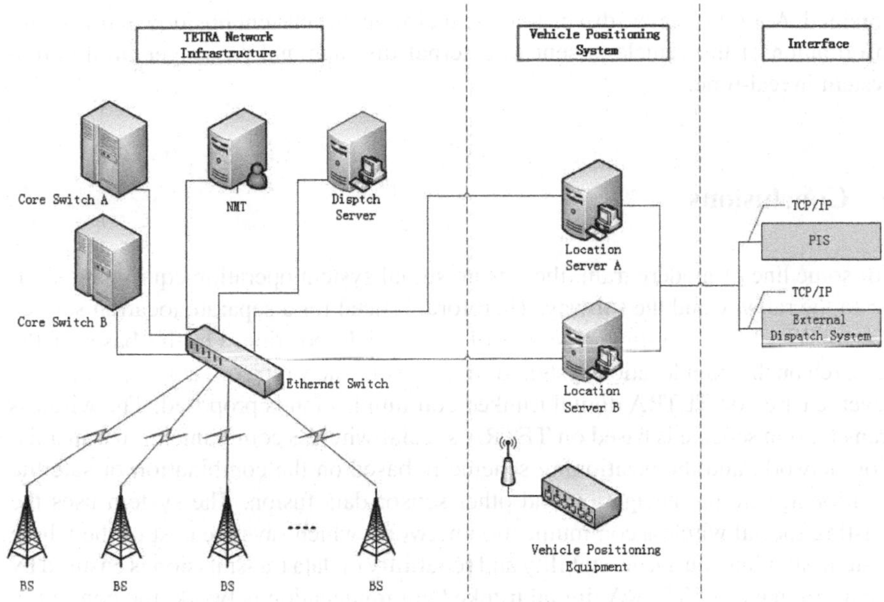

Fig. 1 System architecture of vehicle positioning based on TETRA

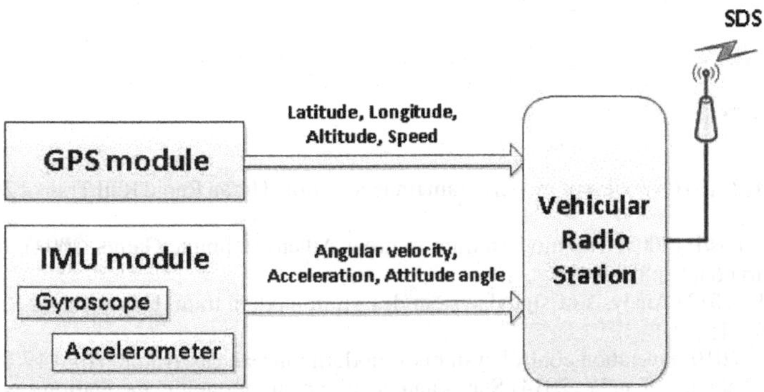

Fig. 2 Schematic diagram of the vehicle positioning device

4.5 Implementation of Location Server

The location server receives the short message data packets sent by vehicle positioning device, then conduct data analysis and location estimation in real-time. In order to ensure the continuity and reliability of the location information, the location server uses a hot standby configuration. The location server requires real-time monitoring power on/off of the vehicle, and according to the received data of sensors to determine the direction of the vehicle. Based on multi-sensor data fusion and dead reckoning algorithm, the prediction and correction of vehicle position and velocity is obtained. After the vehicle drives into the mainline, the current position and velocity information of the vehicle is sent to external dispatch and passenger information system in real-time.

5 Conclusions

For some line of modern tram, there is no signal system operation equipment similar to the railway and the subway. Therefore, it need for a separate location service system to achieve real-time tracking of the vehicle position. On the basis of the research on the train location system of rail transit, a new type of vehicle positioning scheme based on TETRA digital trunked communication is proposed. The wireless transmission scheme is based on TETRA special wireless communication transmission network, and the positioning scheme is based on the combination of satellite positioning, inertial navigation and other sensor data fusion. The system uses the existing special wireless communication network, which save the cost of the whole system, simultaneously, the stability and reliability of data transmission is ensured by the performance of TETRA digital trunked communication network, the continuity and accuracy of the positioning information is ensured by multi-sensor combined positioning technology.

References

1. Qiang D (2013) Overview of modern tram transportation. Urban Rapid Rail Transit 26(6):107–111
2. Konstantinos P (2008) The modern tram in great Athens (Olimpic Games 2004). Ingegneria Ferroviaria 63(11):854–943
3. Haijun L (2013) Analysis of signal system design in modern tram. Urban Rapid Rail Transit 26(6):156–159
4. Jiayan T (2010) Operation control system of modern railroad car. Autom Appl 12:61–65
5. Jiang L, Baigen C, Jian W (2014) Status and development of satellite navigation system based train positioning technology. J Cent South Univ (Sci Technol) 45(11):4033–4042
6. Sahawneh LR, Aijarrah MA, Assaleh K (2011) Real-time implementation of GPS aided low-cost strapdown inertial navigation system. Intel Robot Syst 61:527–544
7. Baoqing G, Tao T, Datian Z (2007) Map aided positioning method used in train locating application. J China Railway Soc 29(6):44–47 (positioning)
8. Qin Y, Bogen C, Jian W, Dawang J (2010) Train position system integrated with GPS and ODO. Mod Electron Tech 19:168–171
9. Xiaochun W, Nanjie L, Bo H (2014) A data fusion algorithm for integrated positioning system based on DR/GPS/MM. Microcomput Appl 33(21):1–3
10. Fuli W, Qinxin W, Yingyan C (2013) Weighting methods in GPS/GLONASS integrated positioning. Sci Surv Mapp 38(1):18–19
11. Liqun G (2010) Research on train integrated location technology based on GSM-R network, Southwest Jiaotong University
12. Xiaojuan L (2009) Research on the key technologies of communication based train control system in urban rail transit, Lanzhou Jiaotong University

Dynamic Path Selection Model Based on Logistic Regression for the Shunt Point of Highway

Bing Chang and Tongyu Zhu

Abstract In the operation management and real-time monitoring of the highway, we always want to know the current position of all vehicles in real time, so that we can find the congestion and accident section timely and make effective treatment. But in reality, only a small part of the highway vehicles equipped with a global positioning system, and can access to the location information in real-time, for the most of the vehicles, we can only access the position point when they are in and out the highway by the toll data, and cannot get access to their specific routing when they are on the highway. Especially when the vehicle is moving to the shunt point, during the current state we cannot know exactly which direction the vehicle will choose next, which leads to the result that we cannot estimate the correct position of vehicles. In order to accurately identify the direction of vehicles in the shunt point, this paper proposes a framework that based on the highway toll data, the A* algorithm and logistic regression were used to predict the direction choose of vehicles on the shunt point of highway. The framework takes the historical toll data as a sample to train the feature weight in logistic regression model, and takes the actual direction of vehicles on the shunt point as a test set to evaluate the effectiveness of the method proposed by this paper.

Keywords Logistic regression · A* algorithm · Shunt point · Toll data

B. Chang (✉) · T. Zhu
State Key Lab of Software Develop Environment,
Beihang University, Haidian, Beijing, China
e-mail: Changbing1402@gmail.com; 776901453@qq.com

T. Zhu
e-mail: zhutongyu@nlsde.buaa.edu.cn

© Springer International Publishing Switzerland 2017

V.E. Balas et al. (eds.), *Information Technology and Intelligent Transportation Systems*, Advances in Intelligent Systems and Computing 454,
DOI 10.1007/978-3-319-38789-5_26

1 Introduction

In recent years, with the continuous development of the transportation infrastructure, the vehicle population continues to increase and people's travel demand is increasing, the management and monitoring of the highway is facing increasing pressure, reasonable and effective management of the highway is a hot research topic in the field of traffic. A feasible solution is if we can get the current position of all vehicles on the highway, then we can monitor the running state of all vehicles in real time, and find out the abnormal traffic phenomena such as congestion and traffic accidents on the freeway timely, then carry on the effective treatment. But at present, only a small part of the highway vehicles equipped with a global positioning system, and can get the running status information such as the position and speed real time and precisely. For most of the vehicle without the global positioning system, the real-time running state information of the vehicle can only be obtained by the fixed detection equipment which is located on the highway, however, due to the high cost of the fixed detection equipment, the layout density is relatively low, resulting in the fact that we cannot obtain the running state information with high accuracy, therefore we need to build a real-time vehicle location estimation model on the whole road network. In the process of building the model, we will encounter a problem: when the vehicle runs to the shunt point, we cannot know exactly the next direction choose of the vehicle in real-time, and will lead to the failure of vehicle position estimation. The main shunt points of the highway are the following: (1) after entering the toll station, the vehicle can choose running up or down; (2) when the vehicle is traveling to the ramp which connects two main line, it may continue to travel along the current main line or turn to another main line on the ramp; (3) when the vehicle runs along the main line near the ramp which links a toll station, it may continue to travel along the main line or through the ramp to the toll station. Analysis of the direction chooses of the vehicle at the shunt point, we find that in the case of different shunt point, vehicles are always faced with two choices: running up or down in case (1) and running on main line or turning to ramp in case (2) (3). Therefore, we can take the problem of direction chooses on the shunt point as a classification problem of the vehicle behavior, namely the vehicles choose the same direction are classified as a class. Logistic regression model is a classical classification algorithm, which can be used to calculate the probability of each category according to the input of a set of independent variables. In this paper, the classification of vehicles on the shunt point is exactly corresponding with logic regression. So this paper proposes a dynamic direction choose model based on logistic regression when the vehicles are facing the shunt point. Firstly the model redistricts the highway section to select the shunt point, then uses A* algorithm to calculate the actual route of the vehicle and to construct the training set based on toll data, the weight of feature vectors in the logistic regression model is obtained by training the training set, finally constructs the logical regression model to predict the vehicle direction selection on shunt point.

The rest of this paper is organized as follows. Section 2 reviews some related works. Section 3 presents the framework of proposed method. Section 4 describes the concrete implementation method in detail. Section 5 evaluates the performance and accuracy of the proposed approach. Finally, we summarize our conclusion and discuss the next step of work.

2 Related Work

In this section, we will review some previous work. Dewen S provided that based on traffic data of the past, the upcoming traffic flows can be estimated. With the help of the improved ant colony algorithm, the dynamic optimal path planning results will meet the need of the travelers according with multiple actual constraints [1]. Bierlaire M proposed a probabilistic path generation algorithm to replace conventional map matching (MM) algorithms. Instead of giving a unique matching result, the proposed algorithm generates a set of potential true paths. Temporal information (speed and time) is used to calculate the likelihood of the data while traveling on a given path [2]. Li investigated morning commute route choice behavior using global positioning systems and multi-day travel data [3]. In Mandirs study of the behavior of the driver path selection in Munich metropolitan area, the vehicle routing model based on GPS data is constructed by using C-logit model to prove the traffic information is helpful to save the total time and total energy consumption [4]. Hood estimated a route choice model with GPS data collected from smartphone users in San Francisco. Traces were automatically filtered for activities and mode transfers, and matched to a network model. Alternatives were extracted using repeated shortest path searches in which both link attributes and generalized cost coefficients were randomized [5]. Broach used the 1449 non-exercise, utilitarian trips to estimate a bicycle route choice model. The model used a choice set generation algorithm based on multiple permutations of path attributes and was formulated to account for overlapping route alternatives [6]. Bierlaire and Frejinger proposed a general modeling framework that reconciles network-free data with a network based model without data manipulations. The concept that bridges the gap between the data and the model is called Domain of Data Relevance and corresponds to a physical area in the network where a given piece of data is relevant [7]. L. Liao introduces a hierarchical Markov model that can learn and infer a users daily movements through an urban community. The model uses multiple levels of abstraction in order to bridge the gap between raw GPS sensor measurements and high level information such as a users destination and mode of transportation [8]. Yang Y proposed a model based on the PSL model, which is based on the influence of road network conditions and traffic conditions on the behavior of the driver path selection. The results show that taxi drivers are more likely to choose the path which has short travel time and less turn, is a trunk road or has more secondary roads [9]. Jia Z presents two new methods based on the shortest-path (SP) approach. These heuristic methodsone centralized, the other distributed are both polynomial. In numerical experiments the proposed algorithms almost always find the optimal paths [10].

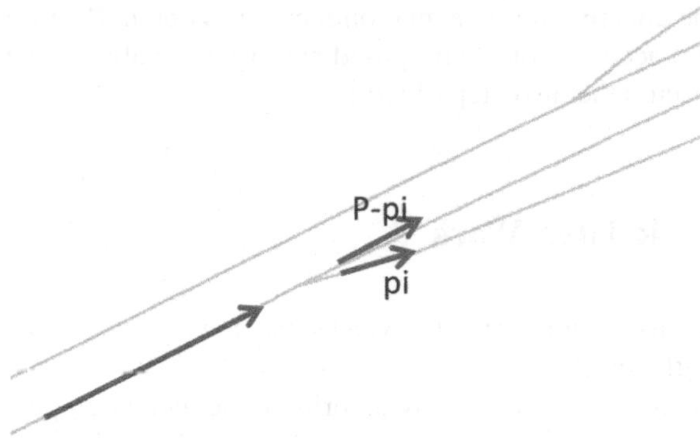

Fig. 1 The framework of the dynamic path selection model

3 The Framework of Dynamic Path Selection Model

As shown in Fig. 1, the path dynamic selection model proposed in this paper consists of three functional modules. The function of the first module is the actual route extraction. In this module, the topology of the highway network is constructed with the shunt point as breakpoint by analyzing the map bottom file, then combined with the location of the toll station, A* algorithm is used to extract the shortest route between the toll station as the input of constructing logistic regression model. The function of the second modules is to construct a logistic regression model. First, we should do preprocessing to remove the invalid toll data. Then with the routing data between stations obtained from the upper module, the historical direction selection data of the vehicle at the shunt point are generated, which can be used to train logistic regression model. The third module is to use the logic regression model to predict the direction selection of the vehicles in the shunt point and evaluate the accuracy of the prediction.

4 The Concrete Implementation Method in Detail

4.1 *Actual Route Extraction*

In this module, it is divided into two steps: the extraction of the highway network topology and the acquisition of the key route of OD. First we define the basic concept of the OD, the continuous process of vehicle enters and leaves the highway as a whole OD, and the entering station called O, the leaving station called D. Highway network topology extraction work mainly uses the map file of the highway to search shunt point and take the road connecting the shunt points as the unit road section, then we save the connecting relationship as the new topological structure of highway network. Combined with subsequent analysis needs, each unit road section sets up

Table 1 The topological structure of the highway network

Road section number	Next road section	Length (km)	Has toll station	Station number
G00G5001738924	G00G5001738766	20.5	N	
G00G5001739859	G00G5001738924	18.34	Y	208
G00G5001740387	G00G5001739859 #J00G42000G4201351964	33.2	N	–
J00G42000G4201351964	G00G4201351964	5.43	Y	3302

Table 2 The actual path selection on shunt point

O	D	Road section before shunt point	Actual selection of next road section
204	3020	G00G5001740387	J00G42000G4201351964
208	220	G00G5001740387	G00G5001739859
5004	5020	GS100200030060	G00G6511644839

a number of key attributes, including the length of the section, the next section, the stations information if there is a toll station in the unit, etc. The topological structure file of the highway network is shown in Table 1.

On the table, if a road section has a plurality of subsequent sections, there is # to split them. The main task of the second step is to extract the main route of OD. Combining the topology structure of the highway network in the first step, taking the shortest route as the filter standard and the A* algorithm as the main filtering algorithm to extract the shortest route between any OD pairs as the actual route selection, further extracting the available shunt nodes in the actual route, and finally obtain the actual path selection on the shunt node (if there is a shunt node in present route). The actual path selection on shunt point is shown in Table 2.

4.2 *Logistic Regression Model Construction*

The main function of this module is to train the logistic regression model using the toll data and the actual path selection data on shunt point.

(1) Data preprocessing: The data quality of the input data has a significant effect on the final construction of the model. So in this module, data preprocessing is first carried out to guarantee the accuracy and reliability of data. The main function of data preprocessing is to delete invalid data, including station information losing and station matching failing records, license plate missing and error records, vehicle type missing and time format error records, and delete the redundant attribute. Table 3 shows the sample data after data preprocessing.

Table 3 The sample data after data preprocessing

Plate	Vehicle type	O	D	Enter time	Out time
AN8360	1	3020	3009	20140901083045	20140901091330
CQ9110	0	220	208	20140901134032	20140901143045
K61630	4	3706	720	20140923183044	20140923171242
...	...	...	...	...	

(2) Extraction of historical path selection on shunt point: Combining data in Tables 2 and 3 according to the same attribute O and D, we can obtain the data of the path selection of all vehicles as well as other attribute data, which can be used as input of feature vector extraction.

(3) Feature vector extraction: In the construction of logistic regression model, the feature vector is used as the direct input. In this paper, the feature vector including the entering station number (S), entering time (T), vehicle type (C) and vehicle license plate (P).

$$(S, T, C, P)$$

Among them, S is the actual number of toll stations, such as 220. The value of T is 1, 2, 3, 4 which represent the 4 time periods of a day: 00:00–06:00, 06:00–12:00, 12:00–18:00 and 18:00–24:00. Vehicle types is the number from 0 to 6, which 0 said the small passenger car, 1 said the large and medium-sized passenger car, 2 said the first kind of truck, 2 said the second kind of truck, 4 said the third kind of truck, 5 said the fourth kind of truck, 6 said the fifth kind of truck. The license plate number is represented by the number which is the sequence number for all vehicles appearing on the highway, such as the number of YUAN8360 is 101.

(4) Logistic regression model construction: On the highway, when vehicle V runs to a designated shunt point, there may be two or more follow-up path selection P=p1p2pn, for any one of the path in the collection P, V either selects the path pi or selects another path from P-pi. So for all vehicles traveling to the shunt point, there are two results: selecting pi or not, as shown in Fig. 2.

In this paper, we use the logistic regression to model each optional path on designated shunt point. Logistic regression model is a classical classification algorithm, which can be used to calculate the probability of each category according to the input of a set of independent variables. Logistic regression which has two categories of 0 and 1 is commonly used, its conditional probability distribution is:

$$\mathrm{P}(Y = 1|x, w) = \frac{e^{w^{\mathrm{T}}x+b}}{1 + e^{w^T x+b}} = \frac{1}{1 + e^{-(w^T x+b)}} \tag{1}$$

$$\mathrm{P}(Y = 0|x, w) = \frac{1}{1 + e^{w^T x+b}} \tag{2}$$

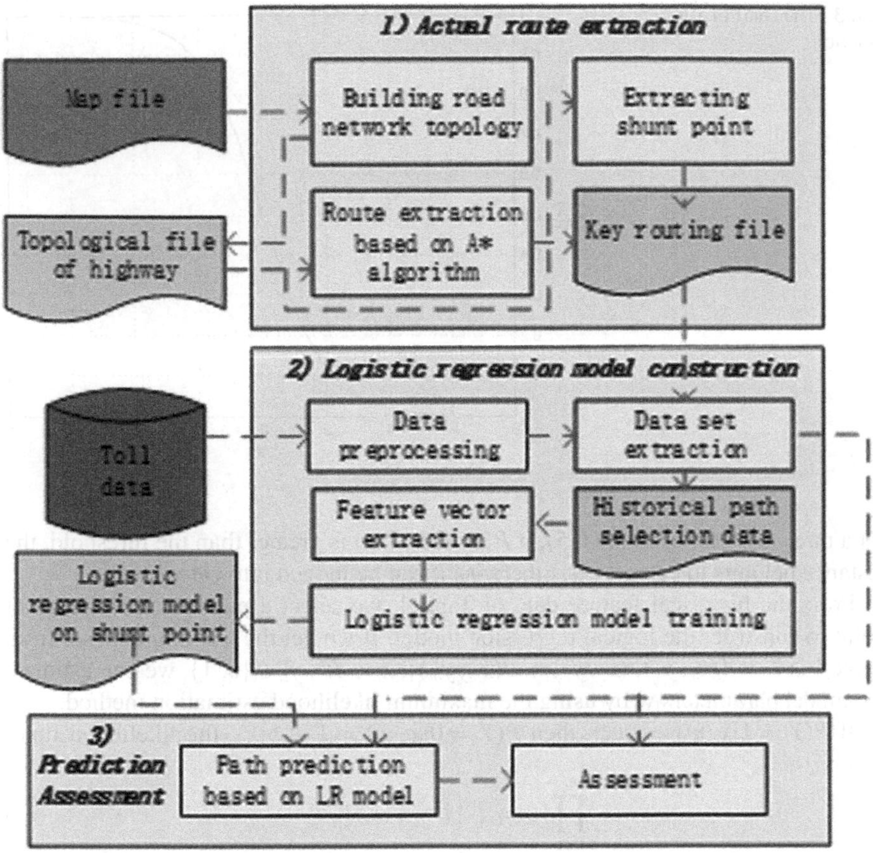

Fig. 2 The map of vehicle selection on shunt point

$x \in R^n$ is the input and called the feature of an instance. $Y \in \{0, 1\}$ is output, 0 and 1 respectively represent selecting pi or not. $w \in R^n$ and $b \in R$ is parameter, w is the weight vector, b is the offset.

Weight vector represents the influence of input on the classification results. The weight is positive suggests that the corresponding input increases the probability of the result; negative weight is just the opposite. The absolute value of the weight is bigger indicates that the impact of the input is stronger; on the contrary, the weight is close to 0, indicating that the input have little effect on the results. The Fig. 3 is the chart of LR function.

For a given training data set $T = \{(x_1, y_1), (x_2, y_2), ..., (x_n, y_n)\}$, We can estimate the model parameters w by using the maximum likelihood estimation method, so as to get a logistic regression model, then for an instance input x, according to the model can calculate the result $P(Y = 1|x, w)$, this is a number between 0 and 1, and we can

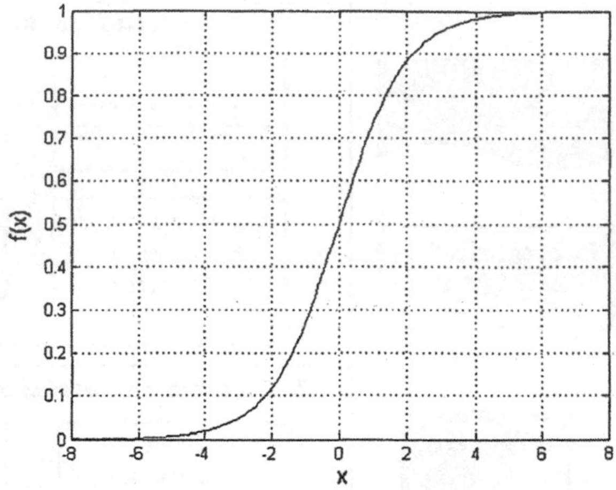

Fig. 3 The chart of LR function

set a threshold (for example 0.5), if $P(Y = 1|x, w)$ is greater than the threshold, the instance belongs to category 1, otherwise it can be judged into category 0.

From the historical feature data of Table 4, we select a path pi on some shunt point to construct the logical regression model. If we get the training data set from Table 4 is $T = \{(x_1, y_1), (x_2, y_2), ..., (x_m, y_m)\}$, $x_i \in R^n$, $y_i \in \{0, 1\}$, we can estimate the model parameters w by using the maximum likelihood estimation method.

If $P(Y = 1|x, w) = \phi(x)$, then $P(Y = 0|x, w) = 1 - \phi(x)$, the likelihood function is

$$\prod_{i=1}^{m} [\phi(x_i)]^{y_i} [1 - \phi(x_i)]^{1-y_i} \tag{3}$$

The log likelihood function is

$$L(W) = \sum_{i=1}^{m} [y_i \log \phi(x_i) + (1 - y_i) \log(1 - \phi(x_i))] \tag{4}$$

Seeking the maximum value of $L(W)$, we can get the estimated value of w.

Table 4 The feature vector data

Station number (S)	Entering time (T)	Vehicle type (C)	Plate (P)
3020	2	1	101
220	3	0	362
3706	4	4	1034
...	...	...	...

Assuming that the maximum likelihood estimate value of w is $\hat{w}$, then the logical regression model is:

$$P(Y = 1|x, w) = \frac{1}{1 + e^{-\hat{w}^T x}} \tag{5}$$

$$P(Y = 0|x, w) = \frac{1}{1 + e^{\hat{w}^T x}} \tag{6}$$

At this point, the problem is transformed into an unconstrained optimization problem with the log likelihood function. In the process of getting a solution, the problem of seeking the maximum value is generally converted to a minimum value seeking problem. So we turn the objective function to the negative log likelihood function, the function is a convex function that can be solved by using the gradient descent method.

Now considering the simple expression of unconstrained optimization problems $\min f(x), x \in R^n$, using x^* to indicate the minimum point of the objective function. Assuming $f(x)$ has two order continuous partial derivatives, the independent variable is x^k in the kth round iteration, the gradient function is $g(x) = \nabla f(x)$, the Hessian matrix of $f(x)$ is $H(x) = [\frac{\partial^2 f}{\partial x_i \partial x_j}]_{n\times n}$, using g_k and H_k represent the value of gradient function and Hessian matrix in x^k respectively, then the two order Taylor expansion of $f(x)$ in x^k is

$$f(x) = f(x^k) + g_k^T(x - x^k) + \frac{1}{2}(x - x^k)^T H_k(x - x^k) \tag{7}$$

We should note the fact that the expansion is ignored in the higher order infinitesimal part. Each iteration starts from the point x^k to seek the minimum point of objective function. x^{k+1} is the value of $k + 1$ iteration, the necessary condition for minimum value is $g(x) = \nabla f(x) = 0$, so $nablaf(x^{k+1}) = 0$ and $g_k + H_k(x^{k+1} - x^k) = 0$, we can get the recursion formula

$$x^{k+1} = x^k - H_k^{-1} g_k \tag{8}$$

p_k represents the search direction of the kth iteration, so

$$p_k = -H_k^{-1} g_k \tag{9}$$

In order to prevent over fitting and to obtain a simple model, we use the L1 regularization term in the training process, the benefits of joining the L1 regularization term are

- The relationship between the L1 regularization term and the sample complexity is logarithmic, which means that the L1 has good adaptability to the independent features.
- L1 can train lots of features which the weight value is zero.

- Compared with L2, L1 can achieve convergence with less times of iterations.

So the original optimization problem into

$$\begin{cases} f(w) = l(w) + r(w) \\ l(w) = -\sum_{i=1}^{m} \log(P(y|x_i, w)) \\ r(w) = C\sum_{j=1}^{n} |w_j| \\ \min_{w} f(w) \end{cases} \tag{10}$$

C is a constant for more than 0. Easily calculated

$$l(w) = \sum_{i=1}^{m} [ch_i \log(1 + e^{-w^T x_i}) + noch_i \log(1 + e^{w^T x_i})] \tag{11}$$

So its gradient is

$$\nabla l(w) = \sum_{i=1}^{m} [-ch_i \frac{e^{-w^T x_i}}{1 + e^{-w^T x_i}} x_i + noch_i \frac{e^{w^T x_i}}{1 + e^{w^T x_i}}] \tag{12}$$

So

$$\nabla l(w) = \sum_{i=1}^{m} [-ch_i P(y = 0|x_i, w)x_i + noch_i P(y = 1|x_i, w)x_i] \tag{13}$$

As for the problem of point zero is not differentiable results from L1, we use the finite memory quasi Newton method (OWLQN) to solve the problem. The simplified form of the objective function containing the L1 regularization term is

$$f(x) = l(x) + C\sum_{i} |x_i| \tag{14}$$

The process of solving logistic regression model based on OWLQN is as follows:

Step 1: Initial value $w_0 = (0, 0, ..., 0)^T$, set iteration convergence parameter ϵ and k = 0;

Step 2: Calculating $\nabla l(w_k)$, thus we can calculate the negative imaginary gradient $v_k = -\Diamond f(x)$ of $f(x)$ in w_k, if $||v_k||_2 < \varepsilon$ then exit, approximate solution to this problem is $x^* = x^k$, otherwise turning Step 3;

Step 3: Calculating $B_k v_k$ using BFGS algorithm, the search direction is $p_k = \pi(B_k v_k; v_k)$;

Step 4: Determining step and linear searching t, making

$$f(x^t + t \bullet p^k) = Min(f(x^k + t \bullet p^k))(t \geq 0)$$

and calculating $x^{k+1} = x^k + t \cdot p^k$, turning Step 3.

After the model is solved, we can get the mapping file of the feature to the weight, which is what we call the model file.

5 Prediction

In this paper, we take the Chongqing highway as the predicting object. We choose the toll data from March 2015 to May 2015 as the sample to train logical regression model and use the toll data from June 2015 to August 2015 testing the model. There are three shunt points are selected to construct the LR model, including the intersection of YuKun highway and the highway round Chongqing (Fig. 4), the intersection of YuSui highway and the highway round Chongqing (Fig. 5), the intersection of BaoMao highway and the ramp connecting JieLong station (Fig. 6).

In the above picture, the green line represents the path to model.

Figures 7, 8 and 9 has shown the prediction of vehicles path selection on shunt point using the LR model. In order to further verify the accuracy of the prediction results, Figs. 10, 11 and 12 compares the predicting results with the actual results in above conditions.

From the above picture, we can see the fact that the predicting result using the LR model has a high fitting with the actual result. But the contrast above is analyzed from a macro point of view, and cannot guarantee the reliability of the predicting results.

Fig. 4 Condition 1

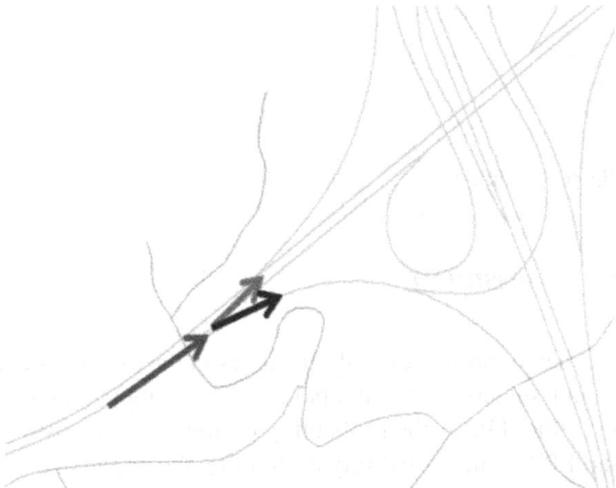

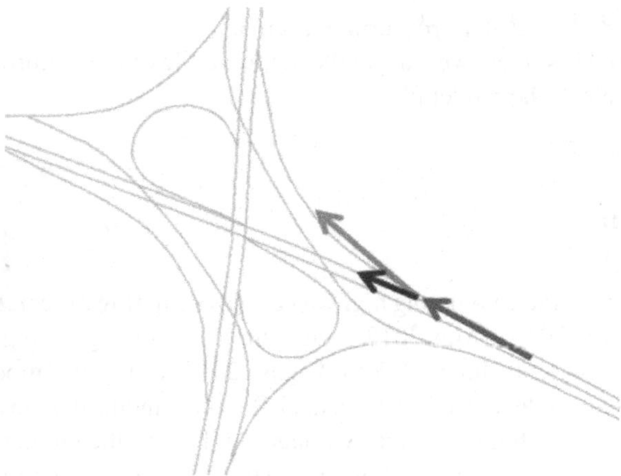

Fig. 5 Condition 2

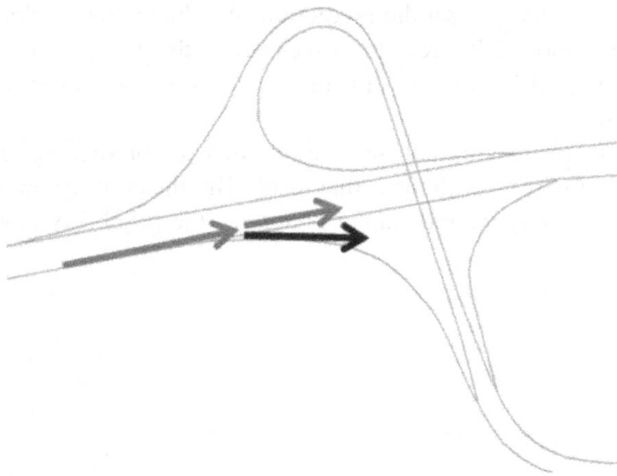

Fig. 6 Condition 3

6 Assessment

In order to further verify the accuracy of the logistic regression model in the prediction of path selection on shunt point from the microscopic, we set up two indicators TPR and FPR. TPR is the probability of dividing the positive cases into the true category, and FPR is the probability of dividing the negative cases into the true category, their calculation method is:

$$TPR = \frac{TP}{TP + FN} \tag{15}$$

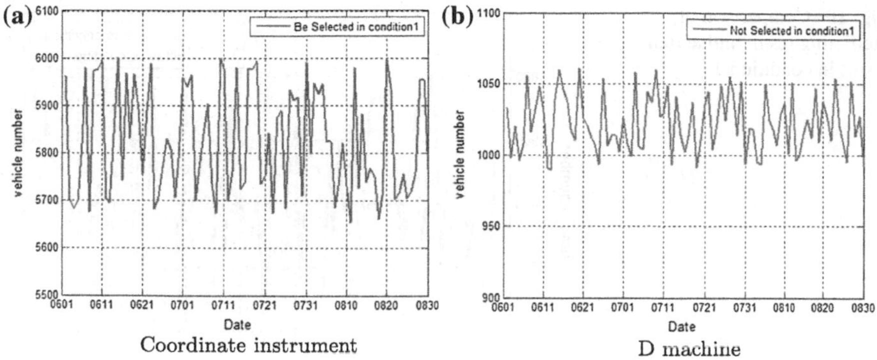

Fig. 7 The result that using the LR model to predict the vehicle numbers which select the green path in Fig. 4 from June 2015 to August 2015 (**a**). The result that using the LR model to predict the vehicle numbers which select the black path in Fig. 4 from June 2015 to August 2015 (**b**)

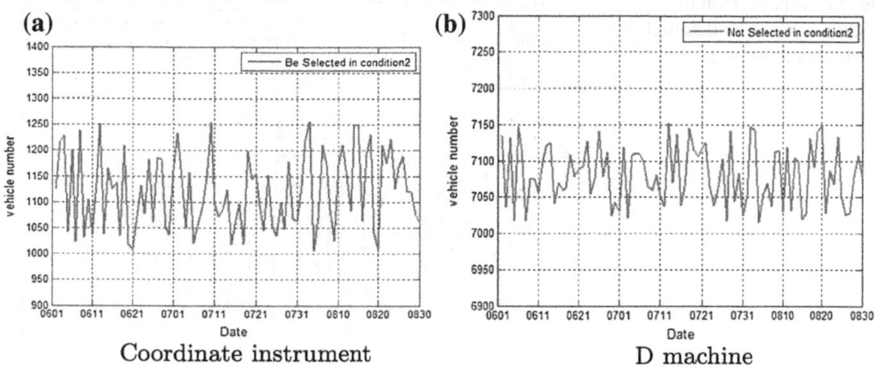

Fig. 8 The result that using the LR model to predict the vehicle numbers which select the green path in Fig. 5 from June 2015 to August 2015 (**a**). The result that using the LR model to predict the vehicle numbers which select the black path in Fig. 5 from June 2015 to August 2015 (**b**)

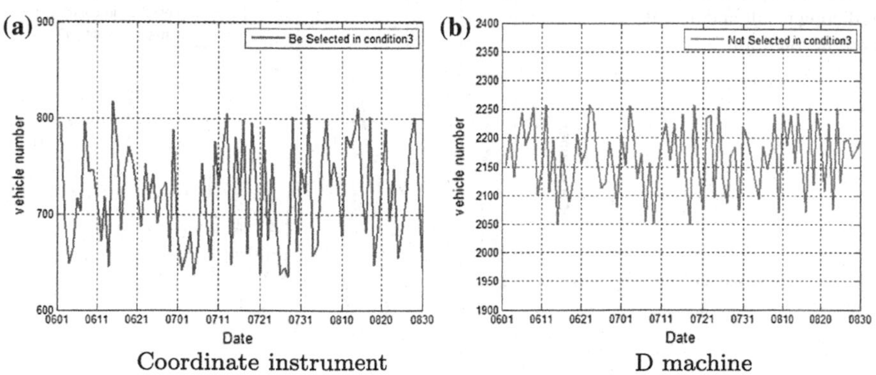

Fig. 9 The result that using the LR model to predict the vehicle numbers which select the green path in Fig. 6 from June 2015 to August 2015 (**a**). The result that using the LR model to predict the vehicle numbers which select the black path in Fig. 6 from June 2015 to August 2015 (**b**)

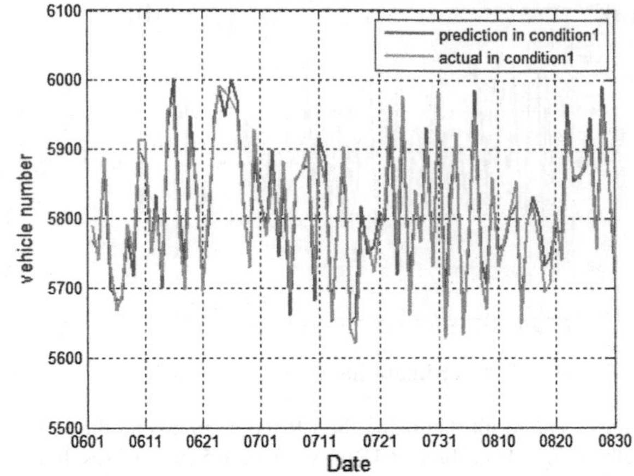

Fig. 10 Comparison of predicting result and actual result in condition 1

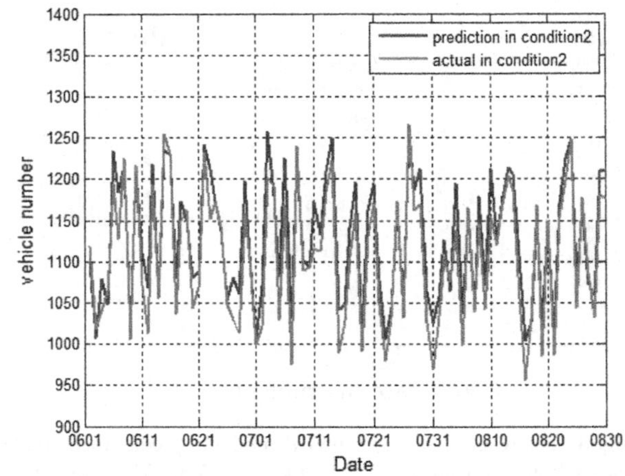

Fig. 11 Comparison of predicting result and actual result in condition 2

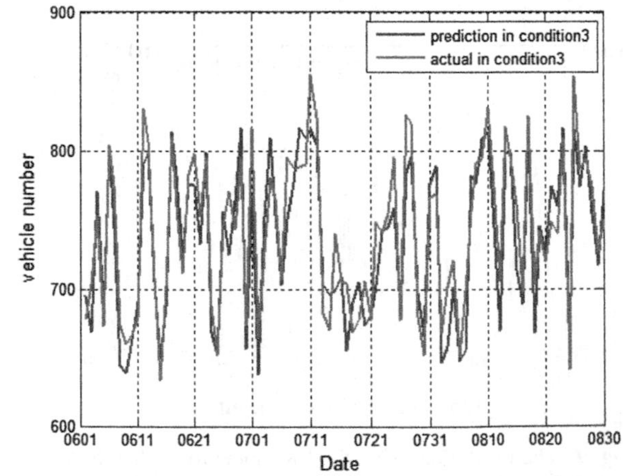

Fig. 12 Comparison of predicting result and actual result in condition 3

$$FPR = \frac{FP}{FP + TN} \tag{16}$$

Each test instance can be divided into one of the four types of the table: TP/FP/TN/FN. Positive represents the test instance is from the positive category. Negative represents the test instance is from the negative category. True represents the true prediction. False represents the false prediction. Figures 13 and 14 show the TPR and FPR of LR model predicting results (Table 5).

Can be seen from the above figure that TPR of the LR model predicting result remains at around 90 %, and FPR remains at around 10 %, so there is no doubt that the logistic regression model has high accuracy and credibility for the prediction of path selection when vehicles are facing shunt point, and the prediction results are relatively stable, which can be used to predict the real-time path selection.

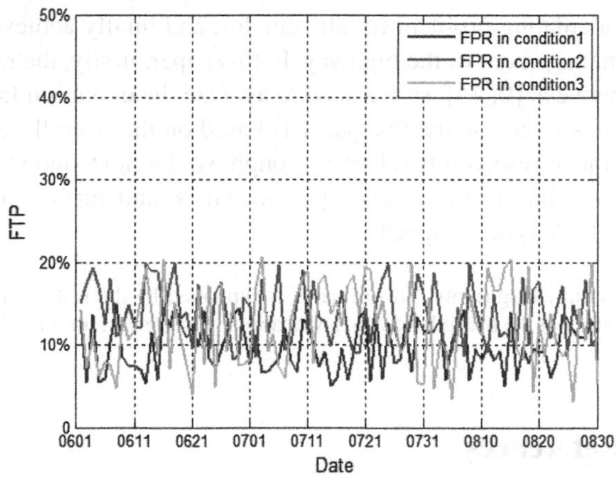

Fig. 13 TPR of predicting result

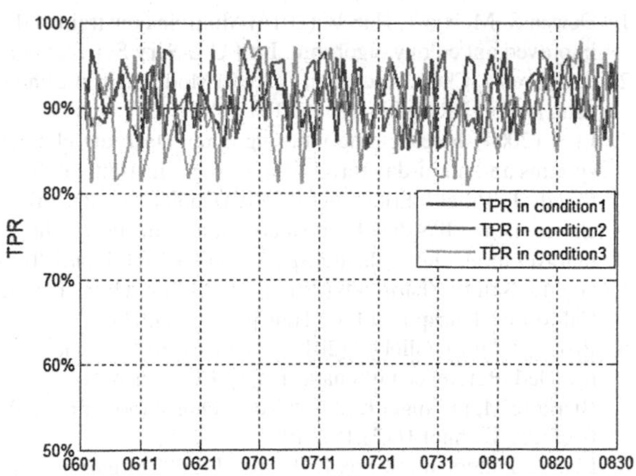

Fig. 14 FPR of predicting result

Table 5 Symbolic interpretation

	Positive	Negative
True	True Positive (TP)	True Negative (TN)
False	False Positive (FP)	False Negative (FN)

7 Conclusion

In this paper, we propose a model for the prediction of the path selection of vehicles with the shunt point and we carried out the actual model construction analysis and have solved all kinds of problems in the process of construction. Finally we take the Chongqing highway as the predicting object and compare the predicting result with the actual result to prove the accuracy and credibility of the proposed model. On the basis of this model, the next step can continue to study the real-time location calculating problem for all vehicles, and finally achieve the overall monitoring and management of the highway. In this paper, firstly, the range of the vehicle density is between [0, 45], so we need to analyze the evolution law of speed in higher vehicle density. Secondly, this paper is based on the overall vehicle density as the research object, research is relatively rough, so the next step is to divide the vehicle density according to the vehicle type attributes, and further analyze the effect of different vehicle type on speed.

Acknowledgments This research is supported by the test environment and demonstration application of the trusted network application software system for vehicle No. F020208.

References

1. Dewen S, Meixia T, Hao W (2014) Multiple constrained dynamic path optimization based on improved ant colony algorithm. Int J U E-Serv Sci Technol 7(6):117–130
2. Bierlaire M, Chen J, Newman J (2010) Modeling route choice behavior from smartphone GPS data. Report TRANSP-OR 101016:2010
3. Li H (2004) Investigating morning commute route choicebehavior using global positioning systems and multi-day travel data. Georgia Institute of Technology, Atlanta
4. Mandir E, Pillat J, Friedrich M (2010) Choice set generation and model identification for route choice using GPS data from smart phones. In: Proceedings of the conference on innovations in travel modeling of the transportation research board, Tempe, Arizona
5. Hood J, Sall E, Charlton B (2011) A GPS-based bicycle route choice model for San Francisco, California. Transp Lett Int J Transp Res 3(1):63–75
6. Broach J, Dill J, Gliebe J (2012) Where do cyclists ride? A route choice model developed with revealed preference GPS data. Transp Res Part A Policy Pract 46(10):1730–1740
7. Bierlaire M, Frejinger E (2008) Route choice modeling with network-free data. Trans Res Part C Emerg Technol 16(2):187–198
8. Liao L, Patterson DJ, Fox D, Kautz H (2007) Learning and inferring transportation routines. Artif Intell 171(5–6):311–331

9. Yang Y, En-jian Y, Long P, Nan Z (2015) Taxi route choice behavior modeling based on GPS data. J Transp Syst Eng Inf Technol 15(1):81–86
10. Jia Z, Varaiya P (2006) Heuristic methods for delay constrained least cost routing using shortest-paths. IEEE Trans Autom Control 51(4):707–712

Research on a Rapid Hierarchical Algorithm for the Wavelet Transform Based Contour Detection of Medium Plates

Ji Li, Xuxiao Hu and Mingmin Yang

Abstract This paper reviews two problems occur in the online real-time detection for cold rolled medium plates-low transmission efficiency and poor performance. The detecting process involves sending collected images to the industrial personal computer (IPC) of a camera where a rapid contour detection scheme of medium plates based on wavelet transform is presented. The overall architecture, contour detection principle and new two graded data transmission structure which are different from traditional ones will be analyzed. In the first place, the wavelet scalar quantization algorithm is used for pre-processing and compressing the picture captured by the image acquisition module. Subsequently, the processed data will be sent to the IPC where it is then uncompressed. Results show that the compressed image size equals 0.09 % of the original one; hence the new method is of capability to improve the overall system effectiveness significantly without affecting the accuracy.

Keywords Laser line · Contour detection · Wavelet transform · Image coding · Data transmission

1 Introduction

The term cold rolling means to roll steel, the raw material at temperature lower than its crystallization point; whereas hot rolling is the process of heating up a piece of steel billet before several rounds of rolling, and subsequently trimming it into a piece

J. Li (✉) · X. Hu · M. Yang
Faculty of Mechanical Engineering and Automation,
Zhejiang Sci-Tech University, Hangzhou 310018, China
e-mail: blueliji@163.com

X. Hu
e-mail: huxuxiao@zju.edu.cn

M. Yang
e-mail: ymingmin@126.com

© Springer International Publishing Switzerland 2017

V.E. Balas et al. (eds.), *Information Technology and Intelligent Transportation Systems*, Advances in Intelligent Systems and Computing 454,
DOI 10.1007/978-3-319-38789-5_27

of steel plate. Moreover, in comparison with the hot rolled ones, cold rolled steel sheets are of finer surface quality, better appearance as well as higher dimension precisions.

According to their thickness, steel plates can be divided into sheets, medium plates, thick plates and super thick plates. For instance, medium plates usually range from 4.5 to 25 mm. Since they have been widely used in the fields of construction engineering, machinery and containers manufacturing, shipbuilding along with bridge construction, etc., their quality is of significant impact on the safety and reliability of them [1]. On the one hand, the contour of medium plates serves as a crucial index for evaluating the performance of the plate mill and the control of their shapes. On the other hand, the accuracy of their detections affects steel losses directly. A research shows that among all losses in the yield of medium plates, cropping ends and trimming has respectively caused 23 and 26 % of the total wastage. Hence, it is clear that the plate contour detector produces great economical values by reducing cutting losses while increasing the yields of medium plates through the improvement of its detecting performance and accuracy [2].

Due to the gradual popularization of automatic product lines in the processing industry, the requirements for cold rolled medium plates tend to become increasingly exigent as they are the main raw materials for manufacturing. Consequently, their shape control has become a problem which the metallurgy industry needs to solve urgently along with shape testing as one of the crucial priorities of reinforcing automatic shape control [1].

To define the contour information obtained from the detector as essential parameters and send it to the Rolling Mill Control System is a way of entitling improved rolling quality to the strip steel in the first place. In addition, transmitting this data into secondary process systems, such as the Optimized Shear System, introduces the possibility of achieving automatic production in the rolling process. At present, although a variety steel strip profile detectors fitted for specific product lines have been successfully developed as well as partially applied in the actual manufacturing of strip steels; certain problems still show up [3]. For example, low transmission efficiency and poor real time effect can occur when the excessive width or high travelling speed of the plates generate a large number of images; and as a consequence, the productivity as well as quality of strip steel products will become less satisfying.

The emphasis of this paper is placed on cold rolled medium plate product lines. A set of rapid hierarchical testing methodologies based on wavelet transform have been developed in order to detect and display the contour data of tested plates for users online. These methods use an industrial plane array CMOS camera to capture the laser line irradiating on the surface of a medium plate at a high speed continuously. Afterwards, the pictures are transmitted into the camera IPC through an image acquisition device. The IPC will then process the received images so that the vertical and end position coordinates of the captured laser lines can be acquired. Next, visual calibration is combined with the speed data of the medium plates on the conveyor collected by pace testing devices, in order to transform the statistics of the vertical and end positions coordinates of laser lines into height information and edge position coordinates of the medium plate. The contour data of cold-rolled medium plates will then be generated after matching the above information and is eventually presented to users in a curve format which shows that the non-contact testing has been achieved.

In order to fulfill the testing purposes while concentrating on improving the system efficiency, the following approaches are included in this paper.

First of all, the global structure of the medium plate contour detecting system is demonstrated. Subsequently, the constitution of the image acquisition module and the data transmitting methods are elaborated. After that is the presentation of the principle of medium plate contour detection. Afterwards, the two-stage-image-compressing algorithm introduced with an emphasis of describing the image compressing arithmetic based on wavelet transform in the second phase. Finally, the advantages of the algorithm in effects and productivity are empirically proved.

2 The Principle of Medium Plate Contour Detection

According to their functions, five modules can be derived from the cold rolled medium plate contour detecting system, including the human-machine interactive host PC module, the front-end IPC module, image acquisition and processing module, medium plate speed and length testing module, as well as the contour graphic work condition and running state real-time monitoring module. Within the whole system, the image acquisition module is the most important but the most time consuming.

2.1 The Structure of Image Acquisition Module and the Statistics Transmission

The structure of image acquisition module The main function of the image acquisition module is to acquire pictorial information of the laser lines crossing the strip surface. Then the collected images will be processed with the results being transmitted into the IPC of the camera.

The main hardware components of the image acquisition module involve plane array CMOS image sensor, linear array semiconductor laser, and image capture card. Being selected as the light source for the image acquisition module, semiconductor laser lines have been installed on flat surfaces at a fixed height. Taking into account the reflection properties of cold rolled medium plate surfaces and the attempt to achieve optimal imaging quality of the camera, diffuse reflection has been chosen as the lighting system; as illustrated in Fig. 1, the axial planes of the laser line and the camera will form an angle named which the light will shine through. In order to eliminate the interference produced by the laser lines which illuminate on non-steel surfaces, and to speed up the image processing by simplifying its algorithm, the system uses plane mirror reflection to filter out the laser lines shining on non-steel

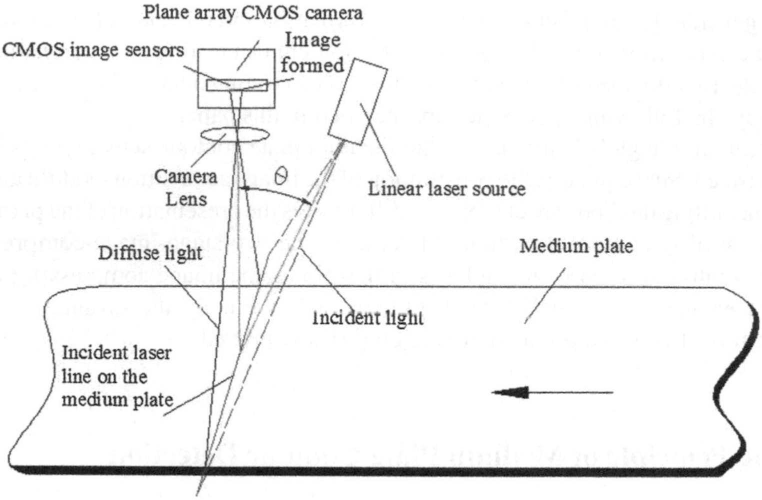

Fig. 1 The illustration of lighting system for contour detection of medium plates

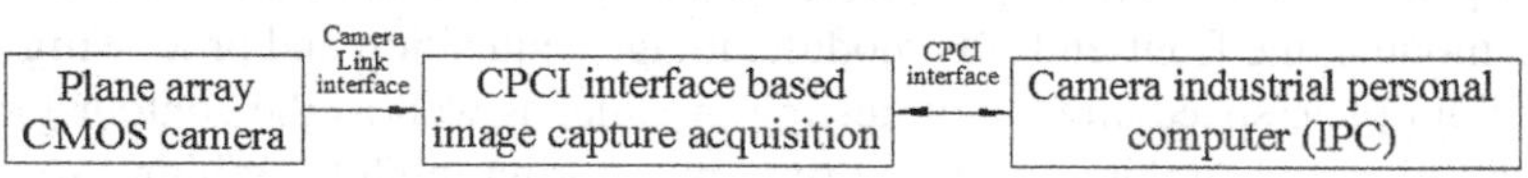

Fig. 2 The data flow diagram of the image acquisition module in a single CMOS camera

plate surfaces. Hence the diffuse reflection line will not be able to enter the image sensors.

Data transmission Figure 2 displays the data flow diagram of the image acquisition module data flow in a single complementary metal oxide semiconductor (CMOS) camera. Every plane array CMOS camera is equipped with a sensor which transforms the acquired optical signals into analog electrical signals; and an inner signal processing circuit where these analog image signals are translated into digital image signals. Subsequently, the Camera Link interface passes the processed digital image information to the image acquisition card based on the compact peripheral component interface (CPCI). The card has a built-in pre-processing algorithm for images with a main purpose of highlighting useful information in the pictures through eliminating noise signals. In addition, such algorithm can speed up the image acquisition procedure of the camera by reducing the amount of pictorial statistics. Afterwards, the card will transport the pre-processed image information to the camera IPC for further treatments through the CPCI. Nonetheless, in the case of testing an excessively thick plate, images captured by a single CMOS camera are not possible to cover all information of the whole plate. As a solution, multiple-CMOS cameras are needed but they will produce enormous images. Therefore, to continue using this transmission method means to incur high costs with low efficiency.

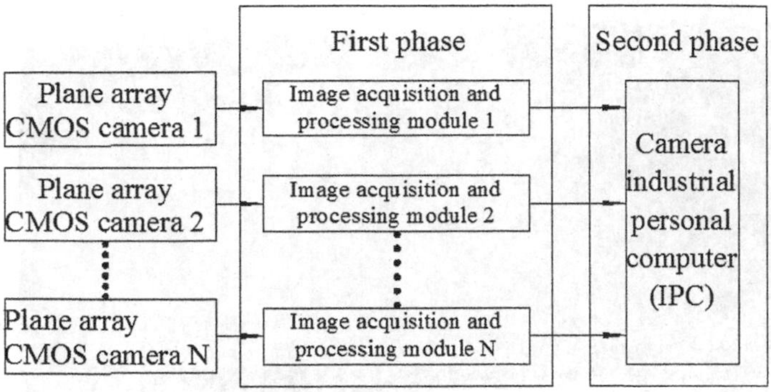

Fig. 3 The new data transmission of image acquisition module

In this paper, a new image acquisition module has been established (Fig. 3). The process of data transmission in this innovative module is divided into two phases. In the first place, pictures captured by multiple CMOS cameras be sent to the first tier of image processing unit, which has been enhanced with an image compressing unit on top of the built-in conventional pre-processing algorithm; whereas in the second stage, instead of having all image information of massive data flushed into the IPC directly, the statistics will be transmitted after being processed in the previous step. As a result, the transmission productivity is significantly improved so that the whole processing duration is shortened to some extent. Eventually, the pictures are decoded by the IPC in order to make subsequent image processing at ease.

In an image acquisition module, the communication between individual hardware components needs coordination and controls. Refer to Reference (2) for detailed communication structures.

2.2 Principle of Contour Detection

According to the demands of techniques, the contour of medium plates mainly involves their sides along with anterior and posterior. Figure 4 shows a picture of laser lines shining on a plate placed on a moving conveyor belt taken by a CMOS camera in proper sequence; whereas Fig. 5 is a diagram of the successful contour detection on cold-rolled medium plates.

When middle part of the medium plate is being inspected, the object of detection is the side contour. In this procedure, laser lines continuously irradiate on the plane of the medium plate, and the number pixels (W) between the upper and lower end points of the laser line will be worked out after the collected images being processed. After the visual calibration by the camera [4], with the help of Eq. (1), the pixel coordinates of the captured laser lines ends can be transformed into space coordinates of the medium plates side edge.

$$L = \lambda W \tag{1}$$

Fig. 4 The image of laser line on the plate captured by CMOS camera

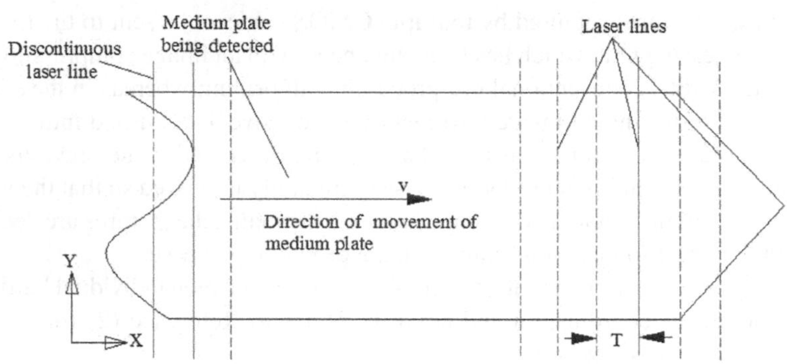

Fig. 5 Diagram of successful contour detection on cold-rolled medium plates

In the above equation, L stands for the width of the medium plate; whereas λ is the actual length represented by a pixel of the image; it is related to the distance (h) between the surface of the tested strip steel and the camera plane. The relationship between and h can be identified through camera calibration.

The anterior and posterior contour of the medium plate is uneven along the width in comparison with its central part. As a consequence, some of the flat laser lines irradiate on these two parts will go across the plate through the grooves instead of remaining on the surface. In this situation, segmentation will occur to the laser lines, as seen from the left end of Fig. 5. According to the principle of side contour detection, likewise, the coordinates of the laser line ends can be worked out with a specific algorithm.

The above procedure can generate coordinate statistics of a medium plate for its side, anterior and posterior end points; by utilizing these data, an overall contour curve of the whole medium plate can be fitted. As we can tell from Fig. 5 that the strip steel moves at a constantly changing speed (v), by observing its pace information with a special speed testing device in a certain cycle (T0), its coordinate on the X direction can be identified. In addition, the image acquisition modules takes pictures of the laser lines on the medium plate in a particular period (T) as well, so that the concurrent coordinate data of its side, anterior and posterior ends can be adopted. In another words, the coordinates of the strip steel edges on the Y direction can be worked out in this way.

To summarize, the contour detection system firstly works out the corresponding space coordinates and speeds of the medium plate edges for the laser line ends in the acquired images, then conducts related processes which mean matching the width statistics to an appropriate position; finally, the contour curve for the medium plate is fitted and shown on the human-machines interactive interface of the host PC.

3 The Description of the Image Compressing Algorithm

Due to the massive space taken up by the pictures captured by multiple CMOS cameras and the low productivity of direct image transmission, compressing has been proposed in the image acquisition module as a means to minimize image sizes and transmitting time without interfering the follow-up processing of laser line coordinate extractions [5].

3.1 *The Algorithm Flow of the Images*

In Fig. 6a, the whole algorithm of the first phase algorithm is illustrated, while the arithmetic flow of decoding and post treatment in the second stage is displayed in Fig. 6b. We can tell that there are also two phases in the compressing procedure, the first step is to extract as-small-as-possible areas in the image that contain the required features. In the second stage, wavelet decomposition will be conducted on the pictures before applying a wavelet scalar quantization to the decomposed coefficient; subsequently, the quantized coefficient will go through run-length and Huffman coding; eventually, the data will be translated into a standard format of document.

3.2 *Image Compressing Process Based on Wavelet Transform*

This paper also works on the image compressing process based on wavelet transform, and precisely studies the wavelet decomposition and reconfiguration of medium plate images.

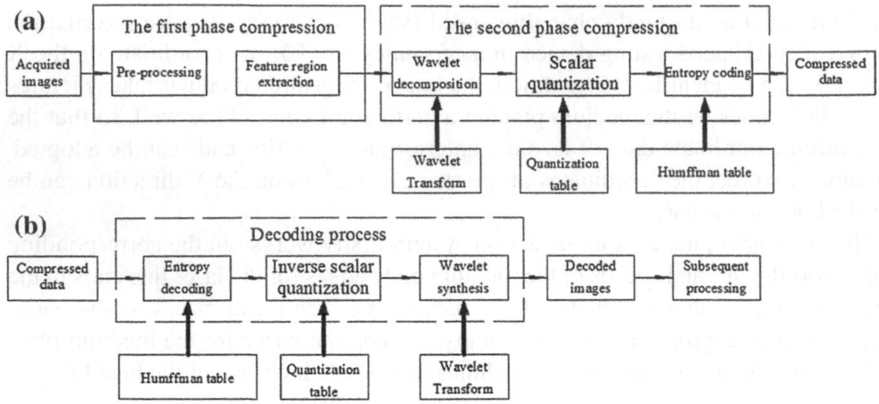

Fig. 6 Image compression and decompression algorithm flow. Flow of the whole first phase algorithm (**a**), flow of the second phase algorithm (**b**)

Table 1 Coefficients of DB97 wavelet decomposition filter group

Coefficients of low-pass filter	Coefficients of high-pass filter
0.037828455506995	
−0.023849465019380	0.064538882628938
−0.11062440441842	−0.040689417609558
0.37740285561265	−0.41809227322221
0.85269867900940	0.78848561640566
0.37740285561265	−0.41809227322221
−0.11062440441842	−0.040689417609558
−0.023849465019380	0.064538882628938
0.037828455506995	

The filter group used for wavelet decomposition in this method is a whole sampling set (WSS). $h0$ is used for representing the low pass filter while $h1$ stands for the high pass one. In response of WSS, the low pass filter needs to symmetric in relation to 0, hence $h0$ runs from $h0(-r0)$ to $h0(r0)$, in another words $h0(-n) = h0(n)$. In contrast, the high pass filter is symmetric in relation to -1, therefore $h1(-1-n) = h1(n-1)$. According to research outcomes, DB97 wavelet decomposition filter set is a group of WSS which has the best compressing effect [6]. Please see its coefficient in Table 1.

As for the wavelet reconfiguration filter group, its coefficient is determined entirely by the anti-alias relativity of Eq. (2).

$$\begin{cases} f_0(n) = (-1)^n h_1(n-1) \\ f_1(n) = (-1)^{n-1} h_0(n-1) \end{cases} \quad (2)$$

Table 2 Coefficients of DB97 wavelet reconfiguration filter group

Coefficients of low-pass filter	Coefficients of high-pass filter
	0.037828455506995
−0.064538882628938	0.023849465019380
−0.040689417609558	−0.11062440441842
0.4180922732222	−0.37740285561265
0.78848561640566	0.85269867900940
0.41809227322221	−0.37740285561265
−0.040689417609558	−0.11062440441842
−0.064538882628938	0.023849465019380
	0.037828455506995

According to this, when the wavelet configuration takes place, the coefficient of the high pass filter depends on the low pass filter which can work reversely as well. The WSS coefficient is worked out through the relativity shown in Table 2.

Thc boundaries of signals are in need of an extension so that the filter set can be helped adapt to signals of diverse lengths. In another word, such extension allows images in random sizes to undergo a multi-tiered wavelet transform. The boundary extension includes cycle extension, symmetry extension and zero extension, to name but a few. Among them, symmetry extension is used in the algorithm of wavelet scalar quantization.

Symmetric wavelet transform indicates that signals are extended before transforming.

Suppose x stands for incoming signals, then x will run from $x(0)$ to $x(N0-1)$, that means the length of signal x is N0. Furthermore, (i, j) represents the coordinates of x. To decide whether the left and right ends need to be extended, we can tell from the figures of the coordinates, if $i = 1$, no extension is needed for the left end; but if $i = 2$, extension for the left end is in need; the same principle applies for j as an indicator of extension needs for the right end. Also, $E(i, j)$ is used for representing extensions [7]. Hence, the relativity can be shown as below,

$$\begin{cases} \text{DirectTransformation} : xy\,\{a_0, a_1\} \\ \text{InverseTransformation} : \{\mathrm{a}_0, \mathrm{a}_1\}\, yx \end{cases} \tag{3}$$

where y stands for the signals acquired though the symmetric extension of x.

During the symmetric wavelet transformation, only half of the signals need to be calculated and saved as the signals are symmetric after the mentioned extension. For instance, after symmetric wavelet transforming, the wavelet coefficient of WSS and its even length signals is shown as Eq. (4),

$$a_i(k) = \sum_{n=0}^{N-1} y(n)h_i(2k-n) \tag{4}$$

The only thing that requires calculation and storage is the coefficient of $k = 0, \ldots, N_0/2 - 1$, therefore,

$$b_i = \downarrow (y * h_i) \tag{5}$$

3.3 The Scalar Quantization of Wavelet Coefficients

Quantization serves as a process of further data discretization. It involves not only scalar quantization which means to quantify a single statistic according to pixels; but also vector quantization in which a data block is quantified. In actual application, the compressing effect of vector quantization tends to be better, but its machinery structure is a lot more complicated hence will much more time is required for coding images [5]. For a compressing system that handles a huge amount of medium plate images, such disadvantage cannot be compensated. Taking this factor into account, scalar quantization has been chosen for the algorithm of this research.

Mathematically, the scalar quantizer can be represented by Eq. (6), with S indicating the quantizer index of discretion of which the value is the data used in entropy coding; meanwhile, E is not reversible, hence the compressed result is damaged. For the coding of this paper, scalar quantizer indicates p, the quantizer index that maps a, the wavelet coefficient, onto the integer value; this index points to the quantized interval where a locates, as shown in Fig. 7, hence

$$E : R \rightarrow S \tag{6}$$

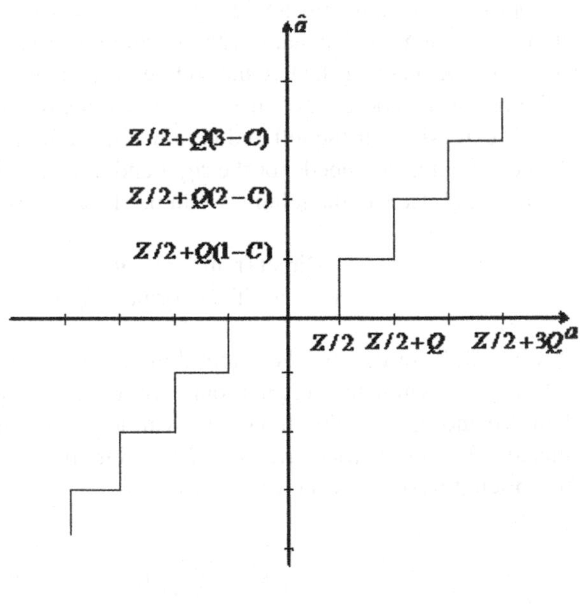

Fig. 7 Scalar quantizer

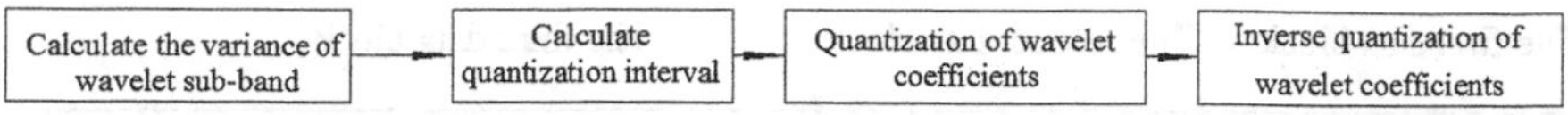

Fig. 8 Quantizer configuration process

During the decompression of image data, the quantizer index requires inverse quantization; inverse quantizer can be mathematically shown as Eq. (7), in which D is not an inverse transformation of quantizer E. Within the decompiler, inverse quantizer maps the quantizer value p onto a discrete reconfiguration coefficient named $\hat{a}$, hence

$$D : S \rightarrow R \tag{7}$$

In accordance to the above description, the priority for quantifying the wavelet coefficient is to construct a quantizer, the constructing process can be found in the Fig. 8 [8].

3.4 *The Entropy Codification of the Quantization Index Value*

The entropy expression of the discrete source and source code First of all, the assumed constitution of the symbol set of a discrete source includes a number of symbols (N) in the range of $\{x_1, x_2, x_N\}$, the occurrence probability of a symbol in the source matches $\{p_1, p_2, , p_N\}$ as well as fulfilling $\sum_{k=1}^{N} p_k = 1$. This paper has chosen memory-less sources for its algorithm so that the current and previous output statistics are detached from each other [9]. The Eq. (8) below decides how much information each symbol contains, and this is the so-called self-information quantity.

$$I(x_k) = -\log_2 p_k \tag{8}$$

Therefore, the average self-information quantity of each symbol in the discrete source set of $\{x_1, x_2, , x_N\}$ is worked out by the relativity of Eq. (9).

$$H(x) = -\sum_{k=1}^{N} p_k \log_2 p_k \tag{9}$$

$H(x)$ represents the entropy of the source.

Source coding implies the representation of a source output sequence by a Binary code sequence. In addition, two conditions must be met in order to apply coding on the memory-less discrete source [10].

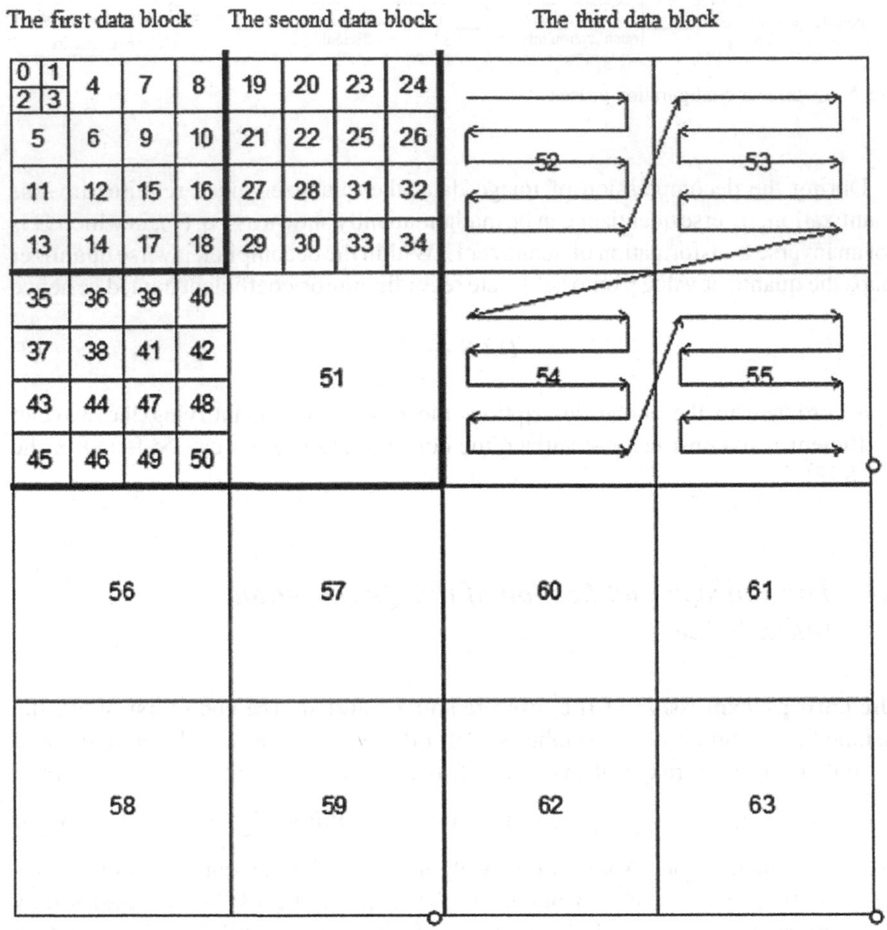

Fig. 9 Divisions of sub-bands and data blocks

Data sequence of entropy coding After the wavelet decomposition, 64 sub-bands can be derived from the image of a medium plate, and their distributions are shown in Fig. 8 [11]. These sub-bands will be divided into 3 data blocks during the entropy coding procedure; these blocks will then be codified individually [12]. Figure 9 illustrates the divisions of sub-bands and data blocks; sub-bands NO.0-18 are in the first data block, NO.19-51 are in the second, while NO.52-63 are in the third block. Scans on each data block is based on the sequence of sub-bands, while within each sub-band, scans are done from left to right and top to bottom.

Zero RLE The continuous occurrence of certain symbols in a signal is absolutely common during the coding process. When encountering this situation, such signals can be represented by the number of occurring times (L) and the symbol (V) [13].

For example, the character string of signal aaabbbbbddddddccc can be simplified as 3a5b6d3c.

In fact, in signals, zero is the only symbol value that will turn up from time to time, whereas others occur randomly. Therefore, RLE is only applicable to zero symbols in this case. In the algorithm of this paper, the quantizer index matrix derived through the scalar quantization of wavelet coefficient is sparse. In another word, the quantizer index matrix has a large number of zero symbols. Relying on the scanning strategies introduced in Sect. 3.4.2, as soon as all statistics are linked together to form a data string, a huge bunch of zero will turn up continuously, while other indexes are distributing randomly. Therefore, RLE has been chosen for initially coding the quantization value matrix.

As for the RLE of zero, results are recorded according to the symbol list in Ref. [14] which is then used as the input symbol table for Huffman coding.

Implementation of Huffman coding In 1952, Huffman put forward the codes of variable length named Huffman coding. Such coding is to allocate code words of different lengths to symbols according to their occurrence probabilities. Generally, symbols of higher probabilities will be allotted shorter code words and vice versa [15–18]. Figure 10 demonstrates the application process of image coding whereas Fig. 11 shows the procedure of the codes being written to files. Huffman coding works in the following sequencing.

First of all, the data distribution probabilities are counted. Secondly, the statistics are ranked from low to high probability. Thirdly, the smallest two nodes are combined into one having the probability value of the combined node equals the total of both nodes. Fourthly, the second and third steps are repeated until the node sum becomes 1 in order to build a binary tree that is named Huffman Tree. Finally, coding takes place based on the route of Huffman Tree, in another words, the 0, 1 code strings on the routes that the terminal node runs through towards each origin node are used as the coding words.

Implementation of Huffman decoding Figure 12 demonstrates the process of Huffman decoding. As shown in Fig. 6b, the first task in Huffman decoding is to read the Huffman table generated by the compression. Hence, there is no necessity

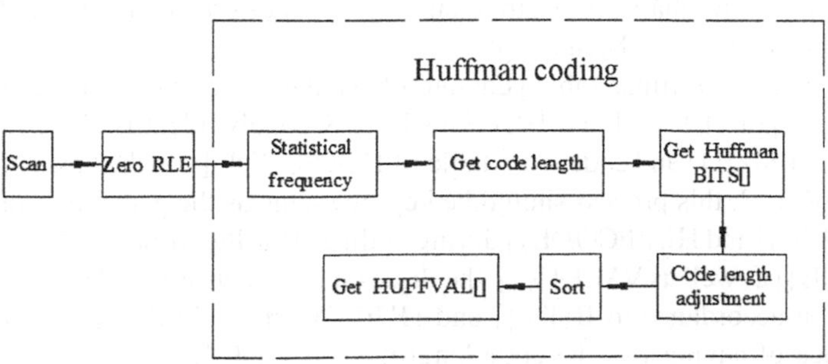

Fig. 10 Implementation of image coding

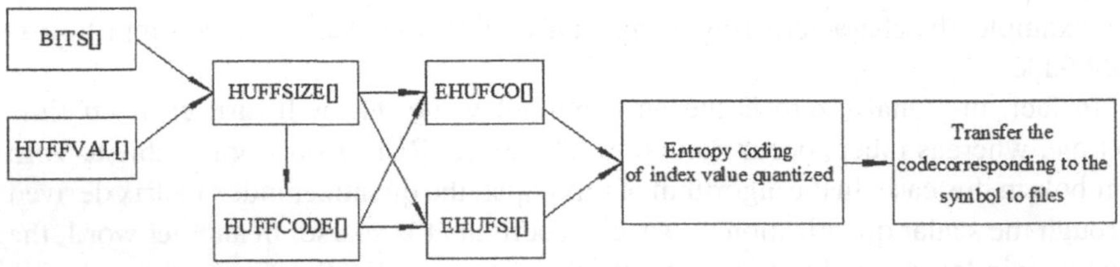

Fig. 11 The process of writing to the file

Fig. 12 Huffman decoding process for this report

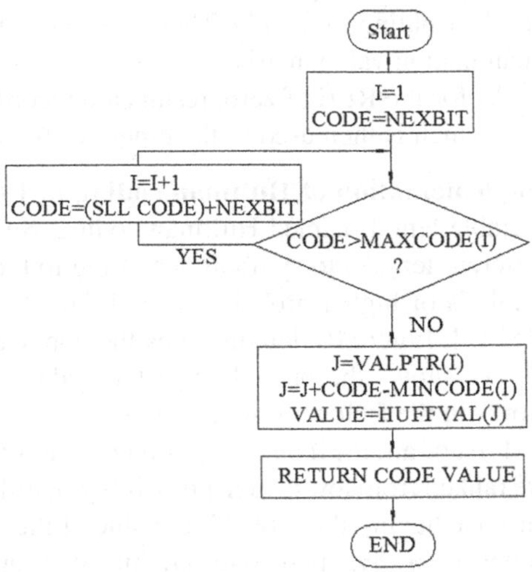

of establishing another Huffman table during decoding. The first basic step of decoding is to read the binary values one by one and store them into internal codes. After that, the data lengths are assessed aiming at verifying whether they equal the code lengths. If both lengths are identical, the code will be inspected in order to search for corresponding code words; if such code words exist, their corresponding symbols will be decoded. The same procedures are repeated in the reading of the next binary value until the whole decoding is completed.

Based on the flow chart demonstrated in Fig. 12, a customized Huffman decoding process for this report has been adopted.

Firstly, read the Huffman table generated from the compressed source document in order to acquire BITS [] and HUFF VAL []. Secondly, identify the code length table of HUFFSIZE [] and code word table of HUFFCODE [] on the basis of BITS [] and HUFFVAL []; this process should be kept the same as the generating procedure of HUFFSIZE [] and HUFFCODE [] during coding. Thirdly, work out MINCODE [], MAXCODE [] as well as VALPTR [], the three tables which are used in the decoding algorithm, in accordance to BITS [] and HUFFCODE []. In this stage, under the circumstance of having I as the code length, MINCODE [I] contains the smallest code word, MAXCODE [I] has the biggest one, whereas VALPTR [I] possesses the first value in the table. Finally, decode the code stream in accordance to the previously acquired tables.

4 Experiments

An experimental prototype for the cold-rolled medium plate contour detection system is built. With the purpose of handling and controlling the acquisition of images, vc++ software is used in the prototype for writing image acquisition processing and controlling programmes. Also, with the goal of displaying the results of contour detections, a human-machine interactive interface application is programmed.

Image compression is used in this paper, and the method used in such procedure is evaluated by the key performance index including compression ratio (CR), peak signal to noise ratio (PSNR) along with the complexity of the algorithm [19]. The main verifying object of the algorithm in this research is to significantly improve the running effectiveness of the whole system by reducing data transmitting time through the compressing algorithm without affecting the accuracy of contour detection. The following pictures illustrate the stages of the algorithm practice.

First of all, Fig. 13a, b are respectively the images of laser lines shining on a medium plate captured simultaneously by the CMOS cameras on the left and right sides, their dimensions are both 718*960*3 and they both have 2067840 bytes; meanwhile, Fig. 14 is a picture which has been through image mosaicking and the extraction of laser-line-only area, it is sized as 54*553*3 with 89434 bytes making it only equals 4.3 % of the original image size.

Secondly, Fig. 15 is the decompressed (with the algorithm of this paper) version of Fig. 14, this image indicates that later acquisitions of laser line edge coordinates are not affected by compression or decompression. The only difference is that the file

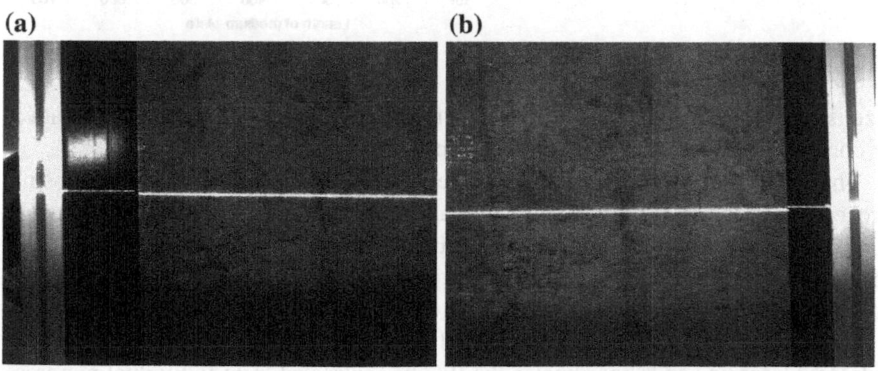

Fig. 13 Images of laser lines shining on a medium plate captured by the CMOS cameras on the *left* (**a**) and *right* (**b**) sides at the same time

Fig. 14 Picture which has been through image mosaicking and extracted the area with only laser line on

Fig. 15 The decompressed (with the algorithm of this essay) image of Fig. 14

Fig. 16 The pixel image generated after a series of image processing work in the IPC

Fig. 17 Image of laser line with 4 ends in it

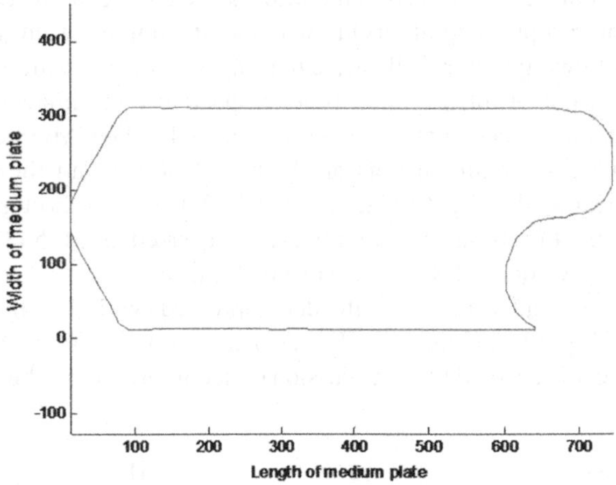

Fig. 18 Outline image of steel plate according to the curve fitting of coordinates of the ends of each laser line

size is reduced to only 2 k with 2*1024 bytes after compression. Also, the duration of compression is verified as 8 ms, while the decompression time is 15 ms.

Thirdly, Fig. 16 is the pixel image generated after a series of image processing procedures in the IPC, through this, the laser line coordinates can be found as (5, 23) and (5, 544). Figure 17 shows the situation of laser line breakage during the measurement of the medium plate ends, the coordinates for the 4 end points in this case are respectively (6, 23), (7, 116), (7, 169), (5, 540).

Finally, Fig. 18 is the contour image of a medium plate fitted according to the extracted laser line ends coordinates the IPC has extracted from individual pictures.

Without interfering subsequent coordinate extractions from the ends of laser lines, the image size will be reduced to as small as 0.09 % of the original picture. This small size will undoubtedly.

5 Conclusion

This paper summarizes the research status of the medium plate detection system which is problematic to some extent. Therefore, an innovative detection system is developed to tackle the problems. In the first place, the detection principle of the system is briefly introduced by emphasizing on the description of the hierarchical data transmission approach in the new system. Afterwards, the theory of the algorithm for medium plate compression is precisely studied, including the basic principles for wavelet decomposition and reconfiguration, wavelet symmetric extension, wavelet scalar quantization as well as quantizer index value coding. In addition, the fundamental theory and application process of Huffman coding arc analyzed. Eventually, a performance test on the algorithm takes place with feedbacks showing that the speed of the system can be improved providing all conditions are met.

Acknowledgments The authors would like to express their gratitude to Professor Hu and Zhidong Zhao for their invaluable help, assistance and guidance throughout this work. This work is supported by Natural Science Foundation of Zhejiang Province under Grant No.LZ14E050003.

References

1. Sun Q, Wang X (2007) Study of flaws in heavy and medium plate. Iron Steel 42(8):41–45. doi:10.3321/j.issn:0449-749x.2007.08.010
2. Haigen Y (2010) Application of plane view pattern control function for 3800 mm plate mill. Metall Equip 2:54–57. doi:10.3969/j.issn.1001-1269.2010.02.014
3. Cheng H (2010) Research on CCD width gauge. Automation Research and Design Institute of Metallurgical Industry
4. RM L (2012) Study on the profile detector of clod-rolled plate based on image processing
5. Taubman DS, Wei J, Michael WM (2004) JPEG2000 image compression fundamentals, standards and practice. Publish House of Electronics Industry, Beijing
6. Wang G, Yuan W (2001) Generic 9-7-tap wavelets filters and their performances studies on image compression. Acta Electronica Sinica (1):130–132. doi:10.3321/j.issn:0372-2112.2001.01.035
7. Wohlberg B, Brislawn CM (2008) Symmetric extension for lifted filter banks and obstructions to reversible implementation. Signal Process 88(1):131–145
8. ZD Z (2013) Research on fingerprint image compression based on wavelet transform
9. Zhu X (2001) Fundamentals of applied information theory. Tsinghua University Press, Beijing
10. Zhang C, Su Y, Zhang J (2005) Digital image compression coding. Tsinghua University Press, Beijing
11. Bradley JN, Brislawn CM (1992) Compression of fingerprint data using the wavelet vector quantization image compression algorithm,". Los Alamos Rep 92:1507
12. Onyshczak R, Youssef A (2004) Fingerprint image compression and the wavelet scalar quantization specification. Autom Fingerpr Recognit Syst, 385–413
13. Huang Z, Ma Z (2001) Wavelet image coding combining zero tree and run length coding. J Image Graph 11:74–80. doi:10.3969/j.issn.1006-8961.2001.11.014
14. Brislawn CM, Bradley JN, Onyshczak RJ et al (1996) The FBI compression standard for digitized fingerprint images. Appl Digit Image Process XIX. In: Proceedings of the SPIE, p 2847

15. Vitter JS (1989) Algorithm 673: dynamic Huffinan coding. ACM Trans Math Softw (TOMS) 15(2):158–167
16. Lakhani G (2004) Optimal Huffinan coding of DCT blocks. IEEE Trans Circuits Syst Video Technol 14(4):522–527
17. Huffinan DA (1952) A method for the construction of minimum-redundancy codes. Proc IRE 40(9):1098–1101
18. DISI. 10918-1. Digital compression and coding of continuous-tone still images (JPEG). CCITT Recommendation T, 1991, 81
19. Jieying Z (2011) Important visual feature-based image quality assessment and image compression, Huazhong University of Science and Technology

Research on Power Dividing Strategy for Hybrid Electric Vehicle

Luming Chen, Zili Liao, Chunguang Liu and Yu Xiang

Abstract In order to eliminate negative effects of transient power demands, configurations and power source characters of series hybrid electric vehicles were analyzed. The power dividing strategy was established based on wavelet transform, and then it was tested in RT-LAB model to compare with the result of constant temperature control strategy. As it was shown in the simulation, the power dividing strategy had an advantage in power following. Meanwhile, power supply quality could be promoted and battery life could be extended.

Keywords Wavelet transform · Power dividing · Instantaneous power

1 Introduction

With the development of science, technology and the advancement of new military revolution, weapons and equipments are becoming the key factor of combat power generation. Due to the limitation of the power system structure and the transmission way in traditional armored vehicles, it is hard to meet firepower, mobility and protective performance with the increasing load demands [1]. New type of energy storage technology is gradually mature, which paves the way for the development

L. Chen (✉) · Z. Liao · C. Liu · Y. Xiang
Department of Control Engineering, Academy of Armored Force Engineering, Beijing 100072, China
e-mail: 18211077415@163.com

Z. Liao
e-mail: 295170692@qq.com

C. Liu
e-mail: 18800130677@163.com

Y. Xiang
e-mail: 13121511200@163.com

© Springer International Publishing Switzerland 2017

V.E. Balas et al. (eds.), *Information Technology and Intelligent Transportation Systems*, Advances in Intelligent Systems and Computing 454,
DOI 10.1007/978-3-319-38789-5_28

of the hybrid power armored vehicles. Integrated power system with multiple power supplies becomes a new development direction of hybrid power armored vehicles [2].

Power supply quality of integrated power system depends not only on topology structures of each power source, but also control strategies. In view of the different integrated power system structures and tasks, there are several control strategies, such as typical biggest charged state of peak power control strategy and constant temperature control strategy, etc. The hybrid control strategy based on electric priority aiming at compound type was established in Literature [3], which improved the efficiency of vehicle braking energy recovery at the UDDS condition. In literature [4], a single point of constant temperature control strategy was applied at power maintain stage to increase the fuel economy. The process of starting, acceleration and braking will be more frequently existed during drive conditions. It may cause negative effects to the durability and reliability of power supplies. Wavelet transform can carry on the decomposition of the power signal in time domain, frequency domain and shunt high-frequency transient power for ultracapacitor, maximizing the advantages of high specific power. The method is suitable for non-steady state and transient signal analysis.

This paper analyzes the hybrid vehicle configuration and the power supply characteristics, selects CYC-HWFET as drive test cycle, uses wavelet transform to realize the diversion of vehicle power. Finally the simulation experiment was carried out on the established RT-LAB vehicle model. Comparing with the result of constant temperature control strategy, it is proved that the power dividing strategy based on wavelet transform can give a full play to different power supplies, extends the service life of power battery and engine/generator set, improves power supply quality of integrated power system.

2 The Basic Situation of the Hybrid Electric Vehicle

2.1 Hybrid Electric Vehicle Configurations

The integrated power system is connected in the form of series [5], and its concrete structure are shown in Fig. 1.

2.2 The Vehicle Power Calculation

According to vehicle dynamics theory, the formula to calculate vehicle power demand is shown as follow [6]:

$$P = \left(mg\sin\alpha + mgf\cos\alpha + \frac{C_d A}{21.15}v^2 + \delta m\frac{dv}{dt} \right)\frac{v}{3600} \tag{1}$$

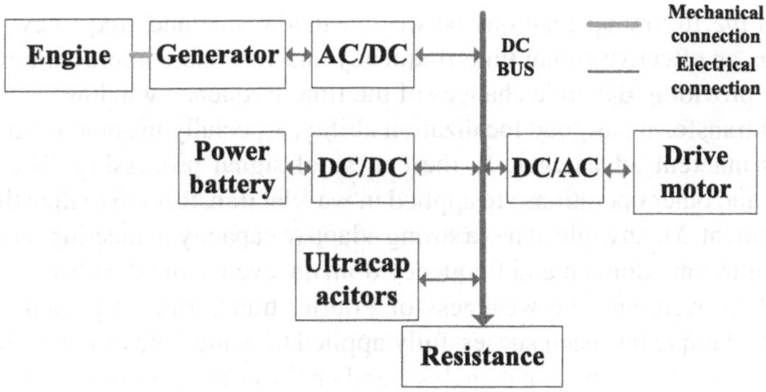

Fig. 1 Structure of integrated power system

In the formula, f is the rolling friction coefficient, α is the slope angle, C_d is the air drag coefficient, A is the windward area, v is speed, m is the quality of vehicle, δ is spinning quality coefficient.

2.3 *Parts Power Characteristic*

Kamaraj [7] Engines: When the dynamic load changes quickly, the change of engines working point will be very intense. While the engine is a big delay inertia component, severe dynamic adjustment process will lead to dynamic losses. It will reduce working efficiency and lead to the deterioration of fuel economy index.

Power batteries: When the amplitude and frequency of the power demand are very high, the stability of the battery and its life will be affected, especially transient peak current flowing through power battery will lead to irreversible damages.

Ultracapacitors: The energy density of ultracapacitor is 10 to 100 times more than other battery powers. In addition, charging and discharging process can be finished in several milliseconds to several seconds because of its excellent high-frequency performance.

3 Wavelet Transform Theory

3.1 *The Characteristics of Wavelet Transform*

Zhang [8] Wavelet analysis, which is originated in the 1980, is the expansion and extension of Fourier analysis. As a function of the wavelet analysis, wavelet transform is mainly used to decompose a given function or time signal into different scales. It absorbs and inherits the localization thought of short-time Fourier Transform, established the follow-up relations between window size and frequency. Wavelet transform is an effective signal time-frequency analysis and compression tool, particularly in providing dynamic changes of the time-frequency window.

Wavelet transform has good localization ability, especially in some local features area it has inherent advantages. In the course of signal processing, The scaling, translation and other operations are applied in wavelet transform to realize the multi-scale refinement. Meanwhile, it has a strong adaptive capacity in meeting the analysis requirement in time domain and frequency domain, even some details of signal can be focused to overcome the weakness of Fourier transform. At present, wavelet transform technique has been successfully applied to many fields such as the signal and graphics compression, signal analysis and engineering technology, etc.

3.2 *Wavelet Transform Process*

The process of wavelet transform generally includes two stages, namely, decomposition and reconstruction. First of all, through the decomposition formula one dimensional time-domain function will be mapped to the two dimensional time-frequency domain. The wavelets with different time-frequency width will be obtained by changing the translation and scaling factors, reaching the purpose of analyzing the signal in time-frequency localization. Finally, the two dimensional signal will be reconstructed back to the original one dimensional signal through the wavelet inverse transformation. Wavelet basis function used by wavelet transform is diverse, so how to choose the function becomes the key factor to outcome result. As one of the most popular wavelet basis function, haar wavelet has the shortest length of filtering in the time domain comparing with other wavelets. As a result, the paper choose the haar wavelet as the wavelet basis function [9], the expression is shown as follows:

$$\psi(t) = \begin{cases} 1 & t \in [0, 1/2] \\ -1 & t \in [1/2, 1] \\ 0 & others \end{cases} \tag{2}$$

For continuous wavelet transform, the decomposition expression is shown as follow:

$$L_\psi f(a,t) = \frac{1}{\sqrt{c_\psi}} \frac{1}{\sqrt{|a|}} \int_{-\infty}^{+\infty} f(u) \overline{\psi\left(\frac{u-t}{a}\right)} \mathrm{d}u (a \neq 0), (t \in R) \tag{3}$$

The expression of continuous wavelet reconstruction transform is shown as follow:

$$f(t) = \frac{2\pi}{\sqrt{c_\psi}} \int_{-\infty}^{+\infty} \int_{-\infty}^{+\infty} L_\psi f(a,t) \frac{1}{\sqrt{|a|}} \omega \left(\frac{t-u}{a}\right) \frac{duda}{a^2} \tag{4}$$

In practical problems of numerical calculation, the continuous wavelet transform is a function of continuous change, which is not convenient for continuous integral of the digital signal. Riesz base is adopted to meet the demand of $f(t)$ transform to achieve discrete results of continuous wavelet transform processing. At this point, decomposition expression of discrete wavelet transform is shown as follow:

$$\begin{aligned} L_\psi f(a,t) &= \frac{1}{\sqrt{c_\psi a}} \int_{-\infty}^{+\infty} f(u)\psi\left(\frac{u-t}{a}\right)\mathrm{d}u \\ (a &= 2^{-j}, t = k2^{-j}, j, k \in) \end{aligned} \tag{5}$$

The reconstruction expression of discrete wavelet transform is shown as follow:

$$f(t) = \frac{2\pi}{\sqrt{c_\psi}} \sum_{j=-\infty}^{+\infty} \sum_{k=-\infty}^{+\infty} 2^{j/2} L_\psi f(a,t)\psi(2^j t - k) \tag{6}$$

3.3 Power Dividing Strategy

The power dividing strategy based on wavelet transform has the advantage in extracting the instantaneous power component from the total power demand in vehicles, and the high frequency components will be allocated to the ultracapacitor. This will make full use of ultracapacitors high specific power characteristics. On the other hand, surplus low-frequency power component is offered by engine\generator set and power battery. Other than one-way characteristic of engine\generator set, power battery allows the bidirectional power flows. Therefore, power battery can not only absorb the low frequency feedback but also act as a capable assistant of the engine/generator set to meet the demand of low-frequency power [10].

In the process of vehicle driving, wavelet decomposition order should be appropriate to ensure simplicity of calculation and meet the demand of the frequency constraints. Hence the paper chooses the third order haar wavelet in decomposition and reconstruction power demand, distribution situation is shown in the Fig. 2.

The expression of power supply power allocation is shown as follow:

$$P_{EG} = \begin{cases} 0.6x_0(n) & x_0(n) > 0 \\ 0 & others \end{cases} \tag{7}$$

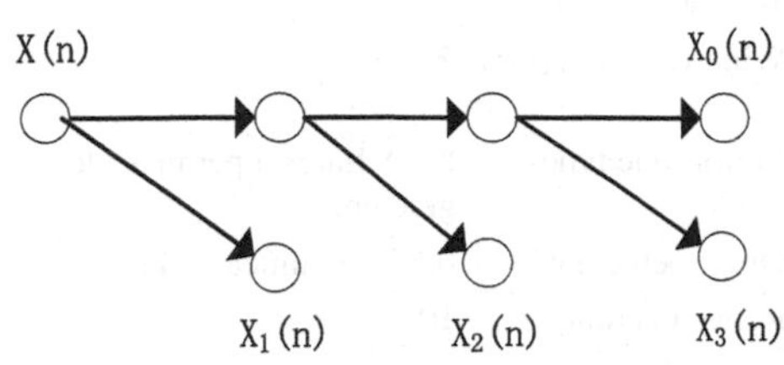

Fig. 2 Power dividing sketch map

$$P_{\text{Batt}} = \begin{cases} 0.4x_0(n) & x_0(n) > 0 \\ x_0(n) & others \end{cases} \quad (8)$$

$$P_{UC} = x_1(n) + x_2(n) + x_3(n). \quad (9)$$

4 Examples of Application

4.1 *Basic Parameters and Performance Indicators*

In order to calculate power demands and performance parameters of the various components, necessary parameters and performance requirements of the hybrid vehicle is listed in Table 1.

4.2 *Drive Test Cycle*

Vehicle driving test cycle is a speed curve which is composed of a large number of data points, reflecting the changing relation between the speed and time. The driving test cycle can realize the simulation of vehicle and it is usually adopted to forecast integrated power system performance at early stage. In this paper, the CYC-HWFET driving test cycle is selected to testing the performance of the hybrid electric vehicle [11, 12]. The speed curve is shown in Fig. 3.

Parts of characteristic values in this driving test condition are shown in Table 2.

4.3 *The Simulation Model*

RT-LAB is an engineering design application platform based on models, which can directly apply a mathematical model of dynamic system, which is established in

Table 1 Basic parameters and design requirements of hybrid electric vehicle

Basic parameters		Design requirements	
Total weight (t)	18	Largest highway speed (km/h)	110
Windward area (m^2)	2.2*3	Largest cross-country speed (km/h)	40
Motorcycle type	8 × 8 Largest permissible gradient	30	
Drag coefficient	0.5 Mute mileage (km)	10	
Reduction ratio	10	0–32 km/h accele ration time (s)	9

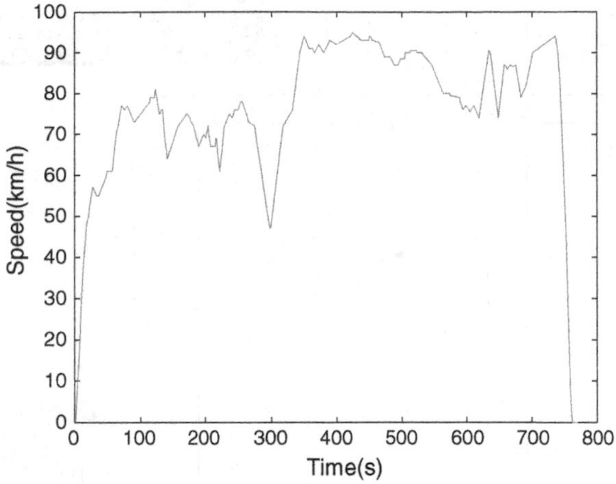

Fig. 3 CYY-HWFET drive test cycle

Table 2 CYC-HWFET characteristics

Types	Values
Time (s)	760
Distance (km)	16.3
Maximum speed (km/h)	95
Average speed (km/h)	77.2
Maximum acceleration (m/s^2)	1.31
Maximum deceleration (m/s^2)	–1.12
Average acceleration (m/s^2)	0.57
Average deceleration (m/s^2)	–0.42

MATLAB/Simulink, to real-time hardware in the loop (HIL) simulation test or other related fields. The hybrid vehicle model built in RT-LAB platform is shown in Fig. 4.

4.4 The Power Dividing Simulation Curve

To compare results of the power dividing strategy based on wavelet transform and constant temperature control strategy, the simulation test was respectively carried out on the vehicle model established in RT-LAB. The power supplys dividing situation is shown in Figs. 5 and 6.

As it is shown in Fig. 5, the peak power of power battery is not more than 50 kW and engine\generator set peak power is 63 kW. While in Fig. 6, peak power demands of power battery and motor\generator set are respectively 76 and 72 kW. Therefore, final conclusion can be drawn that power dividing strategy based on wavelet

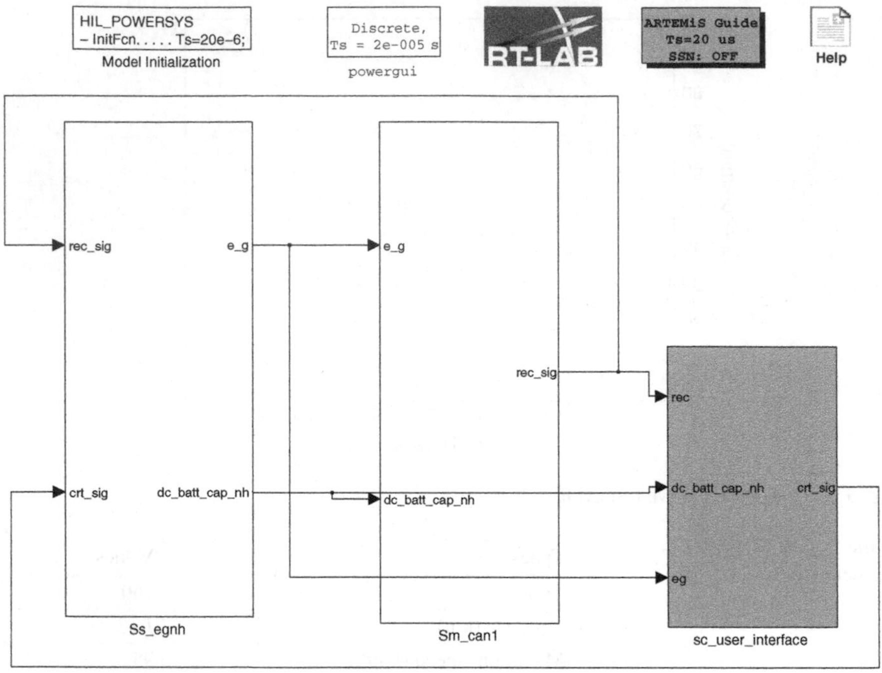

Fig. 4 Simulation model of hybrid electric vehicle

Fig. 5 The power dividing strategy based on wavelet transform

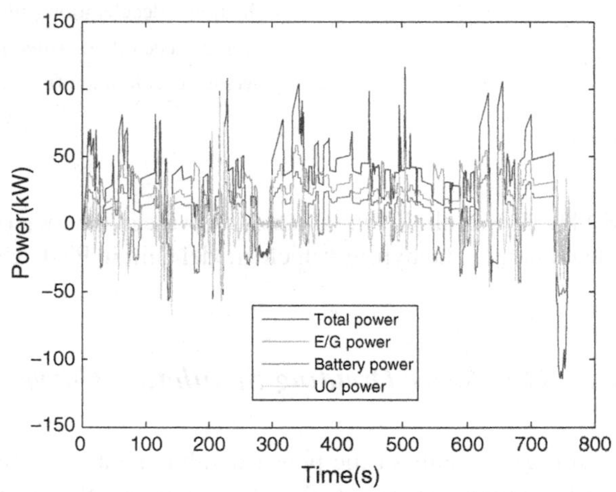

transform has a better performance than constant temperature control strategy. After the wavelet transform process, intensity and frequency of ultra capacitor power increase significantly. It is beneficial to develop its inherent advantage in high specific power. Power flowing to engine/generator and power battery has a lower frequency, a reduced impact on the power supply, and ensures the higher quality of power sup-

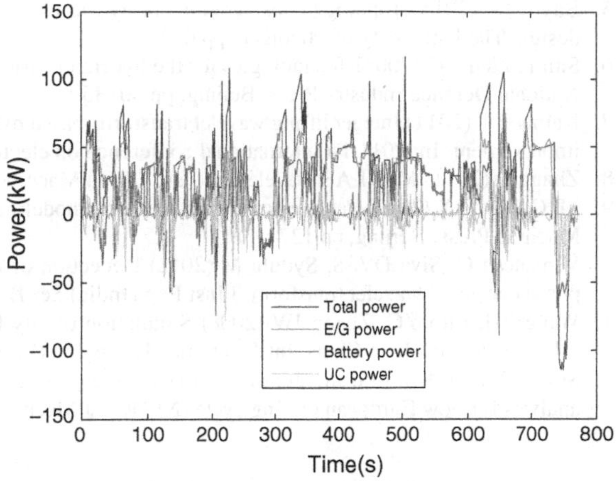

Fig. 6 Constant temperature control strategy

ply coming from integrated power system. The results show that the proposed power dividing strategy based on wavelet transform is superior to the traditional strategy, and the former has the feasibility and adaptive ability.

5 Conclusion

The paper established a power dividing strategy based on the wavelet transform, decomposition and reconstruction of power demand will be fulfilled by time domain and frequency domain transformation. Given the negative impact on power supply from high frequency transient components, the integrated power system assigns high power to ultra capacitor. Meanwhile, the remainder is jointly offered by the power battery and engine\generator set. These measures tend to reduce the power fluctuations on the impact of power sources, extend power battery service life and improve the power supply quality of integrated power system.

References

1. Zili L, Xiaojun M, kemao Z (2008) Research on status quo and key technologies of all-electric combat vehicle. Fire Control Command Control. 33(5):1–4
2. Mao M, Han Z, Liu Y (2014) Research on carload energy management and power management. In: The third BBS on all-electric technology development, pp 162–168
3. Zhang G (2010) Study on parameter matching based on vehicle duty cycle, control strategy and simulation software of hybrid electric vehicle. South China university of technology, 69–86
4. Zhu Wuxi, Sun Liqing (2013) Control strategy research for extended-range electric bus. Automob Technol 4:1–5

5. Bayrak A (2015) Topology considerations in hybrid electric vehicle powertrain architecture design. The University of Michigan, pp 6–9
6. Sun F, Zhang C (2008) Technologies for the hybrid electric drive system of armored vehicle. National Defence Industry Press, Beijing, pp 30–35
7. Kamaraj C (2011) Integer lifting wavelet transform-based hybrid active filter for power quality improvement. In: 2011 1st international conference on electrical energy systems, pp 103–107
8. Zhang D (2011) MATLAB wavelet analysis. China Machine Press, Beijing, pp 53–64
9. Mi C, Zhang X (2013) Vehicle power management: modeling, control and optimization. China Machine Press, Beijing, pp 127–138
10. Venkatesh C, Siva DVSS, Sydulu M (2012) Decection of power quality disturbances using phase corrected wavelet transform. J Inst Eng (India) Ser B, 37–42
11. Wang BH, Luo YG, Zhang JW (2008) Simulation of city bus performance based on actual urban driving cycle in China. Int J Automot Technol 9(4):501–507
12. Shim BJ, Park KS, Koo JM, Jin SH (2014) Work and speed based engine operation condition analysis for new European driving cycle (NEDC). J Mech Sci Technol 28(2):755–761

Driver Compliance Model Under Dynamic Travel Information with ATIS

Ande Chang, Jing Wang and Yi Jin

Abstract This study explores many details of the drivers response to dynamic travel information with variable message signs (VMS) which is the one of the most common advanced traveler information systems (ATIS) deployed in many areas all over the world. A stated preference (SP) survey was conducted in some parts of China to collect various drivers behavior information with VMS. Based on the findings from the surveys, seventeen potential affecting factors for driver compliance with VMS are identified and applied to further study. A binary logistic regression model is adopted to evaluate the significance of these seventeen factors. Gender, age, whether full-time worker, delay ratio of the current route, knowledge of an alternate route, length ratio of an alternate route, and crowded level on an alternate route are proved to be significant variables affecting driver compliance under dynamic travel information with VMS. Classification and regression trees (CART) is adopted to develop the driver compliance model. The CART model reveals the hierarchical structure of driver compliance and produces many interesting findings. The developed model is evaluated with collected data from SP survey and shows a reasonable performance. The CART model explains the behavior data most clearly and maintains the highest prediction rate.

Keywords Intelligent transportation systems · Variable message signs · Driver compliance · Classification and regression trees

A. Chang (✉) · Y. Jin
National Police University of China, Shenyang 110035, China
e-mail: changande@npuc.edu.cn

Y. Jin
e-mail: 18800958@qq.com

J. Wang
College of Engineering, Shenyang Agricultural University, Shenyang 110866, China
e-mail: gzlwangjing0707@163.com

© Springer International Publishing Switzerland 2017

V.E. Balas et al. (eds.), *Information Technology and Intelligent Transportation Systems*, Advances in Intelligent Systems and Computing 454,
DOI 10.1007/978-3-319-38789-5_29

1 Introduction

During the last years, traffic congestion has become one of the most common phrases in the daily news. Traffic congestion is a phenomenon caused when a facility is asked to bear a travel demand that is greater than its capacity. This has been caused, in part, by our rapid economic growth. With our rapid economic growth, we have come to expect more mobility in our modem life.

To mitigate this traffic congestion and improve the efficiency of travel in urban areas, various types of advanced traveler information systems (ATIS) have been studied and implemented during the past several decades. ATIS is a major component of intelligent transportation systems (ITS) that include a wide range of new tools for managing traffic as well as for providing services for travelers. Given the continued increase in travel and the difficulties of building new and expanded roads in urban areas, ATIS is essential to maximize the utilization of existing facilities.

One common ATIS that has been deployed is variable message signs (VMS). They are installed along roadsides and display messages of special events that can affect traffic flow such as football games. VMS usually deliver various types of information such as expected delay, cause of incident, lane closure, etc., which can help drivers make smart decisions regarding their trip. It is expected that by providing real-time information on special events happening on the oncoming road, VMS can improve the drivers route choice, reduce the drivers stress, mitigate the severity and duration of incidents and improve the performance of the transportation network.

Unlike other ITS components such as ramp closure, which is part of the advanced traffic management system (ATMS), VMS are not supported by mandatory regulation. That is, the impact of implemented VMS on the transportation network depends to a large degree on the response of drivers. Therefore, it is very important to find and understand those factors that can affect driver compliance with ATIS. Many past studies were widely applied to the collection of data and the results were analyzed using statistical methods. However, those individual findings are rarely aggregated to explain driver compliance in an effective and efficient manner and there has been no extensive evaluation of those findings [1].

Many researchers, in their attempts to analyze the driver behavior data, have adopted a logistic regression model [2–6] that is based on Logit function and utility models that usually assume the effects of attributes of an alternative are compensatory, which means that increase or decrease of one attribute in a model can be compensated by certain proportional increases or decreases of other variables. In many cases, it is difficult for driver behavior characteristics to meet the assumptions and it is difficult for the researcher to reflect the nonlinear effects of variables or interactions between variables using the logistic regression model.

To overcome this limitation, several new attempts have been made, such as using neural networks [7–9]. These approaches are useful in terms of overcoming many limitations in the traditional logistic regression model by automatically detecting no-compensatory, nonlinear effects of variables and interactions between variables. However, it is difficult to interpret the results of these models to understand the behavior characteristics of the drivers who have been analyzed. In other words, these models would provide good predictions, overcoming many limitations of the traditional approach, but it is very difficult to use the results or findings for any policy implications [1].

This study explores many details of the drivers response to VMS. A SP survey was conducted in China to collect various drivers behavior information with VMS. Based on the findings from the surveys, as much as possible potential affecting factors for driver compliance with ATIS are identified and applied to further study. A binary logistic regression model is adopted to evaluate the significance of these factors. CART model is adopted to develop the driver compliance model. This paper is outlined as follows: The next section depicts the data survey works for driver behavior; Sect. 3 analyses the impacts of VMS on driver behavior; Sect. 4 researches the CART model for driver compliance; Sect. 5 depicts the model validations; Last section presents conclusion.

2 Data Survey for Driver Behavior

The scope of research is usually limited by the availability of data. This study aims to understand the structure of driver compliance with VMS on the urban network. However, driver response associated with VMS is difficult to observe and collect in the field. Thus this study has adopted stated preference (SP) methods involving extensive user surveys. The SP method is a very effective tool to investigate the drivers behavior characteristics under a variety of controlled scenarios. To understand the general drivers response to VMS and to find significant variables that affect the drivers compliance, an online survey was conducted.

Based on the recent enhancement of the internet, online surveys are now generally established as one of the tools for behavior studies. Use of the internet for surveys can improve the reliability of collected data by allowing the researcher to provide more detailed and realistic descriptions to respondents, which allows them to provide answers closer to their behavior under real circumstances. Moreover, it can largely eliminate coding errors and take adaptive designs where the options offered can be modified as a result of the responses to previous answers. However, users of this method should be careful to control the sample population.

The survey was sent to a random sampling of drivers in China based on their travel experiences with a sample size of 1000. A 22-question driver survey questionnaire (see Appendix) was developed to collect the following information: (1) Demographic characteristics of drivers; (2) Perceptions of dynamic travel information; (3) Driver behaviors under dynamic travel information with VMS. Seventeen potential affecting factors are identified as following: city kind (x_1), region (x_2), gender (x_3), age (x_4), marital status (x_5), degree (x_6), job (x_7), whether full-time worker (x_8), monthly income (x_9), crowded level on the current route (x_{10}), vehicle queue length of the current route (x_{11}), delay ratio of the current route (x_{12}), knowledge of an alternate route (x_{13}), length ratio of an alternate route (x_{14}), crowded level on an alternate route (x_{15}), anticipated travel time saving ratio (x_{16}) and quality of dynamic travel information (x_{17}).

540 completed survey forms were returned and used for analysis. Among 490 valid surveys (50 surveys were excluded due to incredulity), more than 50 % of

Table 1 Demographic characteristic of survey respondents

Numbers	Characteristics	Options	Percentages (%)
1	Region	Northeast	55.7
		North China	20.6
		East China	11.4
		Central South	4.5
		Northwest	3.3
		Southwest	4.5
2	County	Village 6.7	5.1
	City Kind	Common City	46.9
	Provincial Municipality	Capital 15.0	26.3
3	Gender	Female	36.9
		Male	63.1
4	Age	Less than 20	4.1
		20–29	51.6
		30–39	27.3
		40–49	11.2
		50–59	4.5
		Greater than 60	1.3
5	Marital Status	No	44.9
		Yes	55.1
6	Degree	Elementary School	12.0
		Middle School	12.7
		University	52.6
		Master or Doctor	22.7
7	Job	Student	16.9
		Company Staffer	31.4
		Civil Servant	6.8
		Teacher or Doctor	11.2
		Freelancer	10.4
		Other	23.3
8	Whether Full-time Worker	No	23.1
		Yes	76.9
9	Monthly Income (Yuan)	Less than 2000	18.4
		2000–4000	40.6
		4000–8000	26.9
		Greater than 8000	14.1

the respondents are somewhat familiar with VMS. The respondents demographic characteristics which present the applicability of research conclusions in this paper are shown in Table 1.

3 Impact Analysis on Driver Behavior

Binary responses occur in many fields of study. Logistic regression analysis has been used to investigate the relationship between these discrete responses and a set of explanatory variables. The response, Y, of an individual or an experimental unit can take on one of two possible values such as Comply ($Y = 1$) or Not Comply ($Y = 0$). Suppose x is a vector of explanatory variables and is the response probability to be modeled. The linear logistic model has the form:

$$\text{logit}\,(p) = \ln\left(\frac{p}{1-p}\right) = a + b_1x_1 + b_2x_2 + \cdots + b_mx_m \tag{1}$$

where a is the intercept parameter and $b_i\,(i = 1, 2n)$ are the slope parameters.

It obtains maximum likelihood estimates of the parameters using an interactive weighted least squares algorithm whereas least squares regression minimizes the sum of squared errors to obtain parameter estimates. The logistic regression has four variable selection methods: forward selection, backward elimination, stepwise selection, and best subset selection. The best subset selection is based on the likelihood score statistic. This method identifies a specified number of best models containing one, two, three variables and so on, up to a single model containing all the explanatory variables.

Total 490 responses were collected and used to develop a binary logistic regression model. The result of binary logistic regression for compliance behavior is summarized in Table 2. This shows that x_3 $(wald = 5.377, sig = 0.02)$, x_4 $(wald = 5.505, sig = 0.019)$, x_8 $(wald = 3.272, sig = 0.07)$, x_{12} $(wald = 4.105, sig = 0.043)$, x_{13} $(wald = 7.652, sig = 0.006)$, x_{14} $(wald = 8.347, sig = 0.004)$ and x_{15} $(wald = 17.666, sig = 0)$ have wald-values are significantly greater than the others indicating that the parameters have the greater impacts on driver behavior than others, and sig-values less than or close to 0.05 indicating that the parameters are not zero using a significance level of 0.05.

It is further proved from the result of the binary logistic regression shown in Table 3 that the driver behaviors under dynamic travel information with VMS depend greatly on their Gender, age, whether full-time worker and the traffic factors, including delay ratio of the current route, knowledge of an alternate route, length ratio of an alternate route and crowded level on an alternate route.

Table 2 Binary logistic regression with complete explanatory variables

Variables	B	S.E.	Wald	DF	Sig	Exp(B)
x_1	−0.092	0.131	0.492	1	0.483	0.912
x_2	−0.069	0.091	0.573	1	0.449	0.933
x_3	0.570	0.246	5.377	1	0.020	1.768
x_4	−0.347	0.148	5.505	1	0.019	0.707
x_5	0.018	0.311	0.003	1	0.953	1.019
x_6	−0.109	0.164	0.439	1	0.507	0.897
x_7	0.010	0.078	0.016	1	0.900	1.010
x_8	0.547	0.303	3.272	1	0.070	1.729
x_9	0.022	0.146	0.023	1	0.879	1.022
x_{10}	−0.046	0.146	0.099	1	0.753	0.955
x_{11}	0.030	0.088	0.116	1	0.733	1.031
x_{12}	0.226	0.112	4.105	1	0.043	1.254
x_{13}	0.416	0.150	7.652	1	0.006	1.516
x_{14}	−0.305	0.106	8.347	1	0.004	0.737
x_{15}	−0.480	0.114	17.666	1	0.000	0.619
x_{16}	0.024	0.095	0.063	1	0.802	1.024
x_{17}	0.019	0.119	0.026	1	0.872	1.019

Table 3 Binary logistic regression with primary explanatory variables

Variables	Transition Variables	B	S.E.	Wald	DF	Sig	Exp(B)
x_3	X_1	0.581	0.241	5.813	1	0.016	1.788
x_4	X_2	−0.317	0.116	7.484	1	0.006	0.729
x_8	X_3	0.516	0.269	3.691	1	0.055	1.675
x_{12}	X_4	0.228	0.084	7.318	1	0.007	1.256
x_{13}	X_5	0.403	0.148	7.426	1	0.006	1.496
x_{14}	X_6	−0.290	0.103	7.972	1	0.005	0.748
x_{15}	X_7	−0.467	0.101	21.343	1	0.000	0.627

4 CART Modeling for Driver Compliance

Using classification and regression trees (CART) is a very effective method when there is a need to classify or makes a prediction from a complicated data set. Especially when traditional statistical methods, which are required to meet many stringent conditions, fail to analyze the given data effectively, trees can be utilized in many circumstances. Trees label records and assign them to discrete classes. They are built through a process known as binary recursive partitioning. The process of computing the CART can be characterized as four basic steps [10]:

(1) Specifying the criteria for predictive accuracy. The CART algorithms are generally aimed at achieving the best possible predictive accuracy. Operationally, the most accurate prediction is defined as the prediction with the minimum costs. The notion of costs is developed as a way to generalize, to a broader range of prediction situations, the idea that the best prediction has the lowest misclassification rate. In most applications, the cost is measured in terms of proportion of misclassified cases. In this context, a prediction would be considered best if it has the lowest misclassification rate. The need for minimizing costs, rather than just the proportion of misclassified cases, arises when some predictions that fail are more catastrophic than others.

(2) Selecting splits. The second basic step in CART is to select the splits on the predictor variables that are used to predict membership in classes of the categorical dependent variables, or to predict values of the continuous dependent variable. In general terms, the split at each node will be found that will generate the greatest improvement in predictive accuracy. This is usually measured with some type of node impurity measure, which provides an indication of the relative homogeneity of cases in the terminal nodes. If all cases in each terminal node show identical values, then node impurity is minimal, homogeneity is maximal, and prediction is perfect.

(3) Determining when to stop splitting. As discussed in Basic Ideas, in principal, splitting could continue until all cases are perfectly classified. However, this wouldnt make much sense since we would likely end up with a tree structure that is as complex and tedious as the original data file, and that would most likely not be very useful or accurate for predicting new observations. What is required is some reasonable stopping rule.

(4) Selecting the right-sized tree. The size of a tree in CART analysis is an important issue, since an unreasonably big tree can only make the interpretation of results more difficult. Some generalizations can be offered about what constitutes the right-sized tree. It should be sufficiently complex to account for the known facts, but at the same time it should be as simple as possible. It should exploit information that increases predictive accuracy and ignore information that does not. It should, if possible, lead to greater understanding of the phenomena it describes. The strategy is to use a set of well-documented, structured procedures developed by Breiman (1984) for selecting the right-sized tree.

CART is an effective and efficient tool that can generate various classification trees. Seven numerical variables are used as predictors. Gender and whether full-time worker are two categorical variables that have binary choice. The other categorical variable is code, which indicates which data belongs to which scenario. The details of the seven variables are summarized in Table 4.

Among total 490 valid surveys, 390 random samples are used to construct a final tree with fourteen intermediate nodes and sixteen terminal nodes. As can be seen in Fig. 1, for example, if a sample has values of X_6 greater than 1, X_5 greater than 1, X_2 less than or equal to 4, X_6 less than or equal to 2 and X_2 greater than 3. It is shows that those who do have an age in 50 59 and know an alternate route well will comply with the given information, and then take an alternate route under the given

Table 4 Summary of seven variables

Variables	Parameters	Value ranges	Digitization
X_1	Gender	{female, male}	{0 1}
X_2	Age	<20, 20 29, 30 39, 40 49, 50 59, >60	{0 1 2 3 4 5}
X_3	Whether full-time worker	{no, yes}	{0 1}
X_4	Delay ratio of the current route	{<0.5, 0.5, 1, 2, >2, not sure}	{0 1 2 3 4 5}
X_5	Knowledge of an alternate route	{lowest, lower, upper, highest}	{0 1 2 3}
X_6	Length ratio of an alternate route	{<0.5, 0.5, 1, 2, >2, not sure}	{0 1 2 3 4 5}
X_7	Crowded level on an alternate route	{lowest, lower, upper, highest, not sure}	{0 1 2 3 4}

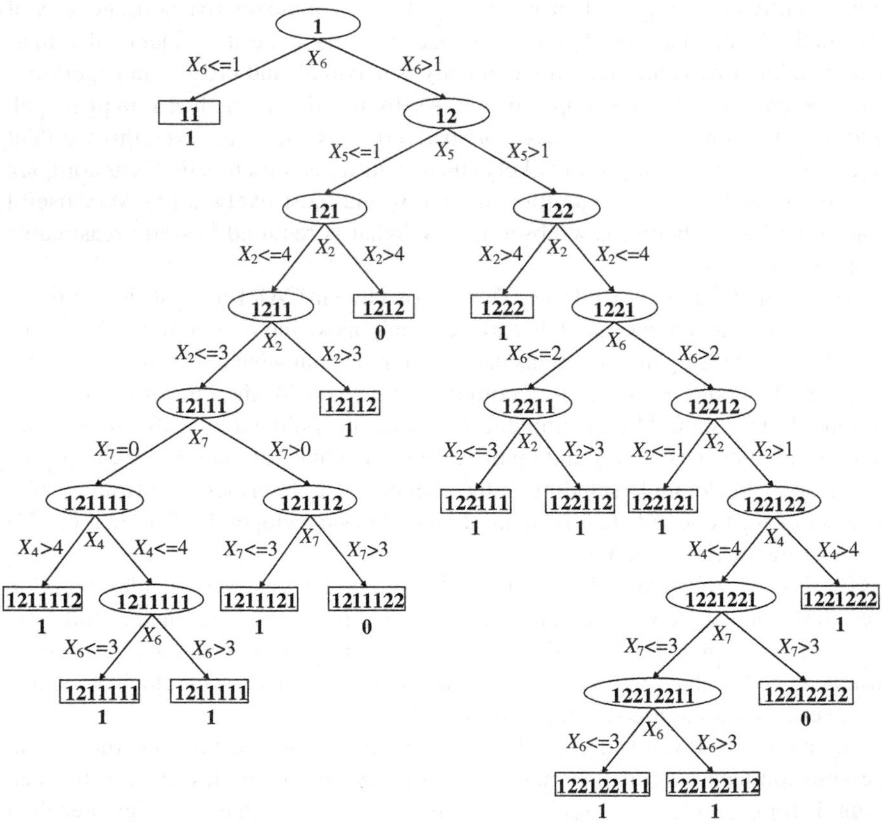

Fig. 1 CART for driver compliance

situation that the length ratio of an alternate route is 1 when a traffic incident has been occurred.

5 Model Validation

The prediction rates of binary logistic regression model and CART are compared to validate the developed model. To compare the result from the models with the other 100 random samples which were collected by the online survey, it is assumed that if p greater than 0.5, a driver would comply with the given dynamic travel information with VMS. Table 5 shows the aggregated error rates for the binary logistic regression model.

As can be seen in Table 6, CART model shows a better performance than the binary logistic regression model. Compared with the binary logistic regression model, the CART model shows a 5 % lower error rate with total validation samples. In particular, the CART model shows a 17 % lower error rate for Comply cases than the binary logistic regression model. According to the developed driver compliance model, driver compliance depends on different variables depending on the characteristics of drivers and trips.

Table 5 Classification matrix for binary logistic regression

		Predicted		
		Y = 0	Y = 1	Percentage Errors (%)
Observed	Y = 0	9	14	25.0
	Y = 1	11	66	

Table 6 Classification matrix for CART

		Predicted		
		Y = 0	Y = 1	Percentage Errors (%)
Observed	Y = 0	8	15	20.0
	Y = 1	5	72	

6 Conclusion

Based on the many findings from the SP surveys, seventeen factors are identified that potentially affect driver compliance with ATIS. With extensive SP approach, the study revealed many details of driver behavior with VMS. Based on online survey, those factors are evaluated and analyzed. To analyze the impact on driver behavior with VMS, a binary logistic regression is adopted. The binary logistic regression approach identifies seven significant variables, which are Gender, age, whether full-time worker, delay ratio of the current route, knowledge of an alternate route, length ratio of an alternate route and crowded level on an alternate route. A CART model is used to analyze driver compliance. It showed great flexibility in analyzing the data and identifying the structure of the given data set. It clearly presents the structure of driver compliance behavior while maintaining a reasonable prediction rate of around 80 %. The study has analyzed driver compliance behavior with the tree approaches which provides more understanding of driver compliance behavior mechanism while maintaining reasonable estimation accuracy.

The study has found the conditional structure of driver compliance with VMS which means non-homogeneous driver would use different factors to make a route choice decision. This finding can be applied to develop implemental operation strategies for ATIS applications including VMS.

The significant variables that affect driver compliance under the provision of travel information are based on the SP study. However, the study collected 490 samples to analyze and construct the models but considering the complicated structures of driver compliance, extensive research with a larger sample size may disclose many answers more clearly. Especially follow-up survey with non-respondent group will be essential to ensure many findings from this research. To enhance the driver behavior study with VMS, ongoing efforts should be made to find more effective uses of available technologies in SP surveys, including the use of interactive and dynamic surveys based on the computer and the internet. This will help in the development of more reliable and cost-effective methods for behavior data collection.

Acknowledgments This paper is supported by Program of Doctoral Scientific Research Foundation of National Police University of China (Traffic Congestion Alarming based on GPS Equipped Vehicles), Training Program of the Major Research Plan of National Po-lice University of China (Study on Road Traffic State Extraction based on Locating Point Group of GPS Equipped Vehicles), General Project of Liaoning Provincial Education Department of China (L2015554), Technology Research Program of Ministry of Public Security of China (Study on Road Traffic State Extraction based on Locating Point Group of GPS Equipped Vehicles) and Project of Natural Science Foundation of Liaoning Province (Dual-mode Traffic Guidance Models and Infor-mation Releasing policies).

Appendix

1. Driver Survey Questionnaire Demographic characteristics of drivers

Where are you living?

(a) Northeast (b) North China (c) East China (d) Central South (e) Northwest (f) Southwest

What kind of city you are living?

(a) Village (b) County (c) Common City (d) Provincial Capital (e) Municipality

What is your gender?

(a) Woman (b) Man

What is your age?

(a) Less than 20 (b) 20 29 (c) 30 39 (d) 40 49 (e) 50 59 (f) Greater than 60

Have you ever get married?

(a) No (b) Yes

What is your degree?

(a) Elementary School (b) Middle School (c) University (d) Master or Doctor

What is your job?

(a) Student (b) Company Staffer (c) Civil Servant (d) Teacher or Doctor (e) Freelancer (f) other

Are you engaged in full-time work?

(a) No (b) Yes

How much is your monthly income (Yuan)?

(a) Less than 2000 (b) 2000 4000 (c) 4000 8000 (d) Greater than 8000

2. Perceptions of dynamic travel information

How serious do you think about traffic congestion around your hometown?

(a) Nothing serious (b) Generally serious (c) Very serious (d) Not sure

Do you feel that the availability of travel information in regards to traffic congestion is important?

(a) Not important (b) Generally important (c) Very important (d) Not sure

3. Driver behaviors under dynamic travel information

Please recall an experience of your travel, and then we will ask you some questions about how you make your decision under traffic congestions and travel information with VMS. If you cant get any experience, please give us your estimation based on the following questions. How do you think the length of current route?

(a) Less than 0.5 km (b) 0.5 1 km (c) 1 2 km (d) 2 4 km (e) 4 8 km (f) Greater than 8 km (g) Not sure

How do you think the travel time of current route?

(a) Less than 5 min (b) 5 10 min (c) 10 20 min (d) 20 40 min (e) Greater than 40 min (f) Not sure

How do you think the crowded level on the current route?

(a) Lowest (b) Lower (c) Upper (d) Highest (e) Not sure

How do you think the vehicle queue length of the current route?

(a) Less than 50 m (b) 50 100 m (c) 100 200 m (d) 200 400 m (e) 400 800 m (f) Greater than 800 m (g) Not sure

How do you think the delay of the current route?

(a) Less than 2 min (b) 2 5 min (c) 5 10 min (d) 10 20 min (e) 20 40 min (f) Greater than 40 min (g) Not sure

Are you familiar with the alternate route?

(a) Lowest (b) Lower (c) Upper (d) Highest (e) Not sure

How do you think the length of alternate route?

(a) Less than 0.5 km (b) 0.5 1 km (c) 1 2 km (d) 2 4 km (e) 4 8 km (f) Greater than 8 km (g) Not sure

How do you think the crowded level on the alternate route?

(a) Lowest (b) Lower (c) Upper (d) Highest (e) Not sure

How do you think the anticipated travel time saving on the alternate route?

(a) Less than 2 min (b) 2 5 min (c) 5 10 min (d) 10 20 min (e) 20 40 min (f) Greater than 40 min (g) Not sure

Based on all given questions above, would you take an alternate route?

(a) Yes (b) No

References

1. Lee C (2004) Developing driver compliance based operations model for ATIS applications. University of Wisconsin, Madison
2. Abdel-Aty MA, Kitamura R, Jovanis PP (1997) Using stated preference data for studying the effect of advanced traffic information on drivers route choice. Transp Res Part C 5(1):39–50
3. Peeta S, Ramos JL, Pasupathy R (2000) Content of variable message signs and on-line driver behavior. Transportation Research Board, Washington D.C
4. Chatterjee K, Hounsell NB, Firmin PE (2002) Driver response to variable message sign information in London. Transp Res Part C 10(2):149–169
5. Tsirimpa A, Polydoropoulou A, Antoniou C (2007) Development of a mixed multi-nomial logit model to capture the impact of information systems on travelers switching behavior. Intell Transp Syst 11(2):79–89
6. Bekhor S, Albert G (2014) Accounting for sensation seeking in route choice behavior with travel time information. Transp Res Part F 22(1):39–49
7. Dia H, Panwai S (2007) Modelling drivers compliance and route choice behaviour in response to travel information. Nonlinear Dyn 49(4):493–509
8. Dia H, Panwai S (2010) Evaluation of discrete choice and neural network approaches for modelling driver compliance with traffic information. Transportmetrica 6(4):249–270
9. Peng JS, Guo YS, Fu R et al (2015) Multi-parameter prediction of drivers lane-changing behaviour with neural network model. Appl Ergon 50(1):207–217
10. Breiman L, Friedman J, Olshen R et al (1984) Classification and regression trees. Chapman & Hall/CRC, London

Binary Probit Model on Drivers Route Choice Behaviors Based on Multiple Factors Analysis

Jing Wang, Ande Chang and Lianxing Gao

Abstract This study explores many details of the drivers response to dynamic travel information with variable message signs (VMS) which is the one of the most common advanced traveler information systems (ATIS) deployed in many areas all over the world. A stated preference (SP) survey was conducted to collect various drivers route choice behavior with VMS. Based on the surveys, seventeen potential affecting factors such as city kind, region, gender, age, marital status, degree, job, whether full-time worker, monthly income, crowded level on the current route, vehicle queue length of the current route, delay ratio of the current route, knowledge of an alternate route, length ratio of an alternate route, crowded level on an alternate route, anticipated travel time saving ratio and quality of dynamic travel information were identified and applied to further study. A binary probit model was adopted to evaluate the significance of these seventeen factors. Gender, age, whether full-time worker, delay ratio of the current route, knowledge of an alternate route, length ratio of an alternate route, and crowded level on an alternate route were proved to be significant variables. Then a model for estimating drivers route choice results was build based on the significant variables. The verification results showed that the model estimating precision could reached 76 %.

This paper is supported by Training Program of the Major Research Plan of National Police University of China (Study on Road Traffic State Extraction based on Locating Point Group of GPS Equipped Vehicles), General Project of Liaoning Provincial Education Department of China (L2015554), Technology Research Program of Ministry of Public Security of China (Study on Road Traffic State Extraction based on Locating Point Group of GPS Equipped Vehicles) and Project of Natural Science Foundation of Liaoning Province (Dual-mode Traffic Guidance Models and Information Releasing policies).

J. Wang · L. Gao
College of Engineering, Shenyang Agricultural University, Shenyang 110866, China
e-mail: gzlwangjing0707@163.com

L. Gao
e-mail: lianxing_gao@126.com

A. Chang (✉)
National Police University of China, Shenyang 110035, China
e-mail: changande@npuc.edu.cn; changande1234@163.com

© Springer International Publishing Switzerland 2017

V.E. Balas et al. (eds.), *Information Technology and Intelligent Transportation Systems*, Advances in Intelligent Systems and Computing 454,
DOI 10.1007/978-3-319-38789-5_30

Keywords Intelligent transportation · Route choice · Probit model · Factors analysis

1 Introduction

During the last years, traffic congestion has become one of the most common phrases in the daily news. Traffic congestion is a phenomenon caused when a facility is asked to bear a travel demand that is greater than its capacity. This has been caused, in part, by our rapid economic growth. With our rapid economic growth, we have come to expect more mobility in our modem life.

To mitigate this traffic congestion and improve the efficiency of travel in urban areas, various types of advanced traveler information systems (ATIS) have been studied and implemented during the past several decades. ATIS is a major component of intelligent transportation systems (ITS) that include a wide range of new tools for managing traffic as well as for providing services for travelers. Given the continued increase in travel and the difficulties of building new and expanded roads in urban areas, ATIS is essential to maximize the utilization of existing facilities.

One common ATIS that has been deployed is variable message signs (VMS). They are installed along roadsides and display messages of special events that can affect traffic flow such as football games. VMS usually deliver various types of information such as expected delay, cause of incident, lane closure, etc., which can help drivers make smart decisions regarding their trip. It is expected that by providing real-time information on special events happening on the oncoming road, VMS can improve the drivers route choice, reduce the drivers stress, mitigate the severity and duration of incidents and improve the performance of the transportation network.

Unlike other ITS components such as ramp closure, which is part of the advanced traffic management system (ATMS), VMS are not supported by mandatory regulation. That is, the impact of implemented VMS on the transportation network depends to a large degree on the response of drivers. Therefore, it is very important to find and understand those factors that can affect drivers route choice behaviors with ATIS. Many past studies were widely applied to the collection of data and the results were analyzed using statistical methods. However, those individual findings are rarely aggregated to explain drivers route choice behaviors in an effective and efficient manner and there has been no extensive evaluation of those findings [5].

Many researchers, in their attempts to analyze the driver behavior data, have adopted logistic regression models [1–4, 6–8] that is based on regression function and utility models that usually assume the effects of attributes of an alternative are compensatory, which means that increase or decrease of one attribute in a model can be compensated by certain proportional increases or decreases of other variables. However, potential affecting factors were considered too small to reflect the drivers route choice behaviors well.

This study explores many details of the drivers response to VMS. A SP survey was conducted in China to collect various drivers behavior information with VMS. Based on the findings from the surveys, as much as possible potential affecting factors for drivers route choice behaviors with ATIS are identified and applied to further study. A probit model is adopted to evaluate the significance of these factors, and then the primary explanatory factors are adopted to develop the driver compliance model.

2 Stated Preference Survey

The scope of research is usually limited by the availability of data. This study aims to understand the structure of driver compliance with VMS on the urban network. However, driver response associated with VMS is difficult to observe and collect in the field. Thus this study has adopted stated preference (SP) methods involving extensive user surveys. The SP method is a very effective tool to investigate the drivers behavior characteristics under a variety of controlled scenarios. To understand the general drivers response to VMS and to find significant variables that affect the drivers compliance, an online survey was conducted.

Based on the recent enhancement of the internet, online surveys are now generally established as one of the tools for behavior studies. Use of the internet for surveys can improve the reliability of collected data by allowing the researcher to provide more detailed and realistic descriptions to respondents, which allows them to provide answers closer to their behavior under real circumstances. Moreover, it can largely eliminate coding errors and take adaptive designs where the options offered can be modified as a result of the responses to previous answers. However, users of this method should be careful to control the sample population.

The survey was sent to a random sampling of drivers in China based on their travel experiences with a sample size of 1000. A 22-question driver survey questionnaire (see Appendix) was developed in June 2011 to collect the following information: (1) Demographic characteristics of drivers; (2) Perceptions of dynamic travel information; (3) Driver behaviors under dynamic travel information with VMS. Seventeen potential affecting factors are identified as following: city kind (x_1), region (x_2), gender (x_3), age (x_4), marital status (x_5), degree (x_6), job (x_7), whether full-time worker (x_8), monthly income (x_9), crowded level on the current route (x_{10}), vehicle queue length of the current route (x_{11}), delay ratio of the current route (x_{12}), knowledge of an alternate route (x_{13}), length ratio of an alternate route (x_{14}), crowded level on an alternate route (x_{15}), anticipated travel time saving ratio (x_{16}) and quality of dynamic travel information (x_{17}). 540 completed survey forms were returned and used for analysis. Among 490 valid surveys, more than 50 percent of the respondents are somewhat familiar with VMS. The respondents demographic characteristics which present the applicability of research conclusions in this paper are shown in Table 1.

Table 1 Demographic characteristic of survey respondents

Numbers	Characteristics	Options	Percentages (%)
1	Region	Northeast	55.7
		North China	20.6
		East China	11.4
		Central South	4.5
		Northwest	3.3
		Southwest	4.5
2	City kind	Village	5.1
		County	6.7
		Common city	46.9
		Provincial capital	26.3
		Municipality	15.0
3	Gender	Female	36.9
		Male	63.1
4	Age	Less than 20	4.1
		20–29	51.6
		30–39	27.3
		40–49	11.2
		50–59	4.5
		Greater than 60	1.3
5	Marital status	No	44.9
		Yes	55.1
6	Degree	Elementary school	12.0
		Middle school	12.7
		University	52.6
		Master or Doctor	22.7
7	Job	Student	16.9
		Company staffer	31.4
		Civil servant	6.8
		Teacher or Doctor	11.2
		Freelancer	10.4
		Other	23.3
8	Whether Full-time Worker	No	23.1
		Yes	76.9
9	Monthly income (Yuan)	Less than 2000	18.4
		2000–4000	40.6
		4000–8000	26.9
		Greater than 8000	14.1

3 Potential Affecting Factors Analyzing

Binary responses occur in many fields of study. Probit model analysis has been used to investigate the relationship between these discrete responses and a set of explanatory variables. The response, y, of an individual or an experimental unit can take on one of two possible values such as Comply ($y = 1$) or Not Comply ($y = 0$). Suppose x is a vector of explanatory variables and is the response probability to be modeled. The probit model has the form:

$$p(y|x) = f(x) \tag{1}$$

If the following formula is established,

$$f(x) = \Phi(\beta_0 + \beta_1 x_1 + \beta_2 x_2 + \cdots + \beta_{\mathrm{m}} x_m) \tag{2}$$

Then,

$$p(y|x) = \Phi(\beta_0 + \beta_1 x_1 + \beta_2 x_2 + \cdots + \beta_{\mathrm{m}} x_m) \tag{3}$$

where $x_i (i = 1, \ldots, m)$ are the arguments, $\beta_i (i = 1, \ldots, m)$ are the correlation coefficients, β_0 are the constant terms and $\Phi(\cdot)$ is standard normal distribution function.

Total 490 responses were collected and used to develop a probit model. The result based on complete 17 variables is summarized in Table 2. This shows that x_3 (Wald = 4.951), x_4 (Wald = 5.726), x_8 (Wald = 3.262), x_{12} (Wald = 3.968), x_{13} (Wald = 7.306), x_{14} (Wald = 9.284) and x_{15} (Wald = 18.318) have wald-values are significantly greater than the others indicating that the parameters have the greater impacts on driver behavior than others. It is further proved from the result of the probit analysis shown in Table 3 that the driver behaviors under dynamic travel information with VMS depend greatly on their x_3 (Gender), x_4 (age), x_8 (whether full-time worker) and the traffic factors, including x_{12} (delay ratio of the current route), x_{13} (knowledge of an alternate route), x_{14} (length ratio of an alternate route) and x_{15} (crowded level on an alternate route).

4 Probit Model for Driver Compliance

The prediction rates of binary probit model between complete explanatory variables and primary explanatory variables are compared to validate the necessity of potential affecting factors analyzing. To compare the result from the models with the other 100 random samples which were collected by the online survey, it is assumed that if p greater than 0.5, a driver would comply with the given dynamic travel information with VMS. Table 4 shows the aggregated error rates for the binary probit model with complete 17 explanatory variables. Table 5 shows the aggregated error rates for the binary probit model with 7 primary explanatory variables.

Table 2 Probit analysis for driver behavior with complete explanatory variables

Variables	B	S.E.	Wald
x_1	–0.044	0.076	0.339
x_2	–0.043	0.053	0.656
x_3	0.315	0.141	4.951
x_4	–0.205	0.086	5.726
x_5	0.028	0.177	0.024
x_6	–0.066	0.094	0.484
x_7	–0.003	0.045	0.005
x_8	0.318	0.176	3.262
x_9	0.005	0.083	0.003
x_{10}	–0.043	0.084	0.259
x_{11}	0.023	0.050	0.310
x_{12}	0.124	0.062	3.968
x_{13}	0.224	0.083	7.306
x_{14}	–0.181	0.059	9.284
x_{15}	–0.285	0.067	18.318
x_{16}	0.013	0.053	0.065
x_{17}	0.014	0.069	0.042

Table 3 Probit analysis for driver behavior with complete explanatory variables

Variables	Transition variables	B	S.E.	Wald
x_3	X_1	0.328	0.138	5.622
x_4	X_2	–0.186	0.068	7.629
x_8	X_3	0.300	0.156	3.686
x_{12}	X_4	0.124	0.047	6.854
x_{13}	X_5	0.219	0.081	7.285
x_{14}	X_6	–0.166	0.058	8.335
x_{15}	X_7	–0.276	0.059	21.893

Table 4 Classification matrix for probit model with complete variables

		Predicted		
		Y = 0	Y = 1	Percentage errors (%)
Observed	Y = 0	7	16	31.0
	Y = 1	15	62	

Table 5 Classification Matrix for probit model with primary variables

		Predicted		
		Y = 0	Y = 1	Percentage errors (%)
Observed	Y = 0	9	14	24.0
	Y = 1	10	67	

5 Conclusion

Based on the many findings from the SP surveys, seventeen factors are identified that potentially affect drivers route choice behaviors with ATIS. With extensive SP approach, the study revealed many details of driver behavior with VMS. Based on online survey, those factors are evaluated and analyzed. To analyze the impact on driver behavior with VMS, a probit model is adopted. The binary probit model identifies seven significant variables, which are Gender, age, whether full-time worker, delay ratio of the current route, knowledge of an alternate route, length ratio of an alternate route and crowded level on an alternate route. Then the primary variables are used to analyze driver compliance. It showed great flexibility in analyzing the data and identifying the structure of the given data set. It clearly presents the structure of driver compliance behavior while maintaining a reasonable prediction rate of around 76 %. The study has analyzed driver compliance through behavior potential affecting factors analyzing which provides more understanding of driver compliance behavior mechanism while maintaining reasonable estimation accuracy.

The study has found the conditional structure of driver compliance with VMS which means non-homogeneous driver would use different factors to make a route choice decision. This finding can be applied to develop implemental operation strategies for ATIS applications including VMS.

The significant variables that affect driver compliance under the provision of travel information are based on the SP study. However, the study collected 490 samples to analyze and construct the models but considering the complicated structures of driver compliance, extensive research with a larger sample size may disclose many answers more clearly. Especially follow-up survey with non-respondent group will be essential to ensure many findings from this research. To enhance the driver behavior study with VMS, ongoing efforts should be made to find more effective uses of available technologies in SP surveys, including the use of interactive and dynamic surveys based on the computer and the internet. This will help in the development of more reliable and cost-effective methods for behavior data collection.

References

1. Khattak AJ, Schofer JL, Koppelman FS (1993) Commuters enroute diversion and return decisions: analysis and implications for advanced traveler information systems. Transp Res Part A 27(2):101–111

2. Abdel-Aty MA, Kitamura R, Jovanis PP (1997) Using stated preference data for studying the effect of advanced traffic information on drivers route choice. Transp Res Part C 5(1):39–50
3. Peeta S, Ramos JL, Pasupathy R (2000) Content of variable message signs and online driver behavior, Transportation Research Board, Washington D.C
4. Chatterjee K, Hounsell NB, Firmin PE (2002) Driver response to variable message sign information In London. Transp Res Part C 10(2):149–169
5. Lee C (2004) Developing driver compliance based operations model for atis applications. University of Wisconsin, Madison
6. Tsirimpa A, Polydoropoulou A, Antoniou C (2007) Development of a mixed multi-nomial logit model to capture the impact of information systems on travelers switching behavior. Intell Transp Syst 11(2):79–89
7. Bekhor S, Albert G (2014) Accounting for sensation seeking in route choice behavior with travel time information. Transp Res Part F 22(1):39–49
8. Manley EJ, Orr SW, Cheng T (2015) A heuristic model of bounded route choice in urban areas. Transp Res Part C 56(1):195–209

Accelerating Reservoir Simulation on Multi-core and Many-Core Architectures with Graph Coloring ILU(k)

Zheng Li, Chunsheng Feng, Shi Shu and Chen-Song Zhang

Abstract Incomplete LU (ILU) methods are widely used in petroleum reservoir simulation and many other applications. However high complexity often makes them the hotspot in the whole simulation due to high complexity when problem size is large. ILU's inherent serial nature also makes them difficult to take full advantage of computing power of multi-core and many-core devices. In this paper, a greedy graph coloring method is applied to the ILU(k) factorization and triangular solution phases. This method increases degree of parallelism and improves load balance. A block-wise storage format is employed in our ILU implementation in order to take advantage of hierarchical memory structures. Moreover, a dual intensive parallel model is proposed to further improve the performance of ILU(k) on GPUs. We test the performance of the proposed parallel ILU(k) with a set of Jacobian systems arising from petroleum reservoir simulation. Numerical results suggest that the proposed parallel ILU(k) method is effective and robust on multi-core and many-core architectures. On an Intel Xeon E5 multi-core CPU, the speedup compared with the serial execution time is 5.6× and 5.4× for factorization and triangular solution, respectively; on an Nvidia K40c GPU card, the speedup can reach 8.6× and 12.7× for factorization and triangular solution, respectively.

Z. Li (✉)
Civil Engineering and Mechanics, Kunming University of Science and Technology, Kunming, China
e-mail: lizhxtu@126.com

C. Feng
Hunan Key Laboratory for Computation and Simulation in Science and Engineering, Xiangtan University, Xiangtan 411105, China
e-mail: spring@xtu.edu.cn

S. Shu
Key Laboratory of Intelligent Computing and Information Processing of Ministry of Education, Xiangtan University, Xiangtan 411105, China
e-mail: shushi@xtu.edu.cn

C.-S. Zhang
NCMIS and LSEC, Academy of Mathematics and Systems Science, Beijing 100190, China
e-mail: zhangcs@multigrid.org

© Springer International Publishing Switzerland 2017

V.E. Balas et al. (eds.), *Information Technology and Intelligent Transportation Systems*, Advances in Intelligent Systems and Computing 454,
DOI 10.1007/978-3-319-38789-5_31

Keywords Graph coloring · ILU · Multi-core · Many-core

1 Introduction

Petroleum reservoir simulation is an important tool to understand residual oil distribution, design development plans, and optimize the recovery processes. High-resolution reservoir models can better characterize the reservoir heterogeneity and describe complex water encroachment [1]. The demand for more accurate simulation results has led to larger and more heterogeneous models, which entail large and difficult-to-solve linear systems. For fully implicit reservoir simulation, the solution of Jacobian systems usually accounts for the majority of simulation time.

The Incomplete LU (ILU) methods are widely used as preconditioners for iterative methods for solving a sparse linear algebraic system $Ax = b$ due to their effective reduction to wide range error components. The ILU methods can be divided into two phases: the setup and solve phases. The setup phase yields a lower triangular sparse matrix L and an upper triangular sparse matrix U such that the residual matrix $R = LU - A$ satisfies certain conditions. Denote $M = LU$ and M can be viewed as an approximation of A. The block-wise ILU (or BILU) methods are often used as a stand-alone preconditioner or a component of multi-stage preconditioners, such as the well-known CPR-type preconditioners [2–4] in reservoir simulation. Since BILU(k) may have high computational complexity, it usually becomes the bottleneck of linear solution when the problem size grows. Wu et al. [5] reported that the BILU factorization and triangular solution sometimes count for more than 50 % of the CPR solver time in their numerical tests.

Computers are now equipped with multi-core CPUs and coprocessors that have tens or even thousands of light cores, such as Intel Xeon Phi coprocessors and graphic processing units (GPUs). This presents new challenges for solving large problems with parallel algorithms. This is especially true for ILU methods because they have strong data dependency and are naturally sequential. Wu et al. [5] and Chen et al. [6] applied a level-scheduling strategy [7] to parallelize ILU(k) in reservoir simulation as well as circuit simulation on multi-core architectures. Klie et al. [8] partitioned the connectivity graph associated with A using METIS, and do ILU(k) or ILUT factorizations on each local matrix and triangular solution in parallel on GPUs. Li and Saad [9] used a multi-color method to parallelize triangular solution of ILU(0) on GPUs.

In this paper, we apply a greedy graph coloring method [7, Sect. 3.3] to parallelize ILU(k) in both factorization and triangular solution phases. The proposed method is implemented on multi-core CPUs and many-core GPUs. We propose a dual intensive parallel model on GPUs, which greatly improve parallel performance of factorization and triangular solution. To test the robustness and effectiveness of our implementation, we apply BILU(k) to a linear solver for fully implicit petroleum reservoir simulation. We shall note that BILU(k) has a wide range of applications in scientific/engineering computing and our method is not restricted in any ways to the petroleum reservoir simulation.

The rest of this paper is organized as follows: In Sect. 2, we firstly profile the running time of our linear solver in reservoir simulation. In Sect. 3, we briefly review the sequential ILU algorithm. We then introduce the greedy graph coloring algorithm in Sect. 4. We give implementation details of ILU(k) on multi-core CPUs and many-core GPUs in Sect. 5. In Sect. 6, we design several numerical tests and show speedup of BILU(0) and BILU(1) in reservoir simulation on CPUs and GPUs.

2 Petroleum Reservoir Simulation

In reservoir simulation, among many possible discretization methods for the black oil model, fully implicit method (FIM) [10] is often used in commercial simulators due to its robustness and stability. However it has a major disadvantage—the computational cost associated with solving the Jacobian systems arising from the Newton linearization is high. Very often, solving such linear systems takes more than 80 % of the computation time in reservoir simulation. In our numerical experiments, we employ a popular two-stage CPR preconditioner [2, 11, 23]. In the first stage, the pressure equations is solved (approximately) by algebraic multigrid methods (AMG) [12]. And, in the second stage, the BILU(0) method is applied to the whole Jacobian systems. Thus, our linear solver mainly consists of BILU(0), AMG, and GMRES.

It is well-known that AMG's solve phase and GMRES are mainly based on sparse matrix vector multiplication (SpMV) and their implementation on CPUs and GPUs is relatively easy. However, BILU(0) and AMG setup phase are inherently sequential and difficult to be parallelized. We profile our linear solver and evaluate the CPU time ratio of BILU(0), AMG setup and SpMV over the linear solver running time in Fig. 1. From the figure, we can see that BILU(0) accounts for more than 30 % of linear solver time for almost test cases. In particular, it mounts to 40 % for Case–8. Furthermore, the ratio trends to increase with the size of problems. These observations indicate that poor scalability of BILU(0) will result in the poor overall parallel performance of linear solver on multi-core and many-core platform. We have discussed the AMG method on GPUs in [11]. In this paper, we focus on the strategies for implementing BILU(k) on CPUs and GPUs.

We choose ten Jacobian matrices from different real reservoir problems, which are listed in Table 1 in detail. These test problems can be categorised into three types due to their origin. Case 1–4, Case 5–8, Case 9–10 are adapted and modified from the benchmark models in SPE10 [13], SPE1 [14], SPE9 [15], respectively. The size of matrices ranges from tens of thousands to several millions; the block size of them are 2 (two-phase) or 3 (three-phase); In each type, the problems are ordered with increasing matrix size. Due to the presence of wells and faults, the band-width of Jacobian matrix will be affected by number of perforations and the number of non-neighboring connections (NNCs). Therefore, the bandwidth (maximal number of row non-zeros) of these matrices varies largely among different sources.

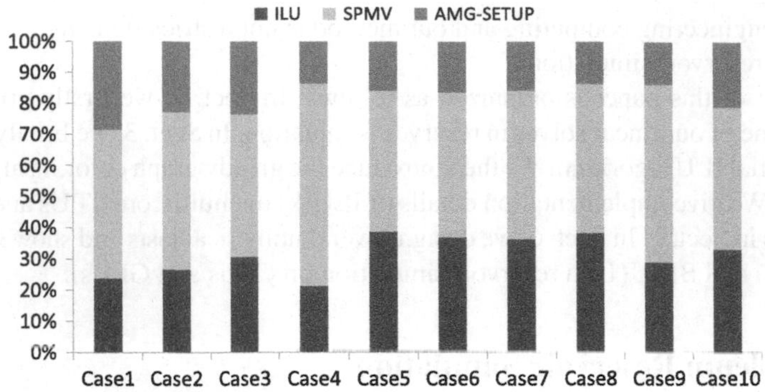

Fig. 1 Time profile for linear solver of reservoir simulation

3 Sequential ILU Algorithm

The setup phase of ILU contains symbolic factorization and numerical factorization, and the solve phase is just triangular solution. The symbolic factorization determines sparsity patterns of L and U; the numerical factorization, on the other hand, is done in place to determine the value of non-zeros. The ILU factorization is similar to the well-known Gauss Elimination process. It introduces non-zero values for some entries of the matrix A whose values are zero initially. These new arrivals often called fill-in's and they can be discarded based on different criteria, such as according to their positions, magnitudes, etc. Many variants of ILU were proposed in the past (see for example [16]) and are widely used in practice. These methods differ in the rules they use to drop fill-in's; see [7] and references therein for details.

In this paper, we only consider the parallel implementation for ILU(k). ILU(k) is obtained based on the concept of "level of fill-in". The initial level of fill-in of a matrix entry a_{ij} is defined as (1). Then level of fill-in is updated for each matrix entry that occurs in the incomplete factorization process following the rules

$$lev(a_{ij}) = \min\{lev(a_{ij}), lev(a_{ik}) + lev(a_{kj}) + 1\}. \tag{1}$$

Fill-in's are dropped if the entry's level is larger than a given integer k after each elimination.

4 A Greedy Graph Coloring

Note that, from the implementation point of view, there is not much difference between BILU and ILU. ILU can be viewed as a special case of BLIU with blocksize 1. For brevity, we will only address algorithms for ILU in this paper. But the following discussed parallel methods can be easily extended to BILU.

Table 1 Summary of test problems

Case	#Rows	Blocksize	#Nonzeros	#MaxRowNZ	#Colors(0)	#Colors(1)	#Levels(0)	#Levels(1)
1	556,402	2	3,832,970	87	4	90	254	531
2	1,094,422	2	7,478,172	85	4	85	364	751
3	2,188,843	2	15,029,123	88	18	94	424	811
4	2,188,843	2	15,026,777	85	4	85	424	811
5	200,002	3	1,372,030	11	3	14	220	357
6	512,002	3	3,520,030	11	3	14	340	537
7	800,002	3	5,504,030	11	3	12	420	657
8	1,200,001	3	7,500,021	11	3	12	429	686
9	540,001	3	3,739,973	12	3	14	245	467
10	900,026	3	6,165,486	8	3	12	504	781

Let $A \in R^{n \times n}$ be a square matrix. The adjacency graph of A is a graph $G(A) = (V, E)$, whose vertices are in the index set $V = \{1, 2, \ldots, n\}$ and whose edges are defined as $(i, j) \in E$ if and only if $a_{ij} \neq 0$. Thus, each vertice corresponds to a row of A. The coming introduced algorithms, level-scheduling algorithm and greedy graph coloring algorithm, are based on matrix graph $G(A)$. A natural parallel method of ILU(k) is to sort the unknowns in a way such that unknowns (except for the i-th variable) involved in the i-th equation must be known at the i-th step. This logic relation can be measured by assigning depth to each unknown $\text{depth}(i) = 1 + \max_j\{\text{depth}(j),\ \forall j \in N_i\}$, where $N_i \subset V$ is the neighboring variables of i.

We sort unknowns with increasing depth and view unknowns with the same depth as in the same level. This way, we form a dependency level sequence $\{\mathcal{L}_i\}_{i=1}^{M}$, M is the number of levels. Then unknowns are solved by increasing depth order. Obviously, those unknowns on the same level $\mathcal{L}_i$ can be simultaneously solved. This method is called level-scheduling [7] and is widely used for parallel implementation. However, this method has two major drawbacks: 1. It sometimes yields too many levels (see rightmost two columns of Table 1); 2. The number of vertices in different levels vary largely. These two disadvantages will result in higher synchronization overhead and serious load imbalance for parallel implementation.

Here we consider a greedy graph coloring method [7], it is based on the graph associated with the matrix A; The graph coloring method aims to divide the $G(A)$ into several independent sets G_i, $i = 0, \ldots, l$, and they satisfy that $G(A) = G_0 \cup G_1 \cup \cdots \cup G_l$. If we apply graph coloring algorithm to the matrix on the left of Fig. 2, three independent sets are generated, $G_0 = \{0, 2\}$, $G_1 = \{1\}$, $G_2 = \{3\}$. The permuted matrix PAP^T, is the one on the right of Fig. 2, where P is permutation matrix. It is easy to see that the first two rows can perform the Gauss Elimination process simultaneously, then the third, and the fourth row at last.

The sparsity pattern after the factorization shows that the non-zero pattern of ILU(k) grows like $|A|^{k+1}$ ($|A|$ denotes the absolute value of A). Inspired by this fact, Heuveline et al. [17] proposed a modify ILU(k) methods by restricting the non-zero pattern of the factorization in the non-zero pattern of $|A|^{k+1}$. He applied the coloring algorithm to $|A|^{k+1}$ to obtain multi-color order, and then permuted A with the order. He proved the fill-in of modified ILU(k) to permuted A only occur on off-diagonal blocks. In reservoir simulation high level of fill-in ILU(k) ($k > 1$) methods are rarely used, we will not consider them in this paper.

We apply the level-scheduling algorithm and the graph coloring algorithm to the Jacobian matrices in Table 1. Note that, due to discretization of petroleum reservoir simulation equations involving upwinding weights of convection terms, the resulted Jacobian matrices are usually not structurally symmetric, i.e.,

$$\begin{pmatrix} 1 & 2 & 0 & 0 \\ 3 & 4 & 0 & 5 \\ 0 & 0 & 6 & 0 \\ 0 & 7 & 0 & 8 \end{pmatrix} \xrightarrow{PAP^T} \begin{pmatrix} 1 & 0 & 2 & 0 \\ 0 & 6 & 0 & 0 \\ 3 & 0 & 4 & 5 \\ 0 & 0 & 7 & 8 \end{pmatrix}$$

Fig. 2 Natural ordering (*Left*) and Multi-color ordering (*Right*)

$a_{ij} \neq 0$ if and only if $a_{ji} \neq 0$, $\forall i, j \in V$. Then the proposed coloring algorithm is applied to the symmetrized matrix $A + A^T$ to obtain the independent sets $\{G_i\}$. Table 1 shows that the number of colors and levels of test problems formed by graph coloring and level-scheduling. In this table, "#Rows" stands for number of rows of block matrix; "#MaxRowNZ" gives the maximal number of nonzeros in each row; #Colors(0) and #Levels(0) are number of colors and levels corresponding to ILU(0) formed by graph coloring and level-scheduling, similarly, #Colors(1) and #Levels(1) are corresponding to ILU(1).

We notice that, for level-scheduling, the number of levels of ILU(0) and ILU(1) are more than 200. On the other hand, for graph coloring, the number of colors are no more than 4 except for Case 3 in ILU(0) and is less than 100 in ILU(1). It is obvious that the degree of parallelism ($\frac{\#\text{Rows}}{\#\text{Colors}}$) of ILU(k) based on the graph coloring is higher than level-scheduling. In addition, since L_0U_0 is more dense than A, the levels (colors) of ILU(1) increase dramatically when compared with ILU(0). Here, L_0 and U_0 are the factors of A for ILU(0). From Table 1, we also find that number colors of ILU(1) are easily affected by the bandwidth of matrix, but it would be bounded by MaxRowNZ(L_0U_0)+1 [7].

5 Parallel ILU Algorithm

In this section, we discuss parallel implementation of BILU(k) based on the greedy graph coloring algorithm on GPUs as well as multi-core CPUs. We begin with parallel implementation on GPUs. GPUs are throughput-oriented processors, and have great computing power and high band-width. Better performance of factorization and triangular solution can be expected from the right combination of data structures and corresponding implementation that best exploit the GPU architecture. Bell and Garland [18] studied several traditional sparse storage formats on GPUs and analysed their performance, and proposed a GPU-suitable Hybrid (ELL/COO) storage format. The advantage of this Hybrid data format lies in that all threads coalesced access to global memory, which reduces cache missing and burden of memory band-width. Li et al. [11] presented a block hybrid (BHYB) format for Jacobian matrices on GPUs and we employ BHYB for ILU(k) implementation on GPUs.

In order to sufficiently expose fine-grained parallelism of GPUs, we now present a new dual intensive parallel model, that's, the outer parallel is that vertices in G_i (rows with same color) can simultaneously process, the inner parallel is that multi-threads are assigned to handle each vertex (each row) together; see Fig. 3. This dual intensive parallel model can be easily applied to numerical factorization and triangular solution. For simplicity, we ignore the irregular part of matrix stored in BCOO format and only show BELL part of matrix in Fig. 3. The column R_i in BELL format denotes the i-th row in A, n_b denotes block size, and m denotes the number of column in BELL format. V_i is global vector index, each vector will handle a column. The number of vectors equal the matrix size.

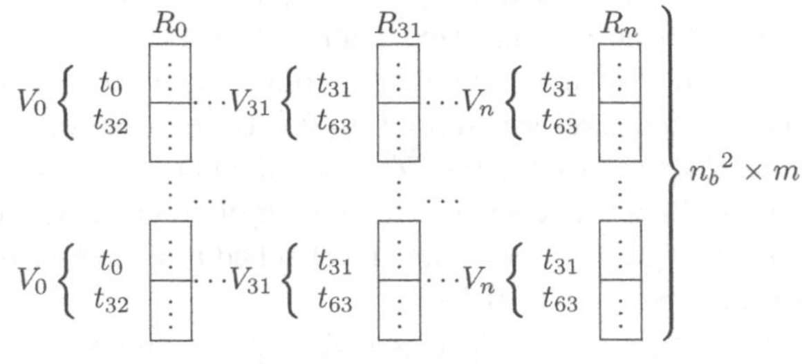

Fig. 3 A dual intensive parallel model

The next is to group all threads in the same block into vectors and also keep thread coalesced access to global memory. Assume 64 threads in each thread block, and each vector has two threads. that's, there are 32 vectors in each thread block. Therefor, each continual 32 columns, e.g. $R_0 \sim R_{31}$, are assigned to each thread block as a task. Some indices are needed to organize threads and vectors:

- the index of thread in global: `id = threadIdx.x + blockIdx.x * blockDim.x`;
- the index of vector in global: `v_i = id / vec_size`, where `vec_size` means the number of threads in each vector;
- the index of vector in thread block: `v_loc = threadIdx.x % #vectors` where `#vectors` is the number of vectors in each block;
- the index of thread in each vector: `tid_loc = threadIdx.x / #vector`.

Threads within the same vector and vectors in the same thread block can communicate with each other via the shared memory as well as the global memory, but vectors in different thread blocks can only communicate via the global memory. Since multiple threads work on a task together, the reduction operation and communication are often needed. A most efficient way to handle this is to cache data in the shared memory, which has higher bandwidth than global memory. Hierarchical memory structure of GPUs are addressed with great details in [19]. Another main factor affecting the performance of this parallel model is the vector size. To determine the vector size, the priority is to select how many threads in each block. Thread configuration of kernel functions is complicated and has much to do with the resources provided by specific GPU cards in use, such as cache size, number of registers, and so on. A lot of research has been done on the issue and interested readers are referred to [20–22] and references therein. In this paper, we just use an empirical value and fix the number of threads in each block to be 128.

In order to investigate performance of the proposed dual parallel model and chose the reasonable vector size, we test the different vector size on Case 7, the execution time of factorization and triangular solution are recorded; see Fig. 4. We can see that time using vector size 4 is considerably shorter than the other cases. When compared with non-vector parallel case (vector size 1), the vector size 4 has 63 and 23 % improvement for factorisation and triangular solution, respectively. Since computa-

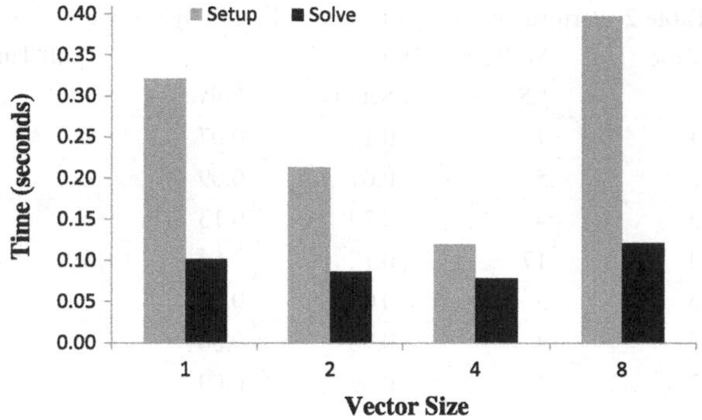

Fig. 4 Performance comparison for vector sizes

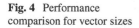

tion of factorization is more intensive than triangular solution, factorization achieves greater improvement than triangular solution. Large improvement is obtained when intensive parallel model is employed.

The graph coloring method discussed here can also be applied for parallel implementation of BILU on multi-core CPUs. Inner loop operations of numerical factorization are small block multiplication, and triangular solution is much like sparse matrix vector multiplication (SpMV). We use block sparse row (BSR) storage format for matrix A, block operations implements explicitly by taking usage of registers and data locality. Since the high degree of parallelism of BILU(k) based on graph coloring, it is easy to implement using OpenMP on multi-core architectures. Here we use the default static task schedule strategy and no extra attentions are paid to the task assignment to each thread.

6 Numerical Experiments

All experiments were run on a work station with 2 Intel Xeon CPU E5-2690 v2 with 20 cores, 175GB of RAM, and an NVIDIA Tesla K40c coprocessor with 2880 cores, 12GB of memory. In order to get better ideas on the efficiency and robustness of the proposed graph coloring based BILU(k) methods, we chose a set of ten Jacobian systems from realistic reservoir simulation, whose main characteristics have been listed in Table 1. The stopping criteria for the preconditioned GMRES method [7] is to reduce the relative residual in the Euclidian norm by 10E-3 or more.

In all tables of numerical experiments part, "NO", "LS", "MC" denote iteration number of sequential linear solver associated with ILU in natural order (NO), level-scheduling (LS) order, and multi-color (MC) order, as preconditioners, respectively; "Wall Time for sequential" records the CPU time of setup and solve phases of BILU run by one core while "Wall Time for LS" and "Wall Time for MC" record the GPU time of setup and solve phases of BILU corresponding to level-scheduling and multi-color order. "Speedup on CPUs" denotes the speedup gained for ILU setup and solve

Table 2 Performance comparison of BILU(0) by LS and MC on GPUs

Case	Wall Time for LS			Wall Time for MC		
	LS	Setup (s)	Solve (s)	MC	Setup (s)	Solve (s)
1	7	0.12	0.07	9	0.03	0.03
2	5	0.07	0.09	6	0.03	0.03
3	4	0.39	0.13	5	0.12	0.06
4	17	0.12	0.54	19	0.04	0.46
5	3	0.03	0.02	5	0.01	0.01
6	4	0.06	0.06	5	0.03	0.03
7	4	0.09	0.09	5	0.05	0.04
8	4	0.13	0.13	6	0.07	0.08
9	7	0.06	0.11	9	0.03	0.05
10	4	0.10	0.11	5	0.05	0.05

phases with using 12 cores. Similarly, "Speedup on GPUs" denotes the speedup of above two parts on GPUs.

We first compare the performance of BILU(0) with level-scheduling and multi-color order on GPUs. From the Table 2, we can see that the iteration number of level-scheduling is less than that of multi-color, this is because level-scheduling method does not change the non-zero pattern of matrix but multi-color method should permute matrix. Besides, the setup and triangular solution time cost twice than multi-color at least. This indicates the poor performance of level-scheduling.

Next, we analysis the performance of multi-color order based BILU. We can see that, from Table 3, the iteration number of linear solver preconditioned by the multi-color ordered BILU(0) is slightly higher than the natural ordered BILU(0). The iteration numbers of parallel linear solver on CPUs and GPUs are exactly the same as the sequential version. This verifies the good parallelism of the multi-color ordered BILU(0).

The speedup corresponding to factorization and triangular solution increases with the problem size regardless on CPUs or GPUs. When compared with Case 4, whose size is same to Case 3, the speedup of factorization of Case 3 is lower than that of Case 4, this is because the number colors in Case 3 is 18, much more than that of Case 4. In our implementation, factorization on each color needs a auxiliary private arrays. Performance usually deteriorates when more private arrays are required. Case 1–4 with blocksize 2 have higher speedup than other test problems with blocksize 3 in terms of factorization and triangular solution. This advantage is more obvious on GPUs. It indicates that speedup does not only depend on problem size, but are also affected by the block size. In most cases on CPUs, the speedup of factorization is higher than that of triangular solution. This is due to computation of factorization is more intensive than triangular solution.

Generally speaking, higher level of fill-in results in more accurate decomposition. However, it results in more fill-in's, which means more memory space and

Table 3 Performance comparison of BILU(0) on CPUs and GPUs

Case	Iteration numbers		Wall time (sequential)		Speedup on CPUs		Speedup on GPUs	
	NO	MC	Setup (s)	Solve (s)	Setup	Solve	Setup	Solve
1	7	9	0.09	0.27	2.56	4.13	2.97	8.04
2	5	6	0.19	0.36	5.14	5.26	6.78	10.79
3	4	5	0.39	0.64	2.99	4.96	3.28	11.11
4	17	19	0.37	3.12	4.89	5.89	8.62	13.84
5	3	5	0.06	0.08	3.82	4.51	4.85	7.18
6	4	5	0.19	0.21	5.17	4.24	6.31	7.59
7	4	5	0.31	0.33	6.34	4.22	6.69	7.62
8	4	6	0.47	0.60	6.17	4.68	6.74	7.73
9	7	9	0.17	0.39	4.82	4.74	5.43	7.44
10	4	5	0.35	0.38	5.55	4.54	5.84	8.91
			Average speedup		4.74	4.65	6.27	9.46

Table 4 Performance comparison of BILU(1) on CPUs and GPUs

Case	Iteration number		Wall time for sequential		Speedup on CPUs		Speedup on GPUs	
	NO	MC	Setup (s)	Solve (s)	Setup	Solve	Setup	Solve
1	6	7	0.86	0.69	3.97	5.71	7.17	11.56
2	4	5	1.69	1.24	4.24	6.84	9.26	18.52
3	3	4	3.59	2.03	4.03	5.75	2.45	17.52
4	13	27	3.64	13.7	4.26	5.89	10.69	18.68
5	2	4	0.46	0.18	6.45	4.69	7.81	7.44
6	3	4	1.35	0.52	6.90	5.16	9.81	9.51
7	3	4	2.07	0.79	6.65	4.56	10.32	9.58
8	3	6	3.02	1.78	6.29	4.72	10.01	14.82
9	6	7	1.41	1.05	6.17	5.79	8.43	9.72
10	4	5	2.34	1.14	6.94	4.91	10.32	9.89
			Average speedup		5.59	5.40	8.63	12.72

more computation for factorization and triangular solution. In the classic CPR-type preconditioners in reservoir simulation, BILU(0) is usually more cost-effective than BILU(1). We apply the graph coloring method to BILU(1) here in order to verify parallel performance of BILU(1) both on CPUs and GPUs. As Table 4 shows, the average speedups of the setup and solve phases on CPUs and GPUs are higher than BILU(0), which is expected due to more intensive computational cost of BILU(1).

Acknowledgments The paper is finished during Li's visit to the State Key Laboratory of Scientific and Engineering Computing (LSEC), Academy of Mathematics and Systems Science. Li is thankful to the kind support from LSEC. The authors would like to thank Prof. Ludmil Zikatanov from the Pennsylvania State University and Dr. Xiang Li from the Peking University for many helpful discussions on ILU methods.

References

1. Dogru AH, Fung LSK, Middya U, Al-shaalan TM, Pita JA, Hemanthkumar K, Su HJ, Tan JCT, Hoy H, Dreiman WT, Hahn WA, Mezghani M (2009) A next-generation parallel reservoir simulator for giant reservoirs. In: SPE/EAGE Reservoir Characterization & Simulation Conferenced, pp 1–29
2. Cao H, Tchelepi HA, Wallis J, Yardumian H, Stanford U (2005) Parallel scalable unstructured CPR-type linear solver for reservoir simulation. In: Society of petroleum engineers, SPE annual technical conference and exhibition, vol 1, pp 1–8
3. Wallis J (1983) Incomplete gaussian elimination as a preconditioning for generalized conjugate gradient acceleration. Paper SPE 12265 presented at the SPE reservoir simulation symposium, San Francisco, California, 15–18 November 1983
4. Wallis J, Kendall R, Little T, Nolen J (1985) Constrained residual acceleration of conjugate residual methods. SPE 13536:10–13
5. Wu W, Li X, He L, Zhang D (2014) Accelerating the iterative linear solver for reservoir simulation on multicore architectures. In: 20th IEEE international conference on parallel and distributed systems (ICPADS), pp 265–272
6. Chen X, Wu W, Wang Y, Yu H, Yang H (2011) An EScheduler-based data dependence analysis and task scheduling for parallel circuit simulation. IEEE Trans Circuits Syst II Express Br 58(10):702–706
7. Saad Y (2003) Iterative methods for sparse linear systems, 2nd edn. SIAM, Philadelphia
8. Klie H, Sudan H, Li R (2011) Exploiting capabilities of many core platforms in reservoir simulation. In: SPE reservoir simulation symposium, Woodlands, Texas, USA
9. Li R, Saad Y (2013) GPU-accelerated preconditioned iterative linear solvers. J Supercomput 63(2):443–466
10. Chen Z, Huan G, Ma Y (2006) Computational methods for multiphase flows in porous media, vol 2. SIAM, Philadelphia
11. Li Z, Wu S, Xu J, Zhang CS, Toward cost-effective reservoir simulation solvers on GPUs. Advances in applied mathematics and mechanics, to applear
12. Stüben K (1999) Algebraic multigrid (AMG): an introduction with applications. GMD Forschungszentrum Informationstechnik
13. Christie M, Blunt M (2001) Tenth SPE comparative solution project: a comparison of upscaling techniques. SPE Reserv Eval Eng 4(4):308–317
14. Odeh AS et al (1981) Comparison of solutions to a three-dimensional black-oil reservoir simulation problem (includes associated paper 9741). J Pet Technol 33(01):13–25
15. Killough J et al (1995) Ninth SPE comparative solution project: a reexamination of black-oil simulation. In: SPE reservoir simulation symposium. Society of Petroleum Engineers
16. Chan T, Van der Vorst HA (1997) Approximate and incomplete factorizations. Springer, Netherlands
17. Heuveline V, Lukarski D, Weiss JP (2011) Enhanced parallel ILU(p)-based preconditioners for multi-core CPUs and GPUs C the power(q)-pattern method. Engineering Mathematics and Computing Lab (EMCL), (08)
18. Bell N, Garland M (2008) Efficient sparse matrix-vector multiplication on CUDA. NVIDIA Technical report, pp 1–32
19. CUDA C Programming Guide, (2015) NVIDIA Corporation
20. Choi JW, Singh A, Vuduc RW (2010) Model-driven autotuning of sparse matrix-vector multiply on GPUs. ACM SIGPLAN Not 45(5):115
21. Monakov A, Lokhmotov A, Avetisyan A (2010) Automatically tuning sparse matrix-vector multiplication for GPU architectures. In: High performance embedded architectures and compilers, pp 111–125. Springer, Heidelberg

22. Zhang Y, Shalabi YH, Jain R, Nagar KK, Bakos JD (2009) FPGA vs. GPU for sparse matrix vector multiply. In: Proceedings of the 2009 international conference on field-programmable technology, FPT'09, pp 255–262
23. Hu X, Xu J, Zhang C (2013) Application of auxiliary space preconditioning in field-scale reservoir simulation. Sci China Math 56(12):2737–2751

Indoor Scene Classification Based on Mid-Level Features

Qiang Zhang, Jinfu Yang and Shanshan Zhang

Abstract In this paper, we use mid-level features to solve the problems of indoor scene image classification. The mid-level patches should satisfy two conditions: (1) representative, they should occur frequently enough in the visual world; (2) discriminative, they need to be different enough from the rest of the visual world. In this paper, we propose a method to select the initial patches. It can eliminate a large number of patches which are mismatch the conditions, and there is no need manual processing. For initial patches we adopt unsupervised cluster algorithm on HOG space. Then, using the purity-discriminative evaluation criteria, the top r clusters were selected to represent each scene. The experimental results on MIT Indoor 67 scene image classification datasets indicate that our method can achieve very promising performance.

Keywords Mid-level features · Scene classification · Initial patches · Unsupervised cluster · Evaluation criteria

1 Introduction

As an important research content of image understanding, scene image classification has become an important research problem in the field of computer vision and pattern recognition. It has been attracted much attention in recent years. It's a difficult task to classify the scene images correctly, because the categories of scene have the nature of variability and ambiguity. Currently, the most popular visual representations in

Q. Zhang (✉) · J. Yang · S. Zhang
Department of Control and Engineering, Beijing University of Technology,
No. 100 Chaoyang District, Beijing 100124, China
e-mail: dkzhang@emails.bjut.edu.cn

J. Yang
e-mail: jfyang@bjut.edu.cn

S. Zhang
e-mail: csustzhang@126.com

© Springer International Publishing Switzerland 2017
V.E. Balas et al. (eds.), *Information Technology and Intelligent Transportation Systems*, Advances in Intelligent Systems and Computing 454,
DOI 10.1007/978-3-319-38789-5_32

computer vision are based on visual words [1] which are obtained by unsupervised clustering of SIFT [2] features. SIFT features is a kind of low-level features. One of the problem is that the low-level features are relatively low-dimensional. So it might not be powerful enough to express anything of higher complexity. In [3], there is an interesting attempt to represent image by high-level semantics. It represents images by pooling the responses of pre-trained object detectors to the image. High level semantics also applied to recognize objects, actions and scene. However, there are also some practical barriers. First, we should require a large amount of labels to train per each semantic entity. Second, many semantic entities are not discriminative enough to act as good features. For example, wall is a well-defined semantic category, but it makes a bad detector because walls are usually plain and not easily discriminable. As a result, most researchers have concerned on using features at an intermediate scale: the mid-level image patches.

For pattern recognition, mid-level features provide a connection between low-level features and high-level semantic concepts. The mid-level features are more informative than visual words and do not require the semantic grounding of the high-level entities. The idea is to search for clusters of image patches that are both (1) representative, i.e. frequently occurring within the dataset, and (2) visually discriminative. The properties means that mid-level patches can be detected in a large number of images with high recall and precision. Mining mid-level patches from a large dataset is difficult for a number of reasons. First, the search space is huge: a typical dataset for visual data mining has tens of thousands of images, and finding something in an image involves searching across tens of thousands of patches at different positions and scales. Second, patch descriptors tend to be thousands of dimensions.

The mid-level features have been used for tasks including image classification [4–6], object detection [7], action recognition [8, 9], visual data mining [10, 11]. Recently, several approaches have proposed mining visual data for discriminative mid-level patches. M. Juneja et al. [4] presented a method to automatic discover the distinctive patches for an object or scene class. In this approach, the author address this problem by learning parts incrementally, starting from a single part occurrence with an Exemplar SVM [12]. And the additional patch instances are discovered and aligned reliably before being considered as training examples. However, it is difficult to train a SVM classifier for every object. In [13], the features are learned using supervised mid-level information in the form of hand drawn contours in images. Patches of human generated contours are clustered to form sketch token classes and a random forest classifier is used for efficient detection in novel images. However, in the real scenario, it's a huge quantity of work to sketch the contours for every training sample.

In this paper, we add a pre-processing step before selecting the patches. Our goal is decrease the number of patches in the initialization step, and reserve the patches which are contain abundant features as far as possible. As we know, the interesting features always contains much more gradient information, so we utilize one of the contour detection methods to obtain the gradient information in the images. As Fig. 1 shows that we can obtain the major objects in the images. Then, we select the

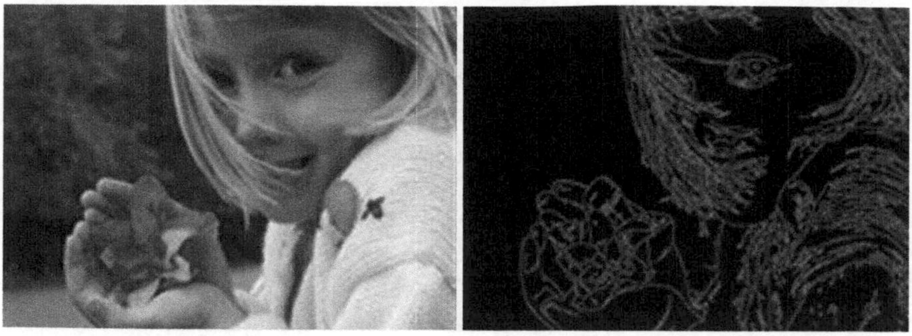

Fig. 1 An example of an original image and its contour image

patches in the region where contain the salient gradient information. The next step is finding good mid-level patches which represent each indoor scene. This procedure is inspired by the work in [14]. In our experiments, we used the original MIT Indoor-67 train/test dataset (80 training and 20 testing images per class) which containing 67 scene categories.

The rest of the paper is organized as follows. We introduce a mid-level patches selecting method in Sect. 2. In Sect. 3, the patches are clustered and ranked. Section 4 describes the results and the analysis. The paper is concluded in Sect. 5.

2 Selecting Initial Patches

In this section, we introduce the method how to select the initial patches in training set. We utilize the Canny edge detector to detect the contour of training images, than we sample the patches in the region of the contour.

2.1 Calculating the Gradient

In this step, we adopt the Canny edge detector [15] to detect the contour of the training images. Comparing with other contour detection methods, Canny operator is the optimal algorithm. Because it use two threshold values to detect weak and strong edges, and the two threshold values are set to clarify the different types of edge pixels, one is called high threshold value and the other is called the low threshold value. If the edge pixels gradient value is higher than the high threshold value, they are marked as strong edge pixels. If the edge pixels gradient value is smaller than the high threshold value and larger than the low threshold value, they are marked as weak edge pixels. If the pixel value is smaller than the low threshold value, they will be suppressed. The two threshold values are empirically determined values, which

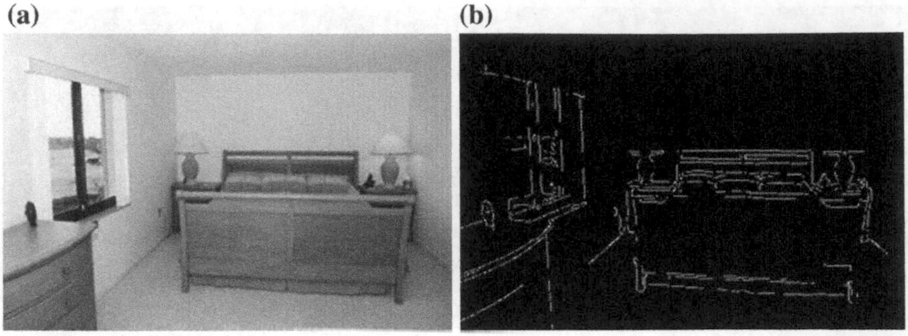

Fig. 2 An image in MIT indoor 67 training dataset and its contour image calculated by Canny detector

will need to be defined when applying to different images. So Canny edge detector is more suitable for detecting the real weak edges.

The process of Canny edge detection algorithm can be broken down to 5 different steps:

1. Apply Gaussian filter to smooth the image in order to remove the noise
2. Find the intensity gradients of the image
3. Apply non-maximum suppression to get rid of spurious response to edge detection
4. Apply double threshold to determine potential edges
5. Track edge by hysteresis: Finalize the detection of edges by suppressing all the other edges that are weak and not connected to strong edges.

Figure 2a is an image in MIT indoor 67 training dataset, which belongs to the scene of bedroom. Figure 2b is the Canny edge detector applied to the color photograph. In Fig. 2b, the main objects' edge which represents this scene have already detected, such as bed, bed lamps, windows. In our experiments, we set the high threshold $k_1 = 0.25$, the low threshold $k_2 = 0.1$.

2.2 *Obtaining the Patches*

In this procedure, we sample the patches on original images in a regular size (64×64 pixels in our experiments) based on the contours. In Fig. 3a, there are many highly overlapping patches. For decreasing the redundant patches, we eliminate the patches which overlapping area ratio greater than 0.5. In Fig. 3b, the region in the red bounding boxes are initial patches for next step. In MIT Indoor-67 train/test dataset (80 training and 20 testing images per class), we discover patches for each scene independently and we select 20 patches per training image, for a total of $107,200$ patches.

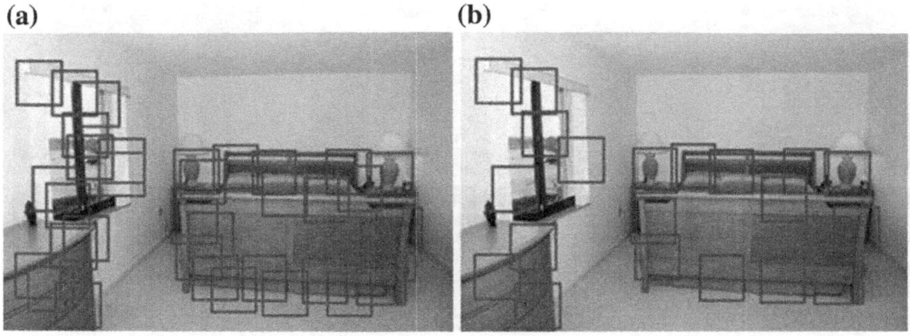

Fig. 3 An example of the overlapping patches in image and eliminate the overlapping image

3 Clustering and Ranking the Patches

In this part, we use HOG features to descript the patches, than run a unsupervised cluster algorithm on HOG space. At last, the clusters were ranked by a evaluation criteria.

3.1 Feature Representation and Clustering

For all selected patches, we compute HOG descriptors [16]. HOG features are based on small spatial regions located in the images. Each 8×8 pixels region, which is called a cell, accumulates a local histogram of gradient directions over the pixels. And each 2×2 cells region, which is called a block, accumulates a measure of local histogram energy over larger spatial regions and uses the results to normalize all of the cells in the block. Firstly, each pixel calculates a weighted vote for an orientation histogram channel based on the orientation of the gradient element centred on it, and the votes are accumulated into orientation bins over cells (i.e., local spatial regions). The orientation bins are evenly spaced over $0°-180°$ to generate 9 contrast insensitive orientations, and $0°-360°$ to generate 18 contrast sensitive orientations. To reduce aliasing, votes are interpolated bilinearly between the neighboring bin centres in both orientation and position. Then, we compute 4 normalization factors for a cell in a measure of its local histogram energy. A HOG feature is originally a 36-dimensional vector. In this procedure, we use an alternative 31-dimensional HOG feature to descript the patches. In Fig. 4, the top row is an example of mid-level patches, the below is their corresponding HOG features. For each scene, we run standard k-means clustering in HOG space. Then, we use patches within the cluster as positive examples and all patches of other clusters as negative examples to train a linear SVM classifier for each cluster.

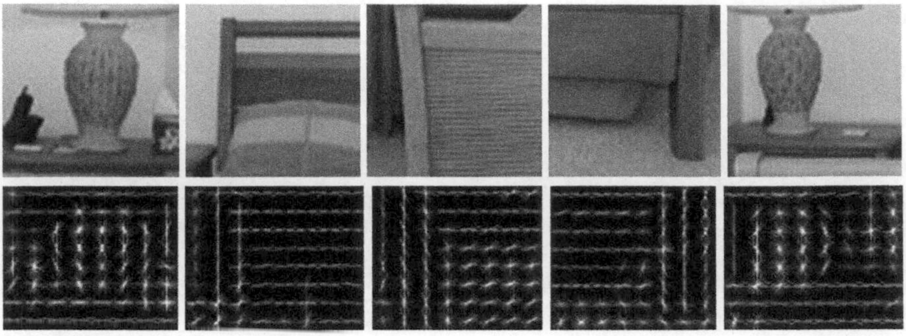

Fig. 4 An example of selected patches and the corresponding HOG features

3.2 *Ranking the Patches*

Our algorithm produces a lot of clusters for each scene. The next task is to rank them, to find a small number of the most discriminative and representative clusters. We utilize the evaluation criteria which proposed in [15]. The evaluation criteria consists of two terms: purity and discriminativeness. A good cluster should have all its member patches come from the same visual concept. We approximate the purity of each cluster in terms of the classifier confidence of the cluster members. Thus, the purity score for a cluster is computed by summing up the SVM detection scores of top r cluster members. For discriminativeness, we can say is that a highly discriminative patch should appear rarely in other scenes. Therefore, the discriminativeness of a patch is defined as the ratio of the number of firings on positive examples to the number of firings on negative examples. All clusters are ranked using a linear combination of the above two scores. For each scene, we select the top 3 clusters to represent their nature. To perform classification, the selected patches in the top 3 clusters of each scene are aggregated into a spatial pyramid [17] using max-pooling over the discriminative patch scores as in [3] use a linear SVM in a one-vs-all classification.

4 Experiment and Analysis

We tested our method on a widely used scene image classification dataset, MIT Indoor-67 dataset. On this dataset, we followed the original splits in [18], which used around 80 training images and 20 testing images for each category.

As shown in Table 1, our method achieved the second highest accuracy. Note that Visual elements utilized numerous patches extracted at scales ranging from 80×80 to the full image size, and the patches were represented by HOG descriptors plus a 8×8 color image in $L * a * b$ space. In our algorithm, we just use HOG features to descript the paths. The HOG features, heavily engineered for both accuracy and speed, are not without issues or limits. They are gradient-based and lack the ability

Table 1 Average classification on MIT Indoor-67 dataset

Method	Accuracy (%)	Method	Accuracy (%)
ROI + Gist [18]	26.05	miSVM [11]	46.40
DPM [19]	30.40	MMDL [5]	50.15
Object Bank [3]	37.60	Blocks that shout [4]	60.77
RBoW [20]	37.93	Visual element [21]	64.03
D-Patches [15]	38.10	Ours	62.30

to represent other features. But an image is not just represented by gradient, it also have other characteristic, for example the color features. So it is suitable for most scenes, not for all.

5 Conclusions

In this paper, we presented a pre-processing method before selecting the patches. Our goal is decrease the number of patches in the selecting step, and reserve the patches which are contain abundant features as far as possible. We utilize the Canny edge detector to detect the contour of training images, than we sample the patches in the region of the contour. Then, we run a unsupervised cluster algorithm on HOG space. At last, the clusters were ranked by a evaluation criteria. Our results shows that the accuracy have much better than most of the method. However, there is still much room for improvement in feature representation. For instance, images were utilizing the multi-feature fusion to represent images.

References

1. Sivic J, Zisserman A (2003) Video google: a text retrieval approach to object matching in videos. In: ICCV
2. Lowe DG (1999) Object recognition from local scale-invariant features. In: Proceedings of ICCV
3. Li L, Su H, Xing E, Fei-Fei L (2010) Object bank: A high-level image representation for scene classification and semantic feature sparsification. In: NIPS
4. Juneja M, Vedaldi A, Jawahar CV, Zisserman A (2013) Blocks that shout: distinctive parts for scene classification. In: CVPR
5. Wang X, Wang B, Bai X, Liu W, Tu Z (2013) Max-margin multiple-instance dictionary learning. In: ICML
6. Fernando B, Fromont E, Tuytelaars T (2014) Mining mid-level features for image classification. Int J Comput Vis
7. Endres I, Shih K, Jiaa J, Hoiem D (2012) Learning collections of part models for object recognition. In: CVPR
8. Jain A, Gupta A, Rodriguez M, Davis L (2013) Representing videos using mid-level discriminative patches. In: CVPR

9. Hu J, Kong Y (2013) Activity recognition by learning structural and pairwise mid-level features using random forest. In: IEEE international conference automatic and face gesture recognition
10. Doersch C, Singh S, Gupta A, Sivic J, Efros AA (2012) What makes Paris look like Paris? In: SIGGRAPH
11. Li Q, Wu J, Tu Z (2013) Harvesting mid-level visual concepts from large-scale internet images. In: CVPR
12. Malisiewicz T, Gupta A, Efros AA (2011) Ensemble of exemplar svms for object detection and beyond. In: Proceedings of ICCV
13. Lim J, Lawrence Zitnick C, Dollar P (2013) Sketch tokens: a learned mid-level representation for contour and object detection. In: CVPR
14. Singh S, Gupta A, Efros AA (2012) Unsupervised discovery of mid-level discriminative patches. In: ECCV
15. Canny J (1986) A computational approach to edge detection, IEEE Trans PAMI 8(6):679–698
16. Dalal N, Triggs B (2005) Histograms of oriented gradients for human detection. In: CVPR
17. Lazebnik S, Schmid C, Ponce J (2006) Beyond bags of features: spatial pyramid matching for recognizing natural scene categories. In: CVPR
18. Quattoni A, Torralba A (2009) Recognizing indoor scenes. In: CVPR
19. Pandey M, Lazebnik S (2011) Scene recognition and weakly supervised object localization with deformable part-based models. In: ICCV
20. Parizi SN, Oberlin JG, Felzenszwalb PF (2012) Reconfigurable models for scene recognition. In: CVPR
21. Doersch C, Gupta A, Efros AA (2013) Mid-level visual element discovery as discriminative mode seeking. In: NIPS

Object Detection Based on Improved Exemplar SVMs Using a Generic Object Measure

Hao Chen, Shanshan Zhang, Jinfu Yang and Qiang Zhang

Abstract Recent years most object detection method tends to use the sliding window fashion, which need to search the whole images entirely, causing the extreme waste of search resources. In this paper we propose an object detection method based on improved Exemplar SVMs (IESVM). Our method mainly includes two steps: (1) Coarse object detection: we use a generic object measure to find a region that may contain objects. In this step we do not care about the category of the objects. (2) Precise object detection: we extract the regions created in last step, in where we use Exemplar SVMs to finish the mission of detection. It is proved that the our method can reduce the search space and improve the accuracy of detection. We evaluate our IESVM method on the PASCAL VOC 2007 dataset and find that our method achieves good results in the PASCAL object detection challenges.

Keywords Object detection · IESVM · Generic object measure · Exemplar

1 Introduction

Over the past decade, object detection has been one of the most fundamental challenges in computer vision. One of the early advancements in statistical object detection came back in 2005 when Dalal and Triggs [1] introduced histograms-of-gradient (HOG) descriptor to represent object templates and coupled it with SVM. Conse-

H. Chen (✉) · S. Zhang · J. Yang · Q. Zhang
Department of Control and Engineering, Beijing University of Technology,
Chaoyang District, Beijing 100124, China
e-mail: chenhao19890807@126.com

S. Zhang
e-mail: 814601344@qq.com

J. Yang
e-mail: jfyang@bjut.edu.cn

Q. Zhang
e-mail: 903155664@qq.com

© Springer International Publishing Switzerland 2017

V.E. Balas et al. (eds.), *Information Technology and Intelligent Transportation Systems*, Advances in Intelligent Systems and Computing 454,
DOI 10.1007/978-3-319-38789-5_33

quently, much subsequent work focused on exploiting the HOG+SVM strategy, in conjunction with exhaustive sliding window search. The most successful have been deformable parts-based models (DPM) [2]. DPM extended these HOG-based templates by adding part templates and allowing deformation between them. The emergence of DPM, and improvements in algorithms to train it, have led to a brisk increase in performance on the PASCAL VOC object detection challenge. Later, numerous works focused on improving the parts themselves, from using strongly supervised parts [3] to using weak 3D supervision [4]. An alternate direction for improvement in performance was to incorporate bottom-up segmentation priors for training DPMs [5].

However, these approaches have a fundamental limitation: given the complexity of exhaustive search, they can only utilize simple features. As a consequence, a major shift in detection paradigm was to bypass the need to exhaustive search completely by generating category-independent candidates for object location and scale [6]. Commonly-used methods propose around 1,000 regions using fast segmentation algorithms, which aim to discard many windows which are unlikely to contain objects [7]. These object proposal methods have resulted in the use of more sophisticated features and learning algorithms.

In recent years sliding-window fashion [8] has been a common approach in object detection research area. While this kind of method need to search the whole images entirely every time which may cause the extreme waste of resources and low efficiency of detection algorithms. In this paper we first use a measure of generic object over classes. It quantifies how likely it is for an image window to cover an object of any class.

We argue that any object has one of three characteristics as follows: a well-defined closed boundary in space; a different appearance from their surroundings [9]; sometimes it is unique within the image [10]. Many objects have several of these characteristics at the same time. We design an generic object measure and explicitly train it to distinguish windows containing an object from background windows. This measure combines in a Bayesian framework several image cues based on the above characteristics. We also show that the combined cues performs better than any cue alone. We demonstrate an algorithm to greatly reduce the number of evaluated windows with only minor loss in detection performance. By this kind of idea, we proposed improved Exemplar SVM based on generic object measure, aiming to reduce the search space and raise the detection accuracy as in Fig. 1 shows.

2 Coarse Object Detection

Recently, several techniques have been proposed to reduce the cost of sliding window algorithms. One approach is to reduce the cost of evaluating a complex classifier on a window. In this step we use a generic object measure, quantifying how likely it is for an image window to contain an object of any class. We explicitly train it to distinguish

Fig. 1 An example of the outputs of the two steps

objects with a well-defined boundary in space, such as cows and telephones, from amorphous background elements, such as grass and road. The measure combines in a Bayesian framework several image cues measuring characteristics of objects, such as appearing different from their surroundings and having a closed boundary.

Objects in an image are characterized by a closed boundary in 3D space, a different appearance from their immediate surrounding and sometimes by uniqueness. These characteristics suggested the four image cues we use in the generic object measure: Multiscale Saliency (MS), Color Contrast (CC), Edge Density (ED), Superpixels Straddling (SS). MS is a global saliency measure based on the spectral residual of the FFT, which favors regions with an unique appearance within the entire image. The CC cue is a local measure of the dissimilarity of a window to its immediate surrounding area. The ED cue measures the density of edges near the window borders. The SS is a different way to capture the closed boundary characteristic of objects rests on using superpixels [11] as features.

We learn the parameters of the generic object cues from a training dataset which consists of 50 images randomly sampled from several well known datasets, including, Pascal VOC 2006 and Caltech 101. These images contain a total of 300 instances of 50 diverse classes including a variety of objects such as animals, buildings, persons and so on. For training, we use annotated object windows, but not their class labels because our aim is to learn a measure generic over classes. There are $8(3 + 1 * 5)$ parameters to be learned because there is 5 scales. The first three are learned in a Bayesian framework, while the last one MS is trained independently by optimizing the localization accuracy of the training object windows at each scale using the efficient technique of Neubeck and Van Gool [12]. Rather than only a single cue can perform very well, instead, CC provides more accurate windows, but sometimes misses objects entirely. ED provides many false positives on textured areas. SS is very distinctive but depends on good superpixels, which are fragile for small objects. So we need to combine the four cues. To achieve this goal, we train a Bayesian classifier

Fig. 2 Some examples of output after coarse object detection

to distinguish between the positive and negative. For each training image, we sample 100,000 windows from the distribution given by the MS cue, and then compute the other cues for them. Windows covering an annotated object are considered as positive examples, all others are considered as negative. After training parameters and combining the cues, we can sample any desired number of windows from computing the cues for them when a test image is given. The posterior probability of a test window is shown below in Fig. 2.

3 Precise Object Detection

After obtaining these regions as shown in Fig. 2 in Sect. 3.1, we use Exemplar SVM (ESVM) to detect just in these regions more precisely. Our main thought can be describe simply as follows: non-parametric when representing the positives, but parametric when representing the negatives. It is the key point of ESVM approach. Our approach can propagate rich annotations from exemplars onto detection windows, with discriminative training, which allows us to learn powerful exemplar-based classifiers from vast amounts of positive and negative data.

Our object detector is based on a very simple idea: to learn a separate classifier for each exemplar in the dataset. We represent each exemplar using a HOG template. Since we use a linear SVM, each classifier can be interpreted as a learned exemplar-specific HOG weight vector. As a result, we have a large collection of simpler individual Exemplar SVM detectors of various shapes and sizes, each of which is highly tuned to the exemplars appearance, instead of a single complex category detector. We perform well on popular dataset such as the PASCAL VOC 2007.

As Fig. 3 shown that given a set of training exemplars, we represent each exemplar E via a rigid HOG template XE. We create a descriptor from the ground truth bounding box of each exemplar with a cell size of 8 pixels using a sizing heuristic which attempts to represent each exemplar with roughly 100 cells. We let each exemplar

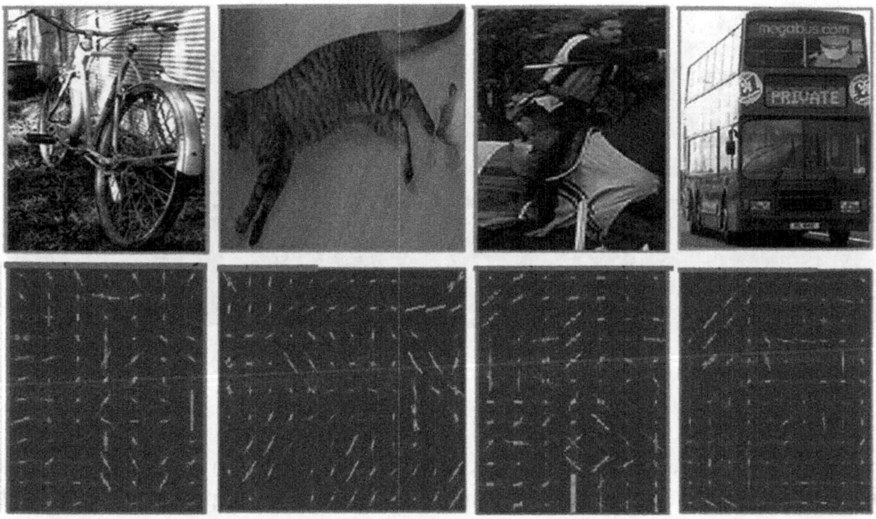

Fig. 3 Some examples of exemplars and the corresponding HOG features

define its own HOG dimensions respecting the aspect ratio of its bounding box. We create negative samples of the same dimensions as XE by extracting negative windows NE, from images not containing any objects from the exemplars category.

Each Exemplar-SVM (WE, bE), tries to separate XE from all windows in NE by the largest possible margin in the HOG feature space. Learning the weight vector WE to optimizing the following convex objective:

$$\Omega_{\mathrm{E}}(\omega, b) = \|\omega\|^2 + C_1 h\left(\omega^T x_{\mathrm{E}} + b\right) + C_2 \sum_{x \in N_{\mathrm{E}}} h\left(-\omega^T x - b\right) \quad (1)$$

Using the procedure above, we train an ensemble of Exemplar SVM. However, due to the independent training procedure, their outputs are not necessarily compatible. In our case, since each exemplar-SVM is supposed to fire only on visually similar examples, we cannot say for sure which of the held-out samples should be considered as positives a priori. Fortunately, what we can be sure is that the classifier should not fire on negative windows. Therefore, we let each exemplar select its own positives and then use the SVM output scores on these positives, in addition to lots of held-out negatives, to calibrate the Exemplar SVM.

To obtain each exemplars calibration positives, we run the Exemplar SVM on the validation set, create a set of non-redundant detections using non maximum suppression, and compute the overlap score between resulting detections and ground-truth bounding boxes. We treat all detections which overlap by more than 0.5 with ground-truth boxes as positives. All detections with an overlap lower than 0.2 are treated as negatives, and we fit a logistic function to these scores.

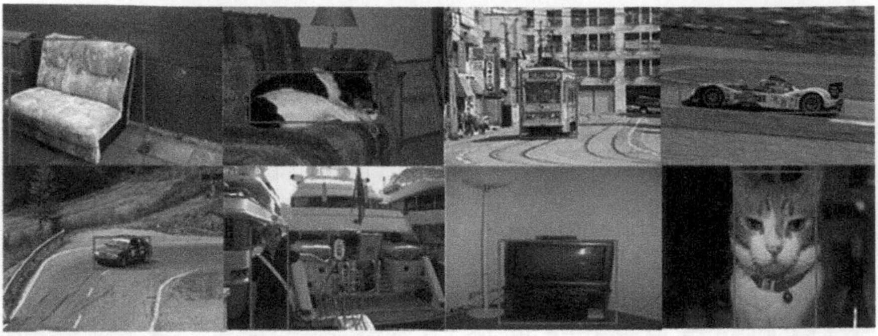

Fig. 4 Some examples of output after precise object detection

After calibration, we can create detection windows for each exemplar in a sliding window fashion. A common and simple mechanism for suppressing redundant responses is non maximum suppression (NMS); however, using NMS directly on exemplars means that multiple exemplars will be competing for detections windows. Instead of just using the raw association score, we propose to augment each detection with an exemplar context score which uses the identities and scores of nearby detections to boost the raw detection score. For each detection we generate a context feature similar to [2] which pools in association scores of nearby and overlapping detections and generates the final detection score by a weighted sum of the local association score and the context score.

At test time, each Exemplar SVM creates detection windows in a sliding window fashion, but instead of using a standard non maxima suppression we use an exemplar context information for suppressing redundant responses. For each detection we generate a context feature similar to [13] which pools in the SVM scores of nearby or overlapping detections and generates the final detection score by a weighted sum of the local SVM score and the context score. Once we obtain the final detection score, we use standard non maximum suppression to create a final, sparse set of detections per image. We choose window corresponding the highest score for the final output as can be seen in Fig. 4.

4 Experiment and Analysis

We test our method on the 20 category PASCAL VOC 2007 dataset. Table 1 shows associated with each detection. Following the protocol of the VOC Challenge, we evaluate our system on a per-category basis on the test set, consisting of 4,952 images. These results have been summarized in Table 1 as Average Precision per class. As we can see that our method IESVM has a better performance in most of the categories.

Table 2 clearly indicates that our method is much better than standard Nearest Neighbor (NN) and a distance function formulation [14]. Not surprisingly, IESVM

Table 1 IESVM object detection results on PASCAL VOC 2007

VOC2007	Areo	Bike	Bird	Boat	Bottle	Bus	Car	Cat	Chair	Cow
ESVM	0.208	0.480	0.077	0.143	0.131	0.397	0.411	0.052	0.116	0.186
IESVM	0.384	0.576	0.207	0.102	0.183	0.581	0.587	0.284	0.190	0.331
DPM	0.332	0.603	0.102	0.161	0.273	0.543	0.582	0.230	0.200	0.241
VOC2007	Table	Dog	Horse	Mbike	Person	Plant	Sheep	Sofa	Train	Tv
ESVM	0.111	0.031	0.447	0.394	0.169	0.112	0.226	0.170	0.369	0.300
IESVM	0.203	0.248	0.563	0.529	0.480	0.128	0.422	0.344	0.495	0.390
DPM	0.267	0.127	0.581	0.482	0.432	0.120	0.211	0.361	0.460	0.435

Table 2 mAP compared with other methods

Method	NN	DFUN	ESVM	DT	LDPM	DPM	IESVM
mAP	0.039	0.155	0.227	0.097	0.266	0.337	0.361

Fig. 5 PASCAL VOC results for varying negative set sizes

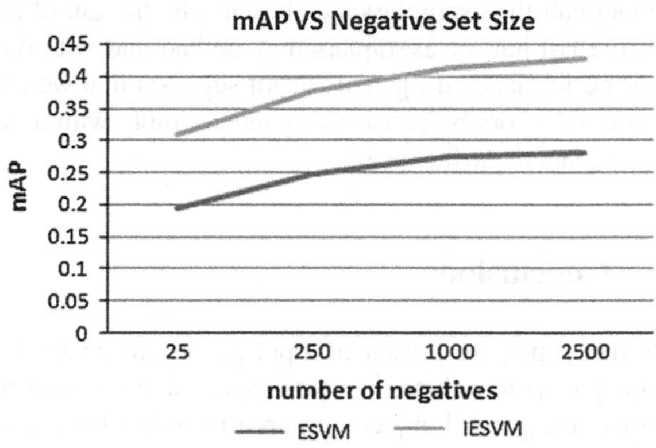

has improved the performance of its original method ESVM, because the reducing the search space and promoting the search accuracy. Even our method edges out the DPM, which has been one of the fundamental method in the area of object detection.

Figure 5 shows the summarized results for the three methods: ESVM and IESVM. We compute the performance of the Exemplar SVM algorithm as a mAP over 6 PASCAL categories: bus, cow, dining table, motorbike, sheep, and train. The x-axis indicates the number of negative images used and the y-axis is the PASCAL VOC 2007 resulting mAP score for each of the 2 methods. We note that increasing the negative set size from 25 to 250 negatives images gives us a much higher boost than increasing it from 1000 to the full 2500 images. This suggests that once each exemplar has seen a sufficient number of negatives, its performance will not significantly improve with more negatives.

The PASCAL VOC mAP versus number of exemplars plots can be seen in Fig. 6. As can be seen from this curve, the performance saturates for these categories when

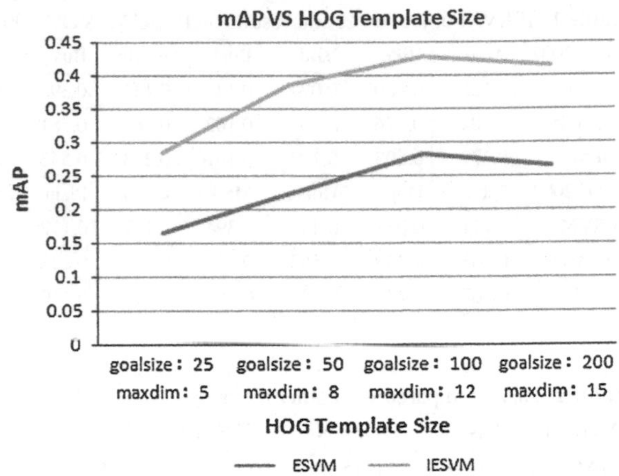

Fig. 6 PASCAL VOC results for varying HOG template sizes

about half the exemplars are chosen. The flat tail of each curve suggests that many of the last half of exemplars may be bad and with the its number added, it makes our performance drop. This result suggests that we can get much faster object category detectors by constructing an ensemble with no more than half the number of exemplars in each category.

5 Conclusions

In this paper, we presented a pre-processing method IESVM for object detection. Our goal is to reduce the search space and improve the accuracy of detection. We utilize the general object measure to conduct the coarse object detection, which we can obtain a region that probably contains the object(regardless of categories), that is we extract the object from the background roughly. then we use the ESVM to finish the whole detection process. We evaluate our IESVM method on the PASCAL VOC 2007 dataset and find that our method achieves good results in the PASCAL object detection challenges. However, there is still much room for improvement in our method. For instance, feature representation for the images such as HOG may need to be improved.

Acknowledgments This work is supported by the National Natural Science Foundation of China under grant Nos. 61201362 and 61273282, and the Beijing Natural Science Foundation under grant No. 7132021, the Scientific Research Project of Beijing Educational Committee under Grant no. KM201410005005.

References

1. Dalal N, Triggs B (2005) Histograms of oriented gradients for human detection. In: CVPR
2. Felzenszwalb P, Girshick R, McAllester D, Ramanan D (2010) Object detection with discriminatively trained partbased models. PAMI 32(9)
3. Bourdev L, Malik J (2009) Poselets: Body part detectors trained using 3d human pose annotations. In: ICCV
4. Shrivastava A, Gupta A (2013) Building parts-based object detectors via 3d geometry. In: ICCV
5. Azizpour H, Laptev I (2012) Object detection using stronglysupervised deformable part models. In: ECCV
6. Alexe B, Deselaers T, Ferrari V (2012) Measuring the objectness of image windows. TPAMI
7. Wang X, Yang M, Zhu S, Lin Y (2013) Regionlets for generic object detection. In: ICCV
8. Arbelaez P, Pont-Tuset J, Barron JT, Marques F, Malik J (2014) Multiscale combinatorial grouping. In: CVPR
9. Fidler S, Mottaghi R, Yuille A, Urtasun R (2013) Bottom-up segmentation for top-down detection. In: CVPR
10. Ahmed E, Shakhnarovich G, Maji S (2014) Knowing a good HOG filter when you see it: efficient selection of filters for detection. In: ECCV
11. Felzenszwalb PF, Huttenlocher DP (2004) Efficient graph-based image segmentation. IJCV, 59(2):167C181
12. Neubeck A, Van Gool L (2006) Efficient non-maximum suppression. In: Proceedings of the 18th international conference pattern recognition
13. Bourdev L, Maji S, Brox T, Malik J (2010) Detecting people using mutually consistent poselet activations. ECCV
14. Malisiewicz T, Efros AA (2008) Recognition by association via learning per-exemplar distances. CVPR

Euler–Maclaurin Expansions and Quadrature Formulas of Hyper-singular Integrals with an Interval Variable

Chong Chen, Jin Huang and Yanying Ma

Abstract In this paper, we present a kind of quadrature rules for evaluating hyper-singular integrals $\int_a^b g(x)q^{\alpha}(x,t)dx$ and $\int_a^b g(x)q^{\alpha}(x,t)\ln|x-t|dx$, where $q(x,t) = |x-t|$ (or $x-t$), $t \in (a,b)$ and $\alpha \leq -1$(or $\alpha < -1$). Since the derivatives of density function $g(x)$ in the quadrature formulas can be eliminated by means of the extrapolation method, the formulas can be easily applied to solving corresponding hyper-singular boundary integral equations. The reliability and efficiency of the proposed formulas in this paper are demonstrated by some numerical examples.

Keywords Hyper-singular integral · Euler–Maclaurin expansions · Quadrature formulas · Hadamard finite part

MSC 45E99 · 65D30 · 65D32 · 41A55

1 Introduction

In recent years, many scientific and engineering problems, such as fracture mechanics, fluid dynamic, and elasticity [1, 2], have attracted attention to boundary integral equations with hyper-singular kernels. To evaluate the hyper-singular integrals [3–10] efficiently is a major problem to solve the hyper-singular integral equations [2, 3, 10–13].

C. Chen (✉) · J. Huang · Y. Ma
School of Mathematical Sciences, University of Electronic Science and Technology of China, Chengdu 611731, China
e-mail: chenchong7890@163.com; ccyaoqing@163.com

J. Huang
e-mail: huangjin12345@163.com

© Springer International Publishing Switzerland 2017

V.E. Balas et al. (eds.), *Information Technology and Intelligent Transportation Systems*, Advances in Intelligent Systems and Computing 454,
DOI 10.1007/978-3-319-38789-5_34

We consider the hyper-singular integral with an interval variable $t \in [a, b]$,

$$(I(g))(t) = \int_a^b g(x)q^\alpha(x,t)dx, \tag{1.1}$$

$$(I(g))(t) = \int_a^b g(x)q^\alpha(x,t)\ln|x-t|dx, \tag{1.2}$$

where $q(x, t) = |x - t|$ (or $x - t$), and $\alpha \le -1$ (or $\alpha < -1$) is a real number. (1.1) denotes the Hadamard finite part [9, 13] of the hyper-singular integral.

The paper is organized as follows: in Sect. 2, we recall the general definition of Hadamard finite-part integrals; in Sect. 3, we present quadrature rules and the Euler–Maclaurin expansions for hyper-singular integrals (1.1) and (1.2); in Sect. 4, some examples are tested. Conclusions are drain in Sect. 5.

2 Definition of Hadamard Finite-Part Integrals

We will recall the definition of Hadamard finite-part integrals at first.

Definition 2.1 The Hadamard finite-part integral [9]: Let $f(x)$ be integrable on (ϵ, b) for any $\epsilon, \epsilon \in (0, b)$ and $b < \infty$. Suppose that there exists a sequence $\alpha_0 < \alpha_1 < \alpha_2 < \ldots$, and a non-negative integral Z such that the expansion

$$\int_\epsilon^b f(x)dx = \sum_{i=0}^{\infty}\sum_{j=0}^{Z} I_{i,j}\epsilon^{\alpha_i} ln^j(\epsilon), \tag{2.1}$$

converges for any $\epsilon \in (0, h)$ with $h > 0$. Then the Hadamard finite-part integral (2.1) is defined as :

$$f.p.\int_0^b f(x)dx = \begin{cases} 0 \ \text{ if } \alpha_k \neq 0 \text{ for all } k, \\ I_{k,0} \text{ if } \alpha_k = 0 \text{ for some } k. \end{cases} \tag{2.2}$$

Based on the Definition 2.1, we recall the following two Lemmas.

Lemma 2.2 ([9]) *For any $b > 0$, we have*

$$f.p.\int_0^b x^\alpha dx = \begin{cases} \ln b, \ for\ \alpha = -1, \\ b^{\alpha+1}/(\alpha+1), \ for\ \alpha \neq -1. \end{cases} \tag{2.3}$$

Lemma 2.3 ([9]) *Assume $\alpha < -1$ and $m > -\alpha - 2$ and $g(x) \in C^{m+1}[0, b)$. Then for any $b > 0$, we have*

$$f.p.\int_0^b g(x)x^\alpha dx = \int_0^b x^\alpha\left[g(x) - \sum_{k=0}^{m} g^k(0)x^k/k!\right]dx + \sum_{k=0}^{m}(g^k(0)/k!)f.p.\int_0^b x^{\alpha+k}dx. \tag{2.4}$$

3 Quadrature Formulas of Hyper-singular Integrals and Their Euler–Maclaurin Expansions

In this section, we major study one of the following integrals

$$I(G) = f.p. \int_a^b G(x,t)dx = f.p. \int_a^b q^{\alpha}(x,t)g(x)dx, \tag{3.1}$$

$$\begin{aligned} I_1(G) &= f.p. \int_a^b G_1(x,t)dx \\ &= f.p. \int_a^b |x-t|^{\alpha}(\ln|x-t|)g(x)dx, \alpha < -1, \end{aligned} \tag{3.2}$$

where $q(x,t) = |x-t|$ (or $x-t$) for $\alpha \le -1$(or $\alpha < -1$), and $g(x)$ is a smooth function on $[a,b]$. $G(x,t)$ and $G_1(x,t)$ are hyper-singular functions about interval variable t as $\alpha < -1$.

The interval $[a,b]$ is divided into n parts, and $h = (b-a)/n$. Let $x_j = a + jh$ $(j = 0, 1, \ldots, n)$ and the singular point t satisfies $t \in \{x_j : 1 \le j \le n-1\}$. In terms of results of Sect. 2 and conclusions of [9], we derive the formulas of integral (3.1) with $q(x,t) = |x-t|(or(x-t))$ in [3].

Lemma 3.1 ([3]) *Let $g(x) \in C^{2m+1}[a,b]$, and $G(x,t) = g(x)(x-t)^{\alpha}$ for $\alpha < -1$. Taking $t = x_i$, $1 \le i \le n-1$, then the following assertions hold.*
(i) When $\alpha = -2l-1$ for $l \in N^+$, the expansion is

$$\begin{aligned} Q(h) &= \sum_{j=0}^{n-1} hG\left(a + \left(j + \frac{1}{2}\right)h\right) \\ &= f.p. \int_a^b G(x,t)dx + \sum_{k=1}^{m+1} \frac{h^{2k}}{(2k)!} B_{2k}\left(\frac{1}{2}\right)[G^{(2k-1)}(b) - G^{(2k-1)}(a)] \\ &\quad + 2\sum_{k=0}^{\min(m,l-1)} \frac{h^{2(k-l)+1}}{(2k+1)!}\zeta\left(2(l-k), \frac{1}{2}\right) g^{(2k+1)}(t) + O(h^{2m+2+\alpha}). \end{aligned} \tag{3.3}$$

(ii) When $\alpha = -2l-1$ with $l = 0$, the form of rule can be written by

$$\begin{aligned} \sum_{j=0}^{n-1} hG\left(a + \left(j + \frac{1}{2}\right)h\right) &= \sum_{k=1}^{m+1} \frac{h^{2k}}{(2k)!} B_{2k}\left(\frac{1}{2}\right)[G^{(2k-1)}(b) - G^{(2k-1)}(a)] \\ &\quad + f.p. \int_a^b G(x,t)dx + O(h^{2m+2+\alpha}). \end{aligned} \tag{3.4}$$

(iii) If α is a negative even integer, the formula is given by

$$\begin{aligned} Q(h) &= \sum_{j=0}^{n-1} hG\left(a + \left(j + \frac{1}{2}\right)h\right) - 2\zeta\left(m+1, \frac{1}{2}\right) g(t) h^{-m} \\ &= f.p. \int_a^b G(x,t)dx + \sum_{k=0}^{m} \zeta\left(-2k-\alpha, \frac{1}{2}\right) \frac{2g^{(2k)}(t)}{(2k)!} h^{2k+\alpha+1} \\ &\quad - \sum_{k=1}^{m} \zeta\left(1-2k, \frac{1}{2}\right) \frac{G^{(2k-1)}(b) - G^{(2k-1)}(a)}{(2k-1)!} h^{2k} + O(h^{2m+\alpha+2}). \end{aligned} \tag{3.5}$$

When we study some boundary integral equations, we notice that, it is demand for calculating the integrals with logarithmic functions to solve the equations. Theorem 3.2 shows the Euler–Maclaurin expansions of (3.2).

Theorem 3.2 *Assume that function $g(x) \in C^{2m}[a,b]$, and $G(x,t) = g(x)|x-t|^{\alpha} \ln |x-t|$, where $\alpha < -1$ and $t \in \{x_j : 1 \le j \le n-1\}$, $l = 1, 2, \ldots$.*
(i) If α is a non-integer, then we have the formulas

$$\begin{aligned} &\sum_{j=0}^{n-1} hG\left(a + \left(j + \frac{1}{2}\right)h\right) - \sum_{k=0}^{l-1} 2\frac{h^{2k+1+\alpha}}{(2k)!}\left[\zeta\left(-2k-\alpha, \frac{1}{2}\right)\ln h \right. \\ &\quad \left. - \zeta'\left(-2k-\alpha, \frac{1}{2}\right)\right] g^{(2k)}(t) \\ &= f.p. \int_a^b G(x,t)dx + \sum_{k=1}^{m} \frac{h^{2k}}{(2k-1)!}\zeta\left(-2k+1, \frac{1}{2}\right)[G^{(2k-1)}(a) \\ &\quad - G^{(2k-1)}(b)] + \sum_{k=l}^{m} 2\frac{h^{2k+1+\alpha}}{(2k)!}\left[\zeta\left(-2k-\alpha, \frac{1}{2}\right)\ln h \right. \\ &\quad \left. - \zeta'\left(-2k-\alpha, \frac{1}{2}\right)\right] g^{(2k)}(t) + O(h^{2m+2+\alpha}). \end{aligned} \tag{3.6}$$

(ii) If α is a negative even integer, then rule has the form of

$$\begin{aligned} &\sum_{j=0}^{n-1} hG\left(a + \left(j + \frac{1}{2}\right)h\right) + 2g(t)h^{\alpha+1}\left(\zeta'\left(-\alpha, \frac{1}{2}\right) - \zeta\left(-\alpha, \frac{1}{2}\right)\ln(h)\right) \\ &= f.p. \int_a^b G(x,t)dx + \sum_{k=1}^{m} \frac{G^{(2k-1)}(a) - G^{(2k-1)}(b)}{(2k-1)!}\zeta\left(1-2k, \frac{1}{2}\right)h^{2k} \\ &\quad + \sum_{k=l}^{m} \frac{2g^{(2k)}(t)}{(2k)!}\left(\zeta\left(-2k-\alpha, \frac{1}{2}\right)\ln(h) - \zeta'\left(-2k-\alpha, \frac{1}{2}\right)\right)h^{2k+\alpha+1} + O(h^{2m+2+\alpha}) \end{aligned} \tag{3.7}$$

Proof Take $t = x_i$, then t is point of division of the interval (a, b). The integral of (3.1) can be decomposed into two parts

$$\begin{aligned} f.p.\int_a^b G(x,t)dx &= f.p.\int_a^b g(x)|x-t|^\alpha dx \\ &= f.p.\int_a^t g(x)(t-x)^\alpha dx + f.p.\int_t^b g(x)(x-t)^\alpha dx. \end{aligned}$$

We consider the two items of the theorem. According to the conclusions of Lemma 2.1 and Lemma 2.3. We have

$$\begin{aligned} \sum_{j=0}^{i-1} hG\left(a+\left(j+\frac{1}{2}\right)h\right) &= f.p.\int_a^t G(x,t)dx \\ &+ \sum_{k=1}^{m} \frac{h^{2k}}{(2k-1)!}\zeta\left(-2k+1,\frac{1}{2}\right)G^{(2k-1)}(a) \\ &+ \sum_{k=0}^{2m+1} \frac{(-1)^k h^{k+1+\alpha}}{k!}\zeta\left(-k-\alpha,\frac{1}{2}\right)g^{(k)}(t) + O(h^{2m+2+\alpha}), \end{aligned} \tag{3.8}$$

$$\begin{aligned} \sum_{j=i}^{n-1} hG\left(a+\left(j+\frac{1}{2}\right)h\right) &= f.p.\int_t^b G(x,t)dx \\ &- \sum_{k=1}^{m+1} \frac{h^{2k}}{(2k-1)!}\zeta\left(-2k+1,\frac{1}{2}\right)G^{(2k-1)}(b) \\ &+ \sum_{k=0}^{2m+1} \frac{h^{k+1+\alpha}}{k!}\zeta\left(-k-\alpha,\frac{1}{2}\right)g^{(k)}(t) + O(h^{2m+2+\alpha}). \end{aligned} \tag{3.9}$$

Combining (3.8) with (3.9), we obtain

$$\begin{aligned} \sum_{j=0}^{n-1} hG\left(a+\left(j+\frac{1}{2}\right)h\right) &= f.p.\int_a^b G(x,t)dx \\ &+ \sum_{k=1}^{m+1} \frac{h^{2k}}{(2k-1)!}\zeta\left(-2k+1,\frac{1}{2}\right)\left[G^{(2k-1)}(a)-G^{(2k-1)}(b)\right] \\ &+ \sum_{k=0}^{m} 2\frac{h^{2k+1+\alpha}}{(2k)!}\zeta\left(-2k-\alpha,\frac{1}{2}\right)g^{(2k)}(t) + O(h^{2m+2+\alpha}). \end{aligned} \tag{3.10}$$

In virtue of α is a negative non-integer, the rule of (3.10) is analysis about α. The conclusion of (i) hold by derivative of (3.10) about parameters α.

The result of (ii) can be obtained by derivative of (3.5) of the Lemma 3.1 about variable α. □

Remark (i) On the bases of the condition that the rules of [3], we can obtain the corresponding quadrature rules of the hyper-singular integral (3.2) as the parameter α of (3.2) changes, by derivative of the rules about parameters α.

(ii) Considering the formulas of Lemma 3.1 and Theorem 3.2, we can obtain a better numerical results by the Richardson extrapolation or Romberg extrapolation method.

(iii) If $G(x)$ is a periodic function, the term $[G^{(2k-1)}(b) - G^{(2k-1)}(a)]$ of error expansions will vanish. Quadrature rules with higher orders accuracy can be achieved by Romberg extrapolation. Furthermore, if $G(x, t)$ is not a periodic function, we can periodizate the integrand by sin^p-transformation [14, 15] to improve the convergence order.

Now, taking Example 4.2 for example, we obtain the quadrature formula and its error expansion. The quadrature rule is $Q(h) = \sum_{j=0}^{n-1} hG(a + (j + 1/2)h)$. Then the kth extrapolation is given by

$$\begin{cases} \bar{Q}^{(0)}(h) = 2Q(h) - Q(\frac{h}{2}), \\ \bar{Q}^{(k)}(h) = [2^{2k}\bar{Q}^{(k-1)}(\frac{h}{2}) - \bar{Q}^{(k-1)}(h)]/(2^{2k} - 1), 1 \le k \le m + 1. \end{cases} \tag{3.11}$$

$\bar{Q}^{(k)}(h)$ has a high convergence order of $O(h^{2(k+1)})$, where $k = 0, 1, \ldots, m + 1$.

4 Numerical Experiments

In this section, numerical experiments illustrate the features of the quadrature rules and verify the theoretical conclusions proposed in this paper. Let $h = \frac{b-a}{n}$ be the step length, and n is the number of nodes. $I(t)$ is the Hadamard finite-part integral of the hyper-singular integral $(I(g))(t)$. $Q^{(i)}(h)$ $(i = 0, 1, 2)$ is the quadrature formula of ith extrapolation. $h^{2k} - ex$ $(k = 0, 1, 2, 3)$ and $|\overline{Q}^{(k-1)}(h) - I|$ $(k = 1, 2)$ are absolute errors of kth extrapolation, and $r_{2n}^{(k)} = \frac{h^{2k}-ex}{(\frac{h}{2})^{2k}-ex}$ (or $\frac{|\overline{Q}^{(k-1)}(h)-I|}{|\overline{Q}^{(k-1)}(\frac{h}{2})-I|}$).

Example 4.1 Calculate the following hyper-singular integral

$$(I(g))(t) = \int_0^1 \frac{g(x)}{(x-t)^2}dx, t \in (0, 1), \tag{4.1}$$

with $g(x) = x^6$ and the exact value of this finite-part integra is

$$(I(g))(t) = \frac{6}{5} + \frac{3t}{2} + 2t^2 + 3t^3 + 6t^4 - \frac{1}{t-1} + 6t^5 ln\frac{1-t}{t}.$$

We use the formulas of (3.5) and corresponding extrapolation of the quadrature formulas to evaluate the hyper-singular integral, and obtain the numerical results in following three Tables 1, 2, and 3. And the quadrature formulas have second order accuracy.

The numerical results the three tables display the fact that $r_n^{(k)} \approx 2^{2k+2}$, $(k = 0, 1, 2)$. The rate $\log_2(r_n^k)$ shows that quadrature rule is $O(h^2)$, and it's kth extrapolation formulas have fourth and sixth order accuracy as $k = 1$ and $k = 2$ respectively.

Example 4.2 ([8]) Calculate the hyper-singular integral

$$(I(g))(t) = \int_0^1 \frac{g(x)}{(x-t)^3}dx, \; g(x) = x^5 + 1, \tag{4.2}$$

where the exact solution is

$$(I(g))(t) = 10t^2 + 5t + \frac{10}{3} + \frac{5t+4}{2t^2} + \frac{t-3}{2t^2(t-1)^2} + 10t^3 \ln \frac{1-t}{t}.$$

We get the numerical results by using the rules (3.3) and (3.11) at $t = 1/4, 7/8, 1/32$ with $\alpha = -3$. On the basis of (3.11), we have the numerical results listed in Tables 4, 5, and 6 for (4.2) by the Romberg extrapolation.

Clearly, the numerical results in Tables 4, 5, and 6 imply that $r_n^{(k)} \approx 2^{2k}$, $(k = 1, 2)$, which meet the formula (3.25) of [3]. The convergence orders of rules $\overline{Q}^{(0)}(h)$ and $\overline{Q}^{(1)}(h)$ are $O(h^2)$ and $O(h^4)$, respectively.

Table 1 The numerical results with $t = 1/8$, $I(t) = 0.283573383$

$ex\backslash e\backslash n$	2^3	2^4	2^5	2^6	2^7	2^8
$Q^{(0)}(h)$	0.280422	0.282784	0.283376	0.283524	0.283561	0.283570
$h^0 - ex$	3.1509e-3	7.8913e-4	1.9737e-4	4.9348e-5	1.2337e-5	3.0844e-6
$r_n^{(0)}$	$2^{1.9974}$	$2^{1.9994}$	$2^{1.9998}$	$2^{2.0000}$	$2^{2.0000}$	$2^{2.0000}$
$Q^{(1)}(h)$		0.283571	0.283573	0.283573	0.283573	0.283573
$h^2 - ex$		1.8751e-6	1.1804e-7	7.3869e-9	4.164e-10	2.8859e-11
$r_n^{(1)}$		$2^{3.9897}$	$2^{3.9981}$	$2^{3.9998}$	$2^{4.0000}$	$2^{3.9999}$
$Q^{(2)}(h)$			0.283573	0.283573	0.283573	0.283573
$h^4 - ex$			8.9780e-10	1.0193e-11	6.3838e-14	2.2204e-16
$r_n^{(2)}$			$2^{6.4607}$	$2^{7.3190}$	$2^{8.1674}$	

Table 2 The numerical results with $t = 1/2 + 1/8$, $I(t) = 1.607735786$

$ex\backslash e\backslash n$	2^3	2^4	2^5	2^6	2^7	2^8
$Q^{(0)}(h)$	1.607736	1.607736	0.283376	1.607736	1.607736	1.607736
$h^0 - ex$	3.1787e-3	7.7770e-4	1.9329e-4	4.8250e-5	1.2057e-5	3.0142e-6
$r_n^{(0)}$	$2^{2.0312}$	$2^{2.0085}$	$2^{2.0022}$	$2^{2.0005}$	$2^{2.0001}$	$2^{2.0000}$
$Q^{(1)}(h)$		1.607736	1.607736	1.607736	1.607736	1.607736
$h^2 - ex$		2.2647e-5	1.5152-6	9.6460e-8	6.0571e-9	3.7900e-10
$r_n^{(1)}$		$2^{3.9018}$	$2^{3.9734}$	$2^{3.9932}$	$2^{3.9983}$	$2^{3.9961}$
$Q^{(2)}(h)$			1.607736	1.607736	1.607736	1.607736
$h^4 - ex$			1.0640e-7	1.8767e-9	3.0286e-11	4.6207e-13
$r_n^{(2)}$			$2^{5.8252}$	$2^{5.9534}$	$2^{6.0344}$	$2^{5.7563}$

Table 3 The numerical results with $t = 1 - 1/8$, $I(t) = -4.417842751$

$ex\backslash e\backslash n$	2^3	2^4	2^5	2^6	2^7	2^8
$Q^{(0)}(h)$	−4.082210	−4.320490	−4.392276	−4.411363	−4.416217	−4.417436
$h^0 - ex$	3.3563e-1	9.7353e-2	2.5566e-2	6.4797e-3	1.6257e-3	4.0678e-4
$r_n^{(0)}$	$2^{1.7856}$	$2^{1.9290}$	$2^{1.9802}$	$2^{1.9949}$	$2^{1.9987}$	$2^{1.9997}$
$Q^{(1)}(h)$		−4.399916	−4.416205	−4.417725	−4.417835	−4.417842
$h^2 - ex$		1.7927e-2	1.6374e-3	1.1753e-4	7.6502e-6	4.8328e-7
$r_n^{(1)}$		$2^{3.4527}$	$2^{3.8003}$	$2^{3.9414}$	$2^{3.9846}$	$2^{3.9960}$
$Q^{(2)}(h)$			−4.417291	−4.417827	−4.417842	−4.417843
$h^4 - ex$			5.5140e-4	1.6206e-5	3.2501e-7	5.4871e-9
$r_n^{(2)}$			$2^{5.0886}$	$2^{5.6398}$	$2^{5.8883}$	$2^{5.9484}$

Table 4 The numerical results for $t = 1/4$

n	2^3	2^4	2^5	2^6	2^7	2^8
I	8.26888	8.26888	8.26888	8.26888	8.26888	8.26888
$Q(h)$	9.37407	11.23356	14.40670	20.59816	32.94096	57.61642
$\overline{Q}^{(0)}(h)$		7.51458	8.06041	8.21524	8.25537	8.26550
$\lvert\overline{Q}^{(0)}(h) - I\rvert$		7.5430e-1	2.0847e-1	5.3640e-2	1.3511e-2	3.3842e-3
$r_n^{(1)}$		$2^{1.9892}$	$2^{1.9973}$	$2^{1.9993}$	$2^{1.9998}$	$2^{1.9999}$
$\overline{Q}^{(1)}(h)$			8.24235	8.26685	8.26875	8.26887
$\overline{Q}^{(1)}(h) - I$			2.6526e-2	2.0292e-3	1.3481e-4	8.5635e-6
$r_n^{(2)}$			$2^{3.9766}$	$2^{3.9938}$	$2^{4.0061}$	$2^{2.6378}$

Table 5 The numerical results for $t = 7/8$

n	2^5	2^6	2^7	2^8	2^9	2^{10}
I	−81.01843	−81.01843	−81.01843	−81.01843	−81.01843	−81.01843
$\overline{Q}^{(0)}(h)$		−79.50641	−80.63022	−80.92070	−80.99396	−81.01231
$\lvert\overline{Q}^{(0)}(h) - I\rvert$		1.51203	3.8821e-1	9.7730e-2	2.4475e-2	6.1215e-3
$r_n^{(1)}$		$2^{1.9616}$	$2^{1.9900}$	$2^{1.9975}$	$2^{1.9994}$	$2^{1.9998}$
$\overline{Q}^{(1)}(h)$			−81.00483	−81.01753	−81.01838	−81.01843
$\overline{Q}^{(1)}(h) - I$			1.3606e-2	9.0230e-4	5.7290e-5	3.5943e-6
$r_n^{(2)}$			$2^{3.9145}$	$2^{3.9772}$	$2^{3.9945}$	$2^{3.9778}$

Table 6 The numerical results for $t = 1/32$

n	2^8	2^9	2^{10}	2^{11}	2^{12}	2^{13}
I	511.85420	511.85420	511.85420	511.85420	511.85420	511.85420
$Q(h)$	509.88417	511.37943	511.77746	511.91933	512.03915	512.3777
$\overline{Q}^{(0)}(h)$		508.38891	510.98139	511.63558	511.79951	511.84052
$\lvert\overline{Q}^{(0)}(h) - I\rvert$		3.4653	8.7280e-1	2.1861e-1	5.4679e-2	1.3671e-2
$r_n^{(1)}$		$2^{1.9892}$	$2^{1.9973}$	$2^{1.9993}$	$2^{1.9998}$	$2^{2.0000}$
$\overline{Q}^{(1)}(h)$			511.84556	511.85365	511.85416	511.85419
$\overline{Q}^{(1)}(h) - I$			8.6399e-3	5.4884e-4	3.4411e-5	2.1946e-6
$r_n^{(2)}$			$2^{3.9766}$	$2^{3.9954}$	$2^{3.9708}$	$2^{6.2235}$

5 Conclusion

From the above results in this paper, we draw conclusions as follows: Evaluating hyper-singular integrals by the quadrature formulas, the algorithms need not calculate any weight. The accuracy order of the algorithms can be improved by extrapolation method. Finally, the examples match with the error analyses.

Acknowledgments Supported by National Science Foundation of China (11371079). The authors in this paper was supported by National Science Foundation of China (11371079).

References

1. Chen JT, Kuo SR, Lin JH (2002) Analytical study and numerical experiments for degenerate scale problems in the boundary element method for two-dimensional elasticity. Int J Numer Method Eng 54:1669–1681
2. Lacerda LAd, Wrobel LC (2001) Hyper-singular boundary integral equation for axisymmetric elasticity. Int J Numer Method Eng 52:1337–1354

3. Chen C, Huang J, Ma YY (2016) Asymptotic expansions of the error for hyper-singular integral with an intervial variable. J Inequal Appl 1:1–16
4. Huang Jin, Wang Zhu, Zhu Rui (2013) Asymptotic error expansions for hyper-singular integrals. Adv Comput Math 38:257–279
5. Hui CY, Shia D (1999) Evaluations of hyper-singular integrals using Gauaaian quadrature. Int J Numer Method Eng 44:205–214
6. Kim P, Choi UJ (2003) Two trigonometric quadrature formulae for evaluating hyper-singular integrals. Int J Numer Method Eng 56:469–486
7. Li J, Yu DH (2010) Rectangle rule for the computation of hyper-singular integral. IOP Conf Ser: Mater Sci Eng 10:012115
8. Li J, Zhang XP, Yu DH (2010) Superconvergence and ultraconvergence of Newton–Cotes rules for supersingular integral. J Comput Appl Math 233:2841–2854
9. Monegato G, Lyness JN (1998) The Euler–Maclaurin expansion and finite-part integrals. Numer Math 81:273–291
10. Tsalamengas JL (2015) Quadrature rules for weakly singular, strongly singular, and hyper-singular integrals in boundary integral equation methods. J Comput Phys 303:498–513
11. Chen YZ (2004) Hyper-singular integral equation method for three-dimensional crack problem in shear mode. Commun Numer Method Eng 20:441–454
12. Kabir H, Madenci E, Ortega A (1998) Numerical solution of integral equations with Logarithmic-, Cauchy- and Hadamard-type singularities. Int J Numer Meth Eng 41:617–638
13. Lifanov IK, Poltavskii LN, Vainikko GM (2004) Hyper-singular Integral Equations and Their Applications. ACRC Press Company, Washington
14. Sidi A (1993) In: Brass H, Hmmerlin G (eds) Numerical Integration, vol IV., A new variable transformation for numerical integration. Birkhäuser, Boston, pp 359–374
15. Sidi A, Israeli M (1988) Quadrature methods for periodic singular and weakly singular fredholm integral equations. J Sci Comput 3:201–231

Optimization Control for Bidirectional Power Converter of DC Micro-grid Power Distribution in Highway Tunnel

Zhou Xiwei, Song Ahua and Xu Hongke

Abstract The characteristic of micro DC-Grid power supply is very valuable to remote mountainous highway tunnel. It can operate not only on parallel mode, but also on isolated mode. Power electronic bidirectional three-level inverter is a key link in the system. A novel three-level inverter SHEPWM algorithm is studied in the paper according to the internal relation between SVPWM and SHEPWM. The switch angle of the algorithm can be achieved by synthesis of the space voltage vector. The method has same Harmonic elimination effect, and can completed the task of AC load in micro DC-Grid system, to achieve stable operation of new energy highway tunnels of DC power supply system.

Keywords Highway tunnel · Micro DC-Grid · Three level inverter · SHEPWM · Voltage vector

1 Introduction

Micro grid research is becoming a hot spot, it will combine micro power, load, energy storage and its control device together. Compared with AC micro-grid, DC micro-grid does not need to track the phase and frequency of the voltage, the controllability and reliability being greatly enhanced, more suitable Distributed Energy Resource (DER) and the load to access.

Z. Xiwei
National Engineering and Research Center for Mountainous Highways, Chongqing, China
e-mail: zhouxiwei@chd.edu.cn

Z. Xiwei · S. Ahua (✉) · X. Hongke
Electronic and Control College, Chang'an University, Xian, China
e-mail: 15202476240@163.com

X. Hongke
e-mail: xuhongke@chd.edu.cn

© Springer International Publishing Switzerland 2017

V.E. Balas et al. (eds.), *Information Technology and Intelligent Transportation Systems*, Advances in Intelligent Systems and Computing 454,
DOI 10.1007/978-3-319-38789-5_35

DC micro-grid has many notable features: dc terminal load for power users, connected to a DC micro-grid can reduce the number of energy conversion, reduce the loss and failure rate. And because no skin effect, the DC power supply cable provides stronger loading capacity. The anti-disturbance of the DC micro-grid is better than AC micro-grid, and the investment on the infrastructure is much lower than AC micro-grid. AC micro-grid has some complex problems such as synchronization of distributed power, Impulse current caused by transformers, the reactive power flow, harmonic currents and phase imbalance, etc., making the control of AC micro-grid is more complex than DC micro-grid.

At present, the highway distribution system still mainly rely on national public AC grid, and the optimal characteristic of DC micro-grid for power supply requirement of highway tunnel is very valuable. For highway tunnels in the remote mountainous areas, the DC micro-grid can not only form a margin of each prepared with public AC grid, improve the security and reliability of the distribution, but also be reasonable and efficient use various forms of energy. The distribution system of the DC micro-grid in the highway tunnel as shown in Fig. 1, DC micro-grid is connected with the large power grid through multiple bidirectional converter. Among them, the distributed DC source includes photovoltaic cells, fuel cells and fuel DC generator sets, distributed AC source has a wind machine, micro turbine and so on.

The mainly use of electricity load in the tunnel is: emergency lighting, general lighting, communications monitoring equipment, fire monitoring, fire control facilities and smoke exhaust fans and other disaster prevention facilities. Among them, the important power load is the first level load, should be set up by two power supply, should also set up separate backup power and continuous emergency power supply. In the distribution system of highway tunnel, the introduction of the DC micro-grid

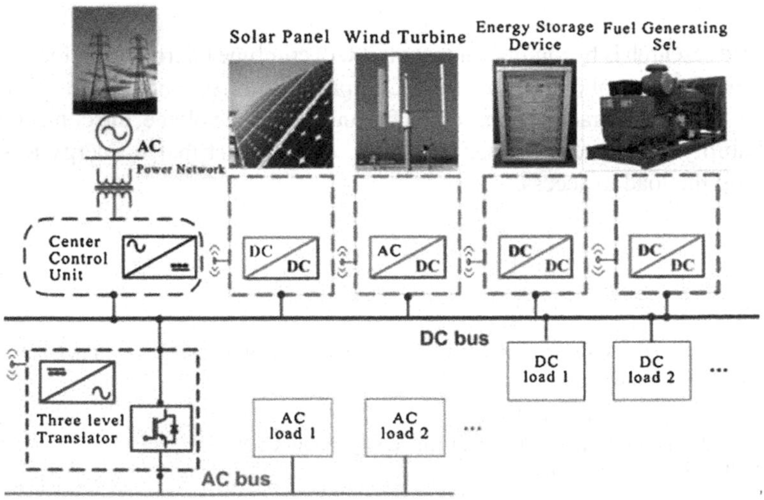

Fig. 1 The distribution system of the DC micro-grid in the highway tunnel

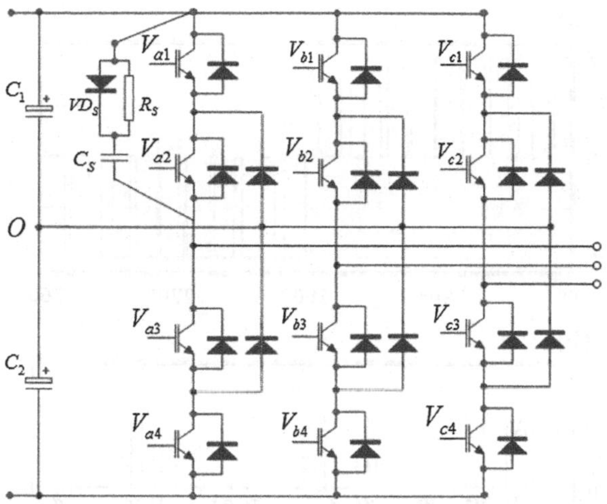

Fig. 2 Main circuit topology structure of the diode clamped three level inverter diode

power distribution, it can not only operation with grid connected but also can operate withan independent island mode operation to reduce dependence on the public grid and provide greater security and reliability of electricity load.

2 Bidirectional Three-Level Converter

Bi-directional power electronic converter is a key link in the system. The converter should be not only can finish the AC–DC transformation, realize feed-in energy received; But also asked to complete a DC–AC inverter, to meet the needs of high quality demand for electricity of AC load.

2.1 Harmonic Elimination Characteristic of SHEPWM

DC micro-grid system can use three-level converter, Fig. 2 shows the main circuit topology structure of the diode clamping, the quality of the output waveform is good, and the switch loss of single IGBT is also small.

Bi-directional power electronic converter is a key link in the system. The converter should be not only can finish the AC–DC transformation, realize feed-in energy received; But also asked to complete a DC–AC inverter, to meet the needs of high quality demand for electricity of AC load.

This converter can use specific harmonic elimination SHEPWM technique. As shown in Fig. 3a is a single phase output voltage waveform for SHEPWM modulation

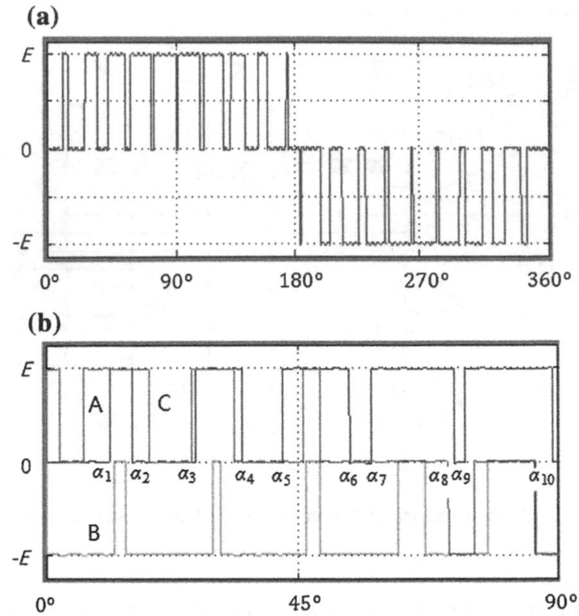

Fig. 3 Phase voltage and switching clamping angle of three level SHEPWM

of the 3 level inverter. It can be seen that this waveform has a 1/4 periodic symmetry, N is the number of switching angles in 1/4 fundamental cycle, Fig. 3b is the inverter switching waveform of three-phase switch angle when the $N = 10$. Phase voltage function can be Fourier analysis:

$$u_A(t) = \frac{a_0}{2} + \sum_{n=1}^{\infty} (a_n \cos n\omega t + b_n \sin n\omega t) \tag{1}$$

where, $a_n = 0$, n = 1, 2, 3, ..., $b_n = 0$, n is an even number. $b_n = \frac{4E}{n\pi} \sum_{k=1}^{N} (-1)^{k+1}$ $cosn\alpha_k$, n is an odd number. Further, there are:

$$0 \le \alpha_0 < \alpha_1 < \cdots < \alpha_j < \cdots \alpha_m < {}^{\pi}/_{2} \tag{2}$$

In order to eliminate kth harmonics, the corresponding coefficient $b_0 = 2$, calculate the switching angle, and the number of $N - 1$ non 3 integer harmonics can be eliminated.

SHEPWM technology has a variety of optimization methods. The typical method of solving the switch angle equation is different, there are: homotopy algorithm, Newton downhill method, genetic algorithm chaos ant colony optimization, bee colony algorithm, etc. In solving the equations, the selection of initial value and optimization of multiple solutions are also studied. In general, the SHEPWM technology

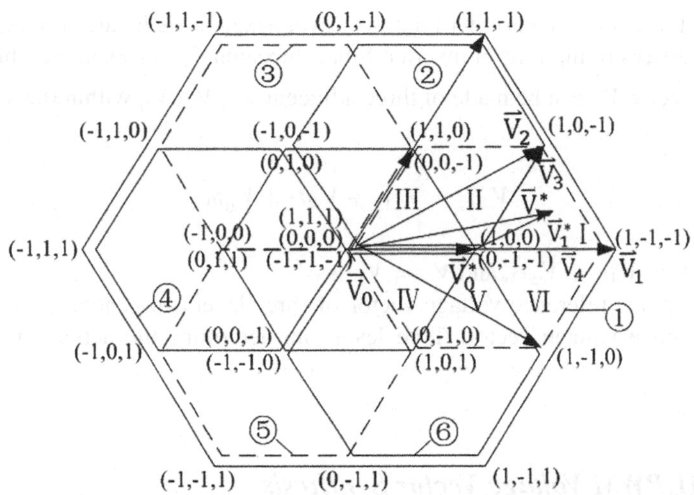

Fig. 4 The space vector hexagonal of three level inverter decomposition and work sector

of the three level inverter is a fundamental bottleneck problem in the solution of the switching angle nonlinear transcendental equations. We can studies a new algorithm on SHEPWM modulation of three level inverter which could suitable for DC micro-grid distribution. The algorithm of switch angle can be directly synthesized by calculating basic space voltage SV vector and have the same harmonic elimination effect, provide AC of high quality waveform for AC load in DC micro-grid.

2.2 *The Voltage Vector Synthesis Algorithm of Three Level SHEPWM*

Due to the three level vector hexagon can be decomposed into two level small hexagon, find the corrected reference voltage space vector, three levels can determine switching sequence and the action time of SV vector as the two-level inverter. And two level SHEPWM modulation can be realization by vector sequence composed of the basic SV vector, function time of basic SV vectors is derived and state switching, SHEPWM instruction of synthetic arm switch, can generate a three-level SHEPWM.

2.3 *The Three Level Space Vector Decomposition*

Each bridge arm switch state combination of three level inverter can produce three output voltage state: $U_{dc}/2$, 0, $-U_{dc}/2$, respectively with a status value of 1, 0, -1. Space voltage vector hexagon of three levels and its decomposition diagram in Fig. 4.

For the three-level, the origin of the two-level hexagon are located on the internal hexagon vertices of three-level inverter. Small hexagon ①, for example, three level reference vector $\vec{V}^*$ can be made of three adjacent $\vec{V}_1$, $\vec{V}_2$, $\vec{V}_0^*$ within the switching period T.

$$\vec{V}^* T = \vec{V}_1 t_1 + \vec{V}_2 t_2 + \vec{V}_0^* t_0$$

while, $\vec{V}_1^* T = \vec{V}_4 t_1 + \vec{V}_3 t_2$, and $\vec{V}^* = \vec{V}_0^* + \vec{V}_1^*$.

Therefore, the reference voltage vector of three level can generated by the two levels reference voltage vector, three level can determine the action time of the switching sequence and the voltage vector as two level inverter.

2.4 SHEPWM Voltage Vector Synthesis

You can analyze the ß/3 interval of SHEPWM modulated when choose $N = 10$, three-phase PWM output pulse waveform can be composed by base vector sequence and the zero vector. As shown in Fig. 3, the two-level small hexagons which are decomposed by the three level, each small hexagonal can be divided into six sectors, the allocation of the work sector could see Table 1.

In each small hexagon, with $k = \mathrm{I}t/(T_S/2) + 1$, I is the under rounding function, $T_s = 2\pi/(3N+1)\, t$ is the instantaneous moment within each $\pi/3$ sector, actually $k_{max} = N + 1$, each sector is divided into k_{max} intervals. Within each k interval, the three-phase output PWM pulse of SHEPWM can be composed of the corresponding three-level basic voltage vectors and the zero vector. Each k range, SHEPWM three-phase output PWM pulse can be made up of three corresponding level of the basic voltage vector and zero vector. The switch state table of all sectors for different modulation M can be classified as shown in Tables 2 and 3, the second line of each table is even number sector, the third line is the odd number sector.

Table 1 The allocation table of work sector

①	②	③	④	⑤	⑥
VI, I	I, II	II, III	III, IV	IV, V	V, VI
II, III	III, IV	IV, V	V, VI	VI, I	I, II

Table 2 Basic vector switching states of SHEPWM I (M < 1)

k=1	k=2	k=odd	k=even	k=N	k=(N+1)
$S_A \to S_B \to \vec{V}_0^*$	$\vec{V}_0^* \to S_B$	$S_B \to S_A \to \vec{V}_0^*$	$\vec{V}_0^* \to S_A \to S_B$	$S_B \to \vec{V}_0^*$	$\vec{V}_0^* \to S_B \to S_A$
$S_B \to S_A \to \vec{V}_0^*$	$\vec{V}_0^* \to S_A$	$S_A \to S_B \to \vec{V}_0^*$	$\vec{V}_0^* \to S_B \to S_A$	$S_A \to \vec{V}_0^*$	$\vec{V}_0^* \to S_A \to S_B$

Table 3 Basic vector switching states of SHEPWM ($1 \le M \le 1.15$)

k=1	k=2	k=odd	k=even	k=N	k=(N+1)
$S_A \to S_B \to \vec{V}_0^*$	$\vec{V}_0^* \to S_B$	$S_B \to S_A \to S_B$	$S_B \to \vec{V}_0^*$	$\vec{V}_0^* \to S_A \to S_B$	
$S_B \to S_A \to \vec{V}_0^*$	$\vec{V}_0^* \to S_A$	$S_A \to S_B \to S_A$	$S_A \to S_B \to S_A$	$S_A \to \vec{V}_0^*$	$\vec{V}_0^* \to S_B \to S_A$

In the table, the SA is the basic voltage vector of the state value of the three-phase bridge arm is zero, the SB is the basic voltage vector of the state value is not zero, $\vec{V}_0^*$ is the zero vector of each small hexagon. And the function of zero vector is the transition of the adjacent basic vector, the choice of zero vector should make three-phase voltage state change single step and only switch one arm bridge at one time.

2.5 The Basic Voltage Vector Function Time of SHEPWM

In the two level small hexagon, the reference voltage vector of three level in the corresponding sector need to correct the bias. Each phase reference voltage vector after correction could see in Table 4.

and $u_{Cs}^{*\prime} = -u_{As}^{*\prime} - u_{Bs}^{*\prime}$ In each small hexagon, the function time of basic SV vectors is different from the parity of k: (1) When $k = is\ odd$

$$\begin{cases} T_A = MT_s \sin\left(\frac{\pi}{3} - \delta\right)/2 \\ T_B = MT_s \sin(\delta)/2 \\ T_0\ or\ T_7 = T_s/2 - T_A - T_B \end{cases} \quad (3)$$

And $k = N,\ T_B = MT_s \sin(\delta)/2,\ T_0\ or\ T_7 = T_s/2 - T_B$.

Table 4 Correction values of the reference voltage vector

Small hexagon	$u_{As}^{*\prime}$	$u_{Bs}^{*\prime}$
①	$u_{As}^* - V_{dc}/3$	$u_{Bs}^* + V_{dc}/6$
②	$u_{As}^* - V_{dc}/6$	$u_{Bs}^* - V_{dc}/6$
③	$u_{As}^* + V_{dc}/6$	$u_{Bs}^* - V_{dc}/3$
④	$u_{As}^* + V_{dc}/3$	$u_{Bs}^* - V_{dc}/6$
⑤	$u_{As}^* + V_{dc}/6$	$u_{Bs}^* - V_{dc}/6$
⑥	$u_{As}^* - V_{dc}/6$	$u_{Bs}^* + V_{dc}/3$

(2) When $k = is\ even$

$$\begin{cases} T_A = MT_s \sin(\delta)/2 \\ T_B = MT_s \sin\left(\frac{\pi}{3} - \delta\right)/2 \\ T_0 or\ T_7 = T_s/2 - T_A - T_B \end{cases} \tag{4}$$

And $k = 2,\ T_0\ or\ T_7 = T_s/2 - T_B, T_B = MT_s \sin\left(\frac{\pi}{3} - \delta\right)/2.$

3 Experiment

Given DC Bus of DC micro-grid is 500 V, used the new synthesis algorithm of three-level SHEPWM voltage vector, modulation coefficient is set to 1, selected the number of switching angles $N = 10$ within 1/4 fundamental period, AC output frequency is 50 Hz, experimental waveforms shown in Figs. 5 and 6 is output line voltage waveform and neutral-point potential waveform, Fig. 7 is FFT spectrum analysis of line voltage.

The results showed that the three level inverters uses SHEPWM algorithm of voltage vector synthesis, the output voltage waveform of the non 3 integer harmonic which is under 30 times with the elimination of the ability, and the output fundamental component achieve modulation factor requirements, meet the requirements of waveform quality of DC micro grid in AC load. But if considering the influence of

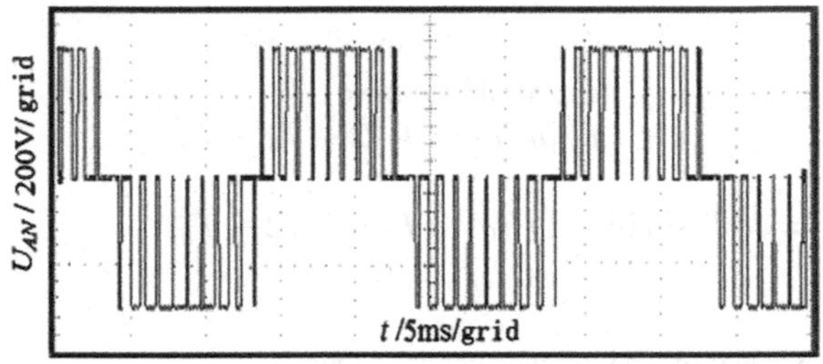

Fig. 5 Output phase voltage of three-level inverter

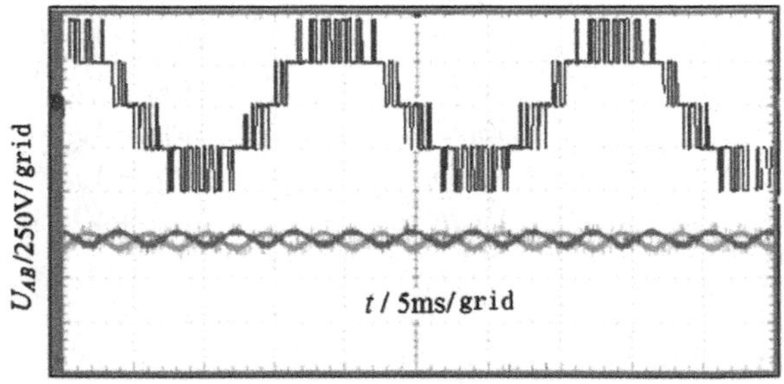

Fig. 6 Output line voltage and neutral-point

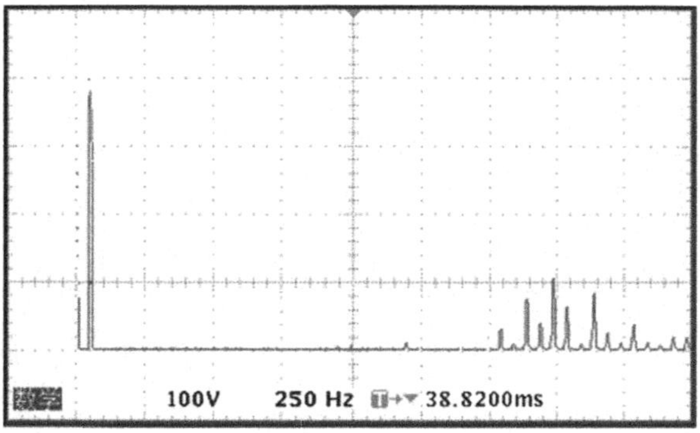

Fig. 7 FFT spectrum potential analysis of line voltage

SHEPWM multiple solutions the neutral-point potential of the three level inverter and the dynamic performance of the algorithm, the method needs to be further studied.

4 Conclusions

The application of DC micro grid technology in the modern Power Supply System is increasingly concerned. But DC micro-grid research in China has just begun. Wind energy and solar energy could be an energy supply unit in DC micro-grid. Thus, the DC micro-grid is not only form the margin of the public AC grid, Improve the security and reliability of the distribution, but also is available for the reasonable and efficient use of clean energy. However, the DC micro grid system requires the coordination and cooperation among the various devices, and the power electronic converter is one of the key point. Only in the rich theoretical basis and experimental study, the DC micro-grid is able to realize the goal of saving energy and user load.

References

1. McGranaghan M, Ortmeyer T, Crudele D et al (2008) Renewable systems interconnectionstudy: advanced grid planning and operations. Sandia National Laboratories
2. Cetin E, Yilanci A, Ozturk HK, Colak M, Kasikci I, Iplikci S (2010) A micro-DC power distribution system for a residential application energized by photovoltaic-wind/fuel cell hybrid energy systems. Energy Build 42(8):1344–1352
3. Behzad V (2010) A new space-vector PWM with optimal switching selection formultilevel coupledinductor inverters. IEEE Trans Indus Elect 57(7):2354–2364
4. Kavousi A (2012) Application of the bee algorithm for selective harmonic elimination strategy in multilevel inverters. IEEE Trans Power Elect 427(4):1689–1696
5. Bowes Sidney R (2007) Optimal regular-sampled PWM inverter control techniques. IEEE Trans Indus Elect 54(3):1547–1559

Bayesian Fault Diagnosis Using Principal Component Analysis Approach with Continuous Evidence

Wenbing Zhu, Zixuan Li, Sun Zhou and Guoli Ji

Abstract For fault diagnosis problems where the historical data is from a number of monitors, conventional likelihood estimation approaches for Bayesian diagnosis are typically independent or lumped approach. However, for most chemical processes the monitor outputs are often not independent, but exhibit correlations to some extent; as for the lumped approach, it is infeasible due to the curse of dimensionality and the limited size of historical dataset. Also there is another limitation to the accuracy of the diagnosis that the continuous monitor readings are commonly discretized to discrete values, therefore information of the continuous data cannot be fully utilized. In this paper principal component analysis (PCA) approach is proposed to transform the evidence into independent pieces, and kernel density estimation is used to improve the diagnosis performance. The application to the Tennessee Eastman Challenge process using the benchmark data demonstrates the effectiveness of the proposed approach.

Keywords Fault diagnosis · Principal component analysis · Kernel density estimation · Tennessee eastman challenge process

1 Introduction

The purpose of fault detection and diagnosis is to detect and isolate in the process the components that have failed. The strong demand for decreased downtime, better performance, safety practices, and energy saving incidents all fuel this active area

W. Zhu · Z. Li · S. Zhou (✉) · G. Ji
Department of Automation, Xiamen University, Xiamen 361005, People's Republic of China
e-mail: zhousun@xmu.edu.cn

Z. Li
e-mail: 2936367051@qq.com

W. Zhu
e-mail: wbzhu@stu.xmu.edu.cn

G. Ji
e-mail: glji@xmu.edu.cn

© Springer International Publishing Switzerland 2017

V.E. Balas et al. (eds.), *Information Technology and Intelligent Transportation Systems*, Advances in Intelligent Systems and Computing 454,
DOI 10.1007/978-3-319-38789-5_36

of research. A Bayesian frame for control performance diagnosis has been proposed in [1]. It takes input from many different fault detection and diagnosis techniques in order to make a synthesize decision [2]. However, the industrial processes often have hundreds of monitors in practice, which results in computation burdens and misdiagnosis. This work proposes the use of principal component analysis in order to handle high dimensional problems, then making use of continuous evidence through kernel density estimation in instead of discrete methods to improve diagnostic performance. The proposed method is applied to the Tennessee Eastman Challenge, simulation results demonstrate the effectiveness.

2 Bayesian Inference

2.1 Bayesian Data-Driven Approach

Bayesian data-driven approach, which is distinct from other fault detection and diagnosis methods, mainly for the reason that the Bayesian approach is a higher-level diagnosis method, other diagnosis methods and even instruments themselves are treated as input sources and are referred to as monitors [3]. Collecting data from monitors for every scenario that one would wish to diagnose, this scenarios which are referred to as operational modes, and the data is called evidence.

According to Bayes Theorem, The Bayesian diagnosis technique ranks each of the modes based on posterior probability, which is calculated

$$p(M|E, D) = \frac{p(E|M, D)p(M|D)}{p(E|D)} \propto p(E|M, D) \tag{1}$$

Given history dataset D, where $p(M|E, D)$ is the posterior probability given evidence E; $p(M|D)$ is the prior probability of the historical mode M; $p(E|D)$ is a normalizing constant; $p(E|M, D)$ is known as likelihood probability, which is the key point of Bayesian inference. Likelihood estimate method include discrete method and continuous method.

2.2 Discrete Method for Likelihood Estimation

For discrete method, the likelihood $p(E|M)$ can be calculated as

$$p(E|M) = \frac{n(E, M)}{n(M)} \tag{2}$$

where $n(E, M)$ is the number of data points where the evidence E and mode M occur simultaneously, and $n(M)$ is the total number of data points were the mode M occurs.

2.3 *Continuous Method for Likelihood Estimation*

Kernel density estimation is the principal technique for the non-parametric estimation of continuous density functions. A kernel density estimate is a summation of kernel functions centered around each data point D [4]. From a multivariate data set D with n entries, a kernel density estimate from the kernel $K(x)$ is given as:

$$f(x) \approx \frac{1}{n}\sum_{i=1}^{n}\frac{1}{|H|^{\frac{1}{2}}}K_H(H^{\frac{1}{2}}(x - D_i)) \tag{3}$$

The kernel function must satisfy

$$\int_{R^d} K(x)dx = 1 \tag{4}$$

Due to the universality of the normal distribution, multivariate Gaussian kernel is a popular choice.

$$K_H(\phi) = \frac{1}{\sqrt{(2\pi)^d}}exp(\phi^T\phi) \tag{5}$$

where d is the data dimension, then

$$f(x) \approx \frac{1}{n}\sum_{i=1}^{n}\frac{1}{\sqrt{(2\pi)^d|H|}}exp([x - D_i]^T H[x - D_i]) \tag{6}$$

3 Dimension Reduction

In practice, Data always has high dimension. It cause two disadvantages: The compute burden and the demand for a large number of samples grows exponentially with the dimensionality of the feature space. This limitation is called the "curse of dimensionality" and severely restricts the practical application of the data-driven Bayesian Fault diagnosis.

3.1 Independent Approach

If the interdependence of each monitor is small, we consider all monitor outputs as a set of single pieces of evidence. All likelihood estimates are one dimensional and then the joint probability could be calculated by multiplying them together. For example, consider a two-dimensional evidence vector $E = (E_1, E_2)$, if the two elements are independent,

$$p(E_1, E_2|M) = p(E_1|M)p(E_2|M),\ E_1|M \perp E_2|M \tag{7}$$

Of course, This assumption depends on the available data.

3.2 PCA Approach

As we all know, each variable of dataset are often not independent, the Hypothesis in Sect. 3.1 will bring large error. Principal component analysis (PCA) serves as the most fundamental technique in high dimensional data preprocessing. PCA is used widely for dimensionality reduction. It transformed the raw data to several principal components, based on the assumption of linear relationship between the variables.

PCA can transform $X = \{x_1, x_2, \ldots, x_d\}$ into $PCS = \{pc_1, pc_2, \ldots, pc_d\}$, where d is the number of dimensions and pc_i denotes the ith principal component. each principal component pc_i is a linear transformation of the variables in the original X and the coefficients defined in this transformation are considered as weight vectors, i.e.,

$$pc_i = \alpha_{1i}x_1 + \alpha_{2i}x_2 + \cdots + \alpha_{di}x_d,\ i = 1, 2, \ldots, d \tag{8}$$

where α is the corresponding weight and x_i denotes the ith variable of X. Moreover, the first principal component sequence pc_1 has the largest variance, it contains most of the information about X. According to the SVD, the covariance matrix can be decomposed by

$$\Omega = ZVZ^T \tag{9}$$

where Z contains the weights for the principal component sequences and the matrix Z has the corresponding variances. The first p principal component are retained,

$$PCS_{n\times p} = X_{n\times d}A_{d\times p} \tag{10}$$

where it is often the case that $p < d$ by set an available contribution rate [5].

In our approach, we set the contribution rate equals one, thus the number of principal components is equal to the number of original variables. The advantage of proposed approach is that Information of raw data will not lost.

An overview of the algorithm is given below:

1. Set the contribution rate equals one, input history data with different modes, output different principal components of history data and the means of history data, coefficient matrix. Each column of coefficient contains coefficients for one principal component.
2. For a validation data, transformed it with the same means and coefficient matrix which came from history data.
3. Now we obtain new history data and a new validation data for Bayesian Fault diagnosis. Making a diagnosis, then joint posterior is obtained by combine multiple posterior.
4. If the new validation data corresponding to one mode which has the largest joint posterior probability, the diagnosis mode is that mode; if the mode equals to the true mode, then we can say it is a correct diagnosis.

4 Application

Although being a rather old process model, the Tennessee Eastman Challenge remains an important tool throughout all disciplines of the system theory for the purpose of comparative studies or validation of algorithms [6]. The benchmark dataset can be downloaded from the website: http://web.mit.edu/braatzgroup/links.html.

4.1 Process Model

There are 15 known preprogrammed major process faults and normal state exist in the process. Extensive reviews of Tennessee Eastman Challenge have been presented in [7]. Flow chart of the process is below (Fig. 1).

4.2 Simulated Modes

Eight types of faults and normal operation status are taken into consideration, except the ones of no significant effect on the process [8], as is shown in Table 1 [9].

The proposed method is applied to Tennessee Eastman Challenge for testing fault detection and diagnosis performance. The diagnostic results in terms of average posterior probability are shown in Figs. 2, 3, 5, 6. A more Intuitive results of validation data be diagnosed mode and the true mode are shown in Figs. 4, 7.

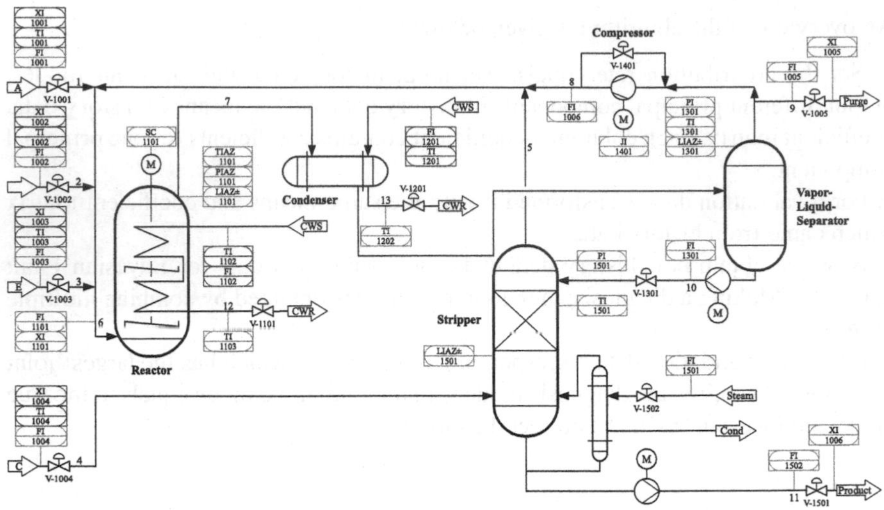

Fig. 1 Process model of the tennessee eastman challenge [7]

Table 1 List of simulated modes for the Tennessee Eastman Challenge

Mode	Process variable	Type
NF	N/A	N/A
IDV1	A/C feed ratio B composition constant (stream 4)	Step
IDV2	B composition, A/C ratio constant (stream 4)	Step
IDV7	C header pressure loss, reduced availability (stream 4)	Step
IDV8	A, B, C feed composition (stream 4)	Variation
IDV10	C feed temperature (stream 2)	Variation
IDV12	Reactor cooling water inlet temperature	Variation
IDV13	Reaction kinetics	Drift
IDV14	Reactor cooling water valve	Sticking

4.3 Comparison of Dimension Reduction Approaches

From the simulation, we can see that independent approach has two wrong diagnosis result which are IDV 7 and IDV 13. It is in sharp contrast with the PCA approach for zero wrong diagnosis. Simultaneously, the misdiagnosed validation data of PCA approach are less than independent approach.

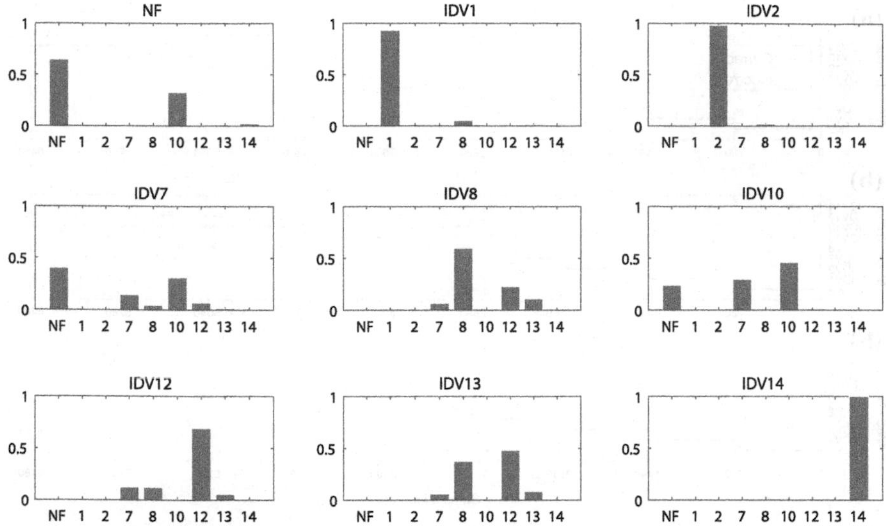

Fig. 2 Posterior probability assigned to each mode for tennessee eastman challenge preprocessing options: independent approach

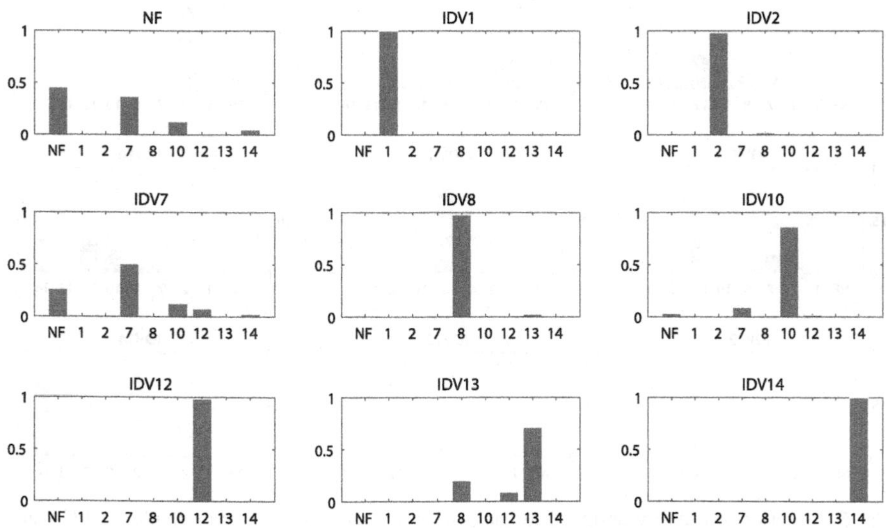

Fig. 3 Posterior probability assigned to each mode for tennessee eastman challenge preprocessing options: PCA approach

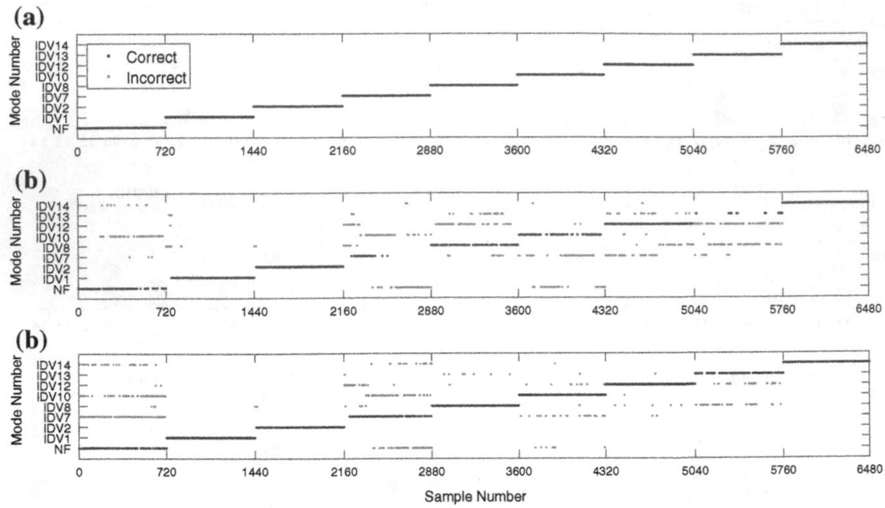

Fig. 4 True mode and diagnosis result of different preprocessing options: **a** True mode; **b** Independent approach Bayesian; **c** PCA approach Bayesian

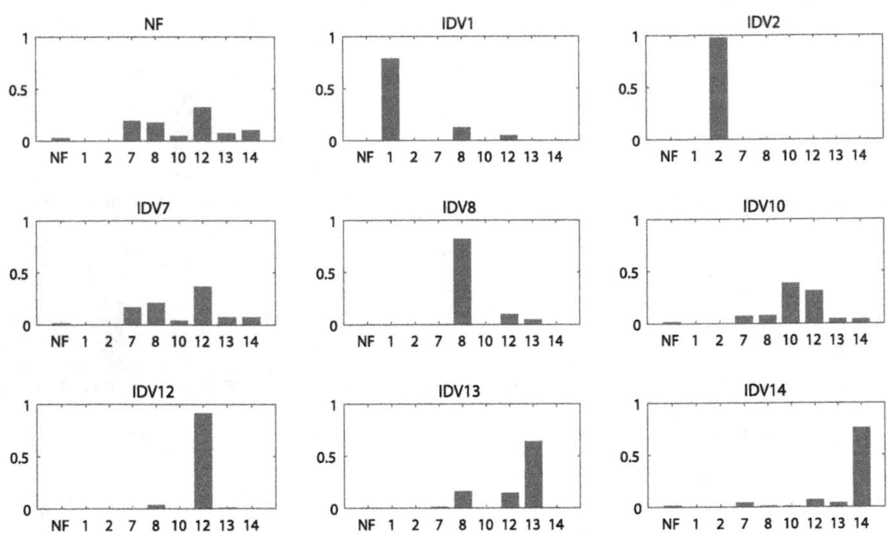

Fig. 5 Posterior probability assigned to each mode for tennessee eastman challenge PCA approaches for discrete method

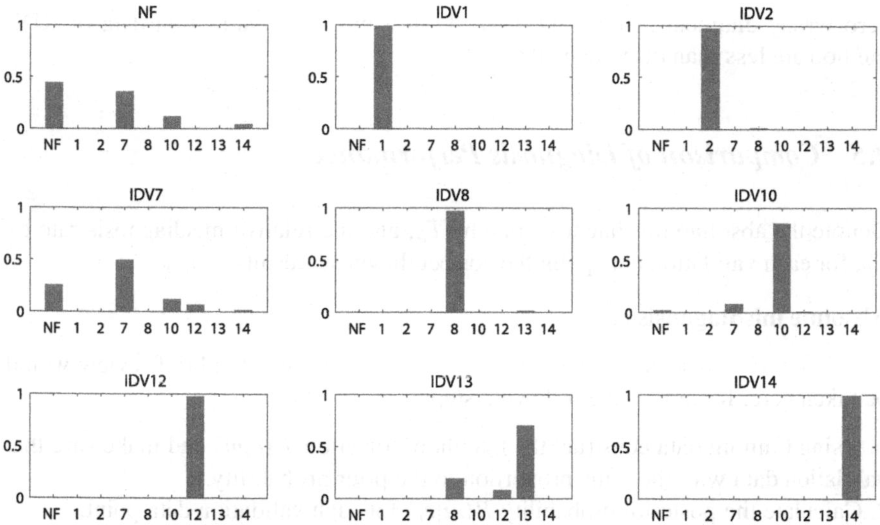

Fig. 6 Posterior probability assigned to each mode for tennessee eastman challenge PCA approaches for kernel density method

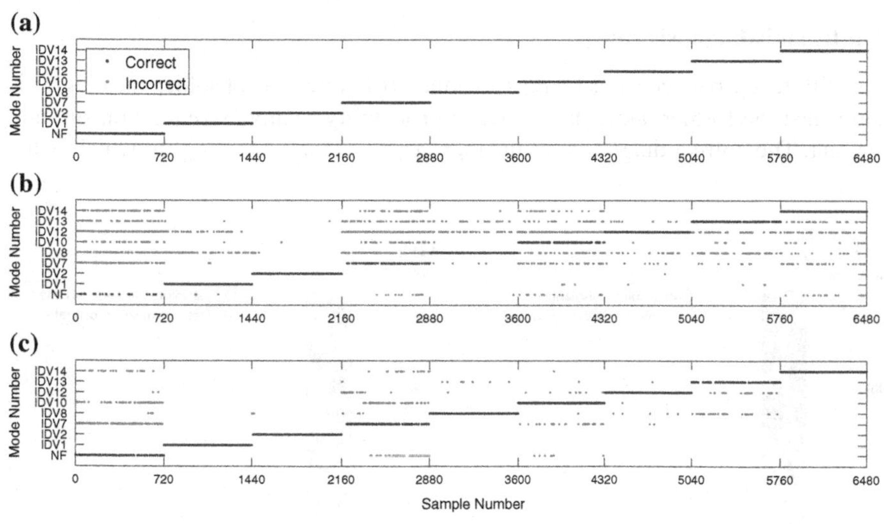

Fig. 7 True mode and diagnosis result of different likelihood estimate method: **a** True mode; **b** Discrete method Bayesian; **c** Continuous method Bayesian

4.4 Comparison of Likelihood Estimate

For the same preprocessing options of PCA, the discrete method has three wrong diagnosis result which are NF and IDV 7. The same with 6.1, the KDE method for zero wrong diagnosis. Simultaneously, the misdiagnosed validation data of KDE method are less than discrete method.

4.5 Comparison of Diagnosis Performance

Denote the absolute misdiagnosis rate by F_a, and the relative misdiagnosis rate by F_b, for each validation data point the correct diagnosis results is n_{cor}.

Absolute misdiagnosis rate

The first performance assessment have been presented in [4], a brief review would be taken here. It requires the following steps:

1. Using training data construct the likelihood function $P(e|m)$ and make sure that validation data with the same proportion to the prior probability.
2. Calculate the posterior probability $P(m|e_i)$ for each validation data point.
3. The mode with the largest posterior $D(m)$ is the one diagnosed, if the diagnosed mode equal true mode we set $n_{cor} = n_{cor} + 1$.
4. F_a can be obtained using $F_a = 1 - \frac{n_{cor}}{no.of validationdata}$.

Relative misdiagnosis rate

The difference between the relative misdiagnosis rate and absolute misdiagnosis rate is that the former taking the posterior probability relative size relationship into account. The correct diagnosis results then set $n_{cor} = n_{cor} + \frac{1}{(m-1)}$, m which is the

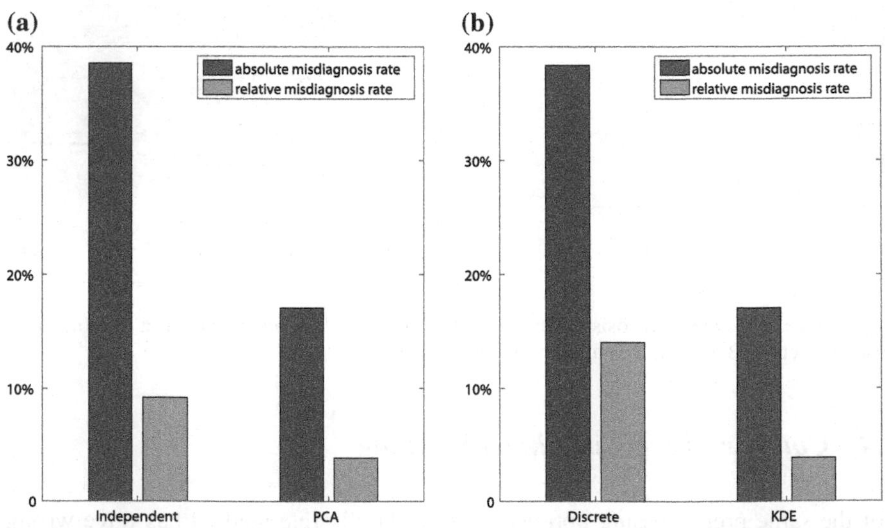

Fig. 8 Absolute misdiagnosis rate and relative misdiagnosis rate: **a** Preprocessing options; **b** Likelihood estimate method

rank of the posterior probability from low to high. Then F_b can be calculated the same with F_a.

To make an integrated evaluation, the proposed assessment criterion are performed. The failed diagnosis rate for each mode is given in Fig. 8.

5 Conclusions

A new data-driven method with consideration of high dimensional problems for Bayesian Fault diagnosis has been proposed. The proposed method uses principal component analysis approach for data preprocessing and utilizes continuous evidence. The proposed method has been applied to the Tennessee Eastman Challenge, where the diagnostic performance of the proposed approaches is demonstrated. One can observe that diagnosis performance is apparently improved.

Acknowledgments This work was supported by the National Natural Science Foundation of China (Nos. 61304141, 61573296), Fujian Province Natural Science Foundation (No. 2014J01252), the Specialized Research Fund for the Doctoral Program of Higher Education of China (No.20130121130004), the Fundamental Research Funds for the Central Universities in China (Xiamen University: Nos. 201412G009, 2014X0217, 201410384090, 2015Y1115) and the China Scholarship Council award.

References

1. Huang B (2008) Bayesian methods for control loop monitoring and diagnosis [J]. J Process Control, 18(9):829–838
2. Qi F, Huang B (2011) Bayesian methods for control loop diagnosis in the presence of temporal dependent evidences [J]. Automatica 47(7):1349–1356
3. Pernestal A (2007) A bayesian approach to fault isolation with application to diesel engines, PhD thesis (2007)
4. Gonzalez R, Huang B (2014) Control-loop diagnosis using continuous evidence through kernel density estimation [J]. J Process Control 24(5):640–651
5. Li H (2016) Accurate and efficient classification based on common principal components analysis for multivariate time series[J]. Neurocomputing 171:744–753
6. Downs JJ, Vogel EF (1993) A plant-wide industrial process control problem [J]. Comput Chem Eng 17(3):245–255
7. Bathelt A, Ricker NL, Jelali M (2015) Revision of the tennessee eastman process model [J]. IFAC-papers online 48(8):309–314
8. Jiang Q, Huang B, Yan X (2016) GMM and optimal principal components-based Bayesian method for multimode fault diagnosis [J]. Comput Chem Eng 84:338–349
9. Namaki-Shoushtari O, Huang B (2014) Bayesian control loop diagnosis by combining historical data and process knowledge of fault signatures [J]. 3696–3704

Behavior Recognition of Scale Adaptive Local Spatio-Temporal Characteristics Harris Algorithm

Zhen Xu, Xiao Li, Fan Wu, FengJun Hu and Lin Wu

Abstract As the accuracy of standard Harris algorithm is not high in the application of behavior recognition, this paper proposed a behavior recognition model based on scale adaptive local spatio-temporal characteristics Harris algorithm, which firstly uses local spatio-temporal characteristics function to achieve fast convolution in order to reduce complexity of the algorithm, and then uses scale adaptive local spatio-temporal characteristics function to replace Gaussian function as the filter of the algorithm, and finally calculates the adaptive matrix and the response value of the angle points in order to improve the accuracy of behavior recognition. Simulation results show that the accuracy of the Harris Algorithm based on scale adaptive local spatio-temporal characteristics, proposed in this paper, is higher than the standard Harris algorithm on behavior recognition.

Keywords Harris algorithm · Behavior recognition · Scale adaption · Local spatio-temporal characteristic function · Gaussian function optimization

Z. Xu · X. Li · F. Wu · F. Hu (✉)
Institute of Information Technology, Zhejiang Shuren University, Hangzhou 310014, China
e-mail: jainism@msn.com; 93966775@qq.com

Z. Xu
e-mail: 61728419@qq.com

X. Li
e-mail: 31798863@qq.com

F. Wu
e-mail: 2846452@qq.com

L. Wu
Hangzhou Maiming Technology Co. Ltd, Hangzhou, China
e-mail: 165600538@qq.com

© Springer International Publishing Switzerland 2017
V.E. Balas et al. (eds.), *Information Technology and Intelligent Transportation Systems*, Advances in Intelligent Systems and Computing 454,
DOI 10.1007/978-3-319-38789-5_37

1 Introduction

The aim of the research on behavior recognition of the athletes in a sport game video is to analyze the image or video, in order to gain the parameters of the posture and movement, and further execute the semantic analysis and the behavioral recognition and understanding [1]. The behavior recognition of the sports game video is a challenging and interdisciplinary research direction, containing characteristics extraction, target tracking and detection, behavior representation, the constructing of the classifier and the behavior recognition etc. specific research contents, which needs use the movement target detection, segmentation, tracking recognition, and semantic representation and reasoning etc. technology, involving mode recognition, image processing, graphics, computer visual, machine learning and artificial intelligence etc. subject [2].

Researchers at home and abroad have already achieved a lot of success in the study of behavior recognition. Park et al. use elliptical structure model to express human head, torso and extremities in the driver behavior analysis [3]. Alexei uses 13 points to represent the human body, and through analyzing the trajectories of feature points recognizes the human behavior [4]. Body contouring appearance and other characteristics can also be used for human action. Liu et al. use the proportion of human head, upper and lower extremities and other main parts of the body to express the body posture [5]. Chung et al. describes the basic movements, such as sit, stand and lie down, through calculating the Gaussian distribution of the target distance projection [6]. Wang et al. use R transformation to study human behavior in the office environment. Although using vision technology to express human behavior has the advantage of perspective invariance, it has a large amount of calculation and high performance requirements of the system, which is not often used [7]. The behavior representation based on motion characteristic is a commonly used method. For example, Zhu et al. use optical flow to study hitting actions in tennis. Furthermore, and using the spatio-temporal feature points to express the behavior is also a commonly used method [8]. Bobick and Davis analyze the human behavior through calculating the motion energy images (MEX) and motion history images (MHI). Veeraraghavan et al. use dynamic time warping (DTW) algorithm to match the behavior sequence. Template matching method does not need extensive sample and has a small calculation, but it is sensitive to the duration and noise of the behavior [9]. Luo et al. use dynamic Bayesian network to study the human behavior recognition and compare with HMM [10]. Buccolieri identifies the posture through using neural network to analyze the body silhouette. In spite of overcoming the drawbacks of template matching, state space method often requires a large number of iterations. The appropriate method should be selected according to the actual situation [11].

Aiming at the characteristics of behavior, this paper proposes a behavior recogni-tion model based on scaling adaptive local spatio-temporal characteristics Harris algorithm, whose accuracy is optimized through scaling adaptive local spatio-temporal characteristic function.

2 Harris Algorithm Analysis

Algorithm is calculated by the pixel gray-scale variance change points in a different direction. The specific steps are as following: select a local window which takes a target pixel as the center; move the window respectively towards up, down, left and right, a small amount of movement; in the four directions, calculate the change volume of average gray value of the pixel points in the window; if the change volume of that response point is greater than scheduled threshold, it is final angle point. Assume that the center point of moving the window is (x, y), the offset in the direction of the X axis is u, the offset in the direction of the Y axis is v, then its calculation formula of regional variation gray is:

$$\begin{aligned} E_{x,y} &= \sum w_{x,y}[I_{x+u,y+v} - I_{x,y}] \\ &= \sum w_{x,y}[uX + vY + o(u^2 + v^2)]^2 \end{aligned} \quad (1)$$

Where are first-order gradient, execute convolution to the image, obtain

$$X = I \otimes (-1, 0, 1) = \frac{\partial I}{\partial x} \quad (2)$$

$$Y = I \otimes (-1, 0, 1)^T = \frac{\partial I}{\partial y} \quad (3)$$

In Eq. (1), $E_{x,y}$ expresses the variations of gray pixels in the window; $I_{x,y}$ expresses the gray of the image in the point (x, y); 0 expresses the higher-order infinitesimals; $w_{x,y}$ expresses the window function, in the form shown below:

$$w_{x,y} = e^{-(x^2+y^2)/2\sigma^2} \quad (4)$$

Removing the infinite events can be expressed as:

$$\begin{aligned} E_{x,y} &= \sum w_{x,y}(uX + vY)^2 \\ &= \sum w_{x,y}(u^2X^2 + v^2Y^2 + 2uXvY) \\ &= Au^2 + 2Cubv + Bv^2 \end{aligned} \quad (5)$$

where:

$$A = X^2 \otimes w_{x,y} \quad (6)$$

$$B = Y^2 \otimes w_{x,y} \quad (7)$$

$$C = (X \cdot Y) \otimes w_{x,y} \quad (8)$$

So, $E_{x,y}$ can be expressed in the form of a matrix

$$E_{x,y} = [\, u \ v \,]M \begin{bmatrix} u \\ v \end{bmatrix} \quad (9)$$

M is a symmetric matrix in the type:

$$M = \begin{bmatrix} A & C \\ C & B \end{bmatrix} \tag{10}$$

Execute similarity diagonalization to M, obtain

$$M = R^{-1} \begin{bmatrix} \lambda_1 & 0 \\ 0 & \lambda_2 \end{bmatrix} R \tag{11}$$

Where, the corner response function $CRF(x, y)$ of the Harris can be expressed in the following type:

$$CRF(x, y) = \det(M) - k(trace(M))^2 \tag{12}$$

In the type, $det(M)$ expresses the determinant of the matrix M, $trace(M)$ expresses the trace of the matrix M, $k = t/(t+1)^2$, k is a variable constant changing with Gaussian function and differential template, and its value usually between 0.04 and 0.06

$$\det(M) = \lambda_1\lambda_2 = A + B \tag{13}$$

$$trace(M) = \lambda_1 + \lambda_2 = AB - C^2 \tag{14}$$

So, the corner response function can be expressed as:

$$CRF(x, y) = AB - C^2 - k(A + B)^2 \tag{15}$$

When the value of the corner response function $CRF(x, y)$ is the local maxi-mum, and it is greater than the given threshold, the corner is regarded as what we have to solve.

Harris algorithm is adaptive to the change of angle and the scale, and it can effi-ciently complete the extraction of angular point, but its accuracy also has some shortcomings.

3 Harris Algorithm Based on Scale Adaptive Local Spatio-Temporal Characteristics

3.1 Scale Adaptive Local Spatio-Temporal Characteristics

According to defect of Harris algorithm, this paper uses scale adaptive local spatio-temporal characteristics to optimize it, and the scale adaptive local spatio-temporal characteristics function is a kind of important spline function, for the node sequence $\{x_i\}$, set:

$$\varphi_m(t) = (t-x)_+^m = \begin{cases} (t-x)^m, (t \geq x) \\ 0, (t < x) \end{cases} \quad (16)$$

The $m+1$ order mean deviation of the function $(x_{j+m+1})\varphi_m(t)$ is:

$$B_{j,m}(x) = (x_{j+m+1} - x_j)\varphi_m(t) \lfloor x_j, x_{j+1}, L, x_{j+m+1} \rfloor \quad (17)$$

The j th of the m time scale adaptive local characteristic function, which can be called local spatio-temporal characteristics function for short. On this basis, the N time local spatio-temporal characteristic function is obtained:

$$B_n(x) = \sum_{j=0}^{n+1} \frac{(-1)^j}{n!} C_{n+1}^j \left(x + \frac{n+1}{2} - h\right)^n u\left(x + \frac{n+1}{2} - j\right) \quad (18)$$

where:

$$u = \begin{cases} 0, x < 0 \\ 1, x > 0 \end{cases} \quad (19)$$

3.2 Improved Harris Algorithm

This paper proposes an improved Harris algorithm, which uses scale adaptive local spatio-temporal characteristics to replace the Gaussian function through introducing scale adaptive local spatio-temporal characteristic function in Harris algorithm.

According to the central limit theorem, when the order of the scale adaptive local spatio-temporal characteristic function approaches infinity, the scale adaptive local spatio-temporal characteristic function infinitely approaches Gaussian function, so the use of scale adaptive local spatio-temporal characteristic function can realize fast convolution and reduce the complexity of the algorithm. degree of scale adaptive local spatio-temporal characteristic function space is a subset of square integrable space. This function is a degree piecewise polynomial and its first order has partial derivative. Its two-dimension form can be expressed as:

$$V^n = \{g^n(x,y) = \sum_{k,l \in z} C(k,l) B^n(x-k, y-l)\} \quad (20)$$

where, $B^n(x)$ can be expressed as:

$$B^n(x) = \frac{(\frac{n+1}{2} + x)B^{n+1}(x+\frac{1}{2}) + (\frac{n+1}{2} + x)B^{n-1}(x-\frac{1}{2})}{n} \quad (21)$$

Formula (20) respectively executes partial derivative to x and y, obtain:

$$\begin{cases} \frac{\partial g^n(x,y)}{\partial x} = \sum\limits_{k,l \in z} C(k,l) \frac{\partial B^n(x-k)}{\partial x} B^n(y-l) \\ \frac{\partial g^n(x,y)}{\partial y} = \sum\limits_{k,l \in z} C(k,l) B^n(x-k) \frac{\partial B^n(y-l)}{\partial x} \end{cases} \tag{22}$$

Formula (21) executes partial derivative to x, obtain:

$$\frac{\partial B^n(x)}{\partial x} = B_{n-1}\left(x + \frac{1}{2}\right) - B_{n-1}\left(x - \frac{1}{2}\right) \tag{23}$$

In conclusion, the differential template of the scale adaptive local spatio-temporal characteristic function in the direction of and are obtained. The steps of the improved Harris algorithm are as follows:

(1) Use the scale adaptive local spatio-temporal characteristic function to take place the Gaussian function as the filter in the algorithm;
(2) According to the formula (20)–(23), calculate the adaptive matrix:

$$M = \begin{bmatrix} I_x^2 & I_x I_y \\ I_x I_y & I_y^2 \end{bmatrix} \tag{24}$$

(3) To avoid selection of k, use the following formula to calculate the response value of angular point:

$$R(x, y) = \frac{I_x^2 \times I_y^2 - (I_x I_y)^2}{I_x^2 + I_y^2} \tag{25}$$

(4) The operation is finished when the value which meets $R(x, y)$ is greater than the given threshold or $R(x, y)$ is the local maximum in particular neighborhood.

4 Algorithm Performance Simulation

In order to verify the validity of this algorithm, execute the simulation experiment. First, apply the Harris algorithm based on adaptive local spatio-temporal characteristics to the behavior recognition, and take a basketball video as example to test the algorithm, as shown below:

From Figs. 1 and 2, it is seen that the athlete behavior in image is obvious passing and blocking, but standard algorithm gives the result of shooting and blocking shot. The improved algorithm recognizes rightly the behavior of athletes.

Then, use five sport videos to test the accuracy of the algorithm in behavior recognition, and compare with the standard Harris algorithm, the results are shown in the following figure.

Fig. 1 Athletes behavior recognition of the results of standard algorithms

Fig. 2 Improved algorithm athlete behavior recognition

As can be seen from the above results, the Harris algorithm based on adaptive local spatio-temporal characteristics, proposed in this paper, greatly improves the accuracy of original algorithm on behavior recognition (Fig. 3).

5 Summarize

The behavior in sports video is a planned, highly collaborative multi-player team behavior. Therefore, sports video behavior recognition is of great economic and social value. This paper presents a behavior recognition model based on scale adaptive local spatio-temporal characteristics Harris algorithm, and seen from the simulation results, the proposed model can effectively improve the recognition accuracy.

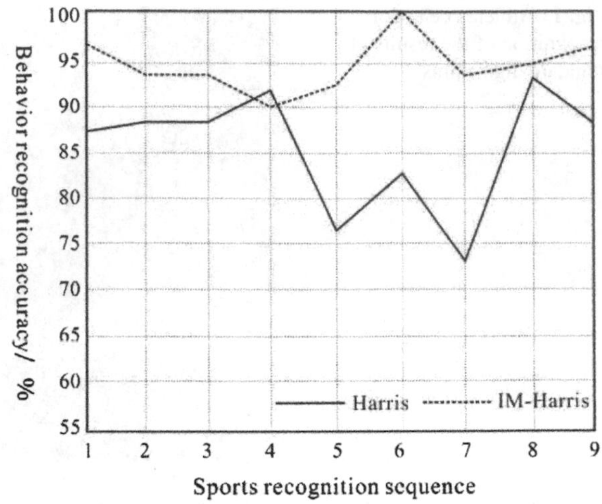

Fig. 3 Compare athlete behavior recognition accuracy

Acknowledgments This work was supported by the Education of Zhejiang Province (Grant No. Y201225667), the Department of Science and Technology of Zhejiang Province (Grant No. 2014C31065), and the Department of Science and Technology of Zhejiang Province (Grant No. 2015C31091).

References

1. Zhao G (2014) Application of improved EM algorithm in recognition of human actions. TV Eng 38(13):196–199
2. Zhou W (2014) Recognizing human actions by particle filter. J Chongqing Norm Univ Nat Sci Edition 31(5):105–109
3. Liu Y (2014) Abnormal behavior recognition based on corner motion history. Comput Eng Sci 36(6):1127–1131
4. Wang M (2014) Time-scale invariant modeling and classifying for object behaviors in 3D space based on monocular vision. Acta Automatica Sinica 40(8):1644–1653
5. Xiao D (2014) KFLD-SIFT with RVM fuzzy integral fusion recognition of human action based on tensor. Pattern Recognit Artif Intell 27(8):713–719
6. Wang H (2014) The research of human behavior recognition based on depth map sequence and space occupancy patterns. Sci Technol Eng 10(24):102–107
7. Li J (2014) Self-adaptive activity recognition method based on CHMMs. Appl Res Comput 31(10):3037–3040
8. Cai J (2014) Human action recognition based on local image contour and random forest. Acta Optica Sinica 10(10):204–213
9. Qin H (2014) Human action recognition based on composite spatio-temporal characteristics. J Comput Aided Des Comput Graphics 26(8):1320–1325
10. Liu H (2014) Human activity recognition based on 3D skeletons and MCRF model. J Univ Sci Technol China 44(4):285–291
11. Wang C (2014) Simulation of movement behavior recognition based on 3D virtual incomplete body image. Comput Simul 31(7):399–402

Calorific Value Prediction of Coal Based on Least Squares Support Vector Regression

Kuaini Wang, Ruiting Zhang, Xujuan Li and Hui Ning

Abstract The calorific value of coal is important in both the direct use and conversion into other fuel forms of coals. Accurate calorific value predicting is essential in ensuring the economic, efficient, and safe operation of thermal power plants. Least squares support vector machine (LSSVM) is a variation of the classical SVM, which has minimal computational complexity and fast calculation. This paper presents Least squares support vector regression (LSSVR) to predict the calorific value of coal in Shanxi Coal Mining Region. The LSSVR model takes full advantage of the calorific value information. It derives excellent prediction accuracy, including the relative errors of less than 3.4 % and relatively high determination coefficients. Experimental results conform the engineering application, and show LSSVR as a promising method for accurate prediction of coal quality.

Keywords Least squares support vector machine · Regression · Calorific value of coal · Prediction

1 Introduction

Coal as a conventional fossil fuel, has played an important role in industrial fields, such as electricity generation, cement making and conversion to coke for the smelting of iron ore. The calorific value of coal refers to the quantity of heat released from unit

K. Wang (✉)
College of Science, Xi'an Shiyou University, Xi'an 710065, Shaanxi, China
e-mail: wangkuaini1219@sina.com.cn

R. Zhang
Canvard College, Beijing Technology and Business University, Beijing 101118, China
e-mail: ruitingzh@163.com

X. Li · H. Ning
China Coal Xi'an Design Engineering Co., Ltd, Xi'an 710054, Shaanxi, China
e-mail: lixujuanhn@163.com

H. Ning
e-mail: ninghui3236@163.com

© Springer International Publishing Switzerland 2017

V.E. Balas et al. (eds.), *Information Technology and Intelligent Transportation Systems*, Advances in Intelligent Systems and Computing 454,
DOI 10.1007/978-3-319-38789-5_38

coal. The calorific value of coal is important in both the direct use and conversion into other fuel forms of coals [1]. It is not only one of the main coal quality evaluations, but also can provide the foundation for the variation of coal and coal classification research. In thermal power plants, the consumption of coal as the main economic indicator must be calculated according to the calorific value. Moreover, The heat balance of the boiler, ratio of mixed fuel, and the theoretical combustion temperature also rely on the calorific value of coal [2]. The measurement of calorific value usually requires sophisticated experimental conditions and experts, and the maintenance of experimental apparatus has many difficulties. It is time-consuming, and the measurement results often lag behind. Therefore, effective and accurate measurement of the calorific value has been a hot topic, which has important practical significance for coal enterprise. In the past, many scholars make use of industrial and elemental analysis of experimental data to establish the empirical formulas [2]. These methods can only approximately estimate the calorific value, because the errors are often big.

In recent years, with the development of artificial intelligence algorithms, such as artificial neural networks (ANNs) [3] and support vector machines (SVMs) [4, 5], which have been widely used to predict the calorific value of coal. These methods like ANNs do not require the knowledge of mathematical relationship between the inputs and corresponding outputs as well as explicit characterization and quantification of physical properties and conditions. However, ANNs need much running time and can not acquire the global minima. This inevitably affects the efficiency and accuracy. SVMs proposed by Vapnik and colleagues [4, 5], have attracted wide attention over the past decades as an elegant machine learning approach. They are based on the structural risk minimization, and have simple structure models, the global optimum and excellent generalization ability, particularly in dealing with high dimensional data. SVMs have been widely applied to pattern recognition, regression analysis, time series prediction and other fields [6–9].

Suykens and Vandewalle [10, 11] proposed least squares support vector regression (LSSVR) which replaces the inequality constraints with equality constraints, and meanwhile adopts sum of squares error loss function instead of epsilon-insensitive loss in the original SVR. In this way, the solution of LSSVR directly follows from solving a set of linear equations, which reduces the computational complexity and storage. More discussions about LSSVR can be found in monograph [11]. Extensive empirical results show that the performance of LSSVR is as comparable as that of SVR whereas the training cost has been cut down remarkably [12, 13].

This paper presents least squares support vector regression (LSSVR) to predict the calorific value of coal in Shanxi Coal Mining Region. Experimental results show that the model has high prediction accuracy. It is applicable to predict the calorific value of coal, and has a certain significance to the industry production.

2 Least Square Support Vector Regression

Consider a regression problem with training samples of input vectors $\{\mathbf{x}_i\}_{i=1}^n$ along with the corresponding targets $\{y_i\}_{i=1}^n$, and the task is to find a regression function that best represents the relation between input vectors and their targets. A nonlinear regressor makes predictions by $f(\mathbf{x}) = w^\top \phi(\mathbf{x}) + b$, where $\phi(\cdot)$ is a mapping which maps the input data into a high-dimensional feature space, w represents the model complexity, and b is the bias. To derive a nonlinear regressor, LSSVR solves the following optimization problem:

$$\min_{w,b,e_i} \frac{1}{2}\|w\|^2 + \frac{C}{2}\sum_{i=1}^{n} e_i^2 \tag{1}$$

$$\text{s.t.} \quad y_i = w^\top \phi(\mathbf{x}_i) + b + e_i, i = 1, \cdots, n \tag{2}$$

where e_i represents the error variables, and $C > 0$ is the regularization parameter that balances the model complexity and empirical risk. We introduce Lagrangian multipliers and construct a Lagrangian function to solve the optimization problem (1)–(2). One can define the Lagrangian function:

$$\mathcal{L}(w, b, e, \alpha) = \frac{1}{2}\|w\|^2 + \frac{C}{2}\sum_{i=1}^{n} e_i^2 + \sum_{i=1}^{n} \alpha_i (y_i - w^\top \phi(x_i) - b - e_i) \tag{3}$$

where $\alpha = (\alpha_1, \alpha_2, \cdots, \alpha_n)^\top$ are Lagrangian multipliers. According to Karush-Kuhn-Tucker (KKT) conditions, we get the following functions

$$\frac{\partial \mathcal{L}}{\partial w} = 0 \Rightarrow w = \sum_{i=1}^{n} \alpha_i \phi(x_i) \tag{4}$$

$$\frac{\partial \mathcal{L}}{\partial b} = 0 \Rightarrow \sum_{i=1}^{n} \alpha_i = 0 \tag{5}$$

$$\frac{\partial \mathcal{L}}{\partial e_i} = 0 \Rightarrow \alpha_i = C e_i \tag{6}$$

$$\frac{\partial \mathcal{L}}{\partial \alpha_i} = 0 \Rightarrow y_i = w^\top \phi(x_i) + b + e_i \tag{7}$$

Eliminating the variable w and e_i, (4)–(7) can be transformed as a set of linear equations:

$$\begin{bmatrix} 0 & \mathbf{1}^\top \\ \mathbf{1} & K + \frac{1}{C} I_n \end{bmatrix} \begin{bmatrix} b \\ \alpha \end{bmatrix} = \begin{bmatrix} 0 \\ y \end{bmatrix} \tag{8}$$

where $\mathbf{1} = (1, 1, \cdots, 1)^\top$, $y = (y_1, y_2, \cdots, y_n)^\top$, I_n denotes identity matrix, and $K = (K_{ij})_{n\times n}$ is the kernel matrix with $K_{ij} = k(x_i, x_j) = \phi(x_i)^\top \phi(x_j)$, $k(\cdot, \cdot)$ is the kernel function, which can be expressed as the inner product calculation in high dimensional feature space.

When (8) is solved, the following regressor is obtained as

$$f(x) = w^\top \phi(x) + b = \sum_{i=1}^{n} \alpha_i k(x_i, x) + b \tag{9}$$

For a new pattern x, we can predict its target by (9).

3 Performance Evaluation Criterions

In the experiments, We adopt the following four popular regression estimation criterions, root mean squared error (RMSE), mean absolute error (MAE), mean relative error (MRE), determination coefficients (R^2) to evaluate the performance of LSSVR for predicting the calorific value of coal.

(1)

$$\mathrm{RMSE} = \sqrt{\frac{1}{N}\sum_{i=1}^{N}(y_i - \widehat{y_i})^2} \tag{10}$$

(2)

$$\mathrm{MAE} = \frac{1}{N}\sum_{i=1}^{N}|y_i - \widehat{y_i}| \tag{11}$$

(3)

$$\mathrm{MRE} = \frac{1}{N}\sum_{i=1}^{N}\left|\frac{y_i - \widehat{y_i}}{y_i}\right| \tag{12}$$

(4)

$$\mathrm{R}^2 = 1 - \frac{\sum_{i=1}^{N}(y_i - \widehat{y_i})^2}{\sum_{i=1}^{N}(y_i - \bar{y})^2} \tag{13}$$

where y_i denotes the actual target, and $\widehat{y_i}$ the corresponding prediction, N refers to the number of test samples. Denote $\bar{y} = \frac{1}{N}\sum_{i=1}^{N} y_i$ as the average value of $y_1, y_2, ..., y_N$. The smaller RMSE is, the better fitting performance is. However, if noises are also used as test samples, too small RMSE probably stands for the over-fitting of the regression. The closer the value of R^2 is to 1, the better the model's regression result is.

4 Experimental Results

In this paper, experimental data of the 38 coal samples were collected from Shanxi Coal Mining Region. All samples were determined and used in the prediction model. The samples cover M_{ad}, A_{ad} and V_{ad} as the input data and $Q_{gr,d}$ as the target. We randomly split the coal samples as the training samples and test samples. The number of training samples is 20, and the rest are test samples. LSSVR are repeated ten times with different partition of training and test samples.

We conduct experiments on the above data to predict the calorific value of coal. Linear kernel $k(x_i, x_j) = x_i^T x_j$ and Radial Basis Function (RBF) kernel $k(x_i, x_j) = \exp(-\|x_i - x_j\|^2/\sigma^2)$ are chosen in the experiments, where $\sigma > 0$ is the RBF kernel parameter. The optimal parameters are through grid search from the search range $\{2^{-9}, 2^{-8}, \ldots, 2^8, 2^9\}$. Table 1 shows the average accuracies of the calorific value

Table 1 Important indexes of calorific value prediction of coal based on LSSVR

Kernel	RMSE	MAE	MRE	R^2
Linear	0.2892	0.2308	0.0054	0.9582
RBF	0.3113	0.2513	0.0058	0.9519

Table 2 The predicted calorific value of coal

		Linear kernel		RBF kernel	
No.	Actual	Predicted	Relative error (%)	Predicted	Relative error (%)
1	24.1900	24.1721	0.0740	24.2352	0.1869
2	23.3000	23.4497	0.6425	23.4637	0.7026
3	21.1000	21.0421	0.2744	20.9007	0.9445
4	21.0400	21.7385	**3.3199**	21.6188	**2.7510**
5	22.2500	22.5341	1.2769	22.4543	0.9182
6	24.7400	24.2078	2.1512	24.2832	1.8464
7	24.5000	24.6721	0.7024	24.7575	1.0510
8	20.6900	20.7408	0.2455	20.5829	0.5176
9	22.9800	22.8833	0.4208	22.8412	0.6040
10	23.8600	23.7512	0.4560	23.7962	0.2674
11	24.6000	24.7376	0.5593	24.8193	0.8915
12	21.0500	21.3183	1.2746	21.1509	0.4793
13	21.2600	20.8900	1.7404	20.7395	2.4483
14	22.4400	22.5768	0.6096	22.4995	0.2652
15	22.6400	22.2399	1.7672	22.1309	2.2487
16	23.1500	23.2518	0.4397	23.2531	0.4454
17	23.3000	23.2020	0.4206	23.1925	0.4614
18	23.1700	23.1459	0.1040	23.1299	0.1731

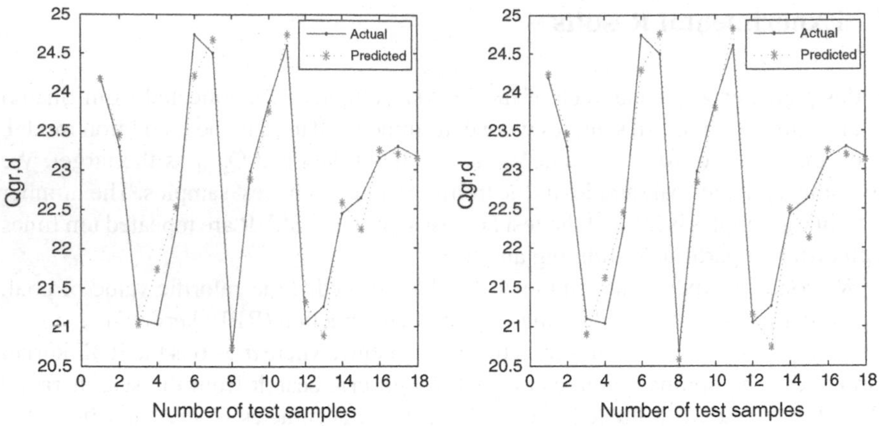

Fig. 1 Prediction results with linear kernel (*left*) and RBF kernel (*right*)

of coal by LSSVR with linear kernel and RBF kernel, respectively. It can be seen that the two determination coefficients R^2 of LSSVR with linear kernel and RBF kernel exceed 0.95, and MRE 0.0054 and 0.0058, which are excellent performance indexes. Compared with RBF kernel, the performance of LSSVR with linear kernel is superior.

Table 2 and Fig. 1 report the prediction results of calorific value. Relative error is usually employed as an engineering requirement [14]. As depicted in Table 2, the maximum relative errors of LSSVR do not exceed 3.4 %, indicating high accuracy, as well as generalization ability of the model. Figure 1 indicates that the predicted values and actual values fit well. Meanwhile, the tendencies are quite close to actual situations.

5 Conclusions

In this paper, the LSSVR model is applied to predict the calorific value of coal incorporating proximate analysis results. It takes full advantage of the information of the calorific value of coal, and derives high determination coefficients. Experimental results show that LSSVR is simple and effective to conduct, and the prediction is satisfactory, which is very important in practical production. It is hoped that these results serve as a reference for the variation of coal and coal classification research in Shanxi Coal Mining Region.

Acknowledgments This work is supported by Xi'an Shiyou University Youth Science and Technology Innovation Fund Project (NO. 2016BS17) and Beijing Higher Education Young Elite Teacher Project (NO. YETP1949).

References

1. Mason DM, Gandhi KN (1983) Formulas for calculating the calorific value of coal and chars. Fuel Process Technol 7:11–22
2. Channiwala SA, Parikh PP (2002) A unified correlation for estimating HHV of solid, liquid and gaseous fuels. Fuel 81(8):1051–1063
3. Patel SU, Kumar BJ, Badhe YP, Sharma BK, Saha S, Biswas S, Chaudhury A, Tambe SS, Kulkarni BD (2007) Estimation of gross calorific value of coals using artificial neural networks. Fuel 86:334–344
4. Cristianini N, Shawe-Taylor J (2000) An introduction to support vector machines. Cambridge University Press, New York
5. Scholkopf B, Smola AJ (2002) Learning with kernels. MIT Press, Cambridge
6. Wang J, Li L, Niu D, Tan Z (2012) An annual load forecasting model based on support vector regression with differential evolution algorithm. Appl Energy 94:65–70
7. Kavaklioglu K (2011) Modeling and prediction of Turkey's electricity consumption using support vector regression. Appl Energy 88(1):368–375
8. Zhou H, Tang Q, Yang L, Yan Y, Lu G, Cen K (2014) Support vector machine based online coal identification through advanced flame monitoring. Fuel 117:944–951
9. Suykens JAK, Vandewalle J (1999) Least squares support vector machines classifiers. Neural Netw Lett 19(3):293–300
10. Suykens JAK, Brabanter JD, Lukas L, Vandewalle J (2002) Weighted least squares support vector machines: robustness and sparse approximation. Neurocomputing 48:85–105
11. Feng Q, Zhang J, Zhang X, Wen S (2015) Proximate analysis based prediction of gross calorific value of coals: a comparison of support vector machine, alternating conditional expectation and artificial neural network. Fuel Process Technol 129:120–129
12. Lv Y, Liu J, Yang T, Zeng D (2013) A novel least squares support vector machine ensemble model for NOx emission prediction of a coal-fired boiler. Energy 55:319–329
13. Zhang W, Niu P, Li G, Li P (2013) Forecasting of turbine heat rate with online least squares support vector machine based on gravitational search algorithm. Knowl Based Syst 39:34–44
14. Li Q, Meng Q, Cai J, Yoshino H, Mochida A (2009) Applying support vector machine to predict hourly cooling load in the building. Appl Energy 86(10):2249–2256

Research on Shortest Path Algorithm in Intelligent Traffic System

Zhang Shui-jian

Abstract The shortest path problem is one of the most critical issues of intelligent transportation systems. Ant colony algorithm is an effective intelligent optimization method to solve the shortest path problem. However, there are some drawbacks when basic ant colony algorithm is used in solving the shortest path problem. So in this paper the basic ant colony algorithm is improved, so that the improved ant colony algorithm is suitable for solving the shortest path problem. In order to improve the efficiency of the algorithm, the search scope is limited, and the transfer rule and the ant colony algorithm pheromone update rule are improved. Simulation results show that the improved ant colony algorithm can efficiently obtain the shortest path.

Keywords Ant colony algorithm · Shortest path · Transportation network

1 Intelligent Transportation Systems

Intelligent Transportation Systems (ITS) effectively integrates a series of advanced information technology, communication technology, control technology, sensor technology and system integration technology on basis of better infrastructure, and uses these services provided by the variety of technologies in ground transportation systems. Information sharing, systems integration and integrated services are essential characteristics of ITS and the fundamental goals the ITS construction [1]. ITS technology enables managers, operators and individual travelers become more informed, more coordinated with each other, and wiser to make decisions. Through the construction and implementation of ITS system, traffic congestion can be alleviated, traffic accidents reduced, transportation costs reduced, and transport efficiency improved. So that the real-time, secure, efficient, comfortable and environmentally friendly intelligent integrated transport system can be established. Shortest path problem is one of the most critical issues of intelligent transportation systems [2, 3].

Z. Shui-jian (✉)
Huzhou Vocational and Technical College, Zhejiang 313000, Huzhou, China
e-mail: zsjsouth@163.com

© Springer International Publishing Switzerland 2017
V.E. Balas et al. (eds.), *Information Technology and Intelligent Transportation Systems*, Advances in Intelligent Systems and Computing 454,
DOI 10.1007/978-3-319-38789-5_39

2 Shortest Path Ant Colony Algorithm

2.1 Basic Ant Colony Algorithm

The rise of intelligent optimization algorithm provides a feasible way for finding the shortest path in a complex transportation network. Italian scholar Colorni A, Dorigo M and Maniezzo V presented a new intelligent optimization algorithm (Ant Colony Optimization, ACO) in 1992 [4–6].

Ants can detect the pheromones released by other ants as they find the path, and use the pheromones to guide their direction of movement, and select the paths with greater amount of pheromones with higher probability. And ants also release pheromones as traveling on this route. Therefore more ants will pass the path with more pheromones, and the amount of pheromones on this path will increase, where by a large number of ants' foraging process constitutes a kind of positive feedback process to the pheromones. Through that way ants can gradually find a shortest foraging path. Ant colony algorithm is a swarm intelligence algorithm devised through simulating the principle of ants how to find the shortest path during the foraging process [7, 8].

The process of basic ant colony algorithm is as follows:

(1) Initializing ant colony;
(2) Computing the fitness of each ant according to the length of the path along that the ant reaches the destination;
(3) Releasing pheromones on the path which the ants pass through according to certain rules;
(4) The ants which depart behind the preceding ants select a path to the destination according to the left pheromones;
(5) The pheromones left on the paths continue to volatilize at a certain rate.

The concentration of the pheromone on each path is equal through initializing ant colony, i.e., as $t = 0$, $\tau_{ij}(0) = C$ (C is a constant), $\tau_{ij}(0)$ denotes the amount of pheromones of section ij at initial moment. Each ant $k(k = 1, 2, \ldots, m)$ (assuming ant scale is m) chooses the next section in accordance with the rules of random proportion according to the concentration of the pheromone on the path in the foraging process, i.e., at time t the ant k that is in position i chooses to move to position j in a certain probability, this probability is called as transition probability, which is calculated by the formula (1):

$$P_{ij}^{k}(t) = \begin{cases} \dfrac{\tau_{ij}^{\alpha}(t)\,\eta_{ij}^{\beta}(t)}{\sum_{s\in notcrossed_k} \tau_{is}^{\alpha}(t)\,\eta_{is}^{\beta}(t)}, & j \in notcrossed_k \\ 0 & otherwise \end{cases} \tag{1}$$

Where $P_{ij}^{k}(t)$ denotes this probability in which at time t the ant k that is in position i chooses to move to position j, η_{ij} denotes the visible coefficient from position i to position j, the parameter α denotes the importance of the pheromone on the path to the ants choose the path, the parameter β denotes the influence degree of the heuristic information to the ants; $notcrossed_k = \{0, 1, \cdots, n-1\}$ denotes the allowed positions that the ant k can reach next. The formula (1) shows that the transition probability is in direct proportion to $\tau_{ij}^{\alpha}\eta_{ij}^{\beta}$, i.e., the more amount of pheromone on the path and the more heuristic information, the bigger probability in which ants choose the path. To prevent ants pass through the passed nodes in one cycle, taboo list is designed to record the positions which the ants has passed through, the taboo list is emptied after a cycle ends.

After all ants complete a process of finding paths, the pheromones of each section are updated with the following formula:

$$\tau_{ij}(t+1) = \rho \cdot \tau_{ij}(t) + \Delta\tau_{ij}(t, t+1) \tag{2}$$

$$\Delta\tau_{ij}(t, t+1) = \sum_{k=1}^{m} \Delta\tau_{ij}^{k}(t, t+1) \tag{3}$$

where $\Delta\tau_{ij}^{k}(t, t+1)$ denotes the amount of pheromones which the kth ant releases on section (i, j) at time $(t, t+1)$. $\Delta\tau_{ij}(t, t+1)$ denotes an incremental of pheromones on section (i, j) during this loop; The pheromones which ants release on the path evaporate at a certain rate, only part of the pheromones is left, $1-\rho$ is the pheromone evaporation coefficient, ρ is the residual coefficient, generally $\rho < 1$.

2.2 Improved Shortest Path Ant Colony Algorithm

Considering the characteristics of shortest path problem of transport network, in this paper the basic ant colony algorithm is improved based on fully taking advantage of the excellent performance of basic ant colony algorithm, so one improved ant colony algorithm is obtained which is suitable for solving the shortest path problem in a transportation network. The improvements which are made for improving the basic ant colony algorithm are the followings:

(1) To speed up the operation and reduce storage space, the search area of the network can be restricted considering the characteristics of space distribution of network, i.e., the amount of data for finding the shortest path can be reduced based on maintaining a certain degree of confidence. The algorithm of restricted area can be used to achieve that purpose; the restricted area can be oval, rectangular or polygonal area. The larger the transportation network, the more obvious the advantages of the algorithm has. In this paper a rectangular area is used to limit the search area. Only the data included in the rectangle are loaded in to find the shortest path. The search efficiency is improved effectively through this way under ensuring a certain degree of confidence of path search.

(2) In a transportation network the shortest path from one origin point to the destination does not deviate from the straight line connection the origin point and destination, so it is reasonable to choose the sections close to the line in the process of finding the shortest path. Shown in Fig. 1, assuming origin point is point A, B is the destination, one ant starts from point A, and arrive at node i, there are two nodes j and E alternatively when it chooses the next node. The Fig. 1 shows that the perpendicular distance from the point j to the line iB is less than the perpendicular distance from the point e to the line iB, then it can be known that the section ij is more inclined to iB, if the amount of pheromones of section ij and iE equal approximately and the traffic situations are roughly similar, ants will travel along thc section ij, which is in line with the usual people's psychology of choosing the travel path. For this reason we have made improvements on the transition rule of ant colony algorithm.
In the improved shortest path ant colony algorithm, the coefficient η_{ij} in the formula (1) is improved by adding heuristic information of direction in the calculation of transition probabilities in which ants choose the next position in the travel process. η_{ij} is calculated as follows:

$$\eta_{ij} = 1/w_{ij} \cdot d \tag{4}$$

where w_{ij} is the traffic impedances of section ij, d is the perpendicular distance from the node to the line iB as shown in Fig. 1. The formula (4) shows that the smaller the impedances of the section and the smaller the perpendicular distance, the more probable the section will be chosen.

(3) In basic ant colony algorithm the pheromones of all sections which each ant passed through must be updated in a loop. The efficiency of this pheromone-update rule is low. To improve the performance of the algorithm,

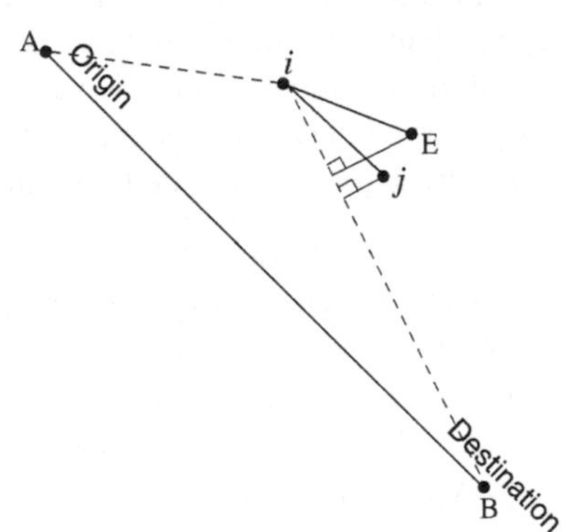

Fig. 1 Search direction

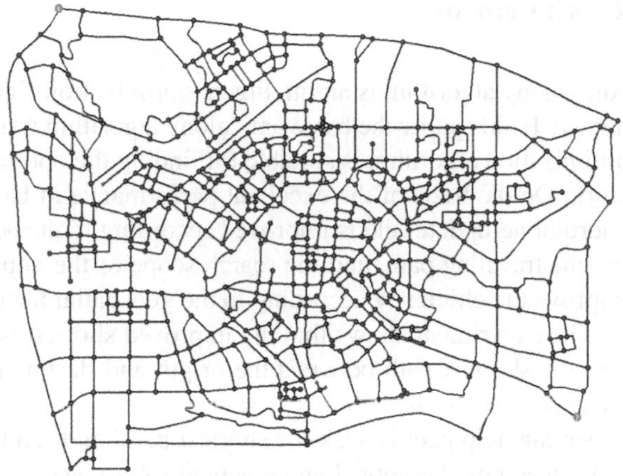

Fig. 2 One city transportation network

the pheromone-update rule of basic ant colony algorithm has been improved. First, the ants whose fitness is high are selected, i.e., the *m* ants are chosen, the paths are the top *m*th shortest path, and then the pheromones of the *m* shortest paths will be updated. The amounts of pheromones released on the paths are determined according to a feature of the paths. The people generally prefer the paths with less traffic lights and intersections, so in this paper the amounts of the pheromones of the top *m* selected paths depend on the number of intersections of the paths, the less the intersections, the more pheromones released on the path. The increment of pheromones is still calculated according to the formula (2) (3), where *m* represents the top *m* shortest paths.

3 Simulation Experiment

The shortest path ant colony algorithm has been simulated on a transport network which has 1015 nodes and 1006 sections (shown in Fig. 2). The program of the improved ant colony algorithm is developed with C# under the ArcGIS10.2 environment. Three pair of origin and destination are selected to test the algorithm. The experimental results show that the improved ant colony algorithm has significant performance than the basic ant colony algorithm.

4 Conclusion

Ant colony algorithm is an intelligent optimization algorithm with excellent performance. In this paper the basic ant colony algorithm was improved so that ant colony optimization algorithm is suitable for finding the optimum path on the transport network. On the basis of the excellent performance of basic ant colony algorithm, the pheromone-update rule is improved according to the people's psychology of choosing the traveling path, and the search scope of the improved algorithm is limited to improve the efficiency according to the geospatial feature of transport network.

The experiment shows that the improved shortest path ant colony algorithm can find the shortest path between the origin and destination more efficiently than the basic ant colony algorithm. The shortest path ant colony algorithm proposed in this paper can help people seek the shortest path, and can be used to assign traffic flow. It has a certain theoretical and practical significance.

Acknowledgments Supported by: Natural Science Research Project of Education Department of Zhejiang Province (Y201432450), and Huzhou Municipal Science and Technology project (2014YZ09).

References

1. Pham, Duc-ThinhHoang, Bao An Mai; Thanh, Son Nguyen; Nguyen, Ha; Duong, Vu. A Constructive Intelligent Transportation System for Urban Traffic Network in Developing Countries via GPS Data from Multiple Transportation Modes, Intelligent Transportation Systems (ITSC), 2015 IEEE 18th International Conference, 1729 – 1734, 2015
2. Szeto WY (2014) Dynamic modeling for intelligent transportation system applications. J Intell Transp Syst: Technol Plan Op 18(4):323–326
3. Dorigo M (1997) IRIDIA, Vrije Univ., Brussels, Belgium; Gambardella, L. M. Ant colony system: a cooperative learning approach to the traveling salesman problem. IEEE Trans Evolut Comput 1(1):53–66
4. Reed Martin, Yiannakou Aliki, Evering Roxanne (2014) An ant colony algorithm for the multi-compartment vehicle routing problem. Appl Soft Comput 15:169–176
5. Dash Madhumita, Balabantaray Madhusmita (2014) Routing problem: manet and ant colony algorithm. Int J Res Comput Commun Technol 3(9):954–960
6. Dorigo M, Birattari M, Sttzle T, Libre U, Bruxelles D, Roosevelt AFD (2006) Ant colony optimization C artificial ants as a computational intelligence technique. IEEE Comput Intell Mag 1:28–39
7. Reed M, Yiannakou A, Evering R (2014) An ant colony algorithm for the multi-compartment vehicle routing problem. Appl Soft Comput 15:169–176
8. Dash Madhumita, Balabantaray Madhusmita (2014) Routing problem: Manet and ant colony algorithm. Int J Res Comput Commun Technol 3(9):954–960

Surface Defect Detection Algorithm Based on Local Neighborhood Analysis

Chengfei Li and Xinhua Chen

Abstract Surface defect detection algorithm based on local neighborhood is proposed to improve the accuracy and real-time of surface defect detection in automation industrial production. A local neighborhood window slides over the entire inspection image, and the coefficient of variation is used as a homogeneity measure. A defect-free region will generate a smaller value of Variation Coefficient than that of a defective region. A simple threshold can thus be used to extract and segment the defective regions. The integral image is introduced to increase the computational efficiency. The proposed algorithm is used to detecting only one single discrimination feature. It could avoid complicated Spectral decomposition and sample learning. Experimental results from material surface detection in the industry, has shown the feasibility and effectiveness.

Keywords Surface inspection · Local neighborhood · Variation coefficient

1 Introduction

In the modern industrial automation production, involving a variety of product quality inspection, requests on product quality become higher and higher in industrial production line. Due to the influence of fatigue and psychological factors, the

---

Corresponding author, research field: industry control, vision detection, smart grid etc. email:car1234566@sina.com; This paper is supported by 2012 the first batch of industrial technology research and Development Fund Project of Guangdong Province (item number:2012B010100016 and supported by 2013 basic theory and science research project of Jiangmen City, and 2015 Guangdong Province Characteristic innovation project (educational research project).

C. Li (✉) · X. Chen
School of Information Engineering, Wuyi University, JiangMen, GuangDong, China
e-mail: car1234566@sina.com

X. Chen
e-mail: chenxhwin@126.com

© Springer International Publishing Switzerland 2017
V.E. Balas et al. (eds.), *Information Technology and Intelligent Transportation Systems*, Advances in Intelligent Systems and Computing 454,
DOI 10.1007/978-3-319-38789-5_40

human eye cannot usually continue to steadily and reliably detect the defects in real time. With the development of signal processing theory and computer technology, based on the image sensor and image processing technology of machine vision automatic detection system has been widely used. Compared with traditional methods, machine vision inspection technology has non-contact, high speed, high precision, low cost advantages [1, 2]. In machine vision systems, surface defect detection is very important, it is widely used in steel balls surface defect detection [3], the rail surface detection [4], the bottle caps surface inspection [5], the slab surface defect detection [6] and solar wafer surface detection [7] and many other fields.

In recent years, surface defect detection technology has developed rapidly. There are many related algorithms, such as histogram statistics, co-occurrence matrices, autocorrelation, the Fourier transform, the Gabor filtering [8]. However surface defect inspection technology is still faced with many problems, such as uneven illumination, defect area and the non-defect area low contrast, large environmental noise, detecting slow, low recognition accuracy. To overcome the problem mentioned above, in this study, we propose a simple and rapid detection method based on local neighborhood, using the coefficient of variation as defect detection and localization of homogeneity measure and using the integral image technology to reduce computing time. The proposed measure has high inspection speed and high accuracy, and insensitive to uneven illumination. It is well applied to defect detection in steel plate, leather in this study.

2 Surface Defect Detection Algorithm Based on Local Neighborhood Analysis

2.1 *The Coefficient of Variation*

The coefficient of variation, also known as dispersion coefficient. It is a reflection of a statistical distribution of the amount of data, which reflects the degree of variation of a set of observed data. The coefficient of variation is the ratio of the standard deviation of the mean. This ratio is a relative variation index, used to reflect the degree of dispersion on mean unit. When you need to compare the size of the dispersion of two or more sets of data, if the difference between the two sets of data measuring scale is too large, the coefficient of variation can eliminate the effects of scale. The coefficient of variation is not only influenced by the value of the discrete variable degree, but also affected by the variable value average size. The coefficient of variation is given by

$$C_v = \frac{\sigma}{\mu} \times 100\,\% \tag{1}$$

where σ is standard deviation and μ is mean.

In general, if the variation coefficient of a group of data is larger, it means that the group data relative to the distribution of discrete, density change is bigger, non-uniform distribution of data. On the contrary, if a group of data of variation coefficient is small, then the group data density change is not big, the data distribution is relatively uniform.

2.2 *The Description of the Local Neighborhood Algorithm*

A defect images usually consists of defective region and defect-free region. The defect-free region of the pixel gray level distribution more uniform, the defective region is abnormal region, its presence pixel gray mutation or unevenly distributed. Homogeneity, also known as homogeneousness, it is largely associated with the image of local information, it reflects the distribution of gray scale [9]. The coefficient of variation described above can be very good reflection of the distribution of data. Therefore, the coefficient of variation can be used as a measure of the value of the image homogeneity, used to test whether there are defects in the image. For an image I of size $M \times N$, it can be seen as an image matrix of N rows and M columns. $I(x, y)$ is a corresponding gray value at the coordinates (x, y), where $x = 1, 2, \ldots, M$, $y = 1, 2, \ldots N$. Define local homogeneity measure (LHM) of the pixel $P(x, y)$. Let $W \times W$ be the neighborhood window size centering on the pixel $P(x, y)$, where $W = 2w + 1$ for some integer w, calculate coefficient of variation of pixel gray value in this neighborhood $C_v(x, y)$ which is given by

$$\mu_{x,y} = \frac{\sum_{i=-w}^{w} \sum_{j=-w}^{w} I(x+i, y+j)}{W \times W} \tag{2}$$

$$\sigma_{x,y} = \sqrt{\frac{\sum_{i=-w}^{w} \sum_{j=-w}^{w} \left(I(x, y) - \mu_{x,y}\right)^2}{W \times W}} \tag{3}$$

$$C_v(x, y) = \frac{\sigma_{x,y}}{\mu_{x,y}} \times 100\,\% \tag{4}$$

where $\mu_{x,y}$ is the gray value of pixels which within the neighborhood window, $\sigma_{x,y}$ is standard deviation, $C_v(x, y)$ is the local homogeneity measure value.

For a homogeneously textured or uniformly non-textured surface image, gray distribution on any local neighborhood uniform, its LHM value will be relatively small. In the image which contains defects, the gray distribution of defective region not uniform, its LHM value will be relatively large and the LHM of defect-free region is small. Using neighborhood window of size 25×25, the LHM value of each pixel can be calculated according to Eq. (4). Figure 1 shows a metal image

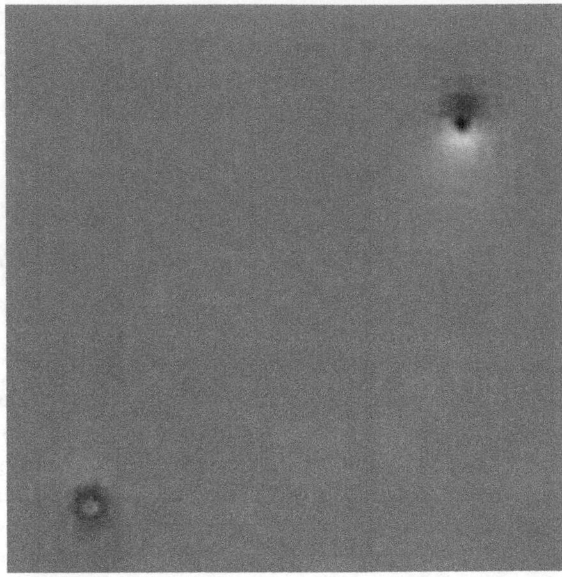

Fig. 1 Defect image

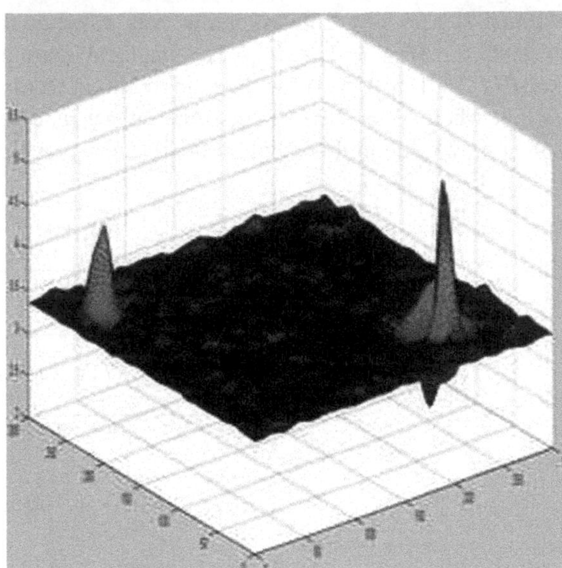

Fig. 2 LHM distribution

of size 300 × 300 which contains the hole defect. Figure 2 shows the distribution of LHM, the small value of LHM at relatively flat area representative defect-free region, the large value of LHM at two peaks representative defective region.

So we can put LHM value as discriminating feature to distinguish the defect region and the defect-free region. The greater the LHM values of pixels, it belongs to the possibility of defective region is larger, through adaptive LHM threshold selection to segment defect region. When LHM value of a pixel is above the threshold, judging it belongs to the defective region. If LHM value is less than the threshold value, it is determined that it is defect-free region. The threshold is given by

$$T_{cv} = \mu_{cv} + \omega \tag{5}$$

$$\mu_{cv} = \frac{\sum_{i=1}^{M} \sum_{j=1}^{N} C_v(i, j)}{M \times N} \tag{6}$$

where μ_{cv} is the mean of LHM values of all the pixels in an image, w is the control variable. As can be seen from the above definitions, different images have different thresholds T_{cv}. That is to say threshold T_{cv} is not fixed, it is self-adaptive variable with different images. Control variable is usually according to different object detection and takes different values, and it can be obtained by offline learning through a lot of defective samples and defect-free samples.

2.3 *Improved Algorithm Basing on Integral Images Technique*

In this paper, improved algorithm base on integral images technique is adopted. The concept of integral image was first proposed by Paul viola, for a grayscale image I, its integral image is a map. The value of arbitrary point $G(x, y)$ in integral image refers to the sum of all the pixel gray value of rectangular area which from the current point to the upper left corner in the gray image. The integral image associated with $G(x, y)$ is constructed by $G(x, y) = \sum_{x' \le x} \sum_{y' \le y} I(x', y')$. So in the image I, the sum of all the pixel gray value within the neighborhood window which centering $P(x, y)$ with size $(2w + 1) \times (2w + 1)$, with integral image can be expressed as

$$\sum_{i=-w}^{w} \sum_{j=-w}^{w} I(x + i, y + j) = G(x + w, y + w) - G(x + w, y - w) - G(x - w, y + w) + G(x - w, y - w) \tag{7}$$

So, how large the size of the neighborhood window, the sum of all pixel gray value within any window can be obtained through just 3 times with integral image arithmetic. In the first step of the local neighborhood algorithm to calculate the value of each pixel LHM, its computational complexity is $O(W \cdot W \cdot M \cdot N)$. Use integral image can greatly reduce the computational complexity, only need to $O(M \cdot N)$. And no matter how big the window size to take, it will not increase the amount of calculation.

2.4 Process of Local Neighborhood Algorithm

(1) Given a neighborhood window of size $W \times W$, and then slide over the entire inspection image in a pixel by pixel basis. Calculate the mean and standard deviation of all pixels in the window with pixel $P(x, y)$ as the center, using Eq. (4) to obtain the LHM of pixel $P(x, y)$. Then get LHM value of each pixel in the image.
(2) Calculating the mean of LHM value of all the pixels in the image, then using the Eq. (5) obtain the adaptive threshold T_{cv}.
(3) Using the threshold T_{cv} to segment defective image. Scanning the entire image, the pixel which LHM value is less than the threshold is determined as a defect-free region, while LHM value is greater than the threshold is determined as a defective region.
(4) According to the definition of surface defect area size, filter the pseudo defect which area is too small. Then get the final defect segmentation image.

3 Experimental Results

To evaluate the accuracy and reliability of the proposed surface detection algorithm, a large set of images has been used in our experiments, including smooth metal defective images, textile defective image, steel defective images. The size of each image is. In defect segmented image, the white represent the defect-free region, black represent defective region. Figure 3 shows the detect result of smooth metal. Figure 4 shows the detect result of textile. Figure 5 shows the detect result of steel.

There are various defect types in the above several sample images. The gray value of some defect regions have obvious difference with the background, some defects and background have low contrast with gray value, some defects are relatively large, some relatively small and some defects in the dark, some bright defects. As can be seen from the experimental results, for the above several kinds of surface defects, by using the method of this paper can accurately identify and locate.

When the light changed in the process of industrial production, uneven illumination is created in the image. But in a small image it can be considered to be uniform illumination. In this paper, the local neighborhood method is proposed, and it can greatly reduce or even eliminate the effects of uneven illumination. As shown in Fig. 5 detect results show that the surface defect detection algorithm based on local neighborhood effects on light has a good robustness.

The choice of Neighborhood window size $W \times W$ and control variable threshold will directly affect the result of detection algorithm. If selected window size is too small, only the edges of the defect is detected, resulting in defect detection incomplete, and even cause false detection and missed detection. If selected window size is too large, it may smooth the image, ignoring small defects, and cause missed detection. Generally, the size of neighborhood window should be larger than the

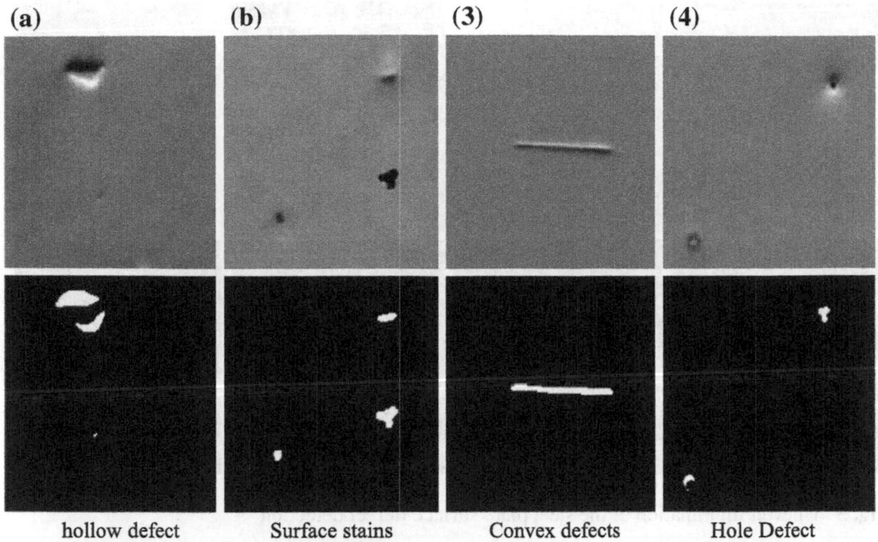

Fig. 3 Different defect types of smooth meta

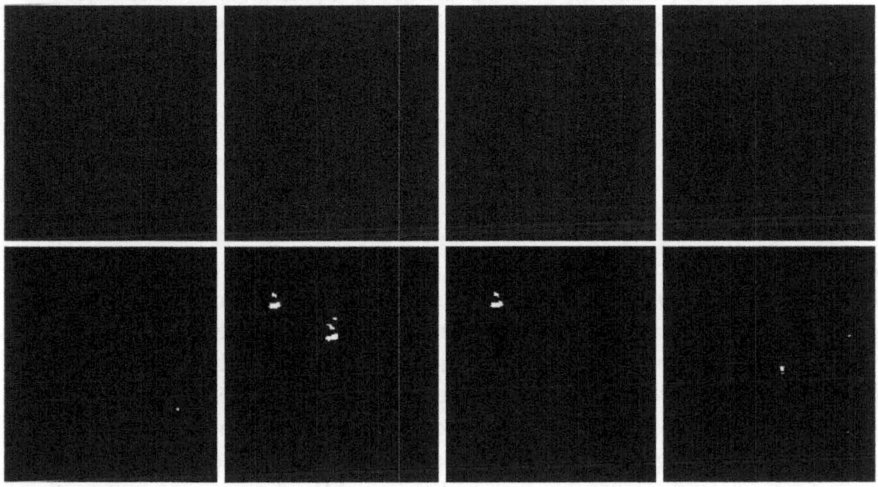

Fig. 4 Textile surface defect detection results

defect. The size of the defect is known in advance in industrial production, so you can confirm the proper neighborhood window size to detect defect. Choosing control variable threshold is very important. If you select a smaller, may generate pseudo defects and noise in the segmented defect image. If you select a larger, it maybe cause defect detection incomplete, and even defect missed detection. Similarly, suitable can be obtained by testing of a large number of defect images. In the experiments, the smooth metal surface, textile surface, steel plate surface defects are detected, their

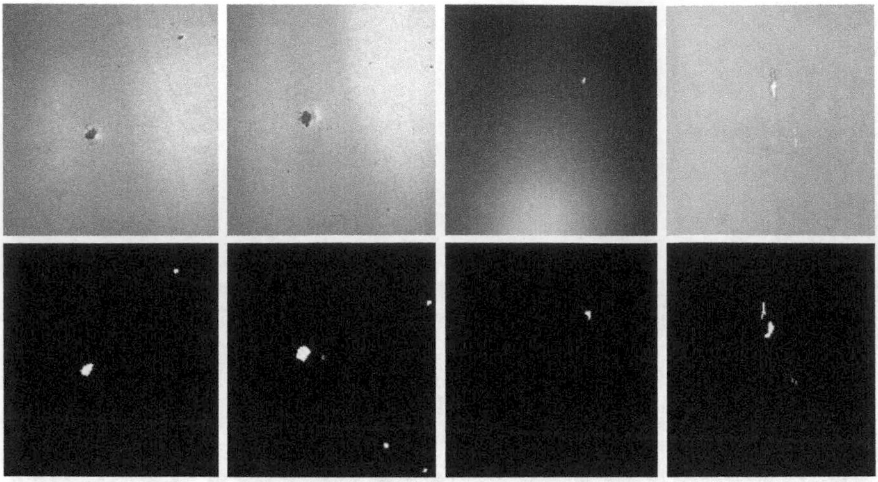

Fig. 5 Uneven illumination of the steel plate surface defect detection

Table 1 Comparison of two kinds of algorithm performance

	The number of unqualified products	Miss detection rate (%)	The number of qualified products	False detection rate (%)
Du-Ming Tsai	74	7.5	37	7.5
Algorithm in this paper	80	0	39	2.5

window size are chose to 13 × 13, 13 × 13, 16 × 16, their control variable threshold are chose to be 0.3, 0.8, 0.3. It is tested to judge the performance of the proposed defect detection algorithm in this paper and to confirm to meet the industrial production line quality control reliability requirements. In the test there are 120 samples of various kinds of steel strip surface defects, including 80 defective samples and 40 defect-free samples. At the same time, compared with another advanced method proposed by Du-Ming Tsai [10], miss detection rate and false detection rate to be as the performance of the defect detection technology. Miss detection rate is the ratio of the number of defective samples which are detected as defect-free samples and the number of all defective testing samples. False detection rate is the ratio of the number of defect-free samples which are detected as defective samples and the number of all defect-free testing samples (Table 1).

From the experiment results, the detection algorithm performed a high success rate of this paper. The experiment completed on the inter Core $i3 - 2348$ M 2.3 GHz CPU, memory for 2 g platform, using Matlab software, the average time is 0.37 s to detect an image. Therefore both in terms of detection effect and running speed can well meet the requirements of industrial production.

4 Conclusion

In this paper, an automated fast surface defect detection method has been proposed using local neighborhood analysis. The variation coefficient is used as a local neighborhood homogeneity measure, using adaptive threshold processing technology, realizing the detection and localization of surface defects. Experiments show that the method can effectively detect different types of surface defects, and is insensitive to uneven illumination, has a good robustness. It is very easy to implement and is computationally fast for online and can meet the high-speed detection requirements on the production assembly lines.

References

1. Yadong L, Peihua G (2004) Free-form surface inspection techniques state of the art review. Comput-Aid Des 36(13):1395–1417
2. Xianghua X (2008) A review of recent advances in surface defect detection using texture analysis techniques. Electron Lett Comput Vis Image Anal 7(3):1–22
3. Yuzhen M, Guoping L (2012) Detection on surface defect of large length scale steel balls. Transducer Microsyst Technol 31(8):144–146
4. Qingyong L, Shengwei R (2012) A visual detection system for rail surface defects. IEEE Trans Syst Man Cybern: Part Appl Rev 42(6):1531–1542
5. Wenju Z, Minrui F, Huiyu Z et al (2014) A sparse representation based fast detection method for surface defect detection of bottle caps. Neuro Comput 123:406–414
6. Jiawei W, Jingqi Y, Zhihong F et al (2012) Surface defect detection of slab based on the improved Adaboost algorithm. J Iron Steel Res 24(9):59–62
7. Li WC, Tsai DM (2012) Wavelet-based defect detection in solar wafer images with inhomogeneous texture. Pattern Recogn 45(2):742–756
8. Jing L, Tingting D, Dan S et al (2014) A review on surface defect detection. J Front Comput Sci Technol 8(9):1041–1048
9. Cheng HD, Sun Y (2000) A hierarchical approach to color image segmentation using homogeneity. IEEE Trans Image Process 9(12):2071C2082
10. Tsai D-M, Chen M-C, Li W-C, Chiu W-Y (2012) A fast regularity measure for surface defect detection. Mach Vis Appl 23(869):886

Part III
Proceedings Papers: Application and Technologies in Intelligent Transportation

Technique of Target Tracking for Ballistic Missile

Xu-hui Lan, Hui Li, Changxi Li and Mao Zheng

Abstract Ballistic missile tracking belongs to the domain of group tracking, and arouses new difficulty for the research of target tracking technique, due to numerous group members. This paper emphasized on the research of group tracking applied in ballistic missile tracking, and concluded the ballistic missile tracking technique from the development process of group tracking, tracking techniques based on data association and non-data association, etc., and analyzed the relative techniques of random finite set (RSF) applied in group tracking, and deduced the recursive formulation and defined the algorithm performance assessment index, and finally pointed out the development trend of ballistic missile tracking technique.

Keywords Ballistic mission · Group tracking · Data association · Random finite set

1 Introduction

In order to improve the missile penetration capability, ballistic missile often takes penetration measures in the middle of the flight, such as launching confusion jamming and simulated warhead, or exploding the end-stage rocket into pieces forming interference fragment cloud. Since there is no air resistance at this stage, warhead,

X.-h. Lan (✉) · H. Li · C. Li · M. Zheng
No. 4 Department, Air Force Early Warning Academy,
Wuhan 430019, China
e-mail: 4517665@qq.com

H. Li
e-mail: 358620937@qq.com

C. Li
e-mail: 1847743223@qq.com

M. Zheng
e-mail: maomao_bear@163.com

© Springer International Publishing Switzerland 2017

V.E. Balas et al. (eds.), *Information Technology and Intelligent Transportation Systems*, Advances in Intelligent Systems and Computing 454,
DOI 10.1007/978-3-319-38789-5_41

decoy, fairing, bus and wreckage move at high speed in the vicinity of trajectory, which formed the group targets that diffuse range up to several kilometers [1].

Group targets refers to multi-objective set that maintain a relatively fixed spatial position over a long period of time and meeting a given space constraint. The literature about the technique of group targets tracking published at home and abroad, mainly aims at aircraft fleet and ship formation and there is a few of literature specific to ballistic missile group targets tracking [2]. Group tracking belongs to multi-target tracking in essence, but the framework for group tracking adds two functional units, clustering detection and combine separation detection, with respect to the framework for multi-target tracking.

Compared ship formation, aircraft fleet and other group targets, the number of ballistic missile group targets is large without certain rules, at the same time, the phenomenon of group member occlusion, coupled with the limited resources and radar resolution, the radar could not distinguish each target in the group. Therefore ballistic missile group targets belong to the target group that can not be distinguished or can be distinguished partially, and we classified this kind of group tracking as dense multi-target tracking.

From the formation of the idea of ballistic missile group targets tracking, the defects of the traditional data association algorithm when applied to track ballistic missile group targets and the technique of non-associated ballistic missile group targets, this paper conducted a comprehensive exposition about ballistic missile group targets, and raised some new ideas on specific issues, and elaborated the trend and difficulty of the technique of ballistic missile group targets tracking finally.

2 The Formation of the Idea of Ballistic Missile Group Targets Tracking

Wax proposed the basic concept of multi-target tracking in 1955 [2], and the basic process and algorithm have matured after decades of development. For now, the single radar has been successfully achieved for sparse multi-target tracking [3], while the multi-target tracking algorithm under coarctation multi-echo becomes more perfect. However, there is no effective way for dense multi-target tracking of group targets such as ballistic missile.

Based on the analysis of the existing tracking algorithm, the entire neighborhood data association algorithm and the formation target tracking algorithm, which are representative of joint probability data association algorithm, are most likely apply to dense multi-target tracking. However, there are still some flaws when the two algorithms apply to track ballistic missile group targets. First, multi-target tracking and other joint probability data association algorithms exist mistracking and lose targets problem when track ballistic dense group. We can improve the dense multi-target tracking capability of the algorithm by reducing its radar tracking gate to some extent, but the radar tracking gate can not be infinitely narrowed. However, the members of the ballistic missile group targets are numerous, and the capability of tracking ballistic missile group targets is limited radar tracking this by reducing the radar tracking gate. Secondly, multi-targets in formation is target collection that target spacing is small, but the speed, the direction are basically the same, which is a special form of dense multi-target, without universal. Meanwhile, there is no law in orientation of the members of the ballistic missile group target, nor in line with the target formation characteristics. Therefore, whether the entire neighborhood data association algorithm, or formation tracking algorithm based on formation target, can not completely solve the problem of dense multi-target tracking.

To solve these problems, given the current sparse multi-target tracking algorithm has matured, and ballistic missile group can be viewed as a whole, and be divided into a number of discernible single target, indiscernible dense multi-target group or both coexist within the allowable range of radar resolution.

After the above analysis, we can make a conclusion that the achievement of ballistic missile target group tracking mainly has the following steps:

(i) Divide the ballistic missile group into a number of discernible single target and indiscernible dense multi-target group according to certain rules;

(ii) View the indiscernible dense multi-target group as a whole, form a number of discernible and partial discernible multi-group target;

(iii) Deal with multi-group target as sparse multi-target tracking.

3 Defects of Traditional Data Associated Multi-target Tracking Method to Track Ballistic Missile Group Target

In the field of multi-target tracking, traditional data association algorithm firstly determines the correspondence between the targets and sensor measurements using data association techniques, and then separately estimates the state of each target, its main typical algorithms including Nearest Neighbor Data Association (NNDA), Probability Data Association (PDA), Joint Probability Data Association (JPDA), Multiple Hypothesis Tracking (MHT), etc.

NNDA algorithm determines the correct measurement based on the principle of minimum statistical distance between measurement and prediction centers achieving multi-target tracking. This method is only applicable to sparse multi-target tracking, and unable to meet need of ballistic missile dense multi-target tracking. PDA algorithm assumes that there is a only correct measurement in each tracking gate, and other measurements are considered as false measurements obeyed uniform distribution. This method is applicable to single target tracking under dense echo. When the distance among targets is large, the measurement of a target falls into its tracking gate with small probability, and the interference disposed as false measurements obeyed uniform distribution can be tolerated. But with regard to ballistic missile group, the spacing between members is not determined, some spacing may be large, meet the conditions that processed as false measurement obey uniform distribution, and some may be close to each other, which will interfere with the other goals, and lead to tracking unstable, even result in the mistracking. Therefore, PDA algorithm can not completely solve the problem of ballistic missile group tracking.

JPDA algorithm taking into account the joint probability data association of all targets within the surveillance zone is one of the accepted and most effective multi-target tracking methods under dense multi-echo currently. JPDA algorithm is based on PDA algorithm, and needs to meet the following two conditions. Firstly, the target number is known, namely the target tracking has already started. Secondly, the relationship between the measurement and the target is unique, i.e., each measurement can only be derived from a target, and there is only one measurement at most for a target. For ballistic missile group, the number of the group is unknown, and as a result of the spacing between a small part of the group members is tiny, it is difficult to guarantee the correspondence between the measurements and the targets.

MHT algorithm assumes that each newly received measurement may comes from new target, clutter or existing target, and establishes a plurality of candidate hypotheses by a sliding window with finite length of time, then achieve the multi-target tracking by techniques of assessment hypothesis, management hypothesis, is a delay logic based method. It not only provides an effective solution to the problem of data association in the process of keeping track, but also can effectively initiate and stop multi-target track. At the same time, the number of targets and clutter presents exponential growth as the increase of the number of feasible joint hypothesis, which leads to incomplete data estimate problem, and constraints its application scope.

It is difficult or not necessary to track members one by one when tracking ballistic missile group. To solve these problems, Feldmann and Franken [6], Koch [7] proposed group tracking algorithm based on Bayes recursion formula, which requires that the motion model and measurement model of the group is linear, and only applies to single group target under no clutter conditions. However, the motion model and measurement model of ballistic missile group generally exhibit non-linear characteristics, and ballistic missile group is divided into a number of relatively independent small groups, which evolved into a multi-group tracking problem. Therefore, Koch's method is not suitable for ballistic missile group tracking.

4 Ballistic Missile Group Tracking Technique Based on Non-association

In the field of non-associated multi-target tracking algorithm, the main algorithm are symmetrical measurement equation algorithm [8] and random finite set (RFS) [9, 10] algorithm.

4.1 The Technique of Ballistic Missile Group Tracking Based on Symmetrical Measurement Equation Algorithm

Embed data association in the target state estimation process is the main issue of symmetry measurement equation method. To solve this problem, we assume that the measurement equation is a symmetric function on the target location, and don't need to consider with the measurement and target association process. However, this algorithm is strict with symmetry demanding, and only for given the number of targets case, can not be directly set when the number of targets is unknown, while embedding association problem in state estimation process also increases the complexity of the state estimation. In terms of ballistic missile group, the measurement equation and the target position is asymmetric, while the number of members in a single group is unknown, and even varies, so these harsh restrictions will limit the application of this method in ballistic missile group targets tracking.

4.2 The Technique of Ballistic Missile Group Tracking Based on RFS

Data association is the core of the whole issue of tracking the problem in traditional in the data association multi-target tracking algorithm, and error association will result in larger tracking error or losing target. Mahler apply the random finite set theory to the field of dense multi-target tracking, and proposed dense multi-target tracking algorithm based on probability hypothesis density filter (PHDF) [11–14], with the goal of avoiding the problem of data association between measurement and target. PHDF is particularly suitable for the group tracking, nevertheless mainly used in conventional individual targets multiple tracking, and rarely applied to the group tracking according to open literature.

Group tracking algorithm based on Gaussian mixture probability hypothesis density filter (GM-PHDF) [15], proposed by Clark, does not track the overall group directly, but rather focuses on the individuals within the group, and is suitable to group tracking under the condition of a linear track, do not apply to ballistic missile group tracking.

Lian-feng et al. proposed a partial distinguished group tracking algorithm based on sequential Monte Carlo probability hypothesis density filter (SMC-PHDF) [16]. It fits resample particle distribution using a Gaussian mixture model (GMM), regards the group as a whole, gets the number and state estimation of group through modeling the evolution process and the measurement equation of the group. Because the use of SMC, it is applied to the condition of nonlinear target motion model and measurement model, while has a good track effect for the issue of group tracking with clutter, the number of group and individual members of the group changed over time.

Time evolution model of ballistic missile group State Assume that state time evolution model of group $g(g = 1, 2, 3, \ldots, N)$ at k meets Markov, then the transition probability density of group center state is [17]:

$$f_{k/k-1}(g_k^g | g_{k-1}^g) = f_{k/k-1}(x_k^g | g_{k-1}^g) f_{k/k-1}(X_k^g | X_{k-1}^g)$$

In the equation, x_k^g represents the center state of group $g(g = 1, 2, 3, \ldots, N)$ at k, x_k^g represents the shape matrix state of group $g(g = 1, 2, 3, \ldots, N)$ at k, g_{k-1}^g represents the state of group $g(g = 1, 2, 3, \ldots, N)$ at $k - 1$, $f_{k/k-1}(X_k^g | X_{k-1}^g)$ represents the group shape transition probability density.

Sensor measurement model Assume that z_k^m represents target m produced by group $g(g = 1, 2, 3, \ldots, N)$ at k, then its likelihood function is [18]:

$$f_k(z_k^m | g_k^g) = \int f_k(z_k^m | p_k^g) f_k(p_k^g | g_k^g) dp_k^g$$

In the equation, p_k^g represents measurement position of group $g(g = 1, 2, 3, \ldots, N)$ at k, $f_k(p_k^g | g_k^g)$ describes the distribution of p_k^g relied on group target state g_k^g.

It should be noted that the equation given above is measurement likelihood function expression derived from group, which influenced by the environment fiercely, and has more clutter, with regard to fleet of aircraft, ships formation and other group targets. Document [16] gives a measurement intensity expression derived from clutter. However, ballistic missile group move outside the atmosphere with little clutter, therefore, there is no need to consider the clutter impact on ballistic missile group.

Ballistic missile group tracking PHDF recurrence formula Assume that multiple groups motion procession meets Poisson distribution, Mahler has get multi-sensor multi-target group Bayes filter under RFS framework, and obtained the PHDF of group by solving its first order approximation [19], as follows :

$$D_{k|k}(g_k | Z_{1:k}) = \int_{g_k \in G_k} f_{k|k}(G_k | Z_{1:k}) \delta G_k$$

In the expression, $G_k = \{g_k^g\}_{g=1}^N$ represents the state set of the group targets, $Z_k = \{z_k^m\}_{m=1}^{M_k}$ represents the sensor measurement at, M_k represents the number of measurement at k, $f_{k|k}(G_k | Z_{1:k})$ represents the posterior probability density of multiple targets state set under RFS framework, $\int \cdot \delta G_k$ represents the set integral of finite set statistic character, the process of PHDF, as follows Forecasting process

$$D_{k|k-1}(g_k | Z_{1:k-1}) = \gamma_k(g_k) + \int \left[p_{S,k|k-1}(g_{k-1}) f_{k|k-1}(g_k | g_{k-1}) + \beta_{k|k-1}(g_k | g_{k-1}) \right] D_{k-1|k-1}(g_k | Z_{1:k-1}) dg_{k-1}$$

Renewal process:

$$D_{k|k}(g_k|Z_{1:k}) = (1 - p_{D,k}(g_k))D_{k|k-1}(g_k|Z_{1:k-1}) + \sum_{z_k \in Z_k} \frac{p_{D,k}(g_k) f_k(z_k|g_k) D_{k|k-1}(g_k|Z_{1:k-1})}{\int p_{D,k}(g_k) f_k(z_k|g_k) D_{k|k-1}(g_k|Z_{1:k-1}) dg_k}$$

In the expression, $\gamma_k(g_k)$ is the intensity of group targets, $p_{S,k|k-1}(g_{k-1})$ represents the survival rate of group targets, $\beta_{k|k-1}(g_k|g_{k-1})$ represents the regeneration or split intensity of group targets, $p_{D,k}(g_k)$ represents detection probability of group targets of sensors.

Performance evaluation index of ballistic missile group tracking algorithm In the field of single target tracking, root mean square error (RMSE) is a good indicator of quantitative evaluation algorithm performance, and its expression is:

$$RMSE_k(x_k, \hat{x}_k) = \sqrt{\frac{1}{N}\sum_{i=1}^{N}(d(x_k^i, \hat{x}_k^i))}$$

In the expression, N is Monte Carlo simulation times, $\hat{x}_k$ is the estimation of x_k, $d(x_k, \hat{x}_k)$ is distance function, and is usually taken as the norm 2, namely:

$$d(x_k, \hat{x}_k) = \left\| x_k - \hat{x}_k \right\|_2$$

Loss track rate and correct association rate are as well as evaluation indicator of multi-target tracking algorithm based on data association commonly used. However, multi-target tracking algorithm based on RFS, on one hand does not perform data association, on the other hand can track multiple targets at the same time. Therefore, RMSE, loss track rate and correct association rate etc. can not be directly used to evaluate the multi-target tracking algorithm based on RFS.

Currently, the evaluation indicator of multi-target tracking algorithm based on RFS includes Hausdorff Distance (HD) [21], Wasserstein Distance (WD) [22] and Circular Position Error Probability (CPEP) [23] etc. Although these indicators are proposed for multi-target tracking algorithm, Lian-feng etc. have applied WD and CPEP to group tracking algorithm and achieved good results. Only if grasp well the distance relationship between the target set, multi-group can be regarded as a multi-target in essence. Therefore, these indicators are still suitable for quantitative evaluation of ballistic missile target tracking algorithm.

(1) Hausdorff Distance (HD)

Assume that real state set and estimation state set of the group at k are defined as G_k and $\hat{G}_k$, the definition of HD is [16]

$$HD_k = \max\left\{ \max_{x_k \in G_k} \min_{\hat{x}_k \in \hat{G}_k} d(x_k, \hat{x}_k),\ \max_{\hat{x}_k \in \hat{G}_k} \min_{x_k \in G_k} d(x_k, \hat{x}_k) \right\}$$

In the equation, x_k is the state of group center. HD is commonly used to measure the distance between two sets, and can reflect estimation results of local performance, but it is not sensitive to the number of ballistic missile group.

(2) Wasserstein Distance (WD)

Assume that real state set and estimation state set of the group at k are defined as G_k and $\hat{G}_k$, the L^2 order WD of group center state is defined as [16]:

$$WD^2 = \min_{C_k} \sqrt{\sum_{i=1}^{\hat{N}_k} \sum_{j=1}^{N_k} C_k^{ij} \left\| \hat{x}_k^j - x_k^j \right\|_2^2}$$

In the equation, N_k is the real number of group at k, $\hat{N}_k$ is the estimate number of group at k, coefficient matrix C_{ij} meets $C_{ij} > 0$, $\sum_{j=1}^{G_k} C_k^{ij} = 1/\hat{G}_k$, $\sum_{i=1}^{\hat{G}_k} C_k^{ij} = 1/G_k$, $\|.\|_2$ represents 2 norm. The advantage of WD is that when the estimation number of group is wrong, the evaluation results will be punished. Koch etc. describe WD using an oval (or ellipsoid). On the basis of this, Lian-feng defined the axis and direction angle WD distance of group shape estimation at k:

$$WD_k^d = \min_{C_k} \sqrt{\sum_{i=1}^{\hat{G}_k} \sum_{j=1}^{G_k} C_k^{ij} \left\| \hat{d}_k^i - d_k^i \right\|_2^2}$$

$$C_k$$

$$WD_k^\gamma = \min_{C_k} \sqrt{\sum_{i=1}^{\hat{G}_k} \sum_{j=1}^{G_k} C_k^{ij} \left\| \hat{\gamma}_k^i - \gamma_k^i \right\|_2^2}$$

$$C_k$$

In the expression, $d_k = [d_k^{\max}, d_k^{\min}]^T$ is the axis of oval, d_k^{max} and $d_k^{\min}$are major axis and minor axis, $\gamma_k (0^0 \leq \gamma_k \leq 90^0)$ is direction angle, and defined as the angle between d_k^{max} and X axis direction.

(3) Circular Position Error Probability (CPEP)

The CPEP of group position at is defined as:

$$CPEP_k = \frac{1}{N_k} \sum_{x_k \in G_k} \Pr ob \left\{ \left\| H_k \hat{x}_k - H_k x_k \right\|_2 > r, \forall \hat{x}_k \in \hat{G}_k \right\}$$

In the equation, r is the radius of the circle, $H_k x_k$ and $H_k \hat{x}_k$ are real position vector and estimation position in the Cartesian coordinate system at k.

5 The Development Trend and Difficulties of Ballistic Missile Target Tracking Technique

With the development of strategic early warning system construction in recent years, the group tracking technique are paid more and more attention, but most algorithm direct at aircraft fleet, ship formation etc. group tracking, and there is a little of literature about ballistic missile tracking technique currently, and a lot of technical difficulties need to be resolved. From a global perspective, how to improve the tracking equipment precision, how to find a suitable and effective tracking algorithm, how to improve the overall situation cognitive ability of the group, are the difficulty of ballistic missile group tracking technique.

(1) Equipping

The excellent equipment is the basis and the premise of target tracking. The so-called group targets have the relevant with the sensor resolution. When sensor resolution is precise enough, and all individuals within the group targets is fully resolved, the targets is the traditional multi-target; when the sensor resolution is not precise enough, and individuals within the group targets are partially or completely indiscernibility because of cluster phenomena, then all the targets are regarded as a whole, namely the group targets. Improving sensor resolution, is one of the effective ways to improve the ballistic missile group clustering detection. Currently, the resolution of multi-sensor tracking system, including radar systems, space surveillance systems, are more and more higher. For the ballistic missile which diffuses range up to several kilometers, with dense targets in the group, nevertheless, the sensor resolution still should be improved.

(2) Algorithm

At present, the traditional tracking algorithms mainly based on data association have matured, and non-associated tracking algorithms are also gradually improving. Nevertheless, the traditional data association algorithms are difficult to adapt to the needs of the ballistic missile group tracking. At the same time, PHDF recursive formula given in text contains multiple points, and can not be resolved for ballistic missile group with non-linear characteristic, and its forms of expression. Therefore, PHDF recursive formula the need for improved. SMC-PHDF tracking algorithm proposed by Feng [16], is the combination of particle filter and PHDF in essence, and achieves the non-linear characteristic target tracking. Mahler proposed extended PHDF on the basis of PHDF [24], and this algorithm can accurately obtain the state and number of multi-targets, but is more complex and time-consuming, if combined with particle filter, can also solve the non-linear characteristic target tracking in theory. With the development of intelligent information fusion technology, multi-target multi-sensor neural networks—fuzzy—expert tracking system [25] attracts more and more attention. The proposition of new track theories riches the existing tracking algorithm. However, how to integrate these new theories and new algorithms with ballistic missile group characteristics, and find a more efficient, more useful ballistic missile group tracking algorithm, is a difficult in the field of ballistic missile group tracking.

(3) Comprehensive cognitive abilities for the group situation

Ballistic missile group is numerous, widely diffuses in extended range, has only one or a few real warhead. Even if effectively split the group, forming a multi-group target, but the information, such as spatial location, identity attributes, threat degree of these group targets, is intertwined together, result in the comprehensive situation is still very complicated. Distinguishing these group targets on the level of threat class according to certain criteria in accordance with the existing recognition technology, focusing on the high level threat target tracking, decreasing the low threat level target tracking standard or do not even track, may be a good choice. On the one hand, we are able to accurately track the group targets, on the other hand, the sensor also saves resources. However, how to set thc right distinction standard of ballistic missile group threat level according to the existing recognition technology is one of the problem we need to solve.

6 Conclusions

Ballistic missile group tracking is an important part of building air and missile defense system, but also an important part of improving the strategic early warning system. Currently, there is a few published literature on ballistic missile group tracking technique. This paper analyzed the advantages and disadvantages of the traditional multi-target tracking and formation tracking techniques in tracking ballistic missile group, and conducted a systematic summary of ballistic missile group tracking technique, and put forward my own views about the development trends and difficulties of ballistic missile group tracking technique from three aspects, namely equipment, algorithms, comprehensive cognitive abilities of situation. But how to solve these problems is the next step of research we need to focus on.

References

1. Wan-xing Z (2011) BMD radar target recognition technology. Publishing House of Electronics Industry, Beijing
2. Pieume CO, Fotso LP, Siarry P (2008) A method for solving bilevel linear progremming problem. J Inf Optim Sci 29(2):335–358
3. Jian-hui G, Rong-tao Z (2012) Group tracking algorithm of ballistic mission defense. Comput Eng Appl 48(35):243–248. (Appl Phys (1955) 26:586–595)
4. Bar-Shalom Y, Li XR, Kirubarajan T (2001) Estimation with applications to tracking and navigation: theory, algorithms and software. New York, Wiley
5. Hong-ren Z, Zhong-liang J, Pei-de W (1991) Tracking of maneuvering targets. National Defense Industry Press, Beijing
6. Feldmann M, Franken D (2008) Tracking of extended objects and group targets using random matrices: A new approach. In: Proceedings of the 11th international conference on information fusion. Cologne, Germany, pp 1–8
7. Koch W (2008) Bayesian approach to extended object and cluster tracking using random matrices. IEEE Trans Aerosp Electron Syst 44(3):1042–1059

8. Kamen EW (1992) Multiple target tracking based on symmetrical measurement equations. IEEE Trans Autom Control 37(3):371–374
9. Hall DL, Llinas J (2001) Handbook of multi-sensor data fusion. CRC Press, Washingon
10. Goodman IR, Mahler R, Nguyen H (1997) Mathematics of data fusion. Kluer Academic, Norwell
11. Mahler R (2001) Detecting, tracking, and classifying group targets: a unified approach. In: Proceedings of SPIE international conference on signal and data processing of small targets. Orlando, FL, pp 217–228
12. Mahler R (2002) An extended first-order Bayes filter for force aggregation. In: Proceedings of SPIE international conference on signal and data processing of small targets. Orlando, FL, pp 196–207
13. Mahler R (2003) Bayesian cluster detection and tracking using a generalized Cheesemen approach. In: Proceedings of SPIE international conference on signal and data processing of small targets. Orlando, FL, pp 334–345
14. Mahler R, Zajic T (2002) Bulk multi-target tracking using a first-order multi-target moment filter. in: Proceedings of SPIE international conference on signal and data processing of small targets. Orlando, FL, pp 175–186
15. Clark D, Godsill S (2007) Group target tracking with the Gaussian mixture probability hypothesis density filter. In: Proceedings of the 2007 international conference on intelligent sensors, sensor networks and information processing. Melbourne, Australia, pp 149–154
16. Feng L (2009) Study on multi-target tracking algorithms based on random finite set. Xi an Jiaotong University, Xi an
17. Koch W (2008) Bayesian Approach to extended object and cluster tracking using random matrices. IEEE Trans Aerosp Electron Syst 44(3):1042–1059
18. Gilholm K, Salmond D (2005) Spatial distribution model for tracking extended objects. IEE Proc Radar, Sonar Navig 152(5):364–371
19. Mahler R (2002) An extended first-order Bayes filter for force aggregation. In: Proceedings of SPIE international conference on signal and data processing of small targets. Orlando, FL, pp 196–207
20. Bar-Shalom Y, Li X-R (1993) Estimation and tracking: principles, techniques, and software. Boston, USA: Artech House
21. Kanungo T, Mount DM, Netanyahu NS et al (2002) An efficient k-means clustering algorithm: analysis and implementation. IEEE Trans Pattern Anal Mach Intell 24(7):881–892
22. Hoffman J, Mahler R (2004) Multi-target miss distance via optimal assignment. IEEE Trans Syst Man Cybern Part A 34(3):327–336
23. Ruan Y, Willett P (2004) The turbo PMHT. IEEE Trans Aerosp Electron Syst 40(4):1388–1398
24. Blackman S, Popoli R (1999) Design and analysis of modern tracking systems. Boston: Artech House
25. Tai-fan Q (2009) Target tracking: advanced theory and techniques. National Defense Industry Press, Beijing

The Trajectory Optimization of Spray Gun for Spraying Painting Robot Based on Surface Curvature Properties

Zhang Peng, Ning Huifeng, Gong Jun and Wei Lina

Abstract Based on the surface curvature feature, the identification scheme is put forward for curvature of the free-form surface and the complex free surface, which is divided into large curvature and small curvature surface, and the graphic example is given to illustrate scheme effectiveness; Aimed at the trajectory optimization problem of spray gun of spraying-robot on complex free surface, the spray gun trajectory optimization model of different curvature surface is put forward to complete the trajectory optimization on the free surface. Combined with the least square approximation and topology element meshing techniques, the trajectory optimization problem of the intersecting area of two patches of different curvature will be turned into the trajectory optimization problem of the intersecting area of a similar natural quadric patch and an approximate plane patch. Then the trajectory optimization evaluation function model is set up, which verifies the proposed scheme effectiveness by simulation results.

Keywords Spray gun · Spray painting robot · Trajectory optimization · Complex free surface spray · Path combination optimization

Chinese Library Classification: TP24 · **Document code**: A

Z. Peng (✉) · N. Huifeng · G. Jun · W. Lina
College of Mechano-Electronic Engineering,
Lanzhou Univ. of Tech., Lanzhou 730050, China
e-mail: 1271498057@qq.com

N. Huifeng
e-mail: 1933844702@qq.com

G. Jun
e-mail: 2118608439@qq.com

W. Lina
e-mail: 143347382@qq.com

© Springer International Publishing Switzerland 2017

V.E. Balas et al. (eds.), *Information Technology and Intelligent Transportation Systems*, Advances in Intelligent Systems and Computing 454,
DOI 10.1007/978-3-319-38789-5_42

1 Introduction

The key technology research of spray gun trajectory optimization [1–3] that faced with the free-form surface is the focus of related scholars at home and abroad. The literature [4–6] according to the geometrical characteristics of the complicated surface to subdivide the surface, and establish the optimization target mathematical model to realize the Trajectory patterns and trends of spray gun, The scheme will not make coating uniformity problem as the main object to discuss. The literature [7–9] by using plane gun model on smaller surface curvature change along a specified space path to optimize spray gun trajectory, and established the curved surface spray model based on the coating accumulation rate model of plane that has deprived. But the scheme for large curvature surface spray gun spraying robot trajectory optimization is relatively inadequate. The literature [10] optimize the spraying gun trajectory of the free surface boundary, but the solution for the curvature change of more complex curved surfaces is not good enough. The 3D gun model [11–15] for different feature surface is established to solve the cylinder and cone surface and plane intersection of trajectory optimization. According to the above problem, this paper made the following content:

(1) Based on the existing 2 D and 3 D trajectory model, the relatively perfect spray gun models are established in order to achieve better spray gun path generation and optimization;
(2) Combining with the curvature features and small curvature surface mesh topology, When spraying free-form surface, on the surface with small curvature surface curvature on the spray gun trajectory optimization respectively;
(3) Discussed the surface features on the border of large curvature surface and small curvature surface, And established coating uniformity evaluation function model on the surface junction according to the different curvature gun model. Implements different curvature surface border gun spraying robot trajectory optimization;
(4) Simulation verification.

2 Gun Model

When the spray gun to the distance of the workpiece is constant, the traditional 2D gun model [7–10] is obtained by analysis of coating accumulated data in Fig. 1:

$$f(r) = A(R^2 - r^2) \tag{1}$$

Where A is constant.

$$R = h \cdot \tan \frac{\varphi}{2}$$

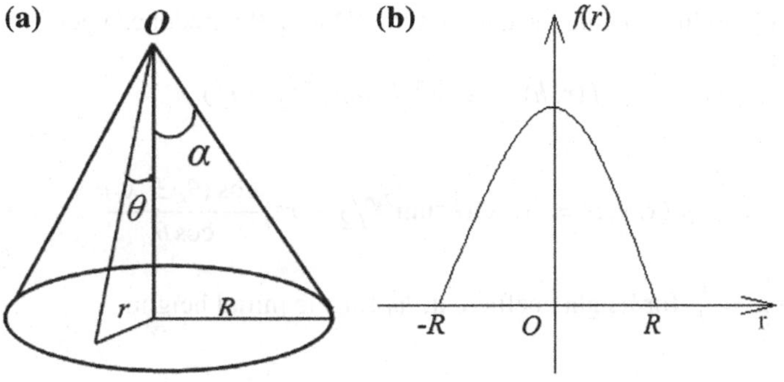

Fig. 1 Plane model of the spray gun and layer growth rate model. **a** Plane model of the spray gun. **b** Layer growth rate model

Fig. 2 Arc coating on the surface of the growth model: **a** Convex arc surface. **b** Concave arc surface

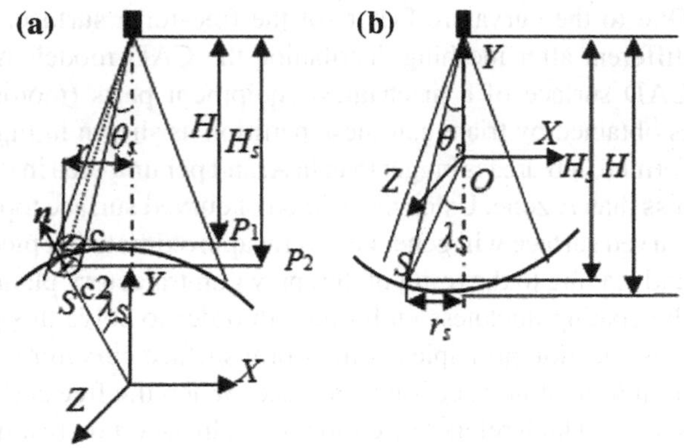

The 3D gun model [11–15] is shown in Fig. 2:

$$q_s(x, y, z) = A\left(R^2 - r^2\right)\frac{\cos(\theta_s \pm \lambda_s)}{\cos\theta_s} \quad (2)$$

where:
$H_s = \frac{r_s}{\tan\theta_s}$, $x^2 + y^2 + z^2 = \rho^2$, $r = H\tan\theta_s$, $r_s = \sqrt{x^2 + z^2}$,
$\lambda_s = \arcsin\left(\frac{r_s}{\rho}\right)$, $\theta_s = \arctan\left(\frac{r_s}{H \pm \rho - y}\right)$

In the process of spraying, the flow is the same, change, but the different paint flow is gotten. According to the Fig. 1, the flow rate of Q can be formulated as:

$$Q = \int_0^R 2\pi r f(r) dr \quad (3)$$

$$R = h \cdot \tan\frac{\varphi}{2}$$

Therefore to modify coating the cumulative 2D and 3D model can get

$$f(r,h) = A \cdot \lambda^4 (h^2 \tan^2 \varphi/_2 - r^2)$$

$$q_s(x,y,z) = A \cdot \lambda^4 (h^2 \tan^2 \varphi/_2 - r^2) \frac{\cos(\theta_s \pm \lambda_s)}{\cos \theta_s} \tag{4}$$

Among $\lambda = \frac{h_0}{h}$ for height coefficient, h_0 for the initial height.

3 The Curvature Division of the Free-Form Surface

Due to the curvature factors of the free-form surface, so the size of triangles are different after meshing distribution the CAD model. As shown in Fig. 3 will be a CAD surface of a mechanical equipment parts (motorcycle wheel cover), which is obtained by triangular mesh partition as shown in Fig. 4, every little triangle area formed in B area is bigger than in A, and per unit area in the form of triangle number is less than A zone. Using the method of curved surface topology principle to model the curved surface will generate several approximate flat piece of area of big differences, and leading to the result of the spray gun trajectory planning is not enough good and the coating thickness uniformity. In order to solve this problem, first to discuss the classification principles of free-form surface curvature. The literature [14] gives the definition of large curvature surface: When the free surface sprayed, large curvature surface which refers to the spraying radius, certain rest if another size is smaller than the length of the spraying radius, says this belongs to the large surface curvature, the method makes spraying radius R as the basis of defining large surface curvature. And in the actual process of spraying, the size of the R influenced by spray gun to spray is the distance, so not intuitive easy to identify.

In order to implement the curvature division of free surface directly and avoid the above problems, this paper puts forward the curvature recognition algorithm:

Fig. 3 The original model of surface

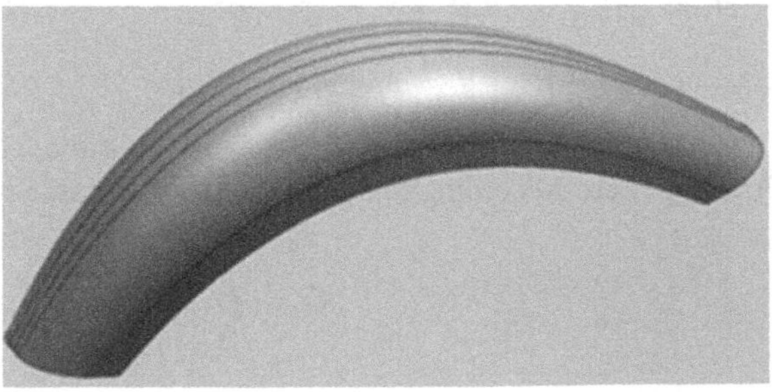

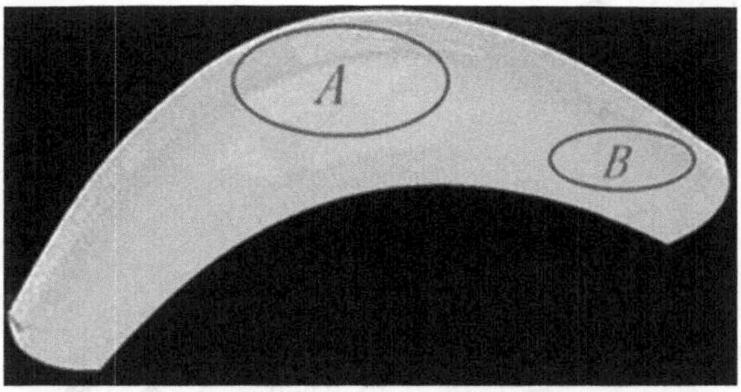

Fig. 4 Meshing figure

Step1: Identify the sprayed surface and establish 3D CAD model;
Step2: Using mesh topology to the 3D model that established in step 1;
Step3: In step 2 the grid model of the part of the triangle density is significantly higher than in other locations, as shown in Fig. 4, B in a single triangle area is almost the same size, every triangle area formed in B area is bigger than in A, and per unit area in the form of triangle number is less than A zone. So we can define B area as large curvature surface directly, area A as small curvature surface.
Step4: Based on the idea of step 3, we can delineate the large curvature surface of the free surface 3D CAD model, which facilitate robot spray gun path generation and optimization.

4 The Junction of Trajectory Optimization for Large and Small Curvature Surfaces

In order to further verify the validity of the model selection here at the junction of large curvature and small curvature surface as the research object. Known from the analysis of the literature [8–10, 15], the problem of small curvature can be divided into several planes to discuss, and the problem of large curvature surface can be translated into the natural quadric surface to discuss. So the junction of trajectory optimization problem of study for large curvature and small curvature surface can be converted to approximate plane and natural quadric surface intersection problem as shown in Fig. 5 to discuss. When natural Quadrille surface is aspheric surface, the boundary line will be circular arc, Select point S, track geometry distribution model as shown in Fig. 5a, we need to optimize the distance h1 that spray gun trajectory to the boundary and corner β that spray gun transition from plane to the sphere. When natural quadric surface is spherical surface, the boundary line will also be circular arc, select point S, track geometry distribution model as shown in Fig. 5b, we need to optimize the distance h1 that spray gun trajectory to the boundary and corner β that spray gun transition from plane to the sphere.

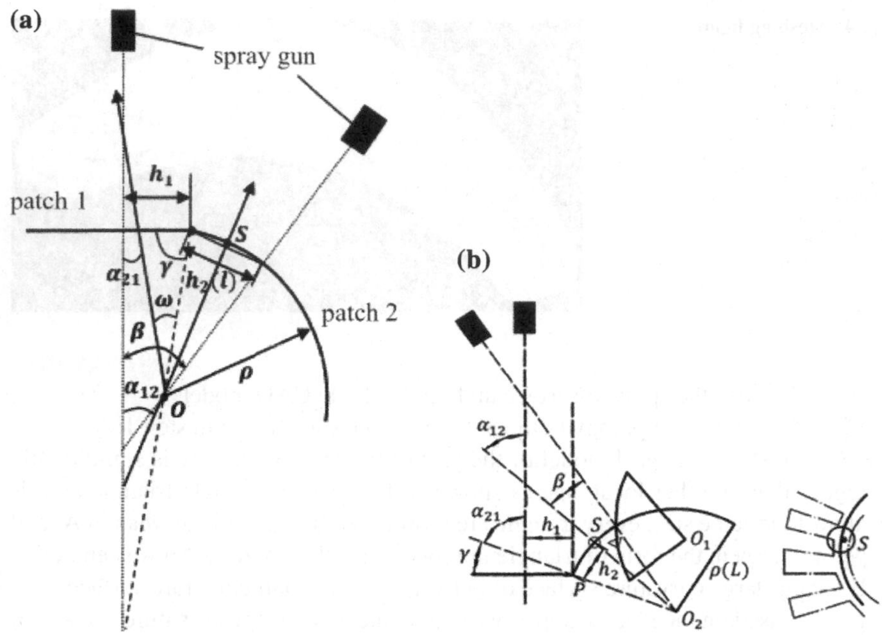

Fig. 5 Intersect model of plane and natural quadric surface. **a** Calculation model at the junction of plane and natural quadric surface (aspherical). **b** Calculation model at the junction of the plane and spherical

Among: a_{12} is the normal Angle of the face 1 relative to face 2 at point S, a_{21} is the normal Angle of the face 1 relative to face 2 at point S, the coating thickness of cylinder and spherical junction at certain point can be represented as type (5):

(1) when $0 \leq x \leq h_1$

$$q_s = q_1(x, h_1) + q_2\,(x, h_2\,(l)) \cos a_{21} \tag{5}$$

(2) When $h_1 \leq x \leq h_1 + h_2\,(l)$

$$q_s = q_2\,(x, h_2\,(l)) + q_1\,(x, h_1) \cos a_{12} \tag{6}$$

The established optimization objective function is:

$$\begin{cases} \min E(h_1, h_2) = \displaystyle\int_0^{h_1+h_2} (T_S(x, h_1, h_2\,(l)) - T_{\rm d})^2 dx \\ s.t.h_1 \geq 0, h_2 \geq 0 \end{cases} \tag{7}$$

Angle that spray gun by plane trajectory to natural quadric trajectory:

$$\beta = \frac{\pi}{2} + 2\arcsin\frac{h_2}{2\rho} - r \tag{8}$$

when the natural quadric surface is aspheric surface (a),

$$\alpha_{12} = \frac{\pi}{2} + 2\arcsin\frac{x - h_1}{2\rho(l)} - \gamma$$

$$\alpha_{21} = \pm\left[\frac{\pi}{2} - \gamma - \omega\right]$$

$$\omega = \arccos\frac{L_{os}^2(l) + \rho^2(l) - (h_1 - x)^2}{2\rho(l)L_{os}(l)}$$

$$L_{os}^2(l) = (h_1 - x)^2 + \rho^2(l) - 2\rho(l)(h_1 - x)\cos\gamma$$

When natural quadric surface is spherical (b),

$$a_{12} = \frac{\pi}{2} - r - \arccos\frac{2\rho^2 - L_{ps}^2}{2\rho^2}$$

$$a_{21} = \pm\left(\frac{\pi}{2} - r\right)$$

$$L_{ps} = h_2 - x$$

5 Simulation Prove

The determination of spray gun parameters: determine the ideal coating thickness is $48\mu m$, spray gun spray cone angle θ=45°, Spraying flow $Q = 200$ ml/s, spraying time of walking by every point is $\Delta t = 0.5$ s, Coating using coefficient is 0.8. Any natural quadric surface and plane will be tangent in junction, means $\gamma = 90°$, arc radius of surface $\rho = 300$ mm, spray gun radius in plane and arc surface are respectively $R = 50$ mm, $\delta = 60.5$ mm, range for coating is $[45.5\mu m, 52.5\mu m]$. Spray gun velocity on the plane $v = 325.6$ mm/s, spraying rate on the arc surface is 322.7 mm/s. Spraying trajectory as shown in Fig. 6.

Stay spray gun rate on the plane and cylinder unchanged, By type [4–8] optimization can get $h_1 = 35.5$ mm, $\beta = 6.9°$. The coating thickness range on the junction is $[51.9\mu m, 47\mu m]$. And the coating thickness distribution of the simulation is shown in Fig. 7. The result shows that the coating thickness range on the junction is in the range of set coating thickness, which shows the effectiveness of optimization plan.

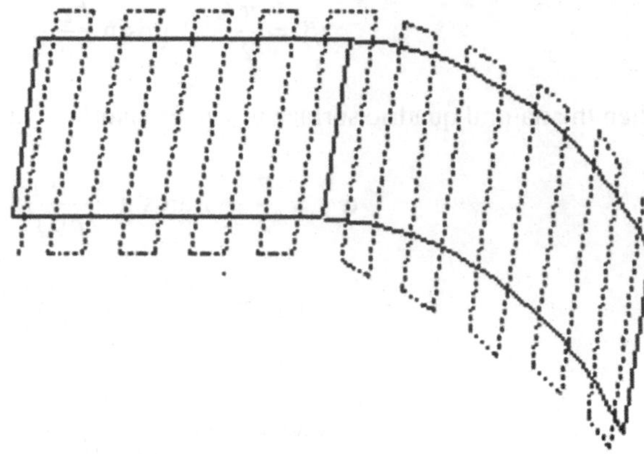

Fig. 6 Spray trajectory schematic diagram

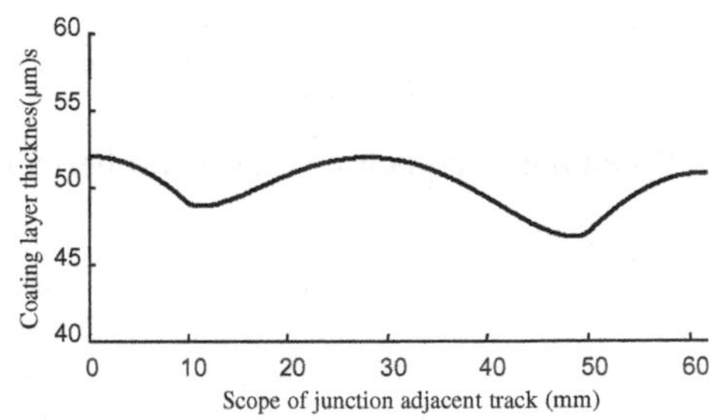

Fig. 7 The coating thickness distribution on the junction of plane and arc surface

6 Conclusion

(1) Aiming at spraying trajectory optimization study that spray the free-form surface do not consider the surface curvature change, on the basis of the CAD model and surface curvature characteristics, depending on grid method, the scheme is put forward for dividing the free surface curvature; and based on features of different curvature of the spray gun spraying model, the spraying trajectory is optimized on the free surface.

(2) According to spray gun trajectory optimization problem on different curvature surface intersect and the theory of least squares and the mesh topology, the junction problems of small curvature and large surface curvature are converted into the junction of the natural quadric and plane intersect trajectory optimization, finally the simulation prove the validity of the scheme, and realize a complex problem simple.

Acknowledgments This project is supported by National Natural Science Foundation of China (Grant no. 51165022) and Natural Science Foundation of Gansu Province (Grant no. 145RJZA028)

References

1. Hyotyiemi H (1990) Minor move-global results: robot trajectory planning. In: Proceeding of the IEEE international conference on decision and control. Honolulu, pp 16–22
2. Suh SH, Woo IK, Noh SK (1991) Development of an automatic trajectory planning system (ATPS) for spray painting robots. In: 1991 IEEE international conference on robotics and automation, 1991. Proceedings. IEEE, pp 1948–1955
3. Tokunaga H, Okano T, Matsuki N et al (2004) A method to solve inverse kinematics problems using lie algebra and its application to robot spray painting simulation. In: ASME international design engineering technical conferences and computers and information in engineering conference. American Society of Mechanical Engineers, pp 85–91
4. Sheng W, Xi N, Song M et al (2000) Automated CAD-guided robot path planning for spray painting of compound surfaces. In: 2000 IEEE/RSJ international conference on intelligent robots and systems, 2000. (IROS 2000). Proceedings, vol 3. IEEE, pp 1918–1923
5. Sheng W, Xi N, Chen H et al (2003) Surface partitioning in automated CAD-guided tool planning for additive manufacturing. In: 2003 IEEE/RSJ international conference on intelligent robots and systems, 2003. (IROS 2003). Proceedings, vol 2. IEEE, pp 2072–2077
6. Sheng W, Chen H, Xi N et al (2005) Tool path planning for compound surfaces in spray forming processes. IEEE Trans Autom Sci Eng 2(3):240–249
7. Chen H, Sheng W, Xi N et al (2002) CAD-based automated robot trajectory planning for spray painting of free-form surfaces. Ind Robot: Int J 29(5):426–433
8. Chen H, Xi N, Sheng W et al (2004) Optimal spray gun trajectory planning with variational distribution for forming process. In: 2004 IEEE international conference on robotics and automation, 2004. Proceedings. ICRA'04, vol 1. IEEE, pp 51–56
9. Chen H, Xi N, Sheng W et al (2005) General framework of optimal tool trajectory planning for free-form surfaces in surface manufacturing. J Manuf Sci Eng 127(1):49–59
10. Wei C, Dean Z (2009) Tool trajectory optimization of robotic spray painting. In: Second international conference on intelligent computation technology and automation, (2009) ICICTA'09, no. 3. IEEE, pp 419–422
11. Zeng Y, Gong J (2010) Optimizing material distribution for complex curved surface in integrated trajectory optimization of spray painting processes. In: 2010 international conference on electrical and control engineering (ICECE). IEEE, pp 1444–1448
12. Zeng Y, Gong J (2010) The trajectory optimization of spray painting robot for conical surface. Adv Mater Res 139:2189–2194
13. Wei C, Dean Z, Yang T (2008) Trajectory optimization on robotic spray painting of free-form surfaces. Trans Chin Soc Agric Mach 39(1):147–150 [in Chinese]
14. Jun G, Yong Z (2009) New models of spray gun and simulation in spray painting. Mach Des Manuf 10:106–108 [in Chinese]
15. Yong Z, Jun G (2011) Trajectory optimization of spray painting robot for natural quadric surfaces. China Mech Eng 22(3):282–290 [in Chinese]
16. Youdong C, Xiaodong L, Hongxing W (2012) Trajectory planning based on latitude for painting on spherical surface. In: 2012 10th IEEE International conference on industrial informatics (INDIN). IEEE, pp 431–435
17. Chen W, Zhao D (2013) Path planning for spray painting robot of workpiece surfaces. Hindawi Publ Corp Math Probl Eng 6:451–457

Characteristics Analysis on Speed Time Series with Empirical Mode Decomposition as Vehicle Driving Towards an Intersection

Liangli Zhang and Bian Pan

Abstract In this paper, we explore the characteristics of vehicle speed time series which described the processes that a driver finishing a specific driving task with different driving operations. Three types of vehicle driving behavior like driving towards an intersection for turn-left, driving for turn-right, and driving for go-straight are designed as a set of real vehicle driving experiments to be carried out on an urban road. Similar to the expected, the collected speed time series of all driving behavior types tend to be non-linear and non-stationary. Therefore, empirical mode decomposition (EMD) is introduced to analyze the characteristic values of speed time series intrinsic mode functions (IMF) and residues. After decomposing, there are 4 levels of IMF with a residue exist existed in the speed time series of turn-left driving behavior, as well as 3 levels in turn-right and 5 levels in go-straight. All the first level IMF of three types of vehicle driving behavior have relatively high frequencies which could be regarded as systematic errors of vehicle speed sensors. As the decomposition continued, subsequent IMF frequencies become lower but average amplitudes have different change trends which could help identifying the driving behavior types. All residue curves are firstly monotone increasing and then monotone decreasing, but the occurrence time of residue maximums are inconsistent. Through this research, we can distinguish the driving behavior type between turn-left, turn-right and go-straight with those vehicle driving behavior time series characteristic values and their changing trend. If all those judgment and statistics of characteristic values be implemented by vehicular industrial control computes, it would improve driving behavior recognition or prediction performances of an advanced driving assistance embedded on vehicle.

L. Zhang (✉) · B. Pan
Wuhan University of Science and Technology, Heping Ave. 947, Wuhan 430081, China
e-mail: zhangliangli@wust.edu.cn

B. Pan
e-mail: 604029680@qq.com

© Springer International Publishing Switzerland 2017

V.E. Balas et al. (eds.), *Information Technology and Intelligent Transportation Systems*, Advances in Intelligent Systems and Computing 454,
DOI 10.1007/978-3-319-38789-5_43

Keywords Driving behavior analysis · Vehicle speed time series · Empirical mode decomposition · Intersection

1 Introduction

Vehicle speed values and time series are used to model subjects like drivers behaviors, vehicle running status and peripheral road traffic environments, which were regarded as the key factors to vehicle driving safety [1–5]. Various forms of vehicle speed information collecting experiments were designed and carried out for those researches. Sato and Akamatsu investigated the influence of the position of a forward vehicle and its following vehicle on the onset of driver preparatory behavior before making a right turn at an intersection [6]. The quantized driving speed values of the vehicles were taken advantage to predict whether a rear-end collision happening. Spek et al. described the process of drivers searching gap acceptance at intersections and proposed a statistical model [7]. It assumed the logarithms of gap time and the speed values of the vehicle closing the gap. The parameter estimation yielded that the speed had a significant effect on gap acceptance behavior.

With the increasing intelligence of modern vehicles, getting and taking advantage of speed information for improving driving safety is becoming common. To get car-following and lane-changing patterns on Swedish roads, Ma and Andreasson collected the motional data of an advanced instrumented vehicle such as acceleration, speed, and position from CAN bus of the vehicle [8]. With the similar data acquisition way, McCall et al. designed an instrumented vehicle test bed [9]. To model driving behavior enabling the simulation of Advanced Driving Assistance Systems (ADAS), Bifulco et al. designed data collecting experiments with various kinds of road-side sensors, vehicular DAQs [10]. In addition, data streams including speed, the pedal positions of the clutch, the brake and the accelerator, the rotation angle of the steering wheel were taken from the CAN network of vehicles. As the On Board Diagnosis-II (OBD-II) data on network of the latest cars widely applied [11], collecting vehicle speed values or engine parameters becomes more and more simple and easy.

Taking a preliminary look at vehicle speed values collected via vehicular experimental devices, it is easily found that speed time series tend to be a non-linear and non-stationary state. Summarizing the literatures, we know that changing trends of time series are affected by drivers intentions and operation habits, vehicle speed data source accuracies, road alignments and speed limits, and uncertain surrounding traffic situations. Such as Shi et al. measured traffic speed time series of multi-fractal behaviors with Legendre singularity spectrum [12]. The proposal approach can reveal significant information that remains hidden. Relative to the above speed time series acquired continuously, some analysis of discrete state time series about vehicle speed were reasoned by the graphical models such as Dynamic Bayesian Network [13] and conditional restricted Boltzmann machines [14].

Before modeling, it is necessary to analyze characteristics of speed time series which are corresponded to a specific driving environment. According to the non-linear and non-stationary state, Wang and Shi forecasted the short-term traffic speed with the model based on Chaos-Wavelet Analysis-Support Vector Machine (C-WASVM) theory [15]. Zheng et al. demonstrated the capabilities of wavelet transform for analyzing the feature data in a systematic congested traffic manner [16]. In addition, fractal theory regarded as multi-scale analyzing method was introduced to decompose non-linear and non-stationary time series [17, 18]. In the past years, empirical mode decomposition was frequently proposed to analyze characteristics of time series in a variety of areas. This method was originally used to find out the residue line which was under the multiple frequency mixed signals [19, 20]. But nowadays, more and more researchers regard that the intrinsic mode functions (IMF) are containing distinct frequency characteristic either [21–23]. From the perspective of recognition modeling, both residue and IMFs are the same important measures of characteristics.

In this paper, we explore the characteristics of a set of vehicle speed time series as a vehicle driving towards an intersection. The speed time series were derived from a real vehicle driving experimental platform operated by a male test driver, and some typical driving behaviors such as turn-left, turn-right and go-straight were asked to be taken into practice. To the general sense, sample values in the vehicle speed time series of those behaviors would be random because of drivers operating habits, complex surrounding driving environments and countdown traffic lights. Driven by different intentions, a driver might control his vehicle with different speed and change them frequently. However, the threshold and the frequencies values can be used as indicators for drivers behavior recognition. Obtaining the above characteristic values, the preliminary work is to adaptively decompose non-linear and non-stationary speed time series. So the selected method is empirical mode decomposition (EMD) which can supply intrinsic mode functions and distinct frequencies of the target time series quickly and accurately.

2 Methodology

In order to analyze nonlinear and non-stationary time series, Huang et al. proposed a new joint time-frequency analysis algorithm called empirical mode decomposition (EMD) [24]. It is a data-driven method which can adaptively decompose the mentioned time series or signals with a set of intrinsic mode functions (IMF). Each IMF corresponds to a specific frequency of the time series, and the specific frequency components during the decomposition are output according to an order from high to low. It is worth to note that the primary function and levels of decomposition are not need to be set in advance. So the method was often used to analyze those time series noises and their trends.

It has been known that the vehicle speed time series are decided by drivers intentions, vehicle running status, road traffic environments and so on. Hypothesize that each factor mentioned may affect the speed fluctuating in an inherent pattern expressed by IMF and a residue which shall be different from each other. The IMF can be considered as a component with a specific frequency, and the residue is the non-periodic part. Both IMF and residue may be linear or may be non-linear. Within each IMF, it has the same number of extremum and zero crossing points, and there must be only one extremum between two zero crossing points. All IMF and residue are independent. Overlaying them along t-axis can obtain the original speed time series overall. Assuming the original speed time series as $x(t)$, $(t = 1, 2, T)$ which can be decomposed to N IMF and a residue, therefore the target IMF are defined as $f_i(t)$, $(i = 1, 2, N)$ and the residue as $r_N(t)$. The followed formula exists.

$$x(t) = \sum_{i=1}^{N} f_i(t) + r_N(t) \tag{1}$$

Steps to obtain $f_i(t)$ and $r_N(t)$ with EMD are as followed.

Step 1: Find out the upper and lower envelope lines of the target time series line. Screen out sectional maximum value points of the target time series line, and then use cubic spline interpolation with the above points to obtain the upper envelope line recorded as $x_u(t)$. With the same way, we get the lower envelope line recorded as $x_l(t)$.

Step 2: Remove the internal low frequency time series from the target time series. Take the original time series for instance. After applying *Step 1* to it, we get $x_u(t)$ and $x_l(t)$ of $x(t)$. Calculate the average line recorded as $x_m(t)$ as

$$x_m(t) = \frac{x_u(t) + x_l(t)}{2} \tag{2}$$

Remove $x_m(t)$ from $x(t)$ as $h(t) = x(t) - x_m(t)$, so we get a new time series without a level of low frequency component. If the number of extremum points in $h(t)$ equals to the number of points on x-axis or their difference does not exceed 1, $h(t)$ is turning to become an IMF. In practice, only one calculation above often cannot come to the requirement. So, replace $x(t)$ with $h(t)$ and execute *Step 1* and *Step 2* again until $x_m(t)$ is gradually closed to the latest $x(t)$.

Step 3: Use a standard deviation of limit to judge whether $h_{ik}(t)$ could be identified as $f_i(t)$. A quantitative judgment threshold can help a computer program automatically cycling or ceasing calculation. Define the standard deviation of limit SD which is usually expressed as

$$SD = \sum_{k=1}^{T} \frac{\left|h_{i(k-1)}(t) - h_{ik}(t)\right|^2}{h_{i(k-1)}^2(t)} \tag{3}$$

where T is the length of the target time series, and i stands for the level number of IMF which has been defined in formula (1). Generally, SD is within (0.2, 0.3).

Step 4: Decompose $f_i(t)$ out from the original time series with the order from high frequency to low frequency. Take $f_1(t)$ for instance. Executing *Step 1* to *Step 3*, we got $f_1(t)$ as the component with the highest frequency. Remove $f_1(t)$ from the original time series as $r_1(t) = x(t) - f_1(t)$, and observe it whether $r_1(t)$ has wave character. If so, replace $x(t)$ with $r_1(t)$ for decomposing $f_2(t)$ which steps are from *Step 1* to *Step 3* again. Finally, as $r_i(t)$ does not have wave character, record it as $r_N(t)$ and $f_i(t)$, $(i = 1, 2, N)$ are identified as IMF of the original time series.

3 Vehicle Speed Time Series Collection

Vehicle speed time series for decomposition and characteristics analysis in this paper were captured by a real vehicle driving experimental platform. And three types of driving behavior were performed on an urban road. In order to make the experimental driving process flow and be out of interferences, an implement procedure of driving test was designed. More details of vehicle speed time series collection are as followed.

3.1 Driving Behavior Types Selection and Data Structure

Driving behaviors of vehicle in urban intersection areas are always be concerned by many experts and scholars of road traffic safety research fields. In order to avoid vehicles collision happening in an intersection, it is necessary to collect some observable time series before vehicles entering an intersection, analyze those time series characteristics and model for accident risk predictions. As a vehicle driving towards an intersection, one of driving behavior types such as turn-left, turn-right and go-straight had been formed in the drivers brain which usually drives this driver operating the steering mechanisms of vehicle. As to vehicle status parameters related to driving behaviors or drivers intentions, vehicle speed should be preferred. We collected the speed values with a real vehicle driving experimental platform. It connected a real car by a CAN bus data analyzer. The vehicle speed values would be intervallic captured per 1.0s and then transmitted to a vehicular industrial control computer. After the outdoor experiments, data would be exported and aligned to be a speed time series form. In order to compare the speed time series of different types of driving behaviors, speed values collected dividedly were artificially re-aligned to the same t-axis.

Fig. 1 The driving area selected for vehicle speed time series collection

3.2 Driving Area and Experimental Participants

We selected a driving area which includes bidirectional six-lanes and an X-shaped intersection in Tuanjie Avenue, Wuhan, China, shown in Fig. 1. The lane length is about 1.6 km and a set of traffic control lights with a count-down timing board is near the intersection. As this area belongs to a new developing zone of the city, few passing vehicles is benefit to the driving test execution.

Two experimental participants were employed in each driving test. One male adult selected as vehicle driver was required to have 3 years of driving experience and no less than 50 km driving per week. The other participant as an assistant was required to operate onboard instruments, give directions to the vehicle driver and monitor real-time data.

3.3 Experimental Procedure and Data Results

The entire driving test lasted 3 days. In each day, the driver should select one type of driving behavior from driving for turn-left, driving for turn-right, and driving for go-straight as his mission to complete. A valid driving section was regarded from the entrance of road (right part in Fig. 1) to the exit of intersection. During this process, the driver should not be disturbed, and he might be asked to implement a single behavior repeatedly. The collected vehicle speed data would be recorded and stored in the hard disk of vehicular industrial control compute. The validity of data was judged by the assistant sitting next to the driver. Screen out the valid speed data of each type of driving behavior and align them with timestamps given by the collection system software. Vehicle speed time series curves of each driving behavior type as driving for turn-left, turn-right and go-straight are shown in Fig. 2. The single line in each coordinate is connected with 140 discrete points of mean arithmetical values of one driving behavior type. Observing directly, those lines in Fig. 2 are obviously found having non-linear and non-stationary characteristics. In fact, we had analyzed the

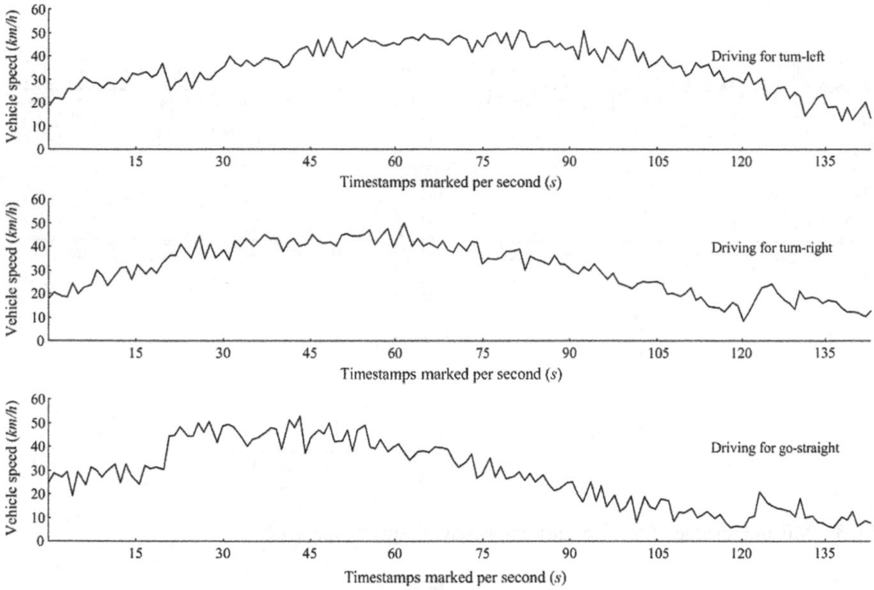

Fig. 2 Speed time series curves collected through the driving test

speed time series with fractal theory, and the results showed that they indeed belong to chaos systems with bias [25]. In this paper, we try to analyze their characteristics with EMD.

4 Decomposition and Characteristics Analysis

4.1 IMF and Residues of Speed Time Series

Decompose the speed time series shown in Fig. 2 with EMD whose computing procedure was encoded by Matlab 2014a. Set the standard deviation of limit SD = 2.5, and IMF and residues of the original time series of 3 types of driving behavior are respectively shown from Figs. 3, 4 and 5, where the abscissa values standing for timestamps marked per second (s) and the ordinate values standing for vehicle speed (km/h). In those figures, we find that 4 levels of IMF with a residue exist in the speed time series of turn-left driving behavior, as well as 3 levels in turn-right and 5 levels in go-straight. All time series of 3 types of driving behavior collected in the experiments have multi-scale decomposition characteristics distinctly.

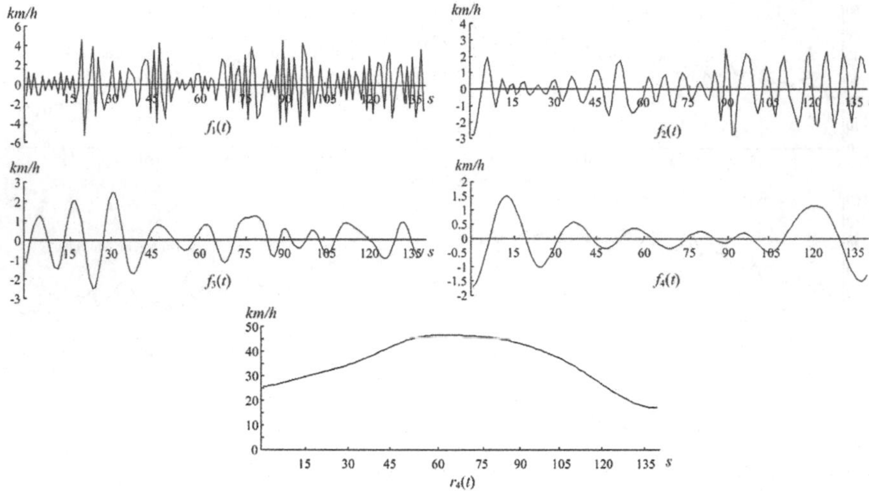

Fig. 3 IMF and residue of turn-left driving behavior speed time series

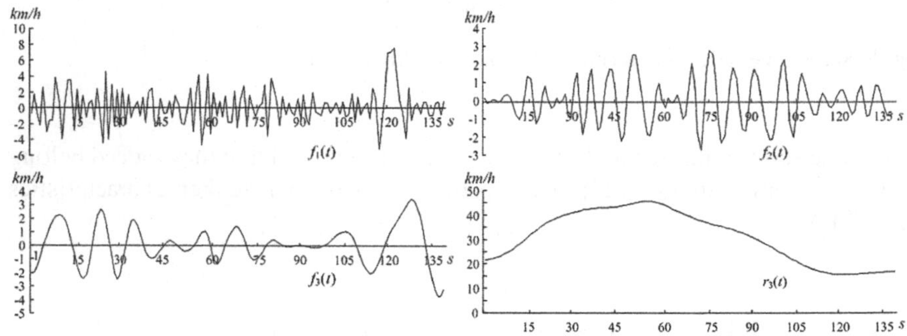

Fig. 4 IMF and residue of turn-right driving behavior speed time series

4.2 *Frequency and Amplitude Characteristics of IMF*

According to the decomposition order of EMD, $f_1(t)$ as IMF with the highest frequency is firstly obtained and often regarded as white noise time series. In this study, $f_1(t)$ of 3 types of driving behavior respectively shown in Figs. 3, 4 and 5 are also regarded as systematic errors of vehicle speed sensors. Due to the same experimental platform and the familiar driving environment, $f_1(t)$ frequency differences between 3 types of driving behavior are not obvious as turn-left 0.364 Hz, turn-right 0.336 Hz and go-straight 0.357 Hz. Meanwhile, $f_1(t)$ amplitude ranges (absolute value) of those behavior types are quite close between each other, where the average amplitude of behavior turn-left is 1.661 km/h (61.6% relative to the upper quartile), turn-right 1.477 km/h(62.9 %), and go-straight 2.295 km/h(64.7 %). All statistics of 3 types of driving behavior $f_1(t)$, such as upper and lower quartiles, maximum,

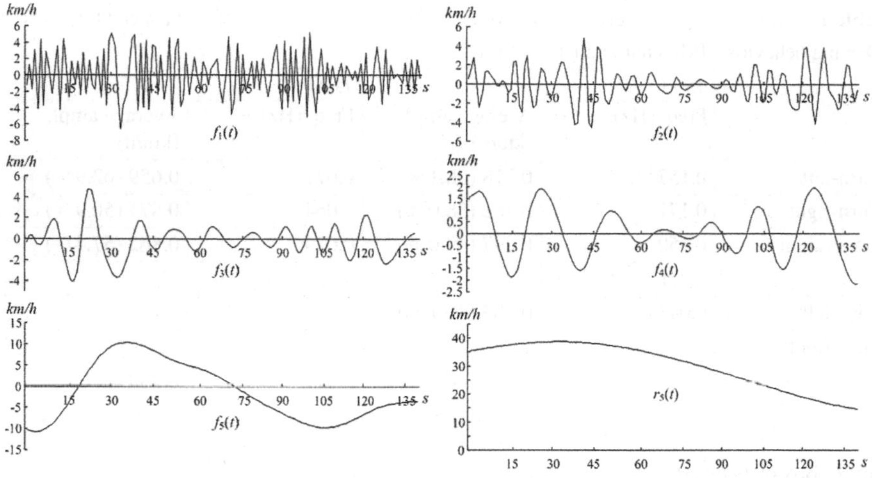

Fig. 5 IMF and residue of go-straight driving behavior speed time series

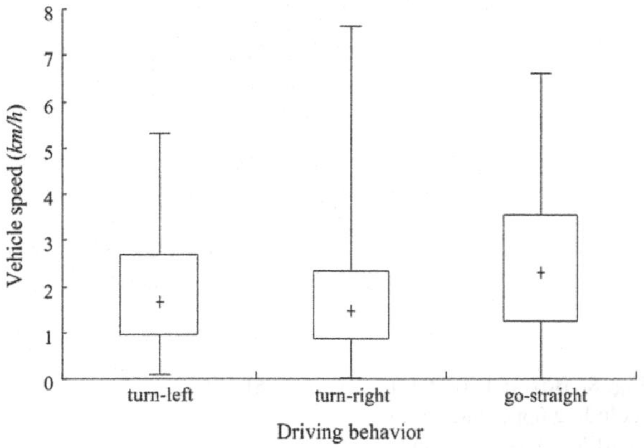

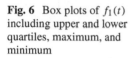

Fig. 6 Box plots of $f_1(t)$ including upper and lower quartiles, maximum, and minimum

and minimum are shown in Fig. 6. Based on the frequencies and amplitude ranges of $f_1(t)$ regarded as sensor errors, we may select an adaptive filter algorithm such as Kalman filter to get rid of them.

When it comes to characteristics of the other IMF defined as $f_i(t)$, $(i > 1)$, we also focus on the differences of frequencies and amplitude ranges. $f_i(t)$, $(i > 1)$ characteristic values of turn-left, turn-right and go-straight speed time series are shown in Table 1, where percents in parentheses followed the average amplitudes stands for the corresponding one related to upper quartile. As all types of driving behavior speed time series have IMF as $f_2(t)$ and $f_3(t)$, we draw box plots to express the upper and lower quartiles, maximums, and minimums of 3 types of driving behavior for the comparative analysis between each other, shown in Figs. 7 and 8.

Table 1 $f_i(t)(i > 1)$ characteristic values of turn-left, turn-right and go-straight speed time series

Driving behavior	IMF characteristic values			
	$f_2(t)$		$f_3(t)$	
	Freq. (Hz)	Average ampl. (km/h)	Freq. (Hz)	Average ampl. (km/h)
Turn-left	0.157	0.778 (56.1 %)	0.071	0.659 (62.9 %)
Turn-right	0.171	0.782 (58.0 %)	0.064	0.871 (50.9 %)
Go-straight	0.150	0.997 (49.0 %)	0.079	0.862 (57.4 %)
	$f_4(t)$		$f_5(t)$	
Turn-left	0.043	0.333 (40.1 %)		
Turn-right				
Go-straight	0.05	0.730 (50.2 %)	0.011	6.28 (70.0 %)

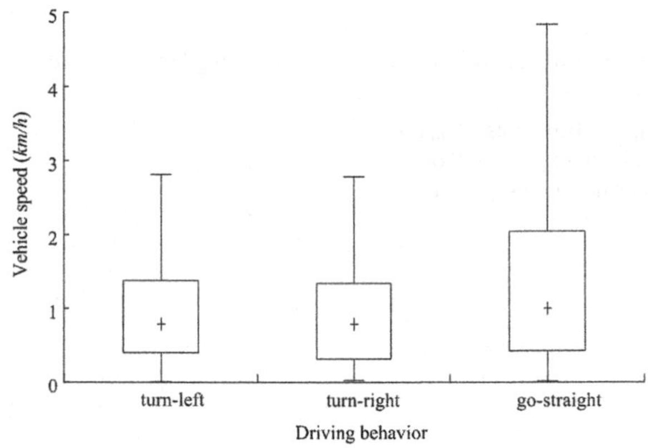

Fig. 7 Box plots of $f_2(t)$ including upper and lower quartiles, maximum, and minimum

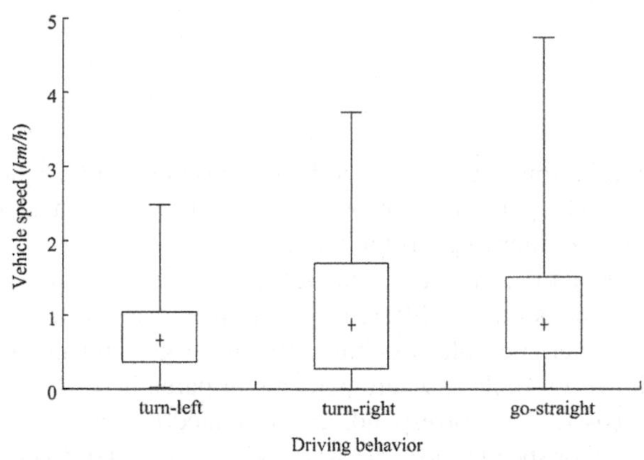

Fig. 8 Box plots of $f_3(t)$ including upper and lower quartiles, maximum, and minimum

4.3 *Trend Characteristics of Residues*

Residue curves of 3 types of driving behavior respectively shown as $r_4(t)$ in Fig. 3, $r_3(t)$ in Fig. 4 and $r_5(t)$ in Fig. 5 are firstly monotone increasing and then monotone decreasing. We focus on the maximums and their timestamps where turn-left 46.67 km/h at 64s in $r_4(t)$ of Fig. 3, turn-right 45.65 at 57s in $r_3(t)$ of Fig. 4 and go-straight 38.74 at 33s in $r_5(t)$ of Fig. 5. In addition, the average speed values during increasing and decreasing are followed by turn-left 35.93 and 35.30 km/h, turn-right 36.62 and 27.30 km/h, and go-straight 37.46 and 28.20 km/h.

5 Conclusions

Through this research, we know that speed time series derived from the vehicle driving towards an intersection with driving behavior as turn-left, turn-right or go-straight are non-linear and non-stationary. So the empirical mode decomposition (EMD) is introduced to analyze the speed time series characteristic values of intrinsic mode functions (IMF) and residues. After decomposing the speed time series captured by a real vehicle driving experimental platform, it is found that 4 levels of IMF with a residue exist existed in the speed time series of turn-left driving behavior, as well as 3 levels in turn-right and 5 levels in go-straight. That means all time series of driving behavior designed in the experiments have multi-scale decomposition characteristics distinctly. As first level IMF, all $f_1(t)$ of speed time series of driving behavior have relatively high frequencies (no less than 0.336 Hz) which can be regarded as systematic errors of the vehicle speed sensor and filtered by many software methods. As the decomposition continued, $f_i(t)(i > 1)$ frequencies become lower as i increasing but the average amplitudes have different change trends. According to the statistical results, it is found that $f_i(t)$ of turn-left and go-straight are increasing but turn-right decreasing. The residue trends of 3 types of driving behavior are similar. All residue curves are firstly monotone increasing and then monotone decreasing, but the occurrence time of residue maximums are inconsistent. Searching the driving behavior time series characteristic values and their changing trend can promote models for driving behavior recognition or prediction. Through this research, we can distinguish the type of driving behavior between turn-left, turn-right and go-straight based on $f_i(t)(1 < i < 4)$ and residues. For instance, comparing $f_2(t)$ and $f_3(t)$ characteristic values of 3 types of driving behavior, it could be found that the turn-left average amplitude in $f_3(t)$, the turn-right frequency in $f_2(t)$, the go-straight average amplitude in $f_2(t)$ respectively has significant differences relatively to the other two driving behavior types. Searching the maximum occurrence among all the residues is a relatively easy way to distinguish the driving behavior type. All those judgment and statistics can be implemented by a vehicular industrial control compute. The premise is that IMF and residues characteristic values, D-values or thresholds must be accurately obtained in advance and stored in it. That process is especially important as the mentioned method applied in a real advanced driving assistance system.

Acknowledgments This paper is supported by the National Natural Science Foundation of China under grant No. 51308426.

References

1. Eluru N, Chakour V, Chamberlain M, Miranda-Moreno LF (2013) Modeling vehicle operating speed on urban roads in Montreal: a panel mixed ordered probit fractional split model. Accid Anal Prev 59:125–134
2. Grumert E, Ma X, Tapani A (2015) Analysis of a cooperative variable speed limit system using microscopic traffic simulation. Transp Res Part C 52:173–186
3. Sun R, Zhuang X, Wu C, Zhao G, Zhang K (2015) The estimation of vehicle speed and stopping distance by pedestrians crossing streets in a naturalistic traffic environment. Transp Res Part F 30:97–106
4. Rossi R, Gastaldi M, Pascucci F (2014) Flow rate effects on vehicle speed at two way-two lane rural roads. Transp Res Procedia 3:932–941
5. Islam MdT, El-Basyouny K, Ibrahim SE (2014) The impact of lowered residential speed limits on vehicle speed behavior. Saf Sci 62:483–494
6. Sato T, Akamatsu M (2007) Influence of traffic conditions on driver behavior before making a right turn at an intersection: analysis of driver behavior based on measured data on an actual road. Transp Res Part F Traffic Psychol Behav 10(5):397–413
7. Spek A, Wieringa P, Janssen W (2006) Intersection approach speed and accident probability. Transp Res Part F Traffic Psychol Behav 9(2):155–171
8. Ma X, Andreasson I (2005) Dynamic car following data collection and noise cancellation based on the Kalman smoothing. In: 2005 IEEE international conference on vehicular electronics and safety. Beijing, China, pp 35–41
9. McCall JC, Achler O, Trivedi MM (2004) Design of an instrumented vehicle test bed for developing a human centered driver support system. In: 2004 Proceedings IEEE intelligent vehicles symposium
10. Bifulco GN, Pariota L, Brackstione M, Mcdonald M (2013) Driving behaviour models enabling the simulation of advanced driving assistance systems: revisiting the action point paradigm. Transp Res Part C 36:352–366
11. Baek S, Jang J (2015) Implementation of integrated OBD-II connector with external network. Inf Syst 50:69–75
12. Shi W, Shang P, Wang J (2015) Large deviations estimates for the multiscale analysis of traffic speed time series. Physica A 421:562–570
13. Sun J, Sun J (2015) A dynamic Bayesian network model for real-time crash prediction using traffic speed conditions data. Transp Res Part C, 54:176–186
14. Fink O, Zio E, Weidmann U (2013) Predicting time series of railway speed restrictions with time-dependent machine learning techniques. Expert Syst Appl 40:6033–6040
15. Wang J, Shi Q (2013) Short-term traffic speed forecasting hybrid model based on chaos-wavelet analysis- support vector machine theory. Transp Res Part C 27:219–232
16. Zheng Z, Ahn S, Chen D, Laval J (2011) Applications of wavelet transform for analysis of freeway traffic: bottlenecks, transient traffic, and traffic occillations. Transp Res Part B 45:372–384
17. Pei Y-L, Li H-P (2006) Research on fractal dimensions of traffic flow time series on expressway. J Highw Transp Res Dev 23(2):115–119, 127

18. Zhang Y, Guan W (2010) Analysis of multifractal characteristic of traffic-flow time series. Comput Eng Appl 46(29):23–25
19. Li X, Ding Z (2008) EMD method for multiple time-scale analysis on fluctuation characteristic of natural annual runoff time series of fen river. Water Resour Power 26(1):30–32
20. Xu T, Li K (2009) Analyzing the dynamic characteristic of the traffic flow using the EMD method. Sci Technol Eng 9(11):3003–3008
21. Molinari F, Martis RJ, Acharya UR, Meiburger KM, Luca RD, Petraroli G, Liboni W (2015) Empirical mode decomposition analysis of near-infrared spectroscopy muscular signals to assess the effect of physical activity in type 2 diabetic patients. Comput Biol Med 59:1–9
22. Mao C, Jiang Y, Wang D, Chen X, Tao J (2015) Modeling and simulation of non-stationary vehicle vibration signals based on Hilbert spectrum. Mech Syst Signal Process 50–51:56–69
23. Kacha A, Grenez F, Schoentgen J (2015) Multiband vocal dysperiodictities analysis using empirical mode decomposition in the log-spectral domain. Biomed Signal Process Control 17:11–20
24. Huang NE, Shen Z, Long SR, Wu MC, Shin HH, Zheng Q, Yen N-C, Tung CC, Liu HH (1998) The empirical mode decomposition and the Hilbert spectrum for nonlinear and non-stationary time series analysis. Proc R Soc A Math Phys Eng Sci 454:899–955
25. Zhang L, Zhu H, Chen L, Zheng A, Chu W (2015) Fractal characteristics analysis on driving behavior time series: example with speed data as vehicle driving towards an intersection. In: 2015 Proceedings of the IEEE transportation information and safety international conference, Wuhan, China, pp 126–132

Low-Complexity and High-Accuracy Carrier Frequency Offset Estimation for MB-OFDM UWB Systems

Xiu-Wen Yin and Hong-Zhou Tan

Abstract Most of carrier frequency offset (CFO) estimation methods for multi-band orthogonal frequency-division multiplexing (MB-OFDM) based ultra-wideband (UWB) systems focus on improving the robustness of CFO estimation. However, as the CFO between the transmitter and the receiver in MB-OFDM UWB systems is small, the accuracy of CFO estimation is also important. In this paper, a CFO estimator targeting at low complexity and high accuracy is proposed. A structure with dual auto-correlation (AC) is designed to improve the accuracy of CFO estimation. To maintain high robustness, a modified AC block, in which the samples that have been greatly affected by noise are removed, is proposed. The evaluation results show that the proposed CFO estimator has low implementation complexity, high accuracy and high robustness.

Keywords MB-OFDM · UWB · CFO estimation

1 Introduction

Ultra-wideband (UWB) technology can offer ideal solution for high-speed wireless personal area network (WPAN) due to the good features such as high spectrum efficiency and low power spectral density. Among the approaches for UWB implementation, multi-band orthogonal frequency-division multiplexing (MB-OFDM) [1, 2] is one of the most promising UWB technique for high-speed transmission and has been adopted by many groups such as ECMA and WUSB. In the newest version of the PHY specification of MB-OFDM proposed by WiMedia, the supported data rate up to 1024 Mbps. However, as normal OFDM systems, MB-OFDM UWB

X.-W. Yin (✉) · H.-Z. Tan
School of Information Science and Technology, Sun Yat-Sen University, Guangzhou, China
e-mail: yinxw08@163.com

H.-Z. Tan
e-mail: issthz@mail.sysu.edu.cn

© Springer International Publishing Switzerland 2017
V.E. Balas et al. (eds.), *Information Technology and Intelligent Transportation Systems*, Advances in Intelligent Systems and Computing 454,
DOI 10.1007/978-3-319-38789-5_44

systems are sensitive to carrier frequency offset (CFO). Carrier frequency synchronization error will destroy the orthogonality between sub-carriers, and lead to severe performance degradation.

Many effective methods have been proposed to improve the performance of CFO estimation, either for normal OFDM systems [3–10] or for MB-OFDM UWB systems [11–15]. However, most of them are implemented at the cost of greatly increased implementation complexity. Furthermore, these methods are effective in improving the robustness of CFO estimation, but most of them do not consider raising the accuracy of CFO estimation. For example, both the BLUE based method in [12] and the method in [15] can achieve robust results, but they do not promote the accuracy of CFO estimation.

According to [1, 2], the CFO between the transmitter and receiver in MB-OFDM systems is very small, the maximum value of the CFO is. Therefore, for MB-OFDM systems, the accuracy of CFO estimation is as important as robust.

By considering that MB-OFDM UWB systems have small CFO and are usually applied in low SNR environments, a CFO estimator, which can achieve both high accuracy and high robustness, is proposed in this paper. The rest of the paper is organized as follows. Section 2 gives a brief description of the system model. The proposed CFO estimator is presented in Sect. 3. Simulation results are shown in Sect. 4. Section 5 concludes the paper.

2 System Model

In MB-OFDM UWB systems, there are a total of $N_s = 165$ samples in a symbol, including $N_g = 32$ null samples, called zero padded (ZP), 5 null guard samples, and $N = 128$ sub-carrier samples. The time domain signals of the lth symbol after IFFT transform at the transmitter can be given as

$$s_{l,n} = \frac{1}{\sqrt{N}} \sum_{k=0}^{N-1} d_{l,k} e^{j\frac{2\pi nk}{N}} u(n) \qquad n < N_s \tag{1}$$

where, $d_{l,k}$ is the data on sub-carrier k. When $n < N$, $u(n) = 1$, and u(n) = 0 when $N \le n < N_s$. A modified S-V channel model is adopted in UWB systems. The impulse response (IR) of the channel can be given by

$$h(t) = X \sum_{l=0}^{L} \sum_{k=0}^{K} a_{k,l} \delta(t - T_l - \tau_{k,l}) \tag{2}$$

where, X is the log-normal shading, T_l is the delay of the lth cluster, $\tau_{k,l}$ is the kth ray delay related to the lth cluster, $a_{k,l}$ is the multipath gain coefficient.

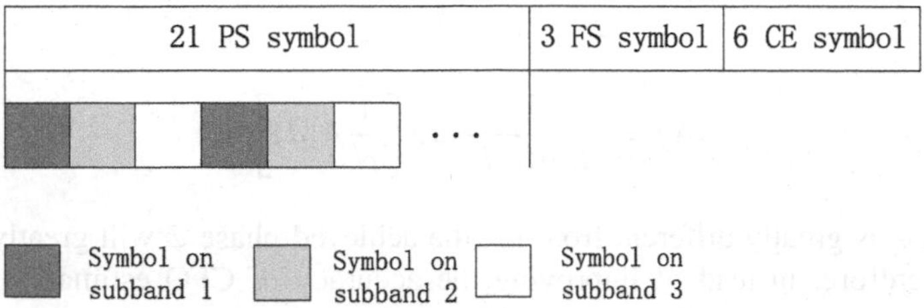

Fig. 1 The sequence of the preamble symbols under TFC 1

By assuming that timing offset has been compensated correctly, the time domain signals with CFO at receiver can be expressed as

$$r_n = \sum_{k=0}^{N-1} d_k H_k e^{j\frac{2\pi}{N}nl} e^{j\frac{2\pi}{N}n\varepsilon} + v_n \tag{3}$$

where, ϵ is the normalized CFO.

In MB-OFDM systems, CFO estimation is implemented based on aided data. There are 30 preamble symbols in the standard preamble structure, including 21 repeated packet synchronization (PS) symbols, which are used for timing and carrier frequency synchronization, 3 frame synchronization (FS) symbols and 6 channel estimation (CE) symbols. In MB-OFDM system, symbols are transmitted on sub-bands of the same band group according to a time frequency code (TFC). The sequence of the transmitted standard preamble symbols under TFC 1 is shown in Fig. 1.

3 The Proposed CFO Estimator

As repeated PS symbols are adopted for timing and frequency synchronization in MB-OFDM systems, CFO can be estimated by adopting auto-correlation (AC). Based on AC, CFO can be estimated as

$$A(d) = \sum_{m=0}^{N-1} r_{n+m}^{*} r_{i+dN_s+m} \tag{4}$$

$$\Delta \hat{f} = \frac{1}{2\pi d N_s T_s} \arg\left[\frac{1}{N} A(d)\right] \tag{5}$$

where, T_s is sampling interval and $d = \{1, 2, 3\}$, which is the symbol interval between two adjacent symbols of the same sub-band. From (5), we can see that the estimated CFO will be more accurate if d is larger. However, it will increase the possibility of the event that channel IR varying within AC operation period.

$$\varphi = \arg(r_n^* r_{n+dN_s})$$

$$\Delta \hat{f} = \frac{1}{2\pi d N_s T_s} \arg\left[\frac{1}{N} A(d)\right] \tag{6}$$

If h_{n+DN_s} is greatly different from h_n, the achieved phase φ will greatly be distorted. Therefore, instead of improving the accuracy of CFO estimation, it may degrade the estimation performance if d is increased simply.

Regardless of the used TFC type, a symbol will be received again on the same sub-band in MB-OFDM UWB systems at time index $n + 6N_s$ if a symbol is received at the nth time index. In addition, the maximum CFO in MB-OFDM UWB systems is ± 40 ppm. To the situation that the carrier frequency of about 4 GHz and the subcarrier spacing of 4.15 MHz, the normalized phase rotation between the two symbols with 6 symbol interval is approximately 0.2. Therefore, $A(d)(d = 6)$ can be used for CFO estimation in MB-OFDM UWB systems without phase ambiguity. In this paper, two AC operations are adopted in the CFO estimation process to eliminate the interference from channel time-varying. The proposed CFO estimation process is shown below. Let

$$A_s(n) = \sum_{m=0}^{N-1} r_{n+m}^* r_{n+m+sN_s} \tag{7}$$

$$\phi_s(n) = \arg\left[\frac{1}{N} A_s(n)\right] \tag{8}$$

$$A_l(n) = \sum_{m=0}^{N-1} r_{n+m}^* r_{n+m+6N_s} \tag{9}$$

$$\phi_l(n) = \arg\left[\frac{1}{N} A_l(n)\right] \tag{10}$$

According to the used TFC type, S is given by

$$s = \begin{cases} 3 & \text{TFC1} - 2 \\ 2 & \text{TFC8} - 10 \\ 1 & \text{TFC3} - 7 \end{cases} \tag{11}$$

Let

$$d_\phi = |\frac{6}{s}[\phi_l(n) - \phi_s(n)]| \tag{12}$$

The normalized CFO can be estimated as

$$\hat{\varepsilon} = \begin{cases} \frac{1}{2\pi s N_s} \phi_s(n) & (d_\phi \geq \gamma) \\ \frac{1}{2\pi \cdot 6N_s} \phi_l(n) & (d_\phi < \gamma) \end{cases} \tag{13}$$

where, γ is the predefined threshold. When $d_\phi < \gamma$, which means the value of $\phi_l(n)$ is close to $\phi_s(n)$. It can be seen as $\phi_l(n)$ is not interfered by channel time-varying. In this case, the estimated CFO based on $\phi_l(n)$ is more accurate than that based on $\phi_l(n)$. When $d_\phi \geq \gamma$, it is much possible that $\phi_l(n)$ is interfered by channel time-varying. In this case, $\Delta \hat{f}$ can be estimated based on $\phi_s(n)$.

The above method can improve the accuracy of CFO estimation while avoiding the interference from channel time-varying. However, it does not promote the noise immunity in low SNRs. As UWB systems are usually applied in low SNR environments, the noise immunity ability of CFO estimator is also important. To improve the robustness of the proposed CFO estimator, the modified $A_s(n)$ is computed as below. Let

$$a_s(n+m) = \begin{cases} r^*_{n+m} r_{n+m+sN_s} & |r^*_{n+m} r_{n+m+sN_s} - c^2_m| < \lambda \\ 0 & |r^*_{n+m} r_{n+m+sN_s} - c^2_m| \geq \lambda \end{cases} \tag{14}$$

where c_m is the m coefficient of the PS symbol, λ is the predefined threshold. Define variable M with the initial value $M = 0$. If $a_{s,n+m} = r^*_{n+m} r_{n+m+sN_s}$, M is set as $M = M + 1$. Then, the modified $A_s(n)$ and $\phi_s(n)$ are given by

$$A^{'}_s(n) = \sum_{m=0}^{N-1} a_s(n+m) \tag{15}$$

$$\phi^{'}_{s,n} = \arg\left[\frac{1}{M}\sum_{m=0}^{N-1} A^{'}_s(n)\right] \tag{16}$$

In the same way, we can achieve the modified $\phi_l(n)$. Let

$$d^{'}_\phi = |\frac{6}{s}[\phi^{'}_l(n) - \phi^{'}_s(n)]| \tag{17}$$

Finally, the estimated CFO is given by

$$\hat{\varepsilon} = \begin{cases} \frac{1}{2\pi s N_s}\phi^{'}_s(n) & (d_\phi \geq \gamma) \\ \frac{1}{2\pi \cdot 6N_s}\phi^{'}_l(n) & (d_\phi < \gamma) \end{cases} \tag{18}$$

Compared to methods in [6, 8], another major advantage of the proposed method is that the implementation complexity is much lower. The structure of the proposed CFO estimator is shown in Fig. 2.

In Fig. 2, arg is used to achieve angular component of AC results. From the figure, we can see that one AC block and one arg block are used. As $\phi^{'}_s(n)$ is achieved based on the AC which is executed by using the ith and the $i + s(s \leq 3)$th PS symbol, the process for achieving $\phi^{'}_s(n)$ has completed before PS symbol $i + 6$ received.

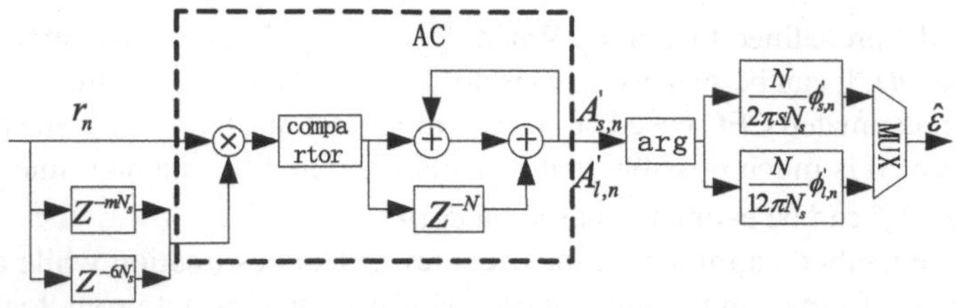

Fig. 2 Structure of the proposed CFO estimator

Therefore, $\phi_l'(n)$ can be computed by reusing the resources that used for computing $\phi_s'(n)$. In this way, only one AC block and one arg are needed in the whole structure of the estimator.

4 Evaluation

In this section, we evaluate the proposed CFO estimator. We first evaluate the complexity of the proposed CFO estimator and then simulate the performance of it. The parameters used in simulations follow the ones in [1]: $N = 128$, $N_s = 165$, the subcarrier spacing $f_i = 4.125\,\text{MHz}$ and the carrier frequencies are $\{3432, 3960, 4488\}$ MHz. The normalized CFO between the transmitter and receiver is set with $\epsilon = 0.033$. In simulations, two UWB channel modes, CM1 and CM4, and the TFC type, TFC 3, are used.

The implementation complexity comparison among CFO estimation methods is shown in Table 1. The CFO estimation method in [15] and the traditional AC based algorithm are included in the comparison. As the AC block and the arctangent operation can be reused, the proposed CFO estimator needs only one AC block and one calculating arctangent block. In the table, the ARG represents arctangent operation. As the real comparator and ABS can be implemented independently and be shared by other operations, 1 ARG, 1 complex multiplier, 1 real division, 2 complex adder, 1 real comparator and 1 ABS are needed to implement the proposed CFO estimator. Compared to the scheme in [15], which needs 3 complex multiplier and 8 complex adder, the proposed needs much fewer complex multiplier and complex adder, which means the proposed scheme has much lower implementation complexity. Of the three methods in the table, the traditional AC based CFO estimator has lowest implementation complexity. Actually, the traditional single AC based method has the lowest implementation complexity among the CFO estimation algorithms for MB-OFDM systems. As to the traditional AC based method, the proposed scheme only needs 1 real comparator and 1 real ABS more than the traditional AC based method. Therefore, the hardware cost of the proposed CFO estimator is almost the same as the traditional AC based method.

Table 1 Implementation complexity comparison among CFO estimators

Hardware resource	Proposed	Method in [15]	Traditional AC method
ARG	1	1	1
Complex multiplier	1	3	1
Real division	1	1	1
Complex adder	2	8	2
Real comparison	1	0	0
Real ABS	1	0	0

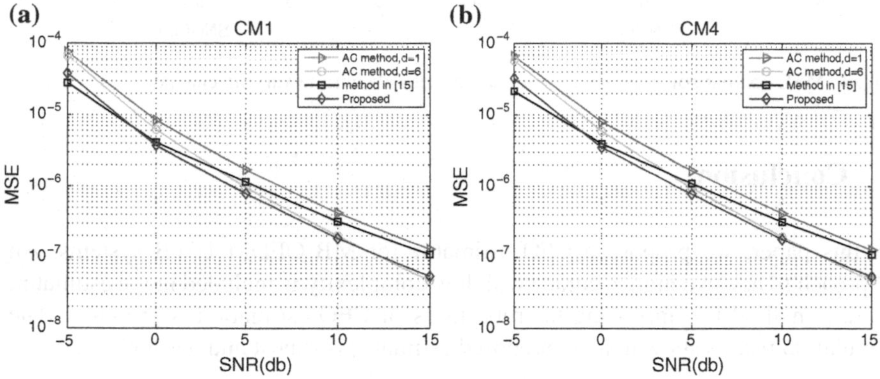

Fig. 3 MSE comparison among CFO estimation methods in time invariant channel

The performance of CFO estimation methods can be evaluated through MSE (mean square error). Figure 3 shows MSE comparison among CFO estimation methods in CM1 and CM4 by assuming the channel IR is invariant with time. The traditional AC based CFO estimation method, in which the symbol intervals are $d = 1$ and $d = 6$ respectively, and the method in [15] are included in the comparison. The BLUE based CFO estimator in [15] is not included in the comparison because the very high complexity of it. From the figure, we can see that the traditional AC method with $d = 6$ has higher accuracy than it with $d = 1$ and that the MSE of proposed CFO estimator is very close to the AC method with $d = 6$ under high SNR environments (SNR > 10 db). The method in [15], of which the CFO is estimated though taking average of the CFOs of all sub-bands, has high robustness. Under SNR = − 5 db, we can see that the proposed CFO estimator has a MSE lower than 4×10^4 in CM1, which is very close to the performance of the method in [15]. This means the proposed method has high robustness.

Figure 4 shows MSE comparison among CFO estimation methods in CM1 and CM4 by assuming that channel IR varying within the time interval between the $(i + 1)$th and the $(i + 6)$ PS symbol with 15 % probability. From the figure, we can see that the performance of the proposed CFO estimator is almost not affected in the time-varying channel while the AC method with $d = 6$ is affected significantly.

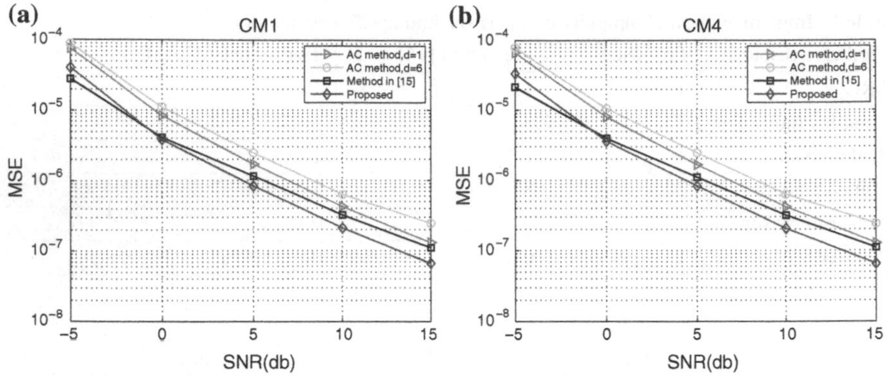

Fig. 4 MSE comparison among CFO estimation methods in time varying channel

5 Conclusion

In this paper, we propose a CFO estimator for MB-OFDM UWB systems. An approach to achieve high accuracy with low complexity is included in the estimator. And, a method for improving the robustness of CFO estimation is proposed. The simulation results show that the proposed estimator is robust and has high accuracy.

References

1. WiMedia Alliance (2009) Multiband OFDM physical layer specification
2. ECMA-368 (2008) High rate ultra wideband PHY and MAC standard, 3rd education
3. Minn H, Bhargava VK, Letaief KB (2003) A robust timing and frequency synchronization for OFDM systems. IEEE Trans Wirel Commun 2(4):822–839
4. Shim E-S, Kim S-T, Song H-K, You Y-H (2007) OFDM carrier frequency offset estimation methods with improved performance. IEEE Trans Broadcast 53(2):567–573
5. Rha HY, Jeon BG, Choi H-W (2012) Simple wide range carrier frequency offset estimator for coherent optical OFDM. IEEE Photonics Technol Lett 24(22):2064–2066
6. Morelli M, Moretti M (2013) Joint maximum likelihood estimation of CFO, noise power, and SNR in OFDM systems. IEEE Wirel Commun Lett 2(1):42–45
7. Zhou X, Yang X, Li R, Long K (2013) Efficient joint carrier frequency offset and phase noise compensation scheme for high-speed coherent optical OFDM systems. J Light Technol 31(1):1755–1761
8. Wang X, Bo H (2014) A low-complexity ML estimator for carrier and sampling frequency offsets in OFDM systems. IEEE Commun Lett 18(3):503–506
9. Qingfeng J, Cheng Ming L, Yuping ZW, Hongwei Y (2014) Pseudo-noise preamble based joint frame and frequency synchronization algorithm in OFDM communication systems. J Syst Eng Electron 25(1):1–9
10. Lmai S, Bourre A, Laot C, Houcke S (2014) An efficient blind estimation of carrier frequency offset in OFDM systems. IEEE Trans Veh Technol 63(4):1945–1950
11. Yinghui L, Minn H, Win MZ (2007) Frequency offset estimation for MB-OFDM-based UWB systems in time-variant channels. WCNC 2007, Kowloon, pp 1019–1024

12. Yinghui L, Minn H, Jacobs T, Win MZ (2008) Frequency offset estimation for MB-OFDM1-based UWB systems. IEEE Trans Commun 56:6
13. Fan W, Choy C-S (2009) Power efficient and high speed frequency synchronizer design for MB-OFDM UWB. ICUWB 2009, Vancouver, pp 669–673
14. Ye Z, Duan C, Orlik PV, Zhang J, Abouzeid AA (2010) A synchronization design for UWB-based wireless multimedia systems. IEEE Trans Broadcast 56(2):211–225
15. Sen D, Chakrabarti S, Kumar RVR (2008) An improved frequency offset estimation algorithm by multi-band averaging method for MB-OFDM based UWB communication for WPAN applications. ANTS 2008, Mumbai

The Research of Coal Mine Underground Rubber Tyred Vehicle Wireless Video Aided Scheduling System

Xiucai Guo and Boli Zhang

Abstract This paper analyzes the existing disadvantages of the rubber tyred vehicle scheduling system of mine, and design a video assistant monitoring system which helps the scheduling center to know the real-time status of the blind area and identify the vehicle license plate according to the video processing technology. Based on the license plate information from the control system database to call out the vehicle information. Provide decision making basis for scheduling.

Keywords Scheduling system · Wireless video · License plate recognition

1 Introduction

Coal mine auxiliary transport system is responsible for the transport of materials, equipment and personnel. Because of the variety of underground material and equipment, transportation sites are scattered. And work on the request in a short time to send personnel to the designated point. Auxiliary transportation is becoming more and more important. The traditional auxiliary transport uses electric motor vehicle, Winch, rail car, monorail and so on. However, due to its many links, with many people, with less weight, accidents, flexibility, low efficiency. Become a key link in production.

With the emergence and development of the non rail transportation in the underground. There is a new form of transportation in China Trackless rubber tyred vehicle. It has the characteristics of large load, strong climbing ability, high speed, high speed, high efficiency and high efficiency. All the large new mines are used to solve the problem of the bottle strength which restricts the production capacity. And achieved good economic and social benefits. Compared with the traditional trackless auxiliary

X. Guo (✉) · B. Zhang (✉)
Xi'an University of Science and Technology, Xian, China
e-mail: 870082058@qq.com

B. Zhang
e-mail: 498911923@qq.com

© Springer International Publishing Switzerland 2017

V.E. Balas et al. (eds.), *Information Technology and Intelligent Transportation Systems*, Advances in Intelligent Systems and Computing 454,
DOI 10.1007/978-3-319-38789-5_45

transportation tools in the past, it has obvious advantages. Rubber tyred vehicle is not restricted by orbit. It saves the cost and investment of laying track. And it is flexible, and it is suitable for various types of vehicles, which is of great capacity, and it is safe and efficient. It is the best choice to ensure the efficient delivery of materials, equipment, personnel. In the fully mechanized mining face, fully mechanized face relocation, the overall delivery and installation support, the performance of trackless rubber tire vehicle is more remarkable [1, 2].

Because the mine is influenced by the investment cost and the geological condition. Some auxiliary transport roadway can only accommodate a single lane. This can easily lead to the occurrence of a vehicle in a certain area of obstruction. Caused by frequent reverse and other issues, even the occurrence of safety accidents. In order to ensure the safety of production and improve production efficiency, the research and design of automatic dispatching system of rubber tire vehicle is very necessary. According to the domestic existing trackless tyred vehicle scheduling system, and put forward the shortcomings. The design of the auxiliary control of trackless tyred vehicle video system, it improves the reliability of the whole control system [3, 4].

2 Present Situation of Dispatching System

For underground trackless tyred vehicle transportation system in our country, there are Wuhan Qi huan Electrical Co., Ltd.s KJ150A moving target monitoring system for mine, and Tiandi (Chanzhou) automation Limited by Share Ltds KGE58 mining vehicle identification card, and related products early to enter the field. But at present, the domestic technology for vehicle scheduling control is still at the initial stage of exploration.

The main working principle of the system of mobile target monitoring system: application of radio frequency identification technology and modern computer communication technology. Use center control computer system in central control room to real-time monitoring of the underground rubber tyred vehicle. The data is transmitted by CAN bus, Ethernet, fiber and other network connected. Moving target of underground rubber tyred vehicle were carrying RFID tag. Wireless communication between the reader and the RFID tag, make sure target location of the object to be identified, and display the information on a large screen or computer screen of the control center in time. It provides an effective means for the production scheduling, emergency control, regional exclusion and so on. System also back up all the target information, and the data is transmitted to the management departments at all levels. It provide important basis for the leadership to make decision and command. It can also provide the interface for other management system in the mine and realize the information sharing. Through the above introduction, we can find the trackless tyred vehicle scheduling system has a drawback existing. Only when the vehicle reaches the radio frequency reader station, the system can know the location of the vehicle running. In other words, between the two radio frequency card reader and the chamber is the monitoring blind area. When there are two directions in the complex situation of multiple vehicles traveling at the same time, must comply with the following principles: The empty truck gives way to the heavy truck, the small truck gives way to the big truck, the truck with goods gives way to the truck with people, the ordinary truck gives way to the special truck. If there is a blockage in the monitoring area, the dispatch center can not know the situation in time. So this paper designs a video assistant monitoring system, it can help dispatch center to monitor the real-time status of the blind area, use video image processing technology to distinguish vehicle license plate. Based on the license plate information, we can use the control system database to call out the vehicle information, provide decision making basis for scheduling [5].

3 System Architecture

Coal mine wireless video transmission system made by the intrinsically safe camera, wireless transmitter, wireless receiver, switches, underground industrial Ethernet ring network, underground dispatching center, the ground monitoring center. As shown in Fig. 1.

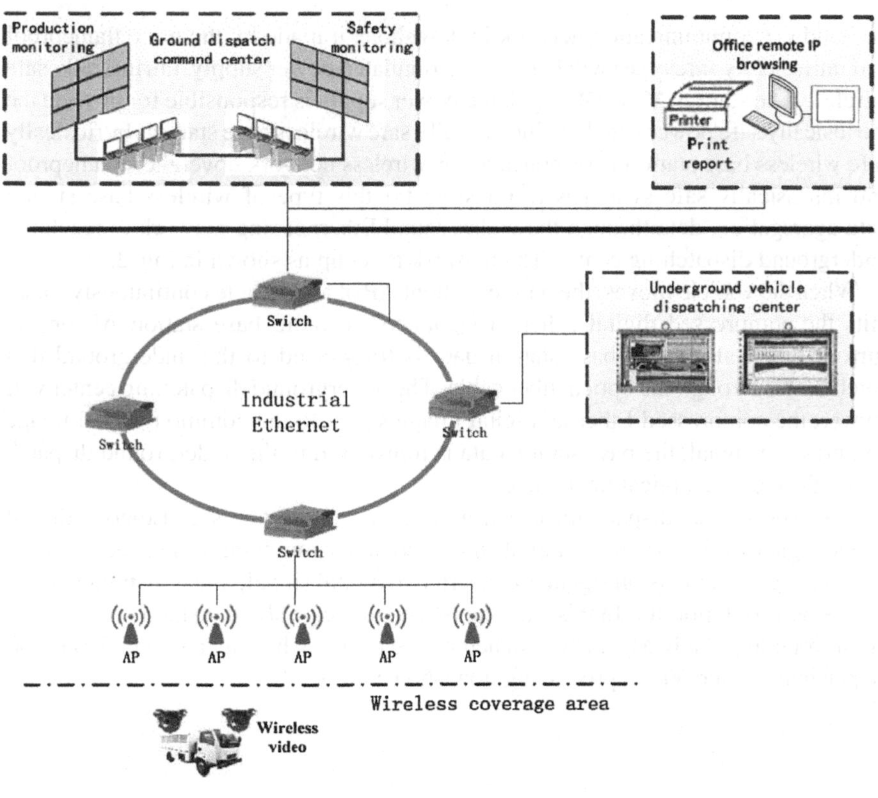

Fig. 1 System architecture

3.1 Underground Wireless Video Terminal

Put wireless safety camera on trackless rubber tyred vehicle front and rear, respectively used to observe the situation in front and rear of the car. According to the characteristics of poor light in underground tunnel, focusing on the use of low illumination, light suppression digital camera technology security camera. The image effect is clear, it overcome the problem of part of the scene camera can not avoid direct light led to a large area of the white screen. It is more conducive to monitoring center observation.

Video image digital signal information is very large, so we must compress the image digital signal. H.264 is a new generation of video compression technology. It has the following advantages: encoding efficiency is high, the ability to adapt to the network is very strong and less encoding options. So H.264 is very suitable for underground video transmission system. After video compression, using the WiFi module to send data to a nearby base station [6, 7].

3.2 Set up the Underground Network

To build the communication network in downhole, it made by the mine flameproof and intrinsically safe type switch, mine DC regulated power supply, intrinsically safe wireless base station. Mine DC regulated power supply is responsible for provide the intrinsically safe power supply to intrinsically safe wireless base station; Intrinsically safe wireless base station is responsible for wireless network coverage; Flameproof and intrinsically safe switch is responsible for this type of wireless base station data aggregation, data through the underground Ethernet ring network to reach the underground dispatching center. The network is set up as shown in Fig. 2.

When the vehicle moves, the wireless client AP of the vehicle continuously transmits the compressed digital Ethernet signal to the radio base station AP on the surrounding wall. These base station data is transferred to the underground dispatch center through the optical fiber cable. The underground dispatching center will convert the compressed Ethernet digital image signal to the common digital image compression signal, the base station data is transferred to the underground dispatch center through the optical fiber cable.

The underground dispatching center will convert the compressed Ethernet digital image signal to the common digital image compression signal, then the ordinary digital image compression signal is converted to digital signal, and transmitted to the explosion-proof monitor. In this way, the real-time image about distance vehicle and the surrounding roadway can be watched by the people who stay in the underground dispatching center. Video processing flow shown in Fig. 3.

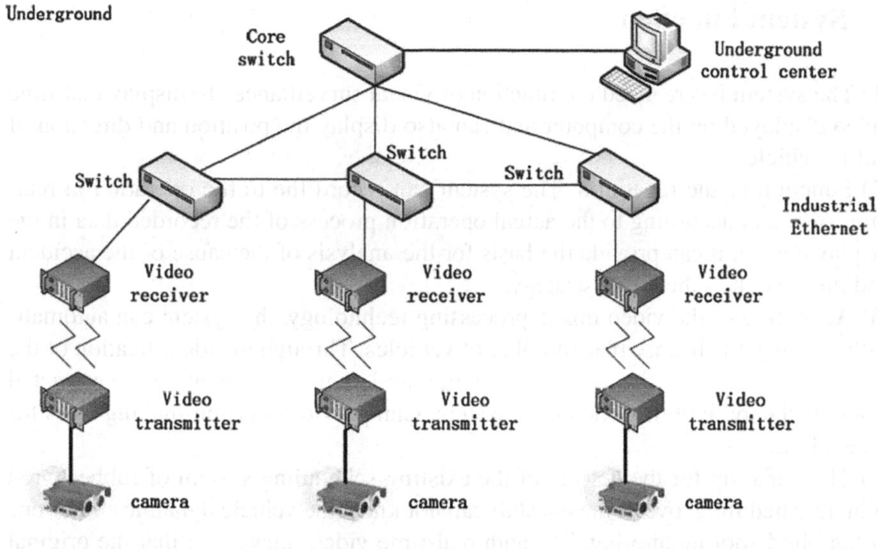

Fig. 2 Network structure

Fig. 3 Video processing

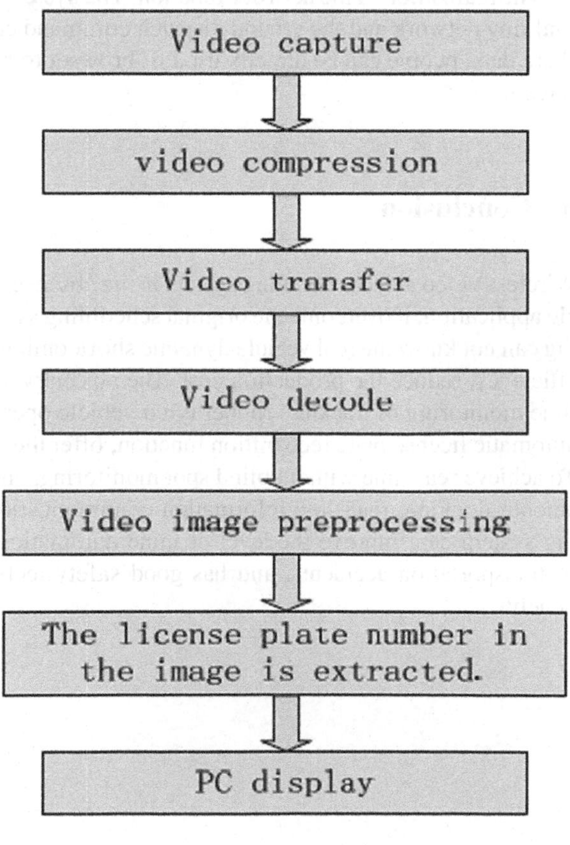

4 System Function

(1) The system has realized the function of visual surveillance. To display real-time video displayed on the computer and can also display the position and direction of traffic vehicle.

(2) Function of the repetition. The system can record the traffic operation in real-time 24 h, and according to the actual operation process of the recorded data in the display device, it can provide the basis for the analysis of the cause of the accident and improve the scheduling strategy.

(3) According to the video image processing technology, the system can automatically identify the license plate number of vehicles. Through the identification of the license plate number, the driver information and vehicle information are extracted from the database of the scheduling system. Can provide decision-making basis for scheduling.

(4) The make up for the defects of the existing scheduling system of rubber tyred vehicle scheduling, overcome the staff can not know the vehicle dynamic conditions in the blind spot monitoring. Through real-time video images, so that the original monitoring blind spots become non blind. Make scheduling more intuitive.

(5) The realization of the network function. The system can directly access the industrial ring network and the ground dispatch command center. Management system to share data, people can be directly used IE browser to browse the system monitoring screen.

5 Conclusion

Wireless video aided scheduling system in the absence of trackless rubber tyred vehicle application, it overcome the original scheduling system in the blind spot monitoring can not know the real vehicle dynamic shortcomings and improve the scheduling efficiency, reduce the production cost. The auxiliary system in addition to the real-time monitoring of trackless rubber tyred vehicle operation, but also supporting the automatic license plate recognition function, offer the decision basis for scheduling. To achieve real-time without blind spot monitoring, and with the mine ring network smooth docking, real-time information communication. The video aided scheduling system can improve the level of mine automation, reduce the occurrence rate of transportation accidents, and has good safety technology and social economic benefits.

References

1. Ma X (2012) Application of monitoring system of KJ150A type underground rubber tyred vehicle in Meihuajing mine. Shenhua Sci Technol 02:40–42
2. Li Y (2013) Rubber-tyred vehicle in coal mine underground traffic control system based on WIFI wireless network, Taiyuan University of Technology
3. Li W (2011) Hardware research on the measurement system of ring laser resonant cavity, Xi'an Technological University
4. Yang F, Zhang J (2013) Research on the video security monitoring system of the underground inclined tunnel transportation. Zhongzhou Coal 08:23–24, 28
5. Zhang Y, Guo D (2010) Principle and technology of wireless mesh network. Publ House Electron Ind 23(3):175–180
6. Romer K, Mattern F (2004) The design space of wireless sensor networks. Wirel Commun IEEE 11(6):54–61
7. Jianfei W (2014) Design and implementation of underground video transmission system based on WiFi technology, Liaoning University of Technology

An Improved Proportional Navigation Guidance Law for Waypoint Navigation of Airships

Ding Han, Xiao-liang Wang, Ming Zhao and Deng-ping Duan

Abstract This article derives an improved proportional navigation guidance law by using the extended gain scheduling law, which can be used in the nonlinear six-degree of freedom (6DOF) model of an airship directly, in which the airship equations of motion based on the Lagrangian approach. Nonlinear simulations for different conditions are performed, including the effect of the wind, and results are discussed. The simulation experiments indicate that, compared with the previous one, the modified guidance law which is used in specific waypoints tracking has better robust and tracking accuracy.

Keywords Lagrangian approach · Improved guidance law · Waypoint navigation

1 Introduction

Airships, also known as lighter-than-air (LTA) aerial vehicles, have some remarkable advantages compared to fixed-wing vehicles (airplanes) and rotary-wing aircrafts (helicopters). A wide range of applications have recently been proposed for airships, such as advertising, surveillance, environmental monitoring, planetary exploration and stratospheric observation [1–6].

Recently, as an autonomy challenge and an important application for a robotic airship, there has been increasing growth in research efforts on the modeling and

D. Han (✉) · X.-l. Wang · M. Zhao · D.-p. Duan
School of Aeronautics and Astronautics, Shanghai Jiao Tong University, Shanghai 200240, People's Republic of China
e-mail: handinghd2006@126.com

X.-l. Wang
e-mail: wangxiaoliang@sjtu.edu.cn

M. Zhao
e-mail: zhaomingpaul@sjtu.edu.cn

D.-p. Duan
e-mail: ddp@sjtu.edu.cn

© Springer International Publishing Switzerland 2017

V.E. Balas et al. (eds.), *Information Technology and Intelligent Transportation Systems*, Advances in Intelligent Systems and Computing 454,
DOI 10.1007/978-3-319-38789-5_46

control of Unmanned Aerial Vehicle (UAV). Practically, airships have more nonlinearities than ordinary aircraft due to the added mass and the time-varying characters, so it is hard to design the heading controller using traditional gain scheduling or feedback linearization, especially when the exact dynamic model is unavailable [1, 7]. Moreover, although in the initial stages of flight dynamics analysis and control design, atmospheric conditions are neglected for the sake of simplicity; when it comes to realistic simulations and practical applications, considering the effects of atmospheric disturbances such as wind is crucially important.

Proportional navigation (PN) is a method of guidance that has been applied to missiles for terminal guidance [8]. It is probably the most popular guidance methods for short-range intercept [9], and it has also been applied to aircraft for purpose of collision avoidance [10].

This study aims to investigate an improved proportional navigation (IPN) law to conduct nonlinear simulations of airship for several flight scenarios. For this purpose, flight dynamics of airship are presented and a nonlinear airship model is generated using Lagrangian approach [11] with a discussion of aero-dynamical characteristics and virtual mass concept. A Proportional navigation (PN) gain scheduling implementation method is stated and the extended guidance law utilized for autonomous flight is described. Finally, nonlinear simulation results for several cases are depicted and discussed. Conclusions of this study are presented and the planned improvements are remarked.

2 The Constitution of an Airship

The airship is consisted of hull, fins, gondola, elevator and rudder (Fig. 1). Every component has its geometric parameters. The general layout airship has been used in this study. For example, the volume of the airship hull $Vol = 5131.013\,\mathrm{m}^3$, the length of the airship hull $L = 50.0\,\mathrm{m}$, the location of the fins on the airship hull $x_f = 36.33\,\mathrm{m}$, the sweepback angle of the fins $\varphi = 41$ deg, the chord length of the fin tip $c_t = 6.0\,\mathrm{m}$, the semi-span of the fin $b = 8.4259$ and the chord length of the elevator or rudder $c_h = 0.25 \cdot c_t$. It should be noted that it is equipped with one forward thruster, which is able to thrust up to 600 N. The two main control surfaces, rudders and elevators, can be deflected from −30 to 30°.

3 Dynamics and Kinematics Modeling of Airship

3.1 Dynamics and Kinematics Modeling of Airship

The weight of airship is balanced by the buoyancy itself, so the displaced air mass is not negligible. In this condition, the equations of motion are usually derived using

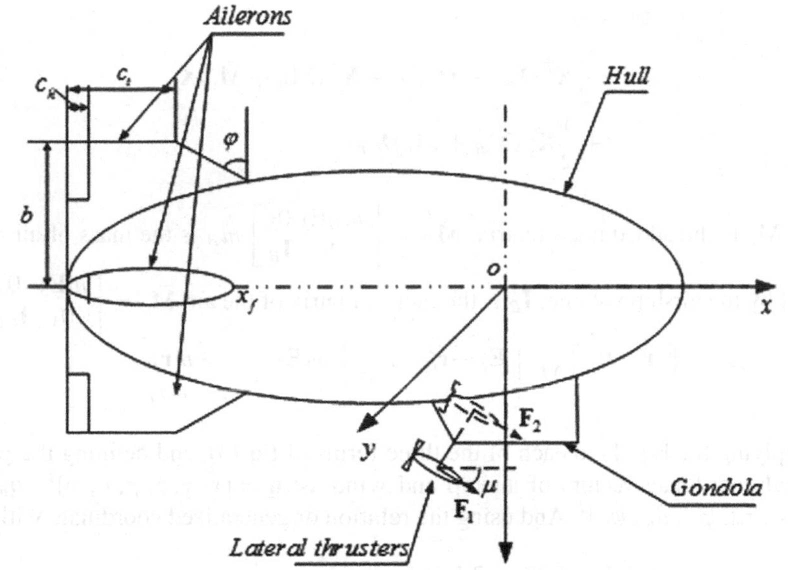

Fig. 1 The structure of airship

the Lagrangian approach. To accommodate the constantly changing CG (center of mass of the airship), the airship motion has to be referenced to a system of orthogonal axes fixed in the vehicle with the origin at the center of volume (CV). The CV is also assumed to coincide with the gross center of buoyancy (CB). The Lagrangian or Euler–Lagrange equations of motion of airship can be expressed as,

$$\bar{\mathbf{F}}(\dot{\mathbf{q}}, \mathbf{q}) = \frac{d}{dt}\left(\frac{\partial W(\dot{\mathbf{q}}, \mathbf{q})}{\partial \dot{\mathbf{q}}}\right) - \frac{\partial W(\dot{\mathbf{q}}, \mathbf{q})}{\partial \mathbf{q}} \tag{1}$$

where $W(\dot{\mathbf{q}}, \mathbf{q})$ the system kinetic energy is expressed as the function of the generalized coordinates $\mathbf{q}$ vector and its time derivative $\dot{\mathbf{q}}$ and $\bar{\mathbf{F}}(\dot{\mathbf{q}}, \mathbf{q})$ is the generalized forces vector.

For the airship, the total system kinetic energy includes the kinetic energy of airship about the center of gravity W^c, the kinetic energy of air displaced by the airship volume about the center of volume W_B^o and the kinetic energy of added mass about the center of volume W_v^o [11].

$$W = W^c + W_B^o + W_v^o \tag{2}$$

Through the deduction, the total system kinetic energy W can be expressed as,

$$
\begin{aligned}
W &= W_1 + W_3 + W_2 \\
&= \frac{1}{2}\mathbf{X}^T(\bar{\mathbf{M}}_o + \bar{\mathbf{M}}_v)\mathbf{X} - \mathbf{X}^T(\bar{\mathbf{M}}_B + \bar{\mathbf{M}}_v)\mathbf{X}_w \\
&+ \frac{1}{2}\mathbf{X}_w^T(\bar{\mathbf{M}}_B + \bar{\mathbf{M}}_v)\mathbf{X}_w
\end{aligned} \tag{3}
$$

With: $\bar{\mathbf{M}}_v$ is the added mass matrix. $\bar{\mathbf{M}}_B = \begin{bmatrix} m_B \mathrm{E}_3 & 0_3 \\ 0_3 & \mathbf{I}_B \end{bmatrix}$ m_B is the mass of air displaced by the airship volume; $\mathbf{I}_B$ is the inertia matrix of the air. $\bar{\mathbf{M}}_c = \begin{bmatrix} m\mathbf{E}_3 & 0_3 \\ 0_3 & \mathbf{I}_{CG} \end{bmatrix}$

$$
\bar{\mathbf{M}}_o = \begin{bmatrix} \mathbf{E}_3 & 0_3 \\ \mathbf{r}_{CG}^{\times} & \mathbf{E}_3 \end{bmatrix} \bar{\mathbf{M}}_c \begin{bmatrix} \mathbf{E}_3 & -\mathbf{r}_{CG}^{\times} \\ 0_3 & \mathbf{E}_3 \end{bmatrix} = \begin{bmatrix} m\mathbf{E}_3 & -m\mathbf{r}_{CG}^{\times} \\ m\mathbf{r}_{CG}^{\times} & \mathbf{I}_{CG} - m\mathbf{r}_{CG}^{\times}\mathbf{r}_{CG}^{\times} \end{bmatrix}
$$

Applying the Eq. (1) to each of the three terms of Eq. (3), and defining the generalized coordinate vectors of airship and wind as: $\mathbf{q} = [x, y, z, \varphi, \vartheta, \psi]^T$, $\mathbf{q}_w = [x_w, y_w, z_w, \varphi_w, \vartheta_w, \psi_w]^T$ And using the relation of generalized coordinate with,

$$
\dot{\mathbf{q}} = \begin{bmatrix} \dot{\mathbf{p}} \\ \dot{\mathbf{\Phi}} \end{bmatrix} = \begin{bmatrix} \mathbf{S}^T & 0_3 \\ 0_3 & \mathbf{R} \end{bmatrix} \begin{bmatrix} \mathbf{V} \\ \Omega \end{bmatrix} \Leftrightarrow \dot{\mathbf{q}} = \mathbf{J}_{\Phi}\mathbf{X} \Rightarrow \mathbf{X} = \mathbf{J}_{\Phi}^{-1}\dot{\mathbf{q}} \tag{4}
$$

$$
\dot{\mathbf{q}}_w = \begin{bmatrix} \dot{\mathbf{p}}_w \\ \dot{\mathbf{\Phi}}_w \end{bmatrix} = \begin{bmatrix} \mathbf{S}^T & 0_3 \\ 0_3 & \mathbf{R} \end{bmatrix} \begin{bmatrix} \mathbf{V}_w \\ \Omega_w \end{bmatrix} \Leftrightarrow \dot{\mathbf{q}}_w = \mathbf{J}_{\Phi}\mathbf{X}_w \Rightarrow \mathbf{X}_w = \mathbf{J}_{\Phi}^{-1}\dot{\mathbf{q}}_w \tag{5}
$$

where the transformation matrix from inertia coordinate system to the body coordinate system $\mathbf{S}$ is

$$
S = \begin{bmatrix} \cos\psi\cos\vartheta & \sin\psi\cos\vartheta & -\sin\vartheta \\ \cos\psi\sin\vartheta\sin\varphi - \sin\psi\cos\varphi & \sin\psi\sin\vartheta\sin\varphi + \cos\psi\cos\varphi & \cos\vartheta\sin\varphi \\ \cos\psi\sin\vartheta\cos\varphi + \sin\psi\sin\varphi & \sin\psi\sin\vartheta\cos\varphi - \cos\psi\sin\varphi & \cos\vartheta\cos\varphi \end{bmatrix}
$$

and the matrix $\mathbf{R}$ is $\mathbf{R} = \begin{bmatrix} 1 & \sin\varphi\tan\vartheta & \cos\varphi\tan\vartheta \\ 0 & \cos\varphi & -\sin\varphi \\ 0 & \frac{\sin\varphi}{\cos\vartheta} & \frac{\cos\varphi}{\cos\vartheta} \end{bmatrix}$ The kinematics equations of airship are:

$$
\begin{bmatrix} \dot{\varphi} \\ \dot{\vartheta} \\ \dot{\psi} \end{bmatrix} = R \begin{bmatrix} p \\ q \\ r \end{bmatrix} \tag{6}
$$

$$
\begin{bmatrix} \dot{x} \\ \dot{y} \\ \dot{z} \end{bmatrix} = S^T \begin{bmatrix} u \\ v \\ w \end{bmatrix} \tag{7}
$$

where, ϑ, ψ, φ are the pitch, yaw and roll angle respectively, x, y, z are the global position coordinates respectively.

From the dynamic equations (3) and the kinematics equations (6 and 7), the state equations of airship motion are:

$$\dot{\mathbf{X}} = \mathbf{f}(\mathbf{X}) \tag{8}$$

where $\mathbf{X} = [u\ v\ w\ p\ q\ r\ \vartheta\ \psi\ \phi\ x\ y\ z]^T$ is the state vector, $\mathbf{f}(\mathbf{X})$ is the nonlinear vector related to the state variables.

3.2 The Aerodynamic Force Vector

The aerodynamic force can be divided into three parts:

$$\begin{aligned}\mathbf{F}_a &= \mathbf{f}(\alpha, \beta, q_\infty, \mathrm{p}, \mathrm{q}, \mathrm{r}, \delta_\mathrm{e}, \delta_\mathrm{r}) \\ &= \mathbf{f}_1(\alpha, \beta, q_\infty) + \mathbf{f}_2(\mathrm{p}, \mathrm{q}, \mathrm{r}, q_\infty) + \mathbf{f}_3(\delta_\mathrm{e}, \delta_\mathrm{r}, q_\infty)\end{aligned} \tag{9}$$

where, q_∞ is the steady-state dynamic pressure, $q_\infty = \rho U_0^2/2$, L_{ref} is the reference length. The aerodynamic coefficients are calculated using the method in Refs. [12, 13].

4 Waypoint Navigation

In straight-line path-following problems, the airship is required to fly through a series of predefined waypoints at a given speed and constant altitude. It is a good application for testing the autonomy level of robotic UAVs. As depicted in Fig. 2, for illustrative purposes, lets name the waypoints A, B, C and D, and assume the airship will visit them in that order. To determine whether the airship has reached an assigned waypoint, a proximity zone of a specified radius is defined. When the distance of the airship to the waypoint is less than the radius (R) of the proximity circle, the airship is considered to have reached the waypoint. When this happens, the next waypoint in the sequence is commanded.

4.1 Proportional Navigation (PN) Guidance Law

In order to fly through a set of given waypoints, the required heading angle, as reference signal r_{com}, is generated by proportional navigation guidance law. The current heading angle, φ_{cur} is calculated from the x and y component of the inertial

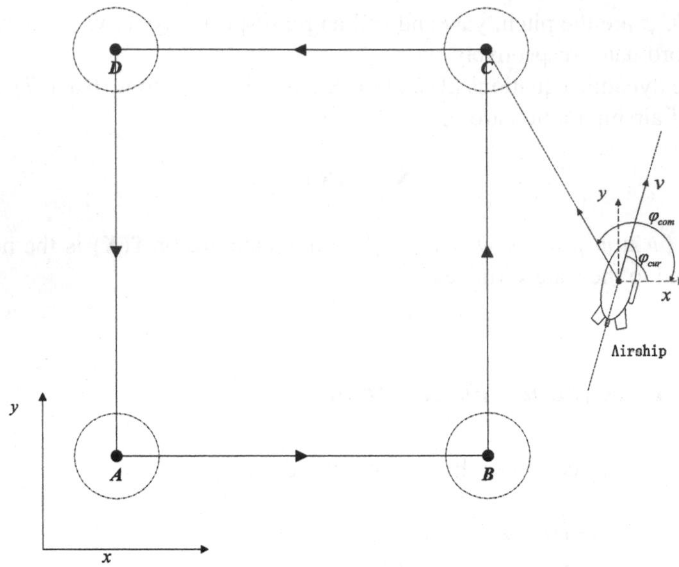

Fig. 2 PN control

velocity vector as

$$\varphi_{cur} = \tan^{-1}\left(\frac{\dot{y}}{\dot{x}}\right) \tag{10}$$

where $\dot{x}$ and $\dot{y}$ are the x and y components of the inertial velocity vector. The commanded heading is calculated by

$$\varphi_{com} = \tan^{-1}\left[\frac{(C_y - P_y)}{(C_x - P_x)}\right] \tag{11}$$

where (P_x, P_y) is the x and y coordinates of airship and (C_x, C_y) is the position of the next waypoint C in the inertial frame, as show in Fig. 2. To turn the velocity vector towards the next waypoint, φ_{cur} should track φ_{com}. This is achieved by generating the commanded yaw rate to reduce the path tracking error, as

$$r_{com} = K_p(\varphi_{com} - \varphi_{cur}) \tag{12}$$

where the proportional gain K_p is a design parameter.

The results of experiments manifest that when the direction of vector V parallels the vector $\overrightarrow{PC}$, it would be falsely considered working well. At the same time, if the normal distance of the airship to the given path (line path connecting waypoints) is more than the radius of the proximity circle, the airship flies along the extension line of the current path and does not turn to the next waypoint.

Only under the control of forwards thrust and rudder and elevator deflection, though the airship can follow the given path, the actual trajectory, which airship past, is not accuracy. To compensate these deficiencies, an improved proportional navigation guidance law is proposed.

5 Improved Proportional Navigation (IPN) Guidance Law

Side slip angle is the angle between a rolling wheel's actual direction of travel and the direction towards which it is pointing. It will determine the vehicle's behavior in a given turn. Consider the relationship between reference line path, vector BC and the current position of the airship; side slip angle β is introduced for tracking error.

As shown in Fig. 3, it can be seen that

$$\mu_{BC} = \tan^{-1}\left(\frac{C_y - B_y}{C_x - B_x}\right) \tag{13}$$

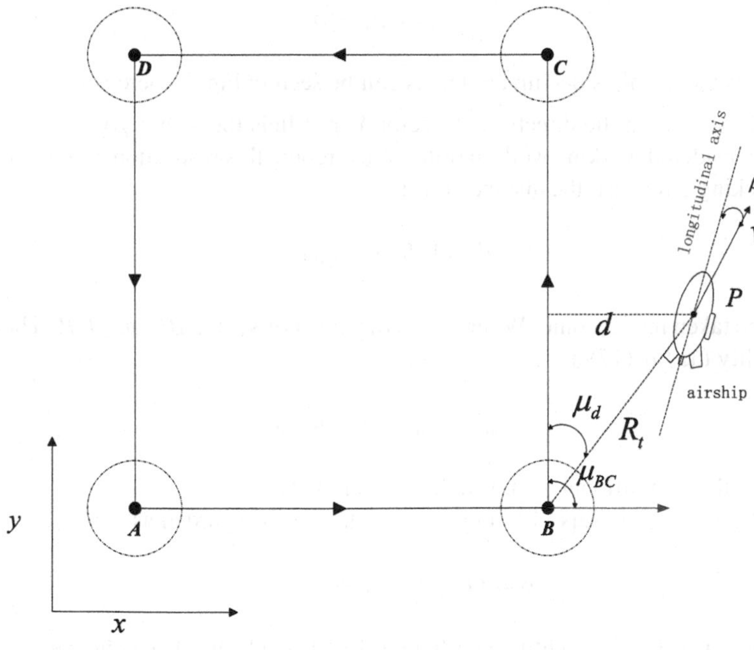

Fig. 3 Improved PN control

where B_x, B_y, C_x and C_y are the x and y coordinates of waypoints B and C respectively. Using the geometric and airship headings a desired heading angle is calculated from the following function

$$\psi_d = \frac{\pi}{2} \tanh \left(\frac{d_v}{L_{des}} \right) \tag{14}$$

where d_v is calculated as follows

$$d_v = \| R_t \| \sin(\mu_{BC} - \varphi_{cur}) \tag{15}$$

which related the direction of velocity vector V and $\| R_t \|$, the distance traveled from waypoint B.

L_{des} is a design parameter which is a function of the airship speed calculated as follows

$$L_{des} = \| V \| \tau \tag{16}$$

with τ being a performance design parameter. The commanded yaw angle provided to the set-point tracking controller is calculated as follows

$$\psi'_{com} = \psi_d - \beta \tag{17}$$

where β is the airship side-slip angle. As can be seen in Fig. 3, there is a restriction. This is to say, when the direction of vector V parallels the vector $\vec{BC}$, it would be falsely considered working well. In order to overcome this restriction and reduce the path tracking error fast, the distance error

$$d = \| R_t \| \sin(\mu_d) \tag{18}$$

should be taken into account. Where μ_d is the angel of vector $\vec{BC}$ and $\vec{BP}$. Then we can modify the Eq. (17) as

$$\psi_{com} = \psi_d - \beta + K \cdot d \tag{19}$$

where K, the proportional gain, is a design parameter.

Finally, combining Eqs. (12) and (19), track error is regarded as

$$e = (1 - \lambda) r_{com} + \lambda \psi_{com} \tag{20}$$

where $0 < \lambda < 1$ is a weighting coefficient. In this study, has been chosen.

6 Results and Discussion

6.1 Flight in the Presence of Wind

In this section, the same tests are repeated, but there are wind disturbances acting on the airship. To conveniently, The paths between waypoints A and B, B and C, C and D, and D and A are called as path1, path2, path3 and path4. In path2, wind is in x-direction (W_x), at speed 1 m/s and in path3 it is in z-direction (W_z), at speed 1 m/s. Wind disturbances exist during all process of path2 and path3. In this section, the trim speed is changed in different paths. In path1, path3 and path4 the trim speed is selected at 6.8 m/s and in path2 the trim speed is 7.5 m/s. The trim altitude of 1002 m is unchanged as before. In Figs. 4 and 5, when the wind is experienced, the altitude changes a little, but effect of on altitude is quite larger than the effect of W_x (nearly zero). Under the improved PN law, linear and angular velocities change differently at the wind components. Moreover, under the IPN law, simulation showed

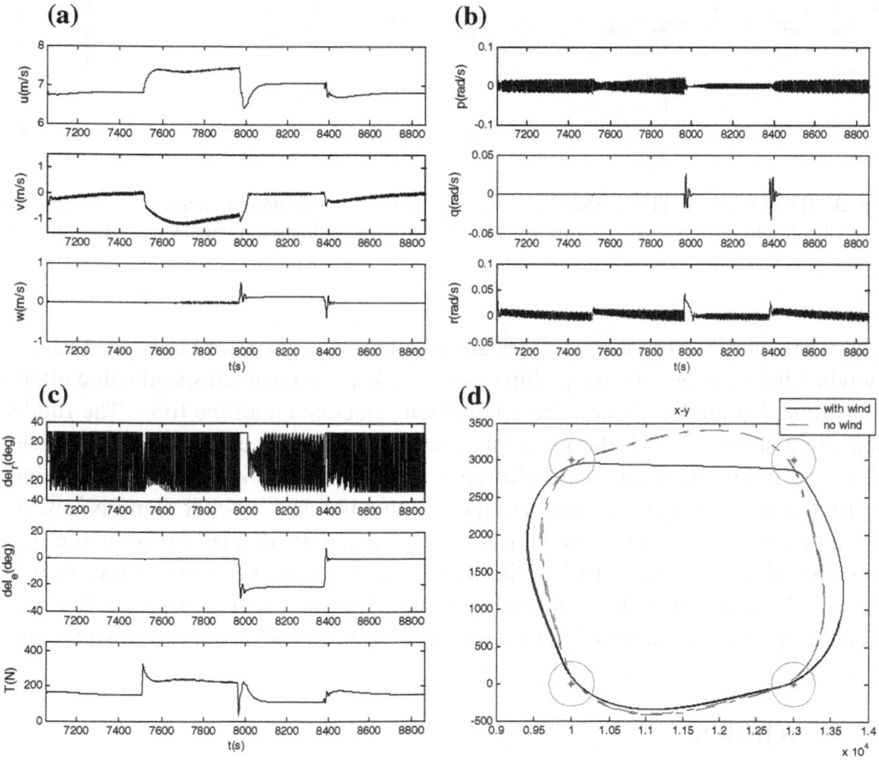

Fig. 4 Time history of airship body linear velocities, angular rates, orientation, control deflections and waypoint graph for Proportional navigation (PN) guidance law under wind disturbance case

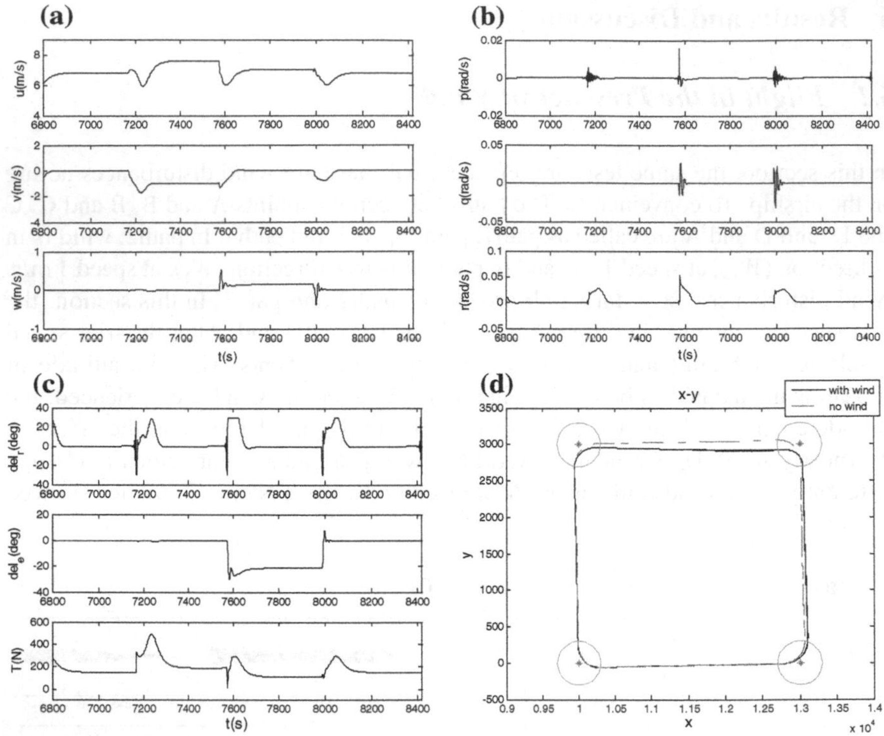

Fig. 5 Time history of airship body velocities, angular rates, orientation, control deflections and waypoint graph for improved Proportional Navigation (IPN) guidance law under wind disturbance case

that, when wind is experienced, there are small overshoots in forward speed caused by wind disturbances and the airship comes back to commanded speed value after a short time. But under PN law, there is velocity fluctuation all the time. The rudder deflection shocks fast all the processing under PN law, as can be seen in Fig. 4e. However, in Fig. 5e, the rudder deflection can overcome the wind disturbances, and return to stability very soon, which is showed that IPN law is better than PN law. As it can be observed from Fig. 5a, b, u, w and q are mainly affected by W_z while v, p and r are affected by W_x, just like that in Fig. 4a, b. All of these above have proved that the IPN law is reliable. Moreover, compared Fig. 4f with Fig. 5f, the IPN law is much better than previous one for waypoint navigation under the wind disturbance.

7 Conclusion

In this article, an Improved Proportional Navigation (IPN) guidance law is derived. The main contribution in IPN is that it makes the airship (only has rudder for turning) much easier to control. The results show that when the speed is constant, the controller can adapt to wind perturbation in or directions. The fundamental difference between the two guidance laws can be clearly seen; the IPN law attempts to track a reference trajectory, therefore flying straight between waypoints and turning in the vicinity of them to navigate towards the next one. The PN law however puts the airship on a continuous arc-like trajectory after it has passed the first waypoint; this is caused by continuously generating acceleration commands to change the heading if the airship towards the next waypoint. In IPN law, the track error, considering β, can reduce the impact of a high lateral acceleration at high speed turns and small turning radius. By comparison, whether under the situation of wind disturbance or not, the proposed algorithm performs better than the previous method in terms of robustness and accuracy. The simulation results agree well with the theoretical analyses.

Acknowledgments This work was partially supported by the National Natural Science Foundation of China (No. 11272205).

References

1. Khoury G, Gillett J (1999) Airship technology. Cambridge University Press, Cambridge, pp 73–106, 141–209, 475–504
2. Elfes A, Bueno SS, Bergerman M, Ramos JG (1998) A semi-autonomous robotics airship for environmental monitoring missions. In: Proceedings of the IEEE international conference on robotics and automation, vol 4, Leuven, Belgium, pp 3449–3455
3. Wilson JR (2004) A new era for airships: lighter-than-air craft, long since mothballed by military planners, may be on the verge of major comeback. Aerosp Am 42(5):27–31
4. Hygounenc E, Jung I, Soueres R, Lacroix S (2004) The autonomous blimp project of LAAS-CNRS: achievements in flight control and terrain mapping. Int J Robot Res 23(4):473–511
5. Kulczycki EA, Joshi SS, Hess RA, Elfes A (2006) Towards controller design for autonomous airship using SLC and LQR methods. AIAA guidance, navigation and control conference and exhibit, Keystone, Colorado, August 21–24
6. Schmidt DK (2007) Modeling and near-space station keeping control of a large high-altitude airship. J Guid, Control Dyn 30(2):540–547
7. Park CS, Lee H, Tahk MJ, Bang H (2003) Airship control using neural network augmented model inversion. Control Appl 1:558–563
8. Shukla US, Mahapatra PR (1990) The proportional navigation dilemma-pure or true? IEEE Trans Aerosp Electron Syst 26(2):382–392
9. Lu P (1998) Intercept of nonmoving targets at arbitrary time-varying velocity. J Guid, Control, Dyn 21(1):176–178
10. Han S-C, Bang H, Yoo C-S (2009) Proportional navigation-based collision avoidance for UAVs. Int J Control Autom Syst 7(4):553–565
11. Azinheira JR, de Paiva EC, Bueno SS (2002) Influence of wind speed on airship dynamics. J Guid, Control Dyn 25(6):1116–1124
12. Jones SP, DeLaurier JD (1983) Aerodynamic estimation techniques for aerostats and airships. J Aircr 20(2):120–126
13. Rajani A, Pant RS, Sudhakar K (2010) Dynamic stability analysis of a tethered aerostat. J Aircr 47(5):1531–1538

Numerical Simulation on Gas Flow Distribution in ESP for Convertor Gas Purification

Lichun Xiao, Zhijiang Ding, Xiaoyuan Yang, Xuyan Liu and Jie Yang

Abstract It is necessary to purify the convertor gas when it is recycled in gas tank. The electrostatic precipitator is widely used for convertor gas dedusting because of its high efficiency and low pressure loss. However, the collection efficiency of electrostatic precipitator is affected by the gas flow distribution uniformity of the inlet and outlet. Based on the computational fluid dynamic (CFD) method, numerical calculation for gas flow distribution is carried out by using the k-ϵ two equation turbulence model and SIMPLE (Semi-Implicit Method for Pressure Linked Equations) algorithm. The results show that the optimal opening ratio of gas flow distribution plate is less than 30 %. The position of inlet gas flow distribution plate has great influence on the gas flow uniformity. Compared with the gas flow field difference between columnar electrostatic precipitator and rectangular one, the columnar electrostatic precipitator has the advantages of high collection efficiency and structure strength. The gas flow distribution uniformity in columnar electrostatic precipitator is also superior to the rectangular one.

Keywords Gas flow distribution · Convertor gas purification · Computational fluid dynamic

---

L. Xiao (✉) · Z. Ding · X. Yang · X. Liu · J. Yang
College of Environmental and Chemical Engineering,
Yanshan University, Qinhuangdao, China
e-mail: xiaolichun2001@sina.com
URL: http://www.ysu.edu.cn

Z. Ding
e-mail: dlyxyj@163.com

X. Yang
e-mail: 18233585619@163.com

X. Liu
e-mail: 15533090562@163.com

J. Yang
e-mail: 18233587442@163.com

© Springer International Publishing Switzerland 2017

V.E. Balas et al. (eds.), *Information Technology and Intelligent Transportation Systems*, Advances in Intelligent Systems and Computing 454,
DOI 10.1007/978-3-319-38789-5_47

1 Introduction

With the rapid progress of industry and development of society, the demand of iron and steel increases quickly. The total outcome of iron and steel keeps the first ten years continuously. The output of by-product gas also increases with steel output and energy consume. The convertor gas is one of important by-product from convertor steel making [1, 2]. It can be recycled as the fuel for power plant. Therefore it is necessary for the iron and steel plant to utilize the convertor gas for reducing the enterprises economic cost. The convertor gas contains large numbers of CO. Its mean caloric is 8000 kJ/m^3. And its dust content is more than 120 mg/m^3 after the purifying by wet method precipitation equipment. As the reason that high concentration dust does harm to the generating equipment, it must be dedusted by a high precision dust removal unit [3].

The wet electrostatic precipitator (ESP) can be used as the high precision dust removal unit to purify the convertor gas [4–6]. However, the gas flow distribution in the electrostatic precipitator has an important influence on its collection efficiency [7]. There is an important method to improve the airflow distribution by adding the gas flow distribution plate in the ESP inlet. It can change the gas flow distribution state by changing the opening ratio. The gas flow will be divided into many small pieces because of the resistance of the orifice flow [8]. It seems that the gas flow is hackled by the gas flow distribution plate. At last the gas flows into the electrostatic field perpendicularly.

The numerical calculation was carried out for the columnar ESP by changing the arrangement mode of the gas flow distribution plate. It has proved that the columnar ESP has many advantages than the rectangular one.

2 Fore Processing of Numerical Calculation and Evaluation Standard

2.1 Geometric Model of ESP

As shown in Fig. 1, the geometric model of ESP is established according to the ESP used in an iron and steel plant which length is 6.5 m, width is 8 m, and highness is 8.5 m. The area of inlet is 1.85 × 1.85 m. The length of inlet is 3.5 m. The geometric model of columnar ESP has the same section area with the rectangular one.

2.2 Computational Grids

Based on comprehensive consideration of the structure and working characteristics of ESP, the node distribution of the grid is simplified for saving the computing

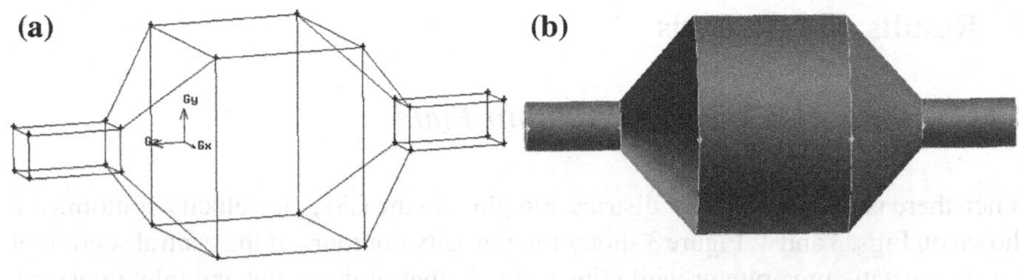

Fig. 1 The geometric model of two kinds of ESP. **a** Model of rectangular ESP. **b** Model of columnar ESP

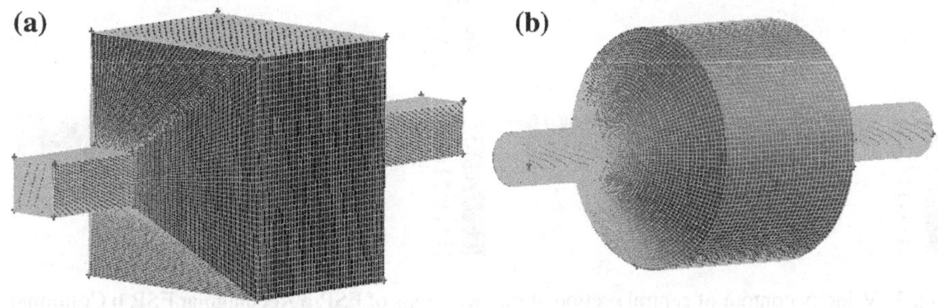

Fig. 2 The computational grid of two kinds of ESP. **a** Computational grid of rectangular ESP. **b** Computational grid of columnar ESP

time. As shown in Fig. 2, the computational grid of two types of ESP are structured processing by using the size control function of soft GAMBIT. The total grid of gas flow distribution plate in the inlet are $8 \times 10^5 \sim 1 \times 10^6$.

2.3 Boundary Conditions

Based on the medium of air the gas flowing through the ESP can be seen as turbulence, low speed, incompressible fluid. The inlet can select the entrance velocity boundary conditions. The outlet can select the free export boundary conditions. Because there are lots of holes in the inlet gas flow distribution plate, the amount of grid from the gas flow distribution plate will be large when the solid model is used. Based on porous medium model, the amount of the computational grid will decrease greatly. The k-turbulence model is used in the paper, a finite volume method is used to solve the discrete problem [9, 10]. The two-step upwind discretization scheme is adopted, and the discrete equation is solved by using the coupled SIMPLE algorithm with the pressure velocity [11–13].

3 Results and Analysis

3.1 Without Gas Flow Distribution Plate

When there is not any gas flow distribution plate in the ESP, the velocity contours are shown on Figs. 3 and 4. Figure 3 shows the velocity contours of the central section of the electrostatic precipitator while Fig. 4 shows that of the outlet and inlet interface.

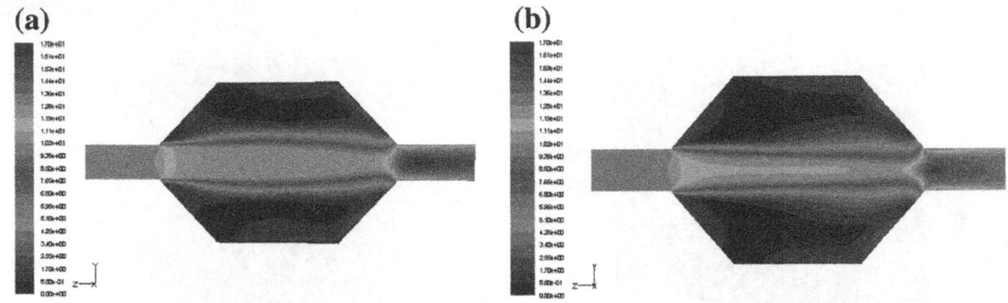

Fig. 3 Velocity contour of central section of the two kinds of ESP. **a** Rectangular ESP. **b** Columnar ESP

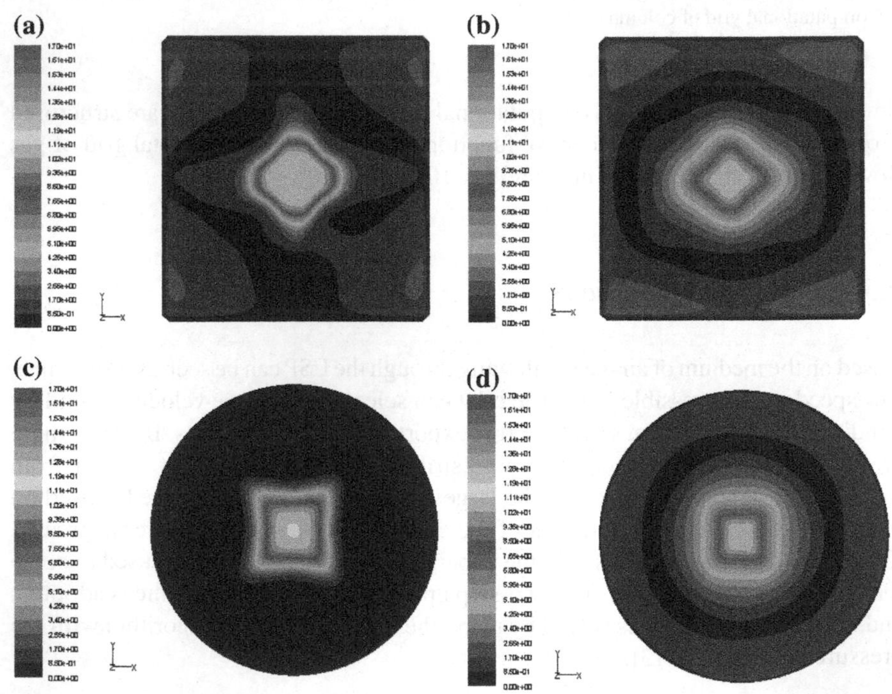

Fig. 4 Velocity contours of the outlet and inlet section of two kinds of ESP. **a** Outlet of the rectangular ESP. **b** Inlet of the rectangular ESP. **c** Outlet of the columnar ESP. **d** Inlet of the columnar ESP

As can be seen from the velocity contours, when there is no gas flow distribution plate the gas mainly flows through the middle part of the dust collector, and a large number of reflux appears in the corner between the inlet and the electric field. So there are both high and low speed areas in the electric field at the same time. The reflux increases disturbance of the gas flow in electric field. This will seriously reduce the dust removal efficiency of electrostatic precipitator. The relative root mean square (RMS) value of square ESP is 1.322 and cylindrical is 1.196 from the statistic results. The gas flow distribution of the columnar ESP is relatively uniform than the rectangular. It also means that the columnar ESP has the superiority that its inner space is smooth.

3.2 *The RMS Values of A Gas Flow Distribution Plate*

From the computational model, the gas flow distribution plate is placed at 1.58 m distance from the inlet. When the percentage of opening area are 20, 25, 30, 35, 40, 45 %, the airflow distribution relative RMS values of two kinds of electrostatic precipitators are shown in Fig. 5.

As can be seen from Fig. 5, gas flow distribution in the electrostatic precipitator turns homogeneous when the gas flow distribution plate is added. The minimum of relative RMS value in rectangular ESP is 0.431 when the percentage of opening area is 40 %, and the minimum of relative RMS value in columnar ESP is 0.358 when the percentage of opening area is 30 %. All of this doesn't accord with the industry standard, it must be improved.

Fig. 5 Comparison of RMS values of different porosities

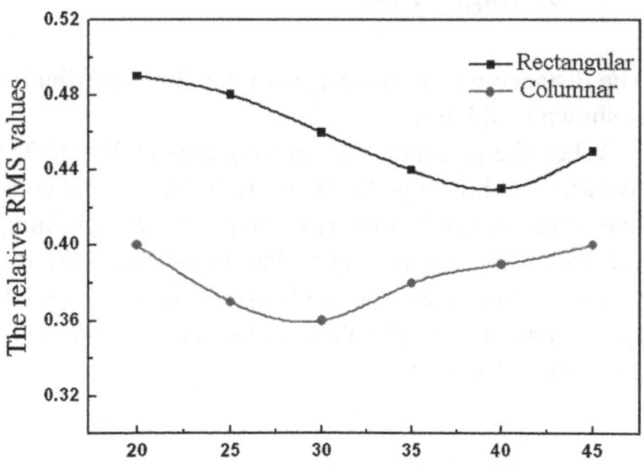

Table 1 The RMS values of three kinds of layout

Percentage of opening area of the first plate	30	30	30
Percentage of opening area of the second plate	25	30	35
The relative RMS value	0.243	0.241	0.237

3.3 Gas Flow Distribution Affected by Percentage of Opening Area

There is each one piece of gas flow distribution plate at the distance of 1.05 and 1 m from the inlet of columnar ESP. The percentage of opening area is selected as Table 1. The relative RMS value of three kinds of gas flow distribution is also given in Table 1. It can be seen that the relative RMS value decreases and the gas flow distribution is obviously improved when there are two pieces of gas flow distribution plate. Compared with three kinds of gas flow distribution, when the percentage of opening area are 30 and 35 % the gas flow distribution state is best. What is more, the relative RMS value = 0.237. The reason is that the gas velocity in electric field is relatively small when the percentage of opening area is small. The pressure loss decreases with the percentage of opening area. Consequently, the gas velocity will increase if the percentage of opening area of the second piece of gas flow distribution plate increases.

3.4 Gas Flow Distribution Affected by the Plates Arrange Mode

The distance which two pieces of gas flow distribution plate set in the inlet of ESP is shown in Table 2.

When the percentage of opening area is 30 and 35 %, the relative RMS value of five kinds of layout is shown in Fig. 6. Horizontal coordinates respectively indicates five layout of gas flow distribution plate. The minimum of relative root mean square value is 0.226. It also denotes that the best gas flow distribution in the ESP is gotten when the plates are respectively arranged at 1.58 and 2.36 m. Therefore, the distance of gas flow distribution plates in the inlet of ESP has great effect on the uniform of gas flow distribution.

Table 2 Distance of the two plates of inlet in the ESP

Number	1	2	3	4	5
Distance of the first plate (m)	1.05	1.58	1.58	1.05	1.05
Distance of the second plate (m)	1.58	2.36	3.15	2.36	2.10

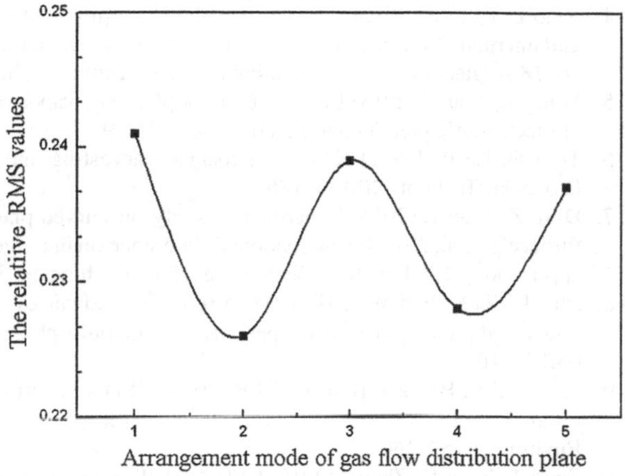

Fig. 6 The RMS values of five locations of layout

4 Conclusions

Through the numerical calculation of the gas flow distribution in the ESP, conclusions are drawn as follows: The gas flow distribution of columnar ESP is better than the rectangular ESP when the percentage of opening area and layout of the gas flow distribution plate are the same. Since there is no sharp edges and corners in the columnar ESP and it has a smooth gas flow field, the uniformity of gas flow distribution has been improved. When there is a piece of gas flow distribution at distance of 1.58 m from the inlet, the minimum of relative root mean square value of rectangular ESP is 0.431 while the columnar ESP is 0.358. The uniformity of gas flow distribution can be also improved by adjusting percentage of the opening area and layout of the gas flow distribution plates. The minimum of the relative root mean square value is 0.226 when the percentage of the opening area first plate is 30 %, the second plate is 35 %, and the tow plates are respectively arranged at 1.58 and 2.36 m.

Acknowledgments This research was supported by the Natural Science Foundation of Hebei province, China (E2015203236).

References

1. Yilmaz SO, Teker T (2015) Experimental research on mechanism and process of direct iron making reduction of mechanically milling scale with coal. J Alloy Compd 650:741–747
2. Zhao X, Bai H, Lu X (2015) A MILP model concerning the optimisation of penalty factors for the short-term distribution of byproduct gases produced in the iron and steel making process. Appl Energy 148:142–158
3. Merritt A, Andersson N, Almqvist C (2013) Cat and house dust mite allergen content is stable in frozen dust over time. Environ Sci Technol 47:3796–3799

4. Xiao L, Ding Z (2012) Coal gas dehydration equipment in combined cycle power plant. In: 2nd international conference on advanced design and manufacturing engineering, ADME 2012, 16–18 August 2012. Trans Tech Publications, Taiyuan, China, pp 554–558
5. Wang X, You C (2013) Effects of thermophoresis, vapor, and water film on particle removal of electrostatic precipitator. J Aerosol Sci 63:1–9
6. Tran D, Le B, Lee D (2013) Microalgae harvesting and subsequent biodiesel conversion. Bioresour Technol 140:179–186
7. Ding Z, Xiao L (2013) Experimental study on curved plate dehydration equipment of blast furnace gas. In: 2013 4th international conference on intelligent systems design and engineering applications, ISDEA 2013. Zhangjiajie, Hunan, China, pp 302–304
8. Zhu J, Xia S, Wei W (2013) PM2.5 removal - advances in wet collection technologies and a novel approach through temperature swing multi-phase flow. Huagong Xuebao/CIESC J 64:155–164
9. Ramponi R, Blocken B, de Coo LB (2015) CFD simulation of outdoor ventilation of generic urban configurations with different urban densities and equal and unequal street widths. Build Environ 92:152–166
10. Song T, Jiang K, Zhou J (2015) CFD modelling of gas liquid flow in an industrial scale gas-stirred leaching tank. Int J Miner Process 142:63–72
11. Asfand F, Stiriba Y, Bourouis M (2015) CFD simulation to investigate heat and mass transfer processes in a membrane-based absorber for water-LiBr absorption cooling systems. Energy 91:517–530
12. Shiehnejadhesar A, Scharler R, Mehrabian R (2015) Development and validation of CFD models for gas phase reactions in biomass grate furnaces considering gas streak formation above the packed bed. Fuel Process Technol 139:142–158
13. Davarnejad R, Jamshidzadeh M (2015) CFD modeling of heat transfer performance of MgO-water nanofluid under turbulent flow. Eng Sci Technol Int J 18:536–542

Optical Characteristics of Antireflection of SiN Layer on the Si Substrate

Haifeng Chen and Duan Xie

Abstract This paper studies the optical characteristics of SiN antireflection coating (ARC) on the Si substrate and mainly focuses on the effects of different wavelength and angle of incident light on the ARC. In this paper the concept of antireflection window (ARW) is proposed and is used to analyze the situation of whole incident angles. It is found that the 3D surface of reflectivity shows a groove which results from the fast change of the light with 400–900 nm wavelength. Additionally, the angle of the incident light with = 616 nm is smaller than 650 for the ARW.

Keywords ARC · SiN · Solar cell · Reflectivity · Si substrate

1 Introduction

Solar cell supplies the clean energy and is used widely. The solar cells are divides into several types and every type has its own characteristic. In this types, the crystal-line silicon solar cell receives the more attentions and applications due to its high photo-electric conversion rate which is reaches 22 % nowadays. One of the key processes improving the photoelectric conversion rate of crystalline silicon solar cell is to make the good antireflection coating (ARC) [1–4]. With the assistant of texturing process, ARC even can make the reflectivity of Si surface decreases below 5 %. The main-stream ARC is the SiN layer which are deposited by the PECVD or PVD [5–8]. Therefore, the study of SiN ARC is very meaningful for the improvement of the solar cell and other optical sensors. Although there are a lot of researches on SiN ARC, all the features of this layer still are not understood completely.

H. Chen (✉) · D. Xie
School of Electronic Engineering, Xian University of Posts
and Telecommunications, Xian 710121, China
e-mail: chenhaifeng@xupt.edu.cn

D. Xie
e-mail: xieduan@xupt.edu.cn

© Springer International Publishing Switzerland 2017

V.E. Balas et al. (eds.), *Information Technology and Intelligent Transportation Systems*, Advances in Intelligent Systems and Computing 454,
DOI 10.1007/978-3-319-38789-5_48

This paper aims to investigate the optical characteristics of SiN ARC and mainly focuses on the effect of the wavelength and incident angle on the surface of reflectivity of Si substrate. With the simulation, the reflectivity curves are shown. Meanwhile, the concept of antireflection window is given.

2 Physical Theory of SiN ARC

When the incident light reaches the Si surface, part of this light steps into the Si and other part is reflected by the Si surface. When the incident angle is 0°, the reflectivity of naked Si is minimum and is about 33 %. Figure 1a shows the reflection illustration of naked Si. This high reflectivity is due to the mutation of the refractive index between the air (n0 = 1) and the Si (nsi = 3.8). Then it is necessary to deposit ARC to decrease the reflectivity. ARC uses the optical phase principle of interference between top and bottom surface of SiN thin film to cancel the reflection light [2, 7].

The refractive index of SiN is about 2, so it can relieve the mutation of the refractive index between the air and the Si. Figure 1b is the reflection illustration of the Si with the SiN ARC. Under the incident angle is 0°, the reflectivity is minimum when the optical thickness equals to the 1/2 of wavelength of incident light.

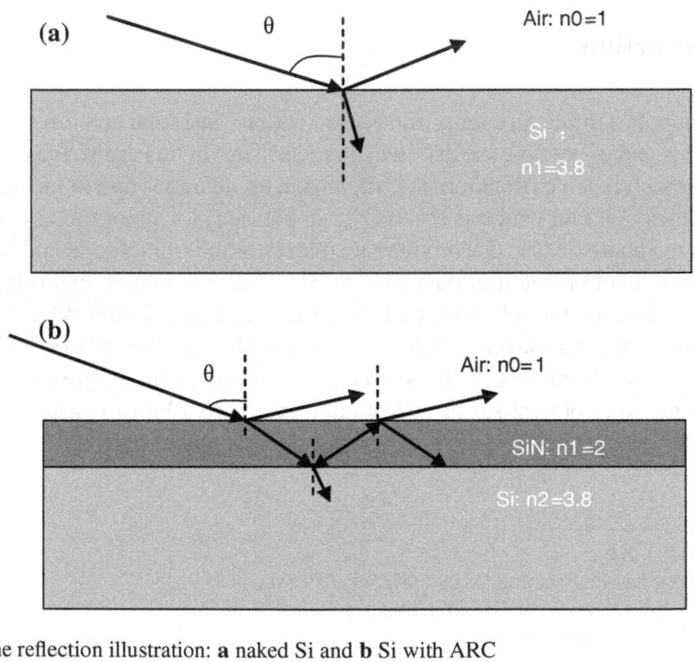

Fig. 1 The reflection illustration: **a** naked Si and **b** Si with ARC

3 Results and Discussion

Figure 2 shows the relationship of reflectivity R with the wavelength with and without SiN ARC at = 0°. In this case, the thickness t of the ARC is 77 nm. In the Fig. 2, the λ of incident light is 616 nm and SiN ARC improves the feature of reflection pretty much. Between the wavelength range of 200–1150 nm, the reflectivity curve of Si with the SiN ARC presents a peak and valley. The whole curve is divided into 3 ranges according to the reflectivity value of 10 %: 250–450, 450–900 and 900–1150 nm. In these ranges, only the range of 450–900 nm has the R below 10 %. Because the spectrum part of strongest intensity of the sunlight just locates in this range, so the SiN ARC plays a good role for the solar cell to decrease the reflection.

For better understanding the ARC, we can call R < 10 % range the anti-reflection window (ARW). Actually, ARW is a filter which lets lights pass through it selectively. In Fig. 2, ARW is the BC. The ARW decides the best range of wavelength which can be absorbed and meanwhile limits this absorb range. However, the refraction index of SiN ARC can be changed by adjusting the constituents of SiN material. Figure 3 gives the comparison of reflectivity between the two cases of nsi = 1.95 and nsi = 2 when the optical thickness of ARC is 1/2 of wavelength. From the Fig. 3, the curve shifts left. This is to say that ARW makes the curve blue shift which means the light with the short wavelength can be absorbed easily.

Since the incident angle of sun light is changed, Fig. 4 shows the 3D surface of R with the angle range of 0–90°. It can be seen that there is a groove at the range of 400–900 nm. As the angle increase, the groove shallows. The formation of the groove is due to the R of the light with the wavelength of 400–900 nm changes faster than that with other range wavelength. The core of this groove has the R below 10 % and actually is ARW field. This ARW is triangle-like. Further, R between 400–900 nm is always very small under the whole θ range.

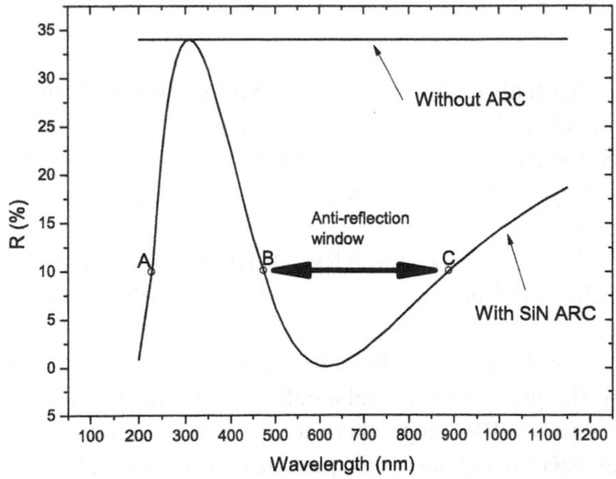

Fig. 2 Relationship of R and With and without SiN ARC at n = 2

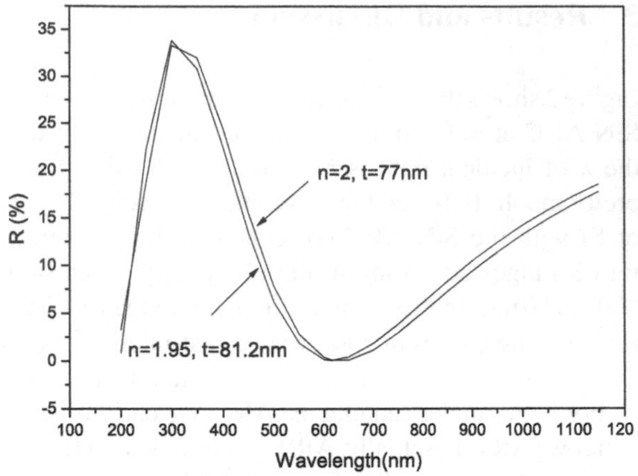

Fig. 3 The blue shift of R curve

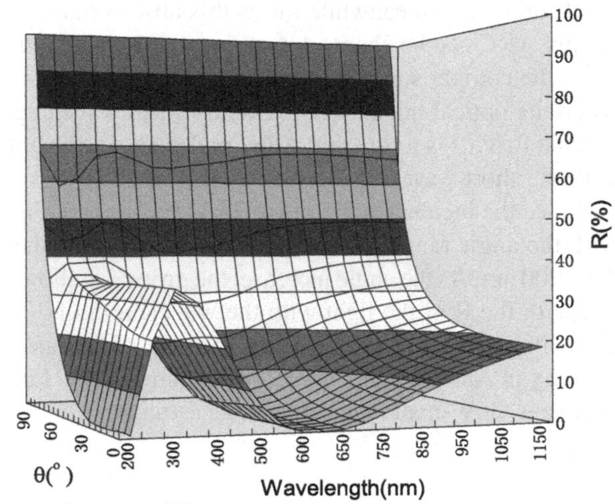

Fig. 4 R under different λ and θ

To analyze the problem, Fig. 5 abstracts the R curves of $\lambda = 200$, 300 and 616 nm wavelength from Fig. 4. It can be shown that R at 616 nm is minimum when $\theta < 80°$. Between $0° < \theta < 40°$, the R at $\lambda = 300$ nm is as small as that at 616 nm. Once θ is larger than 80°, the R of all wavelengths increase rapidly and has the consistent value.

Figure 6 shows the ARW at 616 nm. From this figure, it can be found that R is below 10 % between $0° < \theta < 65°$. R is larger than 10 % and ARW disappears when $\theta > 65°$.

On the basis of above discussion, ARW has the low R and must be considered in the process of the solar cell. In order to decrease R of cell (without texture) and increase ARW range, there are many things that we can do: using multi-coating layer as ARC, using nanoscale particle to transmit high energy light (UV) into visible light and using maximum power point tracking technology on the Solar Cell.

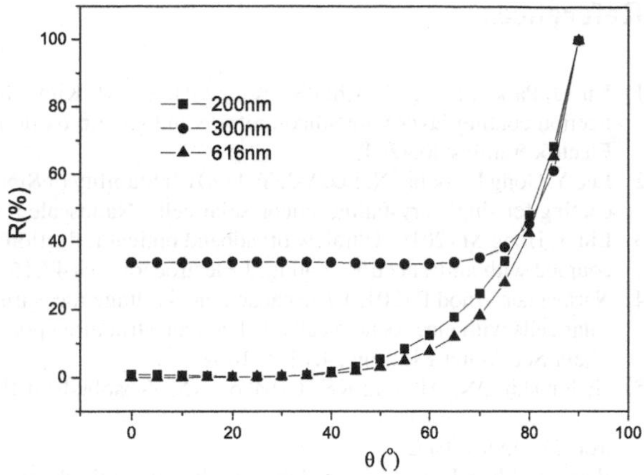

Fig. 5 Relationship of R and θ under the different wavelength

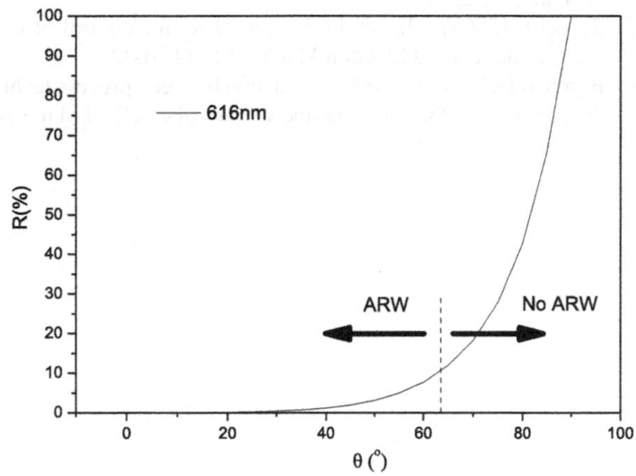

Fig. 6 ARW at the wavelength of 616 nm

4 Conclusion

This paper discusses the optical characteristics of the SiN ARC and mainly focuses on the influence of the wavelength and angle of incident light on the reflectivity. It is proposed the concept of the antireflection window (ARW) and studies the situation of R under the condition of whole incident angles. It is found that the R surface presents a groove and this ascribes to the fast change of R of the incident light with the wavelength of 400–900 nm. Further, the method which increases ARW range is also given.

These results should be helpful to understand the characteristics of the SiN ARC.

Acknowledgments We acknowledge the financial support by the National Natural Science Foundation of China (Grand No. 61306131) and the Science Foundation of Shaanxi Provincial Department of Education (Grant No. 14JK1682).

References

1. Kim J, Park J, Hong JH, Choi SJ, Kang GH, Yu GJ, Kim NS, Song H-e (2013) Double antireflection coating layer with silicon nitride and silicon oxide for crystalline silicon solar cell. J Electroceramics 30:42–45
2. Lee Y, Gong D, Balaji N, Lee Y-J, Yi J (2012) Stability of SiNx/SiNx double stack antireflection coating for single crystalline silicon solar cells. Nanoscale Res Lett 7:50
3. Liu Y, Hong M (2013) Ultralow broadband optical reflection of silicon nanostructured surfaces coupled with antireflection coating. J Electroceramics 47:1594–1597
4. Narayanan Vinod P (2013) The capacitance-voltage measurement of the screen-printed silicon solar cells with electrochemically etched nanostructured porous silicon antireflection coating. J Mater Sci: Mater Electron 24:1395–1404
5. Gadomskii ON, Altumin KK, Ushakov NM, Kosobudskii ID, Podvigalkin VYa (2010) High-efficiency antireflection nanostructructural optical coatings for solar cells. Opt, Quantum Electron 55(7):996–1002
6. Wang X, Shen J (2010) Sol-gel derived durable antireflective coating for solar glass. J Sol-Gel Technol 53:322–327
7. Sopori B (2003) Silicon nitride processing for control of optical and electronic properties of silicon solar cells. J Electron Mater 32:1034–1042
8. Borchert D, Rinio M (2009) Interaction between process technology and material quality during the processing of multicrystalline silicon solar cells. J Mater Sci: Mater Electron 20(1):487–492

The Space Vector Control System for PMSM Implemented by Model-Based Design Techniques

Danping Ma, Ziqiang Tang, Xianwu Gong, Wenjie Hei and Jingfei Yue

Abstract On account of the space vector control system for PMSM (Permanent Magnet Synchronous Motor), adopt MBD (Model-Based Design) techniques, the simulation model of space vector control for PMSM was built with MATLAB/Simulink Embedded Coder toolbox, and the performance of the speed control system is simulated. After verifying the correctness of the simulation model, the fixed-point code model was built for the control system. And used the fixed-point code model to generate C code automatically, and the generated C code was debugged in the designed hardware test platform debugging. The experimental result shows that the space vector control system for PMSM implemented by MBD techniques has good dynamic response performance.

Keywords MBD · PMSM · Space vector control

D. Ma (✉) · X. Gong · W. Hei · J. Yue
School of Electronic and Control Engineering, Chang'an University, Xian 710064, Shaanxi, China
e-mail: danpingma2799@163.com; 838827831@qq.com

X. Gong
e-mail: xwgong@chd.edu.cn

W. Hei
e-mail: 331795864@qq.com

J. Yue
e-mail: 345079535@qq.com

Z. Tang
School of Automobile, Chang'an University, Xian 710064, Shaanxi, China
e-mail: tangzqa@126.com

© Springer International Publishing Switzerland 2017

V.E. Balas et al. (eds.), *Information Technology and Intelligent Transportation Systems*, Advances in Intelligent Systems and Computing 454,
DOI 10.1007/978-3-319-38789-5_49

1 Introduction

Permanent magnet synchronous motor is a kind of synchronous motor, which is widely used in speed control and servo system [1] because of small size, light weight, high efficiency and good dynamic performance. With the integration of power electronics and the constant improvement of the digital, DSP has been widely used in power electronic field and motion control field such as inverter circuit, frequency conversion speed control and so on. The development mode of DSP in the traditional [2] is: first verifying engineers idea by using the simulation software (such as MATLAB); and programming, debugging and commissioning on the self-made hardware platform, in order to realize the demand of engineers idea in the analysis stage; finally testing the whole system design. Because the whole development process is independent each other, the organic unity between the various processes is separated, in addition to the development of more personnel, each person's understanding of the document is inevitably biased, which will result in the function and design of the product is not consistent with the design expectations. The disadvantages of this approach are to reduce the development efficiency, and make the development cycle extended. Compared with controller algorithm of traditional project development is developed with the method of manual programming, Model-Based Design is based on the Coder Embed toolbox developed by TI company and MathWorks company, which is a new way [3] for the development of embedded system. The controller code can be generated automatically by Matlab Real-Time Workshop Embedded Coder in the MBD techniques, which greatly reduces the manpower, material resources and time needed for manual programming, improves the reliability and consistency [4] of the code, accelerates the development of software and hardware, and shortens the development cycle. In this paper, the permanent magnet synchronous motor as the research object, with MBD techniques method to study the space vector control of permanent magnet synchronous motor, to explore the application of the MBD techniques in the permanent magnet synchronous motor control, in order to improve the development efficiency of permanent magnet synchronous motor control system.

2 Model-Based Design

In the process of Model-Based Design system design, all the information transmission and the basis of the work is the system model from the requirements analysis to the system design and implementation, and the testing and verification, which is the core of the development process through the modeling and simulation.

The basic flow chart of Model-Based Design is shown in Fig. 1. It is not dependent on the physical hardware and executable text specification, the Model-Based Design can be implemented as a system model. This executable specification supports design and simulation of system level and hardware level, automatic code generation, and testing and verification throughout the development process.

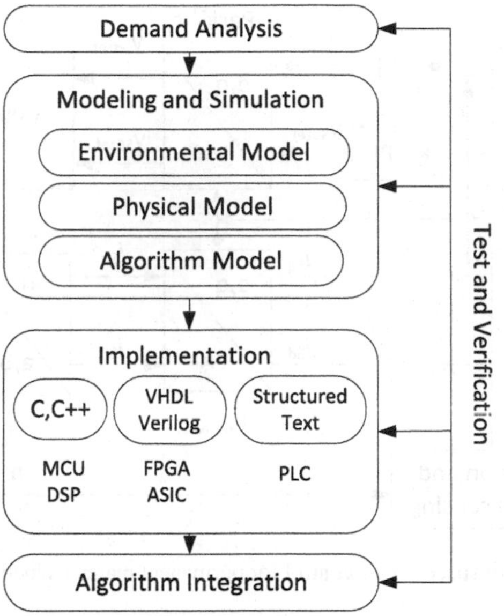

Fig. 1 The basic flow chart of model-based design

3 Modeling and Simulation of PMSM

3.1 The Principle of PMSM Control System

In this paper, the PMSM space vector control block diagram is shown in Fig. 2. i_a, i_b is the PMSM two-phase current detected by current sensor, using the Clark transformation to convert i_a, i_b to i_s, i_s in the two-phase static coordinate system (a, b). Two phase current of Clark coordinate after Park transformation, the output current in the two-phase rotating coordinate system is i_{sd}, i_{sq}, i_{sd}, i_{sq} as the feedback current, comparing with the given current i_{Sdref}, i_{Sqref}, the output v_{Sdref}, v_{Sqref} are adjusted after PID controller. The voltage v_{Sref}, v_{Sref} in the two-phase static coordinate system are obtained by inverse Park transformation. After space vector operation, the control signal of the three-phase inverter bridge is output to drive the inverter bridge.

i_{Sqref} in this paper is the torque current reference value for a given speed and the actual speed difference after the adjustment of speed PID.

3.2 The Simulation Model of PMSM Control System

According to the space vector control structure and performance requirements of the permanent magnet synchronous motor, the PMSM space vector control system

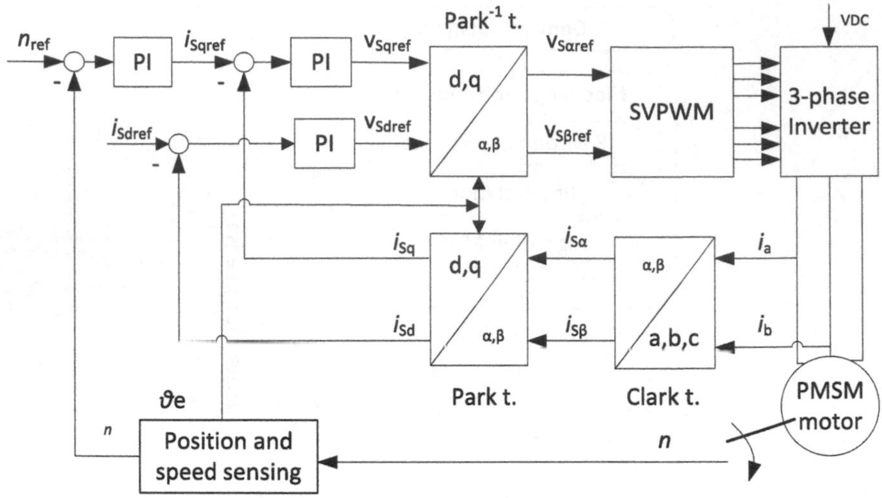

Fig. 2 Block diagram of space vector control for permanent magnet synchronous motor

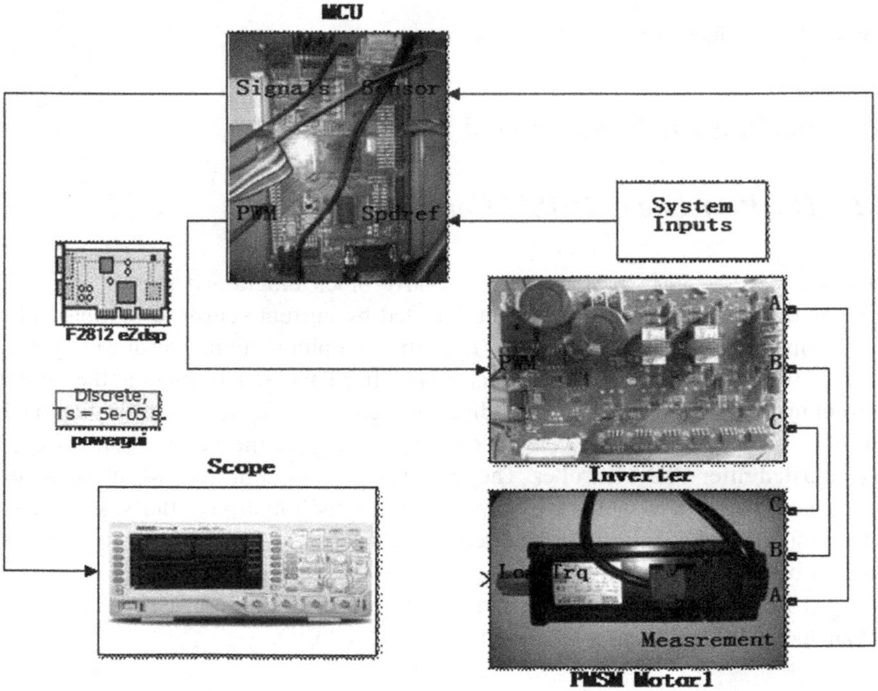

Fig. 3 The simulation model of PMSM space vector control system

Simulink simulation model is built in Fig. 3. The control system mainly includes system input model, the model of controller, the inverter and the model of PMSM, scope is the system signal observation model.

The parameters of the control system simulation model include the speed loop and the current loop PID [5] parameter setting, the parameter setting of the permanent magnet synchronous motor, and the setting of the system current loop and the speed loop.

4 Simulation Result Analysis

In order to verify the correctness and validity of the permanent magnet synchronous motor control system, the simulation and verification of the whole control system model are carried out. Figure 4 shows the speed, current and rotor position waveform when the given speed is $300\,\mathrm{r} \cdot \mathrm{min}^{-1}$ in speed closed loop.

From Fig. 4 it can be seen that the motor can reach a given value at 0.2 s at a given speed, and has no overshoot, which shows that the dynamic response of the motor is good, and it also shows that the speed PID control parameter is reasonable.

5 Experimental Results Analysis of PMSM Control System

5.1 The Code Model of PMSM Control System

The code model of control system is shown in Fig. 5. The basic structure is timing through AD interrupt and calling for the entire control algorithm. According to the requirements of control system, the sampling frequency of current loop is 50 us, while the sampling frequency of speed loop is 5 ms, using PWM underflow to start the AD conversion, after the completion of the AD converter generates an interrupt, the given speed of the motor is achieved by GUI control interface to write RTDX channel.

The implementation part of current loop PID control and SVPWM algorithm are implemented in Function Call Subsystem, as shown in Fig. 6, including PWM module, QEP module, ADC module, Interrupt Hardware module, RTDX module, etc.

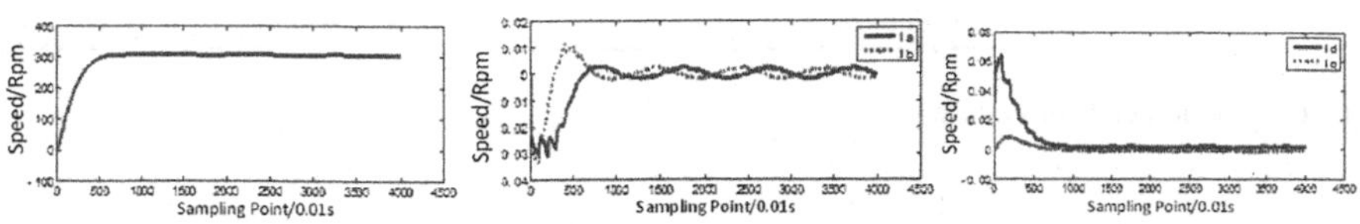

Fig. 4 The simulation model of PMSM space vector control system

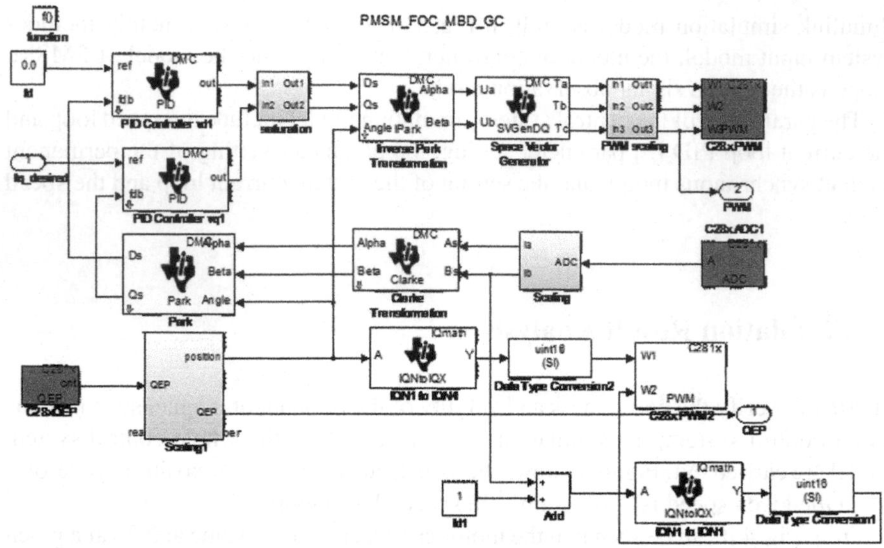

Fig. 5 The code model of PMSM control system

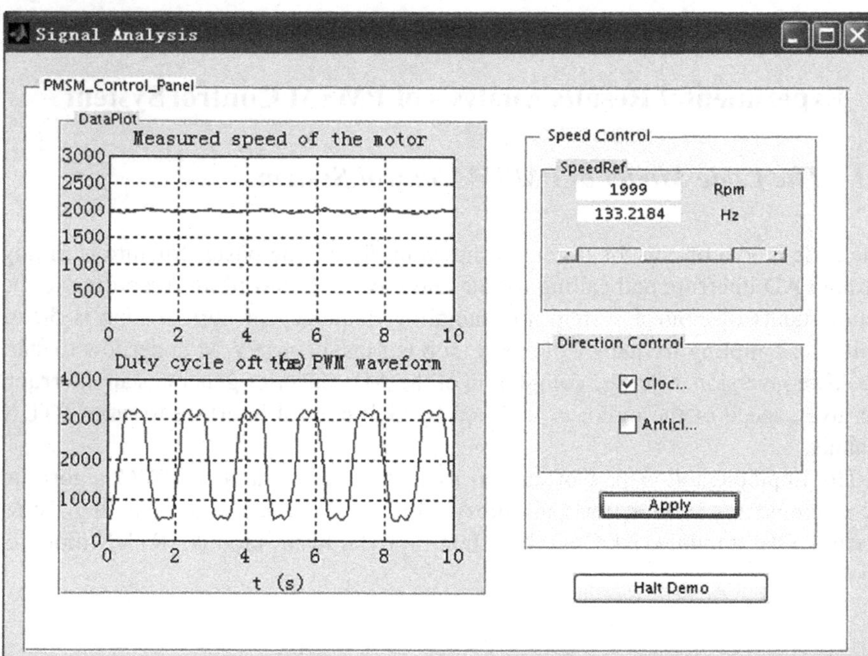

Fig. 6 Function call subsystem

After the completion of the permanent magnet synchronous motor vector control theory analysis, control system simulation and control algorithm code model, combined with a hardware test platform, the control algorithm is automatically generated code download to the hardware test platform for actual debugging and functional verification. The test platform includes control board, power drive board, permanent magnet synchronous motor, DC power supply and oscilloscope.

5.2 *Debug Result Analysis*

GUI real-time control effect is shown in Fig. 7. GUI control interface can be used to modify the given speed of the motor, achieve the motor speed control, but also achieve directional control of motor by select the check box in the GUI interface. The real-time current and speed of the motor are observed by the oscilloscope.

The speed and current response waveform is shown in Figs. 8 and 9. It can be seen that the space vector control system has the characteristics of quick start and fast response.

Comparing the results of experiment and the results of simulation, the current and speed In the speed waveform, the x-coordinate represents the sampling point and the

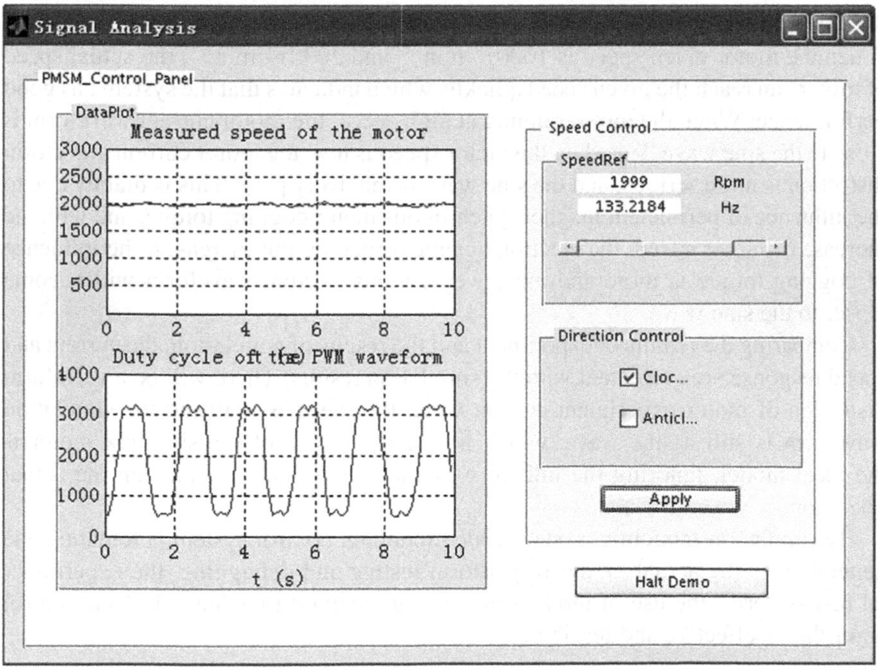

Fig. 7 GUI display and control interface for PMSM

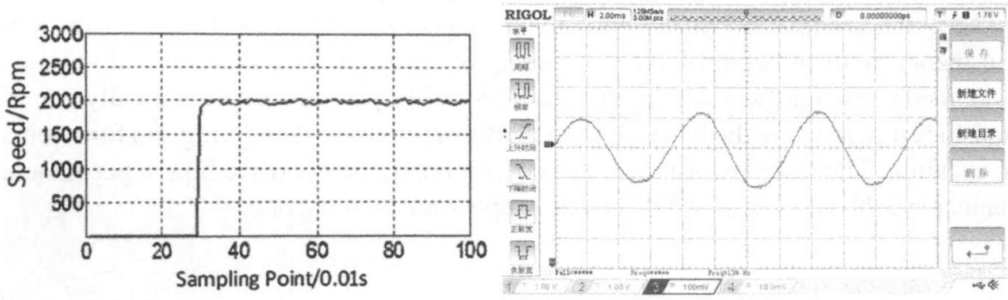

Fig. 8 The motor speed and A phase current waveform when $SpdRef = 2000\,\mathrm{r} \cdot \mathrm{min}^{-1}$

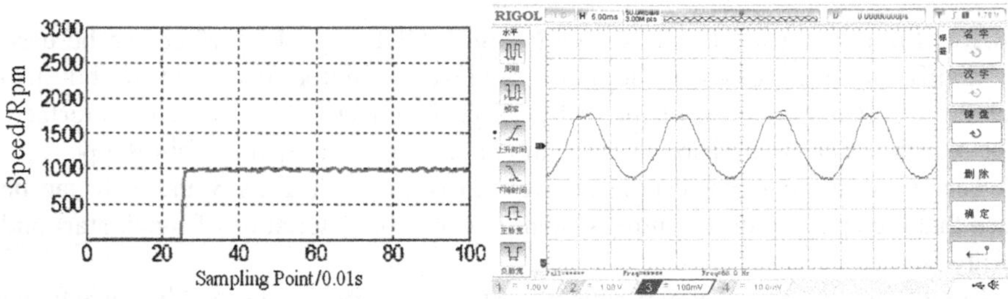

Fig. 9 The motor speed and A phase current waveform when $SpdRef = 1000\,\mathrm{r} \cdot \mathrm{min}^{-1}$

data sampling frequency is 0.01 s, such as the x-coordinate is 20, the time is 0.2 s. When the motor given speed is $1000\mathrm{r} \cdot \mathrm{min}^{-1}$ and $2000\mathrm{r} \cdot \mathrm{min}^{-1}$, the actual speed of motor can reach the given speed quickly, which indicates that the system has good performance; When the motor running at high speed, the motor current waveform is close to the sine wave, but when the motor speed is low, the motor current waveform distortion is more serious, and the sine wave is relatively poor. This is mainly due to the influence of permanent magnet synchronous motor cogging torque, and with the increase of motor speed, the electromagnetic torque output increases, the influence of cogging torque is more and more weak, motor current waveform will become closer to the sine wave.

Comparing the reaults of experiment and the results of simulation, the current and speed response are consistent with the simulation results. There will be a waveform distortion of motor experiment current when the speed is low, and the simulation waveform is still a sine wave, which is due to the use of the simulink motor is the ideal model, ignoring the influence of motor physical structure on the actual operation.

By simulation modeling, system code modeling, control system automatic code generation to the actual hardware platform testing and debugging, the experimental results verify the use of model-based design method to achieve PMSM control algorithm is effective and feasible.

6 Conclusion

The simulation model of the space vector control system of permanent magnet synchronous motor is built by using Coder Embedded toolbox, the speed control performance of the system is simulated. Based on the TMS320F2812 control chip, the hardware test platform of the permanent magnet synchronous motor is designed. The code model of the permanent magnet synchronous motor on the basis of the simulation model is built. Finally, C code is automatically generated by the code model, the actual debugging on the hardware platform is completed. The experimental results show that the speed regulation performance of the permanent magnet synchronous motor based on the model based design method can meet the requirements.

Acknowledgments This work has been supported by the Opening Project of Key Laboratory of Intelligent Transportation Systems Technologies, Research institute of highway ministry of transport and the NSFC (51507013) and Fundamental Research Funds for the Central Universities (2014G1321040, 310822151025).

References

1. Huaji Y, Jiali Q (2014) Based on FPGA and DSP full digital permanent magnet synchronous motor (PMSM) servo system design. J Electric Mach Control Appl 41:28–32 (in Chinese)
2. Zhiqiang L (2008) The DSP program design by Model-Based. J Program Design Micro Comput Appl 29:93–96 (in Chinese)
3. Ting Z, Chaokun X, Weixuan L (2011) Realization and verification of embedded C code based on model. J Microcontroller Embed Syst Appl 12:15–17 (in Chinese)
4. Zhenheng Q, Zhongjie S, Tao L (2010) RTW embedded code automatic generation mechanism and code structure analysis. J Comput Meas Control 18:639–643 (in Chinese)
5. Zhongran J, Wenping C, Bin B (2011) Control system controller parameter optimization design. J. Motor Permanent Magn Synchronous Motor Control Appl 38:26–31 (in Chinese)
6. Mingji L, Jiachi S, Hanjin G (2014) Maximum torque current ratio control of permanent magnet synchronous motor based on fuzzy PI controller. J Motor Control Appl 41:10–16 (in Chinese)

Stability Analysis of Car-Following Model with Uncertainty in Perceiving Velocity

Geng Zhang, Dihua Sun, Min Zhao, Hui Liu, Dong Chen and Yang Li

Abstract To reveal the impact of uncertain traffic information on traffic flow, a new car-following model with consideration of uncertainty in perceiving velocity is proposed. The linear stability criterion of the new model is derived through linear stability theory and it shows that the stable region for a small perceived velocity of the preceding vehicle is larger than that for a large perceived velocity of the preceding vehicle. Numerical simulation is in good agreement with the analytical results, which reveals that the uncertain traffic information would influence the stability of traffic flow importantly.

Keywords Traffic flow · Car-following model · Uncertainty · Linear stability analysis

G. Zhang (✉) · D. Sun · M. Zhao · H. Liu · D. Chen · Y. Li
Key Laboratory of Dependable Service Computing in Cyber Physical Society of Ministry of Education, Chongqing University, Chongqing 400044, China
e-mail: zhanggengfaraday@163.com

D. Sun
e-mail: d3sun@163.com

M. Zhao
e-mail: zhaomin@cqu.edu.cn

H. Liu
e-mail: hly_huiliu@163.com

D. Chen
e-mail: chendong1418@163.com

Y. Li
e-mail: uply@cqu.edu.cn

G. Zhang · D. Sun · M. Zhao · H. Liu · D. Chen · Y. Li
College of Automation, Chongqing University, Chongqing 400044, China

© Springer International Publishing Switzerland 2017

V.E. Balas et al. (eds.), *Information Technology and Intelligent Transportation Systems*, Advances in Intelligent Systems and Computing 454,
DOI 10.1007/978-3-319-38789-5_50

1 Introduction

Traffic congestion is a serious problem in modern society and has attracted extensive research in the last few decades. In order to reveal the mechanism of traffic congestion, many traffic models have been constructed. Among them, the car-following models aiming at revealing the motion of an individual vehicle to a stimulus from its preceding vehicle without lane-changing are widely studied.

The pioneering car-following model is proposed by Pipes [1] and the vehicles motion is controlled by the velocity difference of two successive vehicles. In 1995, Bando et al. [2] assumed that the following vehicles motion is determined by an optimal velocity and put forward the famous optimal velocity (OV) car-following model. After that, the OV model was extended by Helbing and Tilch [3] to the generalized force (GF) model and further extended to the full velocity difference (FVD) model by Jiang et al. [4]. Recently, lots of extended car-following models [5–13] with consideration of various real traffic information have been developed.

However, the information introduced in the aforementioned models are determinate and they are unable to reveal the impact of uncertain traffic condition on traffic flow. Actually, the traffic system is operated under uncertain environment on account of the irregular road surface, bad weather, traffic interruption, equipment failure, and driver's personality and the uncertain traffic condition would influence the stability of traffic flow significantly. But few researches were conducted to explore the car-following behavior with uncertain traffic condition [14].

In order to reveal the effect of uncertain traffic information on traffic flow, a new car-following model is constructed in this paper. The new model is introduced in the following section. In Sect. 3, the linear stability analysis is carried out for the new model. Numerical simulation is carried out to validate the analytic results in Sect. 4 and a conclusion is given in Sect. 5.

2 The New Model

Generally, the car-following theory describes the motion of the following vehicle to the stimulus from its surrounding traffic condition. In 1995, Bando et al. [2] assumed that the following driver always seeks a safe and optimal velocity in the car-following process and proposed the following OV model

$$\frac{dv_n(t)}{dt} = a\left[V(\Delta x_n(t)) - v_n(t)\right] \tag{1}$$

where $x_n(t)$ and $v_n(t)$ are the position and velocity of vehicle n at time t respectively, $\Delta x_n(t) = x_{n+1}(t) - x_n(t)$ denotes the headway between the preceding vehicle $n+1$ and the following one n, a is the driver's sensitivity and V refers to the optimal velocity function. Equation (1) shows that the variation of the following vehicle's speed is determined by the difference of the optimal velocity and its current velocity.

After that, Jiang et al. [4] considered that the preceding vehicle's velocity is also very important for the motion of the following vehicle and developed the famous FVD model.

$$\frac{dv_n(t)}{dt} = a\left[V(\Delta x_n(t)) - v_n(t)\right] + ak\left[v_{n+1}(t) - v_n(t)\right] \tag{2}$$

where $\Delta v_n(t) = v_{n+1}(t) - v_n(t)$ denotes the velocity difference between vehicle $n+1$ and vehicle n, and k represents driver's response coefficient to the velocity difference information.

Recently, the FVD model was widely extended by considering many real traffic factors like multiple information of headway or velocity [5–7], driver's individual behaviors [8–11] and various road condition [12, 13]. It has been found out that these information influence the stability of traffic flow significantly. But to the author's knowledge, these information introduced in the above car-following models are quite deterministic and can not be used to uncover the traffic property caused by uncertain traffic situation. In real traffic, due to the irregular road surface, slushy weather, traffic interruption, and equipment failure, the following driver's perceived velocity of the preceding vehicle is uncertain. In order to investigate the effect of uncertainty of the preceding vehicle's velocity on the stability of traffic flow, a new car-following model is constructed based on the FVD model as follows:

$$\frac{dv_n(t)}{dt} = a\left[V(\Delta x_n(t)) - v_n(t)\right] + ak\left[(1+p)v_{n+1}(t) - v_n(t)\right] \tag{3}$$

where p is the coefficient reflecting the level of uncertainty of the preceding vehicle's velocity and a big absolute value of p indicates a higher level of uncertainty. The optimal velocity function V adopted here is the same as that in Ref. [2]:

$$V(\Delta x_n(t)) = \frac{v_{\max}}{2}\left[\tanh(\Delta x_n(t) - h_c) + \tanh(h_c)\right] \tag{4}$$

where $v_{max} = 2$ is the maximal velocity and $h_c = 4$ represents the safety distance.

3 Linear Stability Analysis

In this section, the linear stability analysis of the new car-following model is carried out to investigate the influence of uncertainty of the preceding vehicle's velocity on traffic flow. We first rewrite Eq. (3) into the following difference form in terms of headway:

$$\frac{d^2\Delta x_n(t)}{dt^2} = a\left[V(\Delta x_{n+1}(t)) - V(\Delta x_n(t)) - \frac{d\Delta x_n(t)}{dt}\right] + ak\left[(1+p)\frac{d\Delta x_{n+1}(t)}{dt} - \frac{d\Delta x_n(t)}{dt}\right] \tag{5}$$

The stable traffic flow is such a state that all vehicles move with the same headway h and the same velocity $V^* = \frac{V(h)}{(1-kp)}$ on a ring road. Then the solution of the stable traffic flow is:

$$x_n^0(t) = hn + V^*t, h = L/N \tag{6}$$

where N is the total car number and L refers to the length of the road. Let $y_n(t)$ be a small deviation from the stable solution $x_n^0(t)$ as below:

$$y_n(t) = x_n(t) - x_n^0(t) \tag{7}$$

Substituting Eqs. (6) and (7) into Eq. (5) and linearizing it, then we have

$$\frac{d^2\Delta y_n(t)}{dt^2} = a\left[V'(\Delta y_{n+1}(t) - \Delta y_n(t)) - \frac{d\Delta y_n(t)}{dt}\right] + ak\left[(1+p)\frac{d\Delta y_{n+1}(t)}{dt} - \frac{d\Delta y_n(t)}{dt}\right] \tag{8}$$

where $\Delta y_n(t) = y_{n+1}(t) - y_n(t)$ and $V' = d(V(\Delta x_n(t)))/d(\Delta x_n(t))|_{\Delta x_n(t)=h}$. By expanding $\Delta y_n(t) = \exp(ikn + zt)$, we obtain the following equation for z

$$z^2 + az - aV'(e^{ik} - 1) - ak\left[(1+p)e^{ik} - 1\right]z = 0 \tag{9}$$

By expanding $z = z_1(ik) + z_2(ik)^2 + \cdots$ and inserting it into Eq. (9), the first-order and second-order terms of ik are obtained

$$z_1 = \frac{V'}{1-kp} \tag{10}$$

$$z_2 = -\frac{(V')^2}{a(1-kp)^3} + \frac{V'}{2(1-kp)} + \frac{k(1+p)V'}{(1-kp)^2} \tag{11}$$

The uniform flow remains stable if z_2 is positive but becomes unstable when z_2 is negative for long wavelength modes. Thus, the neutral stability condition for the new model is:

$$a = \frac{2V'}{(1-kp)\left[1-kp+2k(1+p)\right]} \tag{12}$$

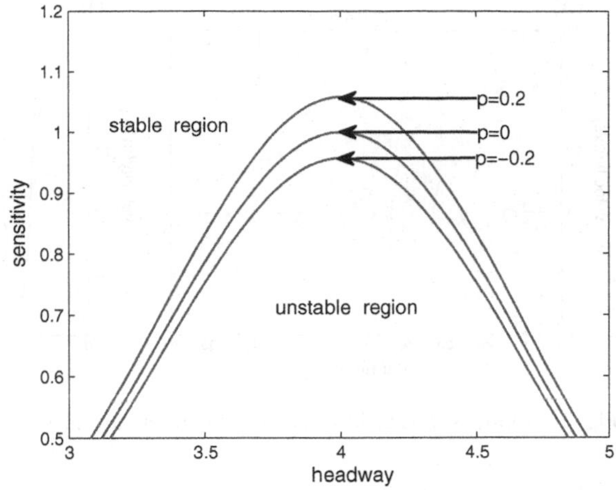

Fig. 1 Phase diagram in headway-sensitivity space

So, the uniform traffic flow is stable if

$$a > \frac{2V'}{(1-kp)\,[1-kp+2k(1+p)]} \tag{13}$$

When $p = 0$, the stable condition reduces to that of the FVD model

$$a > \frac{2V'}{1+2k} \tag{14}$$

From Eq. (13), we can see that the parameter p impacts the stability of traffic flow importantly. Figure 1 shows the neutral stability curves of the new model in the headway-sensitivity space $(\Delta x, a)$ based on Eq. (12) for different values of p when $k = 0.5$. The apex of each curve indicates the critical point (h_c, a_c). In Fig. 1, it is clear that the headway-sensitivity space is divided into the stable region (above the neutral stability curve) and the unstable region (below the neutral stability curve) by each neutral stability curve. In the unstable region, traffic density wave will result from the original stable traffic flow under a small disturbance. To the contrary, the traffic flow is stable in the stable region and traffic jam will not appear. Furthermore, we can see from Fig. 1 that the unstable region expands with the value of p increasing and the stable region for p with negative value is much larger than that for p with positive value. This means that the uncertainty of the preceding vehicle's velocity affects the traffic stability importantly and a smaller perceived preceding vehicle's velocity can enhance the stability of traffic flow but a larger perceived preceding vehicle's velocity can reduce traffic stability and lead to traffic congestion.

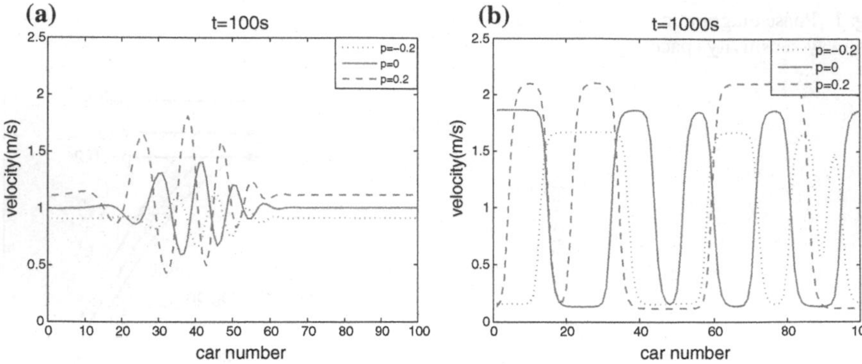

Fig. 2 Snapshots of vehicle's velocities for different values of p when $t = 100\,\mathrm{s}$ and $t = 1000\,\mathrm{s}$

4 Numerical Simulation

In order to check the theoretical results, numerical simulation is carried out for the new model with periodic boundary condition to study the space-time evolution of the velocity when a small perturbation appears. The total car number is $N = 100$ and the length of the ring road is $L = 400\,\mathrm{m}$. Other parameters are $a = 0.64$, $k = 0.5$. The initial condition is taken as $x_1(0) = 0.2\,\mathrm{m}$ and $x_n(0) = (n-1)L/N\mathrm{m}$ for $n \neq 1$, the initial velocity is $x_n'(0) = V(L/N)$, and the simulation time step is $\Delta t = 0.1$.

Figure 2 shows the snapshots of the velocities of all vehicles at $t = 100\,\mathrm{s}$ and $t = 1000\,\mathrm{s}$ for the new model. The result is reduced to the FVD model when $p = 0$. From Fig. 2, we can see that the initial stable flow will develop into traffic congestion under a small disturbance as the stability condition of Eq. (13) is unsatisfied. In addition, the amplitude of the velocity curve for p with negative value is much smaller than that for p with positive value, which means that a smaller perceived preceding vehicle's velocity is more useful than a larger perceived preceding vehicle's velocity in stabilizing traffic flow.

5 Conclusion

In this paper, we propose a new car-following model with consideration of driver's uncertainty in perceiving velocity of the preceding vehicle based on the full velocity difference model. The stability criterion of the new model is obtained through linear stability theory. Also numerical simulation is carried out to verify the analytical results and it is proved that the uncertain traffic information has an important role in the stability of traffic flow.

Acknowledgments This work was supported by the National Natural Science Foundation of China (Grant No. 61573075), the 2015 Chongqing University Postgraduates Innovation Project, the Research Fund for the Doctoral Program of Higher Education of China (Grant No. 20120191110047), and the Fundamental Research Funds for the Central Universities (Grant No. 106112014CDJZR178801).

References

1. Pipes LA (1953) An operational analysis of traffic dynamics. J Appl Phys 24(3):274–281
2. Bando M, Hasebe K, Nakayama A, Shibata A, Sugiyama Y (1995) Dynamical model of traffic congestion and numerical simulation. Phys Rev E 51(2):1035–1042
3. Helbing D, Tilch B (1998) Generalized force model of traffic dynamics. Phys Rev E 58(1):133–138
4. Jiang R, Wu QS, Zhu ZJ (2001) Full velocity difference model for a car-following theory. Phys Rev E 64(1):017101
5. Ge HX, Dai SQ, Dong LY, Xue Y (2004) Stabilization effect of traffic flow in an extended car-following model based on an intelligent transportation system application. Phys Rev E 70(6):066134
6. Peng GH, Sun DH (2010) A dynamical model of car-following with the consideration of the multiple information of preceding cars. Phys Lett A 374(15):1694–1698
7. Li YF, Sun DH, Liu WL, Zhang M, Zhao M, Liao XY, Tang L (2011) Modeling and simulation for microscopic traffic flow based on multiple headway, velocity and acceleration difference. Nonlinear Dyn 66(1–2):15–28
8. Yu L, Li T, Shi ZK (2010) Density waves in a traffic flow model with reaction-time delay. Phys A 389(13):2607–2616
9. Hu YM, Ma TS, Chen JZ (2014) An extended multi-anticipative delay model of traffic flow. Commun Nonlinear Sci Numer Simul 19(9):3128–3135
10. Zheng LJ, Tian C, Sun DH, Liu WN (2012) A new car-following model with consideration of anticipation driving behavior. Nonlinear Dyn 70(2):1205–1211
11. Zhang G, Sun DH, Liu H, Zhao M (2014) Analysis of drivers characteristics in car-following theory. Modern Phys Lett B 28(24):1450191
12. Tang TQ, Huang HJ, Huang SJ, Jiang R (2009) A new car-following model with consideration of the traffic interruption probability. Chin Phys B 18(3):975–983
13. Tang TQ, Wang YP, Yang XB, Wu YH (2012) A new car-following model accounting for varying road condition. Nonlinear Dyn 70(2):1397–1405
14. Li SK, Yang LX, Gao ZY, Li KP (2014) Stabilization strategies of a general nonlinear car-following model with varying reaction-time delay of the drivers. ISA Trans 53(6):1739–1745

Leader-Following Attitude Consensus for Multi-rigid-body Systems with Time-Varying Communication Delays and Jointly Connected Topologies

Long Ma, Shicheng Wang and Haibo Min

Abstract In this paper, we study the leader-following attitude consensus problem for multi-rigid-body systems under jointly connected topologies coupled with time-varying communication delays. By associating an auxiliary vector for each follower, the control algorithm is designed, and the stability of the whole system is proved by using Lyapunov Krasovskii function and contradiction method. In terms of linear matrix inequalities, we derive sufficient conditions that guarantee all followers asymptotically converge to the leader.

Keywords Multi-rigid-body · Attitude consensus control · Leader-following · Communication delay · Jointly connected topologies

1 Introduction

Attitude consensus control for a group of rigid-body systems has received much research interest recently. Since the biggest difference between multi-agent systems and single-agent system lies in the communication network, two interesting topics on attitude consensus problem have been extensively studied. One is about the communication time-delay, while the other one is on the switching topologies.

Time delays inevitably exist in the system and communication links, which may degrade the control performance of the formation and even destabilize the entire system. References [1, 2] consider the case with constant time-delay. References

This work is supported by National Natural Science Foundation of China (Grant No. 61203354).

L. Ma (✉) · S. Wang · H. Min
High-tech Institute of Xi'an, Xian, China
e-mail: mlong_301@163.com

S. Wang
e-mail: wshcheng@vip.163.com

H. Min
e-mail: haibo.min@gmail.com

© Springer International Publishing Switzerland 2017

V.E. Balas et al. (eds.), *Information Technology and Intelligent Transportation Systems*, Advances in Intelligent Systems and Computing 454,
DOI 10.1007/978-3-319-38789-5_51

[3–5] consider the problem with time-varying communication delays under undirected topology.

Communication outage, new member's joining or quitting, radio silence or recovery will cause the change of the communication topology, named switching topologies, which makes it more difficult to design the control laws. References [6–8] consider the case with communication delays and switching topologies. However, all the above literatures don't consider the more challenging jointly connected topologies. References [9, 10] consider jointly connected topologies but without communication delays.

In this paper, we study the leader-following attitude consensus problem for multi-rigid-body systems under jointly connected topologies coupled with time-varying communication delays. This is the biggest difference between this paper and the others. The controller includes two parts, one of which is designed for compensation, while the other one is designed for coordination. By associating an auxiliary signal for each follower, the whole system is proved to be stable and converge to the leader.

2 Problem Statement and Background Information

2.1 *Dynamics of Rigid Spacecraft Attitude*

We consider a multiple rigid spacecraft system consisting of $n+1$ spacecraft. Suppose that there exist n followers, labeled 1 to n, and one single leader labeled 0. The attitude dynamics of the ith follower is described by

$$\begin{aligned} \dot{\sigma}_i &= G(\sigma_i)\omega_i \\ \dot{\omega}_i &= J_i^{-1}\left(-\omega_i^{\times} J_i \omega_i + u_i\right) \end{aligned} \tag{1}$$

where $\sigma_i \in \mathbb{R}^3$ denotes the Modified Rodrigues Parameters that represents the attitude of the ith spacecraft, $\omega_i = \begin{bmatrix} \omega_{i1} \ \omega_{i2} \ \omega_{i3} \end{bmatrix}^{\mathrm{T}} \in R^3$ denotes the angular velocity of the ith spacecraft, J_i and u_i are respectively the inertial matrix and the external input torque of the ith spacecraft. $\omega_i^{\times}$ is the skew-symmetric matrix with the form $\omega_i^{\times} = \begin{bmatrix} 0 & -\omega_{i3} & \omega_{i2} \\ \omega_{i3} & 0 & -\omega_{i1} \\ -\omega_{i2} & \omega_{i1} & 0 \end{bmatrix}$, the matrix $G(\sigma_i)$ is given by

$$G(\sigma_i) = \frac{1}{2}\left[\frac{(1-\sigma_i^{\mathrm{T}}\sigma_i)I_3}{2} + \sigma_i^{\times} + \sigma_i\sigma_i^{\mathrm{T}}\right], \tag{2}$$

which has the following properties [11]

$$G(\sigma_i)G^{\mathrm{T}}(\sigma_i) = \left(\frac{1+\sigma_i^{\mathrm{T}}\sigma_i}{4}\right)^2 I_3 = p_i I_3. \tag{3}$$

2.2 Graph Theory

Graphs can be conveniently used to represent the information flow between agents. Let $\bar{\mathcal{G}} = \{\bar{\mathcal{V}}, \bar{\mathcal{E}}, \bar{\mathcal{A}}\}$ be an undirected graph or directed graph (digraph) of order $n+1$ with the set of nodes $\bar{\mathcal{V}}(\bar{\mathcal{G}}) = \{v_0, v_1, v_2, \cdots, v_n\}$, the set of edges $\bar{\mathcal{E}} \subseteq \bar{\mathcal{V}} \times \bar{\mathcal{V}}$, and a weighted adjacency matrix $\bar{\mathcal{A}} = \{a_{ij}\}$ with non-negative adjacency elements a_{ij}. Given a piecewise constant switching signal $\varrho(t)$, we can define a nonnegative switching matrix $\bar{\mathcal{A}}_{\varrho(t)} = [a_{ij}(t)]$, $i, j = 0, 1, \ldots, n$, where, for $i = 1, \ldots, n$, $a_{i0}(t) > 0$ if and only if the control input u_i can access the information of the leader's at time instant t, and all other elements of $\bar{\mathcal{A}}_{\varrho(t)}$ are arbitrary nonnegative numbers satisfying $a_{ii}(t) = 0$ for any $t \geq 0$, $i = 0, 1, \ldots, n$. Let $\bar{\mathcal{G}}_{\varrho(t)} = (\bar{\mathcal{V}}, \bar{\mathcal{E}}_{\varrho(t)})$ be a dynamic digraph of $\bar{\mathcal{A}}_{\varrho(t)}$. Then, the node set $\bar{\mathcal{V}} = 0, 1, \ldots, n$ with 0 corresponding to the leader system and the integer i, $i = 1, \ldots, n$, corresponding to the ith subsystem of the follower system, and $\bar{\mathcal{E}}_{\varrho(t)} \subseteq \bar{\mathcal{V}} \times \bar{\mathcal{V}}$ and $(i, j) \in \bar{\mathcal{E}}_{\varrho(t)}$ if and only if $a_{ij}(t) > 0$ at time instant t.

To model the jointly-connected topologies, we consider an infinite sequence of continuous, bounded, non-overlapping time intervals $[t_k, t_{k+1})$, $k = 0, 1, 2, \ldots$ with $t_0 = 0$, $T_0 \leq t_{k+1} - t_k \leq T$ for some constants T_0 and T. Assume that each interval $[t_k, t_{k+1})$ is composed of the following non-overlapping subintervals $[t_k^0, t_k^1), \ldots, [t_k^{j-1}, t_k^j), \ldots, [t_k^{m_k-1}, t_k^{m_k})$ with $t_k^0 = t_k$ and $t_k^{m_k} = t_{k+1}$ for some nonnegative integer m_k. The topology switches at time instants $t_k^0, t_k^1, \ldots, t_k^{m_k-1}$, which satisfy $t_k^{j-1} - t_k^j \geq \tau$, $j = 1, \ldots, m_k$, with τ a positive constant, such that during each subinterval $[t_k^{j-1}, t_k^j)$, the interconnection topology $\bar{\mathcal{G}}_{\varrho(t)}$ does not change. Note that in each interval $[t_k, t_{k+1})$, $\bar{\mathcal{G}}_{\varrho(t)}$ is permitted to be disconnected. The graphs are said to be jointly connected across the time interval $[t, t+T]$ with $T > 0$ if the union of graphs $\bar{\mathcal{G}}_{\varrho(t)} : s \in [t, t+T]$ is connected.

Assumption 1 The communication among the followers is bidirectional, and there exists a subsequence $\{i_k\}$ of $\{i : i = 0, 1, \ldots\}$ with $t_{i_{k+1}} - t_{i_k} < T$ for some positive T such that the union graph $\bigcup_{j=i_k}^{i_{k+1}-1} \bar{\mathcal{G}}_{\varrho}(t_j)$ contains a spanning tree with the node 0 as the root.

2.3 Problem Statement

The control objective is said to be reached if

$$\begin{aligned} &\lim_{t\to\infty} |\sigma_i(t) - \sigma_0(t)| \to 0 \\ &\lim_{t\to\infty} |\omega_i(t)| \to 0, \qquad i, j \in l. \end{aligned} \tag{4}$$

3 Main Results

In this section, we deal with the distributed attitude consensus control problem with not only jointly connected topologies but also time-varying delays among agents. Denote $T_{ij}(t)$ as the time-varying delay from the jth agent to the ith agent.

Assumption 2 The time delays are differentiable, bounded and satisfy $\dot{T}_{ij}(t) \leq \Gamma < 1$, where Γ is a nonnegative constant. We define a positive constant gain dependent on the maximum changing rates of delay as $d^2 \leq 1 - \Gamma$, and also assume that $\ddot{T}_{ij}(t)$ is bounded, i.e., $\ddot{T}_{ij}(t) \in \mathcal{L}_\infty$.

Lemma 1 ([12]) *Given constant matrices Ξ_{11}, Ξ_{12}, and Ξ_{22}, where $\Xi_{11} = \Xi_{11}^{\mathrm{T}}$ and $\Xi_{22} = \Xi_{22}^{\mathrm{T}} < 0$. Then $\Xi_{11} - \Xi_{12}\Xi_{22}^{-1}\Xi_{12}^{T} < 0$ if and only if*

$$\begin{bmatrix} \Xi_{11} & \Xi_{12} \\ \Xi_{12}^{\mathrm{T}} & \Xi_{22} \end{bmatrix} < 0, \text{ or } \begin{bmatrix} \Xi_{22} & \Xi_{12}^{\mathrm{T}} \\ \Xi_{12} & \Xi_{11} \end{bmatrix} < 0. \tag{5}$$

3.1 Distributed Controller

We associate each agent with the following auxiliary signal

$$s_i(t) = \dot{\sigma}_i(t) + \gamma(\sigma_i(t) - \sigma_0) \tag{6}$$

where k is a positive constant.

Substituting (6) into (1), the dynamics (1) can be written as

$$\dot{s}_i(t) = G(\sigma_i)J_i^{-1}(-\omega_i^{\times} J_i\omega_i + u_i) + \dot{G}(\sigma_i)\omega_i + \gamma\dot{\sigma}_i \tag{7}$$

We choose the control input for the ith follower agent as

$$u_i = \Upsilon_i(t) + F_i(t) \tag{8}$$

where

$$\Upsilon_i(t) = \omega_i^{\times} J_i\omega_i - \frac{J_i G^{\mathrm{T}}(\sigma_i)}{p_i}\left\{\dot{G}(\sigma_i)\omega_i + (\gamma + c)\dot{\sigma}_i\right\}, \tag{9}$$

$$F_i(t) = -\frac{J_i G^{\mathrm{T}}(\sigma_i)}{p_i}\left\{\sum_{j\in N_i} a_{ij}(t)[\sigma_i(t - T_{ij}(t)) - \sigma_j(t - T_{ij}(t))] + b_i(t)[\sigma_i(t - T_{ij}(t)) - \sigma_0]\right\}. \tag{10}$$

where $j \in \mathcal{N}_i$, $a_{ij}(t)$ is the i, j entry of the weighted adjacency matrix $\mathcal{A}_{\varrho(t)}$, and $T_{ij}(t)$ is the time-varying communication delay.

3.2 Lyapunov Function and Convergence Analysis

Suppose that the time-invariant communication graph $\bar{\mathcal{G}}_{\varrho(t)}$ on interval $[t_k^{j-1}, t_k^j)$ contains $l_\varrho \geq 1$ connected components with the corresponding sets of nodes denoted by $\varphi_k^{j-1}(1), \ldots, \varphi_k^{j-1}(l_\varrho)$, and in each $\varphi_k^{j-1}(i)$ there are $n_k^{j-1}(i)$ nodes with $i = 1, \ldots, l_\varrho$. For example, there is one connected component in $\bar{\mathcal{G}}_1$ of Fig. 1, therefore $l_\varrho = 1$ and $n_k(1) = 3$. To analyze the networks, it is natural to consider the dynamics of agents on each possible connected component.

Define $\tilde{\sigma}_i = \sigma_i - \sigma_0$, $\tilde{\sigma}_i(t-\tau) = \sigma_i(t-\tau) - \sigma_0$, then we can get from (7)–(10) that

$$\dot{s}_i(t) = -c\dot{\sigma}_i - \sum_{j\in N_i} a_{ij}(t)[\hat{\sigma}_i(t - T_{ij}(t)) - \hat{\sigma}_j(t - T_{ij}(t))] - b_i(t)\hat{\sigma}_i(t - T_{i0}(t)). \tag{11}$$

Define

$$l_{kij}(t) = \begin{cases} -a_{ij}(t), & j \neq i, \tau_k = T_{ij}(t) \\ 0, & j \neq i, \tau_k \neq T_{ij}(t) \\ \sum_{j=1}^{n} l_{kij}(t), & j = i \end{cases}, \quad b_{ki}(t) = \begin{cases} b_i(t), & j = i, \tau_k = T_{ij}(t) \\ 0, & j = i, \tau_k \neq T_{ij}(t) \\ 0 & j \neq i \end{cases} \tag{12}$$

then (7) can be written in the matrix form as

$$\dot{s} = -c\dot{\sigma} - \sum_{k=1}^{r} (L_{\varrho k}^i \otimes I_3 + B_{\varrho k}^i \otimes I_3)\tilde{\sigma}(t - \tau_k), \tag{13}$$

where $s = [s_1^{\mathrm{T}}, \ldots, s_n^{\mathrm{T}}]^{\mathrm{T}}$, $\sigma = [\sigma_1^{\mathrm{T}}, \ldots, \sigma_n^{\mathrm{T}}]^{\mathrm{T}}$, $\tilde{\sigma}(t-\tau_k) = [\tilde{\sigma}_1^{\mathrm{T}}(t-\tau_k), \ldots, \tilde{\sigma}_n^{\mathrm{T}}(t-\tau_k)]^{\mathrm{T}}$, $H_{\varrho k}^i = L_{\varrho k}^i + B_{\varrho k}^i$, $L_\varrho = \sum_{i=1}^{r} L_{\varrho k}^i$, $L_{\varrho k}^i = [l_{kij}(t)]$, $B_{\varrho k}^i = [b_{ki}(t)]$.

Define

$$\Xi_\varrho^i = \begin{bmatrix} \Xi_{11} & \Xi_{12} \\ \Xi_{12}^T & \Xi_{22} \end{bmatrix}, \tag{14}$$

where

$$\Xi_{11} = -\gamma \sum_{k=1}^{r} H_{\varrho k}^i + \frac{\gamma^2}{4} \sum_{k=1}^{r} \frac{\tau_k}{(1-d_k)} {H_{\varrho k}^i}^T H_{\varrho k}^i, \tag{15}$$

$$\Xi_{12} = -\frac{c\gamma I_n + \sum_{k=1}^{r} H_{\varrho k}^i}{2} + \frac{\gamma}{4} \sum_{k=1}^{r} \frac{\tau_k}{(1-d_k)} {H_{\varrho k}^i}^T H_{\varrho k}^i \tag{16}$$

$$\Xi_{22} = (-c + \sum_{k=1}^{r} \tau_k) I_n + \frac{1}{4} \sum_{k=1}^{r} \frac{\tau_k}{(1-d_k)} {H_{\varrho k}^i}^T H_{\varrho k}^i, \tag{17}$$

then we can get the following result:

Theorem 1 *Under Assumption 1–2, by choosing the input as* (8), *if there exists positive constants c and* γ *such that* $\Xi^i_\varrho < 0$, *then consensus of system* (1) *is achieved.*

Proof Define a Lyapunov-Krasovskii function as

$$V(t) = \frac{1}{2}s^{\mathrm{T}}s + \sum_{k=1}^{r}\int_{-\tau_k}^{0}\int_{t+\theta}^{t}\dot{\tilde{\sigma}}^{\mathrm{T}}(\varsigma)\dot{\tilde{\sigma}}(\varsigma)\mathrm{d}\varsigma\mathrm{d}\theta. \tag{18}$$

It is worth noting that $V(t)$ is continuously differentiable in spite of the existence of the switching topologies. The derivatives of $V(t)$ can be written as

$$\begin{aligned}
\dot{V}(t) &= s^{\mathrm{T}}\dot{s} + \sum_{k=1}^{r}\tau_k\dot{\tilde{\sigma}}^{\mathrm{T}}(t)\dot{\tilde{\sigma}}(t) - \sum_{k=1}^{r}(1-\dot{\tau}_k)\int_{t-\tau_k}^{t}\dot{\tilde{\sigma}}^{\mathrm{T}}(\varsigma)\dot{\tilde{\sigma}}(\varsigma)\mathrm{d}\varsigma \\
&= \sum_{i=1}^{l_\varrho}\left[\dot{\tilde{\sigma}}^i_\varrho(t) + \gamma\tilde{\sigma}^i_\varrho(t)\right]^{\mathrm{T}}\left[-c\dot{\tilde{\sigma}}^i_\varrho - \sum_{k=1}^{r}(L^i_{\varrho k}\otimes I_3 + B^i_{\varrho k}\otimes I_3)\tilde{\sigma}^i_\varrho(t-\tau_k)\right] \\
&\quad + \sum_{i=1}^{l_\varrho}\sum_{k=1}^{r}\tau_k\dot{\tilde{\sigma}}^{i\,\mathrm{T}}_\varrho(t)\dot{\tilde{\sigma}}^i_\varrho(t) - \sum_{i=1}^{l_\varrho}\sum_{k=1}^{r}(1-\dot{\tau}_k)\int_{t-\tau_k}^{t}\dot{\tilde{\sigma}}^{i\,\mathrm{T}}_\varrho(\varsigma)\dot{\tilde{\sigma}}^i_\varrho(\varsigma)\mathrm{d}\varsigma \\
&\leq \sum_{i=1}^{l_\varrho}\left\{\begin{array}{l}\left[\dot{\tilde{\sigma}}^i_\varrho(t) + \gamma\tilde{\sigma}^i_\varrho(t)\right]^{\mathrm{T}}\left[-c\dot{\tilde{\sigma}}^i_\varrho - \sum_{k=1}^{r}(L^i_{\varrho k}\otimes I_3 + B^i_{\varrho k}\otimes I_3)\tilde{\sigma}^i_\varrho(t-\tau_k)\right] \\ + \sum_{k=1}^{r}\tau_k\dot{\tilde{\sigma}}^{i\,\mathrm{T}}_\varrho(t)\dot{\tilde{\sigma}}^i_\varrho(t) - \sum_{k=1}^{r}(1-d_k)\int_{t-\tau_k}^{t}\dot{\tilde{\sigma}}^{i\,\mathrm{T}}_\varrho(\varsigma)\dot{\tilde{\sigma}}^i_\varrho(\varsigma)\mathrm{d}\varsigma\end{array}\right\}
\end{aligned} \tag{19}$$

Define $\delta^i_{\varrho k} = \tilde{\sigma}^i_\varrho(t) - \tilde{\sigma}^i_\varrho(t-\tau_k)$ and by using Jensen's inequality, we get that

$$\int_{t-\tau_k}^{t}\dot{\tilde{\sigma}}^{i\,\mathrm{T}}_\varrho(\varsigma)\dot{\tilde{\sigma}}^i_\varrho(\varsigma)\mathrm{d}\varsigma \leq \frac{1}{\tau_k}\delta^{i\,\mathrm{T}}_{\varrho k}(t)\delta^i_{\varrho k}(t) \tag{20}$$

then (19) can be written as

$$\begin{aligned}
\dot{V}(t) &\leq \sum_{i=1}^{l_\varrho}\left\{\begin{array}{l}\left[\dot{\tilde{\sigma}}^i_\varrho(t) + \gamma\tilde{\sigma}^i_\varrho(t)\right]^{\mathrm{T}}\left[-c\dot{\tilde{\sigma}}^i_\varrho - \sum_{k=1}^{r}(H^i_{\varrho k}\otimes I_3)\tilde{\sigma}^i_\varrho(t) + \sum_{k=1}^{r}(H^i_{\varrho k}\otimes I_3)\delta^i_{\varrho k}(t)\right] \\ + \sum_{k=1}^{r}\tau_k\dot{\tilde{\sigma}}^{i\,\mathrm{T}}_\varrho(t)\dot{\tilde{\sigma}}^i_\varrho(t) - \sum_{k=1}^{r}\frac{(1-d_k)}{\tau_k}\delta^{i\,\mathrm{T}}_{\varrho k}(t)\delta^i_{\varrho k}(t)\end{array}\right\} \\
&= \sum_{i=1}^{l_\varrho}\zeta_i^{\mathrm{T}}(\Pi^i_\varrho\otimes I_3)\zeta_i
\end{aligned} \tag{21}$$

where $\zeta_i = [\tilde{\sigma}_{\varrho}^{i\,\mathrm{T}}(t), \dot{\tilde{\sigma}}_{\varrho}^{i\,\mathrm{T}}(t), \delta_{\varrho 1}^{i\,\mathrm{T}}(t), \ldots, \delta_{\varrho r}^{i\,\mathrm{T}}(t)]^{\mathrm{T}}$, Π_{ϱ}^{i} can be written as

$$\Pi_{\varrho}^{i} = \begin{bmatrix} -\gamma \sum_{k=1}^{r} H_{\varrho k}^{i} & -\frac{c\gamma I_n + \sum_{k=1}^{r} H_{\varrho k}^{i}}{2} & \frac{\gamma}{2}[H_{\varrho 1}^{i}, \ldots, H_{\varrho r}^{i}] \\ -\frac{c\gamma I_n + \sum_{k=1}^{r} H_{\varrho k}^{i\,\mathrm{T}}}{2} & (-c + \sum_{k=1}^{r} \tau_k) I_n & \frac{1}{2}[H_{\varrho 1}^{i}, \ldots, H_{\varrho r}^{i}] \\ \frac{\gamma}{2}[H_{\varrho 1}^{i}, \ldots, H_{\varrho r}^{i}]^{\mathrm{T}} & \frac{1}{2}[H_{\varrho 1}^{i}, \ldots, H_{\varrho r}^{i}]^{\mathrm{T}} & -\mathrm{diag}\left\{\frac{(1-d_1)}{\tau_1} I_n, \ldots, \frac{(1-d_r)}{\tau_r} I_n\right\} \end{bmatrix}. \tag{22}$$

For convenience of analysis, we define

$$\Pi_{\varrho}^{i} = \begin{bmatrix} \Pi_{11} & \Pi_{12} \\ \Pi_{12}^{\mathrm{T}} & \Pi_{22} \end{bmatrix} \tag{23}$$

where

$$\Pi_{11} = \begin{bmatrix} -\gamma \sum_{k=1}^{r} H_{\varrho k}^{i} & -\frac{c\gamma I_n + \sum_{k=1}^{r} H_{\varrho k}^{i}}{2} \\ -\frac{c\gamma I_n + \sum_{k=1}^{r} H_{\varrho k}^{i\,\mathrm{T}}}{2} & (-c + \sum_{k=1}^{r} \tau_k) I_n \end{bmatrix}, \tag{24}$$

$$\Pi_{12} = \begin{bmatrix} \frac{\gamma}{2}[H_{\varrho 1}^{i}, \ldots, H_{\varrho r}^{i}] \\ \frac{1}{2}[H_{\varrho 1}^{i}, \ldots, H_{\varrho r}^{i}] \end{bmatrix}, \tag{25}$$

$$\Pi_{22} = -\mathrm{diag}\left\{\frac{(1-d_1)}{\tau_1} I_n, \ldots, \frac{(1-d_r)}{\tau_r} I_n\right\}. \tag{26}$$

According to the definition of Schur complement, we get that if and only if $\Pi_{22} < 0$ and $\Pi_{11} - \Pi_{12}\Pi_{22}^{-1}\Pi_{12}^{T} < 0$, then $\Pi_{\varrho}^{i} < 0$. As $d_i < 1$, we get that $\Pi_{22} < 0$. Define

$$\Xi_{\varrho}^{i} = \Pi_{11} - \Pi_{12}\Pi_{22}^{-1}\Pi_{12}^{\mathrm{T}} = \begin{bmatrix} \Xi_{11} & \Xi_{12} \\ \Xi_{12}^{T} & \Xi_{22} \end{bmatrix}, \tag{27}$$

we get that if $\Xi_{\varrho}^{i} < 0$, then $\dot{V}(t) \le 0$.

Next, we prove that $\lim_{t\to\infty} \dot{V}(t) = 0$ by contradiction. First, suppose that $\lim_{t\to\infty} \dot{V}(t) \neq 0$, then there exist infinite time series $\{\mathsf{T}_{s_1}, \mathsf{T}_{s_2}, \cdots\}$, $\mathsf{T}_{s_k} \in [t_{s_k}, t_{s_k+1})$, $\lim_{k\to\infty} \mathsf{T}_{s_k} = +\infty$, such that $|\dot{V}(\mathsf{T}_{s_k})| > \varepsilon_0$, where ε_0 is a positive constant.

As $\lim_{t\to\infty} V(t)$ exists, together with Cauchy limit existence criteria, we get $\forall \varepsilon > 0$, $\exists T > 0$ such that when $T_2 > T_1 > T$, then $|V(T_2) - V(T_1)| < \varepsilon$, i.e.,

$$\left| \int_{T_1}^{T_2} \dot{V}(t)\mathrm{d}t \right| < \varepsilon. \tag{28}$$

Denote $\theta = \min\{\varepsilon_0/2K, \mathsf{T}\}$, then $t_{s_{k+1}} - t_{s_k} > \mathsf{T} \geq \theta$, we get that

$$(\mathsf{T}_{s_k} - \frac{\theta}{2}, \mathsf{T}_{s_k}] \subset [t_{s_k}, t_{s_{k+1}}), \tag{29}$$

or

$$(\mathsf{T}_{s_k}, \mathsf{T}_{s_k} + \frac{\theta}{2}] \subset [t_{s_k}, t_{s_{k+1}}). \tag{30}$$

For (29), according to the mean value theorem, we get that for arbitrary $t \in (\mathsf{T}_{s_k} - \frac{\theta}{2}, \mathsf{T}_{s_k}]$, there exists t_m such that

$$\left|\dot{V}(\mathsf{T}_{s_k})\right| - \left|\dot{V}(t)\right| \leq \left|\dot{V}(\mathsf{T}_{s_k}) - \dot{V}(t)\right| \leq \left|\ddot{V}(t_m)\right| \cdot \left|t - \mathsf{T}_{s_k}\right| \leq \frac{\varepsilon_0}{2}, \tag{31}$$

which implies that

$$\left|\dot{V}(t)\right| \geq \left|\dot{V}(\mathsf{T}_{s_k})\right| - \frac{\varepsilon_0}{2} > \frac{\varepsilon_0}{2}. \tag{32}$$

It can be conclude that

$$\left|\int_{\mathsf{T}_{s_k}-\frac{\theta}{2}}^{\mathsf{T}_{s_k}} \dot{V}(s)\mathrm{d}s\right| = \int_{\mathsf{T}_{s_k}-\frac{\theta}{2}}^{\mathsf{T}_{s_k}} \left|\dot{V}(s)\right| \mathrm{d}s > \int_{\mathsf{T}_{s_k}-\frac{\theta}{2}}^{\mathsf{T}_{s_k}} \frac{\epsilon_0}{2}\mathrm{d}s > \frac{\theta\epsilon_0}{4} > 0, \tag{33}$$

for arbitrary $t \in (\mathsf{T}_{s_k} - \frac{\theta}{2}, \mathsf{T}_{s_k}]$. Similarly, we get that

$$\left|\int_{\mathsf{T}_{s_k}}^{\mathsf{T}_{s_k}+\frac{\theta}{2}} \dot{V}(s)\mathrm{d}s\right| > \frac{\theta\varepsilon_0}{4} > 0. \tag{34}$$

Note that when T_{s_k} is enough large, (33) and (34) are contrary with (28). Then we can conclude that $\lim_{t\to\infty}\dot{V}(t) = 0$, i.e., $\lim_{t\to\infty}\tilde{\sigma}^i_{\varrho}(t) = 0$, $\lim_{t\to\infty}\dot{\tilde{\sigma}}^i_{\varrho}(t) = 0$, which implies that $\lim_{t\to\infty}\sigma_i(t) = \sigma_0$ and $\lim_{t\to\infty}\dot{\sigma}_i(t) = 0$.

4 Numerical Example

In this section, we present a numerical example to illustrate the effectiveness of our protocol (8). Consider the attitude consensus problem for five spacecraft with a single leader and four followers. $J_1 = \cdots = J_4 = \begin{bmatrix} 10 & 0 & 0 \\ 0 & 20 & 0 \\ 0 & 0 & 10 \end{bmatrix}$.

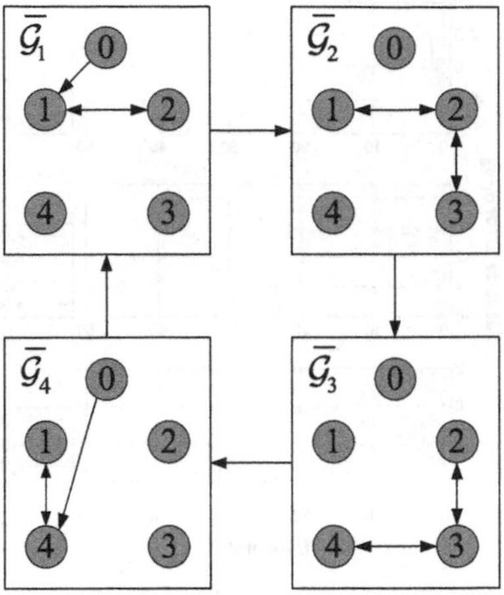

Fig. 1 The switching network topologies

The switching topologies $\bar{\mathcal{G}}_{\varrho(t)}$ associated with the system are shown in Fig. 1, and the switching sequence is $\bar{\mathcal{G}}_1 \rightarrow \bar{\mathcal{G}}_2 \rightarrow \bar{\mathcal{G}}_3 \rightarrow \bar{\mathcal{G}}_4 \rightarrow \bar{\mathcal{G}}_1 \rightarrow \ldots$, which is dictated by the following switching signal:

$$\varrho(t) = \begin{cases} 1, \; m \cdot \tau \leq t < (m+1)\tau \\ 2, \; (m+1)\tau \leq t < (m+2)\tau \\ 3, \; (m+2)\tau \leq t < (m+3)\tau \\ 4, \; (m+3)\tau \leq t < (m+4)\tau \end{cases} \tag{35}$$

where $\tau = 1$ s, $m = 0, 1, 2, \ldots$.

In our simulation, we choose $a_{ij} = 1, i = 1, 2, 3, 4$, if spacecraft j is a neighbor of spacecraft i, and $a_{ij} = 0$ otherwise. We set the initial attitude of the leader to be $\sigma_0 = [2, 5, 6]^{\mathrm{T}}$. The initial attitudes of the three followers are set to be respectively, $\sigma_1 = [1, 6, 3]^{\mathrm{T}}$, $\sigma_2 = [5, 4, 1]^{\mathrm{T}}$, $\sigma_3 = [4, 1, 6]^{\mathrm{T}}$, $\sigma_4 = [1, 5, 2]^{\mathrm{T}}$, and the initial angular velocities of the three followers are set to be respectively, $\omega_1 = [2, 3, 2]^{\mathrm{T}}$, $\omega_2 = [4, 3, 2]^{\mathrm{T}}$, $\omega_3 = [1, 4, 2]^{\mathrm{T}}$, $\omega_4 = [3, 2, 4]^{\mathrm{T}}$. The control parameters are chosen as $c = 2$ and $\gamma = 20$.

Figure 2 show the attitude errors between the leader and the followers as well as the angular velocities of the followers using the control input (8). We see that the attitudes of the followers converge to the dynamic leader, and the angular velocities of the followers converge to zero asymptotically. Figure 3 shows the auxiliary vectors of the followers. We can see that the auxiliary vectors s_i converge to zero, just as what we demonstrate in Theorem 1. Moreover, there exist some vibration in Figs. 2 and 3 because of the switching topologies.

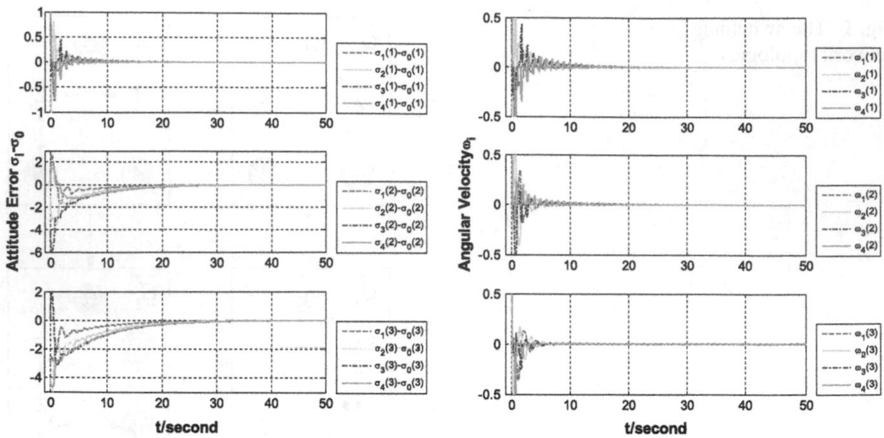

Fig. 2 The attitude errors $\sigma_i - \sigma_0$ and angular velocity ω_i of the followers

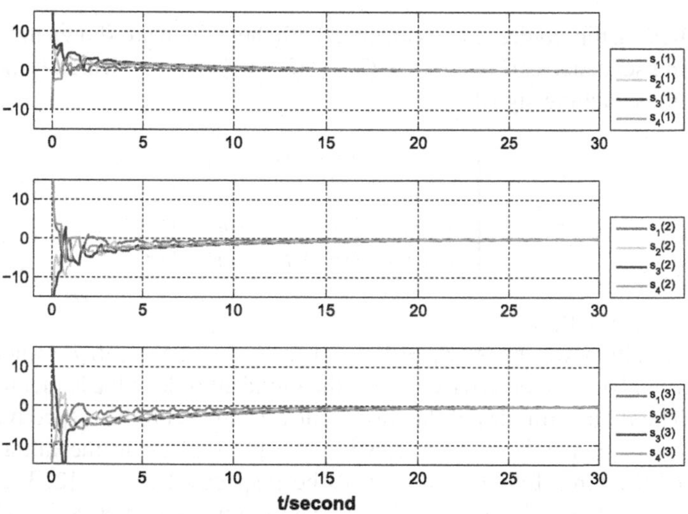

Fig. 3 The auxiliary vector s_i of the followers

5 Conclusion

The leader-following attitude consensus problem is studied for multi-rigid-body systems with time-varying communication delays under jointly connected topologies. The difficulty lies in that how to overcome the bad effect of varying time delay coupled with jointly connected topologies. In terms of linear matrix inequalities, we derive sufficient conditions that guarantee all followers asymptotically converge to the leader. A key idea of our approach is the association of an auxiliary vector for each follower. Moreover, the common Lyapunov Krasovskii function method is employed to prove the stability of the controller. Future work will focus on the dynamic tracking problem for multi-rigid-body systems under time delays and switching topologies.

References

1. Li S, Du H, Shi P (2014) Distributed attitude control for multiple spacecraft with communication delays. IEEE Trans Aerosp Electron Syst 50(3):1765–1773
2. Guo Y, Lu P, Liu X (2015) Attitude coordination for spacecraft formation with multiple communication delays. Chin J Aeronaut 28(1):527–534
3. Abdessameud A, Tayebi A, Polushin IG (2012) Attitude synchronization of multiple rigid bodies with communication delays. IEEE Trans Autom Control 57(9):2405–2411
4. Zhou JK, Ma GF, Hu QL (2012) Delay depending decentralized adaptive attitude synchronization tracking control of spacecraft formation. Chin J Aeronaut 25(3):406–415
5. Li GM, Liu LD (2012) Coordinated multiple spacecraft attitude control with communication time delays and uncertainties. Chin J Aeronaut 25(5):698–708
6. Meng ZY, You Z, Li GH, Fan CS (2010) Cooperative attitude control of multiple rigid bodies with multiple time-varying delays and dynamically changing topologies. Math Probl Eng 2010 Article ID: 621594 :1–19. doi:10.1155/2010/621594
7. Jin ED, Sun ZW (2009) Robust attitude synchronisation controllers design for spacecraft formation. IET Control Theory Appl 3(3):325–339
8. Zhang BQ, Song SM, Chen XL (2012) Robust coordinated control for formation flying satellites with time delays and switching topologies. J Astronaut 33(7):910–919
9. Thunberg J, Song WJ, Montijano E, Hong YG, Hu XM (2014) Distributed attitude synchronization control of multi-agent systems with switching topologies. Automatica 50(3):832–840
10. Chen S, Shi P, Zhang WG, Zhao LD (2014) Finite-time consensus on strongly convex balls of Riemannian manifolds with switching directed communication topologies. J Math Anal Appl 409(2):663–675
11. Du H, Li S, Qian C (2011) Finite-time attitude tracking control of spacecraft with application to attitude synchronization. IEEE Trans Autom Control 56(11):2711–2717
12. Su Y, Huang J (2012) Stability of a class of linear switching systems with applications to two consensus problems. IEEE Trans Autom Control 57:1420–1430

An Urgent Traffic Dispersion and Assignment Model for Urban Road Flooding

Zeshu Zhao, Jiaxian Liang and Guoyuan Li

Abstract Urban road flooding often causes road capacity reduction, traffic congestion and inconvenience in citizens daily travel. This paper proposes an urgent traffic dispersion and assignment model for the case of urban road flooding, with the aim of maximizing the level of service in the road network by controlling the road capacity. First, based on prospect theory and taking the historical travel time average as a reference, the model adopts the BPR (Bureau of Public Roads) function to represent the cost-flow relationship to calculate the prospect value of each route. Second, the travelers route choice behavior is described in logit model with the prospect value as the utility. Third, based on the route choice results, the traffic flow on the congested road sections is dispersed by controlling the road capacity, so the traffic flow to the flooded roads can be adaptively assigned to other roads. Finally, a direct iterative method and genetic algorithm are used to solve the proposed model. The former attempts to implement the traffic assignment based on the travelers route choice behavior, and the latter is used to find the satisfying solution through selection, crossover and mutation. The proposed model is applied to a given road network with an assumption of some road capacity reduction due to road flooding. The results show that when the proposed model is applied, the saturation ratio (or level of service) of the roads in the entire network is more uniform and the distribution of the saturation ratio of main roads is reduced, so the traffic flow in the whole network can remain smooth and the level of service can remain high.

Z. Zhao
Raffles Institution, Singapore, Singapore
e-mail: zzsandrew@gmail.com

J. Liang (✉) · G. Li
Research Centre of Intelligent Transportation System, School of Engineering, Sun Yet-Sen University, 510275 Guangzhou, China
e-mail: 710545808@qq.com

J. Liang · G. Li
Guangdong Provincial Key Laboratory of ITS, 510006 Guangzhou, China
e-mail: 564753569@qq.com

© Springer International Publishing Switzerland 2017

V.E. Balas et al. (eds.), *Information Technology and Intelligent Transportation Systems*, Advances in Intelligent Systems and Computing 454,
DOI 10.1007/978-3-319-38789-5_52

Keywords Urgent traffic assignment · Travelers route choice behavior · Prospect theory · Direct iterative method · Genetic algorithm

1 Introduction

Road flooding belongs to one type of traffic incident, which may disturb road traffic, easily resulting in road capacity reduction, traffic congestion, and even traffic paralysis. These phenomena lead to unnecessarily driving around the congested roads and increasing traffic pressure on other roads, so the level of service and the convenience of vehicular travel for citizens may be decreased. Therefore, urgent traffic dispersion and assignment in the cased of road flooding is becoming increasingly important. Yet, a considerable body of research focuses on traffic dispersion and assignment based on recurring traffic incidents [1]. There is relatively little research on urgent dispersion traffic schemes under non-recurring traffic incidents, such as bad weather, traffic accidents. Zhang [2] explored travelers route choice behavior during periods of traffic accidents and used mixed logit model for simulation based on the travelers psychological traits of traffic accidents, but no solution was put forward for traffic dispersion. Li et al. [3] put forward a generation method for an urgent dispersion assignment scheme by optimizing regional traffic with the goal of minimizing the combination of the total travel time, traffic flows and traffic flow increments. However, most of the studies seeks to find out the traffic dispersion and assignment schemes based on a constant road capacity of each road. Few of them would adaptive induce the drivers route choice by controlling the actual road capacity (such as extending the green time) based on the analysis of travelers route choice behavior.

This study aimed to build an urgent traffic dispersion and assignment model under a road flooding situation. The model controls the capacity of roads and induces travelers route choice behavior to gain an equilibrium traffic assignment on the road network. Therefore, the level of service on the entire road network, especially on main roads, would be stable at a reasonable level to avoid a wide range of road network paralysis due to congestion.

2 Literature Review

This section focuses on establishing the two major themes underpinning our work, namely travelers route choice behavior and a traffic assignment method.

The studies of travelers route choice behavior mainly focus on two aspects. The first one is the selection of the travel cost function. The BPR (Bureau of Public Road) function, which is developed by the Bureau of Public Road, is widely used as the travel cost function [4, 5]. Davidson [6] proposed a gradual model of travel cost based on queuing theory. Wang et al. [7] modified the BPR function by adding the results of the reduction in traffic volumes and road capacity as the traffic flows in the BPR function, which can model the traffic process in which traffic flows first increase and then decrease from smooth traffic to congested traffic. Additionally, more synthetic travel cost functions are proposed by including such factors as travel time, travel fare, degree of congestion, queue length and delay time in the intersection, and etc. [8, 9] in the category of generalized travel cost functions. The second aspect of route choice behavior is the route choice mechanism. Expected utility theory supposes a traveler is reasonable and he/she chooses the alternative with maximum utility in the choice set. Due to the differences in travelers knowledge about the routes, there is a lack of perceived error during the travelers route choice process in expected utility theory. In random utility theory, the perceived utility of each alternative is composed of the actual utility and a random item, where the latter is the travelers perceived error. However, both expected utility theory and random utility theory suppose that all travelers choose the maximum utility and the alternatives are independent of each other in the choice set [10]. Prospect theory was developed by Kahneman and Tversky; in this theory, a decision is made by comparing the differences between the alternatives in the choice set and the reference point. Xia et al. [11] took the arrival time as a reference point in route choice under the framework of prospect theory to analyze the impact of departure time on commuters route choice. Gao et al. [12] analyzed the choice behavior en route using prospect theory with real-time traffic information. Ramos et al. [10] reviewed the development of utility theory, prospect theory and regret theory and attempted to analyze travelers route choice behavior in different theoretical frameworks under uncertain travel time.

There are many studies on traffic assignment methods, which include the all-or-nothing approach, the capacity-restricted approach and the equilibrium traffic assignment model. Li et al. [13] proposed a combination optimization model of traffic assignment with traffic capacity constraints to illustrate travelers decision behaviors and the reasons for traffic congestion on the road network. Sun et al. [14] used an ant colony algorithm to remedy the limitations of the capacity-restricted assignment model and to solve the model by analyzing the patterns of travelers behavior. Jiang et al. [15] introduced crossover and mutation operations from the genetic algorithm into the hybrid artificial fish swarm algorithm to solve the equilibrium traffic assignment model. Luo et al. [16] applied the modified genetic algorithm to the optimization of system traffic assignment.

3 Urgent Traffic Dispersion and Assignment Model

In this section, an urgent traffic dispersion and assignment model is built with the analysis of the travelers route choice behavior. The direct iterative method and genetic algorithm are introduced as the solution algorithms for the proposed model.

3.1 Travelers Route Choice Behavior

The proposed model is based on the utility maximization assumption that travelers act rationally to get the best choice. Both the current travel situation and the historical travel experience are considered when making the proper route choice for travelers. Prospect theory is used as the choice criteria to measure the gains and losses against the historical travel experience (reference point). Travelers route choice behavior can be divided into three stages: (1) generation of the travelers route choice set, (2) determination of the travelers route choice mechanism and (3) travelers route choice.

Travelers Route Choice Set There are many alternative routes between the origin r and the destination s(O-D pair (r, s)). Travelers usually compare those routes based on shortest travel time or distance and ignore the relatively circuitous routes. Thus, the top k shortest routes between O-D pair (r, s) are considered as the route choice set K_n that the drivers may take into account, and the route choice set is generated by using the top shortest route algorithm proposed by Yen [17].

Travelers Route Choice Mechanism The BPR function is adopted to calculate the travel cost, t_k, on route k;

$$t_k = \sum_{a \in k} t_0 \left[1 + \alpha \left(\frac{q_a}{C_a} \right)^{\beta} \right] \tag{1}$$

where $t_k \sim N\left(t_k^{rs}, \sigma\right)$; t_0 is travel time on road section a under free flow conditions; C_a is the actual capacity of the road section a; q_a is the flow of the road section a; α and β are parameters.

Besides the current travel cost, travelers also consider historical travel experience. Prospect theory is introduced to describe the travelers actual route choice behavior by taking the average of the historical travel cost as a reference point. In the prospect theory, a reference point is defined to build the value function is defined as $v(x)$; see formula (2),

$$v\left(x_k\right) = \begin{cases} (x_k)^{\eta}, x_k > 0, \\ -\lambda(-x_k)^{\zeta}, x_k \leq 0. \end{cases} \tag{2}$$

where x_k is the travelers relative gains or losses between the actual travel cost and the historical expected travel cost. η and ζ are sensitivity-decreasing coefficients, $\eta \leq 1$, $\zeta \leq 1$. λ is the loss aversion coefficient, $\lambda > 1$.

The subjective probability weighting function $W(p)$ is defined in the prospect theory to express the subjective probability of occurrence of a given event;

$$\begin{cases} W^{+}\left(p_k\right) = \dfrac{p_k^{\rho}}{\left(p_k^{\rho}+\left(1-p_k^{\rho}\right)^{\rho}\right)^{\frac{1}{\rho}}} \\ W^{-}\left(p_k\right) = \dfrac{p_k^{\delta}}{\left(p_k^{\delta}+\left(1-p_k^{\delta}\right)^{\delta}\right)^{\frac{1}{\delta}}} \end{cases} \tag{3}$$

where $W^{+}(p_k)$ and $W^{-}(p_k)$ represent the probability weighting functions of the gain and the loss, respectively; ρ and δ represent the attitude coefficients of the gain and the loss, respectively; and p_k means the probability of occurrence of x_k, i.e. $p_k = \frac{1}{\sqrt{2\pi}\sigma} e^{-\frac{\left(t-t_k^{rs}\right)^2}{2\sigma^2}}$.

Therefore, the prospect value, Y, of each travel route can be calculated based on the value function and the subjective probability weight function, see formula (4),

$$Y(f) = \int_{-\propto}^{0} W^{-}(p_k) \cdot v(x_k) + \int_{0}^{+\propto} W^{+}(p_k) \cdot v(x_k) \tag{4}$$

where the first term and the second term on the right side of the equals sign represent the prospect value of gain and the loss, respectively.

Travelers Route Choice The logit model is applied in travelers route choice and the prospect value is used as the utility in the logit model to calculate the probability p_k^{rs} of every route in the choice set; see formula (5),

$$p_k^{rs} = \frac{e^{Y_k^{rs}}}{\sum\limits_{k \in K_{rs}} e^{Y_k^{rs}}} \tag{5}$$

3.2 *Urgent Dispersion and Assignment Model*

An urgent dispersion and assignment model is proposed based on the following assumptions:

(1) The top k shortest routes are regarded as a route set that drivers are likely to take;
(2) Travelers consider both their historical travel experience and the current condition;
(3) Travelers are reasonable and choose a travel route by comparing the reference point with the actual travel cost in the route set;
(4) The relationship between the actual road capacity and the depth of the water spot on the road section is not taken into consideration, as the key of the model is to induce thc traffic flows from congested roads to other passable roads;
(5) Traffic flow is not affected by pedestrians, bicycles, and so on.

The objective of the model is to keep the traffic flow smooth in the congested road sections affected by road flooding and make sure the dispersion area and main roads are clear, safe and orderly. Thus, the objective function (formula (6)) includes two parts: the level of service range and the level of service weighted average. The constraints can be defined as follows:

(1) The traffic flow on each route is dependent on the route choice probability and trip distribution between the origin and the destination, see formula (7);

(2) The traffic flow on each road section is based on the traffic flow on each route, see formula (8);
(3) The actual road capacity is between zero and the basic capacity of the corresponding road section, see formulas (10) and (11);
(4) The traffic flow of each road section is within the range of the actual road capacity, see formulas (12) and (13).

Therefore, the model can be describes as:

$$\min \ Z = \max_{a \in A}(q_a / C_a) - \min_{a \in A}(q_a / C_a) + \frac{\sum_{a \in A} \mu_a (q_a / C_a)}{\sum_{a \in A} \mu_a} \quad \forall a \in A \tag{6}$$

subject to

$$f_k^{rs} = q_{rs} \cdot \frac{e^{Y_k^{rs}}}{\sum_{k \in K_{rs}} e^{Y_k^{rs}}} \ , \forall r \in R \ \ \forall s \in S \ , \forall k \in K_{rs} \tag{7}$$

$$q_a = \sum_r \sum_s \sum_k f_k^{rs} \cdot \delta_{a,k}^{rs} \ , \forall a \in A \ , \forall r \in R \ , \forall s \in S \ , \forall k \in K_{rs} \tag{8}$$

$$\delta_{a,k}^{rs} = \begin{cases} 1, \ if \ the \ kth \ route \ from \ the \ origin \ r \ to \ the \ destination \ s \\ \quad includes \ road \ section \ a \\ 0, \ otherwise. \end{cases} \tag{9}$$
$$\forall a \in A \ , \forall r \in R \ , \forall s \in S \ , \forall k \in K_{rs}$$

$$C_a \geq 0 \ , \forall a \in A \tag{10}$$

$$C_a \leq C_a^o \ , \forall a \in A \tag{11}$$

$$v_a \geq 0 \ , \forall a \in A \tag{12}$$

$$v_a \leq C_a, \forall a \in A \tag{13}$$

where Z is the objective function; μ_a is degree of importance of a road section; A is the set of road sections; f_k^{rs} is traffic flow of the kth route between the O-D pair (r, s); q_{rs} is traffic flow between the O-D pair (r, s); R is the set of origins; S is the set of destinations; C_a^0 is the basic capacity of a road section.

3.3 Solution Algorithms

The Solution Algorithm includes the direct iterative method(DIM) and genetic algorithm (GA). The DIM is used to solve the traffic assignment problem considering travelers route choice behavior under the given capacity of each road section in the considered road network. The GA is used to optimize the objective function of the model by controlling the road capacity in the road network, to realize the urgent dispersion assignment in a relatively high level of service.

Direct Iterative Method An equation set can be expressed as $x = Bx + f$ and solved by iterating it in the form $x^{(k)} = Bx^{(k-1)} + f$ with an initialization of $x^{(0)}$. This iterative method is called the direct iterative method, which is a common method in solving equation sets.

The DIM is used to solve the traffic assignment problem considering travelers route choice behavior under the given capacity of each road section. Therefore, the traffic flow q_a can be gained under the given capacity set $\overrightarrow{C}$, as follows (Fig. 1a):

(1) Initialize variables. Set traffic flow $f_k^{rs(0)}$ of the route, traffic flow $q_a^{(0)}$ of road section a to 0 for $\forall k \in K_{rs}$ and $\forall a \in A$. Set the current iterative time to 0.
(2) Calculate the travel cost of each road section. The BPR function is adopted to calculate the current travel cost according to the traffic flow, $q_a^{(i-1)}$.
(3) Calculate the travel cost of each route according to formula (1).
(4) Calculate the route choice probability based on formula (5).
(5) Calculate the traffic flow on each route using formula (7) in the iterative time.
(6) Calculate the traffic flow on each road section using formula (8) in the iteration.
(7) Determine whether the situation satisfies the terminal condition or not. Compare $q_a^{(i-1)}$ with $q_a^{(i)}$; if the difference between them can reach the pre-set convergence precision, finish; if not, then $i++$ and back to step (2).

Genetic Algorithm The GA is a search heuristic algorithm that imitates the biological evolution process. In the GA, after forming the initial group through encoding, the mission of genetic manipulation (including selection, crossover and mutation) is to apply some operations on the units in the group according to their fitness to the environment, thereby selecting the superior and eliminating the inferior.

The algorithm is applied to solve the proposed model by generating a random solution set $C = \{\overrightarrow{c_1}, \overrightarrow{c_2}, \ldots, \overrightarrow{c_m}\}$ (m is the number of solutions in the solution set) and searching the satisfactory solution with the best fitness value through iterations. Each solution (each capacity set of road sections) $\overrightarrow{c} = (c_1, c_2, \ldots, c_n)$ in the solution set is regarded as a unit. c_j refers to the capacity of road section j. n is the number of road sections in the road network. As the objective of the proposed model is to minimize the objective function $Z(C)$ (formula (6)), the fitness function F in the GA is defined in formula (14).

$$F = M - Z(C) \tag{14}$$

where M is a relatively large positive integer. The solution process using the GA is as follows (Fig. 1b):

(1) Initialize variables. Set the number of solutions in the solution set to m and set the maximum iterative time and the current iterative time to G and 0, respectively.

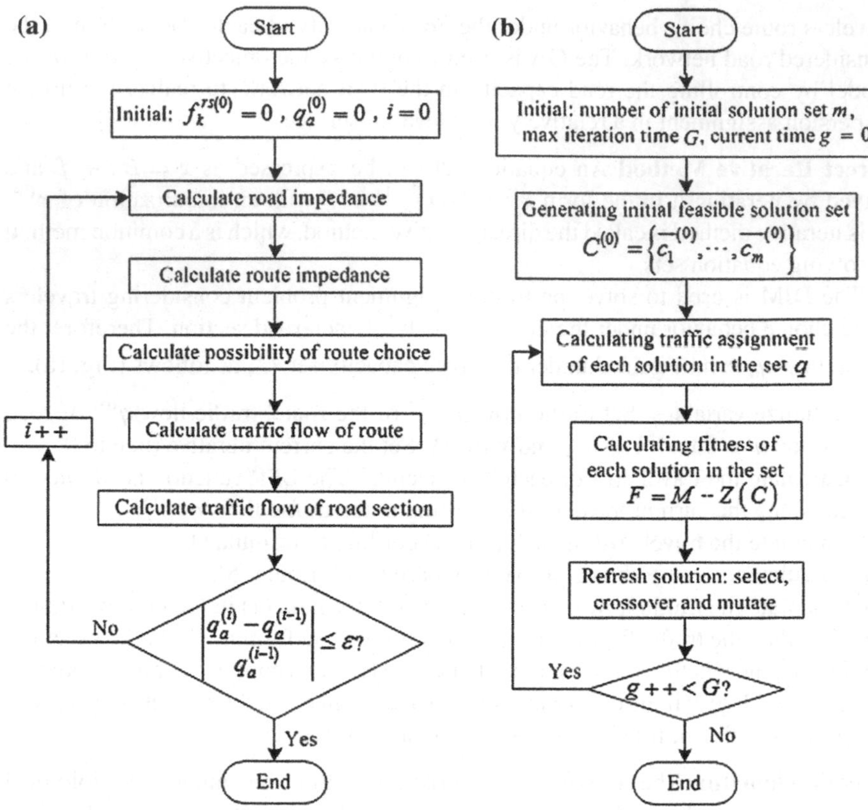

Fig. 1 Flowchart of the solution algorithm: **a** Direct iterative method, **b** Genetic algorithm

(2) Generate the initial feasible solution set. Generate a random solution $\vec{c} = (c_1, c_2, \ldots, c_n)$ and obtain the traffic flow $\vec{q} = (q_1, q_2, \ldots, q_n)$ of road sections using the DIM in Sect. 3.3.1. If this solution meets the constraints, put this solution into the initial feasible solution set. Until the number of solutions in the initial feasible solution set is equal to the pre-set number of solutions, m, repeat step (2).

(3) Calculate the result of urgent traffic dispersion and assignment under each solution in the given set using the DIM in Sect. 3.3.1.

(4) Calculate the fitness value of each solution in the solution set from formula (14) based on the traffic flow, $\vec{q}$, of road sections.

(5) Update the solution in the solution set. The $g + 1$ new solution set can be generated using the operations of selection, crossover and mutation according to the fitness value in the original solution set, $C^{(g)}$. Set $g = g + 1$.

(6) if the current iterative time g is less than the maximum iterative time G, return to step (3); otherwise, finish.

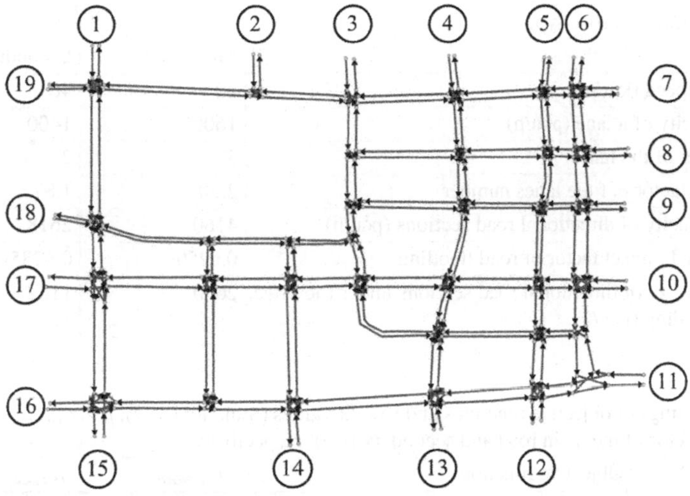

Fig. 2 The assumed road network

4 Case Study

The proposed model is applied to the assumed network, which consists of 53 crossings, 146 directional road sections, and 353 directional vehicle turn arcs at the crossings. All road segments at the boundary of the network are assumed to be both Origins (O) and Destinations (D) (Fig. 2). The roads are divided into two types (Table 1): main road and secondary road. The design capacity and free-flow speed are set according to road type. The assumed travel volume between each Origin and Destination is set in Appendix 1. The point A is assumed to be a place with occasional road flooding, which will decrease the capacity of the nearby bi-directional road sections from 2618 pcu/h to 1122 pcu/h, thus the road congestion causes around point A. The maximum number of iterations of the GA is set to 200 with 20 units. The number of unit gene fragments is set to 146 (equal to the number of road sections), and the probability of crossover is 0.7, while that of mutation is 0.01. The maximum number of iterations of the DIM is set to 50, and the convergence accuracy is 0.01.

To compare with the dispersion results of different optimization objectives, the solutions are generated under different conditions of objective functions and weights (Table 2).

The results are illustrated in Figs. 3 and 4. Figure 3a shows that the saturation rate of the road sections ranges mainly from 0.25 to 0.75, and 0.45 is the most common rate among the road sections in situation 1 (no dispersion). Meanwhile, there are several road sections with saturation rates in the range between 1 and 1.1, which means the traffic flows on the road sections exceed the road capacity and traffic

Table 1 Settings of road type

Road type	Main road	Secondary road
Free-flow speed (km/h)	60	40
Basic capacity of a lane (pcu/h)	1600	1400
The number of the lanes	3	2
Correction factor o, f the lanes number	2.60	1.87
Design capacity of directional road sections (pcu/h)	4160	2618
The assumed impact factor of road flooding	0.6250	0.4285
Actual capacity of directional road sections under the effect of road flooding (pcu/h)	2600	1122

Table 2 Settings of objective functions and weight values (Situations 1–7), $\mu_{a,main}$, $\mu_{a,sec}$ are the weighting factor of the main road and secondary road, respectively

Situation ID	Objective function	$\mu_{a,main}$	$\mu_{a,\text{sec}}$
1	No dispersion	–	–
2	$Z = \max\limits_{a \in A} (q_a/C_a)$	–	–
3	$Z = \max\limits_{a \in A} (q_a/C_a) - \min\limits_{a \in A} (q_a/C_a)$	–	–
4		1	0.5
5	$Z = \max\limits_{a \in A} \left(\frac{q_a}{C_a}\right) - \min\limits_{a \in A} \left(\frac{q_a}{C_a}\right) + \frac{\sum_{a \in A} \mu_a (q_a/C_a)}{\sum_{a \in A} \mu_a}$	2	1
6		3	1.5
7		1	1

congestion is especially prone to occur. After the proposed model is applied using different objective functions (situations 2, 3, and 4), all of the saturation rates of the road sections are less than 1.00, and the rates are mainly distributed from 0.25 to 0.65. Figure 3b shows that the saturation rate of the main roads is the most uniform in situation 4, and the number of road sections at different saturation rates decreases from low rates to high rates. Therefore, the maximum saturation rate decreases about 2.03 % in the entire road network and 0.54 % on the main roads (Table 3). That is, after the proposed model is applied, not only does the entire road network remain smooth, but the traffic congestion on main roads are eased.

Figure 4 shows that the distribution of the saturation rate values of the road sections under different situations is not much different, with most values between 0.3 and 0.7, while Fig. 4b shows a different case. The number of road sections at different saturation values varies greatly under situation 7 because the weight of all of the road sections is set to 1, then the effect of the main road decreased. When the weight of the main roads is set to double that of the secondary road, the maximum number of main roads is at a saturation value of 0.25 (situation 4, 5 and 6), and the mean of

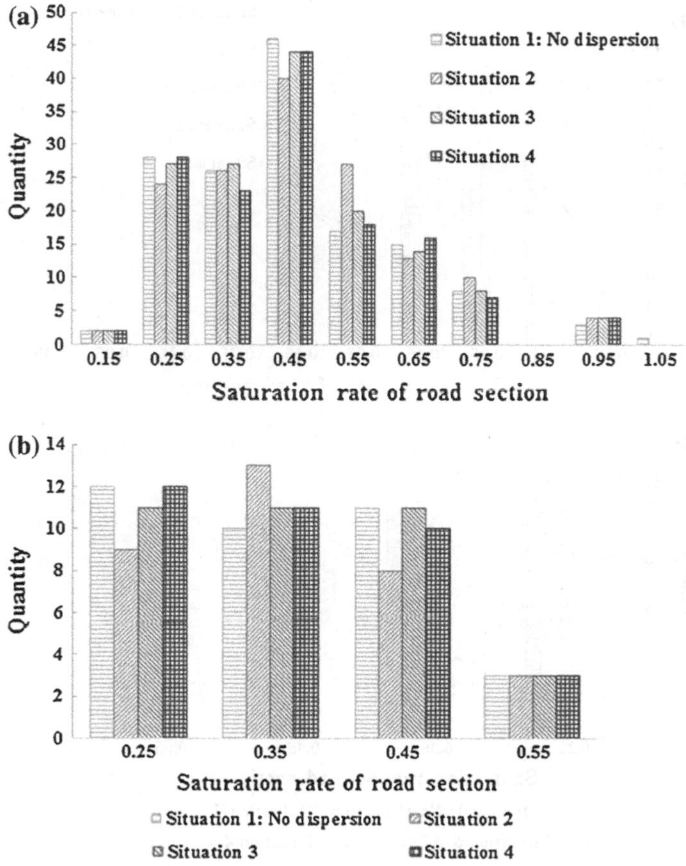

Fig. 3 Saturation rate among different objective functions: **a** saturation rate of road sections, **b** saturation rate of main roads.

the saturation rate increases 0.17 % from situation 4 to situation 1 with no dispersion schemes (Table 3).

Therefore, the results can be summarized as follows: (1) some road sections will exceed their actual road capacity under the case of the road floods, which results in traffic congestion; (2) the saturation rates of flooding roads are improved and all of them are less than 1.00, after the capacity of the road network is controlled in the model; (3) after the proposed model is applied, the average saturation rate of the entire road network is slightly higher (such as –0.07 % in Table 3), but the road network traffic assignment is more uniform and the saturation rate of the main roads will be reduced (0.17 % in Table 3); (4) different schemes bring different results, and all of them has more uniform distribution than that considered the same weight value on different road types.

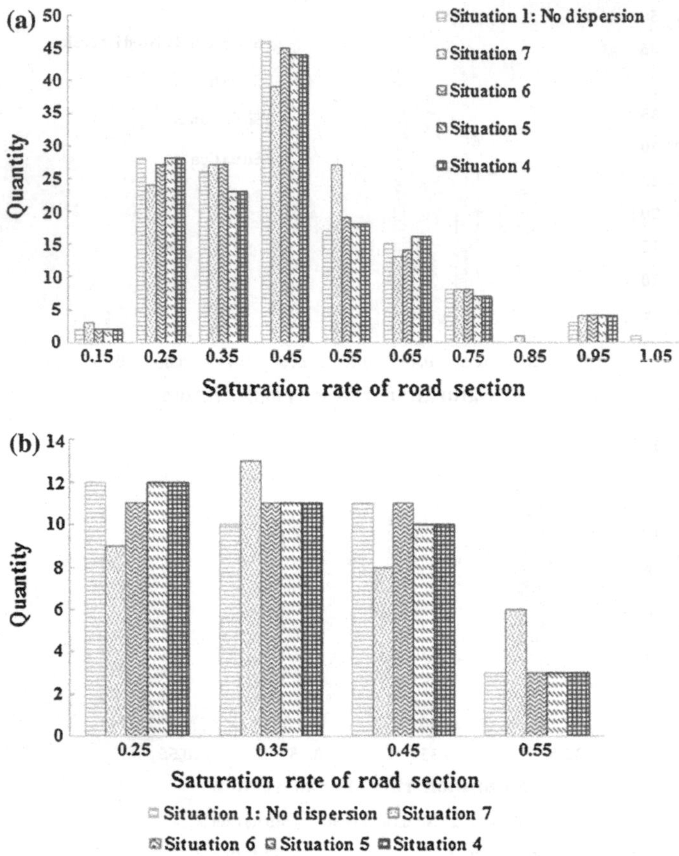

Fig. 4 Saturation rates with different weights: **a** saturation rate of road sections, **b** saturation rate of main roads.

Table 3 Comparison of results between Situation 1 and Situation 4

ID	(All of the road sections			The main road section		
	Max	Min	Average	Max	Min	Average
1	1.0125	0.1314	0.4534	0.5070	0.2036	0.3551
4	0.9920	0.1315	0.4538	0.5042	0.2065	0.3545
Difference	2.03 %	–0.11 %	–0.07 %	0.54 %	–1.42 %	0.17 %

5 Conclusion and Future Work

The paper analyzed travelers route choice behavior based on prospect theory and put forward an urgent traffic dispersion and assignment model that aimed to optimize the saturation rate (level of service) of roads in the road network by controlling the actual road capacity, in view of road flooding. The direct iterative method and genetic algorithm were described for finding the satisfying solution (the set of actual road section capacities). Then the proposed model was applied in the assumed network to optimize the level of service both on the roads surrounding road flooding and the affected road network. However some influences on traveling (such as pedestrians, bicycles) and on travel cost (such as comfort, fares), should be considered in further studies, as well as the effect of various weight of the roads, in order to develop the model.

References

1. Xiaojie T (2005) Research on knowledge-based methodologies and system for decision making of traffic congestion management. Southeast University, Nanjing
2. Yukun Z (2012) Study on travelers route choice behavior during traffic incident. Southwest Jiaotong University, Chengdu
3. Hongwu L, Xiaojian H, Jian L (2007) Research on a method of an urgent traffic dispersion assignment. 2007 3th China Annual Conference on ITS
4. Wenjing W, Ming J, Dongmei L et al (2012) Study on capacity limitation-multipath routing traffic allocation based on logit model and BPR impedance function. J Highw Transp Res Dev 29(1):81–85
5. Suxin W, Li G, Xiao C et al (2008) Particle swarm optimization arithmetic of traffic assignment. J Traffic Transp Eng 7(5):97–100
6. Davidson KB (1966) A flow travel time relationship for use in transportation planning. In: 3rd Conference on Australian Road Research Board (ARRB), Sydney, 3(1)
7. Suxin W, Leizhen W, Li G, et al. (2009) Improvement study on BPR link performance function. J Wuhan Univ Technol (Transp Sci Eng), 33(3):446–449
8. Yuanqing W, Wei Z, Lianen L (2004) Theory and application study of the road traffic impedance function. J Highw Transp Res Dev 21(9):82–85
9. Wei W, Qu D, Zhong Z (2000) On the models of integrated-equilibrium trip assign-ment in urban traffic network. J Southeast Univ (Nat Sci Ed) 30(1):117–120
10. Ramos GM, Daamen W, Hoogendoorn S (2014) A state-of-the-art review: developments in utility theory, prospect theory and regret theory to investigate travellers' behaviour in situations involving travel time uncertainty. Transp Rev 34(1):46–67
11. Jinjiao X, Zhicai J, Jingxin G (2012) Travel routing behaviors based on prospect theory. J High Transp Res Dev 29(4):126–131
12. Gao S, Frejinger E, Ben-Akiva M (2010) Adaptive route choices in risky traffic networks: a prospect theory approach. Transp Res Part C Emerg Technol 18(5):727–740
13. Xinhua L, Guirong H, Huihui M et al (2006) An optimal method for capacity-constrained traffic assignments. Cent South Highw Eng 30(4):116–118
14. Huacan S, Xuhong L, Yanzhong L et al (2009) Ant colony optimization arithmetic of capacity restraint traffic assignment. J Southeast Univ (Nat Sci Ed) 39(1):177–180
15. Shan J, Yefei J (2011) Artificial fish swarm algorithm for solving road network equilibrium traffic assignment problem. Comput Simul 28(6):326–329

16. Wenchang L, Zhijue Z (2002) Application of genetic algorithm in the traffic assignment model with system optimization. J Chang Sha Railw Univ 20(1):10–14
17. Yen JY (1971) Finding the k shortest loopless paths in a network. Manag Sci 17(11):712–716

Simulation of the Effective Highway Capacity Under Unexpected Traffic Incident

Cheng Ding and Chennan Gou

Abstract Because highway is fully sealed road, there is no immediately available alternative road for vehicular evacuation in the event of emergency. The impact on traffic flow and road capacity is far greater than ordinary roads, which will produce more serious than the average delay road loss. In this study, the VISSIM software was used to simulate the "3 off 1" situation on highway. The effective highway capacity under unexpected traffic incident is simulated at different mainline traffic amount, different road configuration and different speed limits in this article.

Keywords Unexpected traffic incident · Highway emergency · Simulation

1 Highway Capacity Overview

Highway capacity refers to the maximum number of vehicles that can pass a section of highway in a unit time. Highway capacity factors are:

(1) Road Condition, which refers to the geometric characteristics of the highway road conditions, including the type of transport facilities, the number of lanes, lane and shoulder width, and design speed.
(2) Traffic Condition, which means the transportation traffic configuration, traffic volume and direction of traffic lanes available for distribution and the distribution of traffic flow.
(3) Control Condition. The highway is running in the form of non-interrupted traffic flow. Unexpected events can interfere with the normal flow of traffic on the highway, causing traffic delay, and even lead to serious traffic jam. For non-intermittent traffic control conditions have lane use control, parking and yield signs.

C. Ding (✉) · C. Gou
School of Economics and Management, Chang'an University, Xian 710061, China
e-mail: 394741854@qq.com; 68053339@qq.com; dc826@163.com

C. Gou
e-mail: 455666@qq.com

© Springer International Publishing Switzerland 2017

V.E. Balas et al. (eds.), *Information Technology and Intelligent Transportation Systems*, Advances in Intelligent Systems and Computing 454,
DOI 10.1007/978-3-319-38789-5_53

2 Traffic System Simulations

System Simulation refers to design a mathematical or logical model for describing the real system of procedures and experiments within the constraints of the system is running a series of guidelines, in order to understand, explain or predict the behavior of the real system performance or evaluation of various strategic purposes [1]. VISSIM is a discrete, random, 10–1S time step micro simulation model developed by PTV Germany. It is currently a great traffic model that integrates urban roads, highways, and public transit, regular of traffic, etc. In the VISSIM model, the vertical movement of the vehicle is calculated with a psycho-physiology model. While, the lateral movement (lane change) is calculated using a rule-based algorithms. The model also provides a graphical interface, using 2D and 3D animation to exhibit the vehicle movement to users, and can use the dynamic traffic assignment for path selection.

3 Establish Factors Should Be Considered in the Simulation Model

(1) Number of lanes and the effective number of lanes

At present, Chinese highway is generally divided into two-way four-lane highway, two-way six-lane, two-way eight-lane three. The highway is a fully closed road, and is divided road, which eliminates the impact on the opposite road, so the design of simulation models is only considered the case in one direction.

(2) Traffic volume

Traffic volume is the traffic flow through a section of the road in a unit of time. Congestion refers to the traffic demand exceeds the capacity of a road traffic, vehicles stranded excess phenomenon occurs. Traffic congestion is caused mainly by the imbalance of supply and demand. The factors causing traffic congestion can be broadly divided into two categories. (a) The frequently excessive traffic demands, such as the peak of the holiday traffic demand is significantly greater than usual demand; (b) The occasionally and temporarily reduce capacity, such as traffic accident or road maintenance and etc. Correspondingly, traffic congestion can be divided into recurrent congestion and occasional congestion. Recurrent congestion means traffic congestion phenomenon occurs frequently in certain periods; sporadic congestion is caused by the reduced capacity resulted from random events such as accidents or other special events. After freeway incident occurred, the incident location sudden decrease capacity. The upstream of incident highway will easily get congestion, causing traffic delays, travel time increases, which have a huge impact on the regional road network.

(3)Traffic configuration

Traffic configuration refers to the various models of vehicles in traffic flow proportion. Different levels of vehicle characteristics are different, have different effects on Highway capacity.

According fees and vehicle characteristics, and refer to "Highway Engineering Technical Standards", the model is divided into small cars, midsize and large car categories [2]: (a) Small cars: general car length less than 5 m, the total weight of the vehicle loaded is less than 4 tons, including passenger cars (sedans), light trucks, sport utility vehicles and light small wagon, etc.; (b) Midsize car: in a car length of 5–8 m, vehicles loaded with a total weight of not more than 10 tons, general load automotive and off-road cars are classified as medium-sized car; (c) Large car: car length is usually greater than 8 m, loaded with more than 10 tons total weight, such as heavy trucks, buses and semi-trailers.

Mentioned below is "carts rate" in the definition of carts for midsize and large cars. Through the survey of China's highway running, we found that the main factors affecting the operation of the highway traffic is the traffic vehicle configuration [3–5], namely "carts rate". It is mainly in two aspects: First, the "Large car" has bigger volume than the "small car", and thus take up more road space; the second is the driving performance of these Large vehicles (such as acceleration, deceleration, and ability to maintain speed, etc.) is much lower than "small car." Different driving performance of the vehicle will result in a gap in the traffic stream, which cannot be utilized as overtake zone, especially for the long-distance continuous uphill. In this case, the "large car" has to reduce speed, causing large gaps in the traffic flow, reducing road capacity.

(4) Limit speed

The goal of traffic management is always to improve traffic safety and to ensure the smooth flow of traffic on freeway. A reasonable highway speed limit value can balance the relationship between speed and safety, and improve transport efficiency.

4 The Simulation Examples of Highway Under Emergency

Currently, the majority of our highways are two-way six-lane highway. After the incident occurred, it will generally occupy a lane. Therefore, the condition of one lane closed in a three-lane road (referred to as "3 off 1") is simulated in this paper. The effective highway capacity under unexpected traffic incident is simulated at different mainline traffic amount, different road configuration and different speed limits in this article.

Establish the following basic model: simulate road is the basic sections of Highways, which does not consider the impact of highway alignment, slope, and weather conditions. The simulation section length is 5 km, lane width is 3.75 m. Normal road

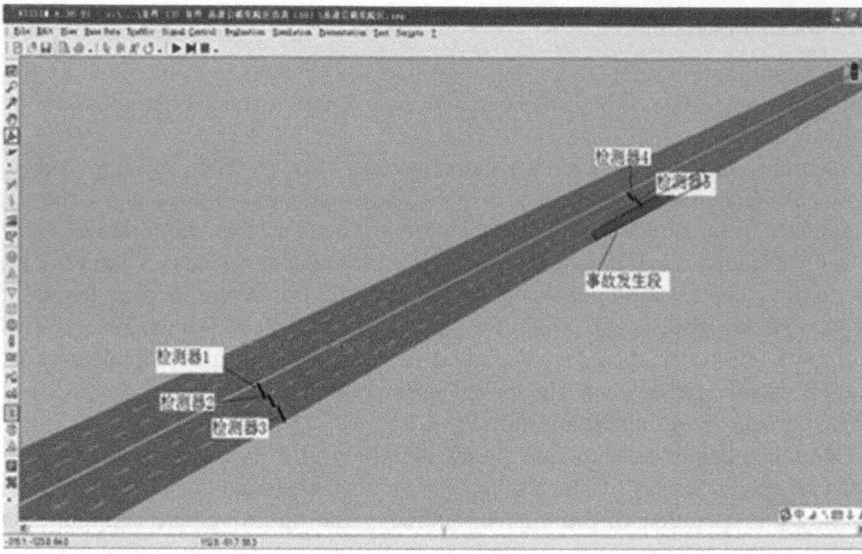

Fig. 1 Diagram of the detectors placement

is one-way road with 3 lanes; while, the incident road is a one-way road with 2 lanes. The road accident is occurred at a distance of 4 km at the starting point. In the upstream far from the accident, detector 1, detector 2 and detector 3 are setup, which can detects the actual capacity of the normal section of Ca. Two other detectors 4 and 5 are placed in the sections where the accident happed, which can detect the actual highway capacity at the accident segment Cb, shown in Fig. 1.

Simulation models are based on the basic model by changing the main stream of traffic volume and traffic configuration. In this experiment, the initial capacity is 1200 vehicles one lane/h capacity and ramp up to 2000 vehicles one lane/h at a 100 vehicles one lane/h rate. Meanwhile, at each traffic conditions, we study the influence of the large car rate (10, 20 and 50 %) on the road section actual capacity of highway accidents.

The highway speed limit values are different for different sections of the highway, which also impacts on the highway capacity. Therefore, we investigate the effect of vehicle speed on the highway capacity by changing the speed characteristics of the vehicle on the basic experiments (20 km/h increments, from 120 km/h decreases to 60 km/h).

According to the design simulation experiments, we build 9*3 simulation models, each simulation model run five times in average, and 1800 s each time. 5 min is set as statistics interval. And then, we take its average as the simulation results.

Input traffic volume to the simulation model. Because the impact of traffic configuration on the traffic flow has already been considered in this experiment, all the traffic inputs are absolute traffic value. The traffic flow before the occurrence of unexpected events is shown in Fig. 2.

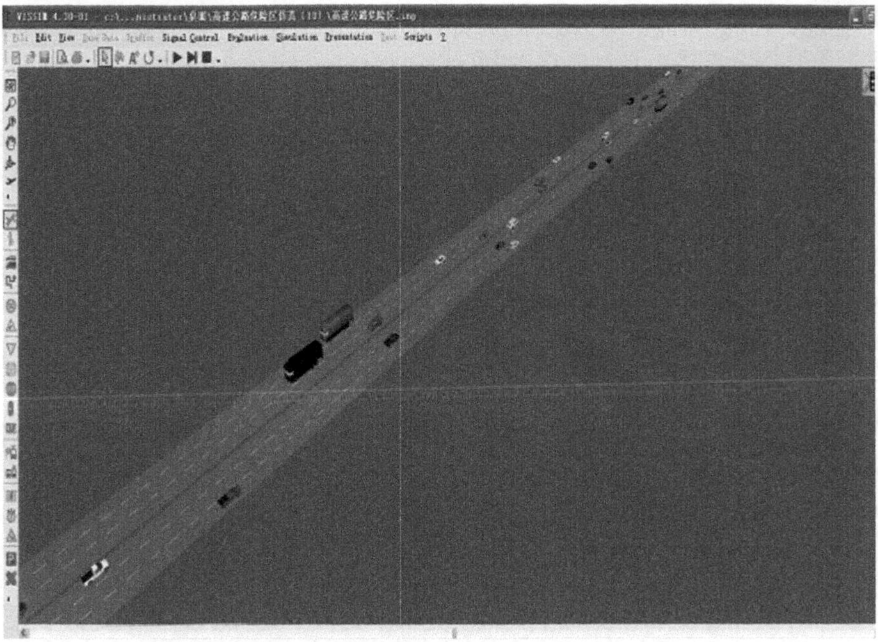

Fig. 2 Traffic flow diagram before the incident

When an unexpected incident happens at a section of the road, the tapered barrels should be placed around the accident and warning signs should be placed about 300 meters upstream of the accident, in order to warn drivers about to change lane. After the lane closed, the upstream of the vehicle lane must change lanes to merge into the other two lanes. Because of lateral clearance and the effect of traffic enforcer on the driver, the driving speed is reduced. When the traffic flow on other two lanes is relatively large, the merging is even more difficult, and resulting in the aggregation of traffic congestion. The traffic flow conditions of unexpected incident are shown in Fig. 3.

After vehicles in the adjacent lane pass through the accident zone, some of those vehicles will change lane. And all the vehicles will accelerate and drive away. The congestion will gradually dissipate after the emergency is finished processing. Experimental data are shown in Tables 1, 2, 3 and 4 and Figs. 4, 5, 6 and 7.

As can be seen from the above data, in the case of “3 off 1” and the design speed of 120 km/h, “3 off 1” and “large car” rate have little effect on the capacity of the highway. By merging into the other two lanes, the vehicles from upstream can successfully pass through the accident segment. At the design speed of 100 km/h, the large car rate of 50 % has more significant impact on highway capacity than the large car rate of 30 and 10 %. Therefore, in this similar situation, emergency management departments should make the necessary traffic control, leading the driver selects unobstructed path to travel, and control the amount of large car on the road.

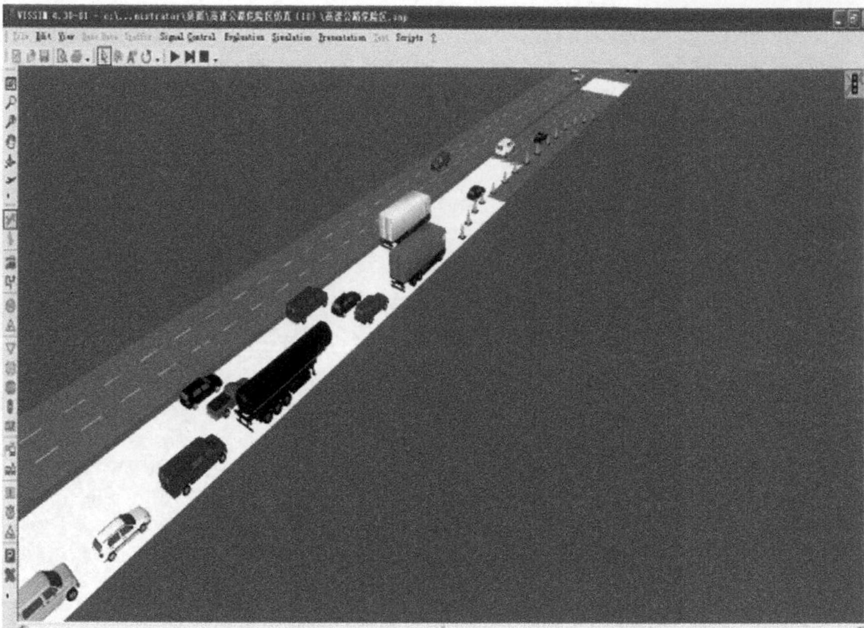

Fig. 3 Traffic flow diagram after the incident

Table 1 120 km/h design speed accident segment capacity

Input traffic	Carts rate		
	10 %	30 %	50 %
1200 vehicles/lane/h	1065	945	910
1300 vehicles/lane/h	1075	1014	980
1400 vehicles/lane/h	1084	1025	997
1500 vehicles/lane/h	1142	1065	920
1600 vehicles/lane/h	1198	1132	924
1700 vehicles/lane/h	1230	1170	1100
1800 vehicles/lane/h	1210	1140	1023
1900 vehicles/lane/h	1280	1189	1033
2000 vehicles/lane/h	1300	1265	982

Table 2 100 km/h design speed accident segment capacity

Input traffic	Carts rate		
	10 %	30 %	50 %
1200 vehicles/lane/h	1022	1004	889
1300 vehicles/lane/h	1120	989	978
1400 vehicles/lane/h	1267	1202	996
1500 vehicles/lane/h	1290	1254	1005
1600 vehicles/lane/h	1200	1102	1002
1700 vehicles/lane/h	1220	1189	1000
1800 vehicles/lane/h	1190	1186	1023
1900 vehicles/lane/h	1300	1265	998
2000 vehicles/lane/h	1345	1222	1100

Table 3 80 km/h design speed accident segment capacity

Input traffic	Carts rate		
	10 %	30 %	50 %
1200 vehicles/lane/h	1189	1107	1003
1300 vehicles/lane/h	1299	1189	980
1400 vehicles/lane/h	1167	1054	997
1500 vehicles/lane/h	1214	1077	1023
1600 vehicles/lane/h	1198	1200	982
1700 vehicles/lane/h	1300	1180	1011
1800 vehicles/lane/h	1312	1166	1006
1900 vehicles/lane/h	1245	1273	–
2000 vehicles/lane/h	1178	–	–

–Indicating that traffic has started in normal road, and vehicles stuck at the bottleneck, unable to pass.

Table 4 60 km/h design speed accident segment capacity

Input traffic	Carts rate		
	10 %	30 %	50 %
1200 vehicles/lane/h	1130	1056	967
1300 vehicles/lane/h	1127	1006	971
1400 vehicles/lane/h	1133	1007	867
1500 vehicles/lane/h	1200	1022	–
1600 vehicles/lane/h	1165	–	–
1700 vehicles/lane/h	–	–	–
1800 vehicles/lane/h	–	–	–
1900 vehicles/lane/h	–	–	–
2000 vehicles/lane/h	–	–	–

–Indicating that traffic has started in normal road, and vehicles stuck at the bottleneck, unable to pass.

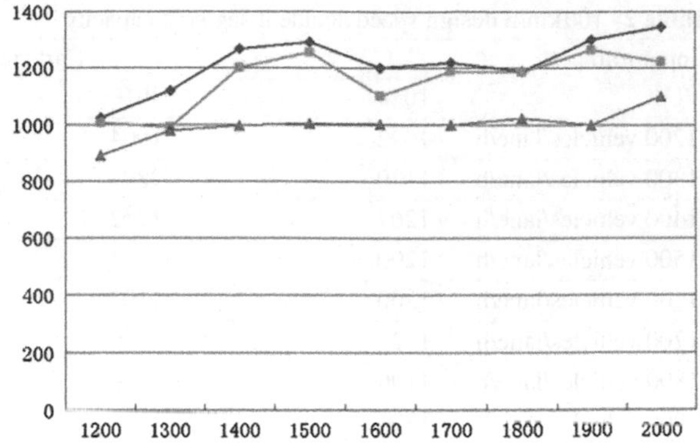

Fig. 4 120 km/h design speed accidentssegment capacity

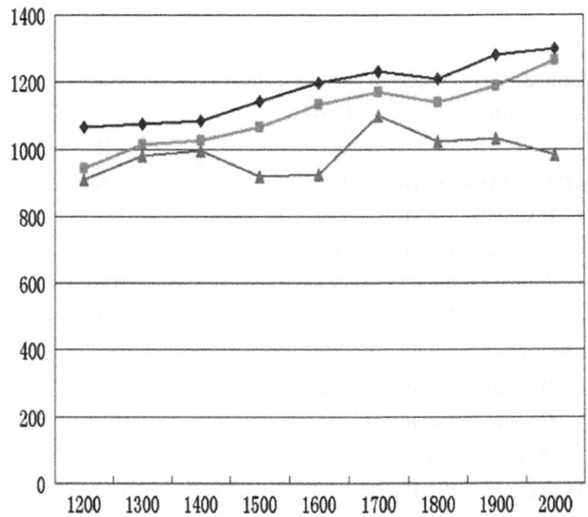

Fig. 5 100 km/h design speed accidents segment capacity

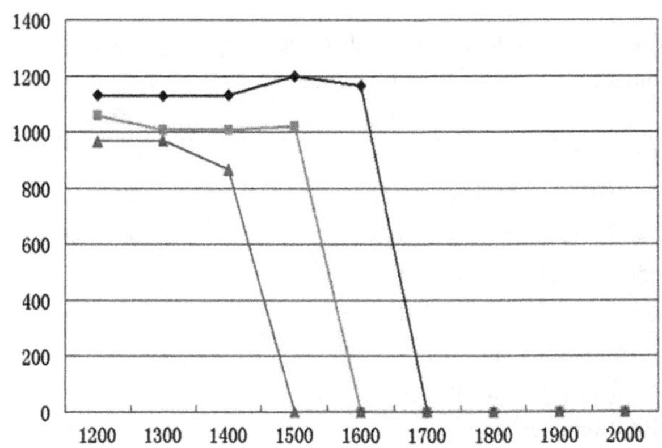

Fig. 6 80 km/h design speed accident segment capacity

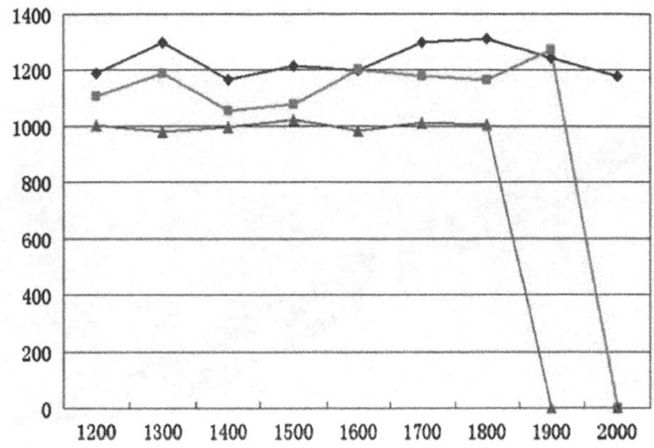

Fig. 7 60 km/h design speed accident segment capacity

As can be seen from the above data, in the case of “3 off 1” and the design speed of 80 km/h, when the input traffic volume reached 1900/lane/h, vehicles will stuck in traffic bottleneck, and congestion happens on the normal road, vehicles cannot pass smoothly. In this situation, the relevant emergency management departments should take the necessary traffic control, control the number of vehicles on the road and the number of large cars. At the design speed of 60 km/h, when the input traffic volume reached 1500/lane/h, there will be vehicle stuck in the traffic bottleneck. Continue to increase the input traffic volume will result in road congestion and an increased length of vehicle queue, as shown in Fig. 8. In this situation, the relevant emergency management departments should immediately take the necessary traffic control, leading the driver to select different travel paths.

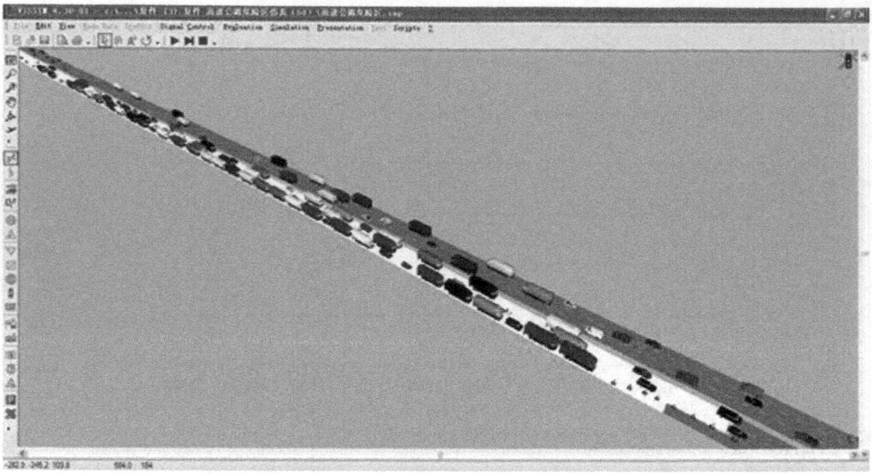

Fig. 8 diagram of vehicle queue at road congestion

References

1. Liu Y, Shi J (2002) Transportation system simulation technology. China Communications Press, Beijing
2. Sa Yuangong, Jianmin Xu, He Jingkai (2002) Model and simulation of freeway dynamic traffic flow. Comput Commun 20(3):221–227
3. Deng S (2008) US emergency management system and its implications. National School of Administration
4. Ma Jianhua (2009) Major disaster supplies discuss security issues. Dep Emerg Manag J 1:233–235
5. Xuan D, Jian L (2010) Estimating the effective capacity of unexpected traffic incident on expressway. J Highw Transp Res Dev 27(11):133–139

The Design and Implementation of Openstreetmap Application System Aimed at Emergency Logistics

Long Li, Renjie Pi and Huiling Zhou

Abstract Compared to traditional logistics, the emergency logistics show the characteristics of sudden, uncertainty, time urgency. The regular commercial maps used in the traditional logistics system (such as Baidu map, Google map), cannot satisfy the need that the emergency logistics will plan the best route autonomouslyin order to ensure the rescue time under the condition of disaster. Openstreetmap (OSM) is an open sourcemap which the greatest strength is the open data and the free secondary development. In this paper, we designed three parts on the Android terminal including the view layer, the logic control layer and data interaction layer based on the OSM data and the REST service interface which the ArcGIS Server provides; The system has achieved the offline maps reading and parsing function and real-time locationfunction. So it will combine the open source OSM with emergency logistics demand and provide a reference for the emergency logistics to plan the bestroute.

Keywords Openstreetmap · ArcGIS · The emergency logistics

1 Introduction

Map service is a branch of a Web service and many Internet companies offer a map service API interface, for example, Baidu map and Google map. These map services are becoming popular because of its simple interface, easy to operate, and get maps

This work is partially supported by the 2013 National Science and Technology support program (2013BAD17B06).

L. Li (✉) · R. Pi · H. Zhou
Automation School, Beijing University of Posts and Telecommunications, Beijing, China
e-mail: lilong8167@163.com

R. Pi
e-mail: pirenjie@gmail.com

H. Zhou
e-mail: huiling@bupt.edu.cn

© Springer International Publishing Switzerland 2017

V.E. Balas et al. (eds.), *Information Technology and Intelligent Transportation Systems*, Advances in Intelligent Systems and Computing 454,
DOI 10.1007/978-3-319-38789-5_54

quickly. Nowadays, most of the Internet maps are using their map service. But it also has some obvious imperfection. The map can't meet the needs of customization and cannot open more permission for the autonomous path planning the post-disaster emergency logistics is different from ordinary logistics. The path planning in ordinary logistics considers the optimal distance or cost as the best path while in the post-disaster emergency path planning, the rescue time is the first element to consider, so there are certain differences in path planning. In addition, in a period of time after the disaster, the disaster area also has some different intensitys secondary disasters, which will affect the traffic network to a certain extent. So the emergency logistics requires overall arrangement, and plan the optimal path to submit the relief goods to the disaster site at first time. Obviously, the business map (Google map, Baidu map, etc.) can't meet the needs of emergency logistics. OSM is a collaborative project to create the global map. The map is produced by the portable satellite navigation equipment, aerial images, and other free data sources (such as the local common sense). Its data is released under the Creative Commons appears—Share Alike 2.0 in order to encourage the free use and distribution of the date. (Commercial and non-commercial). Series of ArcGIS software provides the function of production, editing maps, and ArcGIS Server can publish source OSM data at maps service and provide the REST interface for terminal use, which fully meet the needs of the emergency logistics resources based on GIS.

Currently there are many researches in which the OSM data is used to position and route design. Bonn University of Germany has done a series of case study about traffic networks routing study based on OSM data analysis. The routing service step for walking and cycling in the travel mode of travel route design launched earlier than Google Maps path analysis [1]. Germany's university of Fribourg Gobelbecker has carried out some discussion and implement on how to apply the OSM data to the simulation of the rescue environment after disaster. At the national university of Ireland, OSM is applied to position related to environmental information services as the basic data [2]. But there is little study that combines the OSM data with the demands of the emergency logistics, so this is the purpose of this study.

ArcGIS software platform has provided a better solution for online mobile electronic map application, which can give a general application implement according to the needs of different mobile terminal access to GIS data. These papers called the REST interface that ArcGIS Server has provided, designed three layers including view layer, logic control layer, data interaction layer, realized the maps display, and support the basic map operations, such as the zoom and map roaming function. The system has achieved the offline maps reading and parsing function and real-time location function.

2 Structure Design of the System Function

Accessing to the map service at the Android platform should consider the network problems in the process of emergency logistics. The communication equipment in the disaster-affected areas is generally suffered different degrees of damage which may

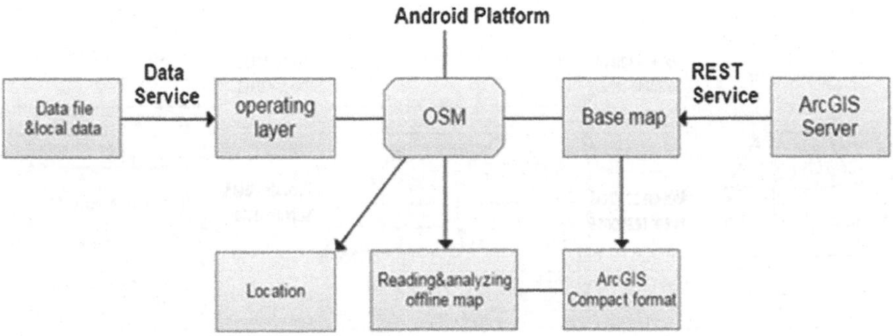

Fig. 1 System function design block diagram

cause the network connection quality and speed of the network to be significantly reduced. Considering these problems, the map is divided into operation layer and base layer in the functions design. The operation layer updates in time and the base map layer uses offline map so that it will reduce the traffic on the network. The concrete proposal of design as shown in Fig. 1.

In the diagram above, the application will regard the map as a superposition of two layer logically. The two layers are layer and operation layer. The reproduction of data on the base layer is generally will not change, its function is to provide users with general reference map location while the date on the operation layer change over time or the user can modify its content. The data on the server side serves for data through the server publishing and get address of the date service on the client service and send the HTTP request data, parse the response returned by the XML data, and the information encapsulated in the ArcGIS for Android corresponding object class, applying to a drawing on the operation layer by ArcGIS. The data on the operation layer composed of the data on the other server rendering, its content can change in time and its layers content can edit and modify. The operating layer data can be stored in the database and can also be stored in a file, using the data from the file server, database, and data encapsulation for the REST of data services. The interactions of the data service and client is that the client sends a POST or GET request, the server can return XML or JSON format of the data, the client receiving, parsing and then use them. Map layer reproduction data source is disaster area city area of the map, so generally, it will not change. Adopting the way of offline maps, and release these date by using the ArcGIS Server which serve for the RESTs map. Cache the map service for local Compact slice file by using ArcGIS Server map service, use the ArcGIS for Android API for offline access and display for base layer, and realize the offline maps navigation: enlarge, shrink, translation.

3 Systems Client Design

Systems client take the general logic, data, and interface design of separation mode, which is divided into the UI view layer, logic control, data communication layer. Concrete scheme design is shown in Fig. 2.

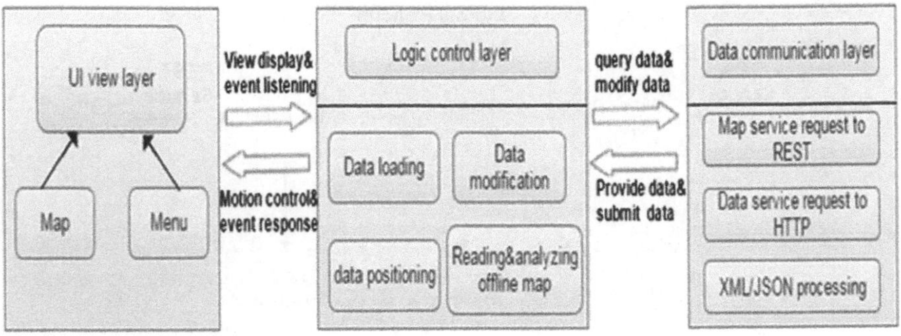

Fig. 2 System clients design block diagram

3.1 The UI View Layer

The view layer conveys the user's commands into the control logic layer. Offline maps reading and analyzing module read maps from the mobile phone memory card by the system directly, and stored in offline map layer. A positioning button was designed to realize positioning function in the orientation module.

3.2 Logic Control Layer

Logic control layers main purpose is to provide the functions about dealing with the interface and dates business logic and make data interaction with data interaction layer, do some data processing and logical combinations in mobile terminal.

3.3 Data Communication Layer

Data communication layer involves both the client and the servers communication problems, using the C/S model based on HTTP protocol. Mobile client sends a request to the server via HTTP protocol and the server sends data encapsulation into XML document to the client. Data interaction layer receive and parse the data, and parsed the XML data encapsulation into the corresponding class and the logic control layer will obtain the required data directly from these classes.

4 The Realization of Functions

4.1 Convert the OSM into Shape File

OSM data cannot be directly analyzed in ArcGIS, it needs to be converted into shape files, and specific steps are as follows:

(1) Use ArcCatalog to build a text geographic database, and name it.
(2) Double-click the loadOSM of the OpenStreetMap Toolbox .tbx under the Toolbox of ArcCatalog. Download the OSM document and choose the geographic database.
(3) The OSM data generated from text geographic database, click export -> export as a shape file.
(4) Add generated data to ArcMap, so it can convert the OSM into shape successfully.
(5) Release a map service.

4.2 The Functions Realization of the Positioning Module

Positioning module The module achieve the function that we can get the coordinate position of the clicks point after clicking the coordinate, and then select the element information that is the minimum distance tolerance with screen coordinate position. Add On Single Tap Listener in the operate layer. When the user clicks on a screen, get the coordinates, and create the parameters of localization task (map service URL and query layer) to perform the positioning. Specific implementation class shown in Fig. 3.

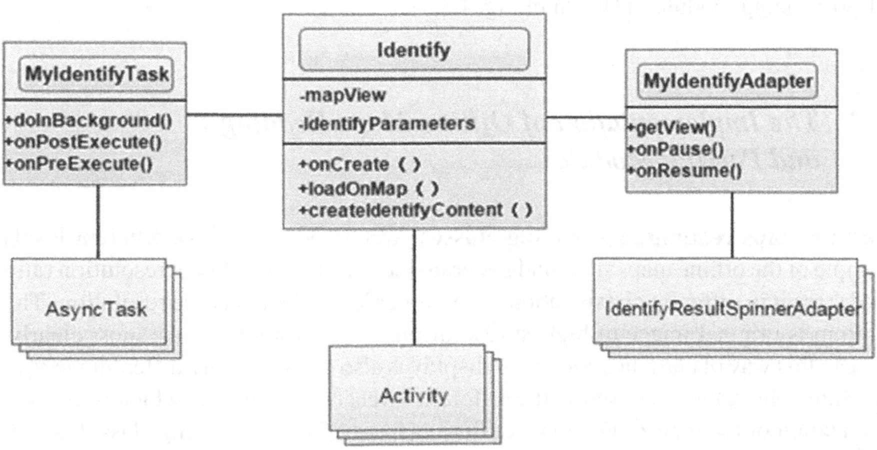

Fig. 3 Main classes diagram in location module

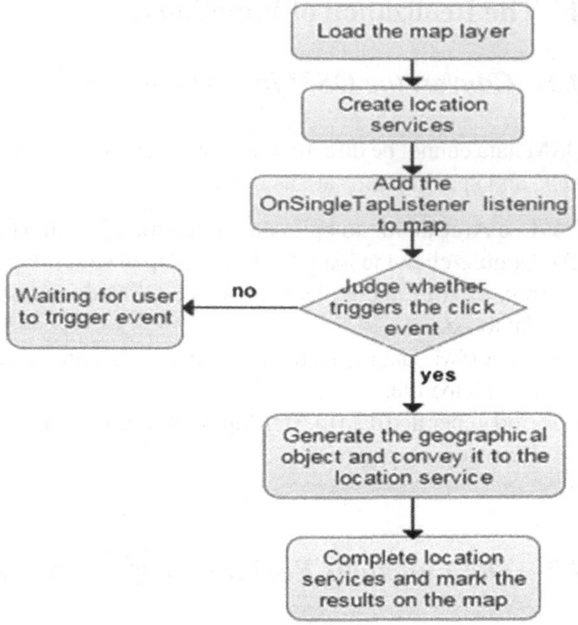

Fig. 4 Logic diagram of the location module

The module performs the query tasks on the map, used to get the position of the mobile client, and using the call out to mark on the map, and find the elements which is the nearest location from mobile client. Use the Identify task to perform the query tasks on the map and display in the call out, add a tiled map service as a map and add a dynamic map service, dynamic map service used to perform the Identify task.

The main logic of positioning module The whole implementation process and logic of positioning module as shown in Fig. 4.

4.3 *The Implementation of Offline Maps Reading and Parsing Module*

Offline maps reading and parsing classes ArcGIS Server adopts different levels sample of the offline maps slice, and generate the slice file of different resolution ratio and store it in different classification. The higher level, the lower the resolution. The bottom is original image of highest resolution, and therefore it is the most clearly. Due to this way of classification, maps display is also classified. In a different view, in the same place, the client shows the different levels of slicing data, which is also the realization of the map zoom mode. Offline maps reading and parsing class diagram as shown in Fig. 5.

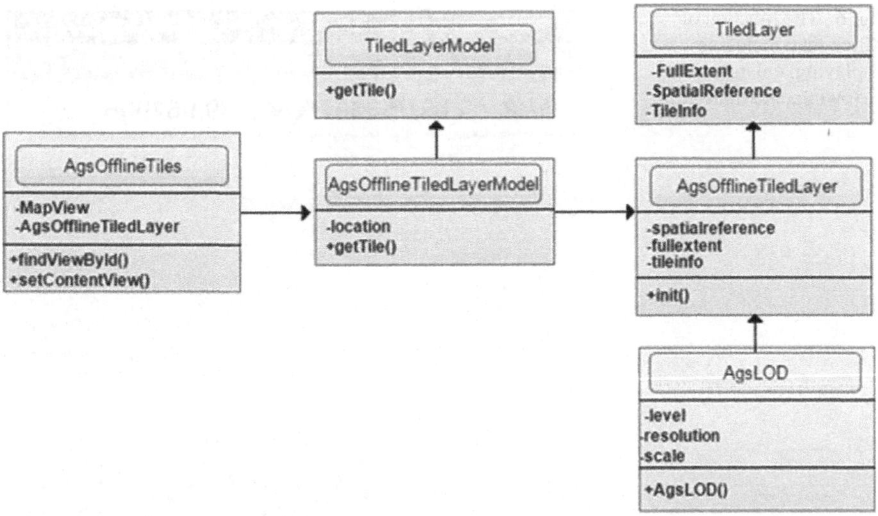

Fig. 5 Main classes diagram in reading and parsing module

The main logic of reading and parsing module Using ArcGIS Server service to slice up the map service into offline map file, ArcGIS Server is taken offline map to different levels of sampling of map service, and generates the slice file of different resolution ratio and makes classification storage. Each level represents a map zoom level; this topic will cut the map into14 levels on behalf of 14 zoom level. Offline map file is stored in the phone's SD card. In the offline map file class diagram, Ags Offline Tiled Layer class analysis offline maps the conf.XML configuration files and configuration files stored the entire cache section information of offline maps, including spatial reference, origin of coordinates, chips level and resolution, etc. And store the information in the tile lnfo object.

5 The System Validation

5.1 *System UI Interface*

After starting the system, enter the systems main interface. The interface is dominated by maps. It can take some basic operation including enlarge, narrow, and translate (Figs. 6 and 7).

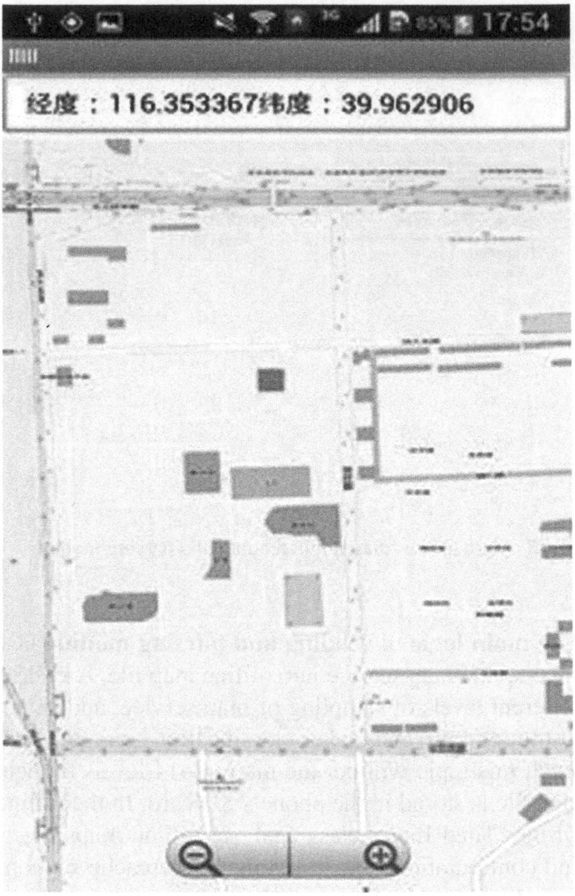

Fig. 6 The function of offline map including displaying, enlarging, narrowing, and translating

5.2 *Positioning Module*

Click on the "localization" menu option, the system mark the location of the client on the map. Positioning module as is shown in the picture.

6 Summary

This paper has studied and implemented the application system based on the OSM date aimed at emergency logistics. It also creatively combined the two together. On the Android platform, the system has called the REST interface provided by ArcGIS Server, designed the UI view layer, logic control layer and data interaction layer, supported basic map operations, such as enlarge, shrink, translation and implemented

Fig. 7 The realization of location services

the reading and parsing of the offline map and real-time positioning function. It has vital point in overcoming the network problems in the emergency logistics and realizing autonomous path planning further.

References

1. Cherdlu E (2007) OSM and the art of bicycle maintenance. In: A presentation at the conference of state of the map. Manchester, UK
2. Ciepluch B, Jacob R, Mooney P, Winstanley AC (2009) Using openstreetmap to deliver location cbased environmental information in Ireland. In: Proceedings of the 7th ACM SIGSPATIAL international conference on advances in geographic information systems, ACM, SIGSPATIAL GIS 2009, November 4–6 2009, Seattle, USA

3. Haklay MM, Weber P (2008) OpenStreetMap: User-generated street maps. IEEE Pervasive Comput 7(4):12–18
4. Haklay MM (2008) How good is OpenStreetMap information? A comparative study of OpenStreetMap and ordnance survey datasets for London and the rest of England, environment and planning B (under review)
5. Zhaobin Z, Qin W, Liang G (2010) Use ArcGIS REST to build high-performance WebGIS service. J Manufact Autom 2010 preceding: + 210, 172–173
6. Jing G, Chenfeng G (2008) Mobile internet mashup applications. J Telecom Technol, the preceding 2008:70–72
7. Skill (2011) 3 D urban system based on ArcGIS development. Shandong university of science and technology
8. Zhao (2009) WebGIS based on ArcGIS server application study. Science and technology plaza, 11:113–117
9. Kuan J, Xiaopeng G (2010) Exposure, etc. Google API to develop a: Google Maps and Google Earth shuangjian combination (second edition) electronic industry publishing house, 2010 practices: 200–201
10. Wang Y-b (2011) The research and application of CAD system based on GIS railway line selection. Lanzhou Jiaotong University
11. ArcGIS server http://en.wikipedia.org/wiki/ArcGIS_Server

Design of UAV Close Formation Controller Based on Sliding Mode Variable Structure

Yibo Li, Guomin Zhou, Wei Chen and Senyue Zhang

Abstract UAV formation flight has many incomparable advantages with single UAV. In this paper, the "lead aircraft C wing aircraft" formation mode was used. The aerodynamic coupling effect of close formation, which was created by the leads vortex, was analyzed. Then the close formation equivalent aerodynamic model was established, which consists of two flying wing layout UAVs. The sliding mode variable structure control theory was introduced to design the UAV close formation controller. The controller was divided into longitudinal channel, lateral channel and vertical channel. These three channels were designed individually, thereby simplify the design of controller, reduce the parameters adjustment difficulty and explore a new method for close formation keeping research. The simulation results show that the formations geometry is maintained in the face of lead aircraft maneuver, and the feasibility and effectiveness of established UAV close formation flight controller was verified.

Y. Li · G. Zhou (✉)
College of Automation, Shenyang Aerospace University,
Shenyang 110136, Liaoning, China
e-mail: qduzgm@163.com

Y. Li
e-mail: liyibo_sau@163.com

W. Chen
AVIC Chengdu Aircraft Industrial (Group) CO, LTD,
Chengdu 610073, Sichuan, China
e-mail: scjjwy@21cn.com

S. Zhang
College of Economics and Management, Shenyang Aerospace University,
Shenyang 110136, Liaoning, China
e-mail: 3310848@qq.com

© Springer International Publishing Switzerland 2017
V.E. Balas et al. (eds.), *Information Technology and Intelligent Transportation Systems*, Advances in Intelligent Systems and Computing 454,
DOI 10.1007/978-3-319-38789-5_55

1 Introduction

UAV (unmanned aerial vehicle) formation flight has many incomparable advantages with single UAV. Especially the UAV close formation flight, which could greatly save power demand, potentially increases the range and endurance of the formation.

The main problem of UAV close formation is to keep the UAVs relative position within a formation when aerodynamic coupling effect was taken into account. Reference [1] proposed the inner and outer loop thinking to research formation flight control problems. Conventional PID approach was used to design a controller in Ref. [2]. The feedback linearization ideas were applied in Ref. [3]. Reference [4] analyzed the aerodynamic coupling effect of close formation, but the paper did not address controller design problem. PID control theory was used in Ref. [5] to design a close formation controller. Reference [6] proposed a complex controller based on CMAC (cerebellar model articulation controller) and PID to maintain close formation, and its effectiveness was verified by simulation results.

Contributions of this paper is that the flying wing layout UAVs are used for the first time and the design of sliding mode variable structure controller based on linear feedback for close formation flight of multiple UAVs. Firstly, the autopilot model and the formation kinematic equations are introduced. Then the aerodynamic coupling effect of close formation is modeled. The upwash and sidewash created by the leads vortices are calculated respectively. The wingman autopilot model is modified accordingly. Thirdly, the complete close formation flight control system is built. Then the sliding mode variable structure control theory based on feedback linearization was used to design the UAV close formation controller. At last, the performance is evaluated using a simulation developed in MATLAB SIMULINK. The feasibility and effectiveness of established UAV close formation flight controller are verified.

2 Kinematic Equations

The research in this paper is primarily focused on flying wing layout UAV in a generic diamond formation. Both the lead aircraft and the wing aircraft are equipped with a flight control system that includes three standard autopilots: Heading-hold, Mach-hold and Altitude-hold autopilots. The Mach-hold autopilot is first order model, the Heading-hold and the Altitude-hold autopilots are second order models. The three standard autopilots are

$$\dot{V}_{\mathrm{i}} = -\frac{1}{\tau_V} V_i + \frac{1}{\tau_V} V_{ic} \tag{1}$$

$$\ddot{\psi}_i = -\left(\frac{1}{\tau_{\psi a}} + \frac{1}{\tau_{\psi b}}\right)\dot{\psi}_i - \frac{1}{\tau_{\psi a}\tau_{\psi b}}\psi_i + \frac{1}{\tau_{\psi a}\tau_{\psi b}}\psi_{ic} \tag{2}$$

$$\ddot{h}_i = -\left(\frac{1}{\tau_{ha}} + \frac{1}{\tau_{hb}}\right)\dot{h}_i - \frac{1}{\tau_{ha}\tau_{hb}}h_i + \frac{1}{\tau_{ha}\tau_{hb}}h_{ic} \tag{3}$$

where the subscript i = W, L, indicates the lead aircraft and the wing aircraft respectively; V is velocity, ψ is heading angle, h is flight height; and τ_{ha} and τ_{hb} are height time parameters, $\tau_{\psi a}$ and $\tau_{\psi b}$ are heading time parameters, τ_v is velocity time parameter.

The analysis of the system kinematics uses two coordinate frames [7]: inertial frame of reference and rotating reference frame centered on the wing aircraft. Equation (4) shows the formation kinematic equations based on the two coordinate frames accordingly.

$$\begin{cases} \dot{x} = V_L\cos(\psi_L - \psi_W) + \dot{\psi}_W y - V_W \\ \dot{y} = V_L\sin(\psi_L - \psi_W) - \dot{\psi}_W x \\ \dot{z} = 0 \end{cases} \tag{4}$$

where the lead aircrafts position in the rotating reference frame is (x, y, z).

3 Aerodynamic Coupling Effect of Close Formation

3.1 Upwash and Sidewash Modeling

The horseshoe vortex model is used in this paper to study the influence of wing aircraft created by lead aircrafts wing tip vortex.

Figure 1 is a view from above the two-aircraft formation horseshoe vortex approximate model [8], filaments A and B are the borders of the horseshoe vortex. $\bar{x}$ and $\bar{y}$ are the longitudinal separation and the lateral separation between the lead aircraft and the wing aircraft respectively. Γ is vortex strength of lead aircraft. Both of the lead aircraft and the wing aircraft are represented by elliptical wing approximation b', that is $b' = \frac{\pi}{4}b$, where b is the wingspan of aircraft.

According to Ref. [8], the average induced upwash $W_{UW_{avg}}$ and sidewash $V_{SW_{avg}}$ on the wingmans wing are

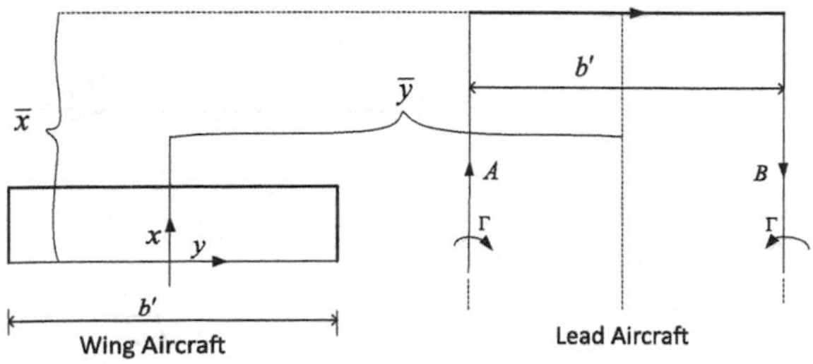

Fig. 1 Top view of the two aircrafts

$$W_{UW_{avg}} = -\frac{\Gamma_L}{4\pi b'}\left\{ln\frac{y'^2 + z'^2 + \mu^2}{(y' - \frac{\pi}{4})^2 + z'^2 + \mu^2} - ln\frac{(y' + \frac{\pi}{4})^2 + z'^2 + \mu^2}{y'^2 + z'^2 + \mu^2}\right\} \quad (5)$$

$$V_{SW_{avg}} = \frac{\Gamma_L}{4\pi h_z}\left\{ln\frac{(y' - \frac{\pi}{8})^2 + z'^2 + \mu^2}{(y' - \frac{\pi}{8})^2 + (z' + \frac{h_z}{b})^2 + \mu^2} - ln\frac{(y' + \frac{\pi}{8})^2 + z'^2 + \mu^2}{(y' + \frac{\pi}{8})^2 + (z' + \frac{h_z}{b})^2 + \mu^2}\right\} \quad (6)$$

where μ^2 is a correction term, $z' = \frac{z}{b}$, $y' = \frac{y}{b}$.

3.2 *Calculation of Change in Wingman's Lift, Drag and Side Force*

The upwash on the wing of the wingman causes a change in the attack angle of the wing. This causes a rotation in the lift and drag vectors of the wing.

Figure 2 is the wing aircraft force change diagram, V is the velocity of the aircraft, W is the upwash, and V' is the composite velocity of the air at the surface of the wing. The original lift and drag vectors are represented by L and D, and the rotated lift and drag are represented by L' and D' respectively.

According to Ref. [9], the non-dimensional coefficient of drag increment is

$$\Delta C_{D_W} = \frac{1}{\pi A_R}C_{L_L}C_{L_W}\frac{2}{\pi^2}\left\{ln\frac{y'^2 + z'^2 + \mu^2}{(y' - \frac{\pi}{4})^2 + z'^2 + \mu^2} - ln\frac{(y' + \frac{\pi}{4})^2 + z'^2 + \mu^2}{y'^2 + z'^2 + \mu^2}\right\} \quad (7)$$

where A_R is the aspect ratio of the wing, and C_{L_L} is the lift coefficient of the lead aircraft. C_{L_W} is the lift coefficient of the wingman. The change in lift coefficient is given by

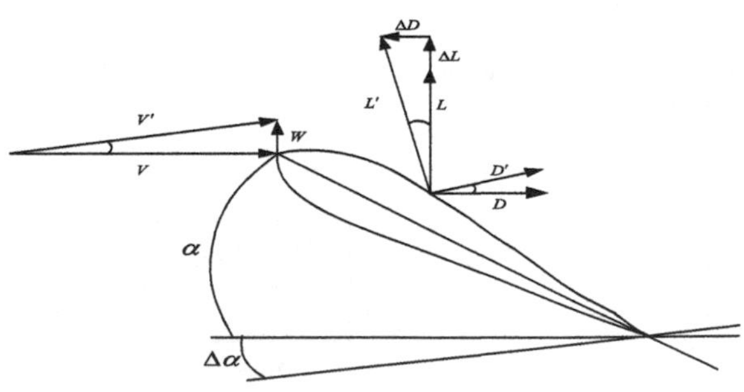

Fig. 2 Wing aircraft force change

$$\Delta C_{L_W} = \frac{a_W}{\pi A_R} C_{L_L} \frac{2}{\pi^2} \left\{ ln \frac{y'^2 + z'^2 + \mu^2}{(y' - \frac{\pi}{4})^2 + z'^2 + \mu^2} - ln \frac{(y' + \frac{\pi}{4})^2 + z'^2 + \mu^2}{y'^2 + z'^2 + \mu^2} \right\} \quad (8)$$

where a_W is the lift curve slope of the wing. The sidewash created by the lead also causes a change in the force on the vertical tail. This change in side force is

$$\Delta C_Y = \eta \frac{S_{Vt}}{S} a_{vt} \frac{|V_{SW}|}{V} \quad (9)$$

and a_{vt} is the lift curve slope of the vertical tail. Because the flying wing layout UAV used in this paper has no vertical tail, so $S_{Vt} = 0$ and the change in side force is $\Delta C_Y = 0$.

3.3 *Perfected Wing Aircraft Autopilot System*

For close formation flight, the wing autopilot needs to be perfected because of the aerodynamic coupling effect created by the upwash and sidewash from the lead aircraft. By calculation, the perfected wing autopilot model for close formation flight is

$$\dot{V}_W = -\frac{1}{\tau_V} V_W + \frac{1}{\tau_V} V_{W_c} + \frac{\bar{q}S}{m} \Delta C_{D_{W_y}} y \quad (10)$$

$$\ddot{\psi}_W = -\frac{1}{\tau_{\psi a}\tau_{\psi b}} \psi_W - \left(\frac{1}{\tau_{\psi a}} + \frac{1}{\tau_{\psi b}}\right) \dot{\psi}_W + \frac{1}{\tau_{\psi a}\tau_{\psi b}} \psi_{Wc} \quad (11)$$

$$\ddot{h}_W = -\left(\frac{1}{\tau_{ha}} + \frac{1}{\tau_{hb}}\right) \dot{h}_W - \frac{1}{\tau_{ha}\tau_{hb}} h_W + \frac{1}{\tau_{ha}\tau_{hb}} h_{Wc} + \frac{\bar{q}S}{m} \Delta C_{L_{W_y}} y \quad (12)$$

where y is the perturbation in the leads y direction position in the formation relative to the wing aircraft from the nominal location.

4 Control System Model of Close Formation

Adjoin equations (4), (10)–(12) yields the complete control system model of close formation. The resulting linear state-space representation of the complete control system is

$$\frac{d}{dt}\begin{bmatrix} x \\ V_W \\ y \\ \psi_W \\ \dot{\psi}_W \\ z \\ \zeta \end{bmatrix} = A\begin{bmatrix} x \\ V_W \\ y \\ \psi_W \\ \dot{\psi}_W \\ z \\ \zeta \end{bmatrix} + B\begin{bmatrix} V_{W_c} \\ \psi_{Wc} \\ h_{Wc} \end{bmatrix} + \Gamma\begin{bmatrix} V_L \\ \psi_L \\ h_{Lc} \end{bmatrix} \tag{13}$$

where $z = h_W - h_L$. The state variables are x, V_W, y, ψ_W, $\dot{\psi}_W$, z and ζ, the three control inputs are V_{Wc}, ψ_{Wc} and h_{Wc}, the disturbance signals are V_L, ψ_L and h_{Lc}. A is dynamic matrix, B is input matrix and Γ is disturbance matrix, their expressions are:

$$A = \begin{bmatrix} 0 & -1 & 0 & 0 & \sin\alpha & 0 & 0 \\ 0 & -\frac{1}{\tau_V} & \frac{\bar{q}S}{m}\Delta C_{D_{Wy}} & 0 & 0 & 0 & 0 \\ 0 & 0 & 0 & -1 & -\cos\alpha & 0 & 0 \\ 0 & 0 & 0 & 0 & 1 & 0 & 0 \\ 0 & 0 & 0 & -\frac{1}{\tau_{\psi a}\tau_{\psi b}} & -(\frac{1}{\tau_{\psi a}} + \frac{1}{\tau_{\psi b}}) & 0 & 0 \\ 0 & 0 & 0 & 0 & 0 & 0 & 1 \\ 0 & 0 & \frac{\bar{q}S}{m}\Delta C_{L_{Wy}} & 0 & 0 & -\frac{1}{\tau_{\psi a}\tau_{\psi b}} & -(\frac{1}{\tau_{\psi a}} + \frac{1}{\tau_{\psi b}}) \end{bmatrix}$$

$$B = \begin{bmatrix} 0 & 0 & 0 \\ \frac{1}{\tau_V} & 0 & 0 \\ 0 & 0 & 0 \\ 0 & 0 & 0 \\ 0 & \frac{1}{\tau_{\psi a}\tau_{\psi b}} & 0 \\ 0 & 0 & 0 \\ 0 & 0 & \frac{1}{\tau_{\psi a}\tau_{\psi b}} \end{bmatrix} \quad \Gamma = \begin{bmatrix} 1 & 0 & 0 \\ 0 & 0 & 0 \\ 0 & 1 & 0 \\ 0 & 0 & 0 \\ 0 & 0 & 0 \\ 0 & 0 & 0 \\ 0 & 0 & -\frac{1}{\tau_{\psi a}\tau_{\psi b}} \end{bmatrix}$$

where α is the formation angle. When the longitudinal separation is x_0, lateral separation is y_0, $\alpha = \arcsin\frac{y_0}{\sqrt{{x_0}^2+{y_0}^2}}$.

5 Sliding Mode Variable Structure Controller Design

The formation flight controller is to make the wingman flying fast track heading, altitude and speed maneuvering of the lead aircraft, while maintaining the formation. Equation (13) states that close formation causes the y channel to be coupled into the x and z channel. The x channel is still decoupled from z channel. So the y channel controller was firstly designed, then the x channel controller, and z channel controller was designed at last.

5.1 Y Channel Controller Design

The goal of designing y channel controller is to solve heading control signal of wing aircraft, that is ψ_{Wc}. It could be seen from Eq. (13) that the y channel state-space equation is

$$\begin{bmatrix} \dot{y} \\ \dot{\psi}_W \\ \ddot{\psi}_W \end{bmatrix} = \begin{bmatrix} 0 & -1 & -\cos\alpha \\ 0 & 0 & 1 \\ 0 & -\frac{1}{\tau_{\psi a}\tau_{\psi b}} & -(\frac{1}{\tau_{\psi a}} + \frac{1}{\tau_{\psi b}}) \end{bmatrix} \begin{bmatrix} y \\ \psi_W \\ \dot{\psi}_W \end{bmatrix} + \begin{bmatrix} 0 \\ 0 \\ \frac{1}{\tau_{\psi a}\tau_{\psi b}} \end{bmatrix} \psi_{Wc} + \begin{bmatrix} 1 \\ 0 \\ 0 \end{bmatrix} \psi_L \tag{14}$$

Set $x_1 = y$, $x_2 = \psi_W$, $x_3 = \dot{\psi}_W$, the Eq. (14) can be written as

$$\begin{bmatrix} \dot{x}_1 \\ \dot{x}_2 \\ \dot{x}_3 \end{bmatrix} = \begin{bmatrix} 0 & -1 & -\cos\alpha \\ 0 & 0 & 1 \\ 0 & -\frac{1}{\tau_{\psi a}\tau_{\psi b}} & -(\frac{1}{\tau_{\psi a}} + \frac{1}{\tau_{\psi b}}) \end{bmatrix} \begin{bmatrix} x_1 \\ x_2 \\ x_3 \end{bmatrix} + \begin{bmatrix} 0 \\ 0 \\ \frac{1}{\tau_{\psi a}\tau_{\psi b}} \end{bmatrix} \psi_{Wc} + \begin{bmatrix} 1 \\ 0 \\ 0 \end{bmatrix} \psi_L \tag{15}$$

$$y = x_1 \tag{16}$$

The first derivative with respect to y is

$$\dot{y} = \dot{x}_1 = -x_2 - \cos\alpha \cdot x_3 + \psi_L$$

The second derivative with respect to y is

$$\begin{aligned} \ddot{y} &= -\dot{x}_2 - \cos\alpha \cdot \dot{x}_3 + \dot{\psi}_L \\ &= -x_3 + \cos\alpha \cdot \frac{1}{\tau_{\psi a}\tau_{\psi b}} x_2 + \cos\alpha \cdot (\frac{1}{\tau_{\psi a}} + \frac{1}{\tau_{\psi b}}) x_3 - \cos\alpha \cdot \frac{1}{\tau_{\psi a}\tau_{\psi b}} \psi_{Wc} + \dot{\psi}_L \end{aligned} \tag{17}$$

Set $f(x) = -x_3 + \cos\alpha \cdot \frac{1}{\tau_{\psi a}\tau_{\psi b}} x_2 + \cos\alpha \cdot (\frac{1}{\tau_{\psi a}} + \frac{1}{\tau_{\psi b}}) x_3 + \dot{\psi}_L$, then Eq. (17) becomes $\ddot{y} = f(x) - \cos\alpha \cdot \frac{1}{\tau_{\psi a}\tau_{\psi b}} \psi_{Wc}$. Define the tracking error as $e = y_d - y$, first and second order derivative of the tracking error is $\dot{e} = \dot{y}_d - \dot{y}$ and $\ddot{e} = \ddot{y}_d - \ddot{y}$ respectively. Where y_d is the nominal lateral separation, y is the real lateral separation. Set the sliding mode surface [10] is $s(y, t) = c_y e + \dot{e}$, where c_y needs to meet the Hurwitz condition, that is $c_y > 0$. Using exponential reaching law, the sliding mode control law based on input-output feedback linearization is designed as

$$\psi_{Wc} = \frac{\tau_{\psi a}\tau_{\psi b}}{\cos\alpha} (f(x) - v - \eta_y \mathrm{sgn} s - k_y s) \tag{18}$$

where v is an auxiliary term of the control law. The Lyapunov function is defined as $V = \frac{1}{2}s^2$, the derivative of V is

$$\dot{V} = s\dot{s} = s(c_y\dot{e} + \ddot{e}) = s(c_y\dot{e} + \ddot{y}_d - \ddot{y}) = s\left(c_y\dot{e} + \ddot{y}_d - f(x) + \cos\alpha \cdot \frac{1}{\tau_{\psi a}\tau_{\psi b}} \psi_{Wc} \right) \tag{19}$$

Substituting Eq. (18) into (19), the derivative of V becomes

$$\dot{V} = s[c_y\dot{e} + \ddot{y}_d - f(x) + (f(x) - v - \eta_y \text{sgn}s - k_y s)] = s(c_y\dot{e} + \ddot{y}_d - v - \eta_y \text{sgn}s - k_y s)$$

Set $v = c_y\dot{e} + \ddot{y}_d$, then $\dot{V} = s\dot{s} = s(-\eta_y \text{sgn}s - k_y s) = -\eta_y |s| - k_y s^2 \le 0$, which means the designed y channel controller has strong robustness.

5.2 X Channel Controller Design

The goal of designing x channel controller is to solve the velocity control signal of wingman, that is V_{Wc}. It could be seen from Eq. (13) that the x channel state-space equation is

$$\begin{bmatrix} \dot{x} \\ \dot{V}_W \end{bmatrix} = \begin{bmatrix} 0 & -1 \\ 0 & -\frac{1}{\tau_V} \end{bmatrix} \begin{bmatrix} x \\ V_W \end{bmatrix} + \begin{bmatrix} 0 \\ \frac{1}{\tau_V} \end{bmatrix} V_{Wc} + \begin{bmatrix} 1 & \sin\alpha \\ 0 & 0 \end{bmatrix} \begin{bmatrix} V_L \\ \dot{\psi}_W \end{bmatrix} + \begin{bmatrix} 0 \\ \frac{\bar{q}S}{m}\Delta C_{D_{Wy}} \end{bmatrix} y \tag{20}$$

Set $x_1 = x$, $x_2 = V_W$, therefore, the Eq. (20) can be written as

$$\begin{bmatrix} \dot{x}_1 \\ x_2 \end{bmatrix} = \begin{bmatrix} 0 & -1 \\ 0 & -\frac{1}{\tau_V} \end{bmatrix} \begin{bmatrix} x_1 \\ x_2 \end{bmatrix} + \begin{bmatrix} 0 \\ \frac{1}{\tau_V} \end{bmatrix} V_{Wc} + \begin{bmatrix} 1 & \sin\alpha \\ 0 & 0 \end{bmatrix} \begin{bmatrix} V_L \\ \dot{\psi}_W \end{bmatrix} + \begin{bmatrix} 0 \\ \frac{\bar{q}S}{m}\Delta C_{D_{Wy}} \end{bmatrix} y \tag{21}$$

$$x = x_1 \tag{22}$$

The first derivative with respect to x is

$$\dot{x} = \dot{x}_1 = -x_2 + V_L + \sin\alpha\dot{\psi}_W$$

The second derivative with respect to x is

$$\ddot{x} = \ddot{x}_1 = -\dot{x}_2 + \dot{V}_L + \sin\alpha\ddot{\psi}_W = \frac{1}{\tau_V}x_2 - \frac{1}{\tau_V}V_{Wc} - \frac{\bar{q}S}{m}\Delta C_{D_{Wy}} y + \dot{V}_L + \sin\alpha\ddot{\psi}_W \tag{23}$$

Set $f(x) = -x_3 + \cos\alpha \cdot \frac{1}{\tau_{\psi a}\tau_{\psi b}}x_2 + \cos\alpha \cdot (\frac{1}{\tau_{\psi a}} + \frac{1}{\tau_{\psi b}})x_3 + \dot{\psi}_L$, then Eq. (23) becomes $\ddot{x} = f(x) - \frac{1}{\tau_V}V_{Wc}$. Define the tracking error as $e = x_d - x$, first and second order derivative of the tracking error is $\dot{e} = \dot{x}_d - \dot{x}$ and $\ddot{e} = \ddot{x}_d - \ddot{x}$ respectively. Where x_d is the nominal lateral separation, x is the real lateral separation. Set the sliding mode surface is $s(x, t) = c_x e + \dot{e}$, where c_x needs to meet the Hurwitz condition, that is $c_x > 0$. Using exponential reaching law, the sliding mode control law based on input-output feedback linearization is designed as

$$V_{Wc} = \tau_V(f(x) - v - \eta_x \text{sgn}s - k_x s) \tag{24}$$

where v is an auxiliary term of the control law. The Lyapunov function is defined as $V = \frac{1}{2}s^2$, the derivative of V is

$$\dot{V} = s\dot{s} = s(c_x\dot{e} + \ddot{e}) = s(c_x\dot{e} + \ddot{x}_d - \ddot{x}) = s(c_x\dot{e} + \ddot{x}_d - f(x) + \frac{1}{\tau_V}V_{Wc}) \tag{25}$$

Substituting Eq. (24) into (25), derivative of V is

$$\dot{V} = s[c_x\dot{e} + \ddot{x}_d - f(x) + (f(x) - v - \eta_x \text{sgn}s - k_x s)] = s(c_x\dot{e} + \ddot{x}_d - v - \eta_x \text{sgn}s - k_x s)$$

Set $v = c_x\dot{e} + \ddot{y}_d$, then $\dot{V} = s\dot{s} = s(-\eta_y \text{sgn}s - k_y s) = -\eta_y |s| - k_y s^2 \leq 0$, which means the designed x channel controller has strong robustness.

5.3 Z Channel Controller Design

The goal of designing z channel controller is to solve height control signal of wingman, that is h_{Wc}. It could be seen from Eq. (13) that the z channel state-space equation is

$$\begin{bmatrix} \dot{z} \\ \dot{\zeta} \end{bmatrix} = \begin{bmatrix} 0 & 1 \\ -\frac{1}{\tau_{ha}\tau_{hb}} & -(\frac{1}{\tau_{ha}} + \frac{1}{\tau_{hb}}) \end{bmatrix} \begin{bmatrix} z \\ \zeta \end{bmatrix} + \begin{bmatrix} 0 \\ \frac{1}{\tau_{ha}\tau_{hb}} \end{bmatrix} h_{Wc} + \begin{bmatrix} 0 \\ -\frac{1}{\tau_{ha}\tau_{hb}} \end{bmatrix} h_{Lc} + \begin{bmatrix} 0 \\ \frac{\bar{q}S}{m}\Delta C_{L_{Wy}} \end{bmatrix} y \tag{26}$$

Set $x_1 = z$, $x_2 = \zeta$, Eq. (26) becomes

$$\begin{bmatrix} \dot{x}_1 \\ \dot{x}_2 \end{bmatrix} = \begin{bmatrix} 0 & 1 \\ -\frac{1}{\tau_{ha}\tau_{hb}} & -(\frac{1}{\tau_{ha}} + \frac{1}{\tau_{hb}}) \end{bmatrix} \begin{bmatrix} x_1 \\ x_2 \end{bmatrix} + \begin{bmatrix} 0 \\ \frac{1}{\tau_{ha}\tau_{hb}} \end{bmatrix} h_{Wc} + \begin{bmatrix} 0 \\ -\frac{1}{\tau_{ha}\tau_{hb}} \end{bmatrix} h_{Lc} + \begin{bmatrix} 0 \\ \frac{\bar{q}S}{m}\Delta C_{L_{Wy}} \end{bmatrix} y \tag{27}$$

$$z = x_1 \tag{28}$$

The first derivative with respect to z is

$$\dot{z} = \dot{x}_1 = x_2$$

The second derivative with respect to z is

$$\ddot{z} = \ddot{x}_1 = \dot{x}_2 = -\frac{1}{\tau_{ha}\tau_{hb}}x_1 - (\frac{1}{\tau_{ha}} + \frac{1}{\tau_{hb}})x_2 + \frac{1}{\tau_{ha}\tau_{hb}}h_{Wc} - \frac{1}{\tau_{ha}\tau_{hb}}h_{Lc} + \frac{\bar{q}S}{m}\Delta C_{L_{Wy}}y \tag{29}$$

Set $f(x) = -\frac{1}{\tau_{ha}\tau_{hb}}x_1 - (\frac{1}{\tau_{ha}} + \frac{1}{\tau_{hb}})x_2 - \frac{1}{\tau_{ha}\tau_{hb}}h_{Lc} + \frac{\bar{q}S}{m}\Delta C_{L_{Wy}}y$, then Eq. (29) becomes $\ddot{z} = f(x) + \frac{1}{\tau_{ha}\tau_{hb}}h_{Wc}$. Define the tracking error as $e = z_d - z$, first and second order derivative of the tracking error is $\dot{e} = \dot{z}_d - \dot{z}$ and $\ddot{e} = \ddot{z}_d - \ddot{z}$ respectively. Where z_d is the nominal lateral separation, z is the real lateral separation. Set the sliding mode surface is $s(z, t) = c_z e + \dot{e}$, where c_z needs to meet the Hurwitz condition, that is $c_z > 0$. Using exponential reaching law, the sliding mode control law based on input-output feedback linearization is designed as

$$h_{Wc} = \tau_{ha}\tau_{hb}(v - f(x) + \eta_z \text{sgn}s + k_z s) \tag{30}$$

where v is an auxiliary term of the control law. The Lyapunov function is defined as $V = \frac{1}{2}s^2$, the derivative of V is

$$\dot{V} = s\dot{s} = s(c_z\dot{e} + \ddot{e}) = s(c_z\dot{e} + \ddot{z}_d - \ddot{z}) \tag{31}$$

Substituting Eq. (30) into (31), derivative of V is

$$\begin{aligned}\dot{V} &= s\dot{s} = s(c_z\dot{e} + \ddot{e}) = s[c_z\dot{e} + \ddot{z}_d - f(x) - (v - f(x) + \eta_z \text{sgn}s + k_z s)] \\ &= s(c_z\dot{e} + \ddot{z}_d - v - \eta_z \text{sgn}s - k_z s)\end{aligned} \tag{32}$$

Set $v = c_z\dot{e} + \ddot{z}_d$, then $\dot{V} = s(-\eta_z \text{sgn}s - k_z s) = -\eta_z|s| - k_z s^2 \le 0$, which means the designed z channel controller has strong robustness.

6 Simulation

Simulation is performed for close formation consisting of two flying wing layout UAVs. The close formations optimal geometry is specified by $\bar{x} = 2b$, $\bar{y} = (\pi/4)b$ and $\bar{z} = 0$, which minimizes the formation drag. The wingspan of the flying wing layout UAV used in this paper is 28.44 ft, the optimal separation of close formation is $\bar{x} = 57.48$ ft, $\bar{y} = 22.57$ ft and $\bar{z} = 0$ accordingly.

The flying wing layout UAV autopilot parameters are listed in Table 1, and the exponential reaching law coefficients of three control channels are listed in Table 2.

Table 1 UAV autopilot parameters

$\tau_{\psi a}$	$\tau_{\psi b}$	τ_v	τ_{ha}	τ_{hb}
1.838	1.838	10	0.615	7.692

Table 2 Exponential reaching law coefficients

η_x	k_x	η_y	k_y	η_z	k_z
3.02	10.06	2.87	8.79	4.61	11.32

The sliding mode surface parameters of three channels are listed in Table 3. The close formation stability derivatives are listed in Table 4.

Two flying wing layout UAVs are used in the simulation, and the simulation time is 30 s. At initial state, the flight heights are $h_{L0} = 10{,}000$ ft, $h_{W0} = 10{,}000$ ft, the heading angles are $y_{L0} = 0$, $y_{W0} = 0$, and the velocities are $V_{L0} = 450$ ft/s, $V_{W0} = 450$ ft/s. The formation geometry is diamond formation, the separation is the nominal value, that is $\bar{x} = 57.48$ ft, $\bar{y} = 22.57$ ft and $\bar{z} = 0$.

The lead aircraft begin maneuvering at 2nd s, the heading changes from 0° to 30°. From 6th to 17th s, the heading maintains 30°. At 17th s, the heading changes from 30° to 0, and then the heading maintains 0° until 30th s. During the heading

Table 3 Sliding mode surface parameters

c_x	c_y	c_z
10.91	15.68	6.53

Table 4 Close formation stability derivatives

$\Delta C_{D_{Wy}}$	$\Delta C_{L_{Wy}}$
−0.000647	−0.0052

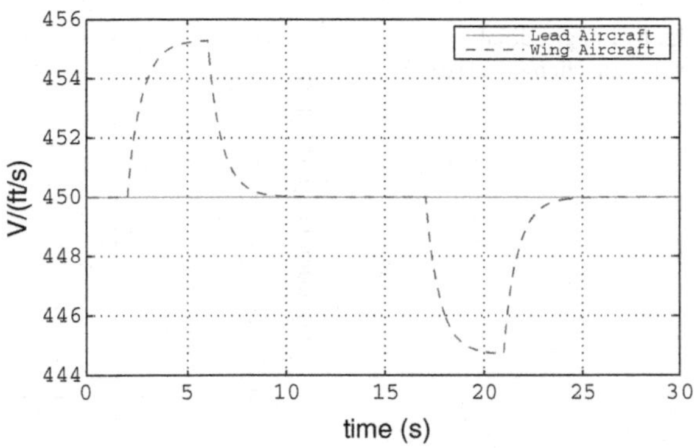

Fig. 3 Velocity change curve

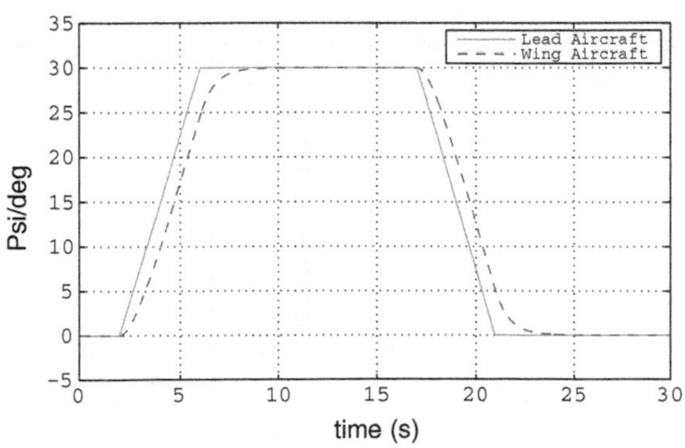

Fig. 4 Heading change curve

maneuvers, the height increases 300 ft, which changes from 10,000 to 10,300 ft. Simulation results are displayed in the following figures.

Figures 3, 4 and 5 show the velocity change curves, heading change curves and height change curves respectively. The solid line represents the lead aircraft and the double-dashed line represents the wing aircraft. Figures 6, 7 and 8 show the longitudinal direction separation change curve, lateral direction separation change

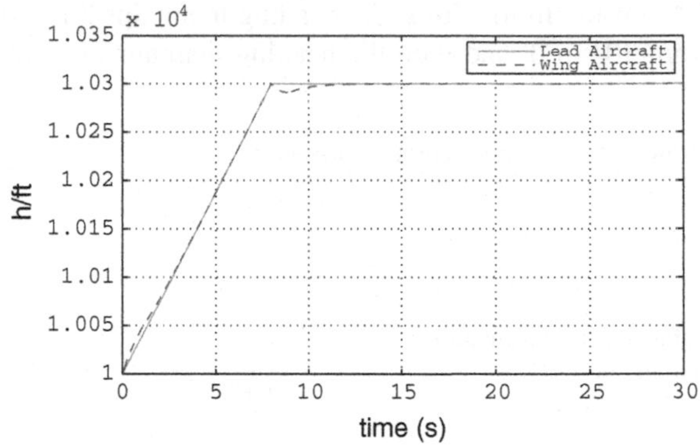

Fig. 5 Height change curve

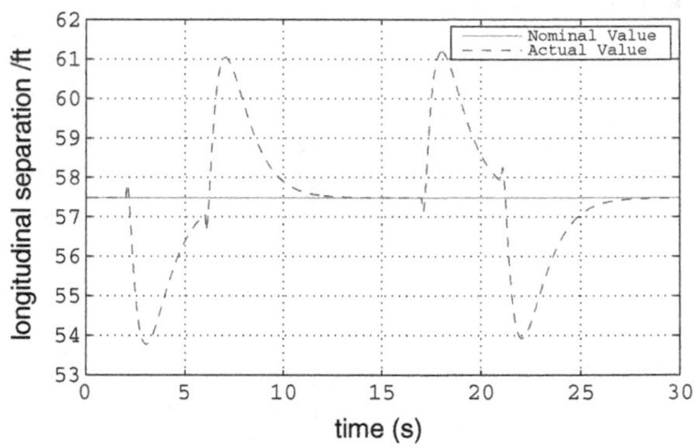

Fig. 6 Longitudinal separation change curve

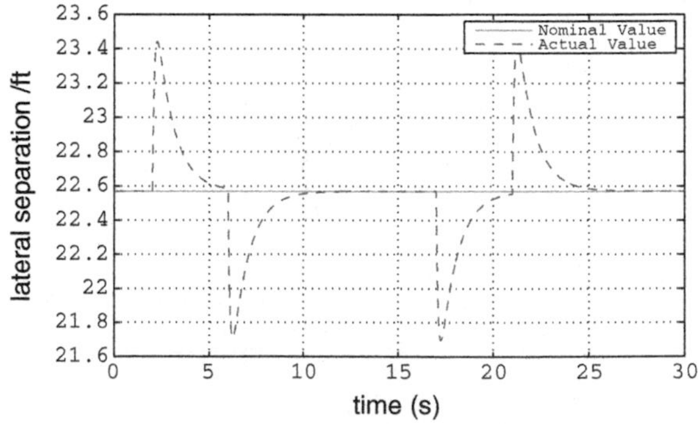

Fig. 7 Lateral separation change curve

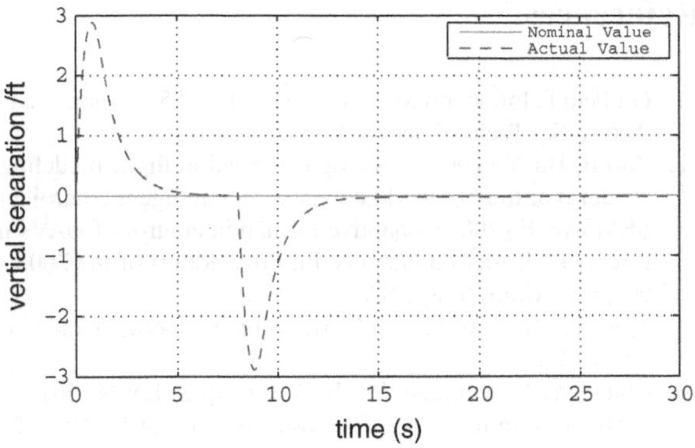

Fig. 8 Vertical separation change curve

curve and vertical direction separation change curve respectively. The solid line represents the nominal value and the double-dashed line represents the actual value.

Figure 3 states that the velocity derivation is less than 6 ft/s, the wing aircraft velocity can reach the nominal value in 8 s. Figure 4 states that the wing aircraft could well follows the heading maneuver of the lead aircraft on the duration of 30 s. The lead aircraft heading maneuver can be completely tracked by the wing aircraft within 2 s. Figures 6 and 7 state that both x direction separation and y direction separation can reach the nominal value in 8 s. The x derivation is less than 1 ft, and the y derivation is less than 4 ft. The closet the wing aircraft comes to the lead aircraft is 53.8 ft in the x direction and 21.7 ft in the y direction, thereby the collision is effectively avoided. Figure 10 states that z direction separation is less than 3 ft during the maneuver, the wingman can track the lead aircraft height in 15 s.

7 Conclusion

The close formation consists of two flying wing layout UAVs was researched in this paper, the aerodynamic coupling effect of the close formation was analyzed from the perspective of flight mechanics, and then three-dimensional formation control mathematical model was established. The sliding mode variable structure control theory was introduced to design the UAV close formation controller, and the controller was divided into longitudinal channel, lateral channel and vertical channel. The simulation results show that the formations geometry is maintained in the face of leads maneuvering. The controller designed in this paper has the advantages of fast response, small overshoot, good robustness and so on. The controller can enable aircraft to take advantage of reduction in induced drag brought by the aerodynamic coupling effect, which will extends the endurance of the formation accordingly.

References

1. Giulietti F, Innocenti M, Napolitano M (2005) Dynamic and control issues of formation flight. Aerosp Sci Technol 36(9):65–71
2. Zuo B, Hu Y (2004) UAV tight formation flight modeling and autopilot designing. In: Proceedings of the 5th world congress on intelligent control and automation. Hangzhou, China
3. SEMSAR E (2006) Adaptive formation control of UAVs in the presence of unknown vortex forces and leader commands. In: Proceedings of the 2006 American control conference minneapolis. Minnesota, USA
4. Yong L, Micui W (2011) Close formation aerodynamic coupling effect analysis. Flight Dyn 19(2):12–16
5. Chundong C, Ruixuan W, Zhi D, Lipeng Z, Lei N (2012) Close formation coordinated flight control design for UAVs. Electron Opt Control 19(7):18–22
6. Chenggong L, Zhong Y, Qiongjian F (2009) Research on multi-UAVs close formation flight control based on CMAC. Transducer Microsyst Technol 28(7):37–40
7. ShengJun Q, Zhe Z (2012) Modeling and simulation of UAV close formation flight control. Adv Aeronaut Sci Eng 3(3):362–366
8. Pather M, Azzo JJ, Proud AW (2001) Tight formation flight control. J Guid Control Dyn 24(2):246–254
9. Andrew WP, Meir P, John JD (1999) Close formation flight control. AIAA, guidance, navigation, and control conference and exhibit
10. Jinkun L (2012) Sliding mode control design and matlab simulation. Tsinghua University Press, Beijing

The Research in the Construction of Aircraft Equipments Visual Warehouse

Shi Yumin, Zhou Bin, Sun Weiqi and Wang Yuhai

Abstract This paper research on how to adopt visual technique such as bar code, RFID, sensor etc. To set up an aircraft equipments warehouse, including storage bar code management system, storage RFID management system, warehouses audio-frequency and video-frequency supervision and drilling system, warehouses temperature and humidity monitoring and controlling visual system, warehouses fire fight visual system are proposed.

Keywords Electronic countermeasures · Resource distribution · Secondary distribution model · Hungarian algorithm · Dynamic programming algorithm

1 Introduction

Currently, the high technique war is an informationize war, battlefield environment is very "transparent", also become higher to the request of aircraft equipments guarantee, if the troops still reserve industrial ages loosing huge and cannot visual warehouse, prepare to place numerous equipment in advance, not only make the system swollen day by day, result in resources maximum wasted, influence national economic development; And result in the guarantee activity dilatoriness, respond the ability and guarantee effects cannot attain anticipant level; It result the operation that

S. Yumin (✉) · Z. Bin · S. Weiqi · W. Yuhai
Naval Aeronautical Engineering Academy Qingdao Branch Qingdao,
266041 Qingdao, China
e-mail: shiyumin1981@126.com

Z. Bin
e-mail: hb_badboy@163.com

S. Weiqi
e-mail: 457361340@qq.com

W. Yuhai
e-mail: ycl.1251@163.com

© Springer International Publishing Switzerland 2017

V.E. Balas et al. (eds.), *Information Technology and Intelligent Transportation Systems*, Advances in Intelligent Systems and Computing 454,
DOI 10.1007/978-3-319-38789-5_56

can cause the sail material manage a low speed degree, low efficiency, and is hard to satisfy the demand of modern war and informationize war [4–8].

Therefore, we have to adopt informationize measure and means, accurate estimate supplies informationize management need, accurate plan and organization military supplies informationize management, carry out the supplies' informationize management changes to "rapid reaction type" from "the amount of resources intensive type", orientation and contented modern wars need to the supplies guarantees. Realization to stock equipments visual management, is be for adapting to modern war to the equipment receiving and supplying quickly and accurately request. Therefore, the construction of visual warehouse which can dynamic and visual follow the equipments receive and supply, control the guarantee condition real time, raise guarantee efficiency and effect, is the inevitable trend that the aircraft equipments guarantee development [9–12].

2 Visual Warehouses Function

2.1 *The Bar Code Management of Storage Equipment*

We will apply the bar code technique to manage the receiving and supplying the equipment and check the warehouse [1]. Bar code technique and aircraft equipments management combine to raise the receiving and supplying efficiency, realize a quickly recording of the database information and goes into, and can manage the batch information and list piece information of equipment, is efficiently means of carrying out stock equipments "the information visual.

2.2 *RFID Management of Stock Equipment*

RFID is a radio frequency identify technique, it can pass wireless radio frequency signal rapid, identifying of batch quantity device. Compared with the bar code management, RFID has advantages, such as "can read and write", "can wear deeply" and "can sweep the batch quantity code"…etc., but manages cost opposite higher, suitable to the important piece manage.

2.3 *Warehouses Audio-Frequency and Video-Frequency Supervision and Drilling System*

System is want to establish audio-frequency and video-frequency supervision system in the statistics room and the storage room, the supervision signal can pass a campus net video frequency to forward platform and carries out the long range supervision in the whole school scope. While carrying on various discrepancy practice, the conductor member broadcast each room to practice a circumstance while canning pass to supervise and control system actually, and issue to direct instruction through an audio frequency. At this time, supervise and control system can be used as drilling system.

2.4 Warehouses Temperature and Humidity Monitoring and Controlling Visual System

System will be builded to display the warehouses temperature and humidity real time, and draw storage rooms temperature and humidity curve of a day and a year. System can choose control measure according to the outside and inside temperature and humidity.

2.5 Warehouses Fire Fight Visual System

Build up a fire fight early warning and automatically extinguish fire system, the fire feeling prognosticate, reports to the police while carrying on actually, and automatically start extinguish fire system.

3 Concrete Construction Project

3.1 The Construction Method of Each System

The bar code management system of stock equipment: The bar code applies systems statistic room only need to add a bar code printer to connect with the aircraft equipment management system, consequently, carry out bar code printing function, and share to the storage room. Bar code applied system need to install a fixed type bar code scanner in each storage room and used to scan bar code to identify the equipment information when receiving and supplying equipment, also, the system need to install a handheld bar code scanner, used to check the storage room. Mainly use in the B database building and the C database building.

RFID of stock device manages system: Be aimed at to the management characteristics of device, adopt RFID to carry on the management that comes in and goes out a database in the A database building, it can pass wireless radio frequency signal rapid, identifying of batch quantity device.

Video frequency in the warehouse supervises and controls system: In each database building install supervision to stretch forward to carry on solid supervise and control, flow to turn to carry on whole distance supervision through a label in the process in the device and the voucher, supervise and control of the result show at the big screen of reference room up, the convenient staff member controls a flowing of database building circumstance and voucher and device to turn a circumstance in time. Is well applied modern technique, adopt existing surveillance supervision and network to transact to wait an advanced equipments, the establishment modernizes of management hardware platform. The network supervises and controls to manage system to integrate each one to supervise and control to order of supervision system, make of ability at unified of management platform up realization with each other connect mutual communication, carry out the share of video frequency data and management information. Carry out supervision to all subordinate supervision points allied the net, centralized management and long range support etc. function. Install an absorbed passive type in the warehouse area red outside locator and red outside laser beam locator, set up a safety alarm system. While discovering danger circus the system quickly connect to the alarm power and report t a sound and light signal to the supervision center; The Supervision and control center receive the alarm, allied with the video-frequency supervision and control system realize function of the video switch (show the police the position of alarm point, real time picture) and automatically record image etc. When the customer confirms the alarm true, the systems automatically carry out a pre-established safety handle emergency instruction;(such as: ring the alarm, dial appointed telephone, show the headwaters apart from a fire to order shortly etc.) If the alarm is true, the system will automatically start ex- record image function, automatic record related information after the alarm [13–15].

Warehouses temperature and humidity monitoring and controlling visual system: The system need to install a monitor, to display the temperature and humidity real time, and can automatically draw storage rooms temperature and humidity curve of a day and a year. The monitor can link intelligence temperature and humidity control system, when the temperature and humidity in the storage room exceed “the 37 lines”, the system can judgment and choose a best control measure according to the temperature and humidity inside and outside the storage room. If suitable and well ventilated, the system will remind the storage-member to carry on airiness, if not, automatically open the dehumidify machine or air- condition according to temperature and humidity, or both open at the same time. The temperature and humidity detect system be used to detect the storage rooms temperature and humidity, the main equipments include temperature and humidity sensor and controller. The system can according to the users fixed time automatically or handlly at any moment collect every sensors data. While discovering alarm, the system can deliver the sound and the light signal to the control center. The control center will automatically show the alarm position when receiving the alarm signal, utilize to allied to the video-frequency system. Via linkage, the system can show every sensors current temperature and humidity, and also show random appointed storage rooms current temperature and humidity on every sensor. The history temperature and humidity can be queried by time, address etc., and make variety curve or report forms and share the temperature and humidity information about the automatic solid warehouse and the temperature invariableness warehouse.

Warehouses fire fight visual system: Build up a fire fight early warning and automatically extinguish fire system, when the smoke density attains certain degree, carries out to report to the police and passes video frequency supervision to tell the preservation member the concrete position of the occurrence a fire, burnable degree etc., thereby, the member can convenient organize to extinguish fire, and automatically start spray water system to carry on extinguishing fire.

The key cabinet management system: The numeral key cabinet management system follow Double person double lock turns the key cabinet management system to observe the management principle of "double person double lock. The system gather to identify technique, gate prohibit, communications technique, picture disposal technique, network technique together, and have four greatly main functions include memory, manage, network, visual. The system can detailly record the person who take away the key and the key cabinet managers identity and the time of open the lock; Via adding chip on the key to identify key is real or false. The system can adopt various mode to manage the user include fingerprint identify, password identify etc. If the key didn't be stipulating time returns or encounters a sabotage, the system will automatically alarm and record, and will automatically report the alarm signal and related information to the supervision and control center at the same time. The system can report the alarm signal and related video frequency signal to network and the network terminal user can share the data such as the supervision and control center. The system can directly link control centre or through a network to control and manage the information. According to the authorization, the system can check the circumstance of the key management and search history record etc. on the network point. When the operator want to open the door, the system will validate the identity, at the same time, take photo towards holding card operator, and collate with the users IC card, write on the key using record. Idiographic constructions projection in following picture (Fig. 1).

3.2 Constructions Difficulty Analysis

No.1 difficulty: bar code labels and RFID labels selection [2, 3].

The aircraft equipment need to be stored a long time, and may experience various complicated external environment during the follow-up conveyance and supply, therefore the label has to enduring using, anti- oil, water tightness etc. characteristic, and make a very high request to the material of the bar code label and the RFID label.

The solution is full investigation, collect the data of various materials, full argument, select the suitable label material via experiment data. No.2 difficulty: The mode of visual technique using in the storage management.

Original aircraft equipment storage management mode is based upon artificial identify and handicraft operate formation. After adopt visual technique in aircraft

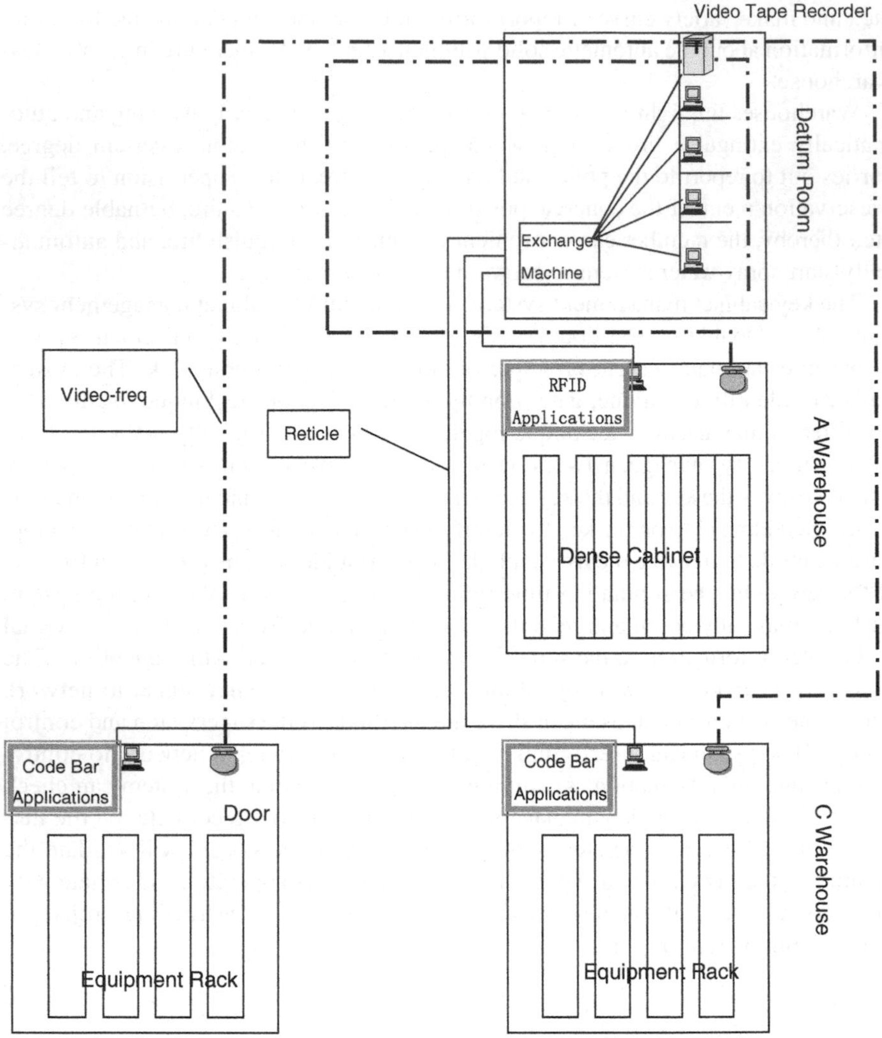

Fig. 1 video tape recorder

equipment storage management, equipments identify and informations inputting and outputting deliver etc. will change into automatic process, therefore, the aircraft equipment storage management mode have to carry on important change. But how adjust current aircraft equipment storage managements various process and rules, not only can develop the advantage of visual technique, promote aircraft equipment storage managements quality and efficiency, but also reserve advantage of original mode, and full permit with original homework mode, this is the item team has to research to solve the great problem which is the key problem of this item. Solution: Promoting storage homework efficiency and equipment management quality is basic purpose, we need to adequately solicit opinion by higher-up machinery and operation member, maximum exert the visual advantage, groping for the applied mode of visual technique using in the storage management.

4 Conclusion

Along with our army information-based level of continuously raising, navys aircraft equipment guarantee request the resources is visual, therefore, navys aircraft equipment department aggressively and gradually push forward the visual resources systems construction. But the visual aircraft equipments construction has not mature experience to be referenced, and need the suggestion that the college provides a viable construction programming and construction mode. The large-scale navys aircraft equipment visual storage construction start, will need to a mass of talented person who have the ability of manage the visual storage, therefore the college has to take preventive measures, advancing to construct a aircraft equipment visual storage to train the student.

References

1. Proon S, Jin M (2011) A genetic algorithm with neighborhood search for the resource-constrained project scheduling problem. Nav Res Logist 58(2):73–82
2. Coello CAC, Gutierrez RLZ, Garcia BM, Aguirre AH (2002) Automated design of combinational logic circuits using the ant System. Eng Opt 34(2):109–127
3. JCordone R, Maffioli F (2001) Coloured ant system and local search to design local telecommunication networks. In: Applications of evolutionary computing proceeding 2001, 2037:60 69
4. Frazelle E (2002) Supply Chain Strategy: The Logistics of Supply Chain Management. McGraw-Hill, NewYork
5. Xiawei Z, Changjia C, Gang Z (2000) A genetic algorithm for multicasting routing problem. In: WCC-ICCT 2000 International conference communication technology proceedings, p 1248–1253
6. Baker B, Yechew A (2003) A genetic algorithm for the vehicle routing problem. Comput Op Res 30(2):787–800
7. Lau HC, Sim M, Teo KM (2003) Vehicle routing problem with time windows and alimited number of vehicles. Eur J Op Res 148:559–569
8. Gutjahr W (2000) J.A graph-based ant system and its convergence. Future Gener Comput Syst 16(8):873–888
9. Gutjahr WJ (2002) ACO algorithms with guaranteed convergence to the optimal solution. Inf Process Lett 82(3):145–153
10. Staezle T, Dorigo M (2002) A short convergence proof for a class of ant colony optimization algorithms. IEEE Trans Evolut Comput 6(4):358–365
11. Yoo JH, La RJ, Makowski AM (2003) Convergence of ant routing algorithms-results for simple parallel network and perspectives [R], Technical Report CSHCN 2003-44, Institute for systems research, University of Maryland, College Park (MD)
12. Hou YH, Wu YW, Lu LJ, et al (2002) Generalized ant colony optimization for economic dispatch of power systems. In: Proceedings of the 2002 international conference on power system technology, 1:225–229

13. Dorigo M, Birattari M, Blum C, Gambardella LM, Mondada F, Statzle T (eds) (2004) Lecture notes in computer science. In: Fourth international workshop on ant colony optimization and swarm intelligence (ANTS), vo1 3172, p 119–130. Berlin, Germany
14. Dorigo M, Gambardella LM, Middendorf M, Statzle T (2002) Special section on ant colony optimization. IEEE Trans Evol Comput 6(4):317–365
15. Isaacs JC, Watkins RK, Foo SY (2002) Evolving ant colony systems in hardware for random number generation. In: Proceedings of the 2002 congress evolutionary computation, p 1450–1455

Calculating Vehicle-to-Vehicle Distance Based on License Plate Detection

Yinan Liu, Yangzhou Chen, Jianqiang Ren and Le Xin

Abstract Vehicle-to-vehicle distance calculation has a great significance to driving assistance and estimation of traffic condition. In this paper, we present an on-board video-based method about calculating distance gap. The method is mainly divided into three major stages. At first stage, an Adaboost cascade classifier using Haar-like features of sample pictures is used to detect preceding vehicles. At second stage, a fusion algorithm combining Maximally Stable Extremal Regions (MSER) for far vehicles with vertical texture method for close vehicles is applied to locate license plate. At the third stage, distance gap is calculated according to the pixel height of plate and the proportion of plate pixel height. Experimental results in this paper showed excellent performance of the method in calculating distance gap.

Keywords License plate detection · Distance gap calculation · Adaboost · MSER

1 Introduction

On-board video-based vehicle detection and distance gap calculation have potential applications in two aspects at least. Firstly, the distance gap obtained by the on-board video, which is defined as the distance from top of front window to preceding vehicle's trail in this paper, can be used for driver-assistance [1]. On the other hand, it can also

Y. Liu · Y. Chen (✉) · J. Ren · L. Xin
Beijing Key Laboratory of Transportation Engineering,
College of Metropolitan Transportation, Beijing University of Technology,
Beijing 100 Pingleyuan Chaoyang District, Beijing 100124, China
e-mail: yzchen@bjut.edu.cn; 283320845@qq.com

Y. Liu
e-mail: LIUYinan@emails.bjut.edu.cn

J. Ren
e-mail: renjianqiang@emails.bjut.edu.cn

L. Xin
e-mail: xinle@bjut.edu.cn

© Springer International Publishing Switzerland 2017

V.E. Balas et al. (eds.), *Information Technology and Intelligent Transportation Systems*, Advances in Intelligent Systems and Computing 454,
DOI 10.1007/978-3-319-38789-5_57

be applied in on-board data collection system, for example, floating vehicle which is equipped with GPS, wireless communication unit and distance gap calculation devices not only can obtain its position coordinates and running speed [2] but also measure distance gap from many preceding vehicles and send the information to traffic surveillance center. Thus the center gathers all the information from a large number of floating vehicles to analyze traffic density and predict traffic status [3].

Various methods which are utilized to detect vehicles and calculate distance gap have been developed such as radar [4], laser [5], infrared [6] or computer vision [7]. Compared to other methods, computer vision has at least two advantages, on the one hand, the cost of camera is lower; on the other hand, it can obtain more information such as pedestrian, bicycle and license plate number. Owing to gradually improved performance of camera and the advantages above, computer vision system finds wide use in intelligent traffic system. The method for vision-based distance measurement includes monocular vision-based distance measurement and binocular vision-based distance measurement. Specifically, Binocular vision needs to use binocular camera and find correspondences of both images. Although binocular vision approach is more accurate than monocular vision to calculate the distance of objects, it is not able to meet the requirement of real-time due to the high algorithm complexity. Thus, in this paper, we propose an on-board monocular vision approach to detect preceding vehicles and calculate distance gap.

The existing literature about monocular vision methods to calculate distance gap can be divided to two categories as follows. The first kind of method is based on trigonometric function. This kind of method needs to detect an angle between camera's vertical line and line connecting camera with preceding vehicle's lamp [8] or bottom [9]. These methods require all the vehicles being in same horizontal plane and utilize interior parameter of camera to calculate the angle. However, the angle is small in general, and it is easy to produce error in detection process by computer vision. The second kind of method is based on proportionality of similar triangles. This kind of method utilizes a length or area ratio between the pixel distance of distance gap and real distance of distance gap [10], pixel distance of lamps and real distance of lamps [11], or pixel width of license plate and real width of license plate [7] to calculate distance gap. Compared to the methods based on trigonometric function, these methods have a better computational accuracy because it is convenient to measure a real object and precise to detect a real object by computer vision. In spite of some improvement of these methods have made, most of them only aim at the calculation in a small detection range and cannot adapt complex traffic scene. Because traffic surveillance center need to acquire large-scale information to analyze traffic condition, the methods will need to calculate vehicles' distance gap as much as possible. Faced with problems above, in this paper, we present a real-time on-board video-based system to detect preceding vehicles, locate license plate and calculate distance gap. Different from other method, the method, proposed in this paper, reduces the image resolution, detects vehicles by using Adaboost and records vehicle's coordinate rather than setting interest-of-region (ROI), then uses a fusion method combining MSER with vertical texture to locate license plate in the vehicles image and acquires the pixel height of license plate to calculate distance gap.

Experimental results showed that the method in this paper has an excellent real-time performance and it can detect vehicles and calculate distance gap accurately in the range of thirty meters.

The rest of this paper is organized as follows. Section 2 details the each part of algorithm in our approach including vehicle detection, plate license location based on vertical texture, plate license location based on MSER, a fusion location method and distance gap calculation method. An experimental study is reported in Sect. 3 that reveals the advantages of proposed method compared to other methods. A conclusion to this paper is given in Sect. 4.

2 Methodology

The proposed distance gap calculation method in this paper can be split into three steps. Firstly, a monocular camera installed on an observation vehicle' s top of front window is utilized to monitor preceding target vehicles and an Adaboost classifier is used for detecting preceding vehicles. In next step, the license plates of the detected preceding vehicles can be located based on a fusion method which combined the MSER with vertical texture algorithm. Thirdly, the distance between the observation and the preceding vehicles, which is called the distance gap, can be calculated according to the height of license plate in the image and the proportion of plate pixel height.

2.1 Detecting Preceding Vehicles

The key points in vehicles detection are accuracy and real-time performance. Due to the traffic congestions and frequent vehicle lane changes, it is hard to use lane lines or road edges to set certain ROI to detect vehicles. Thus, the vehicles detection methods by detecting symmetrical points, vehicle's shadows, and corner points may be not appropriate for on-board video. Therefore, the machine learning approach is employed to complete the detection of targeted vehicles directly. Adaboost algorithm [12] has so high performances on accuracy and real-time performance that it is suitable for changeable traffic scene.

As an iterative algorithm, Adaboost can gather weak classifiers which made up by some weak classification features into a strong classifier. The Haar-like feature presented by Viola and Lienhart [13, 14] is a kind of weak classification feature which can be used as the input of Adaboost.

In the paper, we chose 1000 images with vehicles and 2000 images without vehicles as training samples which is showed in Fig. 1, extracting Haar-like features, generating an 18-layer cascade Adaboost classifier.

In the process of detection, in order to ensure the accuracy of locating license plate in next part, we use 1920 $\times$ 1080 high resolution video to detect preceding vehicles.

Fig. 1 Training samples of adaboost

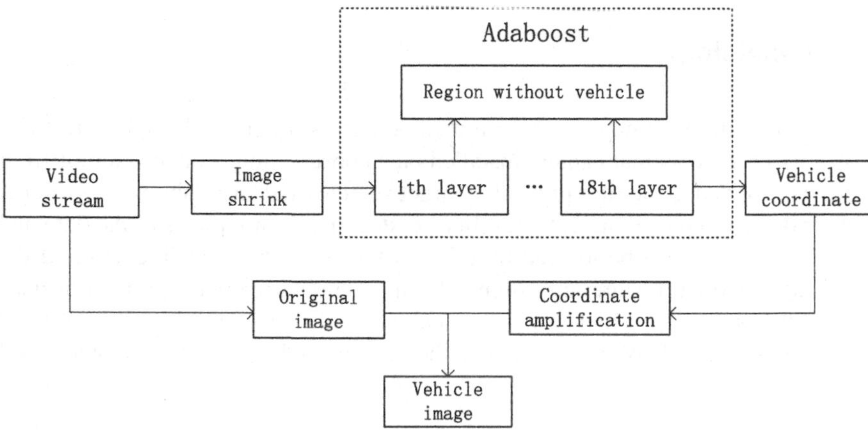

Fig. 2 Adaboost algorithm flow of vehicle detection

So as to reduce the time-consumption of detecting preceding vehicles under the high-resolution video, the system detects preceding vehicles after copying every frame of video and reducing the its size to 576×324. Afterwards, the cascade Adaboost classifier filters out the region without vehicle layer by layer, detects vehicles and records their coordinates Finally the system amplifies vehicles' coordinates and extracts the vehicles images in original image. In Fig. 2 the flow of detecting vehicles is presented.

2.2 Locating License Plate

There have been lots of license plate location methods based on the characters such as texture, fixed size and fixed width-height ratio [15], contours [16], color [17]. However, most of these methods aim at the location in a small detection range and cannot adapt complex traffic scene. The proposed location method focuses on adapting to different scales of vehicles image and locating license plates as much as possible.

In reality, the distance between the different preceding vehicles and the camera may vary, which leads to an apparent differences of the vehicle images' resolution. However, it is not efficient to locate license plates by using a single method. Given the image resolution of detected vehicles, the author will introduce two methods respectively in this paper, and in turn integrate them into a fusion method in order to locate license plates precisely.

License Plate Locating Method Based on Vertical Texture of License Plate

The region of license plate has a rich and inerratic texture in the vertical direction because of regular distribution of characters which is different from other vehicle edge information. Generally, with high-resolution images, the plates' text information can be easily distinguished. Hence, when the vehicles to be detected are closed to the camera, plate images will have rich vertical texture which can be extracted to locate the license plate.

Specifically, we convert the vehicles' color images to gray images. Then, a 3 × 3 window is used for medium filter in the gray image. The pixel value is replaced by medium value in the window to eliminate the noise. In next step, we adopt Sobel operator to calculate the vertical texture.

$$I_s = \begin{bmatrix} -1 & 0 & 1 \\ -2 & 0 & 2 \\ -1 & 0 & 1 \end{bmatrix} * I_m \tag{1}$$

I_s is the plate image processed by Sobel operator, I_m is the plate image processed by median filter and the 3 × 3 window is Sobel operator. The pixels in plate regions execute a convolution with the Sobel operator to eliminate the texture in the horizontal direction and reserve the texture in vertical direction. Because of the influences of illumination, the range of pixel gray-scale value in texture image maybe narrow, it is hard to make binary-conversion. Hence, we project the pixel value of I_s to the gray interval ranged from 0 to 255 and obtain the image I_o which has clear vertical texture by using the following formula.

$$I_o = (I_s(i, j) - \min(I_s)) \cdot 255/(\max(I_s) - \min(I_s)) \tag{2}$$

In formula (2), min is a function to acquire minimum gray-scale value of pixel and max can acquire maximum gray-scale value. Then we inverted the pixel value and generated the image with white connected components.

The brightness of plate images may vary due to the effect of the lighting environment, so a fixed threshold cannot make binary-conversion adaptively to all plate images. In this paper, the adaptive local binary method is chose. The local binary method means that an image is divided into m × n blocks, and then, each block is processed with the binary method Suppose that $I_o(x, y)$ denotes a gray value of point (x, y). Consider a block whose center is a point $(x, y)$and size is $(2w + 1) \times (2w + 1)$. The threshold $T(x, y)$ of $I_o(x, y)$ is computed by

$$T(x, y) = \frac{\max\limits_{-w \le k,l \le w} f(x+l, y+l) + \min\limits_{-w \le k,l \le w} f(x+l, y+l)}{2} \tag{3}$$

Then, the binary image is obtained by

$$I_b(x, y) = \begin{cases} 0, & \text{if } f(x, y) < T(x, y) \\ 255, & \text{else} \end{cases} \tag{4}$$

In formula (4), $I_b(x, y)$ is the image after binarization. Since the local binary method can choose threshold adaptively in every block, it can avoid the effects of patial shadow or uneven illumination to binaryzation.

After a series of operations above, license plate images with a little marge of horizontal texture, disconnected texture of characters, and contour of vehicle were acquired. In the next step, the binary image is processed by morphology to eliminate the texture which is needless and to connect the texture of characters. Specifically, since vertical texture method apply to the vehicle images with high resolution; we use morphological close operation by a fixed 5×1 kernel and open operation by a fixed 5×5 kernel to process the image Afterwards, the narrow gaps and elongated blanks are bridged, and the small holes and cracks are filled up. Therefore, we gain an image with candidate region. Finally, we identify the rectangle contours of white region and select the license plate region based on the conditions of a range of aspect ratio, area, location. By choosing abundant samples including the vehicles with a range of angle of view and conducting experiments to locate license plates the by using vertical texture method, we can draw a conclusion that this method is appropriate for locating license plates which is relatively close to camera. One experiment result is portrayed in Fig. 3.

The method of locating license plate based on the vertical texture has a high efficient performance with high-resolution images. For example, when a preceding vehicle is close to the camera, the texture of license plate's text and edge is clear enough on the image which can be used for locating the license plate. However, when the preceding vehicle is far away from the camera, the plate texture of vehicle image is indistinct, so in this case we need a another way to locate license plate.

License Plate Locating Method Based on Maximally Stable Extremal Regions (MSER)

The MSER algorithm presented by J. Matas [18] aims at segmenting the stable region. The algorithmic process is as follows, a series of thresholds t_n are set and used for binaryzation. For each threshold, a number of black regions $Q_1 Q_2 \ldots Q_k$ can be extracted from the images, which are called extremal regions. If the area of each extremal region maintains stability in a wide range of thresholds it is called MSER.

$$q(i) = \frac{|Q_{i+\Delta} - Q_{i-\Delta}|}{|Q_i|} \tag{5}$$

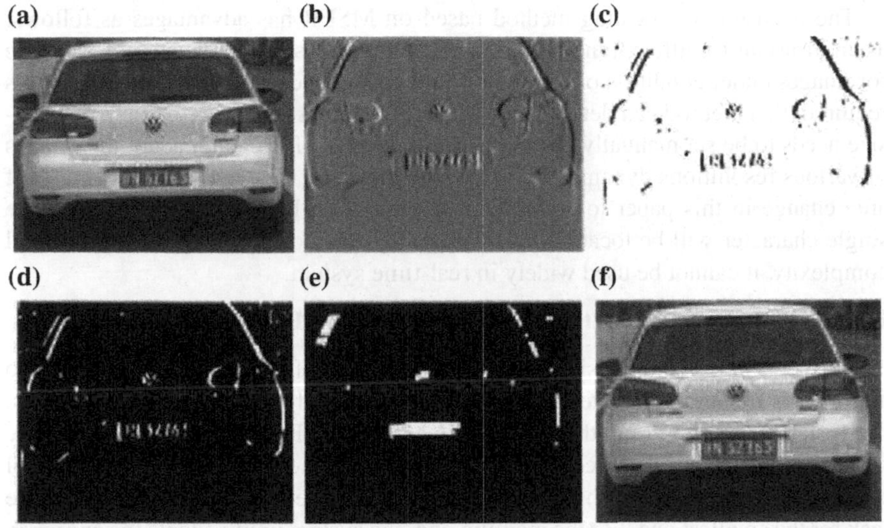

Fig. 3 A Sequence of License Plate Identification by Sobel Method **a** original image **b** image after *gray* processing and sobel operating **c** binary image **d** image after inverting the pixel value **e** image after morphological processing **f** locating license plate by filter condition

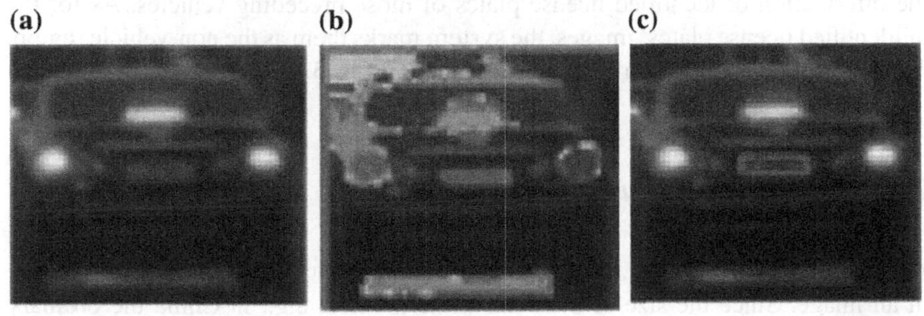

Fig. 4 A Sequence of License Plate Identification by Maximally Stable Extremal Regions (MSER) Method **a** original image **b** image processed by MSER **c** locating license plate by filter condition

In formula (5), Q_i is a connected region when threshold is i, Δ stands for the tiny change of threshold and $q(i)$ is the changing percentage of when threshold is i. When $q(i)$ is local minimum, Q_i is a MSER.

In this paper, the MSER algorithm is used for locating the license plate by detecting the low-resolution vehicle images. In order to endure the high locating rate, we set the step-size as 1, and the threshold of percentage of area change as 0.2. Afterwards, it is crucial to find the rectangle contours of connected components, followed by selecting the region which matches the condition of a range of aspect ratio, area and coordinate. In Fig. 4, a sequence of successful cases of license plate locating by using method with MSER is presented.

The license plate locating method based on MSER has advantages as follows: as an invariant for affine transformation of image gray scale, the method is suitable for images under condition of different illumination; because step-size can be set as required, this method can detect the regions of different fine degree. Since the step-size needs to be set manually, the method based on MSER cannot adapt the images of various resolutions dynamically. If we use the same step-size and percentage of area change in this paper to locate license plate with high-resolution images, the single character will be located. Meanwhile, this method is of high computational complexity, it cannot be used widely in real-time system.

License Plate Locating Method Based on a Fusion Method

This paper will provide a fusion algorithm for license plate locating combining two approaches mentioned above. According to the size of preceding vehicle images, the system is able to group them into three categories. For the largest category, the system could locate their license plates through the method based on the vertical texture of license plates, while it could use MSER method to complete the same task for the smallest one. For the medium one, the system firstly adopts the approach analyzing license plates' vertical texture, and it would stop when the license plates are identified. Otherwise, the MSER approach would be used automatically to conduct the same task until license plates are identified. After that the system records all the information of identified license plates of those preceding vehicles. As for the unidentified license plates' images, the system marks them as the non-vehicle region. The relevant details of algorithm are illustrated in Fig. 5.

2.3 Calculating Distance Gap

After locating license plate, we can obtain a height and width of the license plate in an image. Since the size of a license plate is fixed, e.g., in China the ordinary vehicle rear license plate is 440 mm in width and 140 mm in height, it would be easy to calculate the distance gap according to the pixel height of a license plate and the proportion between pixel height and distance gap. When people drive, turning or lane changing would happen frequently. In these situations, the plane where a license plate stays is not perpendicular to the center axis of camera. Specifically, the width of a license plate's image changes due to angle change of the preceding vehicle, while the height is constant. Hence, we can calculate proportion of plate pixel height and distance gap.

In Fig. 6, the origin of coordinate stands for the camera, whose left side is the plane where license plate stays and the right side is the plane where the images locate. More over the horizontal axis passes through the image plane and license plate plane vertically. In the coordinate system, d is the distance between the camera and the plane of license plate, f is virtual focal length of image's plane, a, b and a', b' are the line connecting between a vertex of license plate and a vertex of license plate

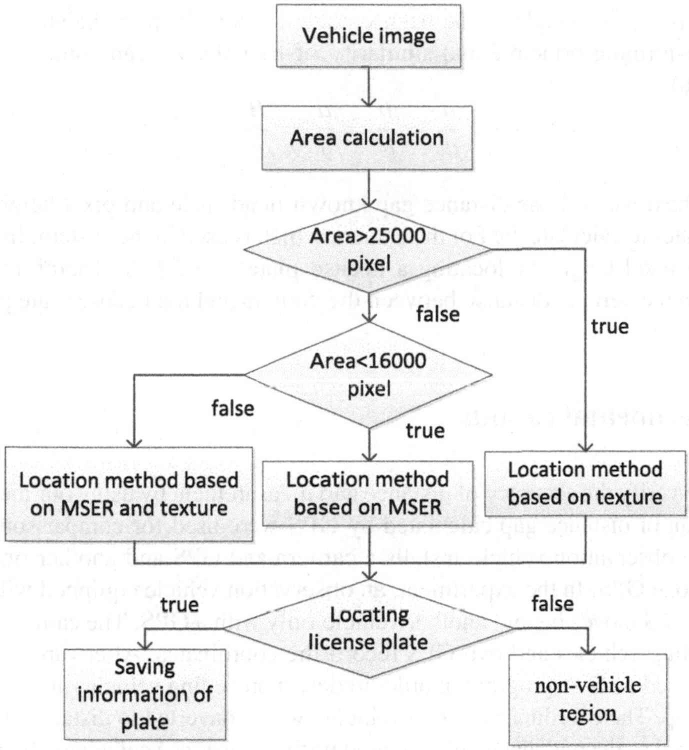

Fig. 5 Algorithm flow of fusion location method

Fig. 6 Camera geometric relationship

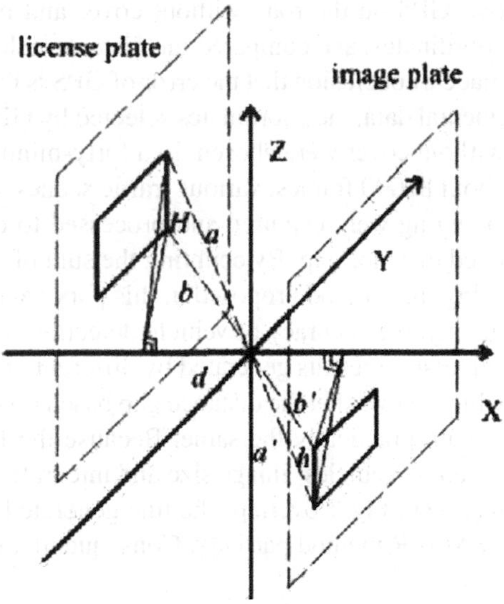

image, H is the real height of the license plate and h is the pixel height. According the image-forming principle and similarity of triangle, we can come up with the formula (6):

$$\frac{a}{a'} = \frac{b}{b'} = \frac{d}{f} = \frac{H}{h} \tag{6}$$

In the next we will use distance gap known in advance and pixel heights of the license plates to calculate the f of the camera which is used in the system. In practice, we obtain pixel height by locating a license plate $d = Ff/h$. Therefore, we can easily gain the vertical distance between the camera and the license plate plane.

3 Experimental Results

In order to verify the accuracy of distance gap measurement by using our method, the information of distance gap calculated by GPS were used for comparison. Specifically, one observation vehicle installs a camera and GPS and another one is only equipped one GPS. In the experiment, an observation vehicle equipped with a camera and a GPS moves behind another vehicle only with a GPS. The camera captures the preceding vehicles and two GPS record the coordinates. Afterwards, the video was processed by C++ program in order to detect preceding vehicles and calculating distance gap. The coordinates of two vehicles were converted to distance gap which is considered as the real distance compared with the distance calculated by program.

Before performing the experiment of the method proposed in this paper, we verified the accuracy of GPS. The researchers selected abundant coordinates by using two GPS on the road without cover and measured the real distance between two coordinates, and compared the distance calculated by GPS coordinates by trial, then made a conclusion that the error of GPS is less than 1 meter. In the selection of experimental data, the coordinates selected by GPS and the video shot by camera on roads without cover were chosen. In a forty-minute vehicle-mounted video which include about 80000 frames, various traffic scenes such as congestion, intersection and lane changing were captured and processed to detect the preceding vehicles and calculated distance gap. By counting the sum of vehicles, the sum of vehicles detected by using the method proposed in this paper and the sum of error vehicles detected artificially, the accuracy of vehicle detection reaches 96 %. As for the distance gap, the calculation results generated by different vision algorithms were compared in Fig. 7, which shows that the distance gap based on GPS and the one based on fusion method remain practically the same. Because the fusion method proposed in this paper is based on vehicles' image size and integrated with Sobel method and MSER method, its line in Fig. 7 overlaps the line generated by Sobel method and the line generated by MSER method partially. Consequently, it can be verified from Fig. 7 that when a

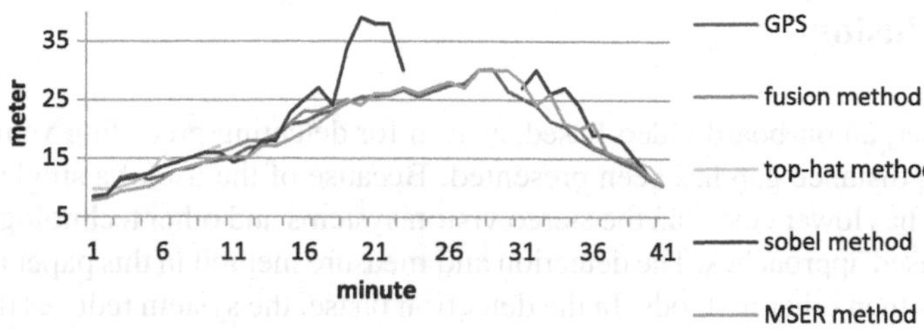

Fig. 7 Comparison of different methods for distance gap calculation

Table 1 Dimensions of scope, error, and running speed of corresponding program detected by GPS and vision method

	Fusion	Sobel	MSER	Top-hat
Scope (m)	0–30	0–22	0–30	0–18
Error (%)	4.3	8.5	5.4	5.5
Frame rate (fps)	30	30	14	30

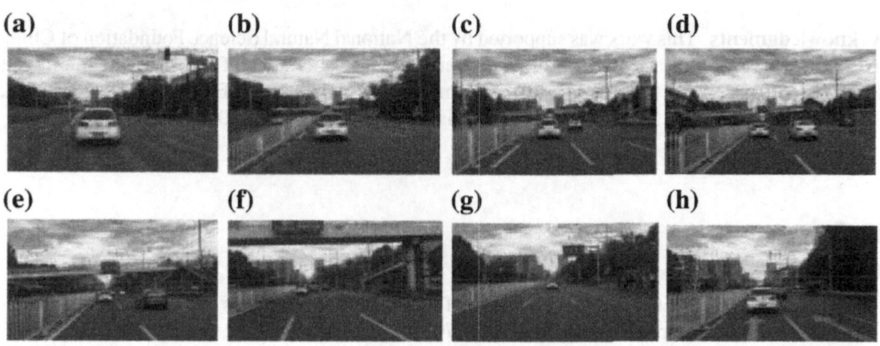

Fig. 8 Computational result of fusion method in a period, **(a)–(f)** were taken in every 5 seconds

preceding vehicle is far away from the camera, the texture of vehicle plate's image is indistinct. In such condition, the Sobel method for distance gap calculation is unstable, and the top-hat method from [7] also has a problem in small detection scope for the same reason. Although the MSER method has a good performance on calculating distance, it is time-consuming compared with other methods. In Table 1, the dimensions of scope, error, and running speed of corresponding program are presented, and the error of vision method is calculated by the absolute difference of distance detected by GPS and vison method divide distance detected by GPS, illustrating that the fusion method proposed in this paper has an excellent performance than other method. Figure 8 describes the experimental result in different timing.

4 Conclusion

In this paper, an on-board video-based system for detecting preceding vehicles and calculating distance gap has been presented. Because of the use of a single camera, the system has lower cost than the stereo vision systems and other technologies, such as radar-based approaches. The detection and measure method in this paper also have advantages than other methods. In the detection phase, the system reduces the image resolution, detects vehicles by using Adaboost and records vehicle's coordinate rather than setting ROI, eliminating errors and reducing time complexity of algorithm. Meanwhile, the system can also detect the vehicles as many as possible in a wider region because of using MSER method, which has a great significance for driving assistance and vehicles information collection. In the step of calculating distance gap, this paper proposed a fusion method which can use different ways to measure distance gap precisely based on a vehicle's image size, and take full advantages of the texture in region of a license plate. Given the excellent robustness and reliability in vehicle detection and high accuracy in distance gap measurement, the system carried out by in-vehicle tests can be used in city traffic.

Acknowledgments This work was supported by the National Natural Science Foundation of China [61273006, 61573030], and Open Project of Beijing Key Laboratory of Urban Road Intelligent Traffic Control [XN070], Scientific Research Foundation of Hebei Higher Education [QN2015209], Science and Technology Support Foundation of Hebei Province [13210807].

References

1. Cherng S (2009) Critical motion detection of nearby moving vehicles in a vision-based driver-assistance system. Intell Transp Syst 10(1):70–82
2. Vandenberghe W (2012) Feasibility of expanding traffic monitoring systems with floating vehicle data technology. IET Intell Transp Syst 6(4):347–354
3. Díaz V, Juan J (2012) Extended floating car data system: experimental results and application for a hybrid route level of service. Intell Transp Syst 13(1):25–35
4. Xun D (2007) Vehicle centroid estimation based on radar multiple detections. Veh Electron Saf, p 1–5
5. Judith S, Velten J, Glasmachers A (2010) Laser phase shift distance meter for vision based driver assistance systems. Imag Syst Tech (IST), p 220–224
6. Lu P-W, Chen R (2014) Infrared-based vehicular positioning with the automatic radiation-strength control. Intell Transp Syst 8(3):273–285
7. Ibarra AM (2014) Monovision-based vehicle detection, distance and relative speed measurement in urban traffic. IET Intell Transp Syst 8(8):655–664
8. Alcantarilla PF, Bergasa LM, Jiménez P (2008) Night time vehicle detection for driving assistance lightbeam controller. Intell Veh Symp, p 291–296
9. Chu J (2004) Study on method of detecting preceding vehicle based on monocular camera. Intell Veh Symp, p 750–755
10. Wu C-F, Lin C-J, Lee C-Y (2012) Applying a functional neurofuzzy network to real-time lane detection and front-vehicle distance measurement Systems. Man Cybern Part C: Appl Rev 42(4):577–589

11. Wang W-Y (2007) Nighttime vehicle distance measuring systems. Circuit Syst II: Express Br 54(1):81–85
12. Freund Y, Schapire RE (1997) A decision-theoretic generalization of on-line learning and an application to boosting[J]. J Comput Syst Sci 55(1):119–139
13. Viola P, Jones M (2001) Rapid object detection using a boosted cascade of simple features. Comput Vis Pattern Recognit Int Conf 1:511–518
14. Lienhart Rainer, Maydt Jochen (2002) An extended set of haar-like features for rapid object detection. Image Process Int Conf 1:900–903
15. Muhammad S, Ahmed MJ, Ghazi SA (2003) Saudi Arabian license plate recognition system. Gom Model Graph Int Conf, p 36–41
16. Tran Duc D, Duc DA, Hong Du TL (2004) Combining hough transform and contour algorithm for detecting vehicles' license-plates Intelligent Multimedia. In: International symposium video and speech processing, p 747–750
17. Haijiao W (2007) Color prior knowledge-based license plate location algorithm. In: Second workshop digital media and its application in museum & heritages, p 47–52
18. Matas J (2004) Robust wide-baseline stereo from maximally stable extremal regions. Image Vis Comput 22(10):761–767

The Improved Weighted Evolution Model of the AS-Level Internet Topology

Xuelian Cai

Abstract By studying the evolution of the Internet behavior, this brief paper introduces a weighted evolution network model based on the Internet Autonomous System (AS) layer. This model considers four kinds of Internet evolutionary behavior: on one hand the AS nodes birth and the links growth based on a preferential attachment rule and on the other hand the AS nodes recession and the links death based on an anti-preferential attachment rule. Through a theoretical analysis with the rate equation method and a simulation experiments on MATLAB environment, the node's degree and strength distribution both follows a power-law distribution. The power-law coefficient is proportional to the ratio of the preferential attachment tendency and anti-preferential attachment tendency. The experimental results were basically consistent with the theoretical analysis value, confirming the correctness of the model.

Keywords Weighted network · Internet · Evolution model · Preferential attachment · Anti-preferential attachment

1 Introduction

The Internet is a large heterogeneous dynamic development of a complex network. The awareness of Internet network topology has experienced random, hierarchy, power-law three stages. In 1999, Faloutsos brothers work [1] reveals the Internet network node distribution follows a power law distribution $P(k) \sim k^{-r}$. Internet network topology produced by the random and hierarchical rule does not have such characteristics. Further later researches [2, 3] showed that the Internet network topology model which is generated on the basis of a power-law rule can reflect more the Internet topological features than a topology model using a random or a hierarchical principle.

X. Cai (✉)
Guangdong Polytechnic of Industry and Commerce, Guangzhou, China
e-mail: xuelian@foxmail.com

© Springer International Publishing Switzerland 2017
V.E. Balas et al. (eds.), *Information Technology and Intelligent Transportation Systems*, Advances in Intelligent Systems and Computing 454,
DOI 10.1007/978-3-319-38789-5_58

Internet power-law topology model is divided into two kinds of static and dynamic models. By using some parameters of specific power-law such as PLOD [4] and PLRG [5], the static model does not consider the growth of the network directly to describe the static topology's properties. The dynamic model explores the internal mechanism of the evolution of the topology growth and reconstructs the topology of the growth process. Compared with the static model, the dynamic model can reflect better the dynamic growth process and the topological characteristics of Internet than the static model. In 1999, Barabasi and Albert give the dynamic model of the network, BA [6] (Barabasi–Albert) model, for the first time. The power-law network models characteristics are attributed to two kinds of evolution mechanisms: growth and preferential attachment. New nodes select the higher degree nodes to connect to them. In 2004, Barrat, Barthelemy and Vespignani proposed a weighted dynamic model called BBV network model [7]. The model studies the dynamic evolution mechanism of the topology and the weight, proposes the new node to join the old nodes based on the preferential attachment mechanism of strength. Subsequent network evolution models of Internet [8–11] were mostly improved on the basis of the BA and BBV models. Reference [8] details a model in which the number of edges between a newly added node and old nodes is allowed to change each time the newly added node enters the network. PFP model [9] simulats the growth of the new nodes and the new internal links combined with the nonlinear preference attachment mechanism. Referring to literature, reference [10] considers the distance between Internet nodes, puts forward a preferential attachment mechanism for which a new node to join an old node is based on the strength of the old and the distance between the two nodes. Reference [11] proposed an evolution network model of Internet based on AS layer with a description of nodes and edges birth and death to characterize the evolution behaviors. Even if most of these already existing models improved model by using some of the characteristics of the evolution behavior of Internet, they rarely consider the node recession rather than the phenomenon of death. In fact, the flow or the capacity limits of some nodes can suppress the priority connectivity to a certain extent, thus the strength will be weakened over time. In order to simulate the real evolution process of the real Internet network, we proposed an improved evolution model. The model does not only consider the growth of nodes and links under a preferential attachment rule, but also considers the decrease of the node strength and the existing links death under an anti-preferential attachment rule.

The second part of the paper introduces the evolution mechanism and the theoretical analysis of the model. The third part studies the node's degree and the strength distribution under different parameters through simulation experiments.

2 Improved Internet Evolution Model

Based on the Euler's theory of graphics, the Internet can form a network on two kinds of meaning: a routing layer and an AS layer. Depending on the topology data acquisition, the approach is different. The AS layer topology data can be directly derived from the Border Gateway Protocol (BGP) routing monitoring system. Data set is relatively complete, and as subsequently mentioned, the Internet is described from the AS layer.

2.1 Network Evolution Model

This paper only considers undirected graphs where the weight of each link is symmetric, i.e. $\omega_{ij} = \omega_{ji}$, where ω_{ij} denotes the weight of a link from the ith node to the jth node. The node strength s_i is defined as $s_i = \sum_{j \in \nu(i)} \omega_{ij}$, where the argument ν (i) for the node i represents its neighbor nodes.

We start with a small number $m_0 (m_0 > 1)$ of fully connected nodes. All links have the same weight for ω_0. The strengh of each node is then $(m_0 - 1)\,\omega_0$. At each step t, one of the following four evolutions is to be performed as random:

(1) With a probability P_1, a new AS node k is added to m_1 old AS nodes, the weight of each new link (k, i) are also ω_0. The node k selects the existing node i to connect to it according to the strength preferential attachment formulas (1).

$$\text{Preferential attachment formula: } \prod (s_i) = \frac{S_i}{\sum S_j} \tag{1}$$

When a new node k will connect an existing node i, this will introduce locally rearrangements of weights between i and its neighbor $j \in V(i)$ according to the simple rule $\omega_{ij}(t) = \omega_{ij}(t-1) + \delta \frac{\omega_{ij}(t-1)}{s_j(t-1)}$, where δ $(\delta > 0)$ is an adjustable parameter. This rule is yielding changes in the strength of node i, $s_i(t) = s_i(t-1) + \delta + \omega_0$.

(2) With a probability P_2, m_2 links is added, the weight of each new link (k, i) is ω_0. Each Link randomly selects a node i on one end, and selects the connection node j according to the formula (1) on the other side. When adding a new edge between i and j nodes, it will cause the strength variation of the network node i and the node j, and the locally adjustment rules in accordance with step (1).

(3) With a probability P_3, m_3 link is deleted. A link arbitrarily selects a node i on one end, and select the node j on the other end side according to the anti-preferential attachment formula (2).

$$\text{Anti-preferential attachment formula: } \prod{}' (s_j) = 1 - \prod s_j \tag{2}$$

(4) With a probability P_4, a node j is selected. The node j is selected according to the anti-preferential attachment formula (2). If the node j is selected, the weight ω_{jk} for all $k \in V(j)$ is decreasing according to the rule $\omega_{jk}(t) = (1-r)\,\omega_{jk}(t-1)$. Therefore, the strength of the node j $s_j(t) = (1-r)\,s_j(1-t)$ $0 \leqslant r \leqslant 1$. When $r = 0$, the node does not have any strength weakened. When $r = 1$, the node strength is reduced to 0, which is equivalent to the node death.

2.2 Theoretical Analysis of the Network Model

We will analysis the node strength distribution with the rate equation method. Assuming $s_i(t)$ is continuous, then there are $M = m_0 + p_1 t$ nodes and $\frac{m_0(m_0+1)}{2} + (p_1 m_1 + p_2 m_2 - p_3 m_3)\,t$ links at time step t in the network. The corresponding rate equation of $s_i(t)$ will change by the four kinds of mechanisms at time step t.

(1) Adding a new AS with the probability P_1 and local rearrangements of strength.

$$\frac{ds_i(t)}{dt} = \sum_{j \in V(i)} P_1 m_1 \delta \frac{\omega_{ij}(t)}{s_j(t)} \prod s_j + P_1 m_1 (\omega_0 + \delta) \prod s_i = (2\delta + \omega_0) m_1 P_1 \prod s_i \tag{1}$$

(2) Adding m_2 links with the probability P_2 and local rearrangements of strength.

$$\frac{ds_i(t)}{dt} = P_2 m_2 \left(\prod s_i + \frac{1}{M} \right) (2\delta + \omega_0) \tag{2}$$

(3) Deleting m_3 links with the probability P_3 and local rearrangements of strength.

$$\frac{ds_i(t)}{dt} = -P_3 m_3 \left(1 - \prod s_i + \frac{1}{M} \right) \omega_0 \tag{3}$$

(4) Decreasing the strength of an AS node with the probability P_4.

$$\frac{ds_i(t)}{dt} = -P_4 r \left(1 - \prod s_i\right) \tag{4}$$

When $t \to \infty$, by combining Eqs. (1)–(4) together, one obtains

$$\begin{aligned}
\frac{ds_i(t)}{dt} &\approx (2\delta + \omega_0) m_1 P_1 \prod s_i + P_2 m_2 \left(\prod s_i + \frac{1}{M} \right) (2\delta + \omega_0) \\
&\quad - P_3 m_3 \left(1 - \prod s_i + \frac{1}{M} \right) \omega_0 - P_4 r \left(1 - \prod s_i \right) \\
&\approx (2\delta m_1 P_1 + \omega_0 m_1 P_1 + 2 P_2 m_2 \delta + P_2 m_2 \omega_0 + P_3 m_3 \omega_0 + P_4 r) \prod s_i \\
&\quad + \frac{2 P_2 m_2 \delta + P_2 m_2 \omega_0 - P_3 m_3 \omega_0}{M} - P_3 m_3 - P_4 r = A \prod s_i + \frac{B}{M}
\end{aligned} \tag{5}$$

In which:

$$\begin{aligned}
A &= 2\delta m_1 P_1 + \omega_0 m_1 P_1 + 2 P_2 m_2 \delta + P_2 m_2 \omega_0 + P_3 m_3 \omega_0 + P_4 r \\
B &= 2 P_2 m_2 \delta + P_2 m_2 \omega_0 - P_3 m_3 \omega_0 - (P_3 m_3 \omega_0 + P_4 r) M
\end{aligned}$$

Note: Time t, have $\frac{m_0(m_0+1)}{2} + (p_1 m_1 + p_2 m_2 - p_3 m_3)\,t$ links in the network,

Get:

$$\sum s_i \approx (2P_1m_1(\delta + \omega_0) + 2P_2m_2(\delta + \omega_0) - 2P_3m_3\omega_0 - P_4r)t$$

Let $D = 2P_1m_1(\omega_0 + \delta) + 2P_2m_2(\omega_0 + \delta) - 2P_3m_3\omega_0 - P_4r$
Equation (5) can be rewritten as

$$\frac{ds_i(t)}{dt} \approx A\frac{s_i}{Dt} + \frac{B}{m_0 + P_1t} \quad (6)$$

We use this approximation in question (6) when $t \to \infty$, Then Eq. (6) can be written then as

$$\frac{ds_i(t)}{dt} \approx \frac{A}{Dt}s_i + \frac{B}{P_1t} \quad (7)$$

Let $X = \frac{A}{D}$, $Y = \frac{B}{P_1}$, Then Eq. (7) can be written as

$$\frac{ds_i(t)}{dt} \approx \frac{X}{t}s_i + \frac{Y}{t} \quad (8)$$

Note that $s_i(t_i) = (m_0 - 1)\omega_0$, Eq. (8) can be solved

$$s_i(t) = \left(\frac{t}{t_i}\right)^X \left((m_0 - 1)\omega_0 + \frac{Y}{X}\right) - \frac{Y}{X} \quad (9)$$

Then we can obtain the strength distribution by using the mean-field method

$$\begin{aligned} p(s_i(t) < s) &= p\left[\left(\frac{t}{t_i}\right)^X \left((m_0 - 1)\omega_0 + \frac{Y}{X}\right) - \frac{Y}{X} < s\right] \\ &= p\left(t_i > t\left(\frac{(m_0 - 1)\omega_0 + \frac{Y}{X}}{s + \frac{Y}{X}}\right)^{\frac{1}{X}}\right) \\ &= 1 - p\left(t_i < t\left(\frac{(m_0 - 1)\omega_0 + \frac{Y}{X}}{s + \frac{Y}{X}}\right)^{\frac{1}{X}}\right) = 1 - \left(\frac{(m_0 - 1)\omega_0 + \frac{Y}{X}}{s + \frac{Y}{X}}\right)^{\frac{1}{X}} \end{aligned}$$

$$p(s, t) = \frac{\partial p(s_i(t) < s)}{\partial s} = \frac{\left((m_0 - 1)\omega_0 + \frac{Y}{X}\right)^{\frac{1}{X}}}{X\left(s + \frac{Y}{X}\right)^{\frac{1}{X}+1}} \quad (10)$$

As it can be seen from the Eq. (10), the strength distribution of the model is subject to a power law distribution, where the power law coefficient is $\gamma = \frac{1}{X} + 1$. In which,

$$X = \frac{A}{D} = \frac{2\delta m_1 P_1 + \omega_0 m_1 P_1 + 2P_2 m_2 \delta + P_2 m_2 \omega_0 + P_3 m_3 \omega_0 + P_4 r}{2P_1 m_1 (\delta + \omega_0) + 2P_2 m_2 (\delta + \omega_0) - 2P_3 m_3 \omega_0 - P_4 r}$$

Let $P_2 = 0$, $P_3 = 0$, $P_4 = 0$, $\delta = 0$, we can get $\gamma = 3$, at this point, the power law coefficient is just the same with BA network. According to the characters of power-law distribution of Internet complex networks, take $2 < \gamma < 3$, then $2 < \frac{1}{X} + 1 < 3$, the solution to $\frac{1}{2} < X < 1$. As long as X satisfies this constraint, the strength distribution of nodes will be in line with the power law characteristics.

3 Simulation Experiment

In order to prove the correctness of the model theoretical derivation, we set up a simulation program in MATLAB for simulating the evolution process of the model. In order to eliminate the randomness of the experimental results, the following data in each group is taken the average of 100 repeated experiments.

Figures 1, 2 and 3 illustrate respectively the degree distribution, the strength distribution and degree-strength correlation diagram of 1000, 2000, 3000 and 5000 nodes. The fixed parameters values are $m_0 = 8$, $\omega_0 = 2$, $\delta = 2$, $P_1 = 0.75$, $P_2 = 0.15$, $P_3 = 0.05$, $P_4 = 0.05$, $r = 0.05$.

As shown in Fig. 1, the model of the distribution of the power law coefficient does not change significantly over time and scale. The power-law coefficient is about 2.2 which is close to the real value for the AS distribution network with the observation in May 15, 2005 [12].

As it can be seen from Fig. 2, most of the strength distribution values follow a power-law distribution pattern. We can also see a heavy tail phenomenon. The

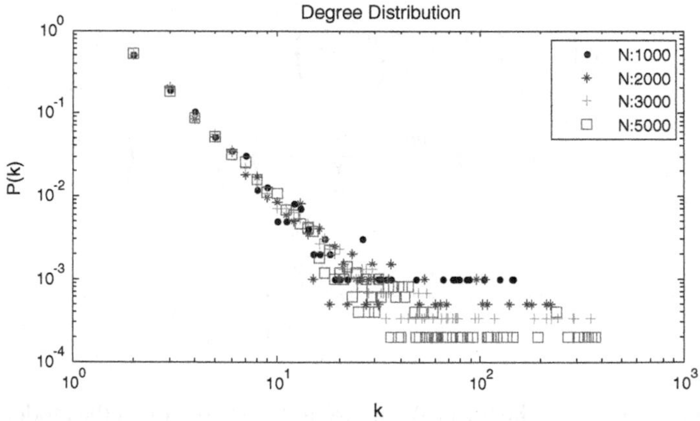

Fig. 1 Network degree distribution of 1000, 2000, 3000, 5000 nodes

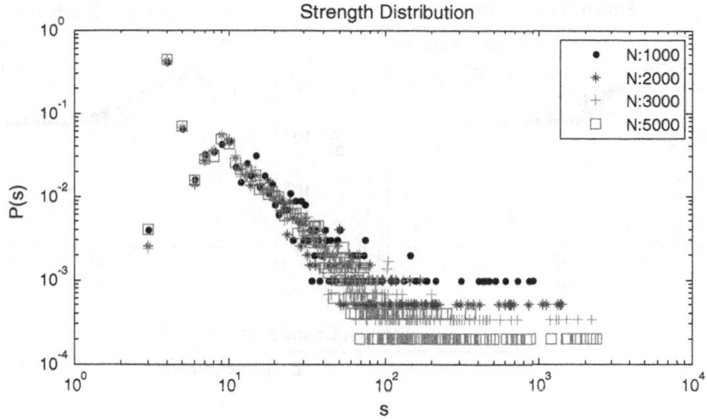

Fig. 2 Network strength distribution of 1000, 2000, 3000, 5000 nodes

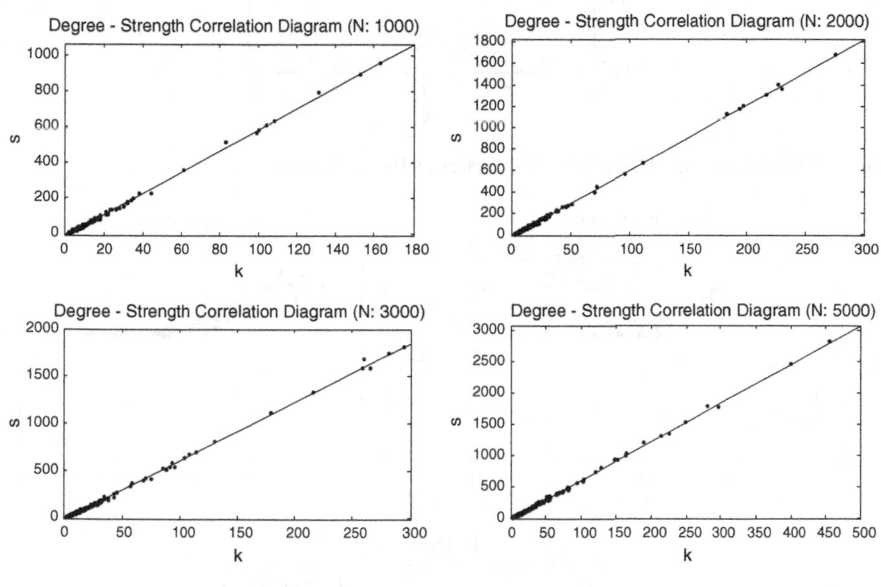

Fig. 3 Degree-strength correlation diagram of 1000, 2000, 3000, 5000 nodes

power-law coefficient value is approximately equal to 2.3. Thus the simulation value is substantially in line with the theoretical value.

Figure 4 plots the relationship between P_2/P_3 proportion and the power-law exponent. The parameters values are $m_0 = 8$, $\omega_0 = 2$, $\delta = 2$, $P_1 = 0.55$, $P_4 = 0.05$, $r = 0.05$. P_2 and P_3 were taken (0.05, 0.35), (0.20, 0.20), (0.35, 0.05), three different sets of values. The power-law coefficient γ is the theoretical value.

Figure 5 is dedicated to plot the relationship between P_2/P_4 proportion and the power-law exponent. This time, the parameters values are $m_0 = 8$, $\omega_0 = 2$, $\delta = 2$,

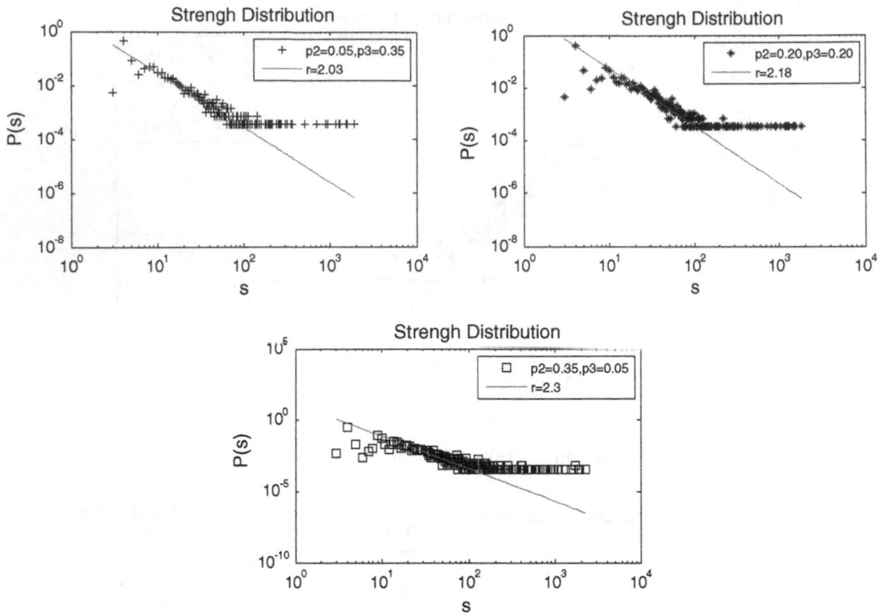

Fig. 4 The relationship of γ to P_2/P_3. γ increases as P_2/P_3 increases

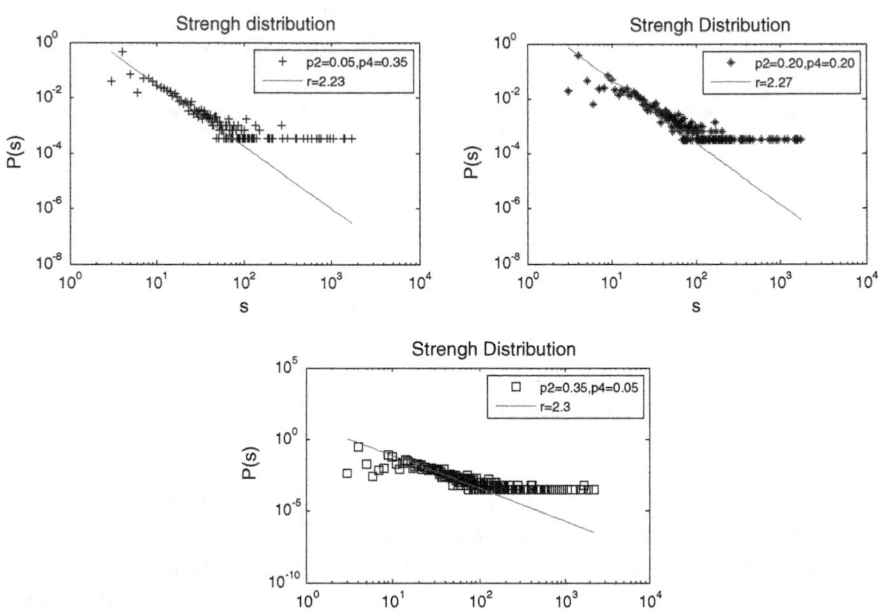

Fig. 5 The relationship of γ to P_2/P_4. γ increases as P_2/P_4 increases

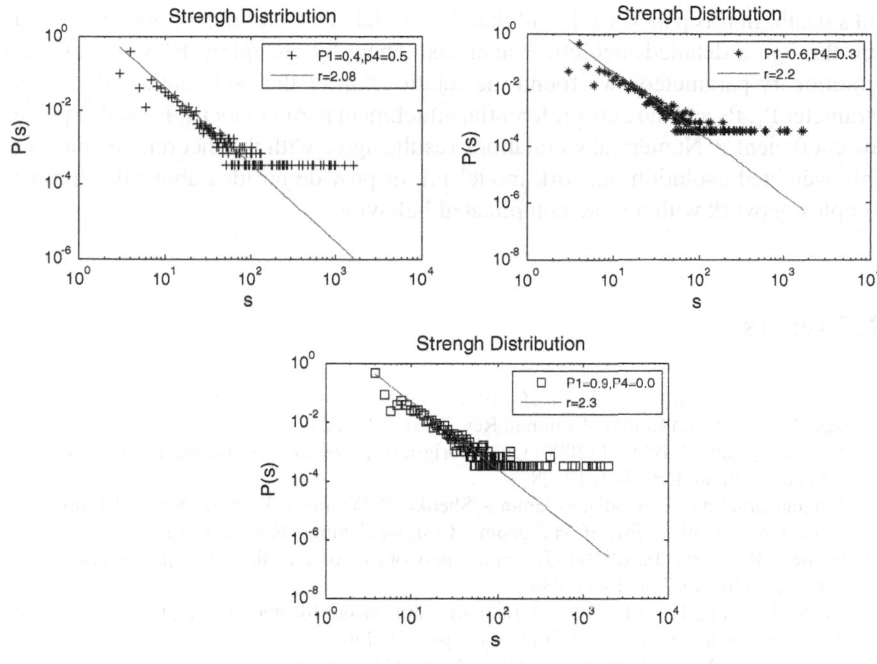

Fig. 6 The relationship of γ to P_1/P_4. γ increases as P_1/P_4 increases

$P_1 = 0.55, P_3 = 0.05, r = 0.05$. P_2 and P_4 were taken(0.05, 0.35), (0.20, 0.20), (0.35, 0.05), again three different sets of values.

Figure 6 plots the relationship between the last ratio P_1/P_4 and the power-law exponent. We taking parameters values as follows: $m_0 = 8$, $\omega_0 = 2$, $\delta = 2, P_2 = 0.55$, $P_3 = 0.05, r = 0.05$. P_1 and P_4 were taken(0.4, 0.5), (0.6, 0.3), (0.9, 0.0) as the two first cases, three different sets of values.

As it can be seen from the three graphs (Figs. 4, 5 and 6), the strength distribution of the experimental values is basically consistent with the theoretical values for the different parameters. The coefficient γ increases along with P_2/P_3, P_2/P_4, P_1/P_4 increase. This proves that the larger the proportion of preferential attachment trends and the anti preferential attachment trends is, the more uneven the network degree distribution becomes. In other words, "the rich are getting richer and the poor are getting poorer."

4 Conclusion

In order to better simulate the behavior of Internet evolution, we proposed an improved evolution model. The model simulated the Internet network's four evolutionary behavior: AS nodes birth, AS links growth, AS nodes recession and AS links death. In this paper, we found that the model had a significant power-law feature through a detailed theoretical analysis. Then, by changing the value of each evolutionary parameters, we found the relationship of the preferential attachment parameter P_1, P_2 and the anti-preferential attachment parameters P_3, P_4 with a power law coefficient γ. Numerical simulations results agree with the theoretical analyses. This weighted evolution network model might provide an idea about the Internet complex network with a more complicated behavior.

References

1. Faloutsos M, Faloutsos P, Faloutsos C (1999) On power-law relationships of the internet topology. Acm Sigcomm Comput Commun Rev 29(3):251–262
2. Medina A, Matta I, Byers J (2000) On the origin of power laws in internet topologies. ACM Comput Commun Rev 30(2):18–28
3. Tangmunarunkit H, Govindan R, Jamin S, Shenker S, Willinger W (2002) Network topologies, power laws, and hierarchy. ACM Sigcomm Comput Commun Rev 32(1):76–76
4. Plamer CR, Steffan JG (2000) Generating network topologies that obey power laws. IEEE Glob Telecommun Conf 1:434–438
5. Aiello W, Chung F, Lu L (2000) A random graph model for massive graphs. Thirty-second Acm Symposium on Theory of Computing, pp 171–180
6. Barabási AL, Albert R, Jeong H (1999) Mean-field theory for scale-free random networks. Phys A Stat Mech Its Appl 272(1):173–187
7. Alain B, Marc B, Alessandro V (2004) Weighted evolving networks: coupling topology and weighted dynamics. Phys Rev Lett 92(22):228701
8. Wang L, Dai GZ, Qin S (2006) A new evolution model of internet. Syst Eng Theor Pract, 102–106
9. Zhou S, Mondragon RJ (2005) A positive-feedback preference model of the as-level internet toplogy. IEEE Int Conf Commun 1:163–167
10. Lan L, Yang H (2009) Internet network model based on fluid and bandwidth. Control Decis Conf 4195–4200
11. Guo H, You-Jun BU, Lan JL , Liu LK (2011) Novel dynamic evolution model for Internet AS-level topology. J Syst Eng 26(2):152–159
12. Oliveira RV, Zhang B, Zhang L (2007) ACM 37(4):313–324

Toward a Green Transport System: A Review of Non-technical Methodologies for Developing Cities

Chen Zhenqi and Lu Weichi

Abstract In this paper we address the question: How should developing cities green their urban transport systems? As a new developing paradigm in the urban transport field, green transport has been advocated and fostered for over 20 years. It is regarded as an effective model to face a series of problems emerging in urban area and support sustainable developments. It desires much further development in the future. However even today most successful practices are found in developed cities and developing cities have performed it poorly. Although many technical solutions were testified useful in developed cities, they may be not so suitable for developing cities for their technology and capital intensive nature. We review non-technical solutions for developing cities, emphasizing on those that contribute to changing the composition of travel modes, i.e. increasing the share of walking, cycling and transit modes. Methodologies working toward a travel style in which walking, cycling and transit modes occupy dominant proportion are studied underlying four strategies: changing land use, giving priority to public transport, promoting walking and cycling and restricting car owning and using. We highlight their effects and limitations and assess their adaptabilities in developing cities. And in the end we summarize some matters need attention for applying the methodologies reviewed.

Keywords Green transport · Methodologies · Non-technical · Developing cities

---

C. Zhenqi (✉) · L. Weichi
Urban Planning Research Center, Yangzhou Urban Planing Bureau,
Yangzhou, Jiangsu Province, People's Republic of China
e-mail: achenery@163.com

L. Weichi
e-mail: 412238096@qq.com

© Springer International Publishing Switzerland 2017
V.E. Balas et al. (eds.), *Information Technology and Intelligent Transportation Systems*, Advances in Intelligent Systems and Computing 454,
DOI 10.1007/978-3-319-38789-5_59

1 Introduction

Green or greening introduced as an innovative term in the field of urban transport has been researched and practiced worldwide for more than 20 years. It was originally raised against the trend of excessive reliance on automobiles globally and with the intention to cope with a series of problems emerged in urban transport, such as congestion, accident, air pollution, significant proportion of energy consumption and inequity among travelers [1]. Although till today it is hard to give a clear definition of green transport, that it is a paradigm supporting sustainable developments and desires further growth and prosperity in the future has been universally agreed [2, 3]. Around the world, western developed cities are pioneers and sophisticated in developing green transport. They have much more successful practices than developing cities. Many developing cities probably are still on the starting point and need to make great efforts to keep up with developed cities. But how should developing cities develop green transport?

As made by and for developed cities, mainstream implementing methodologies focus on reducing transport energy use and greenhouse gas emissions, involved in improvements of energy efficiency and explorations of alternative fuels and power systems [4], which need great advances in technology and thus vast inputs of money. This may be quite difficult for most developing cities. Non-technical solutions somehow inexpensive should be explored.

From the view of the famous green transport hierarchy, more shares of non-motorized or transit modes in total travels surely make the urban transport system greener. Nowadays there are four strategies to increase the share of non-motorized and transit modes: changing land use, giving priority to public transport, promoting walking and cycling and restricting owning and use of private cars. Methodologies underlying the four strategies could be reachable and workable in developing cities. In the following we first study these methodologies within four sections titled by the four strategies respectively. We highlight their effects and limitations and discuss their adaptabilities in developing cities. Then in the end we draw conclusions on some special issues that need attention for employing these methodologies.

2 Land Use Changes

Based on then available data from 46 global cities, Peter Newman and Jeffrey Kenworthy concluded that travel level (measured in passenger-kilometers per capita) had a roughly exponential relationship with urban density [5]. Patrick Moriarty and Damon Honnery found that the share of public transport in overall motorized travel is of much difference between Asian cities with dense population and American or Australian cities that are less densely populated [4]. Thus in general it is believed that increases of urban density will reduce motorized travels and promote shifts to public transport and non-motorized travel modes [6]. Another sound form of land use change lies in optimizations on spatial structure in a certain region of a city or even in the whole city, including redevelopments of utilized lands (such as abandoned dwelling, port and factory area) and constructions of satellite towns around a large city. It is expected to bring out direct changes on the composition of travel modes, especially in new developing area.

However, advantages from land use change are often offset by other factors or there are many difficulties to make changes of land use. There is actually an inevitable problem derived from the incompleteness when concerning urban density. That is in normal conditions density is increased and measured rather locally, for example, in the Central Business District (CBD). Higher density of the CBD could produce within it more short-distance trips, which less depend on automobiles. But on the other hand, more long-distance trips from the periphery to the CBD would also be generated. Then if measured in the entire basis of the city, the average motorized trips between regions would be added. Jeffrey Kenworthy and Christina Inbakaran proved in their research the unexpected result that in some cities, personal travel levels either rose as density rose, or fell as density fell [1].

The government of Shanghai city, P.R.C. has launched an ambitious plan to transfer population and jobs from the central area to the periphery by founding several new towns in the suburb. Under administrative enforcements the withdrawal of residents got great success that while the population of the whole city rose from 18 million in 2000 to 23 million in 2008 (grew nearly 30 %), that within the inner ring region (except for the Pudong District) declined. But efforts to reduce employees in the inner area were frustrated. Very few industries (e.g. only some manufacturing firms) were relocated in the new towns in the past decade. Businesses of high employments and mass activities (such as finance corporations, technology companies, culture and education institutions and large scale retails) were hardly removed from the inner ring region. This is probably explained by the land rent theory and the central place theory under effects of the market selection mechanism.

Since most large developing cities have been tracking the similar path that overly sprawling around a single center and being densely populated, just like what Shanghai and many other Asian metropolises have done, it would take decades to get large rises in density and have to disobey the market law to change spatial structures in these cities, both needing to pay a lot. So for those developing cities which currently are not so large but are free on the way to motorizations, it is probably wise to redefine a small or middle scale of the urban boundary, plan a spatial structure with multi-centers, promote compact, mixed-use developments and take a traffic mode oriented land use strategy, which is related to the concept of transport zoning having been adopted in the urban master plan by a lot of Chinese cities, see Table 1 [7].

Table 1 Transport zoning strategy

Land code	Location	Traffic mode strategy	Land use guidance
A	Conservation areas including natural reserves, scenic spots and historical sites	Restricting motorized travel, especially travel by private cars	Developing lightly and strictly restricting supply of transport facilities
B	Public transport corridors	Giving priority to public transport	Developing densely, restricting parking supply and facilitating transfers between arterial and feeder lines and between public transport and other modes
C	Residential communities in the inner area	Balancing public transport with private cars	Developing moderately and reasoningly increasing supply of transport facilities
D	Residential communities and industrial parks in the peripheral area	Growing motorized travel freely	Developing densely, increasing parking supply and enhancing the connection with the inner area

3 Transit Priority Policies

After the Second World War, public transport use in cities of North America and Western Europe has changed considerably. It first suffered serious decline from the late 1940s to the early 1970s and then stepped into a recovery period since the 1970s [8, 9]. In spite of rising incomes and car ownership and extensive, car-oriented suburban sprawl, these cities have been succeeding in raising and stabilizing the overall passenger level and the market share of public transport [9]. Their successes encouraged other cities globally. Many Asian cities (such as Tokyo, Singapore and Hong Kong) have given continuous priorities to public transport in their overall transport systems and also made significant successes that having increased the proportion of travels taken by public transport up to 60–70 % [10]. Now Transit priority is becoming an important strategy adopted by more and more Asian cities on the way to optimizing the composition of travel modes and greening their urban transport systems.

There are many kinds of policy instrument supporting transit priority, such as the strategy of Transit Oriented Development (TOD), enforcements on supply of transit infrastructures, finance regimes that ensure priority of investments on public transport, traffic rules that guarantee priority of road right of public transport and administrative measures that encourage applications of advanced technologies in public transport [11, 12]. According to lessons from successful cities, it is essential to develop a hierarchy system of public transport that is able to provide differential services to meet demands of diverse groups of people or in different regions. The city must broaden its financing channels and afford continuously funding during the entire life cycle of public transport. Of course it is the most important that public transport is more attractive than other modes. Naturally most cities produce policies of low riding fare to draw passengers, as well as favorable measures of concessionary fares for some particular groups of passengers and subsidies for riding at some certain periods of time [8]. A number of other cities, especially developed cities, also devote to introducing information technologies to improve the overall quality of services provided and boosting advanced public transport systems [12].

However on the whole, public transport is costly and few operators around the world are able to fund it independently [12, 13]. The operators must be compensated by their governments for providing cheap but satisfying services to passengers. It could be a heavy financial burden to the governments. This is one of the most important reasons why Beijing, P.R.C. had to reprice its subway fares in 2014, after having provided fairly cheap services for about seven years since 2007. Nevertheless there is an outstanding exception, Hong Kong. It has successfully made its rail systems self-financing. It is a unique achievement when compared to all other cities in the world that the self-financing policy applies not only to system operation and maintenance, but also to meeting the full cost of rail line construction and the purchase of capital equipment [13].

As being technology, capital and user intensive, rail systems and even bus rapid transits (BRTs) are probably unreachable to many small or middle developing cities. These cities should concentrate their efforts on the provision of quality regular bus services, as well as various other service forms that are easily taken out or withdrawn according to actual needs, such as taxis, school buses, ferries, private hire cars and cableways [14]. It is advisable to plan a binary or tertiary structure of the bus service that includes arterial and secondary lines or arterial, secondary and feeder lines to meet differential demands. It is also suggested that plenty of attentions are paid on priority treatments for bus operations on the road network. Solutions can be made out in forms of preferential signals and exclusive lanes for buses [11, 15, 16]. But it is noted that this kind of solution at the same time is required to balance the improvement of bus operations with the needs of private vehicles and other road users.

4 Tactics to Promote Walking and Cycling

During the past decades, non-motorized transport has experienced significant decline internationally, similar to what public transport has done at the same period. Contemporarily however, development trends of non-motorized transport are different between developed and developing cities and even vary very much among developed cities. All over the world, four development models of non-motorized transport are summarized as below:

Table 2 Walking mode share changes in selected Asian cities (data sources: [19, 20])

City	Year	Before (%)	Year	After (%)	Mode with greatest gain (motorized)
Changzhou	1986	38.24	2006	21.54	Two-wheeler and car
Chennai	2002	47.00	2008	22.00	Two-wheeler
Delhi	2002	39.00	2008	21.00	Two-wheeler and car
Nanchang	2001	44.99	2005	39.11	Car
Shanghai	1986	38.00	2004	10.40	Two-wheeler and bus
Xian	2002	22.94	2006	15.78	Bus
Yangzhou	2007	16.40	2014	15.20	Two-wheeler and car

(1) It is a negligible mode for commuting and has no signs to revive. This model is found in most North American and Australian cities. In 2009, percentage of bicycle travel in each of the American states was not more than 2.34 %; the share of cycling for commuting in Portland was the highest compared to that of all other American cities, but it was only 5.81 % [17]. Currently car-oriented developments remain unchanged in most North American and Australian cities and bicycles are usually used by a minority group of citizens, mainly for sport and leisure activities [18].

(2) It continues to decline but still occupies a large proportion. Cases are found in most developing cities, especially in Asian cities. See Table 2, for example, the change of walking share in some Asian cities. During the past years developing countries accelerated urbanization and motorization. Impact of motorized transport on non-motorized transport has kept rising in developing cities. Now pedestrians and cyclists have become the most vulnerable group in their urban transport systems, in which sources of capital, land, space, infrastructure and management are increasingly favoring car travelers. But non-motorized transport is still substantial in these cities because it was born very dominant. Consequently, most of developing cities are troubled by mixed traffic today.

(3) It first declined and then got some recovery in supporting the development of public transport. Hong Kong, Singapore and Tokyo, as well as other developed cities in the Northeastern Asia are the cases. They have managed to integrate non-motorized transport with rail transits and the share of cycling around rail stations grew much faster than at other sites. Now in Japanese major metropolises the highest percentage of bicycle travel nearby rail stations is 70 % [21].

(4) It first declined sharply and then revived strongly. This fluctuation has occurred in European cities. From 1950 to 1975, bicycle use in the Netherlands, Denmark and Germany decreased by about three quarters and its share glided from 50–85 to 14–35 % [22]. From 1951 to 1972, bicycle travel level per capita in Berlin even reduced by 90 % [18]. But after the middle 1970s, bicycle use in the three countries rose by a

quarter and its share went up to 20–43 %. Especially in Copenhagen, bicycle travels increased by 70 % from 1970 to 2006 [23]. At present, pedestrians, cyclists and drivers in most European cities are treated equally and even non-motorized travelers are given more conveniences than motorized travelers.

The last two development models above prove that it is quite possible to boom non-motorized transport in a highly mobile city. In order to promote walking and cycling, numerous tactics have been put into practices. Overall they belong to three categories: infrastructure supports, program incentives and policy interventions [24].

Appropriate infrastructures supporting walking and cycling include travel paths, bicycle parking facilities and traffic signals, as well as some necessary auxiliaries (such as signs, separators and pavements). Jennifer Dill and Theresa Carr found that the share of workers regularly commuting by bicycle is positively related to the supply level of bike lanes [25] and Anne Vernez Moudon et al. revealed that better accessibility to bike paths will lead to higher probability of cycling [26]. Therefore travel paths are of much importance to fostering walking and cycling. As for travel paths, it is essential to keep pedestrians and cyclists away from motor vehicles. In practices, there are a good number of measures to do it, such as striped or colored bike lanes, raised bike lanes or footpaths, side bike paths, separated bike paths and off-street paths [24], of which off-street paths can carry out thorough separations. Today walking streets and bicycle tracks (also called green ways in Chinese cities) are being appreciated by more and more planners worldwide. Copenhagen has even given birth to a new model of bike track, i.e. so called cycling super highway, which does the most favor to cyclists. However, several stated preference studies will imply that some particular groups of travelers (e.g. women and more experienced cyclists) would choose on-street lanes rather than off-street paths [24]. So planners must take into account differential preferences among travelers when planning travel paths.

Program incentives are formal or informal, long-term or short-term, scheduled or unscheduled campaigns, activities, events, projects and other means hosted by governments or non-governmental organizations. Generally each program incentive is launched upon a specific target, favoring travelers on a certain route, in a certain age group, at a certain time period or by a certain travel mode. There are two kinds of commonly used programs: travel awareness programs and mode access programs. Travel awareness programs aim to arouse public awareness to walking and cycling through education or training schemes, media publicities and public campaigns (such as the Global Car Free Day and the Night of Bicycle in the Europe). Mode access programs are planned to improve the accessibility and convenience of walking or cycling. Nowadays globally, bike sharing (public bicycle) is becoming a well known and widespread bicycle access program and it has been developed for four generations in European cities [27, 28].

Program incentives are helpful to advocate preferences of walking and cycling and cultivate expected traveling habits among travelers. But they need supports from related items, such as improvements of infrastructures, enhancements of traffic managements and establishments of traffic regulations.

Policy interventions are related to legal issues that focus on pedestrians and cyclists safety, such as bicycle helmet laws and lower speed limits for automobiles. But it is

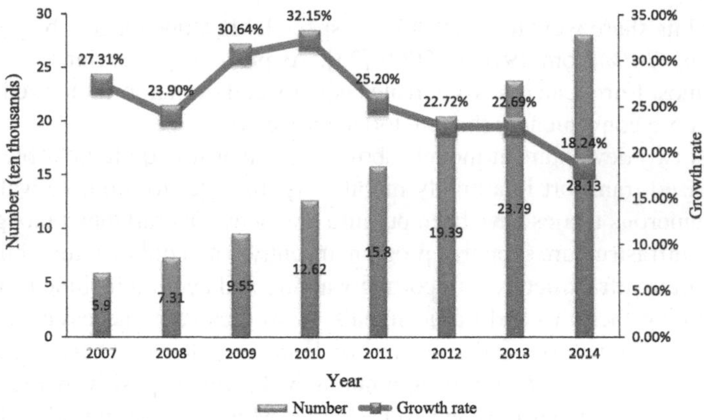

Fig. 1 Growth of private car in Yangzhou city: an instance of constant enthusiasm for motorizaitons (data sources: [20])

noted that bicycle helmet laws could be unwelcome to cyclists while reduced motor speed limits inspiring cycling [29, 30].

It is believed that currently a high level of cycling is beneficial to the development of bicycle mode in the future, for higher cycling level will favor cyclists at a series of aspects, such as further increases, more road right, greater safety, wider support and deeper investments [24, 31–33]. Hence, cities belonging to the second development model of non-motorized transport as summarized before have more endogenous advantages to revive walking and cycling than do those of the first development model. For cities of the second model, there could be a relatively easy way that retaining, advocating and cultivating the traditional travel culture, custom and habit while these factors tend to foster cycling [34, 35]. However it is probably difficult, for those cities that have been feeling free on motorizations, to change again views and preferences of their citizens, who are now strongly interested in owning and using private cars. For example, the annual growth rate of private cars in Yangzhou city has remained no less than 18 % in recent seven years (see Fig. 1).

Furthermore when planning tactics to promote walking and cycling, it is important for planners to make clear the relationship between non-motorized and motorized modes in the current phase of urban transport development. In those developing cities with a small or middle urban area, non-motorized modes would compete with motorized modes for the average distance of trips is short and many trips can be taken by any of them. Under this condition, tactics should pay much attention on temporal and spatial separations between the two. In large developing cities, long-distance trips are dominant and non-motorized modes are usually utilized at the start, transfer or end stage of a trip, as supplements and supports for motorized modes, just like experiences in cities of the third development model of non-motorized transport above. Thus the two kinds of mode are rather of a cooperative relationship and it is important to plan good connections between them.

5 TDM Means to Restrict Car Owning and Using

In order to take immediate effects when deciding to deal with urban transport problems, several developed cities imposed direct restrictions on the ownership and utilization of private cars through powerful economic and policy measures and eventually gave birth to the TDM, the primary objective of which is to reduce motorized traffic volumes in urban area, especially in inner districts of a city. Nowadays TDM means to restrict car owning and use are also abundant, include increasing the cost of holding a car, charging on road use or traffic congestion, lowering vehicular speed limits, cutting parking space supply, reducing wide roads and encouraging car sharing.

High cost on private cars has been maintained in Singapore, Hong Kong and Tokyo and has effectively controlled car growth in the three cities. As early as in 1990, Singapore presented its particular Vehicle Quota System (VQS), by which the urban government could conduct direct control on vehicle growth. The core idea of the VQS is to set a ceiling number of vehicles according to the capacity of road network annually; owners are required to bid for a Certificate of Entitlement (COE) before registering a new car. Since the implementation of the VQS, annual vehicle growth rate in Singapore has dropped from 7.0 to 3.0 % and after 2009 it has even been controlled less than 1.5 % [36]. Congestion charging is normally accepted in very developed cities with heavy motorized traffic or serious congestions in the central area, such as London and other European cities. Experiences of these cities indicate that successful applications of congestion charging are involved into a lot of factors, covering clear goals, public supports, related regulations, charging level adjusting mechanisms, appropriate charging methods and technologies, and reliable assessment systems. Whats more, for a selected city, whether to apply this instrument is determined by some certain preconditions, summarized as: (1) long time traffic jams have frequently struck the central area of the city and likely sprawl to other area; (2) the government has continuously tracked vehicles changing and has accurate vehicle registration records; (3) some other related policies (e.g. parking control) have been implemented; and (4) alternative modes (e.g. public transport) have been well developed.

However even in those cities that have prepared all the related factors and preconditions mentioned above, the effectiveness of congestion charging could fade in the long term for most travelers would be less sensitive to the charging fee as time passed and regard it as the normal travel expense. This is one of the important reasons why London shrank its congestion charging zones in 2011. Moreover for most developing cities, restrictions on vehicle growth may conflict with their economic goals which tend to heavily depend on automobile industries. Therefore this kind of methodology will have the least attractions to developing cities and even meet strong oppositions. Nevertheless, some similar but moderate policies are suggested for large developing cities (e.g. Asian metropolises), such as changing fuel tax, raising parking fares in a certain area and reducing car use with vehicle number plate restrictions.

6 Conclusions

In addition to technical solutions, there are many non-technical methodologies to make urban transport green. They mostly underlie four strategies: changing land use, giving priority to public transport, promoting walking and cycling and restricting owning and use of private cars, aiming to change the automobile travel culture that has flooded developed cities and are spreading to developing ones. Comparing to technical solutions, non-technical ones may be more suitable for most developing cities. However, some difficulties should be noted when applying them. First, although increases of urban density are expected to reduce travel demand, especially motorized travel demand, it is hard to get large density increases in a city. Such a measure would take many decades to work. Secondly, mass rapid transits, such as rails and BRTs could yet be unreachable to many small or middle developing cities as they need investments of advanced technologies and vast capitals and require dense passengers to support their operations. Thirdly, it would be difficult to return to the traditional travel culture that preferring walking and cycling to driving in such a city that has had many freedoms on the way to motorizations for years. Finally, restricting vehicle growth would do harm to some cities economies. Besides, it is also important to: (1) define a small or middle urban area with multi-centers and compact, mixed-use developments, embracing the transport zoning strategy; (2) emphasize on regular bus lines that have road right priorities and are capable to provide differential quality services to meet diverse riding demands; (3) clarify the relationship between non-motorized and motorized modes before planning tactics to promote walking and cycling; and (4) develop some moderate policies to restrict car owning and using, such as changing fuel tax, raising parking fares and introducing vehicle number plate restrictions.

Acknowledgments This work was funded by the Yangzhou science and technology planning program (Program No.: YZ2014235).

References

1. May AD, Zhongming J (2009) European experiences in green transportation development (in Chinese). Urban Transp China 7(6):17–22
2. Huapu L (2009) Approaches towards realization of urban green transportation (in Chinese). Urban Transp China 7(6):23–27
3. Qiaoling L, Martineau D (2010) The urban transportation policies green transition and practice in the U.S. (in Chinese). Planners 26(9):5–10
4. Moriarty P, Honnery D (2013) Greening passenger transport: a review. J Clean Prod 54(9):14–22
5. Newman P, Kenworthy J (1999) Sustainability and cities: overcoming automobile dependence. Island Press, Washington
6. Litman T, Steele R (2013) Land use impacts on transport: how land use factors affect travel behavior, http://wenku.baidu.com/view/71dd3d86f121dd36a22d8236.html

7. The Department of Housing and Urban-Rural Development (2011) Jiangsu Province, P.R.C., Guide on urban comprehensive transportation planning in Jiangsu province
8. White P (2009) Public transport: its planning, management and operation, 5th edn. Routledge, Taylor and Francis Group, London and New York
9. Buehler R, Pucher J (2012) Demand for public transport in Germany and the USA.: an analysis of rider characteristics. Transp Rev 32(5):541–567
10. Linbo L, Chuan C, Cheng S (2006) Comparing study of several cities public traffic. J Shanghai Univ Eng Sci 20(3):259–264
11. Goh KCK (2014) Exploring new methodologies and perspectives on the road safety impacts of bus priority, Thesis submitted in fulfillment of the requirements for the degree of Doctor of Philosophy, Department of Civil Engineering, Monash University, Australia
12. Nakamura K, Hayashi Y (2013) Strategies and instruments for low-carbon urban transport: an international review on trends and effects. Transp Policy 29:264–274
13. Barron B (2002) Simon Kawing Ng. Christine Loh and Richard Gilbert, Sustainable Transport in Hong Kong
14. The Ministry of Construction (2007) P.R.C., Standard for classification of urban public transportation (CJJ/T 114-2007)
15. Hounsell N, Shrestha B (2012) A new approach for co-operative bus priority at traffic signals. IEEE Trans Intell Transp Syst 13(1):6–14
16. Goh KCK, Currie G, Sarvi M, Logan D (2014) Road safety benefits from bus priority: an empirical study. Transp Res Rec 2352:41–49
17. Snyder T (2011) Mapping bicycle mode share where you live, StreetsBlog USA. http://usa.streetsblog.org/2011/03/28/mapping-bicycle-mode-share-where-you-live
18. Wen X (2008) Urban pedestrian and bicycle planning based on human-oriented spatial analysis. Dissertation submitted in conformity with the requirements for the degree of Doctor of Philosophy, School of Transportation Engineering. Tongji University, P.R.C
19. Leather J, Fabian H, Gota S, Mejia A (2011) Walkability and pedestrian facilities in Asian cities: State and issues, ADB Sustainable Development Working Paper Series, Asian Development Bank
20. Yangzhou Urban Planning Bureau (2015) Jiangsu Province, P.R.C., Annual report of Yangzhou urban transportation
21. Andrade K, Kagaya S (2011) Cycling in Japan and Great Britain: A Preliminary Discussion, ERSA conference papers from European Regional Science Association
22. Pucher J, Buelher R (2007) At the frontiers of cycling: policy innovations in the Netherlands, Denmark and Germany. World Transp Policy Pract 13(3):5–56
23. Pucher J, Buehler R (2008) Cycling for everyone: lessons from Europe. In: CD-ROM of 87th TRB annual meeting, Washington, D.C
24. Pucher J, Dill J, Handy S (2010) Infrastructure, programs, and policies to increase bicycling: an international review. Prev Med 50:106–125
25. Dill J, Carr T (1828) Bicycle commuting and facilities in major US cities: if you build them, commuters will use them. Transp Res Rec 116–123:2003
26. Moudon AV, Lee C, Cheadle AD, Collier CW, Johnson D, Schmid TL, Weather RD (2005) Cycling and the built environment, a US perspective. Transp Res Part D Transp Environ 10(3):245–261
27. DeMaio P, MetroBike LLC (2009) Bike-sharing: history, impacts, models of provision, and future. J Public Transp 12(4):41–56
28. Shaheen SA, Zhang H, Martin E, Guzman S (2010) Hangzhou public bicycle: understanding early adoption and behavioral response to bike sharing in Hangzhou, China. Transp Res Rec 2247:33–41
29. Colin C (2006) The case against bicycle helmets and legislation. World Transp Policy Pract 12(3):6–16
30. Robinson DL (2006) No clear evidence from countries that have enforced the wearing of helmets. Br Med J 332(7543):722–725

31. Rune E (2009) The non-linearity of risk and the promotion of environmentally sustainable transport. Accid Anal Prev 41(4):849–855
32. Jacobsen PL (2003) Safety in numbers: more walkers and bicyclists, safer walking and bicycling. Inj Prev 9(3):205–209
33. Robinson DL (2005) Safety in numbers in Australia: more walkers and bicyclists, safer walking and bicycling. Health Promot J Aust 16(1):47–51
34. Pucher J, Komanoff C, Schimek P (1999) Bicycling renaissance in North America: recent trends and alternative policies to promote bicycling. Transp Res Part A Policy Pract 33:625–654
35. de Bruijn GJ, Kremers SPJ, Singh A, van den Putte B, van Mechelen W (2009) Adult active transportation: adding habit strength to the theory of planned behavior. Am J Prev Med 36(3):189–194
36. Kuang LC (2009) Singapore travel demand management: key strategies and characteristics (in Chinese). Urban Transp China 7(6):33–38

Road Network Representation Method Based on Direction Link Division

Ande Chang, Jing Wang and Xiaoshun Zhang

Abstract The present link division methods based on GPS floating vehicles to collect traffic information largely ignored the difference of different directional traffic operating conditions on the intersections, which leads to lower traffic information quality and cannot effectively meet the data demand for dynamic traffic management system. A link division method that is able to distinguish traffic flow directions and to count traffic data is designed. In addition, the corresponding simplified method of road network and the network connected relation expression method are studied, which improves information quality for dynamic traffic management system from the aspects of road network spatial data structure.

Keywords Road network expression · Direction link · Connected relation

1 Introduction

GPS (Global Position System) floating vehicles as an important mean to obtain dynamic traffic data, has short construction cycle, high data precision, wide coverage and strong real-time performance, etc. Since its advent, GPS floating vehicles get rapid worldwide attentions [1]. Network spatial data construction models are the precondition of using GPS floating vehicles to collect dynamic traffic data. The basis of network spatial data model construction is that dividing road network into nodes and links. Therefore, link division methods are closely relating to the quality of traffic data [2].

A. Chang (✉) · X. Zhang
National Police University of China, Shenyang 110035, China
e-mail: changande@npuc.edu.cn

X. Zhang
e-mail: ddzhangxiaoshun@163.com

J. Wang
College of Engineering, Shenyang Agricultural University, Shenyang 110866, China
e-mail: gzlwangjing0707@163.com

© Springer International Publishing Switzerland 2017
V.E. Balas et al. (eds.), *Information Technology and Intelligent Transportation Systems*, Advances in Intelligent Systems and Computing 454,
DOI 10.1007/978-3-319-38789-5_60

Traditional link division methods generally use the intersection point of road centerline, which leads to the fact that the output data of traffic information collecting systems can only reflect the average trends of traffic state of all direction lanes on the intersections, and these methods assumed that the whole road traffic state is equilibrium [3–8]. However, due to the influence of factors such as traffic management measures, traffic states of different direction lanes on the intersections tend to have larger differences. Hence, dynamic traffic management systems have urgent needs to distinguish travel time of vehicles moving direction on the inter-sections.

Therefore, aiming at the information demands of dynamic traffic management systems, according to GPS floating vehicles running characteristics, this article designs a link division method that counts traffic data by different traffic flow moving directions on the intersections. In addition, the corresponding road network simplified method and network connected relation expression method are researched, which will improve the information quality for dynamic traffic management systems from the aspects of road network spatial data structure.

2 Special Arc Processing

In the electronic map, nodes represent the intersections, road end, etc. and arc segments represent the segment between two adjacent nodes. In addition, shape nodes, shape segments and paths are several concepts of extended out nodes and arc. In-stances of network elements show in Fig. 1.

According to the research conclusion from Quiroga, arbitrary arc segment needs not only existing GPS anchor point, but also needs at least two GPS anchor points to guarantee the high quality of link travel time. Therefore, this article draws on Eq. (1) to identify road network special arc segment, which means that the arc segments

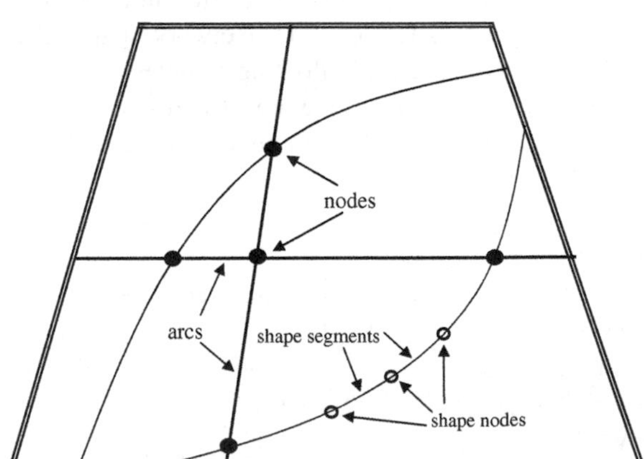

Fig. 1 Network element instances

obey the conditions can be chose as travel time data collection objects, or these arc segments need to be merged.

$$l \geq 2t_c V_c + 2R \qquad (1)$$

where, l is the length of arc segment, t_c is GPS data sampling interval, V_c is design speed of arc segment in the road, R is GPS positioning accuracy. In the electronic map, the close intersection, roundabout and road interchange are the reasons which leads to short arc segment.

(1) The close intersections

The close intersection means that the distance of independent close intersections is too small. This article regards close intersection as common intersection, and uses middle point of internal arc segment as nodes of close intersection, as shown in Fig. 2.

(2) The roundabouts

Based on different scales and structure characteristics of intersections, the roundabout can be dividing into three kinds of circumstances, at the time those nodes marks are also divide into three conditions, as shown in Fig. 3.

(1) The combination decomposed into ordinary intersections. If the length of each mixed link is shown in Eq. (1), roundabout will be seen as many common inter-sections connecting, as shown in Fig. 3a.

(2) The combination decomposed into similar intersections. If the length of each mixed link is not shown in Eq. (1), roundabout will be seen as many common intersections connecting in turn, as shown in Fig. 3b.

(3) The combination decomposed into common intersections and similar intersections. If the length of each mixed link is uneven, the above two cases will appear at the same time and roundabout will be seen as common intersections and similar intersections mixing, as shown in Fig. 3c.

(3) The link interchanges

Due to the link interchange complex internal structure, this article abstractly regards it as common intersection, and uses projection point of road crossing as nodes of link interchange, as shown in Fig. 4.

Fig. 2 Schematic diagrams of marks of close intersection nodes

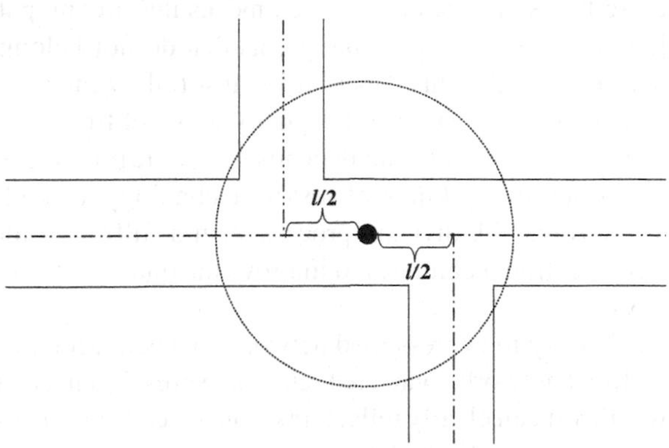

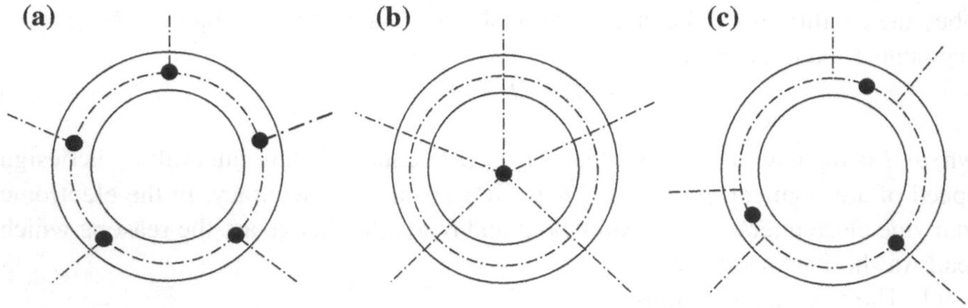

Fig. 3 Schematic diagrams of marks of roundabout nodes

Fig. 4 Schematic diagrams of marks of link interchange nodes

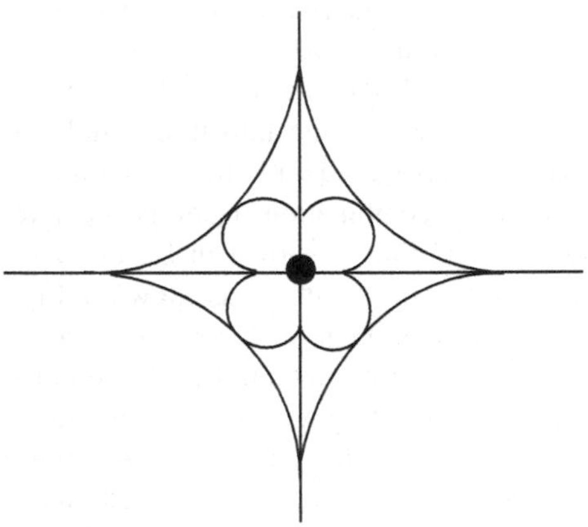

3 Direction Link Division

During link division, two basic principles should be following. First, the close links are independent of each other, which mean that any point belongs to only one link and does not belong to two or more link nodes at the same time in any path. Second, close link s are seamless, which means that in any path any point must belong to a link and there is not exiting points that do not belong to links. In addition, for the research topic of this paper, links must follow the below two special principles.

(1) Easy to guarantee the quality of travel time

Obtaining travel time depends on the traffic detection technology, but the characteristics of all kinds of testing technology to collect traffic data are different. There-fore, link division process should fully consider the characteristics of various detection techniques to improve the quality of travel time data from the view of physical.

(2) Easy to express road network connected relation

Road network connected relation expression needs to consider two aspects. First, whether it can clearly reflect link-connected relation, especially the actual connected relation. Second, whether it has the high efficiency of expression, namely the original size range of the network structure. In view of the fact that natures links do not reflect the difference of travel time from different flows on the intersection, this paper designs a link diversion method. The method divides nature links nodes into several child nodes based on the number of crossing export ways, and takes the child nodes as boundary definition link. Meanwhile, nature links are dividing into several direction links, corresponding to travel time from different flows on the intersection, as shown in Fig. 5.

The GPS technology can realize sample car dynamic positioning, which makes link boundary positions be getting flexibly. However, although this article does not re-search other means which can collect travel time such as loop coil, this article should as far as possible be compatible in link division, otherwise it will reduce the rationality and validity of link dividing method. Therefore, in order to satisfy the first special principle, even in the security of data quality of travel time, this article should consider the characteristics of loop coil technical to determine the location of direction link boundary. In addition, GPS technology is consistent.

According to the different laying location of testing equipment, loop coil detection technology includes two main ways. First, SCATS technology. Testing equipment is located in the natural road downstream near the stop line. Second, SCOOT technology. Testing equipment is located in the reverse extension position of the natural link upstream near the stop line.

Whether SCATS or SCOOT technology, link division should contain the controlled intersection, so that parameters like intersection signal timing are introduced in link travel time estimation model, which reflects the differences between different flow traffic travel time. Therefore, for SCATS technology, this article will choose intersection stop line as direction link defined boundary, as shown in Fig. 6. For SCOOT technology, this paper chooses the reverse extension line of intersection stop line as direction link defined boundary, as shown in Fig. 7.

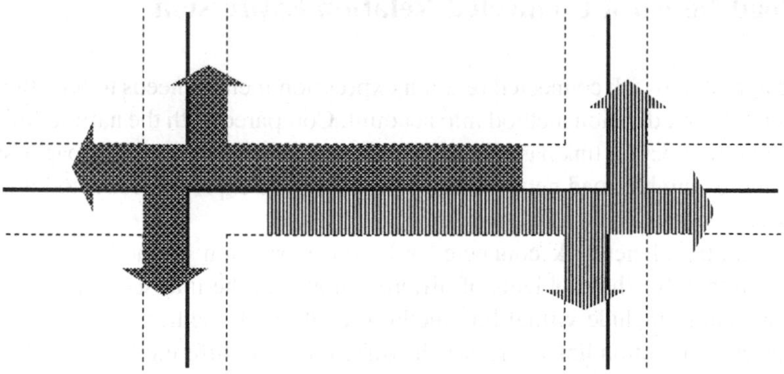

Fig. 5 Schematic diagram of direction link divisions

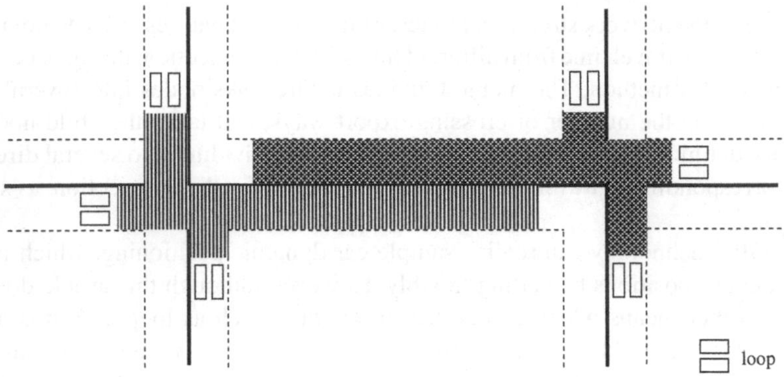

Fig. 6 SCATS signal control system covered link-divided diagram

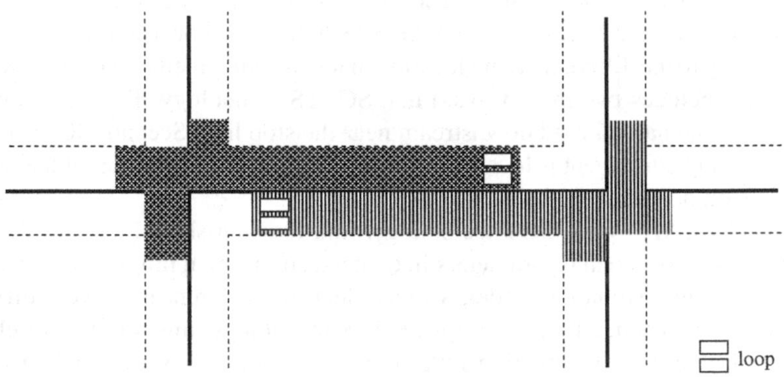

Fig. 7 SCOOT signal control system covered link-divided diagram

4 Road Network Connected Relation Expression

The design of network connected relation expression method needs to take the characteristics of link division method into account. Compared with the natural link division method, direction link has obvious characteristics. Therefore, this paper should design more suitable road network connected relation expression method aimed at the above designing link division method.

The traditional network connected relation expression methods focus on the expression to solve the problems of different flows on the intersection. Traditional methods mainly include virtual line method, dual graph method, labeling method, etc. However, direction link expresses the differences of different flows on the intersection, so the paper just uses ordinary network description method to express the connected relation.

The road network connected relation expression method of SCATS signal control systems show in Fig. 8. The road network connected relation expression method of SCOOT signal control systems show in Fig. 9. In figure, solid line represents

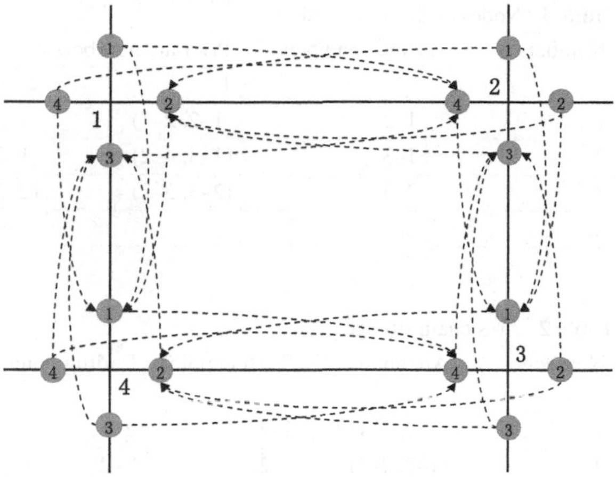

Fig. 8 Road network connected relation schematic diagram of SCATS links

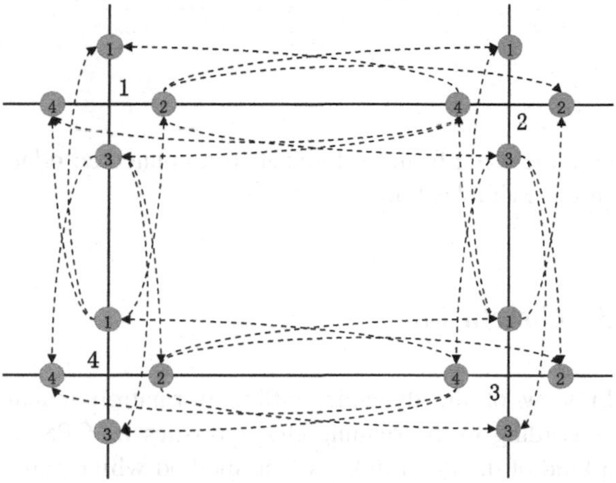

Fig. 9 Road network connected relation schematic diagram of SCOOT links

road centerline, circle represents direction link boundary, and dashed line represents node-connected relation.

Visibly, after dividing direction sections, the above method can clearly express the connected relation between the sections. In addition, it does not as that virtual line method extra adds fictitious domain and fictitious line. Alternatively, it does not like that dual graph method downright changes the original network structure. In addition to expressing method, network storage way of connected relation will affect the efficiency of path optimization algorithm. Therefore, this article should consider the characteristics of road network connected relation expression, and design the storage method of road network connected relation. Direction road division has chosen node weight into the road weight, which means that node, is no longer with the weight. At the same time, the above road network connected relation expression does not change the meaning of the node and arc segment. Therefore, this article

Table 1 Nodes chain storage table

Numbers	Node numbers	Prior arc numbers		
		1	2	3
1	1–2	(1–2, 2–1)	(1–2, 2–2)	(1–2, 2–3)
2	1–3	(1–3, 4–2)	(1–3, 4–3)	(1–3, 4–4)
3	2–3	(2–3, 3–2)	(2–3, 3–3)	(2–3, 3–4)
…	…	…	…	…

Table 2 Arcs chain storage table

Numbers	Arc numbers	Starting point	Ending point	Arc travel time	Length of arc
1	(1–2, 2–1)	1–2	2–1	w(1–2, 2–1)	d(1–2, 2–1)
2	(1–2, 2–2)	1–2	2–2	w(1–2, 2–2)	d(1–2, 2–2)
3	(1–2, 2–3)	1–2	2–3	w(1–2, 2–3)	d(1–2, 2–3)
…	…	…	…	…	…

uses the chain form to store network-connected relation. The chain storage form is shown in Tables 1 and 2.

5 Conclusion

In view of the dynamic traffic information management system requirements, ac-cording to the running characteristics of GPS floating car, this paper designs a kind of direction link division method which can count traffic data according to different directional traffic flows. Then this paper studies the methods of special arc segment processing and network connected relation expression. The results of this research can help to improve the quality of the dynamic traffic information.

Acknowledgments This paper is supported by Program of Doctoral Scientific Research Foundation of National Police University of China (Traffic Congestion Alarming based on GPS Equipped Vehicles), Training Program of the Major Research Plan of National Po-lice University of China (Study on Road Traffic State Extraction based on Locating Point Group of GPS Equipped Vehicles), General Project of Liaoning Provincial Education Department of China (L2015554), Technology Research Program of Ministry of Public Security of China (Study on Road Traffic State Extraction based on Locating Point Group of GPS Equipped Vehicles) and Project of Natural Science Foundation of Liaoning Province (Dual-mode Traffic Guidance Models and Information Releasing policies).

References

1. Rui W, Nakamura H (2003) Validation of an improved method to estimate expressway travel time by the combination of detector and probe data. J East Asia Soc Transp Stud 5(1):2003–2014
2. Li Q (2007) Arterial road travel time study using probe vehicle data. Nagoya University
3. Quiroga CA (1997) An integrated GPS-GIS methodology for performing travel time studies. Louisiana State University
4. Jiang J, Han G, Chen J (2002) Modeling turning restrictions in traffic network for vehicle navigation system. In: Symposium on geospatial theory, processing and application, Ottawa
5. Byon YJ (2005) GPS-GIS integrated system for travel time surveys. University of Toronto
6. Ehmke JF, Meisel S, Mattfeld DC (2012) Floating car based travel times for city logistics. Transp Res Part C Emerg Technol 21(1):338–352
7. Ramezani M, Geroliminis N (2012) On the estimation of arterial route travel time distribution with Markov chains. Transp Res Part B 46(10):1576–1590
8. Cohen S, Christoforou Z (2015) Travel time estimation between loop detectors and FCD: a compatibility study on the Lille network, France. Transp Res Procedia 10(1):245–255

Improved Fuzzy Evaluation for Risk of Explosives in Road Transportation Based on Information Entropy

Lei Wang and Ying Cheng

Abstract According to the complexity and uncertainty characteristics of the explosive road transportation accident, this paper proposed a improved fuzzy synthetic evaluation based on information entropy, established a risk evaluation index system for explosives in road transportation, taken thermal sensitivity, mechanical sensitivity, electrostatic sensitivity, shock sensitivity of explosives as the criterion level. Information entropy method is utilized to determine the index weight coefficient of influence factors, which overcome effectively the subjective influence of index weight caused by experts assigning artificially. Finally the synthetic weight coefficient is obtained by combining information entropy and subjective weighting method of traditional fuzzy evaluation, provided a new effective method for the risk evaluation of explosives in road transportation. The empirical analysis is shown that the transportation risk was relatively large and in line with the actual situation, which shown that the evaluation model is practical and feasible.

Keywords Explosives · Risk level · Road transportation · Information entropy · Improved fuzzy evaluation

1 Introduction

Explosives are widely used in national defense and national economic construction, it is widely used in ore, coal, oil and natural gas exploitation, road construction, pyrotechnics, fuses, fireworks, etc., as well as satellites launch and weapon area. The chemical properties and stability of explosives is poor because of its flammable

L. Wang (✉)
Aviation and Automotive Department,
Tianjin Sino-German University of Applied Sciences, Tianjin, China
e-mail: lei.wang6@aliyun.com; 29729668@qq.com

Y. Cheng
Department of Automobile and Transportation,
Tianjin University of Technology and Education, Tianjin 300222, China
e-mail: chengying1650@aliyun.com

© Springer International Publishing Switzerland 2017

V.E. Balas et al. (eds.), *Information Technology and Intelligent Transportation Systems*, Advances in Intelligent Systems and Computing 454,
DOI 10.1007/978-3-319-38789-5_61

and explosive characteristics. In the transport process, the explosives is extremely easy to explode, poison and cause other serious accidents if meets conditions of light heat, impact, friction and others, resulting in a large number of casualties, property damage and damage to the environment. Explosives transportation is an important part of circulation process of dangerous chemicals, the accident occurrence probability increases greatly due to the changing state in the transport process.

In recent years, influenced by many uncertain reasons, explosives in road transportation accidents occurred frequently and caused serious consequences, the randomness and uncertainty of the risk factors is large during the road transportation process, and the quantitative risk assessment is more complex, so there is no uniform risk assessment model till now. This paper utilizes integrated fuzzy synthetic evaluation method to evaluate the risk of the accident, information entropy method is applied to determine the index weight coefficient of influence factors, to build the fuzzy synthetic evaluation model of the risk evaluation index system, then to determine the risk level of the accident, which should propose protective measures and management approach for achieving explosives safety transportation [1].

2 Evaluation Index System and Structure Model

2.1 Evaluation Index System Establishment

The explosives has minimal explosion possibility if there is no necessary external effect. Any kinds of explosive explosion combustion need to be supplied by the external initiation energy (named initiation energy), the material sensitivity means the minimum initiation energy of a explosives required [2]. Explosive sensitivity includes thermal sensitivity, mechanical sensitivity, electrostatic sensitivity, shock sensitivity. During road transportation process the explosives will be excited by friction, impact and vibration, high fever, electrostatic spark, flame and shock wave and other factors, once external energy exceeds required initiation energy of the explosives, the drastic chemical reaction will be resulted in explosion.

According to experts experience summary in the field of civil explosives work, combined with explosives physicochemical characteristics and road transportation situation, established a risk evaluation index system for explosives in road transportation, taken thermal sensitivity, mechanical sensitivity, electrostatic sensitivity, shock sensitivity of explosives as the criterion level (Fig. 1).

2.2 Hierarchical Structure Model of Evaluation Index Establishment

According to the evaluation index structure, a hierarchical structure model is established, as shown in Fig. 2.

mechanical sensitivity B_1	traffic accident rate C_{11}	explosion reason: fatigue driving, speeding, poor road conditions and climate difference causes crash, impact force exceeds explosives minimum initiation can cause an explosion
	operation standardization C_{12}	explosion reason: non-compatible material transportation; overload and loading, vibration caused by wrong stacking
	driving smoothness C13	explosion reason: squeeze, jitter caused by driver's driving in the emergency brake or vehicle failure
	road condition C14	explosion reason: road bumps, impact and squeeze occurrence at the road bend section
thermal sensitivity B_2	transport vehicle heat source C_{21}	explosion reason: transport vehicle ventilation is not good and temperature rises, smoke exhaust pipe spark, vehicle spontaneous combustion accident
	external kindling C_{22}	explosion reason: the driver took or other unexplained fire into the compartment
	electrical components C_{23}	explosion reason: electrical facilities without explosion-proof function or failure of leakage, insulation / thermal protection device failure occurs
	External temperature C_{24}	explosion reason: close to the heat source, the weather temperature is too high, causing a burning explosion
electrostati-csensitivity B_3	human static electricity C_{31}	explosion reason: operator wearing clothes and shoes do not meet the requirements, generating static, discharge
	vehicle static electricity C_{32}	Explosion reason: the vehicle static electricity do not grounding or grounding device does not work
shock sensitivity B_4	thunder attack C_{41}	explosion reason: vehicles loaded with dangerous goods encounter thunder attack
	RF C_{42}	explosion reason: close to a large power radio transmitting station or operator violating the provisions of using communication equipment

Fig. 1 Risk evaluation index system for explosives in road transportation mechanical sensitivity

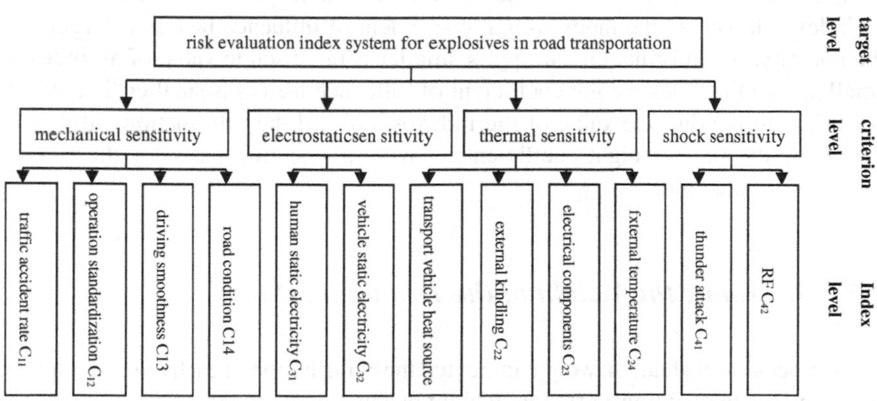

Fig. 2 Hierarchical structure model of evaluation index

3 Establishment of Fuzzy Synthetic Evaluation Model of Risk of Explosives in Road Transportation Based on Information Entropy

Since the fuzzy uncertainty characteristics of risk influence factors for explosive road transport, fuzzy synthetic evaluation method based on information entropy is utilized to conduct the risk evaluation [3]. Information entropy is applied to determine the index weight coefficient of influence factors, then the fuzzy mathematics theory is utilized to analyze and evaluate the evaluating objects [1].

3.1 Principle of Information Entropy

In information theory, entropy is a measure of uncertainty. The larger the amount of information is, the smaller the uncertainty and entropy is; the smaller the amount of information is, the larger the uncertainty and entropy is. According to the characteristics of entropy, the entropy value can be calculated to evaluate the randomness and disorder degree of the event, also can be used to judge a index of discrete degree. The larger index discrete degree is, the larger influence of synthetic evaluation is. Suppose there are n index for the risk of explosives in road transportation, m is evaluating cycle number. Built the original data matrix $x = \left(x_{ij}\right)_{\max}$ for the evaluation system, for an index x_j, it indicated that the index has a larger influence on the sustainable utilization synthetic evaluation if the index value x_{ij} differences is larger, the index is invalid for the sustainable utilization synthetic evaluation if all the index value x_{ij} is equal.

In information theory, utilized function $H(x) = -\sum_{j=1}^{n} g\left(x_j\right) \ln\left(x_j\right)$ to measure the disorder degree which is called information entropy.

Information entropy and information has same absolute value but with opposite sign, the information amount provided by the index is larger if the discrete value of an index is larger, so the index weight coefficient of influence factors is larger. On the contrary, the information entropy is smaller if the discrete value of an index is smaller, then the index weight coefficient of influence factors is smaller. Therefore, according to the discrete value of the indexes, utilized the information entropy to determine the index weight coefficient, provide a scientific basis for the risk of explosives evaluation.

3.2 Determine the Weight of the Evaluation Index

In the process of evaluation, weight indicated the contribution of each index to the risk of explosives in road transportation, weight calculating includes subjective method, objective method and combination method of subjective and objective. The weight calculation steps are as below.

(1) Calculated the index objective weight [4] by utilizing the information entropy method.

(1) There are different dimensions and units for different index, its non measurable caused by the different dimensions and units of the index, therefor the index should be treated in the same degree, the standard formula is as follows. $x'_{ij} = \frac{x_{ij}-\bar{x}_j}{s_j}$, thereinto, $\bar{x}_j = \frac{1}{n}\sum_{i=1}^{n} x_i$, $s_i = \frac{1}{n-1}\sum_{i=1}^{n}\left(x_{ij}-\bar{x}_j\right)^2$. Above, x'_{ij} is the value of same degree, $\bar{x}_j$ is the average of item j.

(2) In order to eliminate the negative values, coordinate can be translated, index x'_{ij} is turned into x''_{ij}, thereinto $x''_{ij} = A + x'_{ij}$. A is the range of coordinate translation.

(3) Calculate the weight R_{ij} for index x''_{ij}, thereinto $R_{\text{ij}} = x''_{ij}/\sum_{i=1}^{m} x''_{ij}$.

(4) Calculate the entropy value e_j for the item j index, thereinto $e_j = -\left(\frac{1}{\ln m}\right)\sum_{i=1}^{m} R_{ij}\cdot \ln R_{ij}$, $e_j \in [0, 1]$.

(5) Calculate the difference coefficient of item j index, thereinto $g_j = 1 - e_j$, the larger the g_j is, the more important of index x_j in the synthetic evaluation system.

(6) Calculate the weight c_j for index x_j, the formula is $c_j = \frac{g_i}{\sum_{j=1}^{n} g_j} = \frac{1-e_j}{\sum_{j=n}^{n}(1-e_j)}$, thereinto $j = 1, 2, \ldots, n$.

(2) Calculate the synthetic weight A, a_i is the component of A and $a_i = c_i d_i/\sum_{i=1}^{n} c_i d_i$, thereinto, d_i is index subjective weight, c_i is information entropy weight, meets $\sum_{i=1}^{n} a_i = 1$, $\sum_{i=1}^{n} c_i = 1$, $\sum_{i=1}^{n} d_i = 1$.

3.3 *Index Set and Evaluation Set Establishment*

The evaluation index system (index set) is a set of all the factors influencing the final evaluation results, expressed by $\text{U} = \{u_1, u_2, \ldots, u_n\}$, $u_i (i = 1, 2, \ldots, n)$ represents each influence factor. The influence of each factor on the evaluation object is divided into m levels, each level can be expressed as fuzzy mathematics language variables, such as "great, better, good, bad, worse", etc. The factor u_i can be expressed as $u_i = \{u_{i1}, u_{i2}, \ldots, u_{im}\}$, $(I = 2, 1, \ldots, n)$.

Evaluation set is a set of all kinds of general evaluation results made from the evaluation object. If it divided into 5 levels, can be expressed as: $V = great, better, good, bad, worse$.

3.4 *Improved Fuzzy Synthetic Evaluation Based on Information Entropy*

(1) Grade evaluation matrix of the influence factors: each factor has a evaluation grade and evaluation index, each factor with different grade has influence of evaluation index, and the influence degree can be expressed by the membership function. The grade evaluation matrix R_i for the item i factor is,

$$\mathrm{R_i} = \begin{pmatrix} r_{i11}r_{i12}\cdots r_{i1k} \\ r_{i21}r_{i22}\cdots r_{i2k} \\ \vdots \\ \mathrm{r_{in1}r_{in2}\cdots r_{ink}} \end{pmatrix} \tag{1}$$

(2) The sub factor level fuzzy evaluation: sub factor level fuzzy evaluation is also called the first level fuzzy synthetic evaluation, synthesized all grade of a factor which influenced the value of the evaluation object, reflected a factors influences on the upper and lower limits of the fuzzy area, it is a single factor judgment. As a single factor evaluation, the first level fuzzy evaluation matrix of the item factor is [5]:

$$\mathrm{B_i} = \mathrm{A_i}\cdot\mathrm{R_i} = (a_{i1}, a_{i2}, \ldots, a_{in}) \begin{pmatrix} r_{i11}r_{i12}\cdots r_{i1k} \\ r_{i21}r_{i22}\cdots r_{i2k} \\ \vdots \\ \mathrm{r_{in1}r_{in2}\cdots r_{ink}} \end{pmatrix} = (b_{i1}, b_{i2}, \ldots, b_{in}) \tag{2}$$

Thereinto means synthesis operation of fuzzy matrix, with $M(.,+)$ algorithm, the formula is $b_{ik} = \sum_{j=0}^{n} a_{ij}rijk$.

(3) Fuzzy evaluation of target level: The first level fuzzy synthetic evaluation matrix reflected single factor influenced the evaluation object value, through the synthesis of fuzzy matrix, the second level fuzzy synthetic evaluation set is obtained as below.

$$\mathrm{C} = \mathrm{A}\cdot\mathrm{B} = (a_1, a_2, \ldots, a_n)\cdot \begin{pmatrix} b_{11}b_{12}\ldots b_{1k} \\ b_{21}b_{22}\ldots b_{2k} \\ \vdots \\ b_{n1}b_{n2}\cdots b_{nk} \end{pmatrix} = (\mathrm{C}_1, \mathrm{C}_2, \ldots, \mathrm{C}_n) \tag{3}$$

4) Processing of the evaluation set: maximum membership principle is usually used to processing of the second level evaluation set, $\mathrm{C}_k = \max\{C_i\}\,(1 \le i \le n)$, then synthetic evaluation result is the grade k.

4 Empirical Analysis

A mining logistics enterprises offer the explosives in road transportation service, making the explosives transportation as an empirical analysis of risk evaluation. Working day average temperature is 20 °C, sunny, dry climate, north wind 315 m/s, via rural, suburban and urban areas road which most is straight, traffic accident rate is 1.03×10^{-6} counts/km.

Table 1 Synthetic weight coefficient of all level evaluation index entropy

Influence factors	Subjective weight coefficient	Synthetic weight coefficient
A1	0.4, 0.3, 0.2, 0.1	0.5010, 0.2750, 0.1791, 0.0449
A2	0.4, 0.1, 0.3, 0.2	0.4358, 0.0815, 0.3415, 0.1412
A3	0.5, 0.5	0.415, 0.585
A4	0.4, 0.6	0.385, 0.615
A	0.4, 0.3, 0.2, 0.1	0.5784, 0.2749, 0.0940, 0.0527

4.1 *Influence Factors Weight Coefficient Determination*

Before the evaluation, five experts consulted and proposed the subjective weight coefficient for all levels, then calculated the synthetic weight based on entropy method, and get the synthetic weight coefficient of all index in the evaluation index system, as shown in Table 1.

4.2 *Establishment of Influence Factors Grade Matrix*

Factors above is used to design the safety evaluation table, selected 20 experts to score according to the actual transport situation, the risk degree of each index was divided into "large, big, general, small, smaller". After processing the statistical data then worked out the membership degree distribution matrix [6].

$$R_1 = \begin{pmatrix} 0.20.50.300 \\ 0.20.50.20.10 \\ 0.30.40.300 \\ 000.50.50 \end{pmatrix}$$

$$R_2 = \begin{pmatrix} 0.40.40.200 \\ 00.50.30.20 \\ 0.40.50.100 \\ 00.10.20.70 \end{pmatrix}$$

$$R_3 = \begin{pmatrix} 00.30.50.20 \\ 00.40.40.20 \end{pmatrix}$$

$$R_4 = \begin{pmatrix} 00001 \\ 000.50.50 \end{pmatrix}$$

4.3 Fuzzy Synthetic Evaluation of Sub Factors

Put the weight coefficient and grade evaluation matrix of the influence factors into formula (2), work out the first level fuzzy synthetic evaluation of all influence factors as below. $B_1 = A_1 \cdot R_1 = (0.2089, 0.4596, 0.2815, 0.0500, 0.0000)$ $B_2 = A_2 \cdot R_2 = (0.3109, 0.3999, 0.1740, 0.1151, 0.0000)$ $B_3 = A_3 \cdot R_3 = (0.0000, 0.3585, 0.4451, 0.2000, 0.0000)$ $B_4 = A_4 \cdot R_4 = (0.0000, 0.0000, 0.3075, 0.3075, 0.3850)$ Then reach

$$\mathrm{B} = \begin{pmatrix} \mathrm{B}_1 \\ \mathrm{B}_2 \\ \mathrm{B}_3 \\ \mathrm{B}_4 \end{pmatrix} = \begin{pmatrix} 0.20890.45960.28150.05000.0000 \\ 0.31090.39990.17400.11510.0000 \\ 0.00000.35850.44510.20000.0000 \\ 0.00000.00000.30750.30750.3850 \end{pmatrix}$$

4.4 Fuzzy Synthetic Evaluation of the Target Layer

$$C = AB = (0.20630.40950.26870.09560.0203)$$

Put the weight coefficient and grade evaluation matrix of the first level into formula (3), work out the second level fuzzy synthetic evaluation as below.

$$C = AB = (0.20630.40950.26870.09560.0203)$$

According to the principle of maximum membership degree, $C_{max} = 0.4095$, the risk of explosive road transportation is second degree, means the risk is bigger.

5 Conclusion

(1) Many influence factors of explosion accident for explosives in road transportation is existent, while there lack of quantitative evaluation model to evaluate the influencing factors with classical mathematical methods. Therefore, its reasonable and feasible to utilize fuzzy synthetic evaluation method to analyze the risk of explosives in road transportation. This paper proposed the combining method of information entropy and fuzzy synthetic evaluation, to evaluate the risk grade of explosives in road transportation. Empirical analysis shown that the evaluation model was practical and feasible.

(2) Based on information entropy fuzzy synthetic evaluation method, weight coefficient of each index was calculated by information entropy, overcomed effectively the subjective influence of index weight caused by experts assigned artificially. In

the fuzzy evaluation process, the same weight coefficient was not adopted by all the enterprises, its more reasonable to calculate the weight index by using information entropy, finally the synthetic weight coefficient is obtained by combining information entropy and traditional subjective weighting method, provided a new effective method for the risk evaluation of explosives in road transportation.

The combined method above reflected different influence factors on the evaluation of different enterprises, and reflected objectively the influence of the evaluation index on the scheme, the evaluation results are more reasonable and scientific.

References

1. Ying C, Xian-ping G (2009) Fuzzy synthetic evaluation on risk of explosives in highway transportation. J Tianjin Univ Technol Educ 19(4):30–33
2. Hua-guang Z, Xi-qin H (2002) Fuzzy adaptive control theory and application. Beihang University Press, Beijing, pp 18–29
3. Guo-feng J (2008) An innovated fuzzy overall evaluation model for liquid-leakage risks in highway transportation. J Saf Environ 8(3):158–161
4. Pei-jie G, Jun-cheng J (2006) Fuzzy multi-criteria modeling and its application in the risk assessment of hazardous chemical materials road transportation. J Nanjing Univ Technol 28(5):57–62
5. Xiao-yan S, Hao-xue L (2008) Safety evaluation on logistics enterprise of hazardous materials based on fuzzy comprehensive evaluation. J Shanghai Marit Univ 29(2):67–71
6. Chen Y, Guang-lu W (2012) Fuzzy comprehensive evaluation of port logistics capacity based on information entropy. Logist Technol 31(5):130–131

Modeling and Simulation Strategies of Cryptographic Protocols Based on Finite State Machine

Ming-qing Zhang, Shu-qin Dong, Hong-shan Kong, Xiao-hu Liu and Hui Guan

Abstract In order to evaluate the influence of cryptographic protocols on the property of communication networks, the method of simulation was adopted. Firstly, a 9-tuples abstract model of cryptographic protocols based on Finite State Machine (FSM) was given, and the process for building FSM models was provided. Secondly, by analyzing the FSM simulation theory of OMNeT++ platform, a dynamic behavior algorithm of simulation models was designed. Thirdly, the validity of the modeling and simulation strategies were tested by taking Internet Key Exchange (IKEv2) protocol as an example and designing a simulation scene of Denial of Service (DoS) attack, and then the usability of cryptographic protocols was analyzed. The simulation results show that, the modeling and simulation strategies are valid, and the average time delay of point to point is approximately increased by 12 % when using IKEv2 protocol in communication networks.

Keywords Cryptographic protocol · Finite state machine · Internet key exchange · Network simulation · Dynamic behavior algorithm

M.-q. Zhang · S.-q. Dong (✉) · H.-s. Kong · X.-h. Liu · H. Guan
Zhengzhou Institute of Information Science and Technology, Zhengzhou 450001, China
e-mail: dongshuqin377@126.com

M.-q. Zhang
e-mail: zmqing0514@163.com

H.-s. Kong
e-mail: konghshan@126.com

X.-h. Liu
e-mail: ganlanlvliu@163.com

H. Guan
e-mail: 83570798@qq.com

© Springer International Publishing Switzerland 2017

V.E. Balas et al. (eds.), *Information Technology and Intelligent Transportation Systems*, Advances in Intelligent Systems and Computing 454,
DOI 10.1007/978-3-319-38789-5_62

1 Introduction

Cryptographic protocols, which are also called secure protocols, are high communication protocols that are running on communication networks and established on the foundation of cryptographic algorithms. They are widely used in many fields to protect private communication messages against divulging. It is vital that whether the cryptographic protocols are really secure. Currently, there are many researches on formal analysis of cryptographic protocols [1–6]. It is a good way to find the problems existing in the design of cryptographic protocols by formally analyzing them, and it can also further improve the security and reliability of cryptographic protocols. However, cryptographic protocols, as a kind of security technology, must be applied to the real network environment if they want to play a role. The degree of their impact on the network performance directly determines the availability of cryptographic protocols, but the analysis of cryptographic protocols on the influence of network performance is rare in existing researches. Simulation, with the advantages of its economy, safety and repeatability, provides an effective means for evaluating the availability of cryptographic protocols and analyzing the performance of cryptographic protocols when used in communication networks.

Before simulating and testing cryptographic protocols based on virtual communication networks, we should firstly build the formal models of cryptographic protocols. The modeling method, which is based on Finite State Machine (FSM) [7, 8], has a great advantage in known formal modeling methods, and quite a number of network protocols are designed and implemented based on FSM, such as TCP protocol, RSVP protocol and so on. Moreover, many simulation tools, such as OPNET and OMNeT++, support the simulation mechanism based on FSM. Therefore, building FSM models of cryptographic protocols can also promote the conversion for cryptographic protocols from system models to simulation models, and effectively reduce the difficulties in developing and maintaining simulation models of cryptographic protocols.

The paper will introduce a modeling method and a simulation algorithm of cryptographic protocols based on FSM, and the rest of the paper is structured as follows: In Sect. 2, a 9-tuples model of cryptographic protocols is proposed based on FSM, and the general steps for modeling cryptographic protocols based on FSM are described. After analyzing the FSM simulation theory of OMNeT++ platform, a dynamic behavior algorithm is designed based on FSM in Sect. 3. In order to verify the effectiveness of the modeling and simulation strategies, a simulation scene is designed by taking International Key Exchange (IKEv2) as an example in Sect. 4, and the validity of the strategies is tested and the influence of cryptographic protocols on communication networks is analyzed. Conclusions and a short outlook on future work are given in Sect. 5.

2 The Modeling Method of Cryptographic Protocols

2.1 The Abstract Model of Cryptographic Protocols Based on FSM

An abstract model of cryptographic protocols can be created from the basic concept of finite state machine. Considering the communication entities and related cryptographic operations in the process of establishment and application of cryptographic protocols, the Finite State Machine of Cryptographic Protocols (CPFSM) can be defined as a 9-tuples model, which is described as following:

$$CPFSM = (Q, E, \Lambda, \Omega, \delta, \phi, \psi, q_0, F) \quad (1)$$

Where, Q is the set of finite states of cryptographic protocols and its elements can be expressed as q_0, q_1, q_2 and so on. E is the set of communication entities of cryptographic protocols and its elements can be expressed as e_1, e_2, e_3 and so on. Λ is the set of finite input events of cryptographic protocols and its elements can be expressed as i_1, i_2, i_3 and so on. Ω is the set of finite output events of cryptographic protocols and its elements can be expressed as o_1, o_2, o_3 and so on. δ is the state transition function of cryptographic protocols, and it is defined as $\delta : Q \times \Lambda \to Q$, for $\forall(q_k, i_l) \in Q \times \Lambda, \delta(q_k, i_l) = q_m$ represents that the state of *CPFSM* will transfer from q_k to q_m when reading the event of i_l, and $k, l, m \in N$. ϕ is the cryptographic operation function of cryptographic protocols, and it is defined as $\phi : E \times \Lambda \to \Lambda$, for $\forall(e_g, i_l) \in E \times \Lambda, \phi(e_g, i_l) = i'_l$ represents that when the communication entity e_g of *CPFSM* reads the input event i_l, it encrypts i_l or hashes i_l, then i_l turns to i'_l, and $l, g \in N$. ψ is the event output function of cryptographic protocols, and it is defined as $\psi : Q \times \Lambda \times E \to \Omega$, for $\forall(q_k, i_l, e_g) \in Q \times \Lambda \times E, \psi(q_k, i_l, e_g) = o_h$ represents that when the communication entity e_g of *CPFSM*, which is at the state of q_k, reads the input event i_l, it outputs event o_h, and $k, l, g, h \in N$. q_0 is the initial state of *CPFSM*, and $q_0 \in Q$. F is the set of final states of *CPFSM*, and $F \subseteq Q$.

2.2 General Steps to Build FSM Models of Cryptographic Protocols

In order to build FSM models of cryptographic protocols, we should carefully analyze the cryptographic protocols with an event viewpoint, so as to determine the main task of the protocol, ascertain the events and conditions which may exist in the running process of the protocol, confirm relevant conditions for the occurrence of an event, and make sure whether the occurrence of an event will change the status of the system and how it behaves. Then, the steps to build state transition diagrams of cryptographic protocols should be determined according to the principle of finite state machine, and FSM models of cryptographic protocols can be built within four steps, just as follows.

Firstly, the protocols should be analyzed carefully. The analysis of protocols, as the first step for building FSM models of cryptographic protocols, is the foundation of the whole process. It directly determines the accuracy of FSM models that whether the analysis of protocols is deep or not. The hierarchy of a cryptographic protocol should be firstly cleared when analyzing it, and then the communication mechanism and communication interfaces for interacting with upper or lower protocols of the cryptographic protocol should be determined according to its operation principle.

Secondly, the events, occurring in the running process of cryptographic protocols, should be listed. In the simulation of cryptographic protocols, events can often be divided into three aspects from the perspective of the event source. One part of them may come from external associated components of the protocols. Another part of them may be produced between different processes when the models of cryptographic protocols are realized by using multiple processes. The third part of them may be internal events that come from the process itself, this kind of situation will often happen when the process executes time-based actions.

Thirdly, the events response table should be established. It is a very important step to determine the transfer situation of the system states when establishing system models of cryptographic protocols, which is the foundation of the state transition diagram. The purpose of establishing the events response table is to determine the state transition situation of cryptographic protocols, and ensure the responses to various events when the states of FSM models transferred. The events response table contains the source state, events, conditions, actions, and an end state. According to the operation principle of the discrete event system simulation, when the simulation runs to a certain state of the FSM model, the simulation clock will not advance until new events come. During this period of time, the FSM model will respond to what has happened in a certain way. The occurrence of new events will lead to the change of the system states, but the master control program of the FSM model will judge relevant conditions before the state transition. Only the transfer conditions are met will the process model be transferred to a new state. Therefore, a state transition condition should be first chosen when determining the state transition, and then the movements that need to be performed after the transition should be determined, finally the end state of the state transition should be determined.

Fourthly, the FSM model should be implemented. The realization of the FSM model is mainly to establish the corresponding state transition diagram based on the above analysis.

3 The Simulation Algorithm of Cryptographic Protocols

3.1 The FSM Simulation Theory of OMNeT++ Platform

OMNeT++ is an object-oriented modular discrete event network simulation framework [9], and it is easy to simulate communication networks on OMNeT++. The structures of the simulated networks can be described with Network Description (NED) language, and the dynamic behaviors of them can be expressed with C++ code in simple modules of the networks. There is an initialization function described as initial(), a function to handle messages described as handleMessage(), a finalization function described as finish() and so on in the C++ code of a simple module. FSM can make life with handleMessage () easier. OMNeT++ provides a class and a set of macros to build FSM, the macros are FSM_Switch(), FSM_Print(), FSM_Transient(), FSM_Steady(), FSM_Enter(), FSM_Exit() and FSM_Goto(). But there are three key points to be noticed, just as follows:

Firstly, the states of FSM in OMNeT++ can be divided into two kinds, one kind of them are transient, and the others are steady. On each event the FSM transfers out of the current steady state, runs through a number of transient states, and finally arrives at another steady state. Thus between two events, the system is always in one of the steady states. Therefore, the transient states are not really a must.

Secondly, all possible states of the system can be defined in an enum type, and each state can be defined as steady state or transient state.

Thirdly, the change of the state must be described in the Exit code.

3.2 The Dynamic Behavior Algorithm of Simulation Models

On the analysis above, the dynamic behavior algorithm of simulation models for cryptographic protocols can be designed below:

Input: CPFSM

Output: The simulation model of the cryptographic protocol

BEGIN

Step1: Declare the object CPFSM of the class cFSM.

Step2: Enumerate all states in CPFSM, set q_0 as INIT, and for $\forall q_k \in Q, k \neq 0$, set q_k as steady state, which can be described as q_k =FSM_Steady(k).

Step3: Initialize the object CPFSM, and enter the stage of message operation.

Step4: CPFSM exits from the current initial state q_0.

Step5: For all $q_k \in Q$, if $\exists i_l \in \Lambda$ makes the equation $\delta(q_0, i_l) = q_k$ valid, then the CPFSM will transfer from the initial state q_0 to the state q_k, and enter the stage of events processing.

Step6: For each $e_g \in E$, if e_g is the receiver of event i_l, then e_g will read-in i_l, and deal with it.

Step7: If e_g is to do some cryptographic operations with i_l, then it will call for the cryptographic operation function $\phi(e_g, i_l) = i'_l$, and then call for the event output function $\psi(q_k, i'_l, e_g) = o'_h$, output the event o'_h. Otherwise, e_g calls for the event output function $\psi(q_k, i_l, e_g) = o_h$ directly, and outputs the event o_h.

Step8: CPFSM exits from the current state q_k.

Step9: For all $q_m \in Q$, if $\exists i_n \in \Lambda$ makes the equation $\delta(q_k, i_n) = q_m$ valid where $i_n = o'_h$ or $i_n = o_h$, then the CPFSM will transfer from state q_k to state q_m, and enter the stage of events processing.

Step10: Turn to Step6 successively until the CPFSM enters its final states set F.

END

4 Modeling and Simulation Tests

In order to verify the effectiveness of the above modeling method and dynamic behavior algorithm, taking IKEv2 protocol as an example, the paper first establishes the FSM model of IKEv2 protocol according to the above modeling method, and then implements it in OMNeT++ using the dynamic behavior algorithm.

4.1 The FSM Model of IKEv2 Protocol

IKEv2 [10], which is an application layer protocol, runs over the UDP protocol. It can negotiate cryptographic algorithms and keys for other cryptographic protocols, such as Authentication Header (AH) protocol, Encapsulating Security Payload (ESP) protocol and so on. There are at least two phases required to establish IKEv2 protocol [11], and the phase of information exchange is designed to send control messages to both communication sides, informing them of the errors occurred in the communication. The whole exchange process is presented in Fig. 1.

The events, occurring in the running process of IKEv2 protocol, are listed in Table 1. The states existed in FSM model of IKEv2 protocol are listed in Table 2. The actions, which need to be executed when the states of FSM model have transferred, are listed in Table 3. Table 4 is the events response table.

The FSM model of IKEv2 protocol, which is also called the state transition diagram of IKEv2 protocol, can be depicted according to the events response table (Table 4). And it is presented in Fig. 2, in which the single circles represent general states of the FSM model, the double circles represent the final states of the FSM model, the virtual arrows represent actions executed by the Initiator, while the real arrows represent actions executed by the Responder, and the bold real arrows represent that the transfers of states do not execute actions.

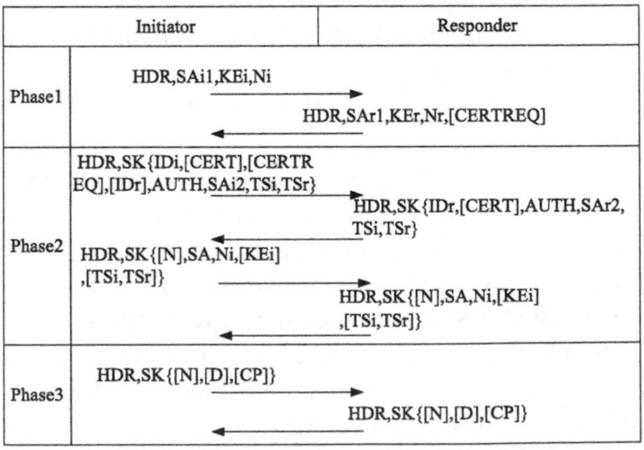

Fig. 1 The exchange phases of IKEv2 protocol

Table 1 The events of IKEv2 protocol

Name	Content
E1	The process of FSM model has started
E2	The Responder has received the message of HDR, SAi1, KEi, Ni from the Initiator
E3	The Initiator has received the message of HDR, SAr1, KEr, Nr, [CERTREQ] from the Responder
E4	The establishment of IKE_SA completes
E5	The Responder has received the message of HDR, SK {IDi, [CERT], [CERTREQ], [IDr], AUTH, SAi2, TSi, TSr} from the Initiator
E6	The Initiator has received the message of HDR, SK {IDr, [CERT], AUTH, SAr2, TSi, TSr} from the Responder
E7	The exchanges of IKE_AUTH complete
E8	The Responder has received the message of HDR, SK {[N], SA, Ni, [KEi], [TSi, TSr]} from the Initiator
E9	The Initiator has received the message of HDR, SK {SA, Nr, [KEr], [TSi, TSr]} from the Responder
E10	Error warning

Table 2 The states of IKEv2 protocol

Name	Content
S1	The Initiator sends a request for establishing IKE_SA to the Responder
S2	The Responder responds to the establishment request of IKE_SA
S3	A shared key is generated, and the IKE_SA is established
S4	The Initiator sends a request of IKE_AUTH to the Responder
S5	The Responder verifies the identity of the Initiator, and responds to the request of IKE_AUTH
S6	The Initiator verifies the identity of the Responder
S7	The Initiator sends a request for establishing IPSec_SA to the Responder
S8	The Responder responds the request of establishing IPSec_SA
S9	The IPSec_SA is established
S10	Exception Interrupt

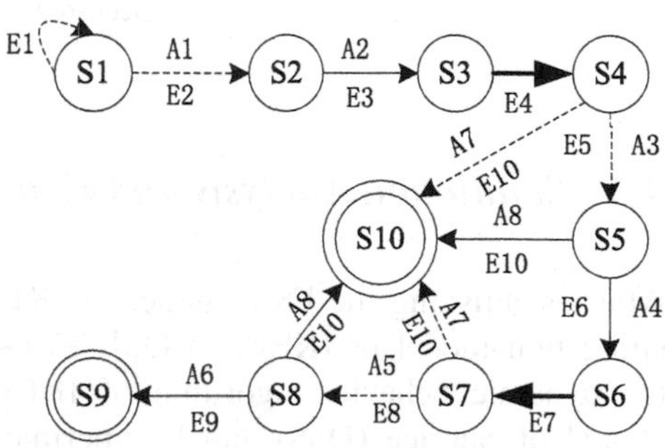

Fig. 2 The FSM model of IKEv2 protocol

Table 3 The actions of IKEv2 protocol

Name	Content
A1	The Initiator sends a message of HDR, SAi1, KEi, Ni to the Responder
A2	The Responder sends a message of HDR, SAr1, KEr, Nr, [CERTREQ] to the Initiator
A3	The Initiator sends a message of HDR, SK {IDi, [CERT], [CERTREQ], [IDr], AUTH, SAi2, TSi, TSr} to the Responder
A4	The Responder sends a message of HDR, SK {IDr, [CERT], AUTH, SAr2, TSi, TSr} to the Initiator
A5	The Initiator sends a message of HDR, SK {[N], SA, Ni, [KEi], [TSi, TSr]} to the Responder
A6	The Responder sends a message of HDR, SK {SA, Nr, [KEr], [TSi, TSr]} to the Initiator
A7	The Initiator sends a warning message of HDR,SK{N, [D], [CP]} to the Responder
A8	The Responder sends a warning message of HDR, SK{N, [D], [CP]} to the Initiator

Table 4 The events response table of IKEv2 protocol

The source state	Events	Conditions	Actions	The end state
S1	E1	No conditions	No actions	S1
	E2	No conditions	A1	S2
S2	E3	No conditions	A2	S3
S3	E4	No conditions	No actions	S4
S4	E5	No conditions	A3	S5
	E10	Exception occurred	A7	S10
S5	E6	No conditions	A4	S6
	E10	Exception occurred	A8	S10
S6	E7	No conditions	No actions	S7
S7	E8	No conditions	A5	S8
	E10	Exception occurred	A7	S10
S8	E9	No conditions	A6	S9
	E10	Exception occurred	A8	S10

4.2 Simulation Analysis and Experiments in OMNeT++

After establishing the FSM model of IKEv2 protocol, the paper implements the simulation model of IKEv2 in OMNeT++. The validity of the FSM model and the dynamic behavior algorithm of IKEv2 protocol are tested by designing a Denial of Service (DoS) attack experiment scene, in which all the clients and

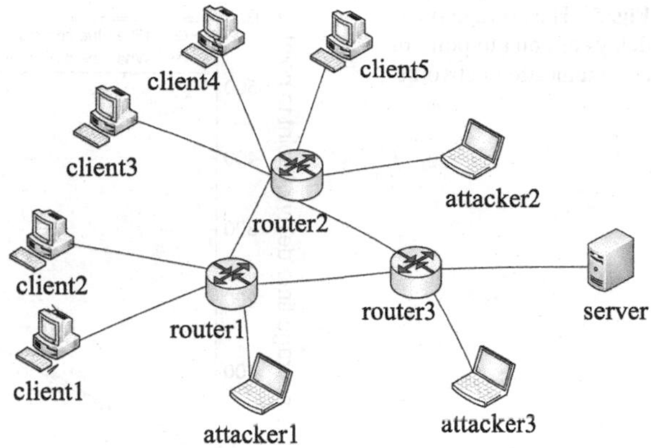

Fig. 3 The DoS attack experiment scene

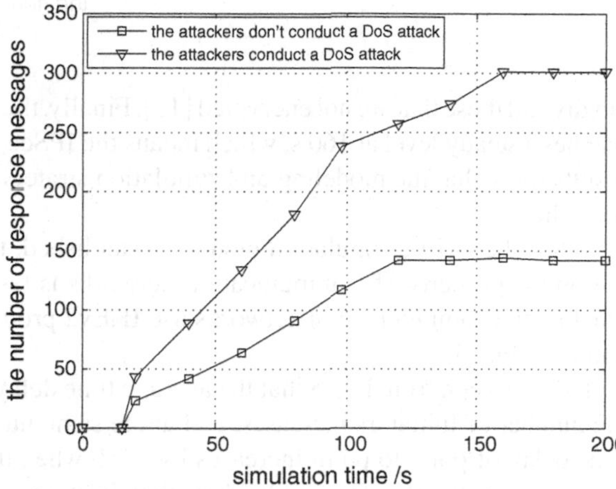

Fig. 4 The number of response messages of the server

attackers are Initiators and the server is the Responder of the communication, and it is shown in Fig. 3.

During the establishment of IKEv2 simulation model, the clients send messages of negotiating parameters for establishing the IPSec_SA to the server, while the attackers send error messages to the server conducting a DoS attack. The server responds to all the messages. Supposing when the IPSec_SA is established, the server will not send response messages. When the simulation time is set to 200 s, the response messages of the server are counted whether the attackers conduct a DoS attack or not, and the results are shown in Fig. 4.

It can be seen from Fig. 4 that the number of response messages sent by the server increases rapidly after 15 s. The number of response messages reaches a steady level at 120 s when the attackers dont conduct a DoS attack, which means the IPSec_SA is established. However, the number of response messages first increases rapidly when the attackers conduct a DoS attack, but the increment speed becomes slow after 98 s, for the IKE_SA is established and it can encrypt the following negotiating messages

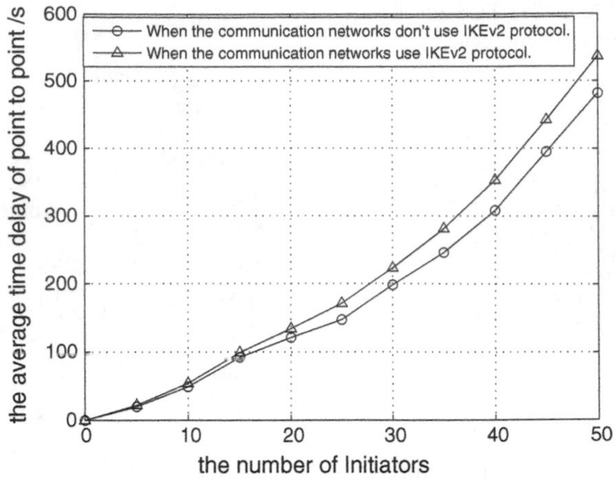

Fig. 5 The average time delays of point to point in communication networks

but discard these that are not encrypted [12]. Finally, the number of response messages reaches a steady level at 160 s, which means the IPSec_SA is established. The above results show that the modeling and simulation strategies of cryptographic protocols are valid.

When the IKEv2 simulation model is established, the influence of IKEv2 protocol on the property of communication networks is tested under the conditions that whether the communication networks use IKEv2 protocol or not, and the results are shown in Fig. 5.

It can be seen from Fig. 5 that the average time delays of point to point increase as the number of Initiators increases, and at the same number of Initiators the average time delay of point to point increases by 12 % when using IKEv2 protocol in communication networks. The results show that the usage of cryptographic protocols has a definite influence on the property of communication networks.

5 Conclusions

The paper proposes a modeling method and a dynamic behavior algorithm based on FSM to simulate cryptographic protocols. The simulation results show that the method is effective in describing cryptographic protocols, and the algorithm is valid in simulating dynamic behaviors of cryptographic protocols. Therefore the paper provides a new effective means to do researches on cryptographic protocols. Mending the modeling and simulation strategies to improve the precision of simulating cryptographic protocols is our future work.

References

1. Meadows C (2011) Formal analysis of cryptographic protocols. Encycl Cryptogr Secur 490–492
2. Liu S, Ye J (2012) Computational rationality of formal analysis on cryptographic protocols. In: Proceedings of the ICCIS. IEEE, pp 904–907
3. Meadows C (2015) Emerging issues and trends in formal methods in cryptographic protocol analysis: twelve years later. Logic, rewriting, and concurrency. Springer International Publishing, New York, pp 475–492
4. Blanchet B (2012) Security protocol verification: symbolic and computational models. In: Proceedings of the POST12. Springer, New York, pp 3–29
5. Meier S, Schmidt B, Cremers C et al (2013) The TAMARIN prover for the symbolic analysis of security protocols. Computer aided verification. Springer, Berlin, pp 696–701
6. Pankova A, Laud P (2012) Symbolic analysis of cryptographic protocols containing bilinear pairings. In: Proceedings of CSF. IEEE
7. Brand D, Zafiropulo P (1983) On communicating finite-state machines. J ACM (JACM) 30(2):323–342
8. Aarts F, Jonsson B, Uijen J et al (2014) Generating models of infinite-state communication protocols using regular inference with abstraction. Form Methods Syst Des 46(1):1–41
9. Varga A (2014) OMNeT++ user manual version 4.6. http://www.omnetpp.org/doc/omnetpp/manual.html
10. Kaufman C, Hoffman P, Nir Y et al (2014) Internet key exchange protocol version 2 (IKEv2). IETF
11. Kasraoui M, Cabani A, Chafouk H (2014) IKEv2 authentication exchange model in NS-2. In: Proceedings of the IS3C. IEEE, pp 1074–1077
12. Patel H, Jinwala D (2013) Modeling and analysis of internet key exchange protocolv2 and a proposal for its variant. In: Proceedings of the 6th ACM India computing convention. ACM

Analysis of Urban Link Travel Time Reliability Based on Odd and Even Limit Period

Wang Zhijian, Li Liang and Hou Zhengying

Abstract In order to analyze urban link travel time reliability on odd or even limit period, firstly, the reliability evaluation index of travel time is established: the 90 % stroke time and the probability of reaching the level of specific service. Then odd or even limit and daily morning peak travel time probability distribution are compared. And we draws a conclusion that for the odd or even limit period, lognormal distribution can better reflect the link travel time probability distribution, while working and rest days are more in line with normal distribution. Finally, travel time reliability is analyzed. And travel time reliability of odd and even limit period is greatly improved compared with daily. The study has guiding significance for odd and even limit period.

1 Introduction

With the development of national economy, the quantity of motor vehicles is increasing rapidly, and road traffic congestion is also increasing seriously. With the concept of time gradually improved, travelers not only hope to reduce travel time, but also hope to reduce the volatility of travel time to ensure the probability of reaching destination on time. Therefore, travel time reliability which can reflect travel time volatility is more and more important to describe the stability of the road, at the same time it has important guiding significance for people to choose the appropriate path.

W. Zhijian · L. Liang (✉) · H. Zhengying
Beijing Key Lab of Urban Intelligent Traffic Control Technology,
North China University of Technology, Beijing 100144, China
e-mail: liliang102018@163.com

W. Zhijian
e-mail: wzjian0722@163.com

H. Zhengying
e-mail: houzhengyingzdh@163.com

© Springer International Publishing Switzerland 2017
V.E. Balas et al. (eds.), *Information Technology and Intelligent Transportation Systems*, Advances in Intelligent Systems and Computing 454,
DOI 10.1007/978-3-319-38789-5_63

2 Status Research

According to the national standard GB3187-825 "Reliability Basic Terms and Definitions", reliability is defined as probability of the product complete specified function under prescribed conditions and provisions time [1]. In 1991, Asakura Y and Kashiwadani M [2] first proposed the concept of travel time reliability: for a given original and destination, the probability of travelers complete the travel within the specified time. The basic expression is:

$$R(t) = P(t_i \leqslant T) = \int_0^T f(t)dt \tag{1}$$

$R(t)$ is link travel time reliability, that is, the probability of successful completion of travel within specified time. And t_i is the time that car i required for the completion, and T is the travel time threshold, which can generally represents the acceptable travel time from original to destination [3]. $f(t)$ is the probability distribution function of link travel time. From the above definition, we can know that the distribution of travel time probability has an important influence on travel time reliability.

Lomax [4] in 2003 proposed buffer time index, which is defined as the difference between the average travel time and 95 % stroke time. Van Lint [5] in 2005 used to the width of travel time distribution, that is the ratio of (90 % bit −50 % bit) and (50 % bit −10 % bit), as well as the partial distribution, that is the ratio of (90 % stroke time −10 % bit) and the median. But the evaluation index proposed in the above two papers can not be directly understood by travelers. The traditional travel time reliability evaluation is usually assumed that link travel time obeys normal distribution [6, 7], but the actual situation is not the case. Arroyo S (2005) [8] and Hesham Rakha (2006) [9] find that log normal distribution is better fitting travel time distribution by using highway data. And Hesham Rakha also think it is also feasible for the calculation of the normal distribution assumption. But the data source is from highway, but the city link is different. In the 2011 year, Takahiro Tsubota [10] analyzed the relationship between traffic accident frequency and travel time reliability.

In 2007, Li Xian [11] proposed unit distance travel time reliability evaluation index and method using taxi IC card data, but the data is from pure taxi data, and it has certain limitations. In the same year, Chen Xiaohong [12] put forward the empirical distribution method of link travel time for a highway road, but the proposed empirical distribution is only suitable for this road, and the characteristics of the highway and urban road section are different. In 2014, Li Changcheng [3] used the data of highway road to analyze the probability distribution of travel time, which showed that using normal distribution to describe link travel time is reasonable. At the same time, some experts and scholars have studied travel time reliability on the abnormal condition, such as Leng Junqiang [13] (2010) analyzed urban road network travel time reliability under the condition of snow.

As China's political center, Beijing has carried out odd and even limit policy sometimes, such as during the 70th anniversary of the victory of the war of resistance against Japan in 2015, and the APEC meeting period, etc. It shows that the research of travel time reliability in odd and even limit period has important significance. But the research of different travel time probability distribution in odd and even limit period and daily is not enough.

3 Establish Evaluation Index

In order to combine travel time reliability and service level, travel time reliability is further defined as the probability of reaching a certain service level in the peak period. So there are two reliability indexes:

(1) 90 % stroke time, which can guide travelers to set aside certain time to ensure the probability of reaching the destination on time, and find the traffic "weak point";

(2) The probability of reaching D service level in each section of the road, which can provide reference for the traffic management and transportation planning. The process of determining the threshold of D service level is as follows: The relationship between road service level and desired speed is shown in Table 1 by the American Road Traffic Capacity Manual (HCM 2000) [14]. The above-mentioned table is converted into the unit distance travel time as in Table 2.

Table 1 The relationship between road service level and desired speed

Road grade	A	B	C	D	E	F
Expressway	72	56	40	32	26	<26
Trunk roads	59	46	33	26	21	<21
Secondary roads	50	39	28	22	17	<17
Branch roads	41	32	23	18	14	<14

Table 2 The relationship between road service level and unit distance travel time

Road grade	A	B	C	D	E	F
Expressway	0.8333	1.0714	1.5	1.875	2.3077	>2.3077
Trunk roads	1.0149	1.3043	1.8182	2.3077	2.8571	>2.8571
Secondary roads	1.2	1.5384	2.1428	2.7273	3.5294	>3.5294
Branch roads	1.4634	1.875	2.0687	3.3333	4.2857	>4.2857

4 Travel Time Data Acquisition and Processing

Because of the periodicity of the travelers' travel and the repeatability of travel time, the time of the road trip is similar in the week. So we collect travel time data on morning rush hour for a total of four classes, respectively Thursday and Saturday on July 1 to July 31 as well as odd and even limit period in Beijing parade (from August 20 to September 3). Thursday represents typical working day, while Saturday is on behalf of typical rest day. At the same time, we can get the difference of travel time probability distribution in odd and even limit period. The selected section is between export of Fushi Lu and Bajiaodong Jie intersection and import of Yangzhuangdong Jie and Jinyuanzhuang Lu. The length of intersection region is determined to import queue length in peak period. And according to the actual investigation, the selected region length of the intersection is 30 m. The collection section is as shown in Fig. 1, which the total length is 1050 m, and reaching D service level is 3.5 min. We can directly collect the data of all kinds of models, mainly for taxis and private cars by using video detection method to collect travel time data.
Some of the data is shown in Table 3:

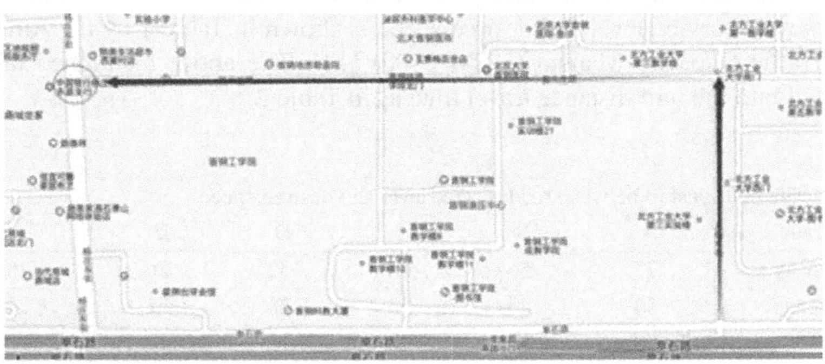

Fig. 1 Diagram of road range

Table 3 Some of the datas

Plate number	Export	Import	Duration	Duration (min)
9Q81	7:22:16	7:23:27	0:01:11	1.1833
2886	7:10:42	7:11:56	0:01:14	1.2334
A519	7:04:46	7:06:31	0:01:45	1.75
FH75	7:53:04	7:54:56	0:01:52	1.867
36Z7	8:06:45	8:08:46	0:02:01	2.0167

5 Travel Time Data Acquisition and Processing

According to the definition of travel time reliability, in order to calculate travel time reliability we should determine the probability distribution of link travel time [15].

5.1 Distribution Diagram

Before determining travel time distribution, the distribution can be estimated, which can validate distribution function purposely to improve the efficiency [16]. In this paper, we use thc histogram estimation method to estimate the distribution. The collected data are transformed into a histogram as shown in Fig. 2. Through the observation of travel time frequency distribution histogram, we can find that the distribution function is generally close to the normal distribution, lognormal distribution and gamma distribution, weibull distribution.

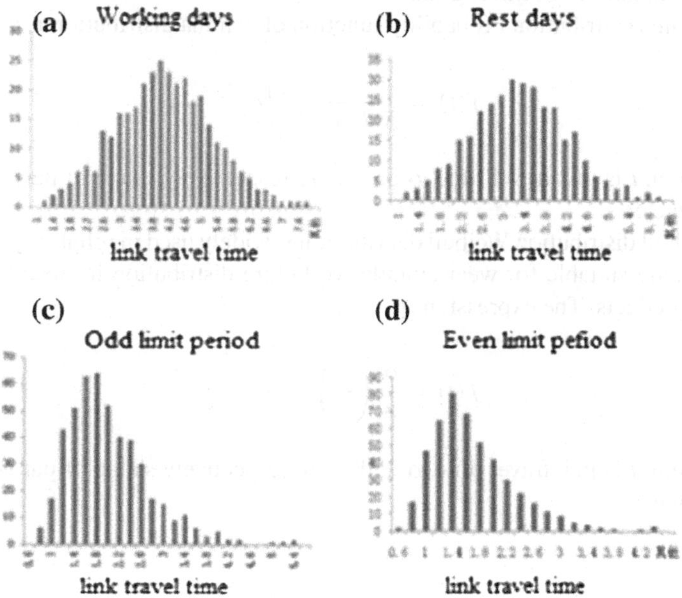

Fig. 2 Diagram of travel time distribution on different days. **a** Working days. **b** Rest days. **c** Odd limit period. **d** Even limit period

5.2 *Common Distribution Function*

(1) Normal distribution At present, the most widely used probability distribution function is normal distribution. The basic expression is:

$$f(t) = \frac{1}{\sqrt{2\pi\sigma}} exp\left[\frac{-(t-\mu))^2}{2\sigma^2}\right] \tag{2}$$

Among them, t is section travel time, μ and σ are respectively the expectation and variance of travel time.

(2) Lognormal distribution If the logarithm of a random variable is normally distributed, then the random variable obeys the lognormal distribution. The expression is:

$$f(t) = \frac{1}{\sigma t\sqrt{2\pi}} exp\left[-\frac{1}{2}\left(\frac{lnt-\mu}{\sigma}\right)^2\right] \tag{3}$$

Among them, t is link travel time, μ and $sigma$ are respectively the expectation and variance of logarithm of travel time.

(3) Gamma distribution Probability function of gamma distribution is as follows:

$$f(t) = \frac{1}{\lambda^a \Gamma(\alpha)} t^{\alpha-1} e^{-\frac{t}{\lambda}} \tag{4}$$

Among them, t is link travel time, α and λ are respectively shape of parameter and scale parameter.

(4) Weibull distribution Weibull distribution is widely used in reliability engineering, especially suitable for wear cumulative failure distribution in mechanical and electrical products. The expression is:

$$f(t) = \frac{\alpha}{\lambda}\left(\frac{t}{\lambda}\right)^{\alpha-1} e^{-\frac{t}{\lambda}^{\alpha}} \tag{5}$$

Among them, t is link travel time, α and λ are respectively shape of parameter and scale parameter.

5.3 *Data Fitting and Test of Goodness of Fit*

The results using dfittool toolbox by matlab are shown in the following figures: From the above curve can visually see for the odd or even limit period, lognormal distribution can better reflect the link travel time probability distribution, while working and rest days are more in line with normal distribution. Then we use sum of squares error and R-square to measure the effect of the curve fitting further [17]. The sum of squares error is the square of difference between the actual value and the fitting value, the calculation formula is (Fig. 3):

$$SSE = \sum_{i=1}^{n} (x_i - \hat{x}_i)^2 \tag{6}$$

Among them, x_i represents the actual value, $\hat{x}_i$ indicates the fitting value. The smaller of the sum of squares error, the better of the fitting effect. R-square is used to characterize the fitting effect by the variation of the fitting data with the actual mean value. Its calculation formula is:

$$R - square = 1 - \frac{\sum_{i=1}^{n} (x_i - \hat{x}_i)^2}{\sum_{i=1}^{n} (x_i - \bar{x}_i)^2} \tag{7}$$

$\bar{x}_i$ is the mean of the original data. Its value is between 0 and 1. The closer R-square to 1, the more close to reality. Goodness of fit test results for each fitting function is shown in Tables 4 and 5. The goodness of fit test results of SSE and R-square above shows that for the odd or even limit period, lognormal distribution can better

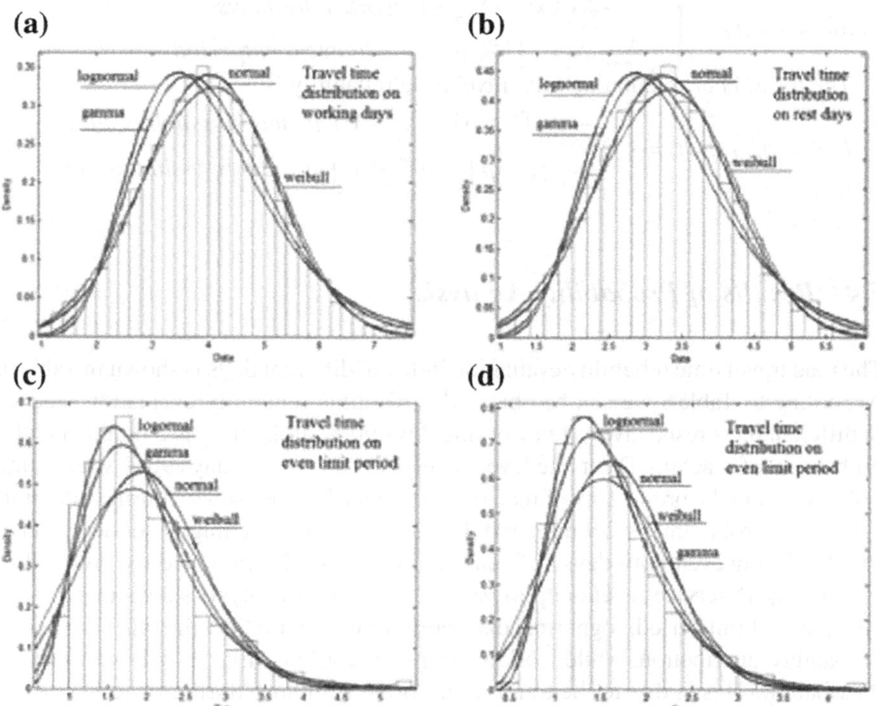

Fig. 3 Diagram of travel time distribution on different days. **a** Working days. **b** Rest days. **c** Odd limit period. **d** Even limit period

Table 4 The SSE of fitting function

	Normal	Lognormal	Gamma	Weibull
Working days	0.001	0.051	0.022	0.005
Rest days	0.005	0.048	0.020	0.013
Odd limit period	0.335	0.018	0.052	0.199
Even limit period	0.202	0.003	0.032	0.176

Table 5 The R-square of fitting function

	Normal	Lognormal	Gamma	Weibull
Working days	0.997	0.872	0.944	0.986
Rest days	0.991	0.916	0.965	0.978
Odd limit period	0.592	0.920	0.910	0.637
Even limit period	0.801	0.996	0.968	0.805

reflect the link travel time probability distribution, while working and rest days are more in line with normal distribution. Therefore, the link travel time distribution is as follows: $f(t) = \begin{cases} \frac{1}{\sqrt{2\pi}\sigma} \exp(-\frac{(t-\mu)^2}{2\sigma^2}), \; workandrestdays \\ \frac{1}{\sqrt{2\pi}\sigma t} \exp(-\frac{1}{2}\frac{(\ln t-\mu)^2}{2\sigma^2}), \; oddorevenlimitperiod \end{cases}$

The formula of calculating link travel time reliability is:

$$R(t) = \int_0^T f(t)dt = \begin{cases} \frac{1}{\sigma\sqrt{2\pi}} \int_0^T exp\left[-\frac{1}{2}(\frac{t-\mu}{\sigma})^2\right] dt, \; workingandrestdays \\ \frac{1}{\sqrt{2\pi}\sigma} \int_0^T exp\left[-\frac{1}{2}(\frac{lnt-\mu}{\sigma})^2\right] dt, \; oddorevenlimitperiod \end{cases}$$

5.4 *Results of Probability Analysis*

The road travel time reliability evaluation index of different days is shown in Table 6. According to Table 6 we can be obtained travel time reliability evaluation indexes in different days respectively. On working days 90 % stroke time is 5.6 min, and the probability of reaching D service level is only 33.0 %; on rest days 90 % stroke time is 4.6 min, and the probability of reaching D service level is 64.9 %; during odd limit days 90 % stroke time is 2.9 min, and the probability of reaching D service level is 96.3 %; during even limit days 90 % stroke time is only 2.4 min, and the probability of reaching D service level is up to 98.9 %. The above analysis shows that for the odd or even limit period, lognormal distribution can better reflect the link t ravel time probability distribution, while working and rest days are more in line with normal distribution. This is due to the traffic in morning rush hour during the odd or even limit period is light, mutual influence between vehicles is also small, most vehicles can travel at high speed, therefore travel time will appear the phenomenon of left side,

Table 6 The R-square of fitting function

	90 % stroke time (min)	The probability of reaching D service level (%)
Working days	5.6	33.0
Rest days	4.6	64.9
Odd limit period	2.9	96.3
Even limit period	2.4	98.9

but due to the drivers' driving habits are different, the speed will be different. While the traffic in morning rush hour during the working days and rest days is heavy, mutual influence between vehicles is large, vehicle speed has a great difference. Making a comparison of travel time reliability evaluation indexes on different days, we can obtain that the travel time reliability during even limit period is highest, odd limit period takes second place, while travel time reliability on working days is the lowest. This is because as a powerful measure nearly 50 % of traffic demand will be restricted during odd or even limit period, which can greatly improve the link travel time reliability. And by the comparison of travel time reliability evaluation index during odd and even limit period, we can conclude that travelers will spend more time on odd limit days than even limit days. As travelers are not concentrated in the morning rush hour on rest days, so travel time reliability on rest days is respectively higher. And because we need set aside a long time to pass this link and the probability of reaching D service level is low on daily, so this link is a weak point of traffic, which is in urgent need of improvement.

6 Summary and Outlook

In this paper, we first establish evaluation indexes of link travel time reliability: the 90 % stroke time and the probability of reaching the level of specific service. Secondly, we analyze link travel time reliability for a specific section And we draws a conclusion that for the odd or even limit period, lognormal distribution can better reflect the link travel time probability distribution, while working and rest days are more in line with normal distribution, which can provide reference for link travel time reliability evaluation. But this paper only analyzes the travel time reliability of a specific road, but also need further study of path travel time reliability, in order to find the "weak point", which can provide reference for traffic management and traffic planning.

Acknowledgments This research is supported by Project of National Natural Science Foundation (Project No.:61503006) and Beijing Municipal Education Commission science and technology Program (Project No.: KM201510009002), the Peoples Republic of China.

References

1. GB3187- 825: Reliability basic terms and definitions (in Chinese)
2. Asakura Y, Kashiwadani M (1991) Road network reliability caused by daily fluctuation of traffic flow. PTRC summer annual meeting, 19th, 1991. University of Sussex, United Kingdom
3. Li C, Wen T, Liu X et al (2014) Study on reliability model of travel time based on expressway toll data. J Highw Transp Res Dev (12) (in Chinese)
4. Lomax T, Schrank D, Turner S et al (2003) Selecting travel reliability measures. Texas Transp Inst Monogr
5. van Lint JWC (2004) Reliable travel time prediction for freeways: bridging artificial neural networks and traffic flow theory. TU Delft, Delft University of Technology
6. Bcll MGH, Iida Y (1997) Transportation network analysis
7. Iida Y (1999) Basic concepts and future directions of road network reliability analysis. J Adv Transp 33(2):125–134
8. Arroyo S, Kornhauser AL (2005) Modeling travel time distributions on a road network. In: Proceedings of the 2005 TRB annual conference
9. Rakha H, El-Shawarby I, Arafeh M et al (2006) Estimating path travel-time reliability. In: Intelligent transportation systems conference, ITSC'06. IEEE, pp 236–241
10. Tsubota T, Kikuchi H, Uchiumi K et al (2011) Benefit of accident reduction considering the improvement of travel time reliability. Int J Intell Transp Syst Res 9(2):64–70
11. Li X, Wen H, Gao Y et al (2007) Beijing road network rate travel time reliability evaluation. J Transp Syst Eng Inf Technol 7(2):72–76 (in Chinese)
12. Chen X, Feng J, Yang C (2007) Research on travel time reliability characteristics based on floating car data. Urban Transp 5:42–45. doi:10.3969/j.issn.1672-5328.200705.008 (in Chinese)
13. Leng J (2010) Travel time reliability of urban road network under ice and snowfall conditions. Harbin Institute of Technology (in Chinese)
14. Manual H C. HCM 2000. Washington, DC: Transportation Research Board, 2000
15. He F (2010) Research on link travel time reliability based on floating car data. Southwest Jiaotong University. doi:10.7666/d.y1687586 (in Chinese)
16. De-feng L, Xiu-cheng G (2011) Study on path travel time reliability model. In: Proceedings of the international conference on intelligent computation technology and automation (ICICTA). IEEE, vol 1, pp 948–951
17. Chen K (2008) Models and algorithms for travel time reliability assessment of Urban road networks based-on moving source data. Beijing Jiaotong University (in Chinese)

Simulation and Analysis of the Performance of a Novel Spread Spectrum Sequence in CDMA System

Lv Hong, Yu Yong-Lin, Qi Peng and Hua Zhi-Xiang

Abstract The selection of spread spectrum sequence in *CDMA* system directly affects the performance of the whole system, which is one of the key technologies of *CDMA* system. The generation model of *m* sequence and *m* sequence is constructed, and the *CDMA* communication simulation system is built by using *MATLAB*/Simulink platform. Then the spread spectrum sequence is introduced into the *CDMA* system to run and analyze the performance of the two sequences in the *CDMA* system with the 6 order and 7 order. The results show that the *m* sub-sequence is rich in resources, and the bit error rate performance of the multi user *CDMA* communication system is better than that of the *m* sequence.

1 Introduction

CDMA technology in anti interference, security, communication capacity, transmission rate, and has a prominent advantages, At first, it is mainly used in the military field and now it has been widely applied in electronic countermeasure, navigation, survey, mobile communication, Internet transmission, intelligent household and other fields to provide users with stable and reliable data transmission as demonstrated [1, 2]. With the development of *CDMA* technology, the use of traditional hardware for communication systems and spread spectrum sequence research will have a variety of defects, such as stability, cost, flexibility, efficiency, etc. The *CDMA* communication system, spread spectrum sequence was simulated based on *MATLAB*/Simulink

L. Hong · Y. Yong-Lin (✉) · Q. Peng · H. Zhi-Xiang
AnHui Jianzhu University, Hefei, China
e-mail: tougao176@163.com; 1065517780@qq.com

L. Hong
e-mail: lvhong176@163.com

Q. Peng
e-mail: yanjiusheng176@163.com

H. Zhi-Xiang
e-mail: ziliaolh@163.come

© Springer International Publishing Switzerland 2017
V.E. Balas et al. (eds.), *Information Technology and Intelligent Transportation Systems*, Advances in Intelligent Systems and Computing 454,
DOI 10.1007/978-3-319-38789-5_64

platform in this paper, and the 6 order and 7 order *m* sequence and the *M* sub sequence generation model are introduced into the construction of *CDMA* system and simulation analysis.

2 Simulation of CDMA Communication System

2.1 CDMA Communication System

CDMA (Code Division Multiple) is a way of communication of using different and the orthogonal (quasi orthogonal) address Code to assign to different users to modulation signals, enabling multiple users to use the same frequency at the same time, same channel access communication network. Due to the use of different, orthogonal (Quasi orthogonal) address code to modulation signal, broadening the original signal spectrum bandwidth, also known as the spread spectrum communication as demonstrated [3, 4].

As shown in Fig. 1, *CDMA* communication principle is: the source of the original signal after channel coding, first carries on the spread spectrum modulation, and then to carry on the digital modulation, to send signals to the channel, the receiver receives signals after demodulation, and it will recover original signal by using the same pseudo random sequences algorithm of the sending end for despreading, then after filtering and sampling judgement process as demonstrated [4, 5].

2.2 The Establishment of the CDMA System Based on SIMULINK

SIMULINK is a tool-kit for *MATLAB*, providing integrated environment for dynamic visual modelling, simulation and analysis. The *SIMULINK* system based on *CDMA*

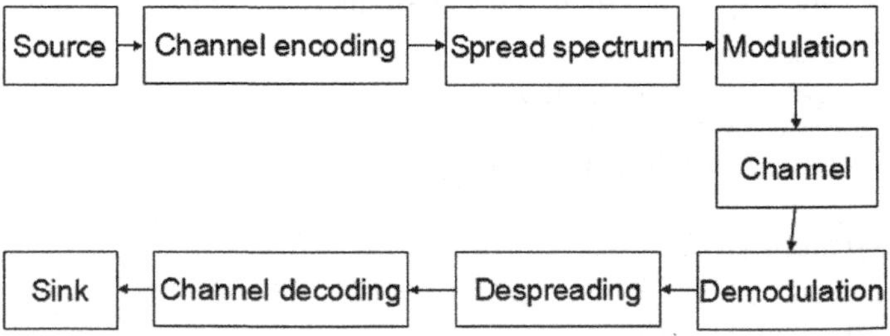

Fig. 1 CDMA communication system schematic diagram

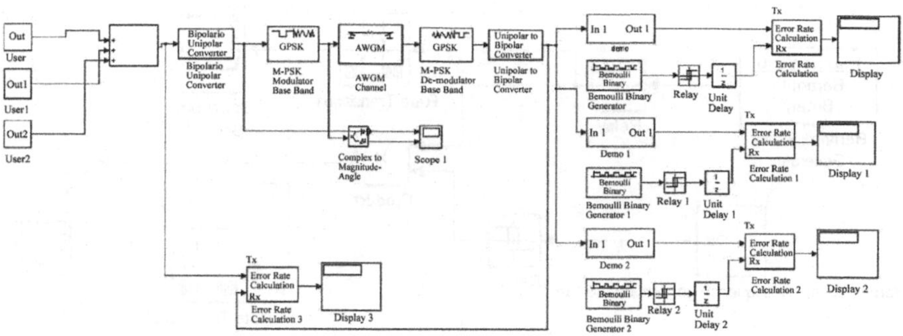

Fig. 2 CDMA simulation system diagram

is built by using Simulink as demonstrated [6, 7], and the system simulation model based on *CDMA* communication principle is shown in Fig. 2.

3 Simulation of Spread Spectrum Sequence

In CDMA systems, spread spectrum uses pseudo-random code with characteristic of very wide and uniform spectrum, the selection of spread spectrum sequence can directly affect the performance of the whole system. The commonly used sequences are m sequences, GOLD codes, etc. In this paper, we will use the m sequence and the m subsequence generated by the m sequence, and then construct the model.

3.1 Construction of Sending Module

Single-user spread spectrum signal generation module Source using binary Bernoulli sequence generator generates a binary random sequence, under the multi user, set each user source of initial state and the binary distribution to simulate the source of each user, and the each are not identical. In the spread spectrum code sequence, the sequence generator is constructed by using the logic operation module and the time delay module. According to the feedback function of m sequence and its sub sequence, the different logical operations are constructed to generate the corresponding pseudo random sequence. In order to make the whole system simple, the source and the spread spectrum part encapsulation in user module, the user is shown in Fig. 3.

The binary information generating by signal source module is the polarity of the binary signal is converted to $(-1, 1)$ through the relay module, spread spectrum code generating module after the polarity of the relay module is converted to $(-1, 1)$. After multiplier module for spread spectrum processing.

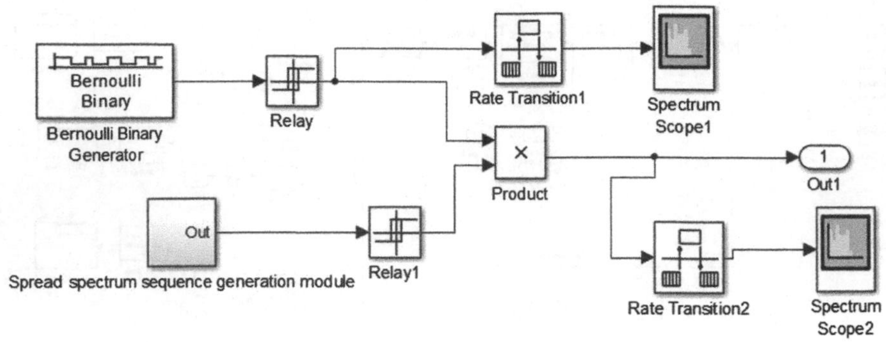

Fig. 3 User module

Fig. 4 *QPSK* signal space constellation

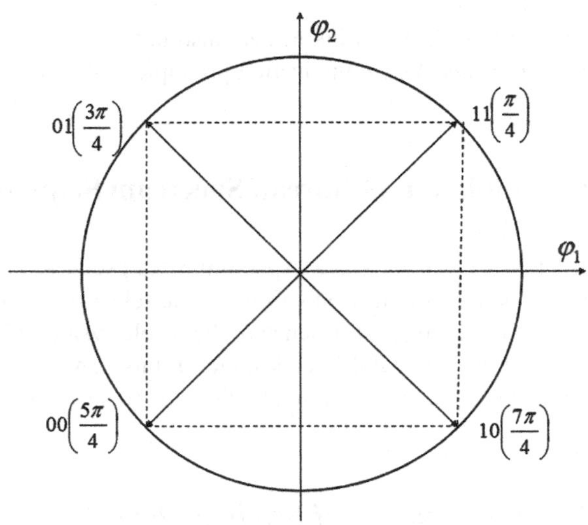

Digital modulation $M - PSK$ modulation (Modulator Baseband $M - PSK$) is used to set the phase modulation, the initial phase is 4, every 2 bits of the input binary signal is divided into a group through *QPSK*, a total of four kinds of combination, namely 11, 01, 00, 10 as demonstrated [8, 9]. To represent each arrangement with one of 4 phases.*QPSK* is the binary signal input into a group of 2 bits, a total of 4 combinations, that is, 11, 01, 00, 10. To represent each arrangement with 4 phases. In the $M - PSK$ (Modulator Baseband $M - PSK$) is set up 4 phase demodulation, *QPSK* signal space constellation, as shown in Fig. 4.

Receiver processing module The sending end of the N user information into the air, information is completely mixed together in the time domain, the frequency domain, each user of the receiving end can receive all signals as demonstrated [10]. The received signal is received by the receiving end of the first subscriber antenna:

$$S(t) = \sum_{i=0}^{N} S_i(t) = \sum_{i=0}^{N} A_i(t) \cdot d_i(t) \cdot c_i(t) \cdot \cos w_c(t) \tag{1}$$

The signal after demodulation is:

$$S_{EX}(t) = \sum_{i=0}^{N} A_i' \cdot d_i(t) \cdot c_i(t) \tag{2}$$

After the correlation detection of the $C1(T)$ with the local spreading code, the signal is amplified:

$$d_1' = \int_0^T S_{EX}(t) \cdot c_1(t)\, dt = \sum_{i=0}^{N} A_i' \cdot \int_0^T d_i(t) \cdot c_i(t) \cdot c_1(t)\, dt \tag{3}$$

Among them, T is the symbol period, which is equal to the source encoding cycle of T_b, so the integral number of citic code $d_i(T)$ is constant,

$$d_1' = \sum_{i=0}^{N} A_i' \cdot d_i(t) \cdot \int_0^T c_i(t) \cdot c_1(t)\, dt \tag{4}$$

Cross-correlation function of the PN code is 0, so:

$$R_{i,j}(\tau) = \int_0^T c_i(t) \cdot c_j(t-\tau)\, dt = 0, \qquad i \neq j \tag{5}$$

Plug in type (4), by orthogonality relation of spread spectrum code, we can get:

$$d_1' = A_1' \cdot d_1(t) \cdot R_1(o) = A_1'' \cdot d_1(t) \tag{6}$$

Among them, t is the peak value of the auto-correlation function of $C_1(t)$. The square wave form $d_1(t)$ is obtained by the sampling of. The receiving end user 1 receives a sequence of 1 of the transmit end user from the sending end N uscr to the air in the time domain and frequency domain as demonstrated [11–13].

As shown in Fig. 5, in the encapsulated module, signal after demodulation multiplied with the corresponding spread spectrum sequence to complete solutions for expansion, the signal after expansion is a narrow-band signal. Other users are still broadband signals, so a low pass filter is designed to extract the narrow band signal. In the system, the filter module (Filter Design Digital) is set the filter for the FIR low pass filter, $F_s = 100\,\text{Hz}$, $F_{pass} = 4\,\text{Hz}$, $F_{stop} = 9\,\text{Hz}$. Finally using the Relay module for sampling judgement, the signal is extracted which is received by the user 1, and the error rate is calculated by comparing with the signal of user 1.

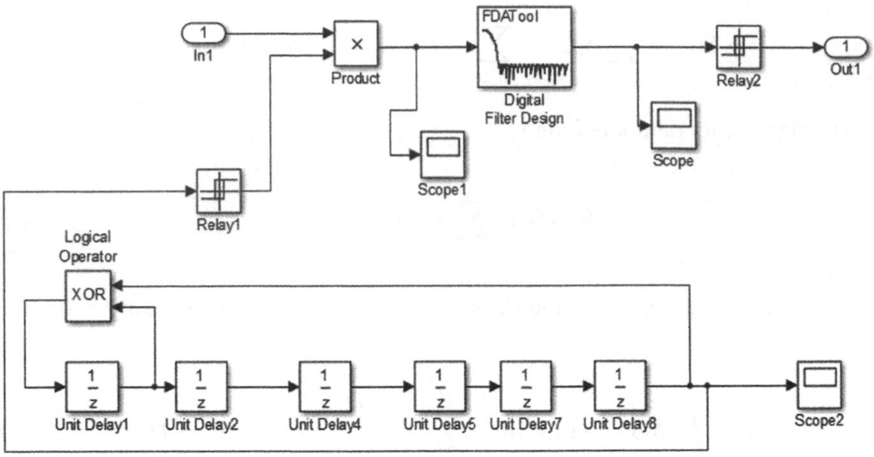

Fig. 5 Receiver processing module

4 The Structure of the Spread Spectrum Sequence

In *CDMA* systems, spread spectrum use pseudo-random code with very wide and uniform spectrum characteristic, the selection of spread spectrum sequence directly impacts on the performance of the whole system, some sequences are frequently used, such as m sequences and *GOLD* code, etc. as demonstrated [14, 15]. In this paper, we will use the m sequence and the *m* sub-sequence generated by the m sequence, and then construct the model.

4.1 m Sequence and the m Sub-sequence

The *m* sequence (the longest linear feedback shift register sequence for short) is the longest cycle sequence produced by linear feedback shift register, a level of *n* linear feedback shift register (as shown in Fig. 6) can produce states, so the maximum period of sequence is, easy to implement, and has good auto-correlation properties as demonstrated [16].

Different from the *m* sequence, the m sub-sequence is generated by the longest period of the non-linear feedback shift register. Because of the *m* sequence of a linear feedback shift register generator, so when the state is full “0”, the state of the generator will not change, it has been maintained all “0”. But m sub-sequence is generated by a non-linear feedback shift register, so the maximum cycle of m sub-sequence can be and can reach and it can achieve the same as the *m* sequence features of run and balance of properties as demonstrated [17].

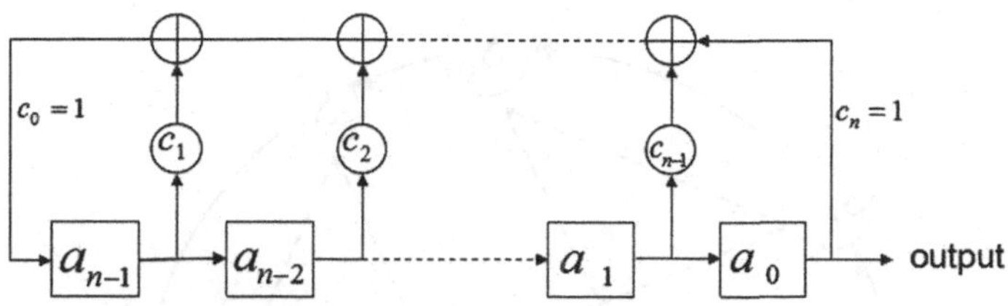

Fig. 6 Principle of linear feedback shift register

4.2 *Structure of the m Sub-sequence*

The *m* sequence is generated by the non-linear feedback function using non-linear feedback shift register, non-linear feedback function is based on primitive trinomials linear function, according to the rules of non-linear sequence generation, with algebraic logic of theoretical, calculated and extracted non-linear features of the sequence function, then the screening function of characteristic function and combined to generate a non-linear feedback function as demonstrated [18]. The feedback function is the key to design the non-linear feedback shift register, a level n *m* sequence generator feedback functional form is:

$$f(x) = c_{n-1}x_{n-1} \oplus c_{n-2}x_{n-2} \oplus \cdots \oplus c_0x_0 \tag{7}$$

where c_i is the feedback coefficient, $c_0 = 1, x_i$ is the register state. The m sequence of state transition of the shift register is determined by the feedback function, According to state transition of linear feedback shift register using to produce the m sequence, we can get m state transitions as shown in Fig. 7 as demonstrated [19]. Changing the state transition, and the length of the cycle is $2^n - 1$, the new state transition graph is a non-linear shift register. According to logic algebra theory,the feedback function can be modified by superposition(mod) of the original feedback function. The structure of the feedback function of the non-linear shift register is:

$$Z = f(x) \oplus y(x) \tag{8}$$

In which $y(x)$ is a feature function, $f(x)$ is a linear feedback function, The corresponding circuit is linear feedback logic circuit, $y(x)$ is a non-linear feedback function, which is corresponding to the non-linear feedback logic circuit. The corresponding logic circuit is a non-linear maximum length shift register, which is a logic circuit for generating a m sub sequence as demonstrated [20], as shown in Fig. 8.

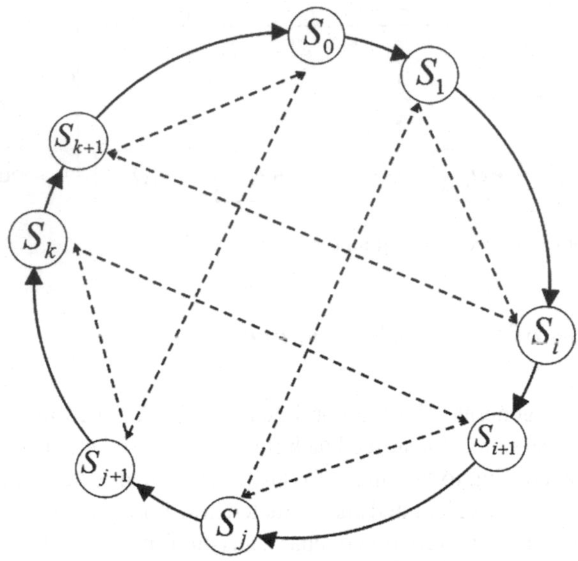

Fig. 7 State diagram of the *m* sequence

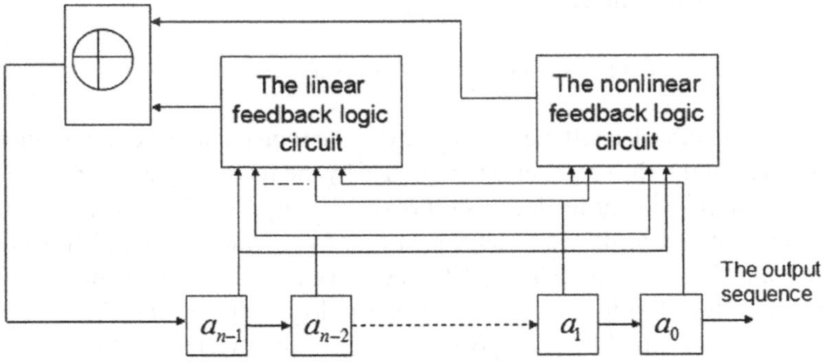

Fig. 8 Logic circuit of generating *m* sub sequence

4.3 Simulation of m Sequence and m Sub-sequence

As shown in Fig. 9, taking the 6 order *m* sequence as an example, the linear feedback shift register is set up in the Simulink platform, which is generated by the *m* sequence. On the basis of *m* sequence generation model, using non-linear feedback function logic to handle, in the condition of characteristic change of the shift register state transition, make *m* sequence generator into *m* sub-sequence generator. Figure 10 is a *m* sub sequence generator based on the *m* sequence generator.

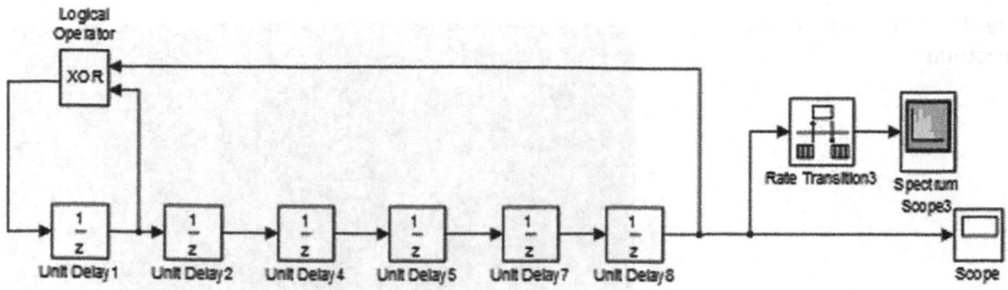

Fig. 9 6 order m sequence generator

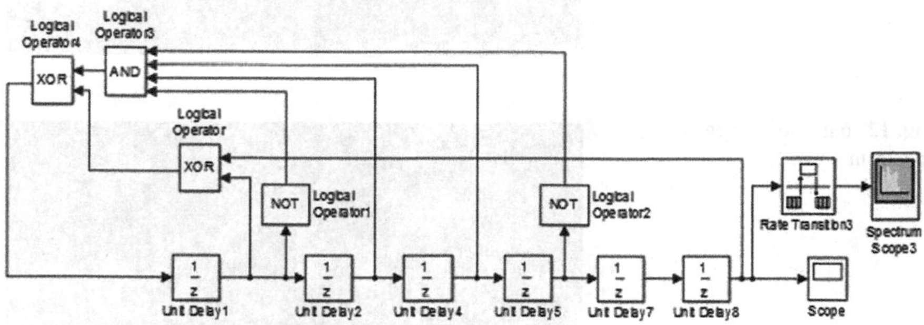

Fig. 10 6 order *m* sub sequence generator

5 Simulation Results and Analysis

The *m* sequence and m sub-sequence generators are introduced into the construction of *CDMA* system and simulation analysis, the article analyzes performance comparison of the 6 and 7 order *m* sequence and *m* sub-sequence in the *CDMA* system To the 6 order sequence, for example, when the system began to run, the source generates a binary code, Converting through the relay to $1-1$), at the same time, *m* sequence or *m* sub-sequence produced by *m* sequence generator(as shown in Figs. 11 and 12) are converted to $(1, 1)$. Multiply and spread spectrum processing (as shown in Figs. 13 and 14). The signals overlay with two other users, then successively through digital modulation and demodulation, channel, solution extender, filtering, and sampling judgement module output signal, compared with the original signal, the bit error rate of the system is obtained.

After many simulation experiments, the bit error rate of 3 system users under the 6 and 7 order of m sequence and m sub-sequence is obtained in the *CDMA* system. Figure 15, respectively, the system bit error rate of 6 and 7 order *m* sequence and m sub-sequence in the *CDMA* system changes to the size of the trend with change of the signal-to-noise ratio in 0–10 db.

From Fig. 15, we can see that system bit error rate declines in a straight line as the *SNR* increases, system bit error rate is almost zero when the signal-to-noise ratio is 0 dB, compared with the literature [20], the bit error rate of change trend and value are

Fig. 11 6 order m sequence spectrum

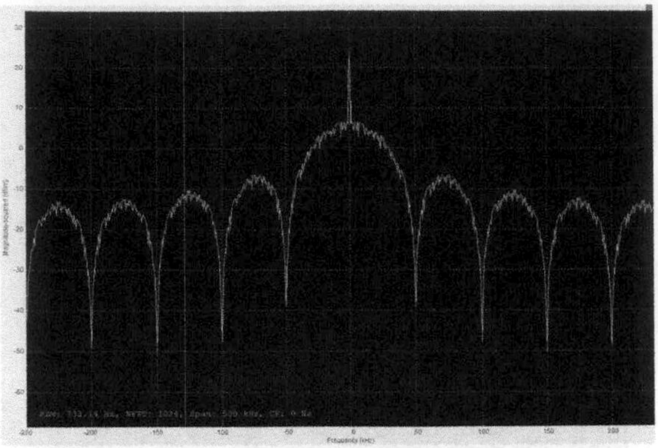

Fig. 12 6 m sub-sequence spectrum

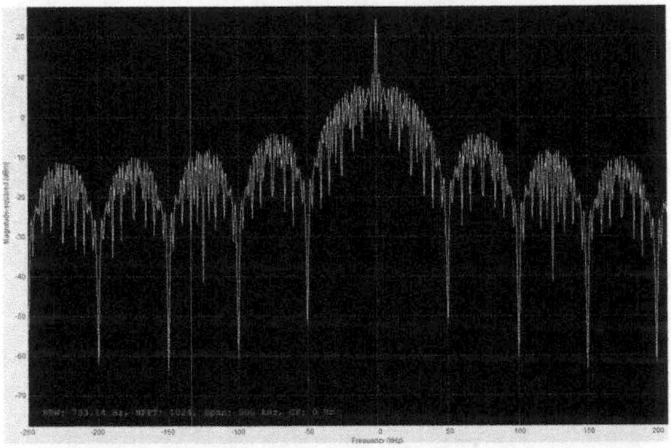

Fig. 13 Spectrum after 6 m sequence spread spectrum

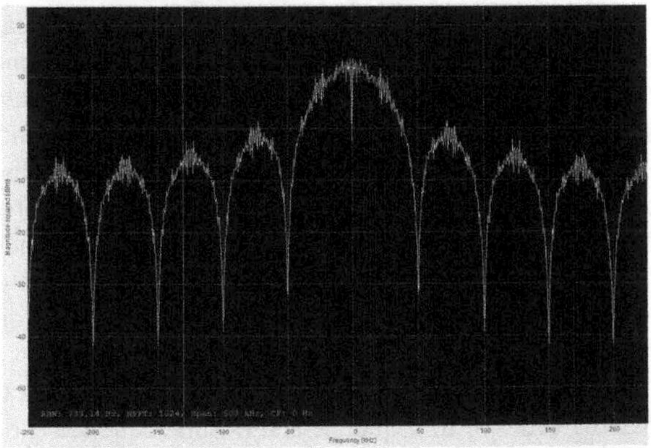

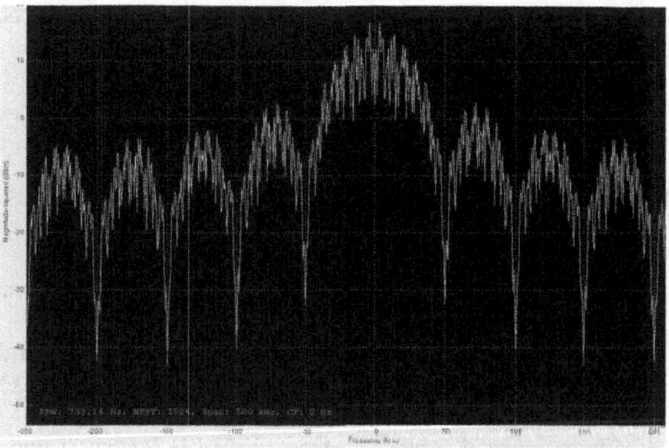

Fig. 14 Spectrum after 6 m sub-sequence spread spectrum

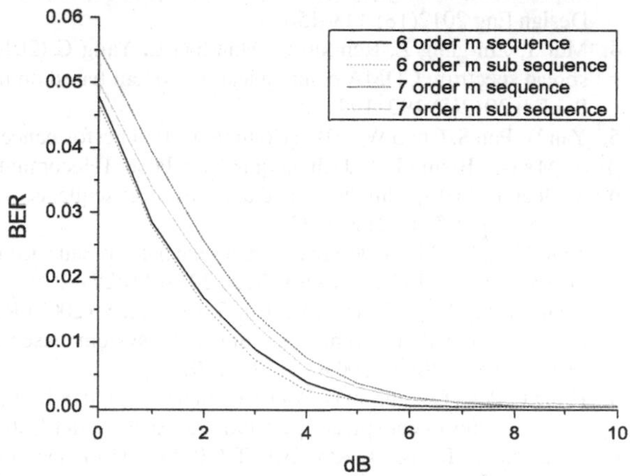

Fig. 15 Bit error rate change trend chart

almost identical, which proves the correctness of the simulation model of *CDMA*. In 3 user *CDMA* system, 6 order *m* sequence and m sub-sequence is obviously superior to the 7 order sequence, in the 6 order and 7 order sequence, performance of *m* sub-sequence in the system is superior to that of *m* sequence from the bit error rate curve trend.

6 Conclusions

In this paper, the *CDMA* system is built and simulated based on the *MATLAB/Simulink*, and the simulation results and the performance of the *CDMA* system are used to prove the correctness of the system. The *m* sequence and its sub sequences are modelled and simulated, and the sequence is fed into the construction of the *CDMA* system. The results show that the performance of m sub-sequence in *CDMA* system is better than that of *m* sequence.

Acknowledgments This work is supported by the National Natural science foundation of China (Grant *NO*. 61372094,61071001).

References

1. Yu-xia L, Hong L (2011) Simulation of m sequence generation and related performance based on MATLAB. J Electron Technol 2011(9):44–47
2. Suchitra G, Valarmathi ML (2013) BER performance of Walsh-Hadamard like Kronecker product codes in a DS-CDMA and cognitive underlay system. J Wirel Pers Commun 2013(3):2023–2043
3. Chong-qiang T, Quan Y, Xiao Y (2012) Research and simulation on m-sequences, gold sequence and orthogonal gold sequence of spread spectrum communication system. J Electron Design Eng 2012(18):148–150
4. Mei Y, Zhi-gang Z, Ren-xiu Q, Man-hua L, Yang C (2012) Simulation of direct sequence spread spectrum CDMA communication system based on m sequences. J Yangtze Univ (Nat Sci Ed) 2012(5):192–199
5. Yan W, Pan S, Chao W (2014) Communication performance of GO-MC-CDMA system based on MATLAB/simulink. J Chongqing Univ Posts Telecommun (Nat Sci Edit) 2014(3):330–333
6. Ya-Juan L (2014) Simulation and analysis of m sequence based on Simulink. J Yangtze Univ (Nat Sci Ed) 2014(22):327–329
7. Xiao-Qing M (2014) Design of direct sequence spread spectrum communication system based on Simulink. J Modern Electron Tech 2014(22):277–279
8. Chang-Geng L, Jia-Ling Z, Ke-Hui S, Li-Yuan S (2009) Research and simulation of chaotic sequence spread-spectrum communication system based on m sequence synchronization method. J Syst Simul 2009(4):1198–1201
9. Hong L, Peng Q, Ying-ni D, Wan-li C, Jian-xia J, Quan-ling S (2013) Construction and analysis of a class non-linear spread spectrum sequence. Acta Electronica Sinica 41(10):1939–1943
10. Takeuchi K, Tanaka T, Kawabata T (2015) Performance improvement of iterative multiuser detection for large sparsely-spread CDMA systems by spatial coupling. IEEE Trans Inf Theory 2015(4):1768–1794
11. Li G-D, Meng W-X, Chen HH (2015) I/Q column-wise complementary codes for interference-resistant CDMA communication systems. J Syst J IEEE 2015(1):4–12
12. Liu Z, Guan YL, Chen H-H (2014) Fractional-delay-resilient receiver design for interference-free MC-CDMA communications based on complete complementary codes. J IEEE Trans Wirel Commun 2014(3):1226–1236
13. Kokate MD, Sontakke TR, Bagul CR (2014) Performance of RAKE-LMMSE receivers in wide-band communication systems. J Wirel Pers Commun 2014(3):317–324
14. Suchitra G, Valarmathi ML (2013) BER performance of Walsh–Hadamard like Kronecker product codes in a DS-CDMA and cognitive underlay system. J Wirel Pers Commun 2013(3):2023–2043
15. Enkoskya T, Stoneb B (2014) A sequence defined by M-sequences. J Discret Math 333:35–38
16. Fernandes MAC, Arantes DS (2014) Spatial and temporal adaptive receiver for DS-CDMA systems. J AEU Int J Electron Commun 2014(3):216–226
17. Zilong L, Yong Liang G, Udaya P (2014) New complete complementary codes for peak-to-mean power control in multi-carrier CDMA. J IEEE Trans Commun 2014(3):1105–1113
18. Suchitra G, Valarmathi ML (2013) BER performance of Walsh–Hadamard like codes based on complementary sequence sets in a CDMA and cognitive underlay system. J Wirel Pers Commun 2013(2):1311–1329

19. He CB, Huang JG, Yan ZH, Zhang QF (2011) M-ary CDMA multiuser underwater acoustic communication and its experimental results. J Sci China Ser F: Inf Sci 2011(8):1747–1755
20. Hong L, Ai-xue Z, Jun-chu F, Jian-xia X, Bing-Rong L, Peng Q (2012) Study of the non-linear feedback functions and a class sub-sequence based on the root-functions. J Acta Electron Sinica 2012(10)

Estimating Initial Guess of Localization by Line Matching in Lidar Intensity Maps

Chongyang Wei, Tao Wu and Hao Fu

Abstract While driving in typical traffic scenes with drastic drift or sudden jump of GPS positions, the localization methods based on wrong initial positions could not select the properly overlapping data from the pre-built map to match with current data, rendering the localizations as not feasible. In this paper, we first propose to estimate an initial position by matching in the infrared reflectivity maps. The maps consists of a highly precise prior map built with offline SLAM technique and a smooth current map built with the integral over velocities. Considering the attributes of the low-texture maps, we adopt the stable, rich line segments to match. A affinity graph to measure the pairwise consistency of the candidate line matches is constructed using the local appearance, pairwise geometric attribute and is efficiently solved with a spectral technique. The initial global position is obtained by converting the structure between current position and matched lines. Experiment on the campus with GPS error of dozens of meters shows that our algorithm can provide an robust initial value with meter-level accuracy.

Keywords Lidar intensity map · Line matching · Initial position · GPS jump · Autonous vehicles

1 Introduction

Much progress in the mapping and localization research community has turned self-driving vehicles into a reality over the past several years. The localization requires firstly to cut out a part of data from the pre-built prior map according to the current

C. Wei (✉) · T. Wu · H. Fu
College of Mechatronic Engineering and Automation,
National University of Defense Technology, Hunan, People's Republic of China
e-mail: cyzq3566@gmail.com

T. Wu
e-mail: wtt09.cs@gmail.com

H. Fu
e-mail: fuhao927@gmail.com

© Springer International Publishing Switzerland 2017

V.E. Balas et al. (eds.), *Information Technology and Intelligent Transportation Systems*, Advances in Intelligent Systems and Computing 454,
DOI 10.1007/978-3-319-38789-5_65

position provided by global position systems (GPS) or GPS-aid inertial navigation systems (INS) to ensure that there exists certain overlapping portions between the current data and the cut data, thus the precise localization can be realized by data registration. Most of the time especially in open space (air, sea, desert), the position of GPS is at least meter-level accuracy and can satisfy the need of providing an initial guess in the localization. State-of-the-art algorithms [9, 15] achieve the decimeter-lever localization with the initial errors of GPS positions being limited within 1.5 m. Unfortunately, in some typical traffic scenes, the GPS could be disturbed owing to the occlusions by vegetation and buildings or multi-path effects due to reflections, resulting in the output producing the error of tens of meters or even more. Thus, there are not enough overlapping portions between the current data and the cut prior data according to the wrong GPS position, render this localization as not feasible.

The key of handling the drift and sudden jump of GPS in the localization lies in that it needs a proper strategy to provide a reliable initial position to limit the error within several meters. After that, the precise localization can be realized by data registering as the common methods [3, 4, 7, 9, 15] do. Unfortunately, to the best of the literature at present, the research on this topic has rarely appeared. We focus on this problem in this paper.

In this paper, we first propose a new algorithm to estimate the initial position by line segment matching in lidar reflectivity maps. We build the prior map by registering overlapping portions with the offline simultaneous localization and mapping (SLAM) technique [5], meanwhile, we also build the current smooth map with the integral over velocities. Considering the special attributes of the low-texture map, rather than using the excellent SIFT feature [10] in camera images, we adopt the stable, rich lines to match the maps. The affinity graph to measure the pairwise consistency of the candidate line matches is constructed using local appearance and pairwise geometric attribute [17] and is efficiently solved with a spectral technique [8]. The initial guess is obtained by converting the relationship between the current vehicle's position and matched lines. Experiment on the campus with GPS error of dozens of meters shows that our algorithm can provide an initial position within 3.5 m, which meets the need of the initial guess of the common localization methods.

2 Related Works

The large-scale, long-term outdoor localization in the field of autonomous driving has attracted considerable enthusiasm and research interest since the DARPA Urban Challenge of 2007. Owing to the lidar's insensitivity of lighting changes and its high frequency precise range measurements, lidar is regarded as a better option and has been widely applied. Hata first proposed to detect curb points [4] in 3D lidar point clouds and later extended to extracting road markings [3] in reflective intensity data, those points were inputted into the monte carlo framework to localize the vehicle. Baldwin [7] proposed to exploit a dual-lidar system, one oriented horizontally for inferring vehicle linear and rotational velocity and one declined for capturing a dense view of the environments, to match data with a prior map. Markus [12] presented to manually label curbs and road markings in the highly accurate lidar intensity map and the vehicle's pose was determined by matching the measurement points in the current camera image and the sampled points from labeled line segments. Jesse used infrared reflectivity from 3D lidar to build an orthographic high-resolution probabilistic map [9], online localization was then performed with the current 3D lidar measurements and an combined GPS/INS systems. Recently, Wolcott [15] localized the vehicle by maximizing the normalized mutual information between the live camera images and the synthetic views within a 3D prior ground map, generated with 3D lidar scanner.

In general, the initial positions are provided by the GPS or GPS-aid INS. Most of methods set the error of initial position to be meter-level accuracy, for example, [7] constraints the error with 3.5 m, [11] thinks the error is 3 m at maximum, [9] limit the error within 1.5 m, [15] also limits the search in the range of 1.5 m, this is feasible when the GPS signal has no severe drift or jump. However, when the GPS drastically jump with error of tens of meters or even more, a meter-order initial position estimation before localization is need. The research on this topic has rarely presented until now.

Existing methods to match lines can be divided into two types: one is matching individual line, one is matching groups of lines. Many methods of matching individual line are based on appearance similarities of the lines, such as [2] where the color histograms of neighboring profiles were used to obtain an original set of candidate matches which grows iteratively, or [14] which proposed a mean-standard deviation line descriptor (MSLD) based on the appearance of the pixel support region. The approaches of matching groups of lines have the advantage that more geometric information is available for disambiguation. Wang [13] used spatial proximity and relative saliency to group lines and the grouped lines are matched by using pairwise geometric configuration and average gradient magnitudes. The approach is shown useful to deal with large viewpoint changes and non-planar scenes. However, it is quite computationally complex because of a large number of line signatures. Recently, Juan [6] proposed an iterative strategy which uses structural information collected through the use of different line neighborhoods, making the set of matched lines grow robustly at each iteration. This method obtains a good performance in the industrial environments, however, the complex calculation of weights in the nine different neighboring structures in each iteration limits its application.

Zhang [17] proposed to combine the local appearance of lines and the pairwise geometric attributes to construct an relational graph and adopt a spectral technique to solve the problem. This approach achieves a better result for large viewpoint and illumination changes in the camera images. In this paper, we follow it to match line pairs in the lidar maps.

3 Building Maps

The first work is to build two maps: an accurate prior map and a smooth local map. The prior map is built with the offline SLAM technique which brings the areas of overlap into alignment and distributes the transformation error over the pose graph when detecting a closed loop. Thus the built prior map is a metrically accurate representation of the observed environment. The local map is created by integrating velocities, invariant to jumps in GPS pose, from INS. Thus we maintain a smooth representation of the local structure.

3.1 Accurate Prior Map

Owing to without a prior knowledge of the environment, we exploit the nonlinear least-squares, pose-graph SLAM technique and measurements from the Velodyne HDL-64E scanner to produce a highly accurate map of the 3D structure. We construct a pose-graph to solve the offline mapping problem, as shown in Fig. 1, where nodes are vehicle poses (X) and edges are either inertial navigation constraints (U), lidar scan-registration constraints (Z), or GPS prior constraints (G). Assumed these constraints are satisfied with Gaussian distributions, the offline mapping is a nonlinear least squares problem. We use the incremental smoothing and mapping (iSAM) [5] algorithm to solve the problem.

Given the prior data containing the global positions from GPS, the velocities from inertial navigation, and the measurements from lidar, we wish to construct a consistent representation of the real environment to put the measurements into the proper positions. We first construct a coarse pose-graph with GPS global positions and lidar scan-registration constraints. When there exist nodes in the pose-graph that are spatially neighboring but temporally separated, the nodes must be refined so that the measurements in the nodes can be matched properly. Briefly, the iSAM [5] is used to solve an optimization problem in which adjacent nodes in the pose-graph are linked by inertial navigation and scan registration, nodes are linked to their estimated global positions, and the transformed error is distributed over the graph once the loop closure is detected.

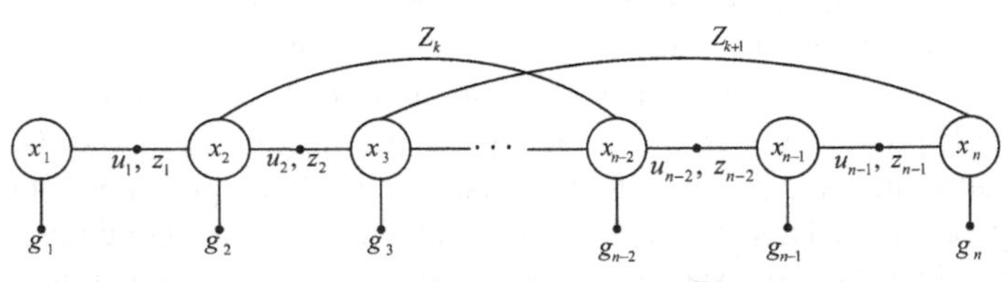

Fig. 1 The pose-graph of the SLAM problem that we construct in the offline mapping process. Here, x_i represents the vehicle's poses, u_i represents inertial navigation measurements, z_k represents lidar scan-registration constraints in adjacent poses and loop closure poses, and $\boldsymbol{g}_i$ is the GPS prior constraints

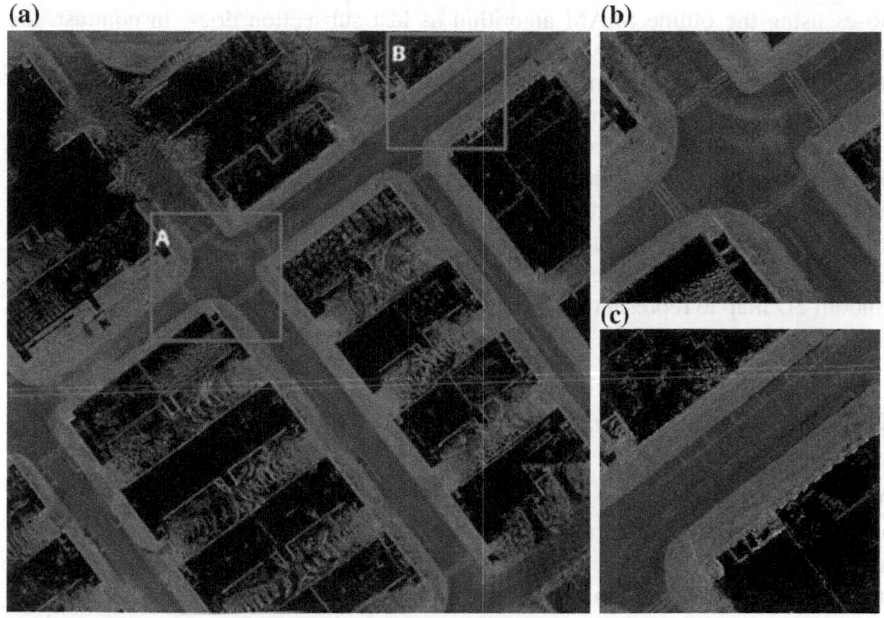

Fig. 2 The built highly accurate intensity average map. In **a**, we show a 200 m × 200 m grid cell map in which each cell shows its average infrared reflectivity. **b** is a zoomed in view of the red rectangular region A in (**a**), we can see clearly the stop lines in the crossroad. **c** shows a zoomed view in region B where the lines of parking space can also be seen clearly

In this implementation, we use the lidar odometry and mapping (LOAM) [16] algorithm to estimate 6-degree of freedom scan-registration constraints between adjacent nodes. Once a series of matches between both general nodes and loop-closure nodes have been estimated, the iSAM objective function is minimized and the nodes, denoting the vehicle's poses, are updated accordingly.

Given the optimized vehicle's poses, the algorithm to build an average intensity map is straightforward. As the vehicle passes through the optimized poses, the lidar points are projected into a x-y grid cell map in which each cell represents a $20 \times 20\,\text{cm}^2$ patch of ground. When new lidar data arrives, each cell updates its intensity average combined with the stored necessary intermediate values. In this way, we build a highly accurate offline 2D prior map, shown in Fig. 2, to represent the known 3D environments.

3.2 Smooth Local Map

The construction of the local map needs to handle the current intensity readings in real-time and it is an online task, thus we can not obtain the well optimized vehicle's poses using the offline SLAM algorithm as last subsection does. In contrast, the inertial updates are smooth and invariant to jumps in GPS positions. We integrate the inertial velocities to form a smooth coordinate system and build the consistent map in this coordinate system.

The construction of the local map is similar with the one of the prior map. Given the intensity readings and the accumulated vehicle's poses, we project the current lidar points onto the 2D grid cell surface ($z = 0$) in which each cell also represents a $20 \times 20\,\text{cm}^2$ patch of ground. Every cell preserves the essential intermediate data to update its average with new readings falling into the cell. In this way, we create a smooth 2D map to represent the current 3D scenes.

3.3 The Map's Attributes

As lidar point cloud is sparse which is especially obvious in the region that is far away from the lidar, in the regions which are close to the vehicle's trajectories, the map is densely covered, whereas in the further regions, the map is sparse with many irregularly shaped "holes" in it, as shown in Fig. 3. The gradients of pixels of the holes vary drastically and results in a powerful disturbance for the construction of robust feature descriptor based on the gradient, for example, the state-of-the-art SIFT [10] which is considered as one of the most descriptive power point features. Figure 3a shows the result of one randomly selected SIFT point and its 50 most similar ones in the map. The selected point is marked with a blue square with white number 1 as its label. The similar points are marked with red stars with green numbers as their labels and the smaller the numbers, the more similar the descriptors.

From Fig. 3a, we can see that most of the points are located in the sparse region of the map and don't come from the special physical positions as expected in the camera image, such as the corner points. These points are unstable and susceptible to changing when new lidar data is accumulated into the map. This is an obvious difference between our generated map and the camera image. In addition, the similar descriptors should, ideally, represent the similar regions, however, the region where the blue square represents is significantly different from the regions where its similar red stars represent. This shows that the point feature descriptor in our low-texture map does not have a sufficient descriptive power. The similarity plot of these descriptors is shown in Fig. 3b. The ratios of near two similarities are in a range of [1.0007, 1.0426], they are so close. So small difference is not enough to distinguish the points.

Another characteristic of the map is its abundance of line segments which consist of the 2D projections of buildings, the boundaries of different reflective attributes, the road markings and so on. Compared with lines in the camera-based image, these lines are immune from shadows and other artifacts caused by changes in lighting, meanwhile, they are stable because the buildings do not move, the reflective attributes of the objects do not vary and the road markings can not disappear. The advantages of lines in the lidar map make it a good option to match maps built in different times

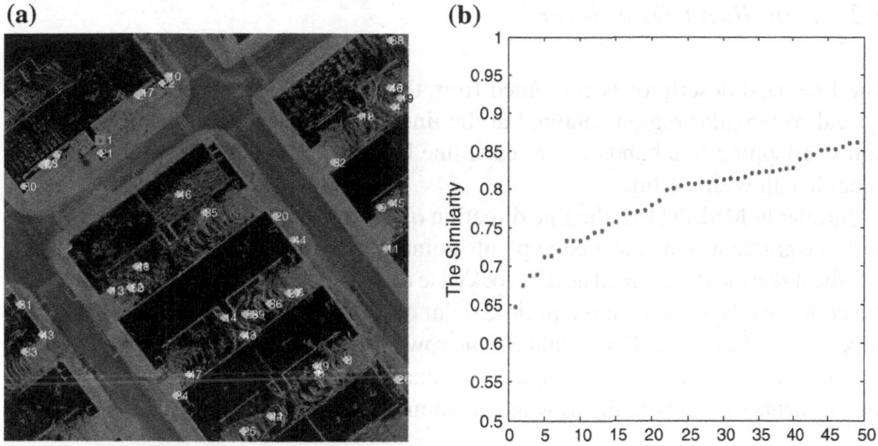

Fig. 3 **a** shows the result of one randomly selected SIFT feature point with white number 1 as its label and its 50 most similar ones with green numbers as their labels. The smaller the numbers, the more similar the descriptors. **b** shows the similarities of descriptors between the number 1 point and its near points. (Best viewed in colour.)

and environments. Thus we firstly propose to adopt line segments to match lidar intensity maps.

4 Line Matching in Maps

In this section, we first depict the method to detect lines [1] in the octave maps, then follow [17] to construct the line band descriptor and the relational graph of the candidate line pairs, finally, the line matching problem is solved efficiently with a spectral technique [8].

4.1 Line Detection in the Octave Maps

To overcome the fragmentation problem of lines, we downsample the original map with a series of scale factors and Gaussian blurring and form a scale-space pyramid consisting of multi-layer maps. In each layer, we adopt the EDLine [1] algorithm to detect line segments. To merge the lines which are related to the same region of the original map, we find them in the scale-space maps, assign them a unique ID and store them into a vector called LineVec [17]. The final detection result is a series of LineVecs. Compared with the method [13] which stores all possible grouping lines, this method reduces the dimension of line matching problem.

4.2 Line Band Descriptor

The line band descriptor is computed from the line support region (LSR) which is a local rectangular region centered at the line. The LSR is consisted of a series of non-overlapping line bands where each line band is parallel with the line and with equal length with the line.

Similar to MSLD [14], the line direction d_L, defined to be parallel to the line and with the gradient of its most edge points pointing from the left side to the right side, and the direction $d_\perp$, defined as the clockwise orthogonal direction of d_L, form a local 2D coordinate system with the middle point of this line as the origin to describe the appearance of this line. For a band B_i, the rows from B_i and its nearest two neighbor bands B_{i-1}, B_{i+1} are used to compute the band descriptor BD_i. For the jth row in B_i, the gradients of pixels in this row are accumulated as:

$$\begin{aligned} \nu 1_i^j &= \alpha \sum_{g^T \cdot d_L > 0} g^T \cdot d_L, \quad \nu 2_i^j = \alpha \sum_{g^T \cdot d_L < 0} -g^T \cdot d_L, \\ \nu 3_i^j &= \alpha \sum_{g^T \cdot d_\perp > 0} g^T \cdot d_\perp, \quad \nu 3_i^j = \alpha \sum_{g^T \cdot d_\perp < 0} -g^T \cdot d_\perp, \end{aligned} \tag{1}$$

where g^T is the gradient of a edge point, α is the coefficient.

Stacking the four accumulated gradient values of each row forms the band descriptor matrix BDM_i, where $BDM_i \in \mathbb{R}^{4 \times n}$ with n denoting the number of rows associated with B_i. BD_i can be simply constructed using the mean vector M_i and the standard deviation vector S_i of the matrix BDM_i: $BD_i = (M_i^T, S_i^T)^T \in \mathbb{R}^{8 \times 1}$. Finally, the LBD is the combination of all BD_i: $LBD = (BD_1^T, BD_2^T, \ldots, BD_m^T) \in \mathbb{R}^{8m}$.

4.3 Constructing the Affinity Graph

An affinity graph is built using the combination of the appearance similarity, the pairwise geometric attributes. The appearance similarity is measured by the distance of the LBD descriptors. The line pairs whose similarities are less than a threshold are taken as the candidate matches. For n candidate matches, the affinity graph is described with an adjacency matrix $M \in \mathbb{R}^{n \times n}$ following the definition in [8]. The value of $M_{i,j}$ is the consistent score of the candidate LineVec (L_p^i, L_c^i) and (L_p^j, L_c^j) where L_p^i, L_p^j are LineVecs in the prior map and L_c^i, L_c^j are LineVecs in the current map. For the lines $l^i(\overrightarrow{s^i e^i})$ and $l^j(\overrightarrow{s^j e^j})$ (the arrows represent their directions), c is their intersection, (s_p^i, e_p^i) are the projections of the endpoints of the line l^i onto the line l^j, and similarly, (s_p^j, e_p^j) are the projections of the endpoints of the line l^j onto the line l^i. The geometric attributes of (l^i, l^j) is described by their intersection ratios (I^i, I^j), projection ratios (P^i, P^j) and relative angle difference θ^{ij},

$$I^i = \frac{\overrightarrow{s^i c} \cdot \overrightarrow{s^i e^i}}{|\overrightarrow{s^i e^i}|^2}, \quad P^i = \frac{|\overrightarrow{s^i s_p^i}| + |\overrightarrow{e^i e_p^i}|}{|\overrightarrow{s^i e^i}|}. \tag{2}$$

I^j and P^j can be computed in the same way.

We use the differences of descriptors and geometric attributes to evaluate the consistent score of pairwise lines and obtain the adjacency matrix M. The diagonal elements of M equal 0 for better results and let $M^{ij} = M^{ji}$ to keep the symmetry. The efficient solving of the line matching problem is realized by the spectral technique [8].

5 Estimating Initial Positions

For a matched line pair (L_p^i, L_c^i), $(P_g^i(x), P_g^i(y))$ is the position of one of the endpoints of the matched line in the prior map, $(P_c^i(x), P_c^i(y))$ is the corresponding position in the current map, (r^i, θ^i) is the location-relation between $(P_c^i(x), P_c^i(y))$ and the vehicle's position $(P_v^c(x), P_v^c(y))$, accumulated with the integral over velocities from INS, the vehicle's global position $(P_v^g(x), P_v^g(y))$ can be simply computed as:

$$P_v^g(x) = P_g^i(x) + r^i cos(\theta^i), \; P_v^g(y) = P_g^i(y) + r^i sin(\theta^i). \tag{3}$$

Ideally, only one point can fix the vehicle's global position, however, owing to inaccurate locations of line endpoints in the low-texture maps, we model the probability density of the neighborhood of the estimated position with a 2D Gaussian distribution $\rho(x, y)$ centered at the position,

$$\rho(x, y) = \frac{1}{2\pi\sigma^2} exp\left\{-\frac{(x - P_v^g(x))^2 + (y - P_v^g(y))^2}{2\sigma^2}\right\}, \tag{4}$$

where the variance σ^2 of the Gaussian represents the uncertainty of the positions of the line endpoints, (x, y) is the neighboring position.

Given the probability of neighboring positions, rather than using directly the center of the probability distribution which will tend to cause the estimated position to be biased too much towards the center, we use the center of mass with enhanced probability with exponent $\lambda > 1$:

$$P_v^g(x) = \frac{\sum_{x,y} \rho(x, y)^\lambda . x}{\sum_{x,y} \rho(x, y)^\lambda}, \quad P_v^g(y) = \frac{\sum_{x,y} \rho(x, y)^\lambda . y}{\sum_{x,y} \rho(x, y)^\lambda}. \tag{5}$$

In this way, the initial guess about the position in the localization is obtained.

6 Experiments

We validate the performance of our algorithm on the data from the campus, which is collected by the autonomous ground vehicle (AGV). The AGV is a modified Toyota equipped with a Velodyne HDL-64E scanner, a NovAtel INS et al. We run the algorithm in C++ on a laptop equipped with a Core i7-2720QM CPU with 2.20 GHZ and 8 GB main memory.

Figure 4a shows the trajectory of the GPS around a closed swimming pool on the campus. The survey vehicle starts from the point *S*, passes through the scene anti-clockwise and finish the test at the end point *E*. In the former testing from point *S* to point *A*, the trajectory is stable and smooth, However, all of a sudden, the GPS happens to jump drastically at the point *A*, which is possible to be caused by occlusions by big trees and/or high buildings, or multi-path effects due to reflections, resulting in its position being transformed to point *B*. After that, the trajectory keeps again stable and smooth. The jump finally produces a error of more than 80 m between the start point and the end point, they should have been closed in the ideal condition. So large error makes the general localization methods based on the initial guess in the meter-level accuracy can not select the properly overlapping portions to match, rending them as not feasible.

(a)

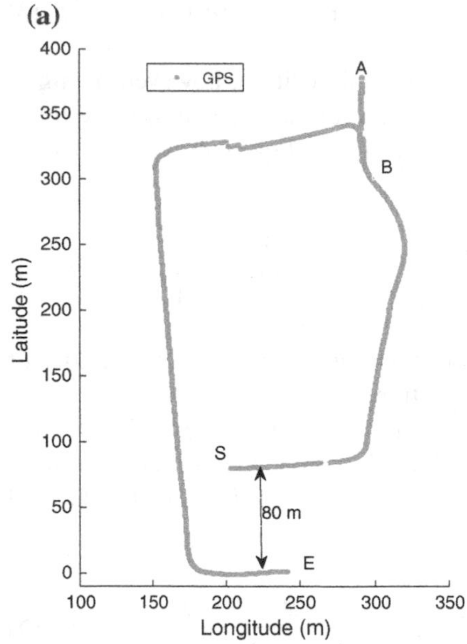

(b)

Fig. 4 Estimated initial positions of the vehicle in the closed environments. The *green*, *blue* and *red points* denote the projections in the prior map of the current GPS readings, the positions of the integral over velocities and our estimated initial values, respectively. The better results should have more points projected in the proper positions in the road and have the smaller closed errors (Best viewed in colour.)

Figure 4b shows the result of our method. The green points are obtained by directly projecting the current GPS positions onto the prior map, the blue points are the projections of the integrated positions only with the start position from GPS. Upon close inspection, the two trajectories are both not in the road with the original GPS positions are wrongly falling in the obstacle regions after the jump position *B* and the blue points having drifted obviously since the first turning in the top of the trajectory. The blue points located within the two turnings in the top fall fully in the obstacle area about 20 m away from the road, the final error between the start point and the end point reaches 24 m. The red trajectory is the estimated result of our method, there are only several points deviating from the road in the top-left of the trajectory, most of them locate in the orthographic positions, and the start point and the end point are well closed. The maximum error in the total trajectory is 3.5 m. The experiment shows that our algorithm can well deal with the drastic jump of GPS signal and provides a reliable initial value for subsequential precise localization.

7 Conclusions

The initial position estimation in lidar intensity maps can be difficult because the maps have the characteristics: low texture, low contrast and great gradient changes in the sparse regions. In this paper, we first propose an algorithm to estimate the initial value for localization by matching line segments in maps. The maps consists of one highly accurate pre-built map with the offline SLAM technique and one smooth accumulated map with the integral over velocities. Line matching is realized by constructing the affinity graph with local appearance, pairwise geometric attributes and solving efficiently with a spectral technique. The initial global position is estimated by transforming the relationship between matched lines and the current pose. Experiment on the campus with GPS error of a few tens of meters shows that our algorithm can provide an initial position with meter-level accuracy, meeting the need of most of localization methods.

References

1. Akinlar C, Topal C (2011) Edlines: a real-time line segment detector with a false detection control. Pattern Recognit Lett 32(13):1633–1642
2. Bay H, Ferrari V, Van Gool L (2005) Wide-baseline stereo matching with line segments. In: International conference on computer vision and pattern recognition, pp 329–336
3. Hata A, Wolf DF (2014) Road marking detection using lidar reflective intensity data and its application to vehicle localization. In: International conference on intelligent transportation systems, pp 584–589
4. Hata AY, Osorio FS, Wolf DF (2014) Robust curb detection and vehicle localization in urban environments. In: IEEE Intelligent Vehicles Symposium, pp 1257–1262
5. Kaess M, Ranganathan A, Dellaert F (2008) iSAM: incremental smoothing and mapping. IEEE Trans Robot 4(6):1365–1378

6. López J, Santos R, Fdez-Vidal XR, Pardo XM (2015) Two-view line matching algorithom based on context and appearance in low-textured images. Pattern Recognit 48(7):2164–2184
7. Lan B, Paul N, Newman P (2012) Laser-only road-vehicle localization with dual 2D push-broom lidars and 3D priors. Intelligent Robots and Systems, pp 2490–2497
8. Leordeanu M, Hebert M (2005) A spectral technique for correspondence problems using pairwise constraints. In: International Conference on Computer Vision, pp 1482–1489
9. Levinson J, Thrun S (2010) Robust vehicle localization in urban environments using probabilistic maps. In: International Conference on Robotics and Automation, pp 4372–4378
10. Lowe DG (2004) Distinctive image features from scale-invariant keypoints. Int J Comput Vis 60(2):91–110
11. Park SY, Choi SI, Moon J, Kim J, Park YW (2011) Localization of an unmanned ground vehicle based on hybrid 3D registration of 360 degree range data and dsm. Int J Control Autom Syst 9(5):875–887
12. Schreiber M, Knöoppel C, Franke U (2013) Laneloc: lane marking based localization using highly accurate maps. In: IEEE Intelligent Vehicles Symposium, pp 449–454
13. Wang L, Neumann U, You S (2009) Wide-baseline image matching using line signatures. In: International Conference on Computer Vision, pp. 1311–1318
14. Wang Z, Wu F, Hu Z (2009) MSLD: a robust descriptor for line matching. Pattern Recognit 42(5):941–953
15. Wolcott RW, Eustice RM (2014) Visual localization within lidar maps for automated urban driving. In: Intelligent Robots and Systems, pp. 176–183
16. Zhang J, Singh S (2014) Loam: lidar odometry and mapping in real-time. In: Robotics: Science and Systems Conference
17. Zhang L, Koch R (2013) An efficient and robust line segment matching approach based on lbd descriptor and pairwise geometric consistency. J Visual Commun Image Represent 24(7):794–805

Correlation Between Respiration and Lung Organ Motion Detected by Optical Flow

Cihui Yang, Lei Wang, Enmin Song, Hong Liu, Shan Gai and Jiehua Zhou

Abstract The method that detects the tumor's position indirectly has played an important role in dynamic radiotherapy for lung cancer because of its non-invasion property. The key problem of this approach, which has not been solved well, is to create a robust correlation model between the respiration and the tumor motion. Aiming at this problem and considering the fact that the lung tumor has similar motion law with the lung organ, we selected the patient's breathing and the fluoroscopic images of lung organ as research objects to investigate the correlation between the respiration and the lung organ motion. The optical flow algorithm is utilized to track the motion of the lung organ in the fluoroscopic images sequence automatically, and the waveform of the lung organ motion is compared with the respiratory signal. Experimental results show that there is a phase discrepancy between these two kinds of waveforms. After phase aligning, a linear relationship can be found between the

C. Yang (✉) · L. Wang · S. Gai · J. Zhou
School of Information Engineering, Nanchang Hangkong University,
Nanchang 330063, China
e-mail: yangcihui@nchu.edu.cn

L. Wang
e-mail: leiwang@nchu.edu.cn

S. Gai
e-mail: gaishan886@163.com

J. Zhou
e-mail: spring19821111@163.com

C. Yang · L. Wang · S. Gai · J. Zhou
Key Laboratory of Image Processing and Pattern Recognition
(Nanchang Hangkong University) of Jiangxi Province, Nanchang 330063, China

E. Song · H. Liu
Center for Biomedical Imaging and Bioinformatics, Huazhong University of Science and Technology, Wuhan 430074, China
e-mail: esong@mail.hust.edu.cn

H. Liu
e-mail: HL.cbib@gmail.com

© Springer International Publishing Switzerland 2017

V.E. Balas et al. (eds.), *Information Technology and Intelligent Transportation Systems*, Advances in Intelligent Systems and Computing 454,
DOI 10.1007/978-3-319-38789-5_66

respiratory signal and the lung organ motion. By using this model, the lung organ motion can be measured with a maximum error of 25 % by means of respiration.

Keywords Optical flow · Motion tracking · Respiratory detecting

1 Introduction

In radiotherapy for lung cancer, the treatment effect is affected by the lung tumor motion due to respiration [1, 2]. There are several methods which can be used to compensate the intrafractional motion, including enlarging target volume [3], respiratory gating [4] and dynamic radiotherapy [5]. Enlarging target volume to encompass the tumor's movement range will deliver extra dose to more healthy tissues surrounding the tumor. Respiratory gating can limit the dose delivered to the healthy tissues by performing radiotherapy during end-exhalation and end-inspiration. However, it requires training the patients and is not applicable for the patients featuring respiratory impairment. Meanwhile, it increases the radiotherapy time [6]. Dynamic radiotherapy is an alternative method to compensate the intrafractional motion, which is achieved by detecting the tumor motion in real time and changing the position of the beam or the patient in real time to align the beam and the tumor [7]. It can minimize the irradiated volume of healthy tissues.

X-ray imaging is the most frequently used method in dynamic radiotherapy to obtain the tumor motion. However, it will cause imaging dose to the patient. Considering this problem, some researchers proposed indirect methods, such as detecting the respiration to acquire the tumor's position. Thus, the problem that the patient receives extra imaging dose can be avoided. The effect of indirect detecting methods relies on the precision of the correlation model between the respiration and the tumor motion created before radiotherapy.

A few researches have been done to investigate the correlation between the respiration and the tumor motion. Chen et al. [8] compared the motion patterns of internal targets and external radio opaque markers on patient's chest, and found that for some patients there are significant motion phase discrepancies between an internal target and an external marker. Tsunashima et al. [9] investigated the correlation between the respiratory waveform and three-dimensional (3D) tumor motion by means of the Fourier transform and a cross-correlation function. Their results showed that there is an evident correlation between the respiratory waveform and the 3D tumor motion. However, the respiratory waveform did not always accurately correspond with the 3D tumor motion. Seppenwoolde et al. [10] used synchronized recordings of both internal tumor motion and external abdominal motion of 8 lung cancer patients to simulate Cybernife Respiratory Tracking System (RTS) treatments, and found that the reduction in treatment error could be reached with a simple linear model. Torshabi et al. [11] selected 20 cases treated with real-time tumor tracking by means of the Cyberknife Synchrony module, and compared three different approaches that infer tumor motion based on external surrogates. All these researches have prompted the development of radiotherapy technology. However, it is still an important problem that should be solved in the dynamic radiotherapy field to find a way to create a robust correlation model between the respiratory signal and the tumor motion.

Aiming at this problem and considering that the lung tumor, as a kind of organ attached in normal organ, has similar motion law as the normal lung organ, we investigate the correlation between the respiration and the lung organ motion in this paper. The fluoroscopic image sequences of the patient's lung organ are obtained by a fluoroscopic imaging system, and the respiratory signal is acquired by a respiratory detecting device based on pressure detecting at the same time. The optical flow algorithm is utilized to track the lung organ motion automatically, and the least square method is adopted to model the correlation between the patient's respiration and the lung organ motion.

2 Methods and Materials

2.1 Fluoroscopic Imaging

The fluoroscopic image sequences of the patient's lung organ were obtained by a Shimadzu AC50-XP fluoroscopic imaging system of the hospital of Huazhong University of Science and Technology, and 12 patients that need to be performed fluoroscopic examination participated in this research. In the process of sampling the images, the doctor performed fluoroscopic examination according to normal procedure, thus it will not deliver extra imaging dose to the patients.

Since the fluoroscopic imaging system uses specific video format, it's necessary to read the imaging system's video signal by customized video capture card. However, it's difficult to buy the video capture card in a short time. Therefore, a normal digital camera was used to capture the fluoroscopic image displayed on the screen of the fluoroscopic imaging system. To avoid affecting the doctor's operation and examination, the camera was fixed on the desk near the fluoroscopic imaging system, with a strabismus angle that can capture the whole fluoroscopic image displayed on the screen. Image capturing speed was set to 15 frames per second (fps).

2.2 Image Preprocessing

It's necessary to adjust the captured image and extract fluoroscopic image from it, for the image was captured from a strabismus angle. Use I_{fs}, I_{fd}, I_{fa} and I_d to denote the fluoroscopic image displayed on the screen, the fluoroscopic image in the image captured by the digital camera, the fluoroscopic image after adjusting and the image captured by the digital camera respectively. The preprocessing can be done by the following steps.

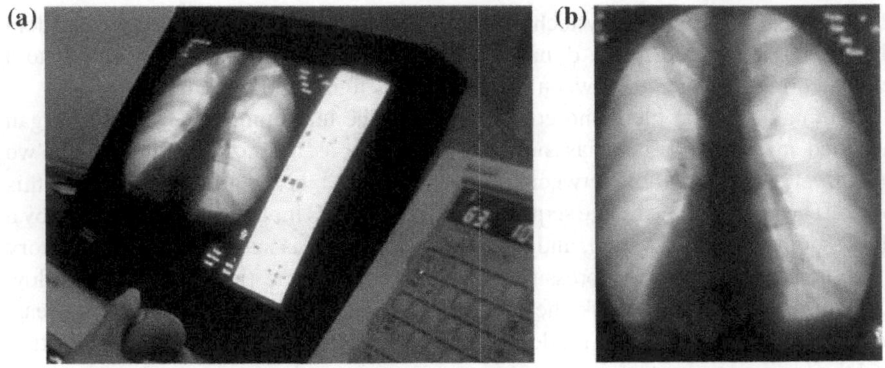

Fig. 1 Original image captured by the camera and adjusted fluoroscopic image extracted from the original image. a Original image captured by the camera. b Adjusted fluoroscopic image

(1) Measure the real size of width w_s and height h_s of I_{fs}, and calculate its width-height ratio r_s.
(2) Calculate width w_d and height h_d of I_{fa} according to r_s, the size of I_d and the ratio r_c that I_{fd} occupied in I_d.
(3) Mark out four vertexes of the fluoroscopic image in the first image captured by the digital camera.
(4) Calculate perspective transformation matrix M according to the coordinates of the four points, w_d and h_d.
(5) Adjust every image by the perspective transformation matrix M, and extracted the fluoroscopic image from the adjusted image.

Figure 1a shows an original image captured by the camera, and Fig. 1b shows an adjusted fluoroscopic image extracted from the original image. Since the original image is captured from a strabismus angle, the fluoroscopic image is not so clear after adjusting.

2.3 Motion Tracking

Optical flow algorithm [12] has good performance on measuring small changes in deformation motion, and can reach sub-pixel accuracy. Thus it is adopted to track the lung organ motion in our research. The main idea of optical flow algorithm is that, for each point (x, y) in frame t, there will be a corresponding point in frame $t + \Delta t$. That is to say, if $I(x(t), y(t), t)$ is used to denote the continuous space-time intensity function of an image, then

$$I(x(t), y(t), t) \approx I(x(t + \Delta t), y(t + \Delta t), (t + \Delta t)). \tag{1}$$

Applying Taylor expansions to the right side of (1), we can get the following optical flow equation

$$\frac{\partial I}{\partial x}\frac{dx}{dt} + \frac{\partial I}{\partial y}\frac{dy}{dt} + \frac{\partial I}{\partial t} = 0. \quad (2)$$

Letting $p = \frac{dx}{dt}, q = \frac{dy}{dt}$, we can get another form of optical flow equation

$$I_x p + I_y q + I_t = 0, \quad (3)$$

where p and q represent the components of the image velocity vector respectively. I_x, I_y are the components of spatial gradient ∇I. I_t is partial differentiation of I with respect to time.

However, the optical flow Eq. (3) is ill-posed because it has 2 variables. There are many techniques for computing optical flow, which differ from each other in their differing assumptions. One of these techniques is Lucas Kanade (LK) [13] which is proposed by Lucas and Kanade in 1981.

The LK optical flow method is adopted to estimate the motion in our research. When using the LK optical flow method, one point is selected first in the lung organ area of the first adjusted fluoroscopic image as reference point. Then its corresponding point in the second adjusted fluoroscopic image is calculated using the LK optical flow method. After obtaining the second point, the second point is adopted as reference point, and its corresponding point is calculated in the third adjusted fluoroscopic image. And so on, all the points detected consist of lung organ's motion trajectory.

2.4 Respiration Detecting

When capturing the fluoroscopic image of the patient's lung, a respiration detecting device based on pressure changing measuring is used to detect the patient's respiratory signal synchronously, as Fig. 2a shows. This respiratory detecting device consists of air sac, data acquisition box, air tube and inflator ball. When the patient breathes, the pressure of his breast or abdomen on the air sac will change. The pressure is a kind of representation of the patient's respiration. By detecting the pressure using the data acquisition box, we can get the patient's respiration signal.

Acquisition speed of respiration signal is set to 100 per second. Figure 2b shows a respiration signal obtained by the respiration detecting device. Considering that there exists some noise during acquisition procedure, Gaussian filter was applied to denoise the respiration signal after acquisition.

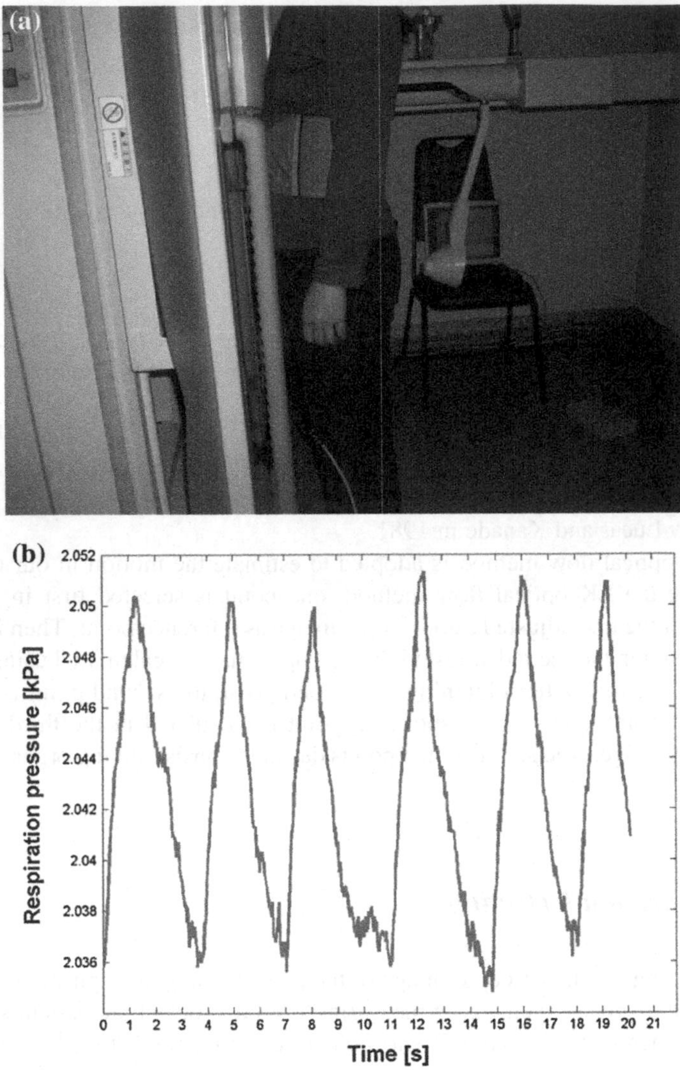

Fig. 2 Respiration detecting and detected respiration signal. **a** Respiration detecting. **b** Detected respiration signal

2.5 Synchronization of Fluoroscopic Imaging and Respiration Detecting

To avoid X ray's damage to the doctor, fluoroscopic imaging machine and observing display device are placed in two different houses respectively, which are isolated by the door and walls with radiation protection function. Thus, two computers were used to capture fluoroscopic image and detect respiration signal respectively. The two computers should have the same time reference to synchronize fluoroscopic image and respiration signal. To solve this problem, specific software was designed to synchronize the two computers' time. After synchronizing, time difference of two computers are limited to 10 ms.

Since the speed of respiration detecting is higher than the one of image capturing, it's necessary to down-sample the respiration signal to make it has the same acquisition frequency as image capturing when analyzing the correlation of respiration and lung organ motion.

3 Experiments

In this section, we'll analyze the correlation between respiration and lung organ motion and try to derive a correlation model. Two criteria are used to validate the correlation model's accuracy. One is root mean squared error (RMSE), which is defined as follows:

$$RMSE = \sqrt{\frac{1}{n}\sum_{i=1}^{n}(F(i) - O(i))}, \tag{4}$$

where $F(i)$ and $O(i)$ are the fitted value calculated by the correlation model and the original value at position i. Another criterion is maximum error (ME), which is defined as:

$$ME = max(|F(i) - O(i)|). \tag{5}$$

The respiration signal and fluoroscopic image of 12 patients were acquired by the methods mentioned above. Of which 7 cases with stable respiration and image data were selected, and the lung organ motion was tracked by the LK optical flow algorithm.

Figure 3 shows No. 3 patient's respiration signal and organ motion waveform. From this figure, we can see that the lung organ motion waveform is very similar to the respiration signal. By observing carefully, we can see that there exists a phase discrepancy between the lung organ motion waveform and respiration signal. In Fig. 3a, the first peak of the curve occurs at about 1.4s, and the second peak occurs at about 6.8 s. While in Fig. 3b, the two peaks occur at about 1.7 and 7.1 s respectively. The lung organ motion lags behind the respiration signal about 0.2 $\sim$ 0.3 s, which occupies 3.7 $\sim$ 5.6 % of the whole breathing cycle. That's because the fluoroscopic images were captured twice and the capturing and transferring of images need more time than the acquisition of respiration signal.

The respiration signal was filtered by a Gaussian filter, which width is 5, to remove the noise. After de-noising, it was down-sampled to make it has the same acquisition

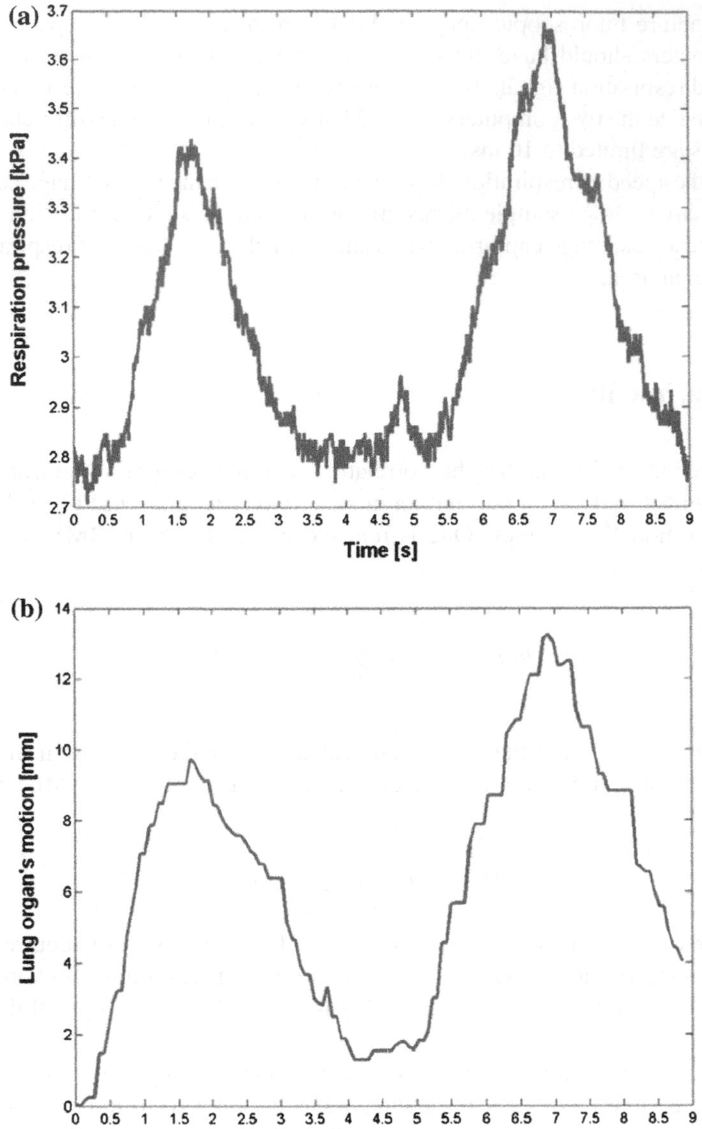

Fig. 3 Respiration signal and lung organ motion waveform of No. 3 patient

frequency as the fluoroscopic image. Thus, it was aligned according to the phase with the lung organ motion, and the least square method was used to fit the relationship between the respiration signal and the lung organ motion waveform. The correlation model between the respiration signal and the lung organ motion of No. 3 patient after phase aligning is

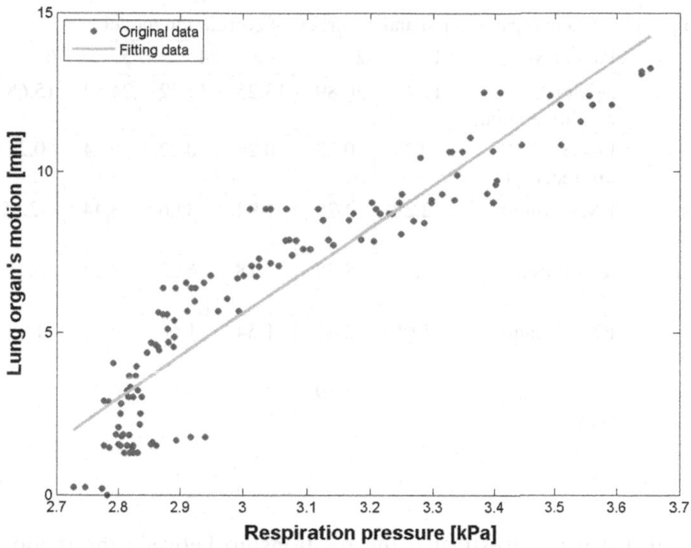

Fig. 4 Correlation curve between the respiration and lung organ motion of No. 3 patient

$$y = 13.24x - 34.10.$$

Figure 4 shows the correlation curve. It's easy for us to see that there is an approximate linear relationship between respiration signal and lung organ motion after phase aligning.

The other 6 patients' data were analyzed by the same way, and the correlation models were created according to these data before and after phase aligning respectively. Thus, motion amplitude of lung organ, RMSE and ME for these correlation models were all calculated. Table 1 shows amplitude of lung organ motion and accuracy of all correlation models. From this table, we can see that the motion amplitude of the patients' lung organ are between 11.47 ~ 24.82 mm. Before phase aligning, the prediction RMSE, RM of the correlation models falls in 1.94 ~ 4.25 mm and 4.18 ~ 8.28 mm respectively. For every patient, the proportion of RMSE in motion amplitude is less than 22 %, and the proportion of ME in motion amplitude is less than 40 %. After phase aligning, the prediction RMSE, RM of the correlation models falls in 1.05 ~ 2.47 mm and 2.85 ~ 6.25 mm respectively. For every patient, the proportion of RMSE in motion amplitude is less than 12 %, and the proportion of ME in motion amplitude is less than 25 %.

The above results show that the lung organ motion can be predicted by respiration with a maximum error of 25 % by means of modeling the correlation between the respiration and the lung organ motion. This will be very helpful to measure the lung tumor motion indirectly in dynamic radiotherapy.

Table 1 Amplitude of lung organ motion and accuracy of correlation model

Phase	Patient No.	1	2	3	4	5	6	7
	Motion amplitude (mm)	11.47	20.89	13.25	15.32	24.82	15.68	18.62
	Phase discrepancy(s)	0.25	0.23	0.26	0.22	0.24	0.24	0.26
Before phase aligning	RMSE (mm)	2.22	4.25	1.94	3.06	3.44	2.55	3.48
Before phase aligning	Max difference (mm)	4.50	8.28	4.18	6.25	8.27	5.41	7.58
After phase aligning	RMSE (mm)	1.05	2.47	1.34	1.68	2.22	1.74	2.15
After phase aligning	Max difference (mm)	2.85	5.79	3.08	3.01	6.25	4.52	5.82

Chen et al. [8] have investigated the relationship between the motion models of the patient's internal organization and the external radio opaque markers on the patient's chest, and found that for some patients there are significant motion phase discrepancies between an internal target and an external marker. The research result is identical to our research. However, Chen etc. just showed the existence of phase discrepancy. In this paper, we further research on the correlation between the respiration and lung organ motion after phase aligning, and derive a concrete correlation model. This method is meaningful for clinic radiotherapy.

4 Conclusion

Correlation between respiration and lung organ motion induced by respiration was presented in this paper. Even though the lung tumor motion can not be obtained accurately by the respiration, its approximate position acquired according to the respiration and correlation model is still helpful to decrease the target volume. Only several data sets are used in this paper to investigate the correlation. More data sets of lung tumor motion and respiration are expected to be obtained and used in further research to investigate more accurate correlation model between respiration and lung tumor motion.

Acknowledgments The authors acknowledge the financial supports of the National Natural Science Foundation of China under Grant Nos. 61402218, 61201439, 61202319, the Jiangxi Provincial Department of Science and Technology under Grant Nos. 20142BAB217013, 20151BAB217022, the Educational Bureau of Jiangxi Province under Grant Nos. GJJ14540, GJJ13487, and the Key Laboratory of Image Processing and Pattern Recognition (Nanchang Hangkong University) under Grant No. TX201304002. The authors also thank Qi Zhang, Qin Li, Zhenhuan Li, doctor Wang and doctor Zhang of Huazhong University of Science and Technology for their help during the procedure of data acquisition and processing.

References

1. Li W, Purdie TG, Taremi M et al (2011) Effect of immobilization and performance status on intrafraction motion for stereotactic lung radiotherapy: analysis of 133 patients. Int J Radiat Oncol 81:1568–1575
2. Cole A, Hanna G, Jain S et al (2014) Motion management for radical radiotherapy in non-small cell lung cancer. Clin Oncol 26:67–80
3. Keall PJ, Kini VR, Vedam SS et al (2001) Motion adaptive x-ray therapy: a feasibility study. Phys Med Biol 46:1–10
4. Giraud P, Yorke E, Ford EC et al (2006) Reduction of organ motion in lung tumors with respiratory gating. Lung Cancer 51:41–51
5. Jang SS, Huh GJ, Park SY et al (2014) The impact of respiratory gating on lung dosimetry in stereotactic body radiotherapy for lung cancer. Phys Med 30:682–689
6. Rosenzweig KE, Hanley J, Mah D et al (2000) The deep inspiration breath-hold technique in the treatment of inoperable non-small-cell lung cancer. Int J Radiat Oncol 48:81–87
7. Hugo GD, Weiss E, Badawi A et al (2011) Localization accuracy of the clinical target volume during image-guided radiotherapy of lung cancer. Int J Radiat Oncol Biol Phys 81:560–567
8. Chen QS, Weinhous MS, Deibel FC et al (2001) Fluoroscopic study of tumor motion due to breathing: facilitating precise radiation therapy for lung cancer patients. Med Phys 28:1850–1856
9. Tsunashima Y, Sakae T, Shioyama Y et al (2004) Correlation between the respiratory waveform measured using a respiratory sensor and 3D tumor motion in gated radiotherapy. Int J Radiat Oncol 60:951C958
10. Seppenwoolde Y, Berbeco RI, Nishioka S et al (2007) Accuracy of tumor motion compensation algorithm from a robotic respiratory tracking system: a simulation study. Med Phys 34:2774–2784
11. Torshabi A, Pella A, Riboldi M et al (2010) Targeting accuracy in real-time tumor tracking via external surrogates: a comparative study. Technol Cancer Res Treat 9:551–561
12. Horn BKP, Rhunck BG (1981) Determining optical-flow. Artif Intell 17:185–203
13. Lucas BD, Kanade T (1981) An iterative image registration technique with an application to stereo vision. In: Proceedings of the international joint conference on articial intelligence, pp 674–679

Joint Optimization of Repeater Gain Setting and Power Allocation in Satellite Communication Systems

Liu Yang, Bangning Zhang, Hangxian Wang and Daoxing Guo

Abstract The power allocation optimization algorithm can improve the efficiency of resource utilization, which is of great significance to the power restrained satellite communication systems. This paper taking both performance analysis of the end-to-end link and influence of repeater gain on power allocation into consideration, and a repeater gain settings based resource allocation model is built. On the basis of analyzing the above model, an algorithm of joint optimization of repeater gain setting and power allocation is proposed. After dual decomposition algorithm searching the optimal solution of power allocation with certain repeater gain, the proposed algorithm searches the optimal repeater gain by applying the steepest descent method. Simulation results show that global optimization solution of system capacity is obtained with detailed results of power allocation and repeater gain.

Keywords Satellite communication · Resource allocation · Repeater gain setting · Optimization algorithm

1 Introduction

The resource optimization algorithm can improve the efficiency of resource utilization, which is of great significance to the power restrained satellite communication systems for enhancing system capacity. That's the reason why resource allocation

---

L. Yang · B. Zhang · H. Wang · D. Guo (✉)
College of Communication Engineering, PLA University of Science and Technology, Nanjing 210007, China
e-mail: rgsc2014@163.com; rgsc@163.com

L. Yang
e-mail: 18795968630@163.com; kongzheyang@gmail.com

B. Zhang
e-mail: bangn_zhang@163.com

H. Wang
e-mail: wanghangxian@163.com

© Springer International Publishing Switzerland 2017
V.E. Balas et al. (eds.), *Information Technology and Intelligent Transportation Systems*, Advances in Intelligent Systems and Computing 454,
DOI 10.1007/978-3-319-38789-5_67

optimization algorithm is a hot issue in the research of satellite communication system. Numerous publications have studied to improve the efficiency of resource utilization in satellite communication systems. References [1–3] took system throughput as the optimization goals. In order to meet the QoS (Quality of Service) for users, an optimal power allocation method was studied for mobile satellite communication system in [1]. Taking physical layer security into consideration, J. Lei et al. [2] pro-posed a joint optimization algorithm of beam-forming and power for meeting safety capacity of the system. A power allocation algorithm was proposed in [3], which was designed for satellite communication systems under fading channels. On the basis of satisfying user requirements, a two-step power allocation scheme was studied in [4], which kept the balance between total power consumption and channel capacity. Considering delay constraints, another joint optimization algorithm of power and bandwidth was researched in [5], so that the throughput of the system can be maximized. Under conditions of user's service requirements and channel conditions, various methods were investigated in [6–8], such that second-order differential system capacity can be minimized.

Previous works mostly concentrated on the study of the downlink in satellite communication systems, and repeater gain was regarded as a fixed value, whose impact on resource allocation of the system was neglected. Nevertheless, in consideration of the satellite communication systems with a transparent repeater, Uplink noise may be amplified and impacts downlink. Though as a trivial factor in resource allocation of general repeater satellite communication, repeater gain on satellites may have severe impacts on resource allocation of transparent repeater satellite communication. Consequently, several factors were analyzed, including effective isotropic radiated power (EIRP), noise spectrum of uplink and downlink, repeater gain on satellite, power on satellite, receiver gain, gain noise-temperature ration. And a resource allocation model suited for transparent repeater in satellite communication system was built. Meanwhile, a joint algorithm of repeater gain setting and satellite communication system resource allocation was proposed in this paper. There must be an optimal solution with certain repeater gain and resource allocation, so that system channel capacity will be global optimal through our corollary. In addition, system channel capacity is a concave function when power allocation is the only factor to be consider. Therefore, dual algorithm was applied to solve optimal power allocation with certain repeater gain. And the steepest descent method was used to evaluate global optimum of repeater gain. In simulation, an ergodic searching algorithm was designed to verify the correctness of the corollary. Finally, simulation results showed that by applying the proposed joint optimization algorithm OPOG (Optimal power allocation and the repeater gain setting algorithm) over the satellite communication system resource allocation, a global optimal solution will be derived with certain repeater gain and power allocation.

2 System Model

2.1 Modeling of Composite Link Capacity

The satellite signal gain consists of three parts, i.e., the receiving antenna gain, the repeater gain and the transmitting antenna gain. Furthermore, repeater gain on satellite is adjustable.

Supposing there are M users in the satellite communication system. Considering the ith user, the power distributed from the back-end of the repeater is defined as p_i, the attenuation factor of weather conditions is α_i^2, and the bandwidth which is pre-distributed is W_i. Free space path loss of uplink and downlink are expressed as L_s and L_r, respectively. Besides, satellite antenna gain and satellite repeater gain can be de-noted as G_s, and G, respectively. In the downlink receiver, the antenna gain of the ith is G_{ri}, and the received power is p_{ri}. The noise spectrum density of uplink and down-link are N_s and N_r, respectively. Therefore, the Shannon channel capacity of ith user is given as:

$$c_i = W_i \log_2 \left(1 + \frac{p_{ri}}{W_i(\frac{N_s G G_s G_{ri}}{L_r} + N_r)} \right), \tag{1}$$

$$p_{ri} = \frac{{\alpha_i}^2 p_i G_s G_{ri}}{L_s}. \tag{2}$$

Supposing the total power on satellite is P. So the *EIRP* from uplink to the transmitter of the ith user is $EIRP_i$. And in this section, the optimization goal is to maximize channel capacity. Consequently, the following mathematical model can be listed as:

$$\max_{\{p_i, G\}} \sum_{i=1}^{M} c_i, \tag{3}$$

$$c_i = W_i \log_2 \left(1 + \frac{{\alpha_i}^2 p_i G_s G_{ri}}{W_i(N_s G G_s G_{ri} + N_r L_r)} \right), \forall i \in \{1, 2, \ldots, M\}, \tag{4}$$

$$P \geq \sum_{i=1}^{M} (p_i + W_i N_s G), \tag{5}$$

$$\frac{EIRP_i G_s G}{L_r} \geq p_i, \forall i \in \{1, 2, \ldots, M\}, \tag{6}$$

Equation (4)means that the distributed capacity cannot exceed the traffic demand; Eq. (5) indicates that the total power of the system is limited; Eq. (6) indicates that the EIRP of user transmitter is limited.

3 Discussion of the Existence of Optimal Value

Since the above optimization problem is a continuous nonlinear optimization problem, we introduce a nonnegative dual variable γ and $\boldsymbol{\beta} = [\beta_1, \beta_2, , \beta_M]$, so that the Lagrange function L of optimization problem is derived as:

$$L(\boldsymbol{p}, G, \gamma, \boldsymbol{\beta}) = \sum_{i=1}^{M} W_i \log_2\left(1 + \frac{{\alpha_i}^2 p_i G_s G_{ri}}{W_i(N_s G G_s G_{ri} + N_r L_r)}\right) + \gamma\left(P - \sum_{i=1}^{M}(p_i + W N_s G)\right) + \sum_{i=1}^{M} \beta_i \left(\frac{EIRP_i G_s G}{L_r} - p_i\right). \tag{7}$$

where Lagrange dual function is

$$g(\gamma, \boldsymbol{\beta}) = \max_{P,G} L(\boldsymbol{P}, G, \gamma, \boldsymbol{\beta}). \tag{8}$$

The dual expression of the original optimization problem can be written as:

$$d = \min_{\gamma \geq 0, \boldsymbol{\beta} \geq 0} g(\gamma, \boldsymbol{\beta}). \tag{9}$$

In dual theory, when the optimal solution of an even function is equal to the optimal solution of the original function, we can obtain that:

$$P = \sum_{i=1}^{M}(p_i^o + W N_s G), \tag{10}$$

$$\beta_i \left(\frac{EIRP_i G_s G}{L_r} - p_i^o\right) = 0. \tag{11}$$

With the results of the second derivative of the objective function, we can derive that:

$$F(P, G) = \max_{\{p_i, G\}} \sum_{i=1}^{M} c_i, \tag{12}$$

$$\nabla_{p_i}^2 F(P,G) = \frac{-\alpha_i{}^2 W}{\ln 2\left(WN_sG + \dfrac{WN_rL_r}{G_sG_{ri}} + \alpha_i{}^2 p_i\right)^2}, \tag{13}$$

$$\nabla_G^2 F(P,G) = \sum_{i=1}^{M} \frac{\alpha_i{}^2 p_i N_s W\left[2N_sW\left(N_sG + \dfrac{N_rL_r}{G_sG_{ri}}\right) + \alpha_i^2 p_i N_s\right]}{\ln 2\left[\left(WN_sG + \dfrac{WN_rL_r}{G_sG_{ri}} + \alpha_i{}^2 p_i\right)\left(N_sG + \dfrac{N_rL_r}{G_sG_{ri}}\right)\right]^2}. \tag{14}$$

From (13) and (14), we can obtain that the objective function is non-concave function, while the objective function is a concave function when the power is the only variable. Due to the concavity of (8), when the solution of the original problem is optimal, the value of the derivative must be 0. Then

$$\frac{W\alpha_i{}^2 G_sG_{ri}}{(W(N_sGG_sG_{ri} + N_rL_r) + \alpha_i{}^2 p_i^o G_sG_{ri})\ln 2} = \gamma + \beta_i. \tag{15}$$

Taking the derivative of (7) with respect to repeater gain G, we obtain that:

$$\nabla_G^2 F(P,G) = \sum_{i=1}^{M} \frac{-\alpha_i{}^2 p_i N_s W}{\ln 2\left(WN_sG + \dfrac{WN_rL_r}{G_sG_{ri}} + \alpha_i{}^2 p_i\right)\left(N_sG + \dfrac{N_rL_r}{G_sG_{ri}}\right)} - \gamma\left(\sum_{i=1}^{M} WN_s\right) + \sum_{i=1}^{M} \beta_i \frac{EIRP_iG_s}{L_r}. \tag{16}$$

Assuming that the repeater gain is G_0 when the value of (16) is 0. Then the repeater gain must be G_0 or the bounds of G. Due to the restrictions of (5) and (6), it is obvious that the objective function cannot be maximized when the repeater gain takes the bound value. So G_0 is the value of the repeater gain when the objective function is maximized.

On the other hand, when the value of the original problem is global optimum, the problem must meet the requirements of (10), (11), (15), and the value of (16) must be 0. Therefore, substituting (10), (11), (15) into (16), and making the value of (16) be 0, we can deduce that:

$$\gamma = \sum_{i=1}^{M} \frac{\alpha_i{}^2 p_i^o W \dfrac{N_rL_r}{G_sG_{ri}}}{\ln 2\left(WN_sG_0 + \dfrac{WN_rL_r}{G_sG_{ri}} + \alpha_i{}^2 p_i^o\right)\left(N_sG_0 + \dfrac{N_rL_r}{G_sG_{ri}}\right)P}. \tag{17}$$

Apparently, dual variable γ will converge to the expression in (17) eventually.

4 Joint Optimization Algorithm of Repeater Gain Setting and Power Allocation

Based on the above reasoning, the objective function of the original problem is a concave function which is derivative with respect to the repeater gain. And the value of the derivative with respect to the repeater gain is 0 when the objective function take the global optimum. That is the reason why a double-layer-nesting algorithm is designed in the paper to search the global optimum of objective function. On the one hand, in the inner layer, we search the optimal solution of power allocation when the repeater gain is certain by applying dual algorithm. On the other hand, in the outer layer, we search the optimum of repeater gain by applying steepest descent method.

It is apparent that power allocation and repeater gain will be optimal if the objective function takes the global maximum when outer algorithm is convergent. Moreover, the dual variable γ is converges to (17). Finally, the repeater gain converges to G_0. We can solve the dual problem of (7) with certain repeater gain by separating it into two parts.

Step 1: Adjustment of power allocation. The value of p_i is expressed by p_i^* when $L(p, \gamma, \boldsymbol{\beta})$ takes its maximum. So the value of the partial derivative of $L(p, \gamma, \boldsymbol{\beta})$ with respect to p_i^* is equal to zero. Then, we obtain that :

$$\frac{W{\alpha_i}^2 G_s G_{ri}}{(W(N_s G G_s G_{ri} + N_r L_r) + {\alpha_i}^2 p_i^* G_s G_{ri}) \ln 2} = \gamma + \beta_i. \tag{18}$$

Therefore the optimum of power assigned to each user is: $p_i^k = max\{0, p_i^*\}$.

Step 2: Updating dual variables. The sub-gradient algorithm is used to update the dual variable.

$$\gamma^{m+1} = \left[\gamma^m - \Delta_\gamma^m \left(P - \sum_{i=1}^{M} (p_i^k + W N_s G)\right)\right]^+, \tag{19}$$

$$\beta_i^{m+1} = \left[\beta_i^m - \Delta_{\beta_i}^m \left(\frac{EIRP_i G_s G}{L_s} - p_i^k\right)\right]^+, \forall i \in \{1, 2, \ldots, M\}. \tag{20}$$

Calculating the optimum of power assigned to each user p_i^k and corresponding value of its dual variable when the repeater gain is certain. Then, substituting the results into the following formula to update the repeater gain:

$$G^{m+1} = \left[G^m - \Delta_G^m \left(-\nabla_G L(\boldsymbol{p}, G^m, \gamma, \boldsymbol{\beta})\right)\right]^+, \tag{21}$$

where $[x]^+ = max\{0, x\}$. m represents the number of iterations, and represents the step size.

The joint optimization algorithm of repeater gain setting and power allocation can be separated into following steps:

Step 1: Initializing the repeater gain G according to its bounds $[A, B]$, and confirming the value of dual variable $\boldsymbol{\beta}$.

Step 2: Under the condition that the power is equally allocated to all users, initialize the value of dual variable γ with the results of calculating (17).

Step 3: Substituting the updated repeater gain and dual variables into (18), solve the results of optimized power allocation.

Step 4: Substituting both the optimized power allocation and the dual variables into (19) and (20), to update the dual variables.

Step 5: If results cannot satisfy (22), go back to Step 3.

$$\left|\gamma^{m+1}(P - \sum_{i=1}^{M}(p_i + WN_sG))\right| < \alpha, \left|\beta_i^{m+1}\left(\frac{EIRP_iG_sG}{L_s} - p_i\right)\right| < \alpha. \quad (22)$$

Step 6: If results satisfy (22), but cannot satisfy (23), update repeater gain by substituting the results of power allocation and dual variables into (18), then go back to Step 3.

$$\left|\nabla_G L(G^m, \boldsymbol{p}, \gamma, \boldsymbol{\beta})\right| < \varepsilon. \quad (23)$$

Step 7: If results satisfy both (22) and (23), output the results of power allocation and values of repeater gain.

5 Analysis of Simulation Results

In order to verify the correctness of the proposed algorithm, a geostationary satellites of Ka band was built. All system parameters are practical for existing Ka band geostationary satellites. It should be noted that we neglect the impact of weather conditions, so the value of α_i^2 is 1 (Table 1).

Figures 1, 2 and 3 shows the convergence of the repeater gain, the dual variable γ, and the system channel capacity, respectively. As depicted in the figures, the convergence of the proposed algorithm is proved.

Numerical results derived by applying OPOG algorithm are recorded in Table 2.

An ergodic searching algorithm was designed to verify the correctness of the proposed OPOG algorithm. The step-length of the values of repeater gain is 0.1 dB. Then we calculate corresponding system channel capacity of repeater gain with different valued by applying dual algorithm.

The maximums of system channel capacity under different repeater gain by applying ergodic searching algorithm are shown in Fig. 4. And numerical results derived by applying ergodic searching algorithm are recorded in Table 3. It should be noted that due to the design of parameters, all the power allocation to users assigned by the system must be restrained by (6) when the repeater gain is smaller than 118 dB. So there is no need to show the repeater gain if its value is smaller than 118 dB.

Table 1 Satellite communication system parameters

Parameter	Value
User number	10
Satellite total power $P(W)$	40
User bandwidth $W(M)$	50
Uplink noise temperature $K_s(k)$	300
Downlink noise temperature $Kr(k)$	50
Satellite antenna gain $G_s(dB)$	38
Link attenuation factor α_i^2	1
Satellite repeater range $G(dB)$	110–130
Ratio of gain and equivalent noise temperature $G_r/T(dB)$	[15 15 15 18 18 18 23 23 25 25]
Effective isotropic radiated power for each user of the sending end $EIRP_s(dB)$	[66 62 55 66 62 59 66 59 62 55]

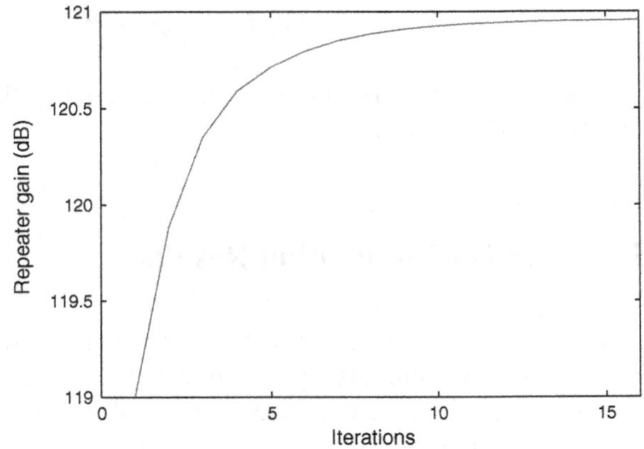

Fig. 1 The convergence of the repeater gain

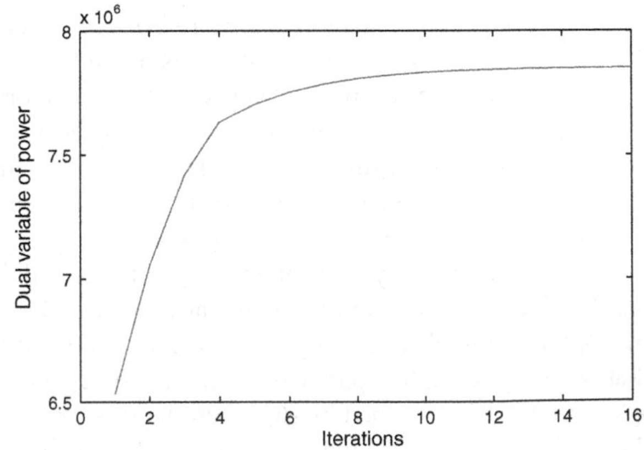

Fig. 2 The convergence of dual variable γ

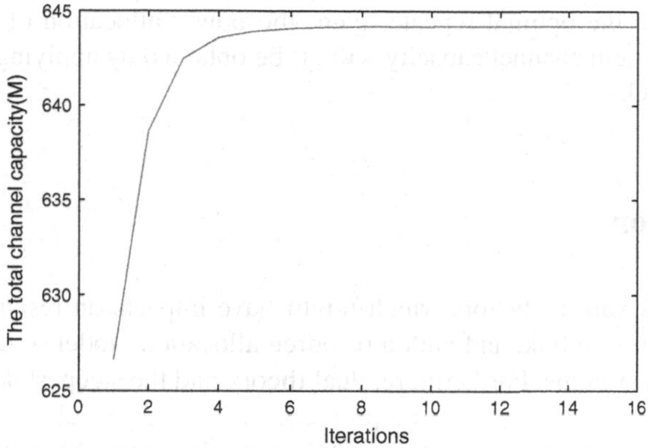

Fig. 3 The convergence of the system channel capacity

Table 2 The optimal repeater gain and corresponding maximum of channel capacity derived by applying OPOG algorithm

Items	Value
Optimal repeater gain (dB)	120.9565
Maximized channel capacity (M)	644.2023
Assigned power to users (W)	[1.9073 1.9073 1.2461 5.4093 5.4093 3.1153 7.8153 3.1153 6.2305 1.2461]

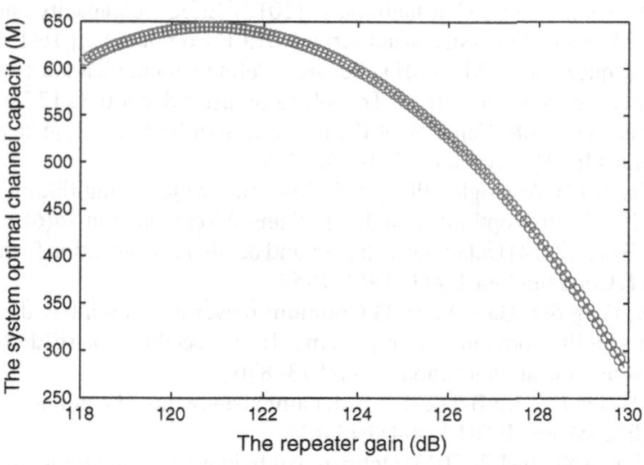

Fig. 4 The maximum of system channel capacity under different repeater gain

Table 3 The optimal repeater gain and corresponding maximum of channel capacity derived by applying ergodic searching algorithm

Items	Value
Optimal repeater gain (dB)	120.9
Maximized channel capacity (M)	643.9

It proves that the optimal repeater gain, the power allocation of users and the maximum of system channel capacity will all be obtained by applying the proposed algorithm OPOG.

6 Conclusion

This paper took various factors which might have impacts on resource allocation in satellite composite link, and built a resource allocation model suited for satellite communication systems. Furthermore, dual theory and the steepest descent method are used to find the optimum of the problem. Simulation results showed that the proposed joint optimization algorithm OPOG can solve the problem of repeater gain and power allocation on transparent transponder satellite communication systems. And channel capacity of the system will be enhanced after applying OPOG algorithm.

Acknowledgments I would like to extend my sincere gratitude to my dear friend Kongzhe Yang, for his instructive advice and useful suggestions on my thesis. I am deeply grateful of his help in the completion of this thesis.

References

1. Vassaki S, Panagopoulos AD, Constantinou P (2012) Effective capacity and optimal power allocation for mobile satellite systems and service. IEEE Commun Lett 16(1):60–63
2. Lei J, Zhu H, Vázquez-Castor MÁ (2011) Secure satellite communication systems design with individual secrecy rate constrains. IEEE Trans Inf Forensics Secur 6(3):127–135
3. Letzepis N, Grant AJ (2008) Capacity of the multiple spot beam satellite channel with Rician fading. IEEE Trans Inf Theory 54(11):5210–5222
4. Aravanis AI, Shankar B, Arapoglou PD (2015) Power allocation in multibeam satellite systems: a two-stage multi-objective optimization. IEEE Trans Wirel Commun 14(6):3171–3182
5. Ji Z, Wang YZ, Wei F (2014) Delay-aware power and bandwidth allocation for multiuser satellite downlinks. IEEE Commun Lett 18(11):1951–1954
6. Feng Q, Li GX, Feng SD, Gao Q (2011) Optimum power allocation based on traffic demand for multi-beam satellite communication systems. In: Proceedings of IEEE 13th international conference on communication technology, pp 873–876
7. Park U, Kim HW, Oh DS, Ku B-J (2012) A dynamic bandwidth allocation scheme for a multi-spot-beam satellite system. ETRI J 34(4):613–616
8. Wang H, Liu AJ, Pan XF, Jia LL (2013) Optimal bandwidth allocation for multi-spot-beam satellite communication systems. In: Proceedings of 2013 international conference on mechatronic sciences, electric engineering and computer, pp. 2794–2798

Cybersecurity Decision Making Mechanism for Defense Strategies in Vehicle Networks

Wei Pan, Mengzi Wang, Yuanyuan Fu and Haobin Shi

Abstract With the development of intelligent transportation system, vehicle networks (VNs) have brought new security challenges due to their mobile and infrastructure-less nature. The existing security solutions lack the comprehensive cybersecurity model for vehicle networks. This paper proposes a combined attack and defense petri net (ADPN) model to describe cybersecurity states and processes of vehicle network. To evaluate security assessment and optimize defense strategies, an attack and defense decision making method for vehicle network (VN-ADDM) is proposed. The method utilizes stochastic games net to analyze and quantify security assessment properties, calculate attack and defense cost, and implement mixed strategy Nash equilibrium. The rationality of the proposed cybersecurity decision making method is well demonstrated by extensive analysis in a detailed case study.

Keywords Cybersecurity · Vehicle network · Decision making · Game theory · Petri net

W. Pan (✉) · M. Wang · Y. Fu · H. Shi
School of Computer Science and Engineering, Northwestern Polytechnical University, Xian 710072, People's Republic of China
e-mail: panwei@nwpu.edu.cn

M. Wang
e-mail: wangmengzi_2013@163.com

Y. Fu
e-mail: fyy186@163.com

H. Shi
e-mail: shihaobin@nwpu.edu.cn

W. Pan
Information Technology Research Base of Civil Aviation Administration of China, Civil Aviation University of China, Tianjin 300300, People's Republic of China

© Springer International Publishing Switzerland 2017
V.E. Balas et al. (eds.), *Information Technology and Intelligent Transportation Systems*, Advances in Intelligent Systems and Computing 454,
DOI 10.1007/978-3-319-38789-5_68

1 Introduction

Vehicle networks (VNs) can be used to improve transportation security, reliability, and management. In VNs, each vehicle is equipped with an On Board Unit (OBU), by which vehicles are able to communicate with each other as well as Road Side Units (or RSUs). VNs are expected to support a wide range of promising applications such as location based services. With the rapid development of VNs, viruses, Trojans and other threats are rapidly spreading to VNs, VN cybersecurity is becoming increasingly prominent. Once VN attacks occur, it may cause potential dangers to information security, transportation security and personal safety. Traditional security methods are passive defense measures and its defense capabilities are limited. VN cybersecurity should pay more attention to the overall operational safety. This paper investigates security aspects of VNs within a game-theoretic framework where defensive measures are optimized with respect to threats posed by malicious attackers.

VNs security is gaining an increased interest from both of academia and industry. Tamer Basar proposed a structured and comprehensive overview of research on cybersecurity and privacy based on game theory [1, 2]. C.W. Ten proposed a quantitative analysis method of attack probabilities which is based on a compromise form and the empowerment of vulnerable trees, and built an network security test bed [3]. Cheminod Manuel used the complementary strengths of two different models and authentication technology to analyze and evaluate the formalizing vulnerabilities of multilayer interconnection of field bus systems [4]. Jaafar Almasizadeh proposed a security metrics model based on stochastic games to quantify and evaluate the reliability of system security [5]. Zhu put forward an information physical integration system security evaluation method, which is based on game theory [6, 7]. Saman proposed a safety assessment model based on Markov decision process for external network attacking the power grid facilities, so that the administrator can decide and response automatically [8]. Prabhakar M used Markov chains to choose the propriate model for security problems in VANET [9].

Although the above methods have made some achievements, they lack uniform quantitation evaluation standards of security properties, and do not fully consider both interact and interdependent strategies of attackers and defenders, which may lead the analysis results mistake. Vehicle network cannot be effectively evaluated of the security situation. To solve the above problems, this paper proposes a security decision making method based on attack and defense model and stochastic games net. This method can reflect the dynamic interaction and state deduction between attackers and defenders.

The paper is organized as follows. In Sect. 2, we introduce attack and defense model for vehicle network security. In Sect. 3, we elaborate cybersecurity decision making method. In Sect. 4, security analysis is given to demonstrate and illustrate the attack-defense game. We conclude this paper in Sect. 5.

2 Attack and Defense Model for Vehicle Network Security

2.1 Vehicle Network Structure

The vehicle network (VN) structure hierarchically consists of three parts, shown as Fig. 1.

(1) OBU: On Board Unit is a vehicular equipment in each vehicle, which can collect traffic information and communicate with other vehicles or communication devices in real time.

(2) RSU: Roadside Unit is a communication base station, and provides communication services for vehicles to connect other communication network or Internet.

(3) TA: Trusted Authority is a security infrastructure owned by government, which provides identity authentication and stores all vehicular information safely.

The communication technologies in vehicle networks can be categorized into V2V (Vehicle to Vehicle) and V2I (Vehicle to Infrastructure). V2V enable vehicles communicate with each other and transfer traffic information directly. V2I enable vehicles access to fixed networks, and can be used to support vehicular services, such as vehicle identification, management and revocation.

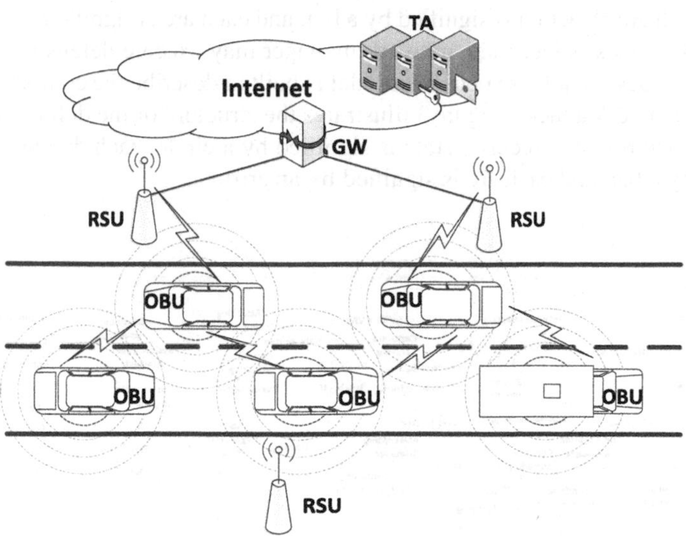

Fig. 1 The structure of vehicle network

2.2 Modeling Attack and Defense Petri Net for Vehicle Networks

Attack and defense modeling is an important part of situation assessment and security decision making for vehicle networks. In this paper, we propose an attack and defense model based on petri net (ADPN) to describe cybersecurity states and processes of vehicle network. Attack behaviors and processes in vehicle networks have the characteristics of diversity, random, concealment, malicious propagation. Petri net has the capability to describe distributed, multi-objective, multi-stage network attack and defense behaviors, especially elaborate concurrent and collaborative attacks and defenses.

A Petri net is a directed bipartite graph, in which the nodes represent transitions (i.e. events that may occur) and places (i.e. states or conditions). The directed arcs describe which places are pre- and/or postconditions for which transitions. In the attack and defense process, an attack or defense action can be described as a transition node. A security state of vehicle networks can be described as a place node. An attack or defense action may trigger another attack or defense action. A trigger condition between these actions can be described as an arc. Generally, vehicle network attacks have three attacking intentions: compromise communication security, compromise location privacy, compromise vehicle safety. We build an attack petri net model to describe the actual process of each attacking intention. Figure 2 illustrates the structure of the attack petri net, in which each possible security state is signified by a circle, each attack action is signified by a bar, and each arc is signified by an arrow.

When an attack is launched, network manager may execute defensive measures to stop the attack. A defense petri net model is built to describe the countermeasures of vehicle network attacks. Figure 3 illustrates the structure of the defense petri net, in which each possible security state is signified by a circle, each defense action is signified by a bar, and each arc is signified by an arrow.

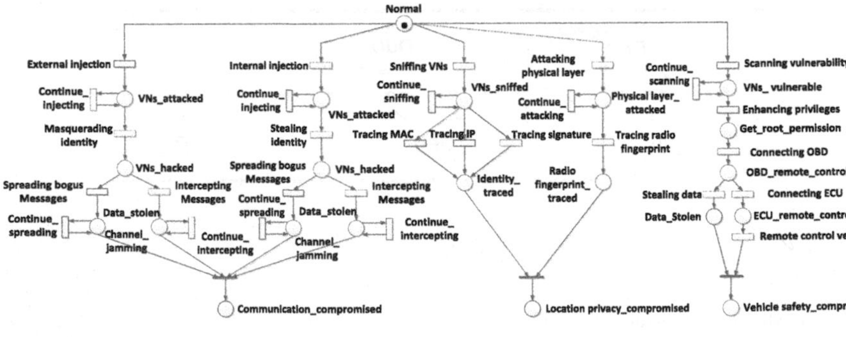

Fig. 2 The attack petri net model

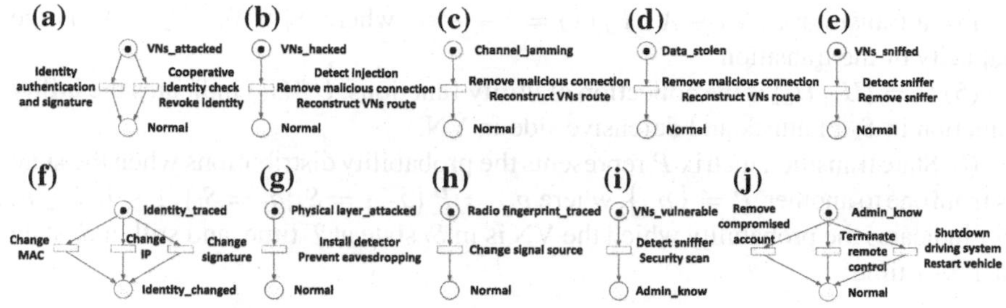

Fig. 3 The defense petri net model

3 Cybersecurity Decision Making Method for VNs

3.1 VN Cybersecurity Game Model Based on Stochastic Game Net

Stochastic Game Net (SGN) is a stochastic modeling and analysis method that combines stochastic game with petri net. It has the capacity to analysis concurrency, asynchronous and uncertainty of a system. It can accurately analyze the state changes of security game between attackers and defenders in vehicle network.

Considering that the attack and defense of VN security is non-cooperative and target opposition between attackers and defenders, the confrontation process of attack and defense is non-cooperative. Because the strategies of attackers and defenders are interdependence, the confrontation process of attack and defense is dynamic. Attackers or defenders cannot obtain all information about counterpart's strategies and VN state changes. The confrontation process of attack and defense is incomplete. Thus, using non-cooperative attack and defense stochastic games network to describe information about VN is suitable for security evaluation and analysis.

Definition 1 The attack and defense decision making model for VN cybersecurity (VN-ADDM) is defined as follows:

(1) $N = (N_1, N_2, \ldots, N_n)$ are players, where N_k represents the kth players. Considering the confrontation of both attack and defensive side, let $n = 2$. N_1 represents attacker, and N_2 represents defender.

(2) $S = (S_1, S_2, \ldots, S_n)$ is a set of security states, where S_i represents an independent state and n indicates the number of states. Different security events will cause a change from one security state to another. The state set in VN-ADDM is consistent with ADPN in the previous section.

(3) $A = A^1 \cup A^2$ is a transition set of player's actions. $A^1 = \left(a_1^1, a_2^1, \ldots, a_m^1\right)$ is a set of attackers actions. $A^2 = \left(a_1^2, a_2^2, \ldots, a_m^2\right)$ is a set of defenders actions.

(4) $\lambda = (\lambda_1, \lambda_2, \ldots, \lambda_n)$ is a set that represents transition firing rate. It depends on the capacity of players in the stochastic games theory and equilibrium strategies selection probabilities. Assume that behavior of players is random and exponential.

For a transition a: $\forall a \in A : F_a(x) = 1 - e^{\lambda_a x}$, where $\lambda_a > 0$, $x \geq 0$. λ_a is the capacity of the transition.

(5) $U = \{U_1, U_2\}$ is the collection of utility functions. $U^m(s_k)$ indicates the utility function in S_k of attack and defensive side in VN.

(6) State transition matrix P represents the probability distributions when the state is from one to another. $P = \{p_{ij}\}$, where $p_{ij} = P(q_{t+1} = S_j | q_t = S_i), 1 \leq i, j \leq n$. p_{ij} indicates the probability which the VN is in S_i state at T time, and still in S_i state at $T + 1$ time.

Mixed strategies are the action rules of players under certain condition. They reflect that the players how to choose specific action in a state. They are described by probabilities.

Definition 2 VN Attack Strategy: π^1 represents the strategy set of VN attacker, where $\pi^1 = (\pi_1^1(a_1^1), \pi_2^1(a_2^1), \ldots, \pi_i^1(a_i^1), \ldots, \pi_n^1(a_n^1))$. $\pi_i^1(a_i^1)$ represents the probability of attacker selecting a_i^1. All VN attackers' actions satisfy $\sum_{i=1}^{N} \pi_i^1(a_i^1) = 1$.

Definition 3 VN Defense Strategy: π^2 represents the strategy set of VN defender, where $\pi^2 = (\pi_1^2(a_1^2), \pi_2^2(a_2^2), \ldots, \pi_i^2(a_i^2), \ldots, \pi_n^2(a_n^2))$. $\pi_i^2(a_i^2)$ represents the probability of defender selecting a_i^2. All VN defenders actions satisfy $\sum_{i=1}^{N} \pi_i^2(a_i^2) = 1$.

According to stochastic game theory, a game model exists mixed strategy Nash equilibrium when state sets and action sets are finite. In this paper, the state sets and action sets in the attack and defense game process are finite. Therefore, VN-ADDM exists mixed strategy Nash equilibrium.

3.2 *The Definition and Quantization of VN-ADDM Properties*

Definition 4 Attack Severity (AS) represents the inherent severity of certain atomic attack. Set the initial value depending on the situation. It is calculated by Eq. (1):

$$AS = \alpha_1 A_k + \alpha_2 \left(\sum_{i=1}^{N} C_r F_r \right) \tag{1}$$

where α_1 and α_2 are respectively the network types impact factor and the attack type's impact factor on AS, which satisfies $\alpha_1 + \alpha_2 = 1$. C_r represents the severitys weight of rth atomic attack, which satisfies $\sum_{r=1}^{N} c_r = 1$. F_r represents for the value correspond to an atomic attack.

Definition 5 Attack Criticality (AC) is the influence caused by atomic attacks in the appropriate network or device importance. Considering network topology (N) and the importance of device (I_m), AC is obtained by the following Eq. (2):

$$AC = \frac{\beta\left(I_{mi} + N_j\right)}{\sum_i \sum_j \alpha_1 I_{mi} + \alpha_2 N_j} \tag{2}$$

where I_{mi} is a discrete value for the importance of a subnet. N_i is a discrete value for the importance of a device. α_1 is the weight of a subnets importance. α_2 is the weight of devices importance. β represents the severity of atomic attack, when it attacks one device in specific subnet.

Definition 6 The cost of system losses (S_c) represents the damage degree of a target resource under attack. The loss of target resource can be defined by attack risk and security property damage. The loss cost of the network system caused by attack (a_i^1) is calculated by the following Eq. (3):

$$S_c(a_i^1) = AD \times AC \times (I_c \times P_i + C_c \times P_c + A_c \times P_a + V_c \times P_v) \tag{3}$$

where the damages of security properties include integrity cost (I_c), confidentiality cost (C_c), availability cost (A_c) and vulnerability cost (V_c). (P_i, P_c, P_a, P_v) represents the proportion of integrity cost, confidentiality cost, availability cost and vulnerability cost, which satisfies $P_i + P_c + P_a + P_v = 1$.

Definition 7 Operation cost (O_c) represents defender's defense operation time and the number of computing resources. It is calculated by the following Eq. (4):

$$O_c = \log\left(O_f + R^2\right) \tag{4}$$

where O_f is the algorithmic complexity and R is the consumption of resources which consist of both human and computer resources.

Definition 8 Negative cost (N_c) represents the loss of network service under defense strategy, such as the decline in service quality or system failure caused by defense strategy. It can be defined by the following Eq. (5):

$$N_c = (A_c \times P_a + V_c \times P_v) \times \theta(a_i^1, a_j^2) \tag{5}$$

where $\theta(a^1, a^2)$ is the negative coefficient, which represents the negative impact on system availability when defense strategy a^2 defends against attack a^1. When a defender adopts a certain action to defend against attacks, $\theta(a^1, a^2)$ is related to the position of VN nodes under attack, vulnerability and services of VN nodes. $\theta(a^1, a^2)$ is a normal random variable. It can be obtained by the following Eq. (6):

$$\theta(a_i^1, a_j^2) = \frac{|P_a - P_v|}{P_a + P_v} e^{-\frac{(x-0.55)^2}{2\times 0.2^2}} \tag{6}$$

where x is obtained by normalizing the weights of attack and defense.

Definition 9 Residual loss (R_c) represents the residual and not eliminated loss caused by attacks when defense strategies are executed. It can be described by the Eq. (7):

$$R_c(a_i^1, a_j^2) = S_c(a_i^1) \times \varepsilon\left(a_i^1, a_j^2\right) \tag{7}$$

where $\varepsilon\left(a_i^1, a_j^2\right)$ represents the residual loss degree when defense strategy a_2 defends against attack a_1. When a defender adopts certain action to defend against the attacks, $\varepsilon\left(a_i^1, a_j^2\right)$ is related to the position of VN nodes under attack, vulnerability and services of VN nodes. $\varepsilon\left(a_i^1, a_j^2\right)$ is a normal random variable. It can be obtained by the following Eq. (8):

$$\varepsilon\left(a_i^1, a_j^2\right) = e^{-\frac{(x-0.3)^2}{2\times 0.15^2}} \tag{8}$$

Definition 10 Defense cost (D_c) is the cost when defense strategies are executed. It is calculated by the following Eq. (9):

$$D_c\left(a_i^1, a_j^2\right) = O_c\left(a_j^2\right) + (A_c \times P_a + V_c \times P_v)\,\theta\left(a_i^1, a_j^2\right) + S_c\left(a_i^1\right)\varepsilon\left(a_i^1, a_j^2\right) \tag{9}$$

Definition 11 Immediate return in a state is represented by r. It can be described by matrix $R^1(S_k)$.

$$R^1(S_k) = \begin{bmatrix} r_{11} & \cdots & r_{1n} \\ \vdots & \ddots & \vdots \\ r_{m1} & \cdots & r_{mn} \end{bmatrix} \tag{10}$$

where r_{ij} is calculated by the following Eq. (11):

$$r_{ij} = \sum S_c\left(a_i^1\right) + 1.4 D_c\left(a_i^1, a_j^2\right) \tag{11}$$

Definition 12 The utility function of VN attackers is described as the following Eq. (12):

$$U^1\left(s_k\right) = \sum_{\forall a_i^1 \in A^1} \sum_{\forall a_j^2 \in A^2} \pi_i^1\left(a_i^1\right) \pi_j^2\left(a_j^2\right) r_{ij} \tag{12}$$

where r_{ij} represents the immediate return when attackers use attack action a_i^1 and defenders use defensive action a_j^2 in the state S_k.

Definition 13 The utility function of VN defenders is described as the following Eq. (13):

$$U^2\left(s_k\right) = -U^1\left(s_k\right) \tag{13}$$

4 Security Analysis

4.1 ADPN Model Analysis

To discern the confrontation process of attacks and defenses, we build a combined model base on the proposed attack model and defense model in the previous section. The combined model takes into consideration the factors about action capability and detection probability of attacks, and solves the probability values by using nonlinear programming method. It is more suitable to analysis the possibility of transition firing.

Figure 4 illustrates the whole combined ADPN model. The gray transitions indicate the attack actions and the white transitions indicate the defense actions. The state set and transition set are detailedly defined in Figs. 2 and 3.

4.2 Implementation of VN-ADDM

The implementation steps of VN-ADDM are as follows:

A. Generating reward matrix

Initialize the VN-ADDM, and construct the state set $S = (S_1, S_2, \ldots, S_n)$.

Build attack actions set $A^1 = \left(a_1^1, a_2^1, \ldots, a_n^1\right)$ through attack petri net. Build defense actions set $A^2 = \left(a_1^2, a_2^2, \ldots, a_m^2\right)$ according to current defense strategies.

Build matrix $r\,[m, n]$.

Get the atomic attacks table of a_i^1 based on attack petri net. Assign values to AS, AC, I_c, C_c and other properties. Enter the related data of the defensive behaviors cost. Assign values to O_c, $\theta\left(a_i^1, a_j^2\right)$ and $\varepsilon\left(a_i^1, a_j^2\right)$.

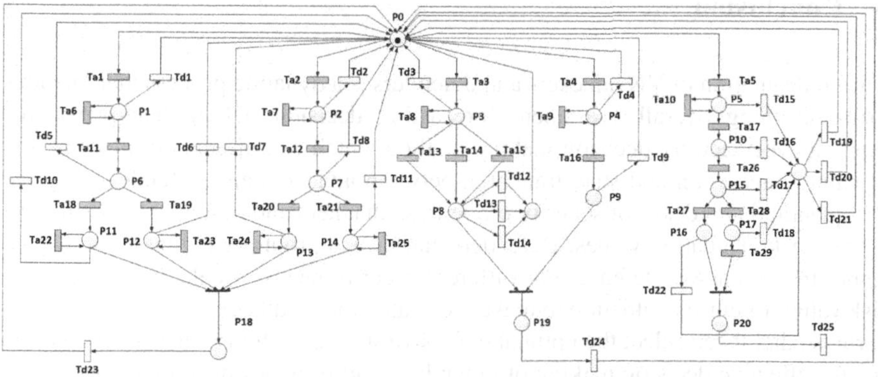

Fig. 4 The ADPN model

Use $r_{ij} = \sum S_c\left(a_i^1\right) + 1.4D_c\left(a_i^1, a_j^2\right)$ to calculate the attack reward that VN defenders use defensive behavior a_j^2 under the attack a_i^1.

B. Generating state transition probability matrix

Build the set of attack strategies $\pi^1 = \left(\pi_1^1\left(a_1^1\right), \ldots, \pi_i^1\left(a_i^1\right), \ldots, \pi_n^1\left(a_n^1\right)\right)$. Build the set of defense strategies $\pi^2 = \left(\pi_1^2\left(a_1^2\right), \ldots, \pi_i^2\left(a_i^2\right), \ldots, \pi_n^2\left(a_n^2\right)\right)$.

Initialize state transition matrix. Calculate the probabilities of every state transition matrix. Calculate correction probability of state transition matrix, then modify state transition probability matrix.

C. Evaluating security situation

Enter the reward matrix and the state information. We can obtain the specific actions of attackers and defenders, get the selecting probabilities of attackers and defenders actions.

Calculate security evaluation values by using: $U = \pi_i^1\left(a_i^1\right)\pi_i^2\left(a_i^2\right) r_{ij}$.

4.3 Security Situation Analysis

Base on the proposed method, we evaluate attack time and attack success probability. The attack probability increases with the increasing VN system time. Moreover, if some attack actions are blocked by defense strategies, then other attack actions are restricted in a controllable way.

The times of intrusion success of different attacks are different. Obviously, external injection attack takes more time. Moreover, there is no relationship between the success rate of attack and the firing rate of attack. In vehicle network, if defense strategies and cybersecurity mechanism are built, attack success rate and attack time are related to attack capability and expected attack returns.

5 Conclusion

The confrontation of VN attackers and defenders is a dynamic process that includes probe scanning, overall assessment, forecasting, decision making. The paper proposes a cybersecurity decision making method which can effectively describe the dynamic interaction and state transition between attackers and defenders. It gives the quantitative indexes of security properties, attackers and defenders benefits and system risk evaluation values. Defenders can take appropriate defensive strategies rather than a passive defense. For different security needs, it calculate the system risk values of current situation to make the evaluation results express in quantitative form, so that it can select the optimal defensive strategies. It provides a strong basis for the effective decision making of defenders, and provides new ideas for vehicle network cybersecurity research.

Acknowledgments This work was supported by Open Project Foundation of Information Technology Research Base of Civil Aviation Administration of China (No. CAAC-ITRB-201402), the Fundamental Research Funds for the Central Universities (No. 3102014JSJ0015), and the Seed Foundation of Innovation and Creation for Graduate Students in Northwestern Polytechnical University (No. Z2015142).

References

1. Nguyen KC, Alpcan T, Başar T (2009) Stochastic games for security in networks with interdependent nodes. In: Proceedings of the 2009 international conference on game theory for networks, gamenets '09, pp 697–703
2. Manshaei MH, Zhu Q, Alpcan T, Başar T, Hubaux J-P (2013) Game theory meets network security and privacy. ACM Comput Surv 45(3):1–39
3. Ten C-W, Liu C-C, Manimaran G (2008) Vulnerability assessment of cybersecurity for SCADA systems. IEEE Trans Power Syst 23(4):1836–1846
4. Manuel C, Alfredo P, Riccardo S (2011) Formal vulnerability analysis of a security system for remote fieldbus access. IEEE Trans Ind Inf 7(1):30–40
5. Almasizadeh J, Azgomi MA (2013) A stochastic model of attack process for the evaluation of security metrics. Comput Netw 57(10):2159–2180
6. Zhu Q, Başar T (2011) Robust and resilient control design for cyber-physical systems with an application to power systems. In: 50th IEEE conference on decision and control and European control conference (CDC-ECC), pp 4066–4071
7. Zhu Q, Basar T (2012) A hierarchical security architecture for smart grid. Smart Grid Commun Netw, pp 413–440
8. Zonouz S, Haghani P (2013) Cyber-physical security metric inference in smart grid critical infrastructures based on system administrators' responsive behavior. Comput Secur 39:190–200
9. Prabhakar M, Singh JN, Mahadevan G (2012) Nash equilibrium and Marcov chains to enhance game theoretic approach for vanet security. In: Proceedings of international conference on advances in computing, pp 191–199

A Storage Method for Large Scale Moving Objects Based on PostGIS

Kai Sheng, Zefang Li and Dechao Zhou

Abstract Storing and managing the large scale moving objects data is one of the research hotspots and difficulties in data mining, trajectory analyzing, location-based services and many other applications. To solve these problems, firstly we design the trajectory point representing model, the trajectory representing model, the moving object data storage model and their relationships based on object-oriented ideology; then we construct a moving objects database using in PostGIS according to the presented models; finally we test the effectiveness of the moving object database with real data. The experimental results show that using the method presented in this paper to store large scale moving objects data can reduce the storage space obviously, meanwhile, it can increase the spatial and temporal querying efficiency effectively.

Keywords Moving objects · Trajectory data · Storage model · PostGIS

1 Introduction

With the fast development and widely application of the space location detecting and the wireless network technologies, large scale trajectory data are generated in the process of positioning and tracking moving objects. Although traditional relation databases can manage non-spatiotemporal data efficiently, but consisted of static non-spatiotemporal data and a large amount of dynamic data such as locations, velocities and azimuths at different times, moving objects data is no easy to be stored and searched in high performance.

K. Sheng (✉) · Z. LiD. Zhou
College of Electronic Engineering, Naval University of Engineering, Wuhan 430033, China
e-mail: shengkai0214@foxmail.com

Z. Li
e-mail: commandern@sohu.com

D. Zhou
e-mail: 2226906254@qq.com

© Springer International Publishing Switzerland 2017
V.E. Balas et al. (eds.), *Information Technology and Intelligent Transportation Systems*, Advances in Intelligent Systems and Computing 454,
DOI 10.1007/978-3-319-38789-5_69

In order to manager moving objects data in a better way, Wolfson et al. presented the concept of Moving-Objects Databases (MODs) in 1997, then propose a Moving-Object Spaito-Temporal (MOST) model [1], which is capable of tracking not only the current but also the near future positions of moving objects, and the query language based on the Future Temporal Logic (FTL). However, this model is unable to support to query historical information. G*ü*ting, a Full Professor of the University of Hagen, presented a data model for managing moving objects based on abstract data types and query operators and built a pototype system of moving objects database named SECONDO [2], which is a foundation of the study in this domain. Recently, the research on spatiotemporal models and indexes is gradually increased in depth. Hadi Hajari and Farshad Hakimpour focus on extending a spatial data model for constrained moving objects [3], Ding Zhiming et al. proposed a network-matched trajectory-based moving-object database (NMTMOD) mechanism and a traffic flow analysis method using in NMTMOD [4], while Chen Nan researched on index and query techniques of moving objects in spatio-temporal databases in his doctoral dissertation in 2010 [5]. Such works has powerfully promoted the development of MODs, but there are still many limits in dealing with the mass moving objects data.

For the characters of moving objects data, in this paper we design the trajectory point representing model, the trajectory representing model, the moving objects data storage model, then we realize these models in PostgreSQL [6] and its spatial data plug-in PostGIS [7] as well as the query methods of spatiotemporal attributes, at last we test the effectiveness of the moving object database with real data.

2 The Advantages of Storing Large Scale Moving Objects Data in Using PostgreSQL/PostGIS

PostgreSQL is an object-relational database management system based on POSTGRES Version 4.2, developed at the University of California at Berkeley Computer Science Department. After years of efforts, it has become the most powerful and the most stable open-source database management system in the world today. PostgreSQL supports a large part of the SQL standard and offers many modern features, such as complex queries, triggers, updatable views, multiversion concurrency control and so on. Meanwhile, PostgreSQL can be extended by users in many ways, for example by adding new data types, functions, operators and index methods.

PostGIS is a spatial database extended for the PostgreSQL database management system. It defines several spatial data types and supports a range of important GIS functions, including full OpenGIS support, advanced topological constructs. Storing and managing moving objects data in Postgre/PostGIS has lots of advantages, such as:

2.1 Clear Hierarchy of Data Organizing

PostgreSQL organizes data by databases, schemas and tables. One database can be include of several schemas, and the names of tables or functions can be absolutely same to each other without any conflicts only if the tables or functions were under different schemas. In PostgrSQL systems, a schema can be understood as a namespace or a catalogue. Using schema allows multi users to connect to one database without any interferences, so in order to manage the large amount of moving objects data from different sources clearly and use it conveniently, we can save it into several schemas according to its sources.

2.2 Abundant Data Types

Other than those ordinary data types in relation databases, PostgreSQL with its spatial plug-in PostGIS provides many special data types such as spatial data types, arrays and range types, which can easily represent the dynamic attributes of moving objects.

Spatial data types consist of geometry types and geography types, both of them can be used to represent spatial information by specifying geospatial coordinate reference system. The difference between them is that the geometry types calculation foundation are plane while geography types are sphere. Generally speaking, if the data is in a small are, using geometry is a better solution in terms of performance and functionality available; if the data is global or covers a continental region, geography types allows users to build a system without having to worry about projection details.

PostgreSQL allows columns of a table to be defined as variable-length multidimensional arrays. Arrays of any built-in or user-defined base type, enum type, or composite type can be created. Arrays are conveniently used to represent dynamic attributes of moving objects. For example, if a user want to store velocity values in each trajectory point, he/she can use double precision arrays, which is defined as double precision[], and if another user want to store the time series, he/she can use time arrays, which is defined as timestamp[].

Range types are new data types after PostgrSQL-9.2. They can represent a range if values of some element type (called the ranges subtype). For instance, ranges of timestamp might be used to represent the ranges of time that a moving object stays in a specific area. In this case the data type is tsrange (short for timestamp range), and timestamp is the subtype. The subtype must have a total order so that it is well-defined whether element values are within, before, or after a range of values. Range types are useful not only because they can represent many element values clearly, also because they support more efficiency indexes such as GiST and SP-GiST, which are mentioned in the next section.

2.3 *Efficient Indexing Mechanism*

PostgreSQL provides several index types: B-tree, Hash, GiST, SP-GiST and GIN. Each index type uses a different algorithm that is best suited to different types of queries. B-tree can handle equality and range queries on data that can be sorted into some order; Hash indexes can only be used to retrieve data in sorted order; GiST (Generalized Search Trees) or SP-GiST (Space-Partitioned GiST) is an infrastructure within which many different indexing strategies can be implemented. GIN indexes are inverted indexes which can handle values that contain more than one key, which can used into full text search. For the static types of moving objects, we can use B-tree, Hash or GIN indexes to accelerate search speed; for spatial or temporal attributes of moving objects, we can use GiST or SP-GiST mechanism, which can not only improve search efficient, also support many complex operators like contain (@>), contained in (<@), overlap (&&), etc.

3 Moving Objects Data Storage Model

Trajectory is an important part of moving objects data. In short, trajectories are space curves generated by moving objects positions change in a continuous time within an area. However, we could not obtain continuous trajectory of moving objects in usual, only can we get are sample positions in some discrete time points. For this reason, we design a kind of trajectory point model, trajectory model and then construct the moving objects data storage model. The relationships among these models are show as Fig. 1.

3.1 *Trajectory Point Model*

Trajectory point is moving objects position at a moment. In PostgreSQL/PostGIS, trajectory point model can be represented by point data type and time data type.

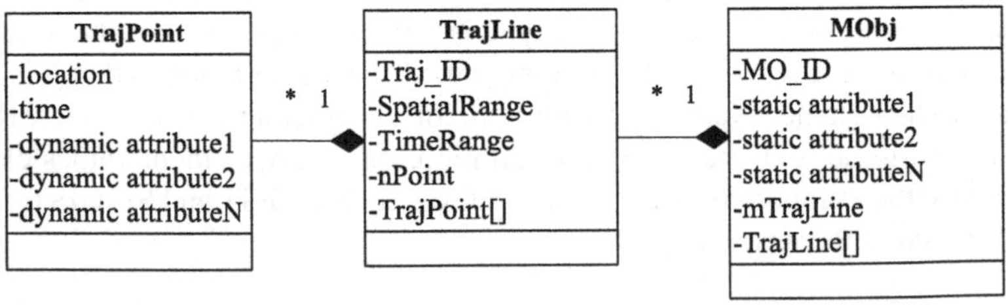

Fig. 1 Moving objects storage model

According to the space dimension, trajectory point model can be divided into 2D-model and 3D-model. The trajectory 2D-model can be represented as:

$$TrajPoint2D = \{Point(x, y), time\} \tag{1}$$

And the 3D-model can be represented as:

$$TrajPoint3D = \{Point(x, y, z), time\} \tag{2}$$

Nowadays many positioning devices can get the instantaneous velocity values, directions and accelerate values besides locations and time points of the moving objects. These information is also related to objects positions, so we can describe them in the trajectory point model, too. Therefore, a common trajectory point model can be represented as:

$$\begin{aligned} TrajPoint =& \{Point(x, y, z), Time, DynamicAttribute\_1, \ldots, \\ & DynamicAttribute\_n\} \end{aligned} \tag{3}$$

3.2 *Trajectory Model*

Trajectory model can be represented as a set of trajectory point ordered by time. Meanwhile, to manage trajectory data conveniently, the quantities of trajectory points, the space scope and time scope of the trajectory can be also defined in the trajectory model explicitly, as follows:

$$TrajLine = \{Traj\_ID, SpatialRange, TimeRange, nPoint, TrajPoint[]\} \tag{4}$$

Traj_ID means the identify code of the trajectory, *SpatialRange* means the space scope, TimeRange means the time scope, nPoint means the quantity of the trajectory points, and *Trajline[]* is the trajectory points set.

3.3 *Moving Objects Storage Model*

In reality, one moving object perhaps contains several trajectories: one reason is that the object is detected by more than one devices at the same time, and each detecting device generates a trajectory about this object; another reason is that the same object maybe appear in an area many times, and each time consist to a single trajectory line. Furthermore, moving objects contain many static attributes like the identity code, object name, object type and so on. Therefore, moving objects storage model can be represented as:

$$MObj = \{MO\_ID, StaticAttribute\_1, \ldots, StaticAttribute\_n, \\ mTrajLine, TrajLine[]\} \quad (5)$$

In this model, MO_ID means the object identity code, StaticAttribute_1 to StaticAttribute_n means the static attributes of the moving object, mTrajLine means the total trajectories quantity, and TrajLine [] store each trajectory information about this object.

4 A Case Study

In order to verify the validity of the data storage model based on PostGIS, this paper selects real data of the ship tracking data in the South China Sea for 3 days as the test data. Compare and analyzed the general data storage method with the method of this paper in storage and query. Test environment: hardware for Core i7 Inter processor, 4G memory, equipped with Win7 64 bit operating system; database system for the PostgreSQL-9.4.4 and PostGIS-2.1.7 extension.

4.1 Ship Track Database Designing Based on PostgreSQL/PostGIS

Database logical structure The ship trajectory tracking of the original data file is dat format and XLS format. File contains messy and irrelevant information. So the first thing to data is cleaning and saving the information which is useful to the database. Then according to the original information database to generate a moving object database. In PostgreSQL/PostGIS database system, in order to facilitate data in management and invocation, different schemas can be used to store the original data information and moving object data. At the same time, because of the large amount of data, it can be used to set up the data table according to time division specific logical structure as shown in Fig. 2.

Trajectory point composite type designing The dynamic properties of the test data include the location, time, instantaneous velocity and instantaneous direction of the ship. According to the trajectory model in third section of this paper, the model of the composite data type can be defined in the database. PostgreSQL/PostGIS supports for the composite type well. The SQL statement to create trajectory point type is as below:

```
    create type trajpoints as (
plocation geography(Point, 4326)[], –point locations
ptime timestamp[], –point times
pvelocity double precision[], –point speeds
pazimuth double precision[], –point directions
);
```

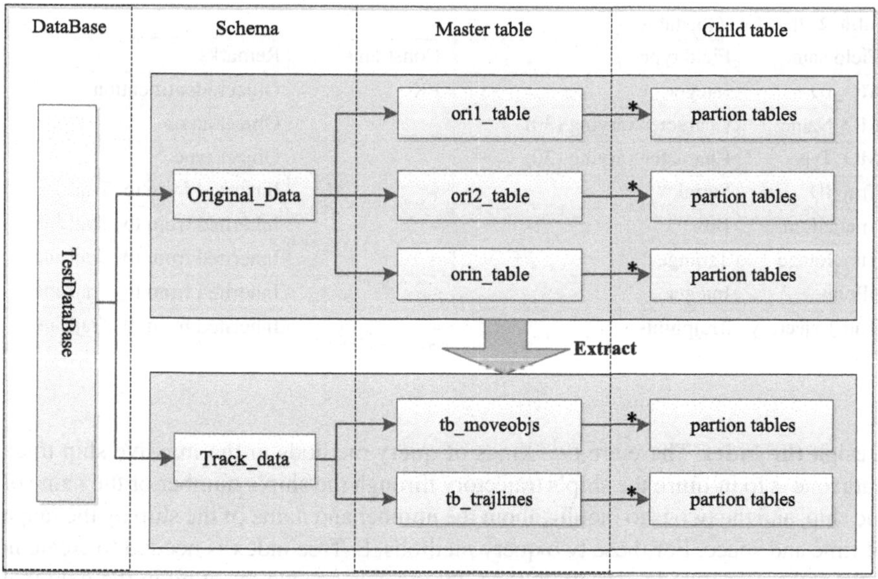

Fig. 2 Logical structure of moving object storage database

Table 1 tb_TrajLinetablee

Field name	Field type	Constraint	Remarks
Traj_ID	Serial	–	Trajectory identification
SpatialRange	Box	–	Trajectory range
TimeRange	Tsrange	–	Time range
nPoint	integer	Not null	Trajectory number
The_trajectory	Trajpoints	Not null	Trajectory sequence

Trajectory table designing In the database, the dynamic properties of the ship are stored in the database. Ac-cording to the third section of the trajectory model, the trajectory table design based on the PostgreSQL/PostGIS is shown in Table 1. It is needed to explain that, the box data type for the storage range can be determined by a pair of point coordinates (lower left and upper right). The tsrange data type for the storage time range is embedded in PostgreSQL, which subtype is timestamp without time zone.

Moving object table designing The static properties of the ship in the test data include the information of the ship identification, the name of the ship, the ship type and so on. As the test data of the ship and the ship's trajectory is one to one, so the dynamic properties of the moving object table can be directly inherited from the trajectory table, while adding the static properties of the ship. Mobile object table design is shown in Table 2.

Table 2 tb_MoveObjstable

Field name	Field type	Constraint	Remarks
MO_ID	Integer	PK	Object identification
MO_Name	Character varying (30)	–	Object name
MO_Type	Character varying (30)	–	Object type
Traj_ID	Serial	–	Inherited from tb_TrajLine
SpatialRange	Box	–	Inherited from tb_TrajLine
TimeRange	Tsrange	–	Inherited from tb_TrajLine
nPoint	Integer	–	Inherited from tb_TrajLine
The_trajectory	Trajpoints	–	Inherited from tb_TrajLine

Design the index There are two kinds of query methods in the massive ship track data: one is to in-quire the ship's trajectory through the ship's number or the name of the ship, and the two is to inquire about the number and name of the ship by the range of time and space. For these two query methods, B-Tree index is needed to create in field of the object's identification, the GIN index is needed to set up in the field that represents the name of the object, and the GiST index is established in the field that represents time and space range, in order to improve the efficiency of the query in the name and time space. SQL statements to achieve the index are: CREATE INDEX ON track_data.tb_moveobjs USING btree (mo_id) TABLESPACE myspace;

CREATE INDEX ON track_data.tb_moveobjs USING gin (mo_name) TABLESPACE myspace;

CREATE INDEX ON track_data.tb_moveobjs USING gist (spatialrange) TABLESPACE myspace;

CREATE INDEX ON track_data.tb_moveobjs USING gist (timerange) TABLESPACE myspace;

4.2 Analysis of Storaging and Querying Moving Objects Data

In this paper, using C# programming language to connected the database. First, the test data is written into the ori_table of the original data table, and then extract the trajectory information from all the object identification which are corresponded, and the moving object table tb_MoveObjs is written by the trajectory information one by one. In order to break through the field of partial trace data in the moving object table exceeds the page size, The TOAST Technology The Oversized-Attribute Storage Technique in PostgreSQL will store these larger fields automatically in the toast table, which is transparent to the users. By comparing the size of the physical space between the original data table and the moving object table, the method of moving object storage based on PostgreSQL/PostGIS can reduce the storage space occupancy obviously (Fig. 3).

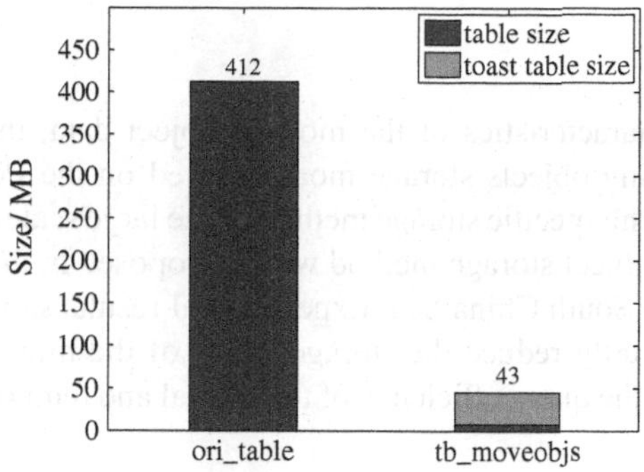

Fig. 3 Contrast of physical space occupancy of moving object data

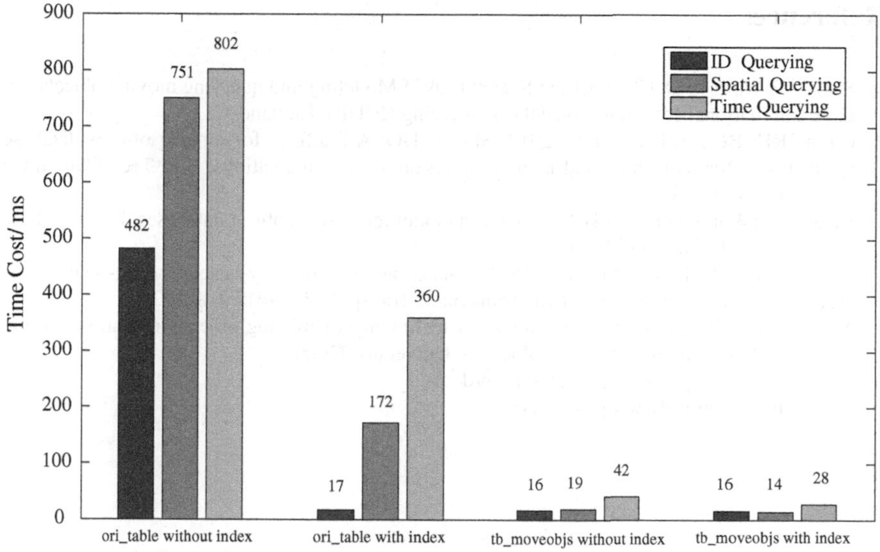

Fig. 4 Time-consuming comparison of moving object data query

The efficiency analysis of the moving objects is carried out in three ways: the object identification, the spatial range query and the time range query. Each query method is carried out in a way that takes the average value of the query. The time consumed by the query is shown in Fig. 4. It can be obtained that in the case of the establishment of the index, the method of moving object storage is not obvious, but the efficiency of spatial query and time query is greatly improved.

5 Summary

According to the characteristics of the moving object data, this paper constructs the large scale moving objects storage model. Based on the PostgreSQL/PostGIS database, we design the specific storage method of the large scale moving object data. And tested moving object storage method which proposed in this paper by the ship detection data of the South China Sea. Experimental results show that the proposed method can significantly reduce the storage space of the moving object database, and greatly improve the query efficiency of the spatial and temporal properties of the moving objccts.

References

1. Sistla AP, Wolfson O, Chamberlain S, et al (1997) Modeling and querying moving objects. In: IEEE international conference on data engineering (ICDE). England
2. Güting RH, Behr T, Düntgen C (2010) SECONDO: A Platform for moving objects database research and for publishing and integrating research implementations. Bull Tech Committee Data Eng. 33(2):1–8
3. Hajari H, Hakimpour F (2014) A spatial data model for moving object databases. Int J Database Manag Syst (IJDMS) 6(1):1–20
4. Ding Z, Yang B, Gting RH et al (2015) Network-mathed trajectory-based moving-object database: models and applications. IEEE Trans Intell Transp Syst 16(4):1918–1928
5. Nan C (2010) Research on index and query techniques of moving objects in spatio-temporal databases. Doctoral Dissertation of Zhejiang University, China
6. Information on http://www.postgresql.org/docs
7. Information on http://www.postgis.net

BPX-Like Preconditioned Conjugate Gradient Solvers for Poisson Problem and Their CUDA Implementations

Jie Peng, Shi Shu, Chunsheng Feng and Xiaoqiang Yue

Abstract In this paper, we firstly introduce two BPX-like preconditioners B_J^1 and B_J^3, and present an equivalent but more robust BPX-like preconditioner B_J^2 for the solution of the linear finite element discretization of Poisson problem. Secondly, we implement these preconditioners and their preconditioned conjugate gradient (PCG) solvers B_l^p-CG($l = 1, 2, 3$) under Compute Unified Device Architecture (CUDA), where we exploit the hierarchical and the overall storage structure, take advantage of the multicolored Gauss–Seidel smoother. Finally, comparisons are made among these PCG solvers and the state-of-the-art SA-AMG preconditioned CG solver (SA-CG) in CUSP library. Numerical results demonstrate that the iteration numbers of B_2^p-CG holds the weakest dependence on the grid size, while B_3^p-CG is the most efficient solver. Furthermore, B_3^p-CG possesses considerable advantages over SA-CG in computational capability and efficiency. In particular, B_3^p-CG runs 3.67 times faster than SA-CG when solving a problem with about one-million unknowns.

Keywords BPX-like preconditioner · Multicolored Gauss–Seidel smoother · CUDA · Poisson equation

J. Peng · S. Shu · C. Feng (✉) · X. Yue
School of Mathematics and Computational Science,
Xiangtan University, Xiangtan, Hunan, China
e-mail: spring@xtu.edu.cn

J. Peng
e-mail: xtu_pengjie@163.com

S. Shu
e-mail: shushi@xtu.edu.cn

X. Yue
e-mail: yuexq1111@163.com

© Springer International Publishing Switzerland 2017

V.E. Balas et al. (eds.), *Information Technology and Intelligent Transportation Systems*, Advances in Intelligent Systems and Computing 454,
DOI 10.1007/978-3-319-38789-5_70

1 Introduction

Poisson equation is one of the fundamental partial differential equations (PDEs). It is widely used in various fields, ranging from environmental science, energy development, hydromechanics to electronic science. The numerical methods to quickly solve Poisson equation have been successfully applied in many practical problems, such as computed tomography (CT), network analysis, quantum chemistry and reservoir simulation, etc. [5, 8]. Therefore, how to design fast algorithm and the corresponding fast solver for solving Poisson equation is a key ingredient to solve many scientific and engineering problems.

The preconditioned conjugate gradient (PCG) method is one of the most commonly used solvers for the discretized system arising from Poisson equation at present. Proper selection of the preconditioner is the key to improve the efficiency of PCG method. The BPX-preconditioner proposed by Bramble, Pasciak and Xu [4] is a typical additive preconditioners which is based on auxiliary variational problem. The key factor which influence the computational efficiency of BPX-preconditioner is the inverse operation of mass matrices in each grid level.

Serial computer programming is usually not sufficient for the vastly increasing problem size and computational complexity in scientific and engineering computation due to the limitation of their memory storage and the computational efficiency. Parallel computing is the affective way to improve computational capability and efficiency. The research based on various CPU environments has made great progress in the last few years [10, 11]. Recently, Graphics processing units (GPUs) burst onto the scientific computing scene as an innovative technology that has demonstrated substantial performance and energy-efficiency improvements for many scientific applications [2, 3]. CUSP is an open source C++ library of generic parallel algorithms for sparse linear algebra and graph computations on CUDA. Till now they have published the SA-AMG and UA-AMG solvers which are suitable for the modern GPUs. These solvers use the weighted Jacobi smoother (WJ) as the default relaxation operator. But it is difficult to select the optimal weight for WJ and the convergence factor is also not high enough for some complex problems. Therefore, how to take the advantage of the CPU-GPU heterogeneous system to reduce the computing complexity of the existing BPX-like preconditioners, with good algorithmic and parallel scalability, is a worthy study work.

In this paper, firstly, we introduce two BPX-like preconditioners [4, 6], denoted by B_J^1 and B_J^3, and an equivalent but more robust BPX-like preconditioner B_J^2 is also presented. Secondly, we design an integrating storage structure oriented to the CUDA environment, and discuss the computational efficiency with hierarchical storage and integrating storage respectively via the basic parallel matrix-vector operations. Then, we implement the parallel programming modules for $B_J^l (l = 1, 2, 3)$ and the corresponding parallel PCG solvers B_l^p-CG(l = 1, 2, 3). The last but not the least, comparisons are made among these PCG solvers and the SA-AMG preconditioned CG solver (SA-CG) in CUSP library.

The numerical results show that, the iteration number of B_2^p-CG depends least on the grid size. All these three parallel PCG solvers yield good parallelism scalability. Extraordinary, when the unknowns reaches 4-million, the acceleration ratio can attain more than eight times. It is particularly worth mentioning here that B_3^p-CG is the most efficient solver. Furthermore, the B_3^p-CG has great advantages over SA-CG in computational capability and efficiency. For example, it runs 3.67 times faster than the state-of-the-art SA-CG when solving a one-million unknowns problem.

The rest of the paper is organized as follows: in the next section, we give the model problem and its corresponding linear finite element discretized system. In Sect. 3, we introduce three BPX-like preconditioners and propose the ways to implement their appropriate parallel solvers. We report our numerical results in Sect. 4 and then summarize the paper in Sect. 5.

2 Model Problem and Linear Finite Element Discretization

Let $\Omega = (a, b) \times (a, b)$, we consider the following 2D Poisson problem

$$\begin{cases} -\Delta u = f, & \text{in } \Omega, \\ u = g, & \text{on } \partial\Omega, \end{cases} \tag{1}$$

where $f \in L^2(\Omega)$, $g \in L^2(\partial\Omega)$.

Set $H_g^1(\Omega) = \{f | f \in L^2(\Omega), f' \in L^2(\Omega), f|_{\partial\Omega} = g\}$, the natural variational problem for (1) reads: find $u \in H_g^1(\Omega)$ such that

$$a(u, v) = (f, v), \quad \forall v \in H_0^1(\Omega), \tag{2}$$

where $a(u, v) = \int_\Omega \nabla u \cdot \nabla v \mathrm{d}x$, $(f, v) = \int_\Omega f v \mathrm{d}x$.

Let $\mathcal{T}_J$ denote a regular and uniform triangulation of Ω (see Fig. 1). The number of partition for each direction is 2^N with mesh size $h_J = \frac{b-a}{2^N}$. Denote V_J as the linear finite element space on $\mathcal{T}_J$ with dimensions $n_J = (2^N - 1)^2$, and M_J is the corresponding mass matrix with the same dimensions.

Using (2) to discrete (1) by linear finite element method, we end up with a system of linear equations

$$A_J u_J = F_J, \tag{3}$$

where $A_J \in \mathcal{R}^{n_J \times n_J}$ is symmetric positive definite, $F_J \in \mathcal{R}^{n_J}$.

PCG method is typically applied to numerically solve symmetric positive definite systems. A key issue for the efficient PCG is the contraction of the preconditioner. In the following section, we discuss three BPX-like preconditioners, and develop the corresponding PCG (BPX-CG) solvers under CUDA.

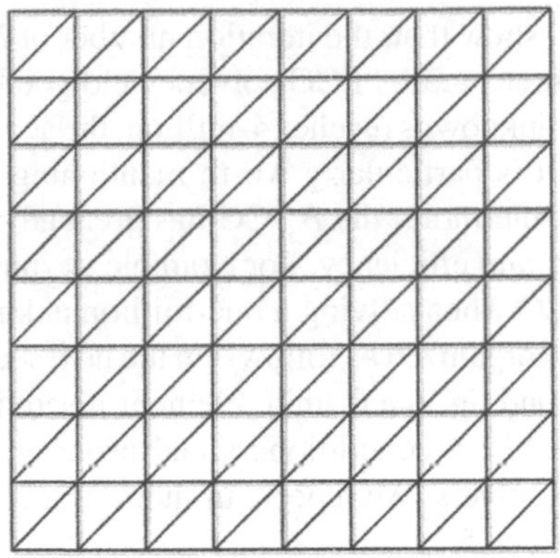

Fig. 1 Illustration for $\mathcal{T}_J$

3 Three BPX-CG Preconditioners and Their CUDA Implementations

3.1 BPX-Like Preconditioners

Before we describe various BPX-like preconditioners that will be used in the remainder of the paper, we define the number of degree of freedoms (DOFs) in the kth grid as $n_k = (2^{N-J+k} - 1)^2 (k = J, \cdots, 0)$, the mesh size as $h_k = \frac{b-a}{2^{N-J+k}}$, and the linear Lagrangian finite element space of the kth grid as V_k.

Let $P^k_{k+1} = (p_{i,j})_{n_k \times n_{k+1}} (k = J-1, \cdots, 0)$ be the restriction operator that maps the $(k+1)$th grid to the kth grid, and details on the generation algorithm of P^k_{k+1} are described in Algorithm 1.

Algorithm 1.

Initialization: $ncx = ncy = 2^{N-J+k} - 1$, $nfx = nfy = 2^{N-J+k+1} - 1$.
for $i_1 = 0$ **to** $ncx - 1$ **do**
  for $i_2 = 0$ **to** $ncy - 1$ **do**
    $i = ncy * i_2 + i_1$; $j_1 = 2 * i_1 + 1$; $j_2 = 2 * i_2 + 1$; $j = j_2 * nfx + j_1$;
    $p_{i,j} = 1.0$;
    $p_{i,j-1} = p_{i,j+1} = p_{i,j+nfx} = p_{i,j+nfx+1} = p_{i,j-nfx} = p_{i,j-nfx-1} = 0.5$.

By using the stiffness matrix A_J, the mass matrix M_J and the restriction operator P^k_{k+1}, combing with the Galerkin condition

$$A_k = P^k_{k+1} A_{k+1} (P^k_{k+1})^T, \quad k = J-1, \cdots, 0, \tag{4}$$

and

$$M_k = P^k_{k+1} M_{k+1} (P^k_{k+1})^T, \quad k = J-1, \cdots, 0, \tag{5}$$

we can obtain the stiffness matrix A_k and the mass matrix M_k in each grid level.

Assume that $g_J \in \mathcal{R}^{n_J}$ is an arbitrary given vector. At first, we can give the classical BPX-like preconditioner [4]

$$B_J^1 = \sum_{k=0}^{J} (P_J^k)^T (h_k^2 M_k^{-1}) P_J^k, \tag{6}$$

where P_J^k is the restriction operator from Jth grid to kth grid, computed by $P_J^k = \prod_{l=k}^{J-1} P_{l+1}^l (0 \le k < J)$.

Remark 1 In this paper, in order to improve the efficiency of BPX-CG solver with B_J^1 as the preconditioner, we invoke m times Gauss–seidel as the smoother to compute M_k^{-1} approximately, where m is independent to the grid size and less than 5 in general. It's worth mentioning that the specified Gauss–Seidel smoother is equivalent to M_k^{-1}.

Then, through theoretical and numerical analysis we can obtain a preconditioner with lower operator complexity and better efficiency,

$$B_J^2 = \sum_{k=0}^{J} (P_J^k)^T B_k^{-1} P_J^k, \tag{7}$$

where $w_k = B_k^{-1} \hat{f}_k$ ($\hat{f}_k \in \mathcal{R}^{n_k}$ is a given vector) is the iterative solution of the discretized system $A_k u_k = \hat{f}_k$, via m times Gauss–Seidel smoother with zero vector as the initial iteration vector.

Last, another BPX-like preconditioner is proposed by AFEM Package [6], which is defined by

$$B_J^3 = \left[\left((P_J^{J-1})^T + \frac{1}{4} I_J \right) \left((P_{J-1}^{J-2})^T + \frac{1}{4} I_{J-1} \right) \left((P_1^0)^T + \frac{1}{4} I_1 \right) A_0^{-1} P_J^0 \right]. \tag{8}$$

In the following section, we will design these three BPX-CG solvers under CPU-GPU architecture.

3.2 Parallel BPX-CG Solvers

The preconditioner B_J^1 and B_J^2 need to store the smoothing operator of each grid level, such as the stiffness matrix $A_k (k = J, \cdots, 0)$ and the mass matrix $M_k (k = J, \cdots, 0)$. In general, these matrices can be stored by two strategies:

Hierarchical Storage (M1): Store the matrix of each grid level separately.
Integrating Storage (M2): Store the matrix of each grid level integrally.

Similar to the matrix, the vector also can be stored by the previous two strategies.

In order to investigate the efficiency of these two storage methods under CUDA environment, we design a corresponding numerical experiment. More specifically, the idea is as follows: first of all, we store the stiffness matrices $A_k(k = J, \ldots, 0)$ and vectors $g_k = P_J^k F_J(k = J, \ldots, 0)$ by these two models. And then we repeatedly compute $A_k g_k(k = J, \ldots, 0)$ for 100 times.

The numerical results are shown in Table 1, where $n_J = (2^N - 1)^2$ is the number of DOFs, T(M1) and T(M2) is the wall times (in second) for these two storage models, and ratio =T(M1)/T(M2).

Compared T(M1) with T(M2), one can see that when n_J is small, M2 has the significant advantages. Whereas, when n_J is becoming larger, the advantage of M2 is gradually unobvious.

The reason for the aforementioned phenomenon is that the CPU wall times of the matrix-vector multiplications on GPU, consist of the computation time and the time of kernel function's distribution and task-scheduling. However, the time of distributing and scheduling the kernel function once is a fixed constant. Therefore, as the number of DOFs is small, the time of distributing and scheduling the kernel function occupies the main part of the total time. Note that if we use M2 there only need to have this distribution and schedule once, but J times in the other way. As a result, M2 is better than M1. On the contrary, when the number of DOFs is large, this part of time would no longer maintain the dominant position of the total time, so the advantage of M2 becoming more and more unobvious.

In the following numerical experiments, we use M2 as the default strategy to store the matrix and vector in each grid level.

It is popular to use Jacobi and Gauss–Seidel method as the smoother in each grid level. Comparing with the Jacobi method, Gauss–Seidel method can exploit more refresh solution into computations. Therefore, the later holds a better approximation and have been widely used. However, Gauss–Seidel method bears the nature of serial communication, which is unfavorable for parallel implementation. There upon, we use a multicolor Gauss–Seidel (MC-GS) method [9] which is based on the strength matrix. It is to be observed that its parallel version is equivalent to the corresponding serial version.

According to the above constraints, we now explain the three critical steps to design BPX-CG solvers based on $B_J^l(l = 1, 2, 3)$ preconditioners.

Table 1 Efficiencies of matrix-vector multiplication based on M1 and M2

N	n_J	T(M1)	T(M2)	Ratio
5	961	1.62E–3	4.16E–4	3.89
6	3969	2.39E–3	1.08E–3	2.22
7	16129	5.63E–3	3.59E–3	1.57
8	65025	1.56E–2	1.30E–2	1.20
9	261121	5.51E–2	5.20E–2	1.06
10	1046529	2.09E–1	2.02E–1	1.03

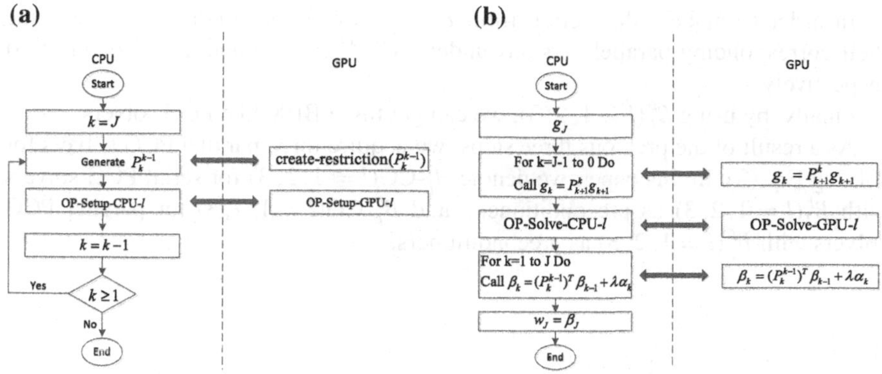

Fig. 2 **a** The flowchart for the setup stage of parallel $B_J^l(l = 1, 2, 3)$; **b** The flowchart for the solve stage of parallel $B_J^l(l = 1, 2, 3)$

Firstly, Fig. 2a gives the detailed flowchart for the setup stage of $B_J^l(l = 1, 2, 3)$ under the CPU-GPU heterogeneous system.

Unless otherwise stated, the black one-way arrows in flowcharts indicate the main process line of the host, and the blue double arrows express invoking the kernel function and transferring data between host and device.

The most important part in the setup stage is OP-Setup-l. Its specific functions are given below.

1. If $l = 1$, it refers to B_J^1. By using Eq. (5) to obtain $h_k^{-2}M_k$ in each grid level at device, and then setup the MC-GS smoother at host.
2. If $l = 2$. By using Eq. (4) to obtain A_k in each grid level at device, and then setup the MC-GS smoother at host.
3. If $l = 3$, there are two situations: when the number of DOFs at the coarsest grid is only one, we needn't to do anything, while the number of DOFs is more than one, we must use Eq. (4) to obtain A_0 at device.

Next, Fig. 2b displays the detailed flowchart for the solve stage of $B_J^l(l = 1, 2, 3)$.

Similarly, the most significant part in the solve stage is OP-Solve-l, which we divide into three cases to describe intrinsic functions.

1. If $l = 1$, we gain $\alpha_k(k = J, \ldots, 0)$ by solving $h_k^{-2}M_k\alpha_k = g_k$ with MC-GS for 4 times (MC-GS(4)) and zero as the initial guess, and then let $\beta_0 = \alpha_0$ at device.
2. If $l = 2$, we gain $\alpha_k(k = J, \ldots, 0)$ by solving $A_k\alpha_k = g_k$ with MC-GS(4) and zero as the initial guess, and then let $\beta_0 = \alpha_0$ at device.
3. If $l = 3$, there are two situations: when the number of DOFs at the coarsest grid is only one, let $\beta_0 = g_0$ at device, while the number of DOFs is more than one, we gain β_0 by using the direct method to solve $A_0\beta_0 = g_0$ at host.

For $\lambda\alpha_k$, let $\lambda\alpha_k = \alpha_k$ when $l = 1, 2$, and let $\lambda\alpha_k = \frac{1}{4}\alpha_k$ when $l = 3$.

In order to make a distinction from $B_J^l (l = 1, 2, 3)$ preconditioners, we name their corresponding parallel versions under CUDA environment as $B_l^p (l = 1, 2, 3)$ respectively.

Finally, by using $B_l^p (l = 1, 2, 3)$, we can get three BPX-like PCG solvers.

As a result of the previous three steps, we acquire three parallel PCG solvers for solving Eq. (3). In this paper, we denote B_l^s-CG$(l = 1, 2, 3)$ for serial PCG solvers with $B_J^l (l = 1, 2, 3)$ as preconditioners, and B_l^p-CG$(l = 1, 2, 3)$ for parallel PCG solvers with $B_l^p (l = 1, 2, 3)$ as preconditioners.

4 Numerical Experiments

Example 1 In model problem (1), let $\Omega = (0, 1)^2, f = 2\pi^2 \sin(\pi x) \cos(\pi y), \quad g = \sin(\pi x) \cos(\pi y)$.

In our numerical experiments, The PCG algorithm is halted when a convergence tolerance of 10^{-6} is reached.

4.1 *The Parallel Scalability for Three BPX-Like PCG Solvers*

In the following subsection, the experimental results for algorithmic and parallel scalability are given respectively.

4.1.1 Algorithm Scalability

Tables 2 and 3 shows the number of iterations for B_l^s-CG$(l = 1, 2, 3)$ and B_l^p-CG$(l = 1, 2, 3)$ with different $N(n_J = (2^N - 1)^2)$, while numbers of DOFs at the coarsest grid are 225, 49, 9 and 1.

Table 2 Serial tests

N	B_1^s-CG				B_2^s-CG				B_3^s-CG			
	225	49	9	1	225	49	9	1	225	49	9	1
7	82	48	36	32	24	16	17	17	21	24	25	25
8	89	51	38	35	24	18	18	18	24	26	27	27
9	94	54	41	37	26	19	18	18	26	28	29	29
10	101	57	43	39	27	20	19	19	28	30	31	31
11	104	60	45	41	28	21	20	20	30	32	33	33

Table 3 Parallel tests

N	B_1^p-CG				B_2^p-CG				B_3^p-CG			
	225	49	9	1	225	49	9	1	225	49	9	1
7	82	48	36	32	24	16	17	17	21	24	25	25
8	89	51	38	35	24	18	18	18	24	26	27	27
9	94	54	41	37	26	19	18	18	26	28	29	29
10	101	57	43	39	27	20	19	19	28	30	31	31
11	104	60	45	41	28	21	20	20	30	32	33	33

It can be seen from Tables 2 and 3:

1. The number of iterations for B_l^s-CG$(l = 1, 2, 3)$ is independent on the problem scale. Specifically, when the number of DOFs at the coarsest grid is only one, B_2^s-CG has the best convergence rate. Compared with B_l^s-CG$(l = 1, 2, 3)$, B_l^p-CG$(l = 1, 2, 3)$ have absolutely the same iteration numbers. Therefore, the good scalability of these parallel BPX-CG solvers is testified.
2. The numbers of iterations for B_1^s-CG and B_2^s-CG monotonically decrease as the number of coarse layers increasing (or the number of DOFs at the coarsest grid decreasing), when the scale of the problem is the same. Note that the number of iterations for B_2^s-CG is dependent least on the scale of the grid.
3. The number of iterations for B_3^s-CG monotonically increases as the number of coarse layers increasing, when the scale of the problem is the same. Notably when the number of DOFs at the coarsest is 1, B_3^s-CG does not need to generate A_0 and solve the system of the coarsest grid.

The parallel scalability of these solvers will be considered and we set 1 as the number of DOFs at the coarsest grid in the following experiments.

Parallel Scalability Table 4 shows the wall time of the solve phase and the parallel speedup ratio of B_l^s-CG and B_l^p-CG$(l = 1, 2, 3)$ for different n_J, where T(B_l^k-CG) $(k = s, p; l = 1, 2, 3)$ denotes the wall time of solve phase of B_l^k-CG, ratio i = T(B_i^s-CG)/T(B_i^p-CG) $(i = 1, 2, 3)$.

As shown in Table 4: as the scale of the problem is becoming larger, there is an increase in parallel speedup ratios of these three BPX-like PCG solvers, which

Table 4 The contrast experiment between B_l^s-CG and B_l^p-CG

n_J	T(B_1^s-CG)	T(B_2^s-CG)	T(B_3^s-CG)	Ratio1	Ratio2	Ratio3
16129	6.00E−2	3.00E−2	2.00E−2	2.61	3.64	1.87
65025	3.70E−1	1.60E−1	6.00E−2	5.45	6.87	2.81
261121	2.01E+0	8.80E−1	3.20E−1	8.04	10.49	5.50
1046529	8.50E+0	3.73E+0	1.35E+0	8.46	11.00	6.75
4190209	3.47E+1	1.54E+1	5.68E+0	8.35	11.06	7.23

Table 5 The contrast experiment between SA-CG and B_3^p-CG

n_J	SA-CG		B_3^p-CG		
	Time	Iter	Time	Iter	Ratio
16129	8.58E-2	19	1.07E-2	25	8.00
65025	1.21E-1	24	2.13E-2	27	5.69
261121	2.58E-1	26	5.82E-2	29	4.43
1046529	7.34E-1	35	2.00E-1	31	3.67
4190209	–	–	7.86E-1	33	–

states clearly that these three solvers have good parallel scalability. Extraordinary, when $n_J = 4190209$, the ratio is up to 11 for B_2^p-CG. In addition, B_3^p-CG has the obvious advantage over other solvers in the wall time of the solve phase.

In the next subsection, we will make a comparison between B_3^p-CG and the PCG solver from CUSP library with its built-in SA-AMG as preconditioner (SA-CG).

4.2 *The Contrast Experiment Between B_3^p-CG and SA-CG*

Let the number of DOFs at the coarsest grid of B_3^p-CG be 1, V-cycle and the default parameters for SA-AMG in CUSP library.

Table 5 shows the total solve time (Time) and the iteration number (Iter) of SA-CG and B_3^p-CG, as well as the ratio between the total time of these two solvers.

One can see from Table 5:

1. As the number of DOFs increasing, the iteration number of B_3^p-CG is more robust than SA-CG. Moreover B_3^p-CG has obvious advantages in total solve time. Extraordinary, when the number of DOFs is 1046529, its total solve time is about a quarter of SA-CG.
2. When the number of DOFs is 4190209, it can be run normally with B_3^p-CG, but it fail with SA-CG due to the lack of memory storage. It means that B_3^p-CG has much more advantages on the computing scale.

5 Conclusion

In this work, we generated a new BPX-like preconditioner for the linear finite element discreted algebraic system of Poisson equation. A kind of storage structure named M2, which takes the matrices in all grid levels as a whole, is also designed for CPU-GPU heterogeneous computer system. Naturally, we discussed the computational efficiency of the matrix-vector multiplication by using hierarchical storage M1 and integrating storage M2. Based on these two storage structures and combined with MC-GS method, we designed the parallel module for $B_J^l (l = 1, 2, 3)$ and their corresponding parallel PCG solvers B_l^p-CG (1 = 1, 2, 3). Numerical results show that the iteration number of B_2^p-CG depend least on the grid size. These three parallel PCG solvers all have good scalability. Even more noteworthy is that B_3^p-CG possesses the best computational efficiency, and it has obvious advantages over the best available SA-CG in CUSP library on the computing scale and computational efficiency.

Acknowledgments The authors would like to thank Dr. Chensong Zhang from NCMIS, Academy of Mathematics and System Sciences of China for his helpful comments and suggestions. They are also grateful for the assistance provided by Dr. Zheng Li from Kunming University of Science and Technology in regard in their numerical experiments. Peng is grateful for the financial support by Hunan Provincial Innovation Foundation for Postgraduate (CX2014B260) and Specialized Research Fund for the Doctoral Program of Higher Education of China Grant 20124301110003. Shu is partially supported by NSFC Grant 91130002 and 11171281. Feng is partially supported by Hunan Provincial Natural Science Foundation of China (2016JJ2129).

References

1. Bell N, Garland M (2013) Cusp: generic parallel algorithms for sparse matrix and graph computations, Version 0.4.0, http://cusp-library.googlecode.com
2. Feng C, Shu S, Xu J, Zhang C (2014) Numerical study of geometric multigrid methods on CPU-GPU heterogeneous computers. Adv. Appl. Math. Mech. 6(1):1–23
3. Top500 list June (2015), http://top500.org/lists/2015/06/
4. Bramble J, Pasciak J, Xu J (1990) Parallel multilevel preconditioners. Math Comput 55(191):1–22
5. Jin J, Shu S, Xu J (2006) A two-grid discretization method for decoupling systems of partial differential equations. Math Comput 75(256):1617–1626
6. Chen L, Zhang C (2006) AFEM@MATLAB: a MATLAB package of adaptive finite element methods. Technical report, University of Maryland, http://math.umd.edu/zhangcs/paper/AFEM40matlab.pdf
7. Xu J (2010) In: Bhatia R (ed) Proceedings of the International Congress of Mathematics, vol 4., Fast Poisson-based solvers for linear and nonlinear PDEsWorld Scientific, Singapore, pp 2886–2912
8. Yang J, Cai Y, Zhou Q (2011) Fast poisson solver preconditioned method for robust power grid analysis. In: 2011 IEEE/ACM international conference on computer-aided design (ICCAD). pp 531–536
9. Feng C (2014) Multilevel iterative methods and solvers for reservoir simulation on CPU-GPU heterogenous computers. Xiangtan university doctoral thesis
10. Xu X, Mo Z, Cao X (2009) Parallel scalability analysis for multigrid solvers in HYPRE. J Softw 20:8–14
11. Zhang Y, Sun J, Chi X, Tang Z (2000) Memory complexity analysis on numerical programs. Chin J Comput 23(4):363–373

Estimation of Travel Time Based on Bluetooth MAC Address Identification

Qiangwei Li, Lufeng Chen, Yun Huang, Yingzhi Wang and Yongbo Mao

Abstract A method to estimate the time of vehicles traveling in roads is presented using Bluetooth sensors. The unique Bluetooth device address of Bluetooth-enabled device inside vehicle can be monitored by some spatially separated Bluetooth sensors. Correspondingly, the travel time can be calculated by the collecting data of these sensors. The reliability and accuracy of our proposed method are demonstrated using field testing data that were collected in an 1863 m long segment of Jiangnan Road in Hangzhou. The experimental results show the consistent detection with the manual recording reference data when the distance between a pair of sensors is bigger than the value of two times of the one by dividing the identification radius of sensor by the system error.

Keywords Bluetooth sensor · Travel time estimation · Intelligent transportation

1 Introduction

Intelligent Transportation System (ITS) is regarded as one of the best solutions for traffic safety and congestion. By means of the new technologies and methods of ITS, accurate traffic information, such as velocity, density, flow and travel time can

Q. Li (✉) · L. Chen · Y. Huang · Y. Wang · Y. Mao
Department of Traffic Management Engineering, Zhejiang Police College, Hangzhou, People's Republic of China
e-mail: liqiangwei@zjjcxy.cn

L. Chen
e-mail: chenlufeng@zjjcxy.cn

Y. Huang
e-mail: huangyun@zjjcxy.cn

Y. Wang
e-mail: wangyingzhi@zjjcxy.cn

Y. Mao
e-mail: maoyongbo@zjjcxy.cn

© Springer International Publishing Switzerland 2017

V.E. Balas et al. (eds.), *Information Technology and Intelligent Transportation Systems*, Advances in Intelligent Systems and Computing 454,
DOI 10.1007/978-3-319-38789-5_71

be obtained for traffic monitoring and ITS management. Travel time over different road segments reflecting the real-time traffic condition is valuable for transportation management and performance measure, which can be published timely to road users.

Traditionally, inductive loop detector, magnetometer, video camera, floating car are used to traffic monitoring. In recent years, many innovative technologies for the travel time estimation are evolved. Bluetooth sensor is an alternative technology for traffic data collection, which is more cost-effective, less influence of weather compared with camera sensor [1–3]. Bluetooth is a short-range, wireless communication technology between devices, such as cellular phones, computers, and digital devices. The most important characteristic of Bluetooth device is that each one has a unique 48-bit MAC (medium access control) address. In another words, if the MAC address can be identified, we can capture the corresponding device.

It is encouraging that the Bluetooth-equipped devices in personal consumer electronics and in car systems are more and more popularly applied, which provides a foundation to develop Bluetooth sensor technologies and methods for application in intelligent transportation system [2, 4]). In this study, Bluetooth sensor technology applied to approximate the travel time is evaluated and discussed. Field test is carried out to demonstrate the effectiveness of the proposed method.

2 Estimation Principle

It is a typical application that the drivers mobile phone is interconnected to a wireless earpiece of on-board vehicle. Observations of multiple vehicles may provide accurate estimates of traffic conditions, such as a number of travel time samples between a pair of Bluetooth sensors [5, 6]. However until 2010 Bluetooth used as a non-intrusive traffic detection technology was reported and obtained more and more attention. The reliability of estimation method based on the Bluetooth sensor depends on widespread penetration of Bluetooth-equipped devices in vehicles. In china, there is no reported proportion of Bluetooth-enabled device in vehicles in the available literature or reports, while the Bluetooth penetration was estimated at between 27 and 29 % [7, 8]. Fortunately, the penetration of Bluetooth-enabled cellular phone is quite higher in China, which results in a possibility of 24 h and 7 days easy collecting traffic information by Bluetooth sensors.

Each Bluetooth device has a unique identification number (MAC address) which results in the possibility of monitoring the behaviour of Bluetooth devices. The basic principle is that a vehicle is recognized at one point by scanning the MAC address, then re-recognized of the same vehicle at another point of the analyzed section. Thus, the travel time can be calculated. Throughout the rest of the section, we refer to these Bluetooth built-in vehicle or carrying Bluetooth enabled device in vehicle (mostly Bluetooth-enabled mobile phone) simply as Bluetooth vehicle.

Figure 1 shows the overall experimental setup of identification points. The Bluetooth sensors deployed at the identification point alongside road scan periodically the short-range medium to collect the MAC addresses of passing Bluetooth vehicles in

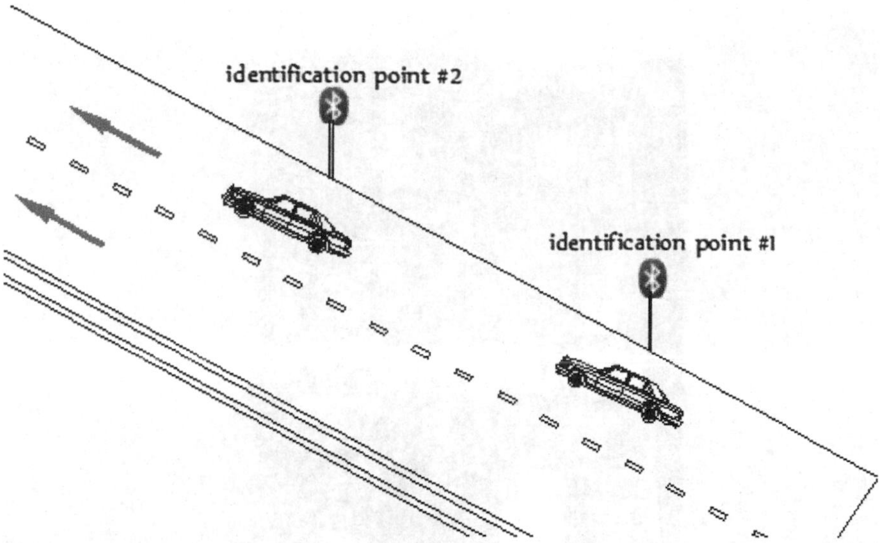

Fig. 1 Overall experimental setup of identification point

its detection neighborhood and corresponding detected time. The travel time is just the passing time difference. And according to the known distance between two sensors, we can calculate the individual traveling speed by dividing the distance by the time difference. If we have a proportion of vehicles traveling time, an approximation of average travel time in this road segment can be extracted.

3 Field Test

We designed a portable Bluetooth sensor (showed in Fig. 2) powered by battery which has 5 h duration and be recharged by USB. We conduct a field test in an 1863 m long segment of Jiangnan Road in Hangzhou. As described in Fig. 3, we deployed two vehicle recognition points (corresponding Bluetooth sensor 1 and sensor 2). In the driving direction, as Bluetooth vehicle passing the sensor 1, the corresponding scanned MAC address and recording time will be stored in the SD card which can be read and analyzed lately. And the second identification of the same MAC address is implemented at recognition point 2. Consequently, the traffic information such as travel time, speed can be calculated.

Travel time estimation requires a reliable reference system, while manual time re-cording by mobile was chosen to provide the reference travel times because of natural high accuracy (almost 100 %) based on the enough long distance of a pair of sensors.

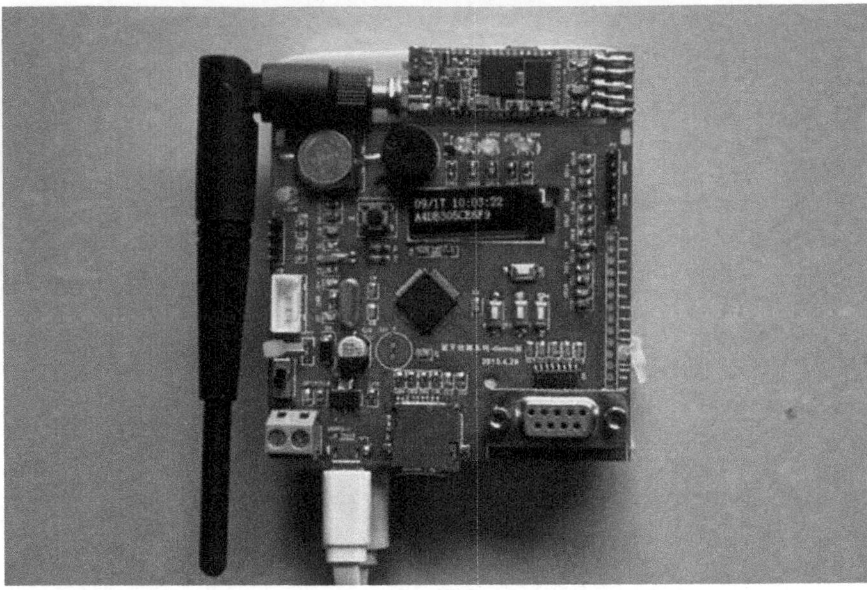

Fig. 2 Portable bluetooth sensor

Fig. 3 Locations of bluetooth sensors on the Jiangnan road

4 Analysis and Discussion

4.1 Analysis Model

Usually, travel time has strong relationship with vehicle speed. To illustrate more explicitly, we simultaneously evaluate two parameters. is defined as the detection radius of the Bluetooth sensor detection neighbor-hood. Some tests were conducted to determine the exact value that is 45 m in this work. should be controlled in the future re-design of the Bluetooth sensor. t_0 is the identification time interval, is the Bluetooth device identified time by the sensor. It is necessarily to point out that the objective of the sensor is to obtain the information about vehicle presence. We dont need to build the connection between sensor and device, so is less than one second. δt is the scanning time period. In the original design, we have no control circuit which will be added a delay control unit. Here equals zero.

$$t = t_0 + \Delta t \tag{1}$$

D is described as the distance of a pair of Bluetooth sensor. Let δt_{cal} and δ_{ref} denote the calculated value and reference value of the identification time difference of Bluetooth vehicle. To evaluate the effectiveness in travel time data collected by the Bluetooth sensors, we convert the travel time into the vehicle speed. Correspond-ing, $V_{cal} = D/\Delta T_{cal}$ and $V_{ref} = D/\Delta T_{ref}$ denote the average vehicle speed values. δ is the tolerant system error. In this study, let δ be 5 %. It is easily understood that practical traveling distance of Bluetooth vehicle is between $D - \lambda$ and $D + \lambda$. Correspondingly, it exits following relationship

$$\frac{2\lambda}{D} \leq \delta \ or \ D \geq \tfrac{2\lambda}{\delta} \tag{2}$$

In this work, it can be inferred the condition of has bigger value than 1800 m.

It is obvious that a Bluetooth vehicle will be continuous detected on the condition of staying in the detection neighborhood of the sensor. Based on the fact that the neighborhood radius of our designed Bluetooth sensor is 45 m and the normal vehicle speed is less than 80 km per hour. The shortest staying time is at least 2 s, which is greater than the identification time interval of Bluetooth sensor. It means that the Bluetooth vehicle is definitely discovered unless the sensor fails to function. Furthermore, in our case the Bluetooth vehicle will be detected at least twice due to fact that the total identification time interval is within 1 s under the absence of the delay control unit condition. Hence, we can define t_{ij} is the detected time of Bluetooth vehicle passing the Bluetooth sensor neighborhood. Subscript means the number of the Bluetooth sensor, while $j = 1$ and $j = 2$ respectively denotes the first and last detected time of Bluetooth vehicle passing the Bluetooth sensor is neighborhood. For one pair of Bluetooth sensors, $\Delta T_{cal} = t_{22} - t_{11}$ is the measured travel time of

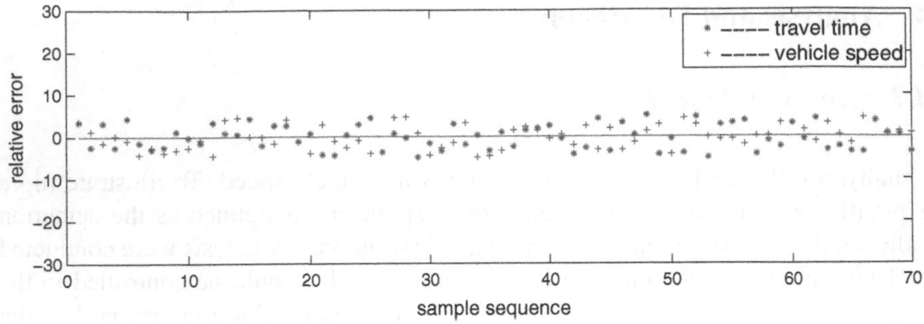

Fig. 4 Relative error of travel time and speed using bluetooth sensors

the individual Bluetooth vehicle. Relative error of travel time e_{time} and speed e_{spped} are chosen as evaluation indices, shown as Eqs. (3) and (4).

$$e_{time} = \frac{\Delta T_{cal} - \Delta T_{ref}}{\Delta T_{ref}} \quad (3)$$

$$e_{speed} = \frac{V_{cal} - V_{ref}}{V_{ref}} \quad (4)$$

Figure 4 illustrated that both relative errors are within 5 %. The accuracy of travel estimation method using Bluetooth tooth seem more consistent with the ground true data collected by the manual operation. In consideration of the enough long distance between a pair Bluetooth sensors mentioned in Eq. (2), the absolute error of travel time collected by manual operation can be ignored.

4.2 *Discussion*

The above results are encouraging for providing a low-cost, non-invasive and non-influence of weather method of estimating the average travel time of the vehicles on different road segments. However, the experiment only shows the feasibility of this Bluetooth technology. While it is transferred the commercial applications, more details should be solved, such as power supply, clock synchronization, message storage and wireless transmission. How many vehicles equipped with Bluetooth device population is sufficient to provide accurate description of travel time is also a challenging work.

By the ways of big data, besides estimation of vehicle travelling time, more potential applications to estimation of pedestrian and bicycle travel times, congestion analyze and origin-destination estimation of urban network are investigated. And in future, the data analyzed quality will be improved by data fusion using Bluetooth data in conjunction with data collected by loop, camera.

5 Conclusions

Currently, the attraction of Bluetooth sensor application is promising, since there are more and more on-board vehicular Bluetooth devices and personal consumer Bluetooth-equipped devices. We investigated the estimation method of travel time using Bluetooth sensors, and some field tests were carried out in order to evaluate the performance of our designed sensor and proposed method. The standard data were synchronous collected by drivers clock recording. On the condition of the distance of a pair of sensors be greater than $\frac{2\lambda}{\delta}$, the gap between estimated travel time and reference travel time show well consistence, and the relative error can be controlled within δ (in this study it is 5 %).

Acknowledgments This work was supported by the Key Scientific Projects of Department of Public Security of Zhejiang Province (No. 20130204) and the Program for Key Science and Technology Innovation Team of Zhejiang Province (No. 2013TD09).

References

1. Araghi B, Pederson KS, Christenson LT, Krishnan R, Lahnmann H (2012) Accuracy of travel time estimation using bluetooth technology: case study limfjord tunnel Aalborg, ITS world congress, Vienna, Austria. Accessed 22–26 Oct 2012
2. Purson E, Klein E, Bacelar A, Reclus F, Levilly B (2014) Simultaneous assessments of innovative traffic data collection technologies for travel times calculation on the East ring road of Lyon, transportation research procedia. In: 4th international symposium of transport simulation of transport simulation, Corsica, France, pp 79–89. Accessed 1–4 June 2014
3. Zoto Jorgos, La Richard J, Hamedi Masoud, Haghani Ali (2012) Estimation of average vehicle speeds traveling on heterogeneous lanes using bluetooth sensors. IEEE Veh Technol Conf 48(2):1–5
4. Bakula C, Schneider WH IV, Roth J (2012) Probabilistic model based on the effective range and vehicle speed to determine bluetooth MAC address matches from roadside traffic monitoring. J Trans Eng 137(1):43–49
5. Liu X, Chien S, Kim K (2012) Evaluation of floating car technologies for travel time estimation. J Modern Trans 1(1):49–56
6. Wasson JS, Sturdevant JR, Bullock DM (2008) Real-time travel time estimates using MAC ad-dress matching. Inst Trans Eng J 78(6):20–23
7. Haghani A, Hamedi M, Sadabadi K, Young S, Tarnoff P (2010) Data collection of freeway travel time ground truth with bluetooth sensors. J Trans Res Board 2175:19–27
8. Friesen MR, McLcod RD (2015) Bluetooth in intelligent transportation systems: a survey. Int J ITS Res 13:143–153

Design and Implementation of the Key Technology for Calling Taxi

Zhang Yongmei, Wang Youwei, Liu Mengmeng, Xing Kuo and Guo Sha

Abstract Since taxis are rather hard to find during rush hours, an intelligent system for calling taxi is proposed. Its main functions include calling taxi, picking up passengers, managing the user information and dispatching the centralized taxi. This paper introduces the Baidu map API for intelligent route planning, avoids going the congested road and shows the traffic route to users vividly. Changing the status of the order timely can ensure the users, the drivers and the administrators get the order's changes in the shortest time. Another feature of this paper is the road information mining by using taxi GPS data of the Beijing. It can dig out where the passengers often appear in a certain period of time. Therefore, it can reach the optimization of resource allocation.

Keywords Calling taxi · Baidu map API · Road information mining

Z. Yongmei (✉) · W. Youwei · G. Sha
School of Computer Science, North China University of Technology, 100144 Beijing, China
e-mail: zhangym@ncut.edu.cn

W. Youwei
e-mail: wyw1018752647@vip.qq.com

G. Sha
e-mail: 2314171316@qq.com

L. Mengmeng · X. Kuo
School of Electronic Information Engineering, North China University of Technology, 100144 Beijing, China
e-mail: 985767102@qq.com

X. Kuo
e-mail: 1430692220@qq.com

Z. Yongmei
Guangdong Key Laboratory of Popular High Performance Computers, Shenzhen Key Laboratory of Service Computing and Applications, 518060 Shenzhen, China

© Springer International Publishing Switzerland 2017

V.E. Balas et al. (eds.), *Information Technology and Intelligent Transportation Systems*, Advances in Intelligent Systems and Computing 454,
DOI 10.1007/978-3-319-38789-5_72

1 Introduction

With the development of economy and the improvement of people's living standard, the process of urbanization is speeding up and the requirement about the public transportation is proposed, especial for the quickness and convenience of taxi scheduling system. Due to the advantages of convenience and comfort, the taxi plays an important role in urban traffic. Today, lots of taxis provides the 24 h service, and stop just by a roadside waving. With the increasing the number of taxis, the extensive management mode led to lower efficiency, high no-load rate, traffic congestion, energy waste and environmental pollution. Therefore, the research of automatic dispatching system for taxi is imperative in the current situation. In the research about the method and structure of the taxi dispatch system, Z. Liao analyzed the application of GPS technology in the real-time intelligent dispatch of taxi [1]. Der-Horng Lee proposed a kind of intelligent dispatching system based on real-time demand and road traffic information [2]. The system focused on how the taxis arrives at the location of the passengers in the shortest time, according to the real-time traffic information.

The GPS technology, GIS technology, GSM communication technology and automation technology had gradually matured and the research of the taxi scheduling system integrated with these technologies had entered a period of rapid development. Centralized taxi scheduling method was that when passengers needs to use a taxi, they would call or send text messages to the dispatch center, telling the center that the arriving time and the target place, and then the dispatch center dispatched a taxi according to the requirement. This kind of method needs more investments in the dispatching center.

In 2010, Yu Bin proposed the GPS/GIS/GSM based taxi dispatch system [3], which effectively solved the vehicle monitoring and scheduling problem. Although this transport vehicle monitoring and dispatching system was suitable for the traffic department in achieving intelligent traffic management, the short message mode of integrated performance was poor and the order information was not detailed which made the driver hard to properly understand the users need, thus impeding the comprehensive promotion of the system.

In 2011, Liu Tang, Peng Jian and other researchers combined Internet of Things technologies with intelligent transportation system, and put forward a kind of intelligent traffic detection system based on Internet of Things (IoT) [4]. The traffic flow information provided a basis for the selection of vehicle route. These studies focused on the management and control of traffic flow, but failed to provide real-time traffic information to the people who are about to go out.

In recent years, with the gradual maturity of the smart phone and 4G communications, taxi-hailing apps are gradually emerging. Drivers can use their mobile phones to deal with orders instead of the vehicle terminals, and passengers can also book taxis. Slowly but surely, it crept into our lives. But according to the survey, most of these taxi-hailing apps are still exist many drawbacks listed as follows:

1. The efficiency of taxis is low, the passengers need wait for a long time on the traffic peak, taxi allocation is unsuitable, and the no-load rate is high in part time.
2. Most of the taxi software only provides quick and convenient services for those who have installed them. So those who are not installed any taxi-hailing apps, or cannot afford the smart phone could not enjoy these services.
3. It is not clear about some passengers in unfamiliar areas to know the distance and route planning from the destination to the starting point. So it may lead the passengers to spend more money. Most of the users are plagued by this problem.

Aiming at the existing problems, a system for booking taxi is designed and realized. The biggest features are the use of Baidu maps API and the second development based on it. It can analysis real-time traffic information, then select the optimal path that does not take the congested road to users. In this way the drivers can arrive at the destination timely. When the drivers accept an order or the passenger arrive at the destination, the orders status will be changed as soon as possible. Thus make the administrators get the latest information in time. This paper meets the needs of passengers without install any taxi-hailing apps, and provides print tickets, checks the weather and other human services. Based on GPS (Global Positioning System) data and the change of the unloaded information, the location of the passengers can be drawn. It can dig out that where the passengers often appear in a certain period of time. Also it can realize the reasonable allocation of taxi resources, improve the efficiency and reduce no-load rates.

2 Taxi Online Call Taxi Overall Function

This paper presents a platform that achieving the function of calling, positioning and querying taxi. The drivers can query and deal with orders conveniently. The administrators can achieve the management and maintenance of the information about passengers and drivers. The platform includes user operation module, driver operation modules and management module.

The functions of passenger operation module include: query and modify personal information, modify personal password, online call taxi. When calling taxi online, the passengers enter detailed information including the title, the starting place, and the destination. The passengers can still get the planning route by Baidu map and search historical orders.

The functions of the driver operation modules include: query personal information, modify personal password and query the current order request. The drivers can receive the order when it meets the requirements. After received the order, the drivers can obtain the optimal planning route from his position to the passenger's, reducing the waiting time of the passenger. Drivers can query the processed orders and select the date to settle accounts.

The functions of background management modules include: user information and account management, order information management and scheduling center management. User's information management includes to add, delete, find and modify user. Order information management includes order's management, set up the weekend mode, set up the usual mode and deal expiration of the order. Scheduling center management includes: the dispatch center uses the server to search the appropriate vehicle for the passenger, then sent the passenger's position information to the taxi. Drivers can browse the details of the order information, and make timely response. Finally the dispatch center will send the information to the passenger.

3 The Key Technology

3.1 Baidu Maps API Interface Technology

Baidu maps API is a set of application interface, which is free for developers, based on Baidu map service, including Java Script API Web server API, Android SDK, IOS SDK, positioning SDK, internet of vehicles API, LSB cloud and many other development tools and services. It can provide with basic map's display, search, positioning, inverse Geographic Encoding and route planning, LBS cloud storage and retrieval, and other functions. It's applied to many kinds of equipment, such as PC, mobile terminal and server and so on.

Based on integrated ArcGIS services of Baidu maps API, a second development is applied and many humanized services are added. After analyzing the real-time traffic situation and avoiding the crowded road, the user can get the best planning path. API is the application programming interface. It is a kind of interface serving for the second party provided by software or websites. Using this interface, you can direct use the function without having to understand the internal mechanism.

Baidu maps API is a set of application programming interface written by Java Script language. The map service API binds this platform and the geographic information data, encapsulating the complex GIS underlying logic, providing itself freely to the users in an intuitive way, reducing the application threshold in both map service and developing layers. It can help users construct feature rich and strongly interactive website application, providing the developers for rich methods, events and encapsulated classes. To use Baidu Maps API the developers only need some HTML and Java Script programming basis. Baidu maps API not only contains the basic interface to build a map, but also provides data services such as local search, route planning. Users do not need to download and install any maps, software or controls. All functions required are returned to the client after the operation on the Baidu server and the developers can simply connect all the services to their own web pages using the API.

To operate the map, it must have tools that can translation, zoom and other abilities. The developer can achieve many functions just added the control or event provided by Baidu maps API.

3.2 The Method of Avoidance Congestion Road

Research on urban traffic optimal path planning is the shortest time of the taxi arrived rather than the shortest path. Due to the complexity of traffic network, the traffic restrictions in urban road network and the actual problems of the vehicle such as delay in intersections, all should be considered. In the dynamic traffic network the selected result of the optimal path is the shortest time. Therefore, the optimal path algorithm must be effective, real-time and rapidity. Only in this way can we quickly find the optimal path in the dynamic traffic network.

In this paper, Traffic Control provided by Baidu maps API can get the real-time traffic information. Dijkstra algorithm is adopted to avoid congested section, and get the path of the shortest time. When the user enters the location and destination, the optimal route planning and the mileage can provided and the blocked is avoided according to the current traffic situation.

The key idea of Dijstra algorithm is to calculate the shortest path from a node to the rest of nodes. The obvious characteristic of it is that the starting center point will extend to the outer until the end. Taking the intersection as the node, each road as the boundary, the network of the city transportation is abstract a positive weight of directed graph. Searching for the shortest path in directed graph, this paper sets the weight of each sides, calculates the shortest time to the destination and combines the situation of urban road and traffic regulations based on the Dijkstra algorithm.

3.3 Mining Road Information

Mining road information: As taxi almost occur every corner of city's streets, you can depict the road network of the city based on taxi GPS data. By using of the spatial-temporal data and the information whether the taxi is no-load or not, a lot of other taxi-related information can be dug out according to the prior-knowledge. For example, according to whether the load have changed, the position that the passengers on or off can be drawn. Then it statistics these data, can described within a certain period of time, passengers often appear in which position. Accordingly, it can later guide the taxis to wait passengers in those moments.

This paper uses the temporal data of Beijing taxi as the example. Define a record as the taxi's data <id, Latitude, Longitude, occupy, time>, where id represents a unique identifier for a taxi, occupy indicates whether the current time the taxi has passengers in it, 1 expresses yes, 0 indicates no. Here use the UNIX time stamp as the standard. It gives a taxi message: (new_abboip latitude40.76212117.2522012 13084687). This message represents the taxi which id is new_abboipat latitude 40.76212 N 117.25242 has no passenger at 1213084687. This data includes time, on the basis of the coordinates, the unique addition of a taxi whether it is no-load information. According to this data, you can dig out a lot of potential for other information. The flow chart is shown in Fig. 1.

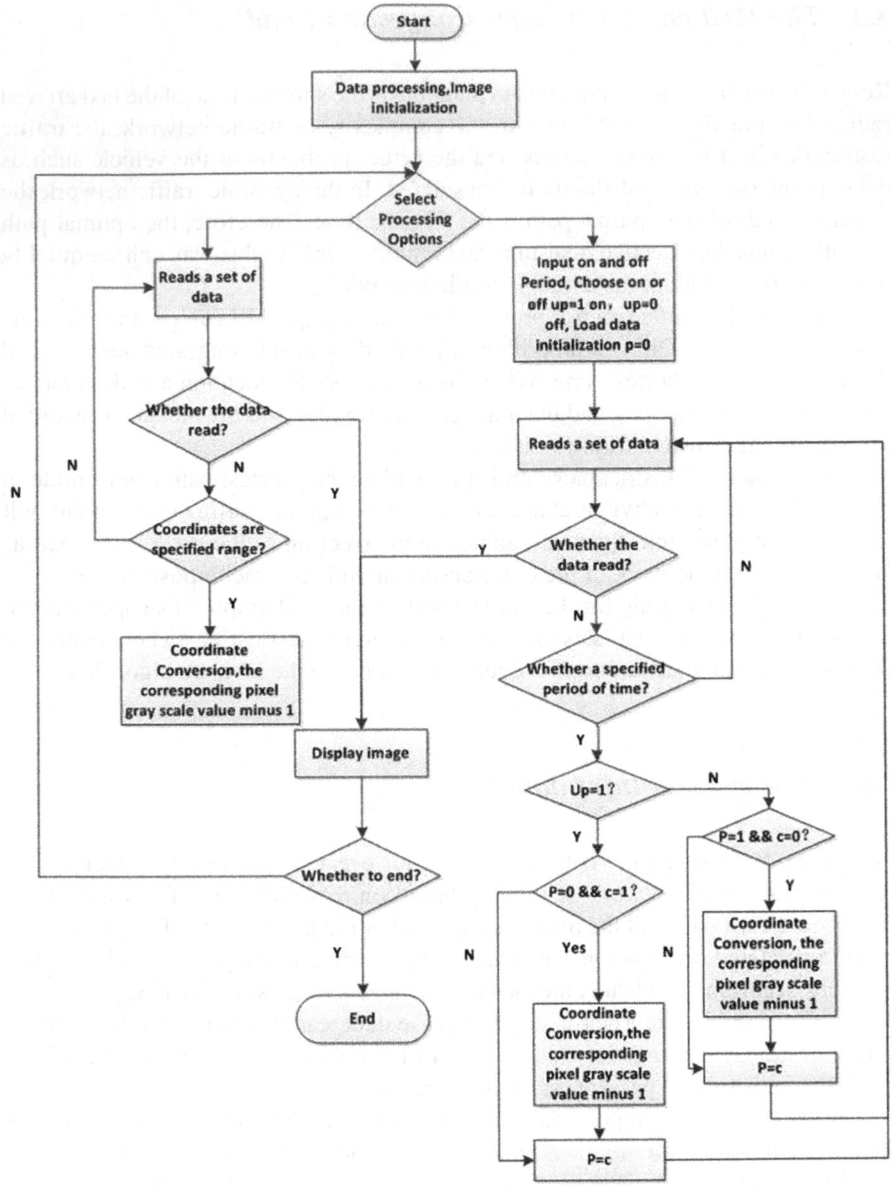

Fig. 1 Flowchart of the mining process

1. Data processing will begin in advance before selecting data mining, and it includes two aspects: the approximate time, position data transfer. The position coordinates of the focus are in latitude 39″26′ to 41″03′ longitude 115″25′ to 117″30′. So it can omit the integer bits and convert to decimal integers (such as 39.75134 converts to 75134). The double-precision coordinates point data into an integer can improve operational efficiency.
2. Select the processing option, here are the road network and on and off options.
3. The program will read all data in turn after the users select the "road network". It will determine whether the data coordinates in a predetermined area. And then the program selects the data within the prescribed time and transforms coordinate and changes the point corresponding to the image gray values. The road network image will be displaying, when all the data read is completed.
4. When selecting "on or off" the program will pop up dialog box that prompts the operator to enter the appropriate start and stop times and choose the on or off the track. The program reads all data sequentially after filling out and confirms the above information and determines whether within the specified time. Selecting the data within the specified time, it will unify the taxi passenger information about the two groups. If around the time information is different, it will coordinate transformation data and changes the gray value corresponding to the point; otherwise unchanged. This allows more of a place on or off the bus number, the lower the dot gradation value, the darker the image. It will display the image after it traverses all the data. You can dig out the off position, statistical data on and off the time and place, and can be described within a certain period of time. Passengers often are seen in which position. This can guide the drivers to pick up the passengers at this time.

4 Experimental Results and Analysis

4.1 Experimental Results

In this paper, online calling taxi system including planning specific route, online payment and so on is carried out after the user login.

The most innovative place is that it can figure out the most specific and convenient route according to the specific requirements of the passengers. Here use the Beijing West Railway Station as the origin and the Beijing Tiananmen Square as the destination. Expect results: Given optimal planning route from Beijing West Railway Station to Beijing's Tiananmen Square, as shown in Fig. 2. Recommended planning route would display in the user's interface. Thus these problems can be solved that it is likely for the passenger to take a detour and spend more money. Especially the passenger in an unfamiliar area is not known the distance and planning route from the origin to the destination.

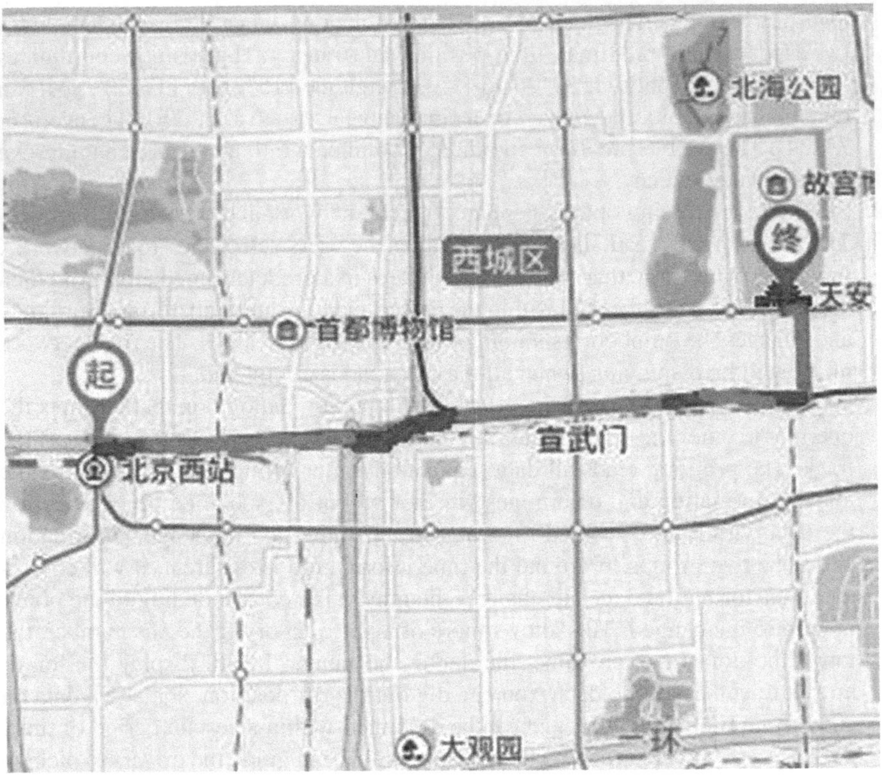

Fig. 2 Road maps

It can avoid congestion road and select the optimal path according to real-time traffic conditions in this paper. Although the starting point and destination are same, the recommended optimal path of different traffic conditions may be different. Figures 3 and 4 show that when traffic flow freely, it recommends a variety of options for users to choose. But in traffic jams, especially during rush hour, it provides a scheme. The green lines represents the recommended route and the red line marks congestion points. It can avoid congested road and choose a smooth path for users. The recommended optimal path is also shown in the driver's interface. After the driver accepts an order, the system will recommend optimal path from current location to the position that the passenger gets on the taxi. So it can avoid crowded road, reduce passenger's waiting time, and avoid miss event. When the driver picks up the passenger, the system will update the optimal path from current location to the passenger's destinations.

This paper also adds a lot of user-friendly designs. Because many passengers forget or lost ticket issue, passengers can print their ticket after the order finished. The driver can select the starting and ending dates for the amount settlement. The above is the results. The running of the entire software supports all the features of the original design.

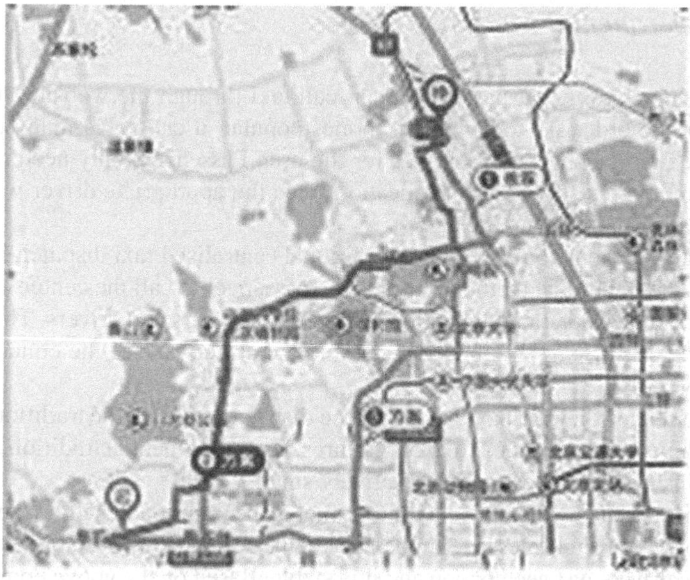

Fig. 3 The recommended path when traffic is clear

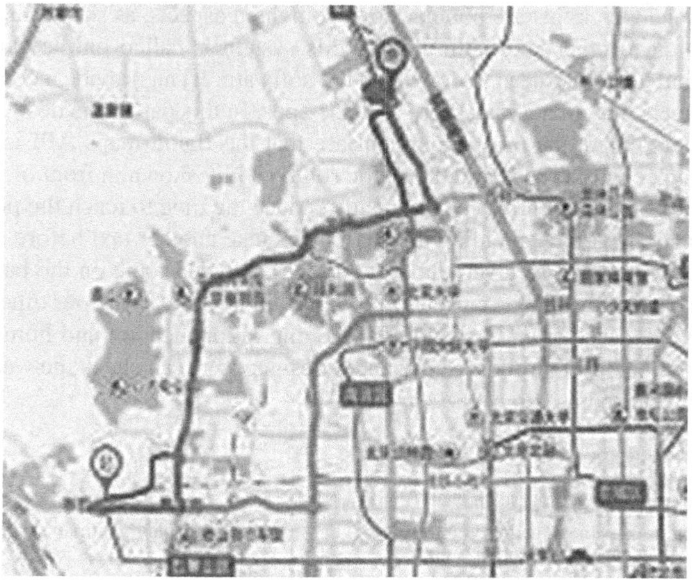

Fig. 4 The recommended path when traffic is heavy

4.2 Experimental Analysis

This system achieves that passengers can call taxi through the website, the driver can accept the order. As today smart phones popular, it can solve many problems and provide more convenient services for the user. Passengers only need to turn on this system and put forward an application, then the appropriate driver will accept the application and pick up them.

This paper system compared with traditional centralized taxi dispatch from four parts [3]. The traditional method requires the passenger to call the center when they need a taxi. It lacks of communication between passengers and drivers. This system only needs the passengers submit orders, and the driver can see the contact details of passengers.

The passengers have to pay the cost of the dispatch center in the traditional methods, and this system only needs to pay taxi fares. The investment of traditional method is large, this systems investment is relatively small.

The drivers accept orders by the car terminal in the traditional method, and it often affected by GSM short message mode and the performance is unstable. But in this system the drivers just need phone and it is stable. Based on the above comparisons, this system has the high efficiency, and it can ensure that the passengers can get a taxi service timely. It meets the increasing demand for urban taxi. It also has compared with Didi and Quick that are popular now from three aspects, as shown in Table 1.

Compared to other taxi-hailing apps, this system is called online calling taxi websites and you don't need to download the software. It can provide a convenience for passengers that do not install or can't use apps. In this paper, the design method has been simplified, the most prominent are that the Baidu maps API is intruded for avoiding congested road and the traffic route vividly shown in front of the users, making travel more scientific. It can greatly reduce the time to reach the position of the passenger departure, avoiding the passengers take another taxi before the driver arrived. The users can be given the best and less crowded route on the basis of the entire route planning and the current time. This will not only save your time, but also eliminates the need for extra spending, reaching rationalization and humane goal. Furthermore, this system adds user-friendly design; users can check the weather and print the invoice copy.

Table 1 Taxi software comparison chart

Taxi software	Whether you need to download APP	Passenger-side display optimal path	Human services
Didi	Yes	No	No
Quick	Yes	No	No
This system	No	Yes	Yes

5 Conclusions

The system is designed and implemented that the passenger can call car. The driver can pick up passengers, easing traffic difficulties. Baidu Maps API is conducted on the basis of secondary development, to complete the intelligent planning route and avoid congested route. The traffic route will visually present to the user. It can ensure users and administrators get the status of the order timely. Using GPS data onto the Beijing's taxi can dig out the information such as which place can get more passengers and which period of time people want to call a taxi, thus guiding a rational allocation of the taxis. It also can help travelers master traffic information and the way to travel timely, while guiding the drivers to choose the most environmental and friendly travel path, reducing empty rates and the consumption of fuel.

Acknowledgments This paper is supported by National Natural Science Foundation of China (61371143), Guangdong Key Laboratory of Popular High Performance Computers, Shenzhen Key Laboratory of Service Computing and Applications (SZU-GDPHPC L2014), Beijing Municipal Education Commission project for the Multi source image registration and recognition research platform of virtual reality fusion (fusion(PXM2015_014212_000024), Multi-source remote sensing image registration and recognition research platform (XN081).)

References

1. Liao Z (2010) Taxi dispatching via global positioningsystems. IEEE Trans Eng Manag 48(3):342–347
2. Lee D-H, Wang H, Cheu RL, Teo SH (2004) A taxi dispatch system basedon current demands and real- time traffic information. Transp Res Rec 1882:193–200
3. Liu T, Jian P, Jin Y, Xiaofen W (2011) The intelligent detection technology of traffic flow based on networking. Comput Sci 9:20–35
4. Hongsong W, Haiyan Z (2014) Tourism geographic information system based on Baidumaps API development [J]. Mod Comput 23:60–63

Optimization of Single Train Operations

Lingna Li, Xinxin Wang, Yuxin Liu and Chao Chen

Abstract Energy efficiency is paid more and more attention in railway systems for reducing the cost of operation companies and emissions to the environment. We both consider the optimizations on timetable and driving strategy are two important and closely dependent parts of energy-efficient operations. It not only regulates the fleet size and the trip time at infestations, but also determines the control sequences of traction and braking force during the trip. In this paper, we analysis and establish the dynamic model of train and build single train optimization Model between two stations and three stations, then propose an iterative search algorithm to get the optimal speed, which can get better energy-saving performance. The simulation results based on a Subway Line in China illustrate that the iterative search algorithm can provide good performance with energy saving.

Keywords Energy consumption · Dynamics principle · Multi-objective optimization · Iterative search algorithm

L. Li (✉) · X. Wang · Y. Liu
School of Science, Southwest Petroleum University, Chengdu 610500, China
e-mail: lilingna@swpu.edu.cn; 1319952@qq.com

X. Wang
e-mail: xin-xin.wang@qq.com

Y. Liu
e-mail: 791815233@qq.com

C. Chen
College of Petroleum Engineering, Southwest Petroleum University,
Chengdu 610500, China
e-mail: chenchaoswpu@qq.com

© Springer International Publishing Switzerland 2017

V.E. Balas et al. (eds.), *Information Technology and Intelligent Transportation Systems*, Advances in Intelligent Systems and Computing 454,
DOI 10.1007/978-3-319-38789-5_73

1 Introduction

Energy consumption of rail transit system refers to the energy consumption of equipment. With energy conservation and emissions reduction is paid more and more attention in low carbon environmental, it is an important research direction in the field of rail transit to reduce the energy consumption of train traction and control train operation optimization in recent years [1]. There is a speed limit according to the line condition, train characteristics, status in front of the line when the train runs between the stations. Under the constraints of it, trains usually contain four operating conditions: traction, cruising, coasting and braking [2]. There are more than speed distance curve to choose from when the train running between stations. Trains can walk the same distance with different operation time and energy consumptions [3]. This paper presents an approach of integrating the train control and distribution of the trip time when considering the variable traction force, braking force, and running resistance [4]. We propose a method called Single objective optimization model to optimize the operation for a single train. In addition, an iterative search algorithm is pro-posed to optimize the timetable and driving strategy, which can achieve global optimization of train operation on energy saving. The rest of this paper is organized as follows. In Sect. 2, we introduce Assumptions and Symbols about models. In Sect. 3, we analysis and establish the dynamic model of train. In Sect. 4, we build single train optimization Model. In Sect. 5, we present the algorithms for calculating the energy-efficient driving strategy, distributing the cycle time, and generating the train timetable. In Sect. 6, a case study is presented based on the infrastructure and operation data from a Subway Line in China, which illustrates that the proposed approach can provide good performance with energy saving.

2 Assumptions and Symbols

For a better understanding of this paper, the assumptions, parameters, and decision variables are introduced as follows.

2.1 Assumptions

(1) Neglect smaller power consumption equipment in the train, such as air conditioning energy consumption.

(2) All trains that run in the same direction share a common timetable, except for the headway time, which means that they are assigned the same dwell time on each station and the same trip time on each section. The only difference is that the latter train is operated at a certain time later than the former train.

(3) In all sections, the accelerating and braking times of trains are known parameters, which can be calculated by using the train energy-efficient operation algorithm based on the current operation timetable.

2.2 Parameters

n	number of service trains
T	cycle time (in seconds)
M	weight of the train (kg))
F_{max}	maximum traction force per unit mass (in meters per square second)
B_{max}	maximum braking force per unit mass (in meters per square second)
V_{max}	maximum velocity of the train in meters per second
μ	energy consumption per unit mass (in kilowatt hours)
S	train position (in meters)
W	total resistance (kN)

3 Dynamic Model of Train

In the process of operation, the actual stress state of the train is very complicated [5]. So we use the elemental point model, let train as the elemental point. According to the principle of dynamics [6, 7], the relationship between energy and consumption of the train shown as

$$E = \int F(t)v(t)dt \tag{1}$$

in which the traction of the train F is equal to the sum of total resistance W and the external force of train operation $F' = F + W$. where total resistance W can be further formulated as [6]

$$W = \frac{Mg(w_0 + w_1)}{1000} \tag{2}$$

Hence, we get relationship of energy consumption with four kinds of train operations.

A. Traction Train speed up, the engine is in a state of energy dissipation, and the traction force, acceleration and energy dissipation formula is as follows [8]

$$F = Mg(\frac{dv}{gdt} + \frac{w_0 + w_1}{1000}), E = \int Mg(\frac{dv}{gdt} + \frac{w_0 + w_1}{1000})vdt \tag{3}$$

B. Cruising Train at a constant speed, the traction force, acceleration and energy dissipation shown as [8]

$$F = \frac{Mg(w_0 + w_1)}{1000}, E = \int (\frac{Mg(w_0 + w_1}{1000})vdt \tag{4}$$

C. Coasting Train Runs slowly down and it does not need traction. So this phase train engine does not waste energy [9]

$$dE = 0, a = \frac{dv}{dt} = \frac{g(w_0 + w_1)}{1000} \tag{5}$$

D. Braking Train runs slowly down and it doesn't need traction. So this phase train engine does not waste energy [9].

$$dE = 0, a = \frac{dv}{dt} = [\frac{B(v)}{Mg} + \frac{(w_0 + w_1)}{1000}]g \tag{6}$$

4 Optimization Model

4.1 Two Stations Optimization Model

According to train dynamics model and the relationship of energy conversion, energy consumed for the appointed time T [10]

$$E(T) = \int_0^T F[v(t)]v(t)dt \tag{7}$$

Requirements for distance and speed of the train operation

$$S = \int_0^T v(t)dt, v(t) \leq V_{max} \tag{8}$$

Requirements for acceleration, the traction and braking in the process of train running and boundary conditions of the train operation

$$|a| \leq 1; F[v(t)] = \mu F_{max}, \mu \in [0, 1]; B(v) \leq B_{max}; v(0) = v(T) = 0 \tag{9}$$

Hence, given the train circle time and distance, we build Single objective optimization model between two stations.

$$\begin{aligned} \min \quad & E(T) = \int_0^T F[v(t)]v(t)dt \\ \text{s.t.} \quad & S = \int_0^T v(t)dt \\ & F[v(t)] = \mu F_{max}, \mu \in [0, 1] \\ & B[v(t)] \leq \mu B_{max}, v(t) \leq V_{max}, |a| \leq a_{max} \\ & v(0) = v(T) = 0. \end{aligned} \tag{4-1}$$

4.2 Three Stations Optimization Model

Among three stations, to make the total minimum energy consumption, so the energy dissipations of the train pass by each station is optimal, namely [11]

$$\begin{aligned} \min E_1 &= \int_0^{T_1} F[v(t)]v(t)dt \\ \min E_2 &= \int_{T_1+T_b}^{T_2} F[v(t)]v(t)dt \end{aligned} \tag{10}$$

where T_1 means train running time of first two stations, T_b is dwell time during the stations, then $T_2 - (T_1 + T_b)$ means train running time of last two stations. So the objective function is shown as

$$\min E(T) = \int_0^{T_1} F[v(t)]v(t)dt + \int_{T_1+T_b}^{T_2} F[v(t)]v(t)dt \tag{11}$$

When the train on the station, the speed of the train is 0, the boundary conditions change as

$$v(0) = v(T_1) = v(T_1 + T_b) = v(T_2) = 0 \tag{12}$$

Hence, given the train circle time, dwell time and distance, we build Single objective optimization model between three stations based on model (4–1).

$$\begin{aligned} \min \quad & E(T) = \int_0^{T_1} F[v(t)]v(t)dt + \int_{T_1+T_b}^{T_2} F[v(t)]v(t)dt \\ \text{s.t.} \quad & S = \int_0^{T_1} v(t)dt + \int_{T_1+T_b}^{T_2} v(t)dt \\ & F[v(t)] = \mu F_{max}, \mu \in [0, 1] \\ & T_2 - T_b = T \\ & B[v(t)] \le \mu B_{max}, v(t) \le V_{max}, |a| \le a_{max} \\ & v(0) = v(T_1) = v(T_1 + T_b) = v(T_2) = 0. \end{aligned} \tag{4–2}$$

5 Algorithm

In this section, we present the iterative search algorithm to optimize the timetable and driving strategy [12]. There has been much research concentrating on how to get the optimal driving strategy such that the energy consumption is minimized. We described the problem with optimal train-control models, which has been presented in (4–1) and (4–2). In general, the energy consumption will decrease as the trip time increases. The minimum energy consumption is uniquely determined by the trip time and vice versa. As we all known, the speed sequences in the traction phase are first calculated with the given energy consumption [13]. If the speed reaches the speed limit during the acceleration process, the rest energy will be used for generating the speed sequences of the cruising phase. For the rest journey, a coasting speed sequence p_k from the end speed of the cruising phase and a braking speed sequence q_j from the end of the braking phase are calculated; then, the minimum speed value of the two sequences is obtained as the optimal speed with given energy constraint [14]. In conclusion, the optimal speed sequences can be solved with the following algorithm.

Algorithm 5.1Iterative Search Algorithm

Step1: Initialize initial speed v_0, final speed v_t and circle time T

Step2: Divide the accelerate weight μ into N parts

$$\mu = [\mu_1, \mu_2, \dots, \mu_N] \in (0, 1]$$

Step3: Let $m = 1\ \mu = \mu_m$

Step4: Set energy consumption E

Step5: Set $i = 0 v_i = v_0\ i = i + 1$ .Calculate

$$a = (\mu F(v_{i-1}) - w_0(v_{i-1}) - w_1(v_{i-1}))/M$$

$$v - i^2 = v - i - 1^2 + 2a(s_i - s_{i-1})$$

$$E = E - (F(v - i - 1)(S_i - S_{i-1})/M$$

Step6: If $v_i < V(s_i)_{max}$ and $E > 0$ return Step5

Step7: If $E > 0$ let $v_{i+1} = v_i$ and

$$E = E - (w_0(v_i) + w_1(s_i))(s_{i+1} - s_i)$$

let $i = i + 1$, return Step6

Step8: Let $k = i\ p_k = v_k$

Step9: If $k \leq l$, calculate

$$p_{k+1}^2 = p_k^2 - 2(s_{k+1} - s_k)(w_0(p_k) + w_1(s_k))$$

let $k = k + 1$, return Step8

Step10: Let $j = l$, $q_j = v_t$

Step11: If $j > i$, calculate

$$q_{j+1}^2 = q_j^2 + 2(s_j - s_{j-1})(\mu B(q_j) + w_0(q_j) + w_1(s_j))$$

let $j = j - 1$, return Step10

Step12: Let $i = i + 1\ v_i = minq_i, p_i$, if $i \leq l$ return step11

Step13: Return the optimal circle time $t = \prod_{i=1}^{l} \frac{s_{i+1} - s_i}{v_i}$

Step14: If $t \neq T, E = E + \Delta E$, return Step4

if t=T, output the optimal speed sequencesv_i^m, $0 \leq i \leq l$

and energy E_m, let $m = m + 1$, if $m \leq N$, return Step3

Step15: Output the optimal energy $E_{min} = minE_1, E_2, \dots, E_m$

corresponding accelerate weight and μ and speed sequences

$v_i, 0 \leq i \leq l$

---

6 Case Study

In this section, a case study is presented based on the infrastructure and operation data from a Subway Line in China. It covers a length of 22.73 km and consists of 14 stations.

6.1 Single Train Between Two Stations

Given the train circle time $T = 110\,\text{s}$ and distance $S = 1355\,\text{m}$. The maximum traction and braking characteristics of expression as shown below

$$F_{max} = \begin{cases} 203, & 0 \le v \le 51.5\,\text{km/h} \\ -0.002032v^3 + 0.4928v^2 - 42.13v + 1343, & 51.5 \le v \le 80\,\text{km/h} \end{cases}$$

$$B_{max} = \begin{cases} 166, & 0 \le v \le 77\,\text{km/h} \\ 0.1343v^2 - 25.07v + 1300, & 77 \le v \le 80\,\text{km/h} \end{cases}$$

According to Algorithm 5.1, we can get the curve of speed and distance and curve of speed and time at two stations as Fig. 1.

From Fig. 1a, we can see the read line means the maximum limit speed at different location, the blue line means the optimize speed at different location. The trains movement firstly speeds up with traction, and then slows down by coasting, finally slows down to stop by braking. Whole process of the train is without cruising phase, and it proves that distance between two stations is shorter. From Fig. 1b, we can

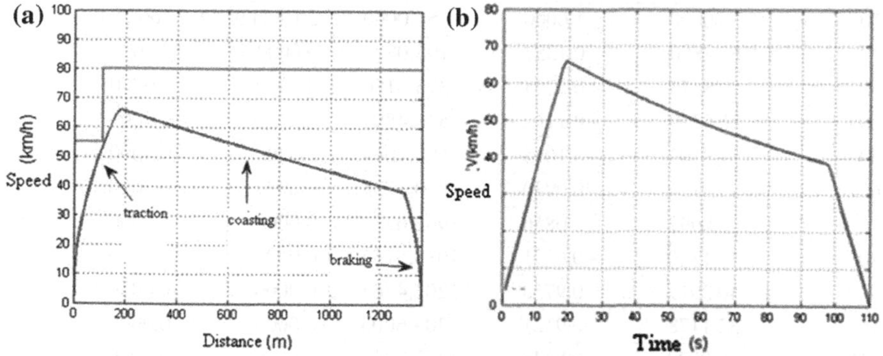

Fig. 1 *Curve* of speed and distance and *curve* of speed and time at two stations. **a** *Curve* of speed and distance. **b** *Curve* of speed and time

see the train began to rapid acceleration, and then slows down, finally the rapid deceleration until the train stops. It satisfies train movement rule. The relation-ship of circle time and energy consumption between two stations and the influence of energy consumption with weigh are shown as Fig. 2.

From Fig. 2a, the relationship of circle time and energy consumption around for inverse. When circle time $T = 110$ s, the optimal energy $E_{min} = 3.571 \times 10^7$ J. From Fig. 2b, with the increasing of weigh μ, energy consumption is decreasing. And when $\mu = 1$, that is traction acceleration or braking deceleration has reached maximum acceleration and deceleration, the energy consumption is the minimum. Record relative parameters at two stations with the interval of 10 s, is seen as Table 1.

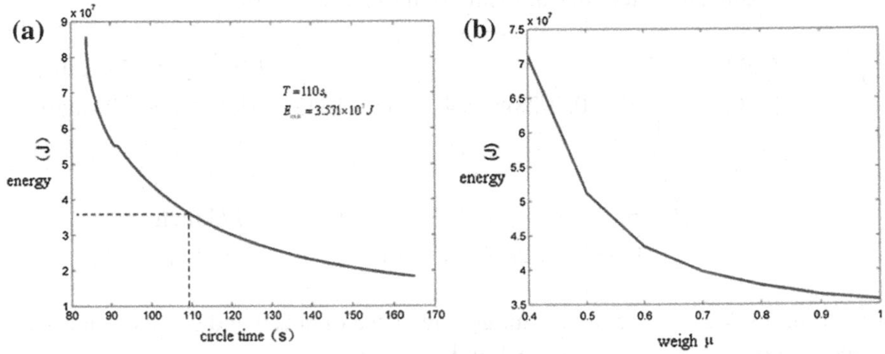

Fig. 2 The relationship of *circle* time and energy consumption with weigh μ. **a** Time and energy consumption. **b** Energy consumption with μ

Table 1 Relative parameters at two stations with the interval of 10 s

Time (s)	Actual speed (km/h)	Acceleration (m/s^2)	Distance (m)	Traction (N)	Traction power (kW)
0	0.0000	1.0000	0.0000	198162.2088	0.0000
10	36.0000	1.0000	50.0000	206884.9829	2068.8492
20	65.4800	−0.1357	198.0880	0.0000	0.0000
30	60.8373	−0.1225	373.4186	0.0000	0.0000
40	56.6349	−0.1112	536.4804	0.0000	0.0000
50	52.8103	−0.1015	688.4064	0.0000	0.0000
60	49.3127	−0.0930	830.1735	0.0000	0.0000
70	46.0997	−0.0856	962.6291	0.0000	0.0000
80	43.1360	−0.0791	1086.5135	0.0000	0.0000
90	40.3919	−0.9758	1202.4767	0.0000	0.0000
100	32.1178	−0.9121	1309.6006	0.0000	0.0000
110	0.0000	−0.8743	1353.9997	0.0000	0.0000

6.2 Single Train Between Three Stations

Given the train circle time $T = 220\,\mathrm{s}$ (without dwell time), dwell time $T_b = 45\,\mathrm{s}$ and distance $S = 2634\,\mathrm{m}$. Use the same algorithm, we can get the curve of speed and distance and curve of speed and time at three stations as Fig. 3.

From Fig. 3a, the read line also means the maximum limit speed at different location, while the blue line means the optimize speed at different location. The trains movement between two stations firstly speeds up with traction, and then slows down by coasting, finally slows down to stop by braking. Whole process of the train is without cruising phase, and it proves that distance between two stations is shorter. From Fig. 3b, with the total time $T = 220\,\mathrm{s}$, the optimal circle time at first two stations $T_1 = 108\,\mathrm{s}$, and the optimal circle time at last two stations $T_2 = 112\,\mathrm{s}$. And the dwell time $T_b = 45\,\mathrm{s}$, so the speed of train is 0 from 108 to 153 s. The relationships of circle time and energy consumption between first and last two stations are shown as Fig. 4.

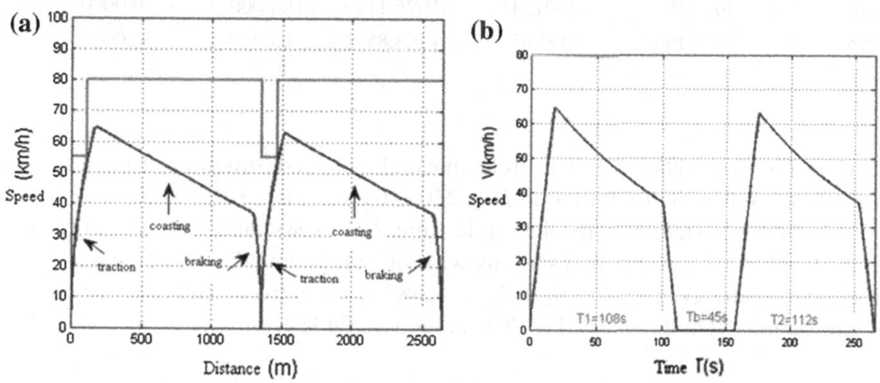

Fig. 3 *Curve* of speed and distance and *curve* of speed and time at three stations. **a** *Curve* of speed and distance. **b** *Curve* of speed and time

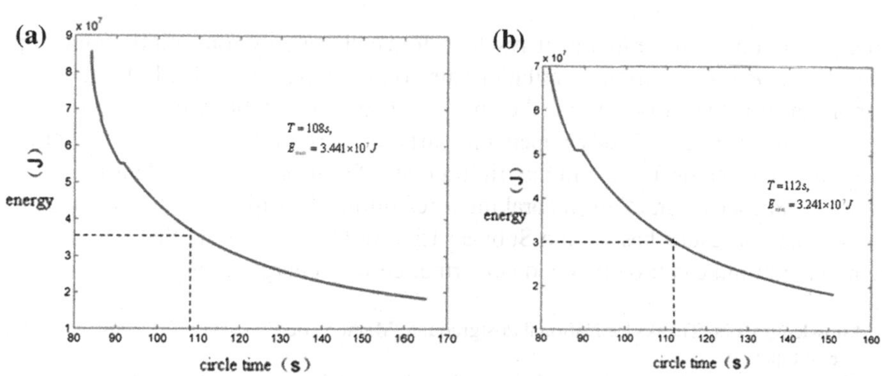

Fig. 4 The relationship of *circle* time and energy consumption at two stations. **a** Relationship at first two stations. **b** Relationship at last two stations

Table 2 Relative parameters at three stations with the interval of 20 s

Time (s)	Actual speed (km/h)	Acceleration (m/s^2)	Distance (m)	Traction (N)	Traction power (kW)
0	0.0000	1.0000	0.0000	198162.2088	0.0000
20	64.1808	−0.1320	197.4974	0.0000	0.0000
40	55.5688	−0.1085	529.3532	0.0000	0.0000
60	48.4200	−0.0909	817.6279	0.0000	0.0000
80	42.3756	−0.0775	1069.3916	0.0000	0.0000
100	37.1852	−0.9244	1290.0452	0.0000	0.0000
120	0.0000	0.0000	0.0000	0.0000	0.0000
140	0.0000	0.0000	0.0000	0.0000	0.0000
160	10.8000	1.0000	4.5000	199842.6160	599.5257
180	60.8092	−0.1225	247.4186	0.0000	0.0000
200	52.7870	−0.1014	562.2641	0.0000	0.0000
220	46.0800	−0.0856	836.3677	0.0000	0.0000
240	40.3750	−0.0734	1076.1140	0.0000	0.0000
260	15.9719	−0.8885	1268.8541	0.0000	0.0000

From Fig. 4, the relationship of circle time and energy consumption also are around for inverse. When the total time is $T = 220$ s, the optimal energy is $E_1 = 3.441 \times 10^7$ J at first two stations with the circle time $T_1 = 108$ s, and the optimal energy $E_2 = 3.241 \times 10^7$ J at last two stations with the circle time $T_2 = 112$ s. Hence, the total optimal energy is $E_{min} = E_1 + E_2 = 6.682 \times 10^7$ J. Record relative parameters at two stations with the interval of 20 s, is seen as Table 2.

7 Conclusion

Energy consumption of rail transit system refers to the energy consumption of equipment, such as train traction, ventilation, air conditioning, elevators, lighting, water supply and drainage. Considering the optimizations on timetable and driving strategy are two important and closely dependent parts of energy-efficient operations. It not only regulates the fleet size and the trip time at infestations, but also determines the control sequences of traction and braking force during the trip. According to the case, the simulation results based on a Subway Line in China illustrate that the iterative search algorithm can provide good performance with energy saving.

Acknowledgments Thanks for National Postgraduate Mathematic Contest in Modeling to provide source of topic.

Research supported by the Young Scholars Development Fund of Southwest Petroleum University (No. 200331010065) and Innovation Foundation of Southwest Petroleum University (No. CXJJ2015027).

References

1. Howllet P (1990) An optimal strategy for the control of a train. J Aust Math Soc Ser B Appl Math 31(04):454–471
2. Rongfang (Rachel) L, Golovitcher LM (2003) Energy-efficient operation of rail vehicles. Transp Res Part A: Policy Pract 37(10):917–932
3. Su S, Tang T, Li X et al (2014) Optimization of multitrain operations in a subway system [J]. IEEE Trans Intell Transp Syst 15(2):673–684
4. Albrecht T, Oettich S (2002) A new integrated approach to dynamic schedule synchronization and energy-saving train control [J]. Publication of WIT Press, Southampton
5. Xin Yang, Xiang Li, Ziyou Gao, Hongwei Wang, Tao Tang (2013) A cooperative scheduling model for timetable optimization in subway systems [J]. Trans Intell Transp Syst 14(1):438–447
6. Shuai Su, Tao Tang, Xiang Li, Ziyou Gao (2014) Optimization of multitrain operations in a subway system [J]. Trans Intell Transp Syst 15(2):673–684
7. Shafia MA, Aghaee MP, Sadjadi SJ, Jamili A (2012) Robust train timetabling problem: mathematical model and branch and bound algorithm. IEEE Trans Intell Transp Syst 13(1):307C317
8. Gao SG, Dong HR, Chen Y, Ning B, Chen GR, Yang XX Approximation-based robust adaptive automatic train control:an approach for actuator saturation, IEEE Trans Intell Transp Syst (to be published)
9. Khmelnitsky E (2000) On an optimal control problem of train operation. IEEE Trans Autom Control 45(7):1257–1266
10. Cacchiani V, Caprara A, Toth P (2008) A column generation approach to train timetabling on a corridor, 4OR. Quart. J. Oper. Res. 6(2):125–142
11. Das D, Zhou S, Lee JD (2012) Differentiating alcohol-induced driving behavior using steering wheel signals. IEEE Trans Intell Transp Syst 13(3):1355–1368
12. Wang K, Shen Z (2012) A GPU-based parallel genetic algorithm for generating daily activity plans. IEEE Trans Intell Transp Syst 13(3):1474–1480
13. Mahdavi M, Fesanghary M, Damangir E (2007) An improved harmony search algorithm for solving optimization problems. Appl Math Comput 188(2):1567–1579
14. Wu Y, Close TJ, Lonardi S (2011) Accurate construction of consensus genetic maps via integer linear programming. IEEE/ACM Trans Comput Biol Bioinf 8(2):381–394

Correlation Analysis Between Risky Driving Behaviors and Characteristics of Commercial Vehicle Drivers

Niu Zengliang, Lin Miao, Chen Qiang and Bai Lixia

Abstract In order to define the factors of commercial vehicles risky driving behaviors and develop driver assistant system about commercial vehicles, this article began to screen risky driving behaviors and its factors for the questionnaire in terms of actual accident cases. We determine the sample size by statistical methods, and then we carry on the actual investigation. Based on the survey data, from perspectives of frequency and cause of accident probability, we researched the correlation between risky driving behaviors and their factors by ordinal polytomous logistic regression model. Researches show that driver characteristics have a significant impact to the judgment on the frequency and cause of accident probability of risky driving behaviors. This article stated the importance of risky driving behavior research in commercial vehicles operation from the perspective of human biology.

Keywords Commercial vehicles · Driving behavior · Correlation analysis

1 Introduction

Risky driving behavior is one of the most main reasons of the commercial vehicles fatal traffic accidents. From 2010 to 2012, our commercial vehicles have a total of

N. Zengliang (✉) · L. Miao · C. Qiang
China Automotive Technology and Research Center, No. 3 Boxing Six Rd, Economic-Technological Development Area, Beijing, China
e-mail: niuzengliang@catarc.ac.cn

L. Miao
e-mail: linmiao@catarc.ac.cn

C. Qiang
e-mail: chenqiang@catarc.ac.cn

B. Lixia
Shanghai Municipal Engineering Design General Institute (Group) CO. LTD, Shanghai, China
e-mail: 851007854@qq.com

© Springer International Publishing Switzerland 2017

V.E. Balas et al. (eds.), *Information Technology and Intelligent Transportation Systems*, Advances in Intelligent Systems and Computing 454,
DOI 10.1007/978-3-319-38789-5_74

742 cases of death more than 3 people traffic accident, which killed 3476 people and caused bad influence [1]. In 2012, in the 8?26 heavy traffic accident in Yan'an, sleeper commercial vehicle rear-end crashed a tanker, which killed 36 people, improper driving behavior is one of the main reasons for the accident [2]. Some research results show that 90.3 % of all traffic accidents related to human factors, and risky driving behavior is one of the most main factors that lead to traffic accidents. So the risky driving behavior research for the commercial vehicles has a great significance.

Around the world research scholars are studying in-depth risky driving behavior, but there are a lot of problems now. Some research institutions developed software and hardware equipment for risky driving behavior, but at the beginning of the development do not aim at the demand, which lead to the development of equipment does not play a good role. Secondly the risky driving behavior research is directed at the ordinary non-professional drivers, for the commercial driver to carry out the effective research are relatively few.

The 100-Car research at Virginia tech used five kinds of on-board channels to collect data driving behavior. Finally it is determined relationship between the driving behavior and the accident by the chi-square test. And the major effect of the special driving behavior is determined by the logistic regression model [3]. The NHTSA analyzed driving behavior difference of the high and almost no collision frequency between different drivers by quantitative and qualitative, the results showed that unsafe drivers for more driving behavior such as a sharp turn [4]. Although these two studies adopted the instruments and equipment for risky driving behavior data analysis, they are not in-depth research to the driver factor.

It is determined that driver whether is risky by the "km fault rate" in the NSTSCE. The research distinguished between individuals and demographic characteristics by these drivers completing the questionnaire and clustering [5]. While this study used the questionnaire to analysis the accident of cause driver factors, but the study did not from the behavior of the commercial vehicle drivers.

This paper listed all possible risky driving behaviors of commercial drivers by refer to relevant information. This paper carried out the plan of questionnaire, screened risky driving behavior by comprehensive factors, and determined the sample size on the basis of statistical methods. Questionnaire survey was conducted in several large passenger transport companies, and it analyzed the validity and reliability of the result of the questionnaire. Finally we sorted comprehensively importance of the risky driving behavior. According to the survey data, the regression model is adopted to risky driving behavior for correlation analysis.

1.1 Questionnaire Design

Questionnaire design principles There are several principles we should pay attention to several principles when make questionnaire. First, we will try our best to describe each kind of risky driving behavior accessibly, and the questionnaire is too. Second, keep questionnaire to a size that limits impact on drivers mood. Last, the options of questionnaire should be made by Likert, and they are based on the frequency of the smallest (or potential) [6]. According to DBQ questionnaire, we drive the questionnaire into two parts [7]: the first part is the basic information of the driver; and the second part is the core part of the questionnaire, which mainly research the frequency of dangerous driving behavior and the possibility of accident it leads to.

Table 1 Statistical analysis of the drivers characteristic (continuous data)

Code	Characteristic attribute	Mean	Variance	Minimum value	Maximum value
G25	Age	34.48	8.15	23.00	59.50
G26	Diving experience	10.74	6.71	0.50	32.00
G27	Diving years of existing car	5.26	4.58	0.25	29.00
G28	The time of accidents	1.28	2.04	0.00	10.00
G29	The time of primary liability	0.52	0.87	0.00	6.00
G30	Running hour	9.18	1.22	1.00	15.00
G31	Continuous driving time	1.67	1.02	0.50	10.00
G32	Interval of time to rest	7.56	4.66	0.00	30.00

Table 2 Statistical analysis of the drivers characteristic (categorical data)

Code	Attribute	Percentage (%)	Code	Attribute	Percentage (%)
G19-1	Male	90.00	G22-2	10–50 thousand km	12.69
G19-2	Female	10.00	G22-3	50–100 thousand km	18.46
G20-1	Junior high school degree and below	13.46	G22-4	100–300 thousand km	36.92
G20-2	High school and technical secondary school	58.08	G22-5	300 thousand km	26.54
G20-3	Junior college and undergraduate	28.08	G23-1	Weekly	66.15
G20-4	Master's and above	0.38	G23-2	Semimonthly	17.31
G21-1	Taxi	0.00	G23-3	Monthly	14.62
G21-2	Bus	96.54	G23-4	Quarterly	1.15
G21-3	Passenger vehicles	2.31	G23-5	Annually	0.38
G21-4	Freight vehicles	0.38	G23-6	Else	0.38
G21-5	Other vehicles	0.77	G24-1	Yes	82.69
G22-1	Ten thousand km	5.38	G24-2	No	17.34

Table 3 Risky driving behaviors statistics

Code	Risky driving behavior	Code	Risky driving behavior
X1	Improper parking	X10	Right turn too fast in the intersection
X2	Not driving on prescribed routes	X11	Overtaking from right side
X3	Don't give way in the no signal intersection	X12	Don't play a turn signal when turning
X4	Driving on line for a long time	X13	Don't play a turn signal when overtaking
X5	Calling when driving	X14	Speeding
X6	Overtaking is not suitable	X15	Changing lane frequently
X7	Severe brake	X16	Bus open the door without parking
X8	Following distance is too close	X17	Speeding when bus getting in
X9	Quick turning steer wheel	X18	Side by side to dock when getting in

The content of questionnaire (1) Drivers characteristic parameters

According to the drivers information in questionnaire, we regard every option as drivers characteristic attribute, which have 14 driver characteristics and 32 attribute in total after numbered [8]. The survey data of drivers characteristic can be divided into two types, one type is continuous data as Table 1, the other is ordered categorical data as Table 2.

(2) Questionnaire core content

We summarized 53 kinds of risky driving behavior through the literature research, finally we selected a total of 18 kinds of risky driving behavior by expert scoring method, as shown in Table 3. In the part of the inspection for the frequency of risky driving behaviors, the option rank from 5 to 1 is with the frequency. In the part of the inspection for the possibility of accident which risky driving behaviors lead to, the option rank from 5 to 1 is with the possibility.

2 Sample Size Calculation

2.1 Computing Method

According to the statistics, in some practical problems, if we choose properly sample size, it can effectively reduce the second category of error. (The error which accept the null hypothesis which it is fake.) Error can be divided into system error and random error. Although random error can not avoid, the more measuring, the less random error, the higher accuracy. Sample size calculation formula

$$n = \left(\frac{t^2\sigma^2}{E^2}\right)$$

t—When the confidence level is 90 %, it values 1.645;

σ^2 Population variance from the sample variance s^2;

E Absolute error, usually values around 5 %; n Minimum sample size.

Sample size selection process is as follows. In the process of the research, firstly we selected n_1 samples, calculated the variance s^2, and estimated the population variance σ^2. According to the error value and the confidence level which we needed, we calculated the required sample size n_2 by the above formula. If the n_1 is greater than or equal to n_2, it will prove the existing enough sample size, or we will still need to continue to research to improve the sample size [9].

2.2 *Calculation Process and Results*

Firstly we calculated total square deviation 2 in the two parts of questionnaire. The more variance 2, the worse consistency whose score results in this question. Then we confirmed E = 5.5 % and $t = 1.645$ by weighing research costs. Ultimately we determined the required sample size, and we calculated the sample variance by $D(X) = EX^2 + (EX)^2$. We Calculated the minimum amount of research which is $n = 251$, and had received 260 valid questionnaires now. The above conditions were determined.

3 Survey Results Analysis

Firstly, in order to ensure the accuracy of the survey data, we need for the reliability and validity analysis of the questionnaire. In the analysis of questionnaire survey, the effective questionnaire is the important condition. The reliability of the questionnaire is the requirement condition of the validity of the questionnaire.

3.1 *Inspection Survey Data*

Reliability test Research took back a total of 285 questionnaires, filtered questionnaires by the logical analysis, and got 260 valid questionnaires finally. We worked out the reliability of the questionnaire for risky driving behavior by SPSS 18.0 software, Cronbach's alpha coefficient of two parts in questionnaire was 0.959 and 0.976, and it showed that the reliability of the questionnaire is very reliable.

Validity test First it is determined whether do factor analysis by applying KMO and Bartlett's spherical test. As shown in Table 4, the KMO value of the part of the frequency is 0.945, and the Sig. of the Bartlett's test is 0.000, which is far less than 0.01. The KMO value of the part of the possibility is 0.961, and the Sig. of the

Table 4 Validity test results

	Sampling enough KMO	Inspection frequency parts 0.95	Investigate possibility parts 0.96
	Approximate chi-square	3892.20	5687.96
Bartlett	df	153	153
Sphericity test	Sig.	0.000	0.000

Bartlett's test is 0.000, which is far less than 0.01. They reached extremely significant level, and they was suitable for application of factor analysis. We tested samples by the structure validity, that is to say that we used factor analysis to identify the potential common factor, and make it orthogonal rotation. The significant correlation factors is the item of the factor loading which is more than 0.4. Table 4 shows the test results of the validity of the two parts questionnaires, and all items of questionnaire passed the validity test.

3.2 Ranking Importance

According to the result of the questionnaire, the ranking importance of the risky driving behaviors, which combined the frequency and possibility of the risky driving behaviors. These two parts are the same important, but the scores of the second part of the questionnaire are generally higher than the first part. So we made the two parts into a uniform weights so that it reach the result of comprehensive. According to the analysis, we ranked synthesis the importance of risky driving behaviors, as shown in Table 5.

Table 5 Ranking importance of risky driving behaviors

No.	Code	Comprehensive weight	No.	Code	Comprehensive weight
1	X14	0.2047	10	X13	0.1053
2	X8	0.1813	11	X16	0.1053
3	X5	0.1579	12	X4	0.0936
4	X6	0.1579	13	X17	0.0760
5	X1	0.1462	14	X9	0.0702
6	X3	0.1462	15	X11	0.0643
7	X15	0.1404	16	X12	0.0643
8	X7	0.1170	17	X18	0.0526
9	X10	0.1053	18	X2	0.0117

4 The Correlation Analysis Based on Probit Regression Model

4.1 Correlation Analysis Method

Ordinal polytomous probit regression After regression model is established, we test its parameter for Wald and eliminate little correlated explanatory variable. According to the data type of explained variable and explanatory variable, we select ordinal polytomous Probit regression [10]. The join function is Probit function, which illustrates the inverse function of explained variable following the normal distribution. Ordinal polytomous Probit regression as follows:

$$\phi^{-1}\gamma_i = \beta_j + \sum_{i=1}^{p} \beta_i x_i$$

Among: β_i—The coefficient of explanatory variables;
β_j—Constant term; x_i—explanatory variable;
γ_j—The former J category cumulative probability of explained variable;
Φ_1—Inverse function of the cumulative standard normal distribution function.

4.2 Analysis Results

Correlation analysis is based on ordinal polytomous Probit regression Wald in the process of inspection. If the results Sig value is less than 0.05, it will mean that they have obvious correlation [11]. Meanwhile, if the parameter estimation is a continuous variable, it will indicate that algebraic values plus or minus corresponds to the plus or minus of correlation. If it is classified variable, the algebraic value change rule of attribute parameter estimates corresponds to the plus or minus of classified variable. Besides, if the topic and a single attribute of classified variable have classified variable, the positive and negative correlation will be ignored. The absolute value of estimate of parameter indicates the incidence of drivers feature to significant correlation topic. Risky driving behaviors leaded to traffic accidents can reflect drivers safety consciousness.

(1) The frequency of the behavior As shown in Fig. 1, there are 17 items of the positive correlation between the first part of the questionnaire and the driver features, which involves 5 kinds of driver feature attributes; there are 9 items of the negative correlation between the first part of the questionnaire and the driver features, which involves 5 kinds of driver feature attributes.

(2) The possibility of the accident As shown in Fig. 2, there are 18 items of the positive correlation between the first part of the questionnaire and the driver features, which involves 5 kinds of driver feature attributes; there are 17 items of the negative

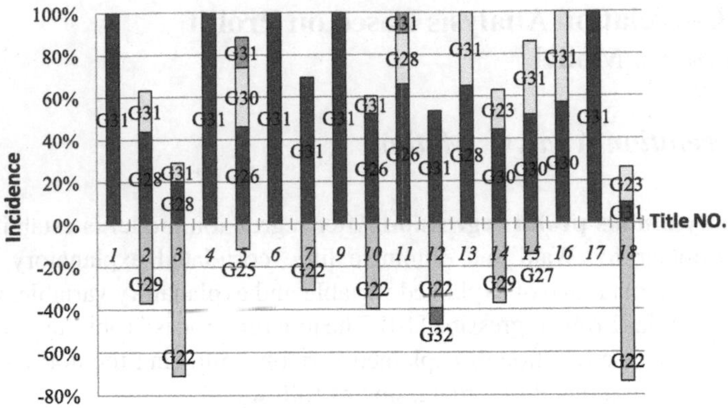

Fig. 1 Related features of the influencing frequency

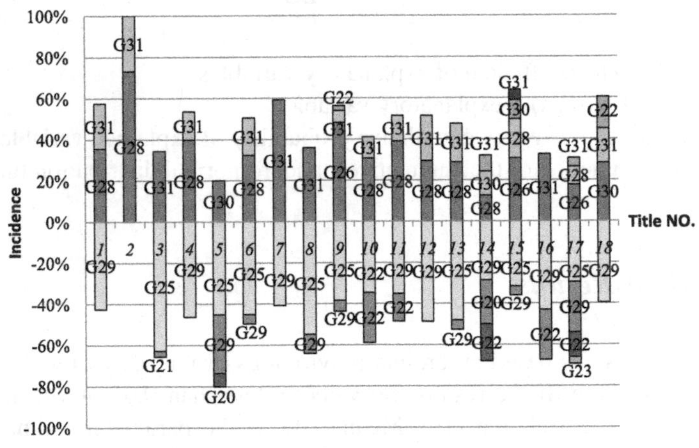

Fig. 2 Related features of the influencing accident probability

correlation between the first part of the questionnaire and the driver features, which involves 6 kinds of driver feature attributes.

5 Conclusion

The main content of this thesis is based on a survey of commercial vehicle drivers. First of all, we screen risky driving behavior, make investigation plan according to the principles of questionnaire, and then determine the minimum amount of research in terms of statistical method. Finally, we made investigation and finish correlation analyses.

(1) We studied a lot of materials in domestic and overseas and collect 53 risky driving behaviors of commercial vehicle drivers. In order to make the survey simply, we determined 18 risky driving behaviors in the survey everything together. According to statistic learning theory, we measured the cost of research, and confirmed absolute error and degree of confidence respectively are 5.5 % and 1.645. At last, we calculated sample capacity of 251, and have received 260 valid questionnaires by now, which are basic conditions for the research.

(2) The survey is highly credible and effective by testing reliability validity of the result. According to the occurrence frequency and the possibility of traffic accidents of risky driving behaviors, we ranked the importance order for risky driving behaviors.

(3) By means of studying the correlation analyses and the questionnaire data types, we decided to use ordinal polytomous Probit regression to analyze own feature dependency and factors of commercial vehicle drivers. The study found that many drivers feature are related to options of the questionnaire, which provides theoretical basis for subsequent research.

References

1. The Ministry of Public Security (2009–2012) Road Traffic Accident Statistics Report in China. The Ministry of Public Security, Beijing
2. Yaqian L (2012) Yan'an commercial vehicle rear-ends accident caused 36 deaths. http://www.jinbaonet.com/topic/2012/august/yakczw. Accessed 17 Sept 2012
3. Klauer SG (2006) Risky driving report, Virginia Polytechnic Institute and State University
4. Klauer SG, Dingus TA, Neale VL et al Comparing real-world behaviors of drivers with high versus low rates of crashes and near-crashes. National Highway Traffic, Washington, DC
5. Soccolich S, Hickman JS, Hanowski RJ (2011) Identifying high-risk commercial truck drivers using a naturalistic approach. Virginia Tech Transportation Institute
6. Qiong Z (2010) Research on the assessment of drivers' safety awareness. Changan University, Xian
7. Dongpeng Z (2013) Study of driver hazard perception and its influencing factors. Shanghai Jiao Tong University
8. Yanli M (2007) Study on characteristics of driving and its countermeasures to road safety. Harbin Institute of Technology, Harbin
9. Zhou S, Shiqian X, Chengyi P (2006) Probability theory and mathematical statistics. Higher Education Press, Beijing
10. Chunqin Z (2011) Research on traffic dynamic information dissemination strategies. Jilin University
11. Wei X (2006) SPSS data analysis

Design of Intelligent Laboratory Based on IOT

Jiya Tian, Xiaoguang Li, Dongfang Wan, Nan Li and Yulan Wang

Abstract Wisdom laboratory based on the internet of things is designed. Cloud computing is the core of this system. Campus laboratory information should be intelligent collection and transmission, intelligent processing and control, intelligent display and push. For the three aspects of "safety, energy saving and high efficiency", paper gives a lot of analysis. Give the flow chart of the subsystem. This system should effectively solve the problem of experimental course.

Keywords Wisdom laboratory · Internet of things · RFID

1 Introduction

"Internet of Things" (IOT) is a new cutting-edge technology. The core and foundation of IOT is "Internet technology" [1]. IOT is based on extension and expansion of a network technology [2, 3]. The client of IOT is extended to any goods and articles to Information exchange and communication [4]. Therefore, the definition

J. Tian (✉) · X. Li · D. Wan
Electronic Information College of Changchun Guanghua University,
Changchun 130031, China
e-mail: tianjiya@sohu.com; 42683387@qq.com

X. Li
e-mail: 843973603@qq.com

D. Wan
e-mail: 14525451@qq.com

N. Li
Electronic Engineering Department of Armored Force Technology College,
Changchun 130031, China
e-mail: 572210786@qq.com

Y. Wang
Scientific Research Department of Changchun University, Changchun 130031, China
e-mail: 14525451@qq.com

© Springer International Publishing Switzerland 2017

V.E. Balas et al. (eds.), *Information Technology and Intelligent Transportation Systems*, Advances in Intelligent Systems and Computing 454,
DOI 10.1007/978-3-319-38789-5_75

of IOT is: through information sensing equipment such as radio frequency identification (RFID), infra-red sensors, global positioning systems, laser scanners and so on to achieve intelligent identification, positioning, tracking, monitoring and management of a network technology. The development of IOT, break the people's traditional way of thinking, brought great changes to human life [5, 6].

This paper presents a wisdom laboratory design method based on IOT. Laboratory is a cluster of instruments and equipment. The respect fire, theft, water, sunscreen of wisdom laboratory projects is very important. This design aims to improve the quality of information services. RFID technology, wireless sensor networks, Zigbee technology, is used to laboratory. Construct an open, common communications laboratory platform.

2 Design of the Laboratory System

Information intelligence is the core goal of wisdom laboratory. This design is composed of information gathering intelligence, intelligent information processing, information display and push intelligent. The core technology of wisdom laboratory is information gathering intelligence, which is different with digital laboratory. The intelligence information gathering of wisdom laboratory has self-network, self-diagnosis, self-healing, uninterrupted operation ability and has high stability and high reliability. It can automatically collect laboratory information, reduce the manpower in information collection. System has accuracy and reliability of the collected information. Intelligent information processing relies on cloud computing processing power, information analysis, processing, provide strong data protection decisions [7] (Fig. 1).

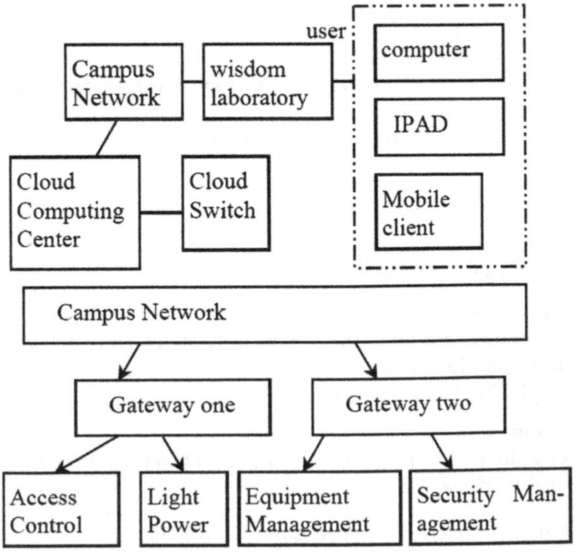

Fig. 1 Management system of wisdom laboratory based on IOT

Information display and push intelligence, using computer, tablet, mobile phones and other platforms, cross-platform, multi-channel implementation, the useful information in a timely, accurate information in the hands of managers need to push to achieve effective use of information. Using the wisdom laboratory to achieve standardization and information of management of university laboratories, laboratory equipment and laboratory supplies management. It could improve the management level and service level of experiment teaching [8, 9].

3 System Function

Wisdom laboratory is composed by the four functional systems and a comprehensive networking experience center platform. The diagram of system function is shown in Fig. 2. There are four subsystems in this system. The four subsystems are: Curriculum Experiment Management Subsystem, Open/innovative experiment management subsystem, Wisdom Laboratory Safety Management subsystem, Wisdom laboratory system log inquiry [10].

3.1 *Curriculum Experiment Management Subsystem*

Before the experiment course, students should make an appointment with this system, including the classroom and the course time. Teachers should upload the preview content and experimental procedures, then students prepare the experiment content. Administrators set the seat for each student. Students should inquire the seat information so he can find his seat accurately before the experiment course. During the experiment course, entrance guard should be open by a card has been authorized or remote control. System should statistic the students and automatically generates a

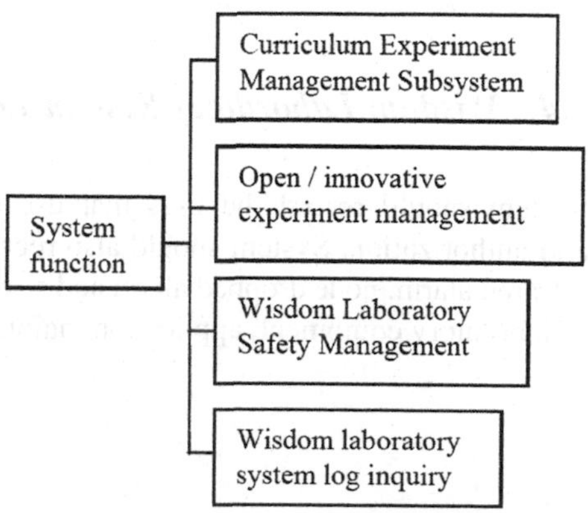

Fig. 2 Qualified parts and defective parts

time sheet for attendance students. System should monitor and record the video real timely. Control the light in the laboratory and feedback the light status. Control the on and off status of equipment power socket and feedback the status.

After the experiment course, system should submit experiment report, download the template and evaluate the report. The curriculum experiment management system could collect information intelligently, and record students word real timely.

3.2 Open/Innovative Experiment Management Subsystem

This system should arrange the seating of classroom. Make an appointment with the open laboratory which the students will use including the time, place and the experiment content. Make a card for students who will go to this laboratory. The students could use this card to using the open laboratory. The system should manage the experimental device and materials.

Students should use the lab after course. First of all, to view the need to use the lab free information, combined with their own actual situation, to submit an open experiment can be selected to the experimental classroom and time, waiting for the system audit. According to the principle of priority, the system automatically review the application submitted by students, and to the relevant students to issue a notice, the way to inform the notice, including the notice in the station letter, notice, etc. If the system cannot automatically resolve the conflict or application changes, the experimental administrator adjust manually.

3.3 Wisdom Laboratory Safety Management Subsystem

This system should manage video. Monitor the security, fire and so on. When danger happens, system should tell people the case. The flow char of this system is shown in Fig. 3.

3.4 Wisdom Laboratory System Log Inquiry

System should record the uses that use this system including login, operation and authorization. System should also record the alarm information. For example, infrared alarm, node dropped alarm and so on. System should record the use process of laboratory equipment, application maintenance and other information.

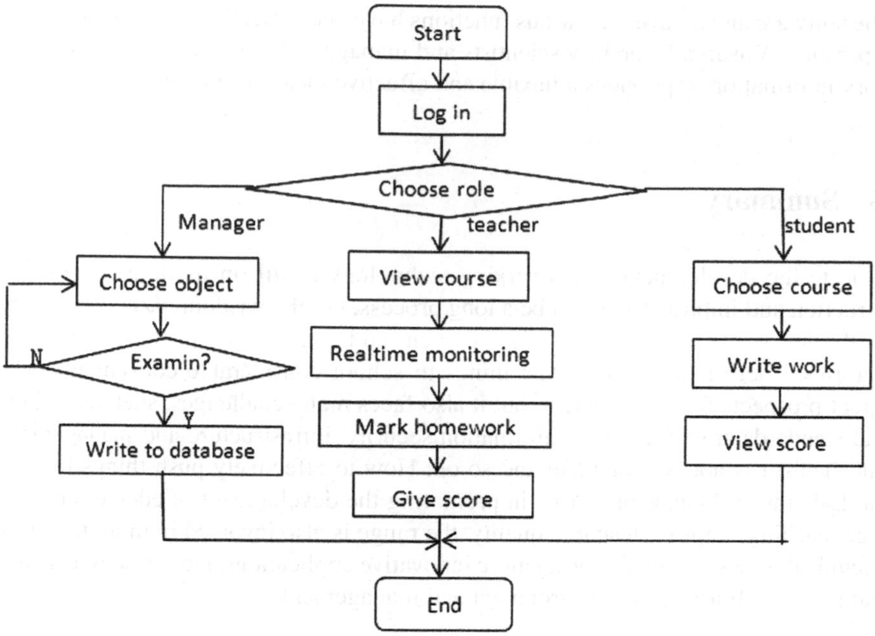

Fig. 3 Flow chart of management subsystem

4 System Deployment and Operation

Figure 4 is a monitoring subsystem interface, including the current equipment running status, environmental information, operating records and so on. The menu bar can be achieved by setting the system parameters, historical data query. Complete

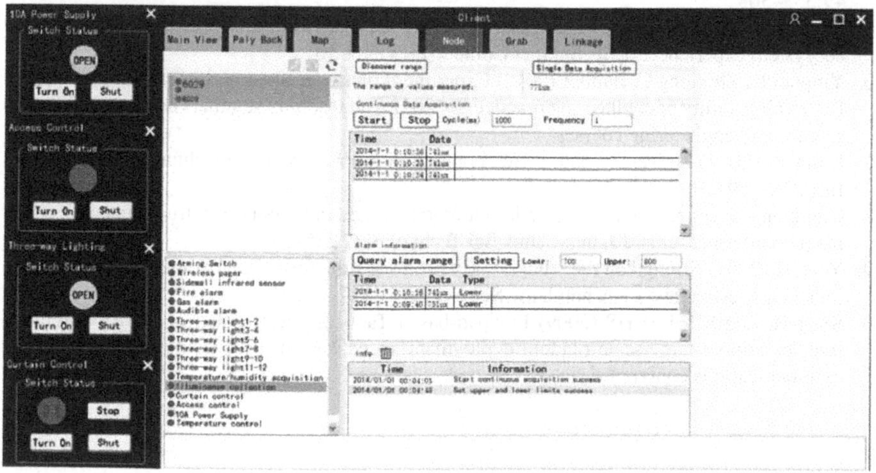

Fig. 4 Monitoring subsystem interface

the software and hardware. Various functions have been effectively tested and stable operation. Wisdom laboratory scientists and managers should easily access laboratory information. It provides a flexible and effective means of monitoring.

5 Summary

Due to the development of networking technology is still in its infancy, its construction and improvement will be a long process, but the wisdom of the laboratory is the inevitable trend of future development and construction of university laboratories. There is no doubt, bring things to school considerable convenience and great prospects for development, but it also faces many challenges, such as technical standards are not unified, information security, infrastructure and management mechanism is not yet complete and so on. How to effectively push things to play on Laboratory Management role in promoting the development of education practice teaching, improve teaching quality, the range is also involved in more than just technical issues, it will also bring more innovative applications and services, causing campus Teaching Reform environment and management.

References

1. Qiang L, Li C, Haiming C (2010) Key technologies and applications of internet of things. Comput Sci
2. Kranz M, Holleis P, Schmidt A (2010) Embedded interaction: interacting with the internet of things. In: IEEE internet computing
3. Weber RH (2009) Internet of things-need for a new legal environment. Comput Law Secur Rev
4. Amardeo C, Sarma JG (2009) Identities in the future internet of things. Wirel Pers Commun 49:353–363
5. Evan W, Leilani B, Garret C et al (2009) Building the internet of things using RFID: the RFID ecosystem experience. IEEE Internet Comput 13(3):48–50
6. Yinghui K, Zhixiong C, Jianli Z (2012) Experiment platform construction of internet of things for training ability of engineering practice. In: 2012 2nd teaching seminars on higher education science and engineering courses
7. Lijun W (2011) University-enterprise cooperation laboratory. J Shijiazhuang Vocat Technol Inst 23(6):69–70
8. Chundong X, Jun W (2011) University-enterprise united laboratory construction and development countermeasures. J Jiangxi Univ Sci Technol 32(2):35–37
9. Yuan H (2001) Strengthen the study jointly promote transformation of scientific and technological achievements. J lanzhou Univ Soc Sci 29(5):147–152
10. Song H, Lee SH, Lee HS (2009) Lowpan-based Tactical wireless sensor network architecture for remote large-scale random deployment scenarios. In: Proceeding of IEEE military communications conference

Research and Implementation for Quadrature Digital Down Converter of Low Intermediate Frequency Signal Based on FPGA

Kang Jun-min

Abstract For the receiver of Global Navigation Satellite System, quadrature digital down converter of low intermediate frequency signal can reduce the data rate and improve the real-time processing ability of baseband. But the phase quantization and amplitude quantization of existing methods may introduce new errors which could have a serious impact on the receiver's ultimate accuracy. According to the result of extensive research in the processing method for quadrature digital down converter, optimization the structure of polyphase filter, Integrated planning for the receiver with the unified optimization of RF chip and baseband circuit. Design a quadrature digital down converter structure which shared with capture mode and tracking mode, and does not introduce gain error and phase error. Simulation and application test results showed that compared with classic quadrature digital down converters, the design put forward in this pater can avoid data and phase truncation with a simple structure and faster processing speed. Under the same image frequency rejection ratio, its circuit is smaller. Practice has proved that the proposed FPGA structure is very suitable for the BeiDou Navigation Satellite System.

Keywords Satellite navigation · BeiDou Navigation Satellite System (BDS) · Digital signal processing · Quadrature down converter · FPGA

1 Introduction

With commercial commercialization of the BeiDou navigation system of China deepening in the Asia-pacific region, three kinds of commercial global navigation satellite systems coexist in the field of global positioning system, that is GPS, GLONASS and BDS. The development direction of the satellite navigation receiver emerged the three systems compatible with trend of fusion and requirements. Current research

K. Jun-min (✉)
School of Information Engineering, Chang'an University, Xi'an 710064, China
e-mail: 9578577@qq.com

© Springer International Publishing Switzerland 2017

V.E. Balas et al. (eds.), *Information Technology and Intelligent Transportation Systems*, Advances in Intelligent Systems and Computing 454,
DOI 10.1007/978-3-319-38789-5_76

focus is on how to design the three systems compatible with the receiver whose system noise is smaller and resource consumption is less.

Global navigation satellite system receiver belongs to low intermediate frequency receiver. The receiving process can be divided into two modes, capturing and tracking.

In capture mode, to evaluate the possibility of the existence of the satellite. Under low dynamic conditions, the requirements for the accuracy of the Doppler frequency is approximately 100 Hz; The precision of ranging code requirements is for half to one code; With the stronger signal energy and reduce the accuracy requirements. In tracking mode, accurate frequency of satellite tracking, parse the data code information for satellite ephemeris and satellite ephemeris resolution and the calculation of position, velocity and time. Low dynamic condition, the accuracy of Doppler frequency in micro-Hz level.

In the navigation receiver design, capture mode to tracking mode is the transformation of the signal frequency, ranging code alignment accuracy of precise process further. The most direct way is in the limited time, for limited long satellite data through multiple iterations, realize the signal frequency and ranging code terminal two-dimensional search. Until the signal frequency and the precision of ranging code terminal track conditions, the receiver into track mode.

The navigation receiver process the digital baseband signal. So RF front-end chip output signal to the baseband signal conversion, as a first step to digital baseband signal processing of the data accuracy and the precision of navigation receiver output has a direct relationship.

In the present of baseband signal processing adopts the baseband signal conversion method emphasis has two: first, is to use the high speed clock of the RF output signal sampling, through a large amount of data compensation data precision. Second, is to use the multilevel filter to filter, plastic and compensation data. Also the treatment effect and disadvantages of the two methods. The first direct consequences of using high speed clock data is required to store huge amount of data, integrated circuit caused by the waste of resources. The second by using the method of multi-stage filters need high order FIR filter, CIC filter and half band filter, contains a large number of multiplication operation, make the data flows through a long and very large scale integrated circuit logic design difficulty. Thus, to seek a balance of performance and resource consumption, and is suitable for large scale integrated circuit logic design implementation method of baseband signal, the navigation receiver has always been a hot spot in the design of integrated circuit.

Under the baseband signal contains a orthogonal frequency conversion, filter and plastic. Under orthogonal frequency conversion processing, solve the uneven IQ signal is always a difficulty. IQ imbalance is the direct source of the tracking precision of the signal frequency error. Predecessors have done a lot of work to study IQ equilibrium degree to improve the problem. Predecessors in analog signal processing, put forward the plural least mean square algorithm [1], the parameter evaluation algorithm based on signal separation feedback [2] and other methods to improve the IQ of RF output balance. In digital signal processing, the improved least mean square algorithm [3], fanaticism, separation interference elimination algorithm [4–7] interface signal estimation method, a priori information parameter extraction correction method [8, 9, 13] adaptive equalization algorithm and so on, most of these methods are using DSP implementation, good adaptability to the signal, but belong to the category of software correction, is not conducive to integrated circuit implementation.

Predecessors in order to be helpful for hardware implementation, also proposed the multi-angle resolution of multi-stage CORDIC algorithm [10, 11] look-up table method combined with CORDIC algorithm and Taylor series with CORDIC algorithm [12], the Angle of the CORDIC algorithm using fixed resolution, produce high precision by using the method of fixed number of rotating iterative mixing part of the carrier to be used under orthogonal frequency, to reduce small value quantization error and phase quantization error. CORDIC the advantage of precise produce fixed frequency signals, but for frequent changes in the frequency of the helpless. To solve the complexity of the mixer, based on polyphase filtering structure under orthogonal frequency [14, 15] have been proposed. In the polyphase filter structure, the use of symbols transformation under the simplified the orthogonal mixing part of inverter. At the same time also by optimizing the FIR filter [16, 17] and streamline FIR filter structure method to speed up the processing speed of the orthogonal separation. However, these methods still can not solve with fuzzy degree of signal sampling, filtering and signal gain error such as contradiction, even can take advantage of the compensation for the loss of half band filter to the filter gain [18].

Any of these orthogonal IQ two output signals in the frequency conversion method under balance and orthogonality depends on the degree of ideal mixer and filter, balance requires the higher requires much higher levels of local mixing the precision of frequency and amplitude, phase and precision filter order number, but the more complex on the logic implementation of integrated circuit, the demand of hardware resources.

Through research under the digital orthogonal frequency conversion algorithm, as well as unified consider RF chip baseband circuit and working mode, optimize the polyphase filter method of polyphase filter, focus from implementation put forward a kind of high-speed digital circuit, itself does not produce phase error and gain error, do not change the RF output signal IQ equilibrium degree under the baseband signal of frequency conversion method.

2 The Improved Quadrature Digital Frequency Down Conversion

Currently using polyphase filtering method under the condition of low intermediate frequency sampling rate, compared with the classical low pass filtering under quadrature frequency conversion method has had the great promotion on the speed and scale, but the polyphase filter must meet with the requirement of time delay filter branch at the same time of filter for delay correction. Although filtering char-

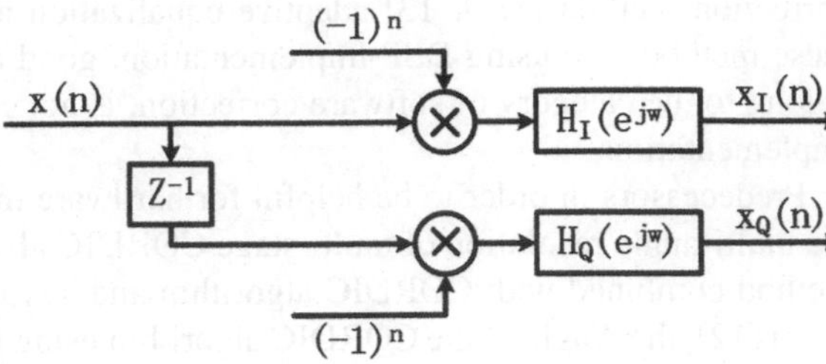

Fig. 1 One level one stage comb filter amplitude frequency response

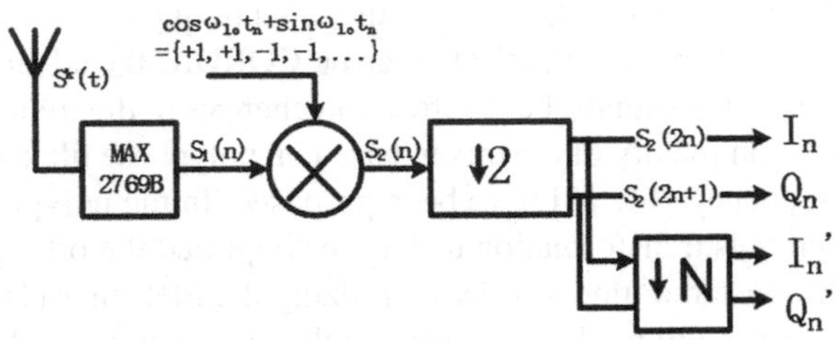

Fig. 2 Single-stage CIC decimation filter

acteristics completely consistent, to ensure the filter after the IQ balanced degree of consistency, but compared with before filter, the balance of IQ gone badly (Fig. 1).

Improved under the digital quadrature sequence and frequency conversion method by adopting the appropriate symbol transform filter structure, on the basis of the guarantee performance and processing speed and simplify the baseband signal under orthogonal frequency hardware logic design. Improve the difficulty lies in the symbol transformation sequence determination and choose the right means of filtering. After the system analysis of receiver system structure, the improved orthogonal frequency conversion principle block diagram is shown in Fig. 2.

The application of the RF front-end chip MAX2769B for global navigation system produced by MAXIM Company. Mixing part by symbol transform in the middle. Quadrature signal separation and filtering performed by odd and even sampling. Even number sequence as the in-phase branch, odd number sequence is quadrature branch. Tracking mode can be used directly parity quadrature sampling output signal. Capture mode to use N times down sampling of quadrature signal after sampling.

2.1 The Sign Changing Sequence

The BeiDou satellite navigation system of China, its civilian CA code rate is 2.046 MHz, B is 4.096 MHz bandwidth. BeiDou second generation satellite launch B1 frequency signal can be expressed in formula 1.

$$
\begin{aligned}
S^i(t) &= A_{B1I} C^k_{B1I}(t) D^k_{B1I}(t) \cos(2\pi f_1 t + \Phi^k_{B1I}) \\
&\quad + A_{B1Q} C^k_{B1Q}(t) D^k_{B1Q}(t) \sin(2\pi f_1 t + \Phi^k_{B1Q})
\end{aligned}
\tag{1}
$$

The superscript k represent different satellite number, A_{B1I}, A_{B1Q}, C^k_{B1I}, C^k_{B1Q}, D^k_{B1I}, D^k_{B1Q}, f_1, Φ^k_{B1I}, Φ^k_{B1Q} represent the B1I signal amplitude; B1Q signal amplitude; B1I signal ranging code; B1Q signal ranging code; The data code modulate on the B1I ranging code; The data code modulation B1Q ranging code; B1 signal carrier frequency; The initial phase of B1I signal carrier frequency; The initial phase of B1Q signal carrier frequency [19].

Satellite signals through the RF front-end chip for many times mixing and AD sampling, could be show like the formula 2.

$$S_1(n) = C(n)D(n)\cos\left(\frac{2\pi(f_{IF}+f_d)n}{F_s} + \phi(n)\right) + e(n) \tag{2}$$

which f_{IF} is the intermediate frequency, $\frac{n}{F_s}$ is the first n point sampling frequency sampling time. f_d is the Doppler frequency drift caused by relative motion, $\phi(n)$ is the signal of the original phase, $e(n)$ for the signal to noise;

The signal after mixing with local carrier signal expressions such as formula 3

$$\begin{aligned} S_2(n) &= C(n)D(n)\cos\left(\frac{2\pi(f_{IF}+f_d)n}{F_s} + \phi(n)\right)\cos\left(\frac{2\pi f_{LO} n}{F_s}\right) \\ &= \frac{1}{2}C(n)D(n)\left(\cos\left(\frac{2\pi(f_{IF}+f_d+f_{LO})}{F_s} + \phi(n)\right)\right. \\ &\quad \left. + \cos\left(\frac{2\pi(f_{IF}+f_d-f_{LO})}{F_s} + \phi(n)\right)\right) \\ &= \frac{1}{2}C(n)D(n)\left(\cos\left(\frac{2\pi(f_{IF}+f_d)n}{F_s} + \phi(n)\right)\cos\left(\frac{2\pi(f_{LO})n}{F_s}\right)\right. \\ &\quad \left. + \sin\left(\frac{2\pi(f_{IF}+f_d)n}{F_s} + \phi(n)\right)\sin\left(\frac{2\pi(f_{LO})n}{F_s}\right)\right) \end{aligned} \tag{3}$$

f_{LO} is the one for local produce mixed frequency; $e(n)$ to contain the high frequency component of noise.

Assuming that

$$\begin{aligned} I_n &= \frac{1}{2}C(n)D(n)\left(\cos\left(\frac{2\pi(f_{IF}+f_d)n}{F_s} + \phi(n)\right)\right) \\ Q_n &= \frac{1}{2}C(n)D(n)\left(\sin\left(\frac{2\pi(f_{IF}+f_d)n}{F_s} + \phi(n)\right)\right) \\ SignI &= \cos\left(\frac{2\pi(f_{LO})n}{F_s}\right) \\ SignQ &= \sin\left(\frac{2\pi(f_{LO})n}{F_s}\right) \end{aligned} \tag{4}$$

formula 3 can be rewritten as formula 5

$$S_2(n) = I_n \cdot SignI + Q_n \cdot SignQ + e(n) \tag{5}$$

According to the bandpass sampling theorem and Nyquist sampling theorem, when F_S, f_{LO} meet $F_S = \frac{4}{(2m+1)f_{LO}}$, at the same time $m = 0$, $SignI$, $SignQ$ respectively

$$\begin{aligned} SignI = \quad \cos\left(\frac{\pi n}{2}\right) &= \begin{cases} 0 \ ; \ n = 1, 3, 5, \ldots (2m+1) \ ; \ m \in N \\ 1 \ ; \ n = 0, 4, 8, \ldots (4m + 0) \ ; \ m \in N \\ -1 \ ; \ n = 2, 6, 10, \ldots (4m+2) \ ; \ m \in N \end{cases} \\ SignQ = \quad \sin\left(\frac{\pi n}{2}\right) &= \begin{cases} 0 \ ; \ n = 0, 2, 4, \ldots, 2m \ ; \ m \in N \\ 1 \ ; \ n = 1, 5, 9, \ldots (4m+1) \ ; \ m \in N \\ -1 \ ; \ n = 3, 7, 11, \ldots (4m+3) \ ; \ m \in N \end{cases} \end{aligned} \tag{6}$$

This formula can be drawn from formula 3 as

$$S_2(n) = \begin{cases} I_n SignI + e(n) \ ; \ n = 0, 2, 4, \ldots (2m) \ ; \ m \in N \\ Q_n SignQ + e(n) \ ; \ n = 1, 3, 5, \ldots (2m+1) \ ; \ m \in N \end{cases} \tag{7}$$

Continue to simplify the type, and in accordance with the sample order. You can get

$$S_2(n) = \begin{cases} +1 \cdot I_n + e(n) \ ; \ n = 0, 4, 8, \ldots (4m) \ ; \ m \in N \\ +1 \cdot Q_n + e(n) \ ; \ n = 1, 5, 9, \ldots (4m+1) \ ; \ m \in N \\ -1 \cdot I_n + e(n) \ ; \ n = 2, 6, 10, \ldots (4m+2) \ ; \ m \in N \\ -1 \cdot Q_n + e(n) \ ; \ n = 3, 7, 11, \ldots (4m+3) \ ; \ m \in N \end{cases} \tag{8}$$

After mixing, the signal is equivalent to the results of sequence $[+1, +1, -1, -1]$ multiplied with the RF output signal. That is:

$$S_2(n) = S_1(n) \cdot M \ ; \ M \in [+1, +1, -1, -1] \tag{9}$$

M for periodic sequence $[+1, +1, -1, -1]$, $e(n)$ is noise signal. $S_2(n)$is still contains the high frequency and low frequency components, the high frequency part contained in $e(n)$ as part of the noise.

Symbol transform implement mixer by changing the sign bit of RF signal, don't operations signal amplitude, so don't introduce new phase error and new gain error. At the same time, the Symbol transformation method is very suitable for very large scale integrated circuit logic design, replaced adding operation module by the NAND gate and inverter, to increase the processing speed and reliability.

2.2 The Improved Filter

The previous section get the symbolic transformation sequence, and now, we would be discuss how to design the filter which after the mixer.

The formula 3 told us that the high part of image frequency is in the Nyquist frequency range of the sample frequency, when both of the intermediate frequency and the local carrier frequency equal the quarter of sample frequency, and the sample frequency is more than double of bandwidth.

After the Symbols transformation, the in-phase branch signals with quadrature branch has one sample point difference. In the poly phase filter method, the poly phase filter is to compensate the time delay difference, filtering the high frequency component of mirror image. According to the theory of poly phase filter, poly phase filter branch can be seen as fractional delay FIR filter. Compare with CIC filter, FIR filter need more large scale integrated circuit resource, so we implement the image frequency filter use the CIC filter which suitable the very large integrate circuit better.

CIC's transfer function contained the integrator transfer function and comb filter transfer function

$$H(z) = H_I^N(z) \cdot H_C^N(z) = \frac{1}{M}\left(\frac{1 - z^{-M}}{1 - z^{-1}}\right)^N = \left(\frac{1}{M}\sum_{k=0}^{M-1} z^{-k}\right)^N \tag{10}$$

The corresponding frequency response

$$H(e^{j\omega}) = \left(\frac{\sin\frac{\omega M}{2}}{M\sin\frac{\omega}{2}} \cdot e^{-j\omega\left(\frac{M-1}{2}\right)}\right)^N \tag{11}$$

N for CIC filter series, M is sampling coefficient; equal to $2f; f$ is the signal frequency. From frequency response expressions can clear see CIC filter has obvious low-pass characteristic. At the same time, the frequency of integer times Fs/M has the minimum value in frequency response curve, the Fs is the sampling frequency. As a result, the simplest CIC filter that could satisfy the requirement of low pass filtering performance is 1 series, the sampling rate M is 2, the frequency response curve as Fig. 3.

Therefore, after the symbol transform the high frequency component of image frequency close to the minimum field of CIC filter frequency response, can achieve maximum attenuation.

One level of CIC filter principle diagram is as Fig. 4.

Figure 4 show the single-stage CIC filter for filtering the Nyquist frequency. Single-stage CIC filter contains two adder and a double sampling. In very large scale integrated circuit logic design, the maximum value is unchanged, so, after each level adder need a gain for 1/2 of the amplifier, to keep the data range does not

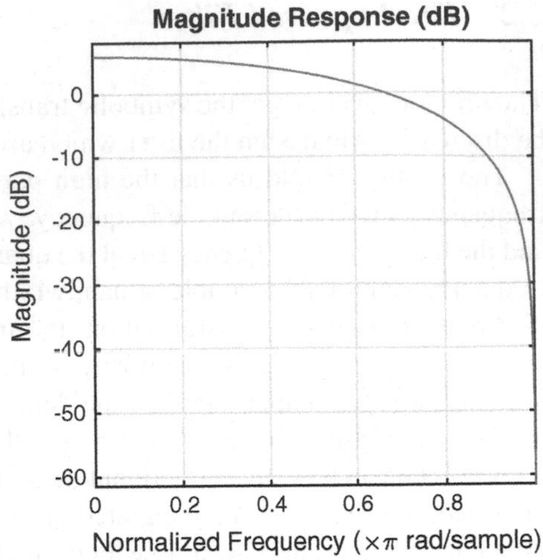

Fig. 3 One level one stage comb filter amplitude frequency response

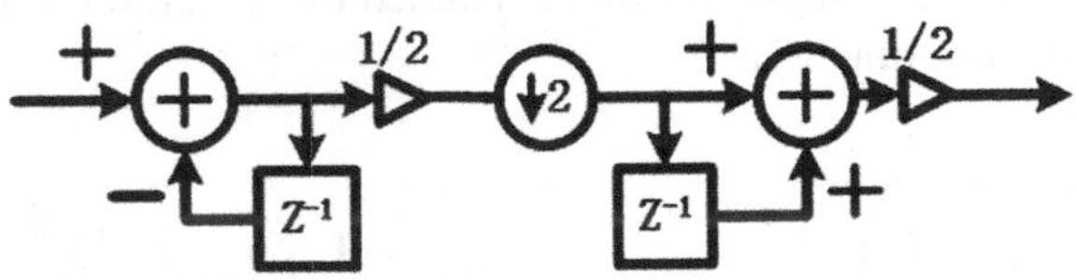

Fig. 4 Single-stage CIC decimation filter

change. In very large scale integrated circuit logic design, 1/2 gain amplifier using data moves to the right a bit.

The signal of symbol transformation by formula 8 is the in-phase branch and quadrature branch alternate combination. In classic of poly phase filter, the first step is signal separation, then after sign changing in-phase and orthogonal branch should be

$$\begin{aligned} S_{2I}(n) &= S_1 \cdot m \ ; \ m \in [+1, 0, -1, 0] \\ S_{2Q}(n) &= S_1 \cdot m \ ; \ m \in [0, +1, 0, -1] \end{aligned} \tag{12}$$

Processing by poly phase filter, the final output signal sampling frequency changed to half of the original signal, the signal energy reduced to a quarter.

As we know, signal accumulation and intercept of poly phase filter must introduce the amplitude and phase errors. In other words, if no accumulation and intercept, then should no amplitude and phase errors.

Now, we have the symbolic transformation sequence, we can use it to mix the signal before the parity separation, and then, we can use double sampling directly extracted the in-phase branch and orthogonal branch from the signal, at the same time, we also implement the low pass filter for two branch signals, so, there are only Doppler frequency include in them.

The final structure diagram of down conversion is shown in Fig. 2. Replace the processing order of classic poly phase filter by mix and then parity separation. Using parity separation as the low pass filter is for in-phase branch and orthogonal branch. The simplified quadrature frequency completely avoids the addition, multiplication and intercept. Thus, data achieved without changing the original signal amplitude, without introducing the phase and amplitude errors.

3 The Model of Improved Quadrature Digital Down Conversion

The image frequency of the quadrature signal processed by quadrature digital down conversion consist of the system noise of RF front end, the Doppler frequency, the amplitude error and phase errors introduced by the processing procedure of quadrature digital down conversion.

Phase error and amplitude error, respectively show as

$$\begin{aligned} \Delta\omega &= \tfrac{2\pi f_d}{F_s} \\ \Delta A &= \sin(\Delta\omega) \end{aligned} \tag{13}$$

The largest image frequency rejection ratio

$$ImageFreqRejectRatio = \left| \frac{1 + (1 + \Delta A) \cdot e^{i\Delta\omega}}{1 - (1 + \Delta A) \cdot e^{i\Delta\omega}} \right| \tag{14}$$

When the RF front-end ADC sampling frequency is 16.368 MHz, the change of Doppler frequency range of plus or minus 10 KHZ, phase error and amplitude error on the theory of mirror frequency inhibition is shown in Fig. 5.

In order to verify the theoretical calculation results, the test is studied by using the method described in this article: 5 KHZ frequency offset load to the BeiDou satellite navigation system of B1 frequency point, after acquisition of RF front-end chip output signal, after dealing with the quadrature digital down converter design in this

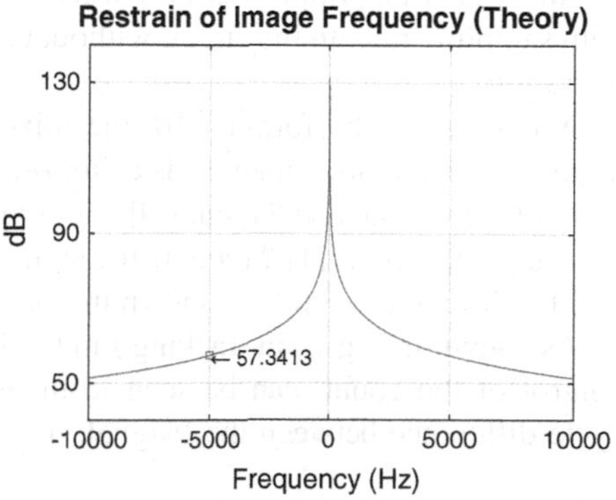

Fig. 5 Image-rejection spectrum diagram

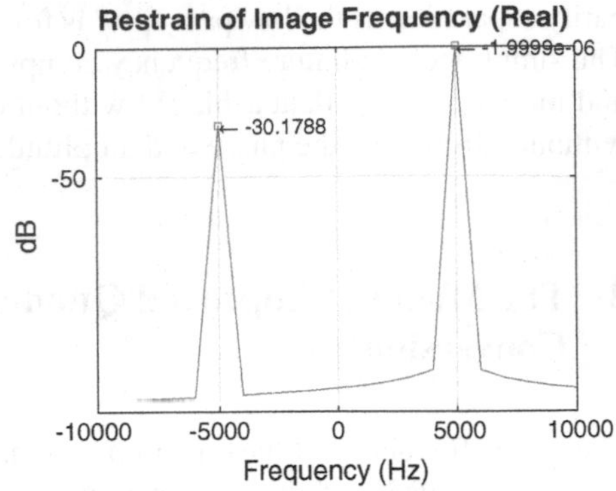

Fig. 6 Image and real spectrum diagram

paper, the results are FFT operation, after the analysis of synthetic signal spectrum, calculation image rejection ratio. Figure 6 for FFT spectrum after operation.

From Fig. 5, Fig. 6 of mirror frequency rejection ratio of 57.34 and 60.36 dB respectively, the actual simulation of mirror frequency rejection ratio is superior to the theoretical calculation of the mirror frequency rejection ratio, analyzing the reason for this is that in theory of mirror frequency rejection ratio and phase error in the calculation is the maximum gain error, are not the maximum value in the practical work.

4 FPGA Implementation of the Improved Quadrature Digital Down Conversion

The result of the improved quadrature digital down conversion circuit synthesized by the VIVADO of the XILINX company which version is 2014.4, as shown in Fig. 7. When the input signal is 1 bit quantitative sign, magnitude, there has nine registers (pictured with characters of FD device) and a small amount of combinational logic. Resource size less than one percent with a same wide classic down conversion circuit, but the signal amplitude can be lossless transmission, and the entire circuit work with sampling rate. In the circuit without using multiplication and addition and shift operation.

According to the formula 10, the mixer function implemented by the Verilog hardware description language is as follows:

iout[1:0] = count[1] ? (count[0] ? iout : sign,mag): sign,mag;

qout[1:0] = count[1] ? (count[0] ? sign,mag : qout): sign,mag;

The simulation results as shown in Fig. 8.

As shown in Fig. 8, in tracking mode, the input signal signs and mag under the control of the count, can be seen from the diagram, there has a sampling clock cycle difference between the 'isignal, qsignal' on the time. After synchronization,

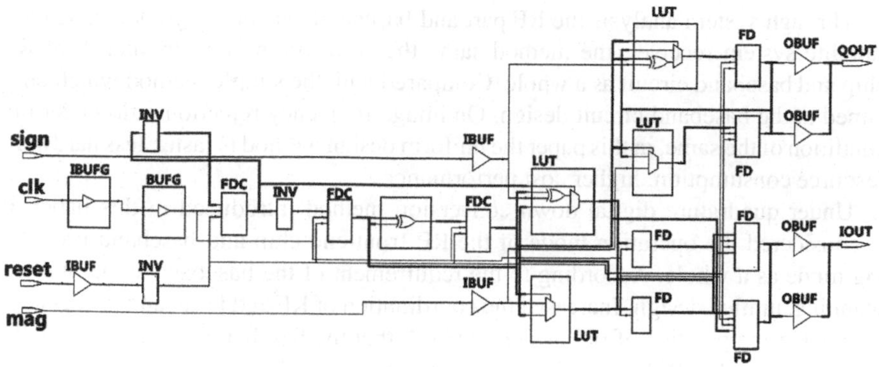

Fig. 7 Image-rejection spectrum diagram

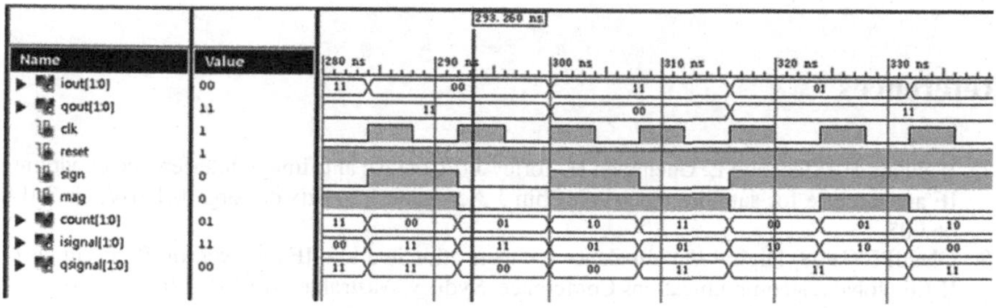

Fig. 8 Image and real spectrum diagram

the output signal 'iout, qout' at a rate of Fs/2. Pseudo-random code signal resolution of the satellite is Fs/2. 046e3/2 per code. When using 16.368 MHz as sampling frequency, the resolution of the satellite pseudo random code can achieve 4 percode.

5 Conclusion

The improved quadrature digital down conversion, making the global navigation satellite system receiver using the same method with capture mode and tracking mode, when the working mode from capture to track the data accuracy is higher, reducing the workload of capture the results of validation, eventually reduce the receiver positioning time. The improved quadrature digital down conversion, do not use multiplication, addition, no introduce new amplitude error; do not use the direct digital frequency produces, no introduce the new phase error. To ensure the frequency consistency between the baseband signal and the satellite signal, the frequency tracking precision is improved. At the same time, the improved quadrature digital down conversion can suitable for all kind of the global navigation satellite system, such as the GPS, the GLONESS and the BDS of China.

Through system analysis the RF part and baseband part of the global navigation satellite system receiver, the method same the work timing and frequency of RF chip and baseband circuit as a whole. Compared with the simple method which only aimed at the baseband circuit design. On image frequency rejection ratio under the condition of the same, in this paper the uniform design method is faster, less hardware resource consumption, higher cost performance.

Under quadrature digital down conversion method introduced in this paper, it is considered the operation mode of the RF front-end chip and baseband processing mode as a whole. According to the requirement of the baseband signal unified planning, unified design. Increased the coordination of RF and baseband processing, avoids the malpractice of old design method, that the baseband processing blindly accommodate the RF front-end, processing efficiency is low, the circuit of the existence of large scale.

References

1. Weidong H, Xiaochun L, Chengyan H, Torley J (2014) An anti-image interference quadrature IF architecture for satellite receivers. Chin J Aeronaut. http://dx.doi.org/10.1016/j.cja.2014.06.005
2. Glas JPF (1998) Digital I/Q imbalance compensation in a low-IF receiver. In: Proceedings of IEEE global telecommunications Conference, Sydney, Australia, pp 1461–1466. doi:10.1109/GLOCOM.1998.776582
3. Valkama M, Salminen K, Renfors M (2002) Digital I/Q imbalance compensation in low-IF receivers: principles and practice. Proc Int Conf Digit Signal Process 2:1179–1182. doi:10.1109/ICDSP.2002.1028303
4. Valkama M, Renfors M, Koivunen V (2001) Advanced methods for I/Q imbalance compensation in communication receivers. IEEE Trans Signal Process 49:2335–2344. doi:10.1109/78.950789
5. Valkama M (2001) Advanced I/Q signal processing for wideband receivers: models and algorithms, Ph. D. dissertation, Tampere Univ. Technology, Tampere, Finland
6. Elmala MAI, Embabi SHK (2004) Calibration of phase and gain mismatches in Weaver image-reject receiver. IEEE J Solid-state Circuits 39(2):283–289. doi:10.1109/JSSC.2003.821779
7. Windisch M, Fettweis G (2004) Blind I/Q imbalance parameter estimation and compensation in low-IF receivers. In: Proceedings of IEEE ISCCSP'04, pp 75–78. doi:10.1109/ISCCSP.2004.1296222
8. Windisch M, Fettweis G (2004) Performance analysis for blind I/Q imbalance compensation in low-IF receivers. In: Proceedings of IEEE ISCCSP'04, pp 323–326. doi:10.1109/ISCCSP.2004.1296292
9. Yu L, Snelgrove WM (1999) A novel adaptive mismatch cancellation system for quadrature IF radio receivers. IEEE Trans Circuits Syst II 46:789–801. doi:10.1109/82.769786
10. Curticapean F, Niittylahti J (2001) An improved digital quadrature frequency down-converter architecture. In: Proceedings of 35th Asilomar conference on signals, systems, and computers, pp 1318–1321. doi:10.1109/ACSSC.2001.987704
11. Nahm S, Han K, Sung W (1998) A CORDIC-based digital quadrature mixer: comparizon with a ROM-based architecture. In: Proceedings of IEEE international symposium on circuits and systems, pp 385–388. doi:10.1109/ISCAS.1998.698871
12. Fu D, Wilson AN (1998) A high-speed processor for digital sine/cosine generation and angle rotation. In: Proceedings of 42nd asilomar conference on signals, systems and computers, pp 177-181. doi:10.1109/ACSSC.1998.750849

13. Gerlach K, Steiner MJ (1997) An adaptive matched filter that compensates for IQ mismatch errors. IEEE Trans Signal Process 45:3104–3107. doi:10.1109/78.650274
14. Bin Z, Hui-min L (2008) Secondary frequency conversion design for wideband digital receiver based on polyphase filtering. Telecommun Eng 48(9)
15. Shao-long P, Zhi-gang M (2013) A new high-speed variable-bandwise digital down-converter based on FPGA. Electronic Sci Tech 9:106–106
16. Po L, Shao-ying S (2008) Implementation of high-speed fir filter in quadrature digital down-convert 8:136–139
17. He-zhong Y, Bei-sheng L et al (2010) An efficiently improved structure for quadrature digital down converter. Electron Des Eng 9:186–189
18. Xiang L, Dong-yi L (2006) Research on decimation technique of digital downconversion. J Univ Electron Sci Tech China 4:471–473
19. BeiDou/Global Navigation Satellite System (GNSS) receiver indepedent exchange format, China satellite navigation systems management office, vol 12 (2013). http://www.beidou.gov.cn/attach/2015/10/30/5.pdf

A Novel Meta-Heuristic Combinatory Method for Solving Capacitated Vehicle Location-Routing Problem with Hard Time Windows

Ali Asghar Rahmani Hosseinabadi, Fataneh Alavipour, Shahaboddin Shamshirbnd and Valentina E. Balas

Abstract Capacitated location-routing problem (CLRP), is one of the new research areas in distribution management. This topic combines two problems: locating the facilities and vehicle routing. The goal of CLRP is to open a set of depots, allocating the costumers to depots and then to design the vehicle tours in order to minimize the overall cost. The limitations of time windows has many applications in the real world, however it has not been noticed enough in the CLRP problem. This article considers the capacitated location-routing problem with hard time windows (CLRPHTW). In this paper, first a mixed-integer linear programming model for CLRPHTW problem is presented and then in order to solve this problem a meta-heuristic method based on variable neighborhood search algorithm is proposed. To assess the performance of the proposed method, this framework is examined with a set of examples. The computational tests demonstrate the efficiency of the proposed method.

Keywords Location · Vehicle routing · Time window · Meta-heuristic method · Variable neighborhood search · Optimization

---

A.A.R. Hosseinabadi
Department of Computer, Technical and Vocational University, Behshahr Branch, Behshahr, Iran
e-mail: A.R.Hosseinabadi@iaubeh.ac.ir

F. Alavipour
Department of Electrical, Computer and Biomedical Engineering, Qazvin Islamic Azad University, Qazvin Branch, Qazvin, Iran
e-mail: F.Alavipour@qiau.ac.ir

S. Shamshirbnd (✉)
Faculty of Computer Science and Information Technology, Department of Computer System and Technology, University of Malaya, 50603 Kuala Lumpur, Malaysia
e-mail: Shahab1396@gmail.com; Shamshirband@um.edu.my

V.E. Balas
Department of Automation and Applied Informatics, Aurel Vlaicu University of Arad, Arad, Romania
e-mail: Valentina.balas@uav.ro

© Springer International Publishing Switzerland 2017

V.E. Balas et al. (eds.), *Information Technology and Intelligent Transportation Systems*, Advances in Intelligent Systems and Computing 454,
DOI 10.1007/978-3-319-38789-5_77

1 Introduction

In the last two decades, the competition between firms for supply of goods and services, has become a serious fact in their progress path. Today firms and factories require unified and flexible production activities from preparation of raw material to delivery of products to the consumers. Distribution and support systems are part of the supply chain, and the producers need to plan, run and control the efficient flow of goods' storage, services and the related information from the start point to consumption [1].

Location-routing problem (LRP), is one of the important and practical problems in distribution and logistics management. LRP is a combination of two problems: locating the facilities and vehicle routing that considers these two problems at the same time. If the problems of locating and routing are not considered at the same time, costs of supply chain will increase. Both of these problems involve the complexities of NP-hard problems, thus LRP is a problem which involves time complexities in the form of NP-hard problems. Therefore solving the LRP problem in large scale using exact methods is difficult and almost impossible [1], and in most of the conducted research, heuristic and meta-heuristic methods are developed to solve these this kind of problems.

In LRP problem, the capacity constraint includes depot or the vehicle (not both of them). Recently some of the researchers have studied the LRP problem with capacity constraint both in depot and in the vehicle [2]. This subject is called capacitated location-routing problem (CLRP). Prins et al. [2] in their paper, have proposed a mathematical model for CLRP. They have proposed a solution for CLRP by combining Greedy Randomized Adaptive Search Procedure (GRASP), a learning procedure and unifying the routs mechanisms. In other paper, they also have used a Memetic algorithm with population management. In this algorithm, the solution is improved by using local search methods and replacing the conventional mutation operator with a dynamic population management technique based on distance to population [3].

Barreto et al. [4] used a heuristic method based on customer classification to solve CLRP. They used several hierarchical and non-hierarchical methods for classification. Prins et al. [5] used Lagrangean Relaxation and Tabu Search methods to develop a two phase approach for solving CLRP. This algorithm alternatively exchange the information between a locating phase and a routing phase. In the first phase, the routs and their customers are integrated in the form of a super customer and the problem is transformed into a facility locating problem. Lagrangean relaxation is used to deal with the allocation constraint in the facility locating problem. In the second phase, Tabu Search method is used to develop the routing of the several obtained depots from the first phase.

In [6] a combined method is used based on particle swarm optimization algorithm to solve the problem. In [7] (GRASP+ELS) algorithm which is a combination of greedy randomized adaptive search procedure and evolutionary local search is employed for CLRP problem. Yu et al. [8] used heuristic simulated annealing algorithm to solve CLRP problem. In order to improve the performance of the simulated annealing algorithm they used three neighborhood structures. They claimed that using this structure will improve the simulated annealing algorithm to solve CLRP problem. They used three neighborhood structures with probable selection to improve the performance of the simulated annealing algorithm. Recently Nguyen et al. [9] in their paper used greedy randomized adaptive search procedure with learning process to solve CLRO in two levels. In other research, they also solved the same problem using a method based on iterative neighborhood search algorithm and improved the results of their previous research [10].

Although time window constraint has numerous applications in real world, little attention has been paid to it in CLRP problem. Jabal-Ameli and Ghaffari-Nasab [11] has modeled the location-routing problem in both hard and soft time windows and by using a simple example has shown the credibility of their model. Nikhbakhsh and Zegardi has studied the two level location-routing problem with soft time window constraint. The authors of this article have employed a four-index mathematical planning model and a heuristic method in order to solve this problem. By using the relaxation Lagrangean method, they also obtained a lower bound for the problem and compared it with the results of their heuristic algorithm.

The main purpose of this article is to add the time window constraint to the CLRP problem in order to obtain a more practical solution. The time window constraint is taken from the vehicle routing problem with time window constraint (VRPTW). This subject is called capacitated location-routing problem with hard time window (CLR-PHTW). In this article first the mathematical modeling of the problem is discussed. Next, considering the complexity level of the problem solution and inability of the exact methods in solving problems with average and large dimensions in reasonable time, a meta-heuristic algorithm based on variable neighborhood search is proposed to solve the problem. The novel features of this article, is addition of waiting time to the mathematical model in [11] for the capacitated location-routing problem with hard time windows and designing a meta-heuristic algorithm to solve this problem.

The rest of the article is organized as it follows: in the second section, the problem is defined and its mathematical model is presented. In the third section, the proposed algorithm and its details are explained. In the fourth section, the results of the computational experiments are illustrated and finally the proposed algorithm and the conclusion are described.

2 Problem Definition and Its Mathematical Model

In the capacitated location-routing problem with hard time windows, a set of customers and a set of candidate points for construction of depots with predetermined geographical coordination exist. The customer demand and capacity of depots is known. All the transportation vehicles have the same capacity. Every customer is assigned only one depot and one vehicle in order to receive his order, therefore constraints for depots capacity and vehicle capacity should be considered.

Every vehicle starts its tour from the depot and after delivering the service to the assigned costumers, returns to the same depot. Every costumer has a predetermined time interval $[e_i, l_i]$, where e_i is the soonest time and l_i is the latest time to receive the service from the vehicle. In the case of hard time windows, the customer receive the service only in the determined interval and if the vehicle arrives sooner that the start time, it should wait until the time window opens. The purpose is to minimize the costs and to determine the suitable location for construction of depots and planning the tours of the vehicles for each depot; such that the all customers' demands is provided [5].

In order to explain the mixed integer linear programming (MILP) of the problem, first the symbols and parameters of the mathematical model are explained, and then by adding the considerations about the interval of the time windows to the capacitated location-routing problem of Prins et al. [5], the problem under study is modeled.

Sets:

I The set of depot candidate points
J The set of customers
V The set of all points $I \cup J = V$
K The set of vehicles

Model Parameters:

d_j Demand of Jth customer
O_i Cost of opening the candidate depot i
F_i Cost of using a vehicle in candidate depot i
C_{ij} Cost of traveling from point I to point j
t_{ij} Time of travelling from point i to j
Cap_i Capacity of candidate depot i
Q Capacity of each transportation vehile
T Maximum time during a tour
e_i Soonest time to deliver service to point i
l_i latest time to deliver service to point i
St_i Time duration of delivering service the customer j
M A large number

Decision variables:

At_i^k Time required for vehicle k to arrive to point i
W_i Waiting time in point i
y_i If candidate depot I is opened, one and otherwise zero
f_{ij} If the demand of customer j is provided by depot I, one and otherwise zero
X_{ij}^k If vehicle K goes directly from point I to point j, one and otherwise zero

Considering the above mentioned points, the mixed integer linear programming of CLRPHTW is presented as below:

$$\text{Minimize} \quad \sum\nolimits_{i\in I} Oi + yi \sum\nolimits_{i\in V} + \sum\nolimits_{j\in V} + \sum\nolimits_{k\in K} + CijX_{ij}^{k} + \sum\nolimits_{i\in I}\sum\nolimits_{j\in I}\sum\nolimits_{k\in K} FiX_{ij}^{k} \quad (1)$$

Subject to:

$$\sum\nolimits_{i\in V}\sum\nolimits_{k\in K} X_{ij}^{k} = 1, \qquad \forall j \in J \quad (2)$$

$$\sum\nolimits_{i\in V}\sum\nolimits_{j\in J} d_j X_{ij}^{k} \leq Q, \qquad \forall k \in K \quad (3)$$

$$\sum\nolimits_{j\in J} d_j Y_{ij} \leq cap_i y_i, \qquad \forall i \in I \quad (4)$$

$$\sum\nolimits_{i\in V}\sum\nolimits_{j\in v} t_{ij} X_{ij}^{k} \leq T, \qquad \forall k \in K \quad (5)$$

$$\sum\nolimits_{i\in s}\sum\nolimits_{j\in s} X_{ij}^{k} \leq |S| - 1, \qquad \forall S \subseteq J, k \in K \quad (6)$$

$$\sum\nolimits_{j\in V} X_{ij}^{k} - \sum\nolimits_{j\in v} X_{ji}^{k} = 0, \qquad \forall i \in V, k \in K \quad (7)$$

$$\sum\nolimits_{i\in I}\sum\nolimits_{j\in J} X_{ij}^{k} \leq 1, \qquad \forall k \in K \quad (8)$$

$$\sum\nolimits_{m\in V} X_{im}^{k} + \sum\nolimits_{h\in V} X_{ih}^{k} \leq 1 f_{ij}, \qquad \forall i \in I, j \in J, k \in K \quad (9)$$

$$At_i^k + St_i + W_i + t_{ij} - At_j^k \leq (1 - X_{ij}^k)M \qquad \forall i \in I, j \in J, k \in K \quad (10)$$

$$e_j \leq At_j^k + W_j \leq l_i, \qquad \forall j \in J, k \in K \quad (11)$$

$$At_i^k + W_j + St_i + t_{ij} - l_i \leq (2 - y_1 - X_{ij}^k) \times M, \qquad \forall i \in I, j \in J, k \in K \quad (12)$$

$$At_i^k = 0, \qquad \forall i \in I, k \in K \quad (13)$$

$$W_i = 0, \qquad \forall i \in I \quad (14)$$

$$At_i \geq 0, \qquad \forall i \in V \quad (15)$$

$$W_i \geq 0, \qquad \forall i \in V \quad (16)$$

$$f_{ij} \in \{0, 1\}, \qquad \forall i \in I, j \in J \quad (17)$$

$$y_i \in \{0, 1\}, \qquad \forall i \in I \quad (18)$$

$$X_{ij}^k \in \{0, 1\}, \qquad \forall i \in I, j \in V, k \in K \quad (19)$$

In this model, the objective function (1), minimizes the sum of depot costs and routing costs which routing costs include vehicle and travelling costs. Constraint (2) guarantees that every customer belongs to only one rout.

Constraints (3) and (4) are vehicle and depot capacity constraints. The maximum time for travelling a tour is shown by expression (5). Constraint (6) is for eliminating the sub tours. Constraint (7) guarantee the tours are continuous and constraint (8) makes sure that every vehicle returns to the same depot from which it started the tour. Expression (9) states that every customer is assigned to one depot, if there is a tour that connects these two together.

Expression (10) shows the arrival time to one customer and to the next customer. Constraints (11) and (12) guarantee that time windows of customers and depots are not violated. Constraints (13) and (14) determine the initial conditions of the waiting time and arrival time variables for the depots. Constraints (15) and (16) make sure that decision variables of arrival time and waiting time are positive and finally constraints (17)–(19) state that other decision variables are binary. After defining the problem and presenting its mathematical model, we propose a meta-heuristic method for the solution.

3 The Proposed Solution Method

The proposed approach for solving CLRPHTW is a method based on variable neighborhood search meta-heuristic algorithm. In the proposed algorithm, first generalized Push-Forward Insertion Heuristic algorithm is used to generate the initial solution for start of the proposed algorithm. The initial solution (S_0) is used as the current solution for the algorithm. Now the main algorithm starts based on the variable search neighborhood algorithm. In order to have a better search of the solution space, the simulated annealing algorithm is used in the framework of the variable neighborhood algorithm. In the following the components of the new method are explained in detail and in the end, the structure of the proposed algorithm is presented.

3.1 Variable Neighborhood Search Algorithm

Variable Neighborhood Search (VNS) algorithm first was introduced by Mladenovi'c Hansen [12] in year 1997. Its pseudo code is provided in Fig. 1. The main idea of this algorithm is to change the neighborhood structure during the search in order to prevent trapping into local minimum. Because of simple implementation and acceptable quality of the obtained results, this algorithm was soon recognized as a good method for solving optimization problems. VSN algorithm starts the optimization procedure by generating initial solution, defining neighborhood structures and using a method for searching the neighborhood. The neighborhood structures of the algorithm are shown by N_l, $l = \{1, 2, \ldots, l_{max}\}$, where N_l is the lth neighborhood. After defining the possible neighborhood structures, the order is determined. Two important points at this stage are "selection of appropriate neighborhood structures and determination of suitable structure order" (for example order based on size of the neighborhood structures). Generating initial high quality solution, defining neighborhood structure, determining neighborhood order and using the suitable method for local search, are important factors which affects the quality of the solution results. The algorithm starts searching by using the initial solution (S_0) and until the termination criteria is met, the main loop of the algorithm is iterated. The main loop of the VSN algorithm includes two phases: moving phase and local search phase.

In the shaking phase, using the lth neighborhood structure, the algorithm moves from the current solution to the neighborhood solution (S'). In the local search phase

Fig. 1 Pseudo code for variable neighborhood search algorithm

```
Input: a set of neighborhood structures N_l, l = 1,2,...,l_max
S=generate initial solution ( );
Repeat
        l = 1 ;
    While (l ≤ l_max)
    {
            S' =Shaking (S, N_l)
            S'* =Local search ( S' )
         if f(S'*) < (f(S)
                S ← S'*
                l = 1 ;
         else
                l = l + 1;
    }
Until stopping condition are met ;
Output : The best solution ;
```

by applying local search methods, searching is performed on the S' solution until local optimum is obtained. Now in the moving phase, if the optimum solution obtained is better than the current solution S, will replace it and the search returns to N_l, otherwise the next neighborhood structure $(N_l + 1)$ is used to continue the search. The search is continued until $<l_{max}$. Figure 1 shows the pseudo code for variable neighborhood search algorithm.

3.2 Simulated Annealing Algorithm

Simulated annealing algorithm (SA) is a heuristic method based on local search. This algorithm is able to prevent trapping in a local minimum by accepting worse solution with a lower probability. The acceptable quality of simulated annealing solutions has motivated the researcher community to apply it for solving complex combinatorial optimization problems in real world applications. The usage of simulated annealing algorithm was made popular by Kirkpatrick et al. [13]. The basic idea of simulated annealing is derived from annealing process of metals in metallurgy industry.

The process of optimization in simulated annealing, is the search for a solution (near) of global minimum. The algorithm starts from a random solution as the initial solution and the system temperature will be equal to this initial temperature ($T = T_0$). In every iteration, a solution in neighborhood of the current solution is obtained. The objective function of the new solution is compared to the current solution. If the solution is better it will replace the current solution and if it is worse, it will replace the current solution by a probability which is obtained from the Boltzmann

Fig. 2 The pseudo code of simulated annealing algorithm

```
Initialize parameters;
S=generate initial solution ( );
T = T_0;
While (T < T_final)
{
   Until (N ≤ I − Iter)
   {
         Generate solution S' in the neighborhood of S
               if f(S'') < (f(S)
                     S ← S'
               else
                     Δ = f(S') − f(S)
                     r = random( );
                     if(r < exp(−Δ / k * T))
                     S ← S'
   }
      T = a×T;
}
Return the best solution found;
```

function $\exp(-\frac{\Delta}{KT})$. This mechanism prevent the algorithm from being trapped in a local minimum. In this relation Δ is the difference in objective function between the current solution and the new solution, K is the Boltzmann constant which is determined beforehand and T is the current temperature. The process of neighborhood search is continued until the number of iterations reaches to a predetermined value. After this step, the system temperature is reduced. This process is continued until the termination criteria is met [13]. The pseudo code of this algorithm is shown in Fig. 2.

3.3 *Generating the Initial Solution*

The algorithm requires an initial high quality solution to start the search, therefore the generalized PFIH algorithm for VRPTW in Solomon [14] paper is used to generate the initial solution. The PFIH algorithm is one of the usual heuristic algorithms for generating initial feasible solutions for solving the routing problem with time window constraint. This algorithm, by considering time window and vehicle capacity constraints, generates feasible solutions. The required steps for generating initial solution are explained below:

1. First all the customers are included in the list of uncounted customers.
2. Every customer in the list of uncounted customers is assigned to the nearest depot. Now every depot which has the most customers in the nearby, will be opened.
3. From the uncounted list, considering the distance to the opened depot, customers are assigned to that depot until the depot capacity is full.
4. After assigning the customers, the assigned customers will be omitted from the uncounted list and the depot will be omitted from the candidate depot list.
5. If there is still any customer in the list of uncounted customers, return to step 2, otherwise go to final step.
6. In this step, the proposed method by Yu et al. [8] for producing the initial solution for vehicle location-routing problem is employed. The initial solution of the proposed algorithm will be used in a reasonable time for achieving high quality solutions.

3.4 *Representing the Solution and Calculating the Objective Function*

In order to implement the computer code of the algorithm, first a structure is required to represent the solution of the problem. The structure for representing the solution, makes the problem understandable for the computer. This structure should be able to represent all the details of the solution clearly.

Designing the suitable structure has considerable role in performance of the algorithm. The structure for representing the solution in CLRPHTW is similar to LPR. If we have n customers, m construction sites for depots and k vehicles, then the solution will be represented by an array with the following order: n customers which are represented by 1, 2, …, n numbers, m depots which are represented by n + 1, n + 2, …, n + m numbers and k − 2 zero elements. Figure 3 illustrates a small CLRP with 10 customers and 3 candidate construction sites for depots.

In this figure depots number 11 and 13 are opened. In this method for solution representation, a depot is open when between this depot and the other depot there is at least one customer in the solution representation. As the figure shows, customers 3, 5, 7, 2 and 1 are assigned to depot 1 and other customers receive service from depot number 13. Also in this method of solution representation, every two depots have two vehicle tours. For instance in the first tour of depot 11, the tour starts from depot 11 and after delivery to customers 3, 5 and 7 respectively, the vehicle returns to depot 11.

11	3	5	7	0	2	1	12	13	4	8	6	0	10	0	0	0

Fig. 3 Illustration of the employed solution for the proposed algorithm

The important point is the violation of depot capacity constraint, vehicle capacity and time windows that may occur in this kind of solution representation. In this paper in order to resolve this problem, a penalty term is added to the objective function to eliminate the infeasible solutions. In this strategy, the penalty is proportional to the extent that constraints are violated. In this paper, the parameters of P-vehicle, P-depot and P-tw are defined for violating the vehicle capacity constraint, depot constraint and time window constraint respectively.

3.5 Neighborhood Structures

In the proposed algorithm, four neighborhood structures are employed. The structures used in this paper are the ordinary neighborhood structures for solving vehicle routing problem with time windows. This four structures include Relocation operator, Swap operator, Or-opt operator and 2-opt operator. The order of the applied neighborhood structures is described below.

In the relocation operator, first two i and j elements in the solution array are selected randomly, then the i element is moved and is placed in the location next to element j. In the swap operator, after selecting two elements, they are replaced in the solution array. Or-opt operator selects a tour and delivers service to three customers and then places this chain in other location of the tour for improvement. 2-opt operator selects two tours from the solution array randomly, then divides each tour in a randomly selected location, next replaces the customers from the first and second parts of the first tour respectively with second and first parts of the second tour in the current solution array.

3.6 The Proposed Algorithm for Solving CLRPHTW

After explaining the main components, in this section we completely introduce the proposed algorithm for solving the location-routing problem with hard time windows. First the algorithm starts the work with the initial solution S_0.

The initial solution is considered as the current solution of the system ($S = S_0$). For solving the problem four neighborhood structures are employed. The algorithm starts with the first neighborhood structure, $l = 1$. First in the shaking phase based on the lth neighborhood structure, the algorithm moves from the current solution S to the neighborhood solution S'. Now in the local search phase, a local search is performed on the neighborhood solution S' so that the local optimum S'^{*} is obtained. Neighborhood search in the proposed algorithm is performed employing the simulated annealing algorithm by using the lth neighborhood structure as the movement operator in order to move to the neighborhood points. Here the length of the Markov chain for neighborhood search with a certain neighborhood structure is equal to N successive iterations without improvement in the objective function. In the

Fig. 4 The pseudo code of the proposed algorithm for solving the CLRPHTW problem

```
Input: a set of neighborhood structures N_l, l = 1,2,...,l_max
S=generate initial solution ();
T = T_0;
While (T < T_final)
{
        l = 1 ;
    While (l ≤ l_max)
      {
          S' =Shaking (S, N_l)
          S'* =Local search ( S' )
          if f(S'*) < (f(S)
                   S ← S'*, l = 1
          else
                   l = l + 1;
      }
    T = a×T;
    Update the best obtained solution
}
Output : The best solution ;
```

movement or non-movement phase if the value of the objective function for local optimum S'^{*} is better than S, *we set* $S = S'^{*}$ and $l = 1$, otherwise the nest neighborhood structure $l = l + 1$ is used. As long as $l \leq l_{max}$ the loop for variable neighborhood search algorithm in a certain temperature is continued. Here $l_{max} = 4$, and after that the temperature is updated according to the geometric reduction law ($T = \alpha \times T$). The proposed algorithm is continued until the final temperature is achieved. The flow chart of the proposed algorithm is illustrated in Fig. 4.

4 Computational Experiments

The proposed algorithm was programmed in C++ language on a Laptop with hardware features of Core i5 2.53 GHz CPU and 4 GB RAM memory. Suitable initialization of algorithm parameters have an important role on the quality of the obtained results. Therefore some values are recommended for each parameter, then according to Table 1 the value of each parameter is varied while other parameters are kept constant at their lowest possible value, and the quantity of the varying parameter is adjusted based on the objective function and the required time to achieve the solution. After performing the above experiment on one of the generated examples, the

Table 1 The proposed values for adjusting the algorithm parameters

Number	Parameters	Recommended values
1	Penalties	50, 100, 200, 500
2	N	0.5L (L – 1), L(L – 1), 2L(L – 1), 3L(L – 1)
3	T_0	30, 50, 70, 100
4	α	0.9, 0.93, 0.95, 0.97, 0.99
5	T_{final}	0.001, 0.01, 0.1,1
6	K	0.1, 0.2, 0.5, 1

algorithm parameters were adjusted as it follows: all the penalty factors (every unit of violating the depot capacity, vehicle capacity and time window) are considered equal to 200 unit, the initial temperature (T_0) was considered 50 °C and the final temperature T_{final} was 0.001 °C. The reduction temperature factor (α) is 0.95 and the Boltzmann constant is 0.2. The number of successive iterations without improvement in the local search (N) is considered L(L – 1) where L is the length of the solution array.

Since there is no benchmark example for the this algorithm, we need to generate a number of examples. In this paper, assessment of the proposed algorithm is done in two parts: in the first part some small size examples are employed to assess the proper performance of the model and the algorithm and in the second part average and large size examples are applied for evaluation.

4.1 Evaluation of Algorithm Performance in Small Size Examples

In order to evaluate the validity of the mathematical model and the performance of the proposed algorithm in small size examples, first some examples should be generated. In this paper 10 small size examples including maximum 15 customers and 5 candidate depot sites are generated. The examples are produced according to the following procedure: 20 points are generated randomly in an area having 50×50 dimensions. The first 15 points are the customers and the other 5 points are candidate sites to construct the depots. To generate the demands of each customer, a number is generated randomly and with a uniform distribution in the interval [10, 20]. For the 15 customers, first a number in the interval of [0, 20] is generated randomly as the soonest delivery time and then a number in the interval of [30, 35] is generated randomly as the latest delivery time. The time window of each depot is considered as [0, 100]. There are 10 transportation vehicles, all of them are the same with the capacity of 80 products. Each depot have a capacity of 200 units and the cost of using a depot is 200 unit. The maximum traveling time to destination for every vehicle is considered 200 time units. The traveling cost for every time unit is considered 1 monetary unit. Table 2 shows the location coordinates, time windows and the customer demands in the above generated examples.

The generated examples in this part are different in the number of customers and number of depot construction sites. For example in a problem with n customers and m candidate depot sites (n and m are maximum 15 and 5 respectively) the n first customers from the customers generated list and m first depots from the depot candidate list are selected. All the other parameters for the examples are considered equal.

The mathematical model for each of the examples are solved by the optimization software GAMS 23.6 with CPLEX solver in a laptop with hardware specifications of Core i5 2.53 GHz processor and 4 GM RAM memory in the time period of maximum 18,000 s (5 h).

In Table 3, a comparison between the solutions of the proposed algorithm in a single run and the solutions resulted from solving the mathematical model using the software is presented. In Table 3, the specifications of the 10 small size generated examples are presented in the first three columns; the examples differ in number of customers and number of candidate depot sites. In the three next columns, solution, solution time and the status of the obtained solution by GAMS software are shown respectively. In the status column, ***opt*** indicates the optimality of the solution and ***time*** indicates the time constraint to achieve the solution. The next two columns, the solution and time to achieve the solution by the proposed algorithm in a single run are shown. The last column shows the deviation of the algorithm solution with respect to the optimal solution in percent.

As it can be seen, the algorithm has reached to the optimal solution in the first 7 examples. In examples 8 and 9, the proposed algorithm has obtained solutions similar to GAMS, but solutions in GAMS are obtained with the 5 h' time constraint; however there is no guarantee that the obtained solutions are optimal. In the last example during the defined time constraint, GAMS software even did not reach to a solution equal to the proposed algorithm. considering Table 3, it can be concluded that: "as the problem size increases, the required time to achieve the optimal solution increases intensely, while the proposed algorithm has the ability to achieve a satisfactory solution in a very short time". The obtained solutions show the validity of the presented mathematical model. In the following part, the performance of the algorithm for average and large size examples will be evaluated.

4.2 *Evaluation of Algorithm Performance for Average and Large Size Examples*

Exact solution and obtaining the optimal results usually is not feasible expect for some small size examples. In addition, because there is no method for solving the capacitated location-routing problem with hard time window constraints, it is not

Table 2 The specifications of the generated small size example

Points	Location coordinates		Demand	Time window		Points	Location coordinates		Demand	Time window	
	X	Y		E	L		X	Y		E	L
1	27	4	18	1	37	11	22	33	11	18	49
2	16	30	16	8	48	12	26	20	20	12	50
3	5	24	1	5	37	13	23	41	20	11	42
4	31	35	20	16	32	14	44	36	15	3	31
5	39	35	16	9	46	15	26	49	18	17	34
6	21	32	11	19	38	16	36	9	0	0	100
7	4	1	13	3	35	17	24	37	0	0	100
8	13	3	16	5	38	18	30	13	0	0	100
9	7	16	20	3	32	19	18	21	0	0	100
10	14	27	20	2	32	20	31	46	0	0	100

Table 3 Comparison between GAMS software results and the proposed algorithm results

No Problem	Number of customers	No depot	Games			Proposed Algorithm		GAP (%)
			Answer	Time (s)	Status	Answer	Time (s)	
1	5	2	342.867	1.016	opt	342.867	0	0
2	7	2	457.845	16.782	opt	457.845	0	0
3	10	2	520.602	1801.34	opt	520.602	1	0
4	10	3	425.199	824.43	opt	425.199	1	0
5	12	2	524.147	12400	opt	524.147	1	0
6	12	3	428.981	14812.4	opt	428.981	2	0
7	12	5	387.127	17563	opt	387.127	2	0
8	15	2	631.029	18000	time	631.029	2	–
9	15	3	612.264	18000	time	612.264	2	–
10	15	5	626.015	18000	time	605.089	2	–

possible to compare the obtained results with previous method. Because of this, in the present paper, the two proposed algorithms are compared to each other. The first method which is called VSN algorithm, is the variable neighborhood search method without applying the simulation annealing algorithm in its neighborhood search structure. The second algorithm is called VNS+SA and employs the proposed variable neighborhood structure and applies simulated annealing algorithm in its neighborhood search structure.

Up to our knowledge, there is no benchmark example for capacitated location-routing problem with hard time constraints. Therefore in this paper, some of the benchmark examples in Solomon's work [14] after applying some variations are used as benchmark examples for CLRPHTW problem.

Solomon's benchmark examples for vehicle routing problem with time window constraint are available in [15]. Solomon [14] have proposed some benchmark examples for VRPTW problem as well. Their database is the most well-known database for assessment of proposed algorithms' performance for VRPTW problem and similar problems. In their proposed examples, number of customers is equal to 25, 50 and 100. They considered 100 customers and 1 depot to generate the examples and generated smaller size examples by considering the first 25 and 50 customers. They classified their examples in three main categories: C, R and RC.

The geographical locations which are considered for examples of C category are classified, for R category are random and for RC category are a combination of both classified and random and are generated in 2 dimensions. In each category, the customer and depot locations and customers' demands are considered the same and they only differ in time windows. Each category has consisted of 1 and 2 subcategories and examples in subcategory 2 has wider time windows. The transportation vehicles are the same with equal capacity and their quantity is known and limited.

To edit Solomon's benchmark examples, 4 depots are added to examples with 25 customers and also 9 depots are added to other examples. All the depots have the same capacity and costs. Generated examples for CLRPHTW are named by adding letter M to the beginning of Solomon's codes.

Every algorithm for every example is run 10 times and the best obtained answer together with the run time is shown. In these tables, the example code, the best obtained answer and the run time of both algorithms is presented. The last column is the deviation percentage between the best obtained answer by VNS+SA algorithm compare to VSN algorithm. If the obtained solution of VNS algorithm is S and that of VNS+SA algorithm is S', then the deviation percentage is calculated from the following relation: $GAP = \frac{s'-s}{s} \times 100$.

As it is indicated in Table 4, in 7 examples out of 18 examples including 25 customers and 5 candidate depot sites, the solutions of the proposed algorithm are equal to the solutions of the variable neighborhood search algorithm and in other cases VNS+SA algorithm has resulted in better solutions compared to VNS algorithm. In all the examples with 50 customers and 10 candidate depot sites except MC203 example, VNS+SA algorithm has resulted in better solutions compared to VNS algorithm (Table 5).

In order to improve the validity of the proposed algorithm, in examples with 100 customers and 10 candidate depot sites, in addition to comparing the solutions of the proposed algorithms from the view point of the best obtained solution, Paired-sampled T test is also used for statistical analysis. Because for every example and every algorithm 10 runs are available, therefore the sample size for every example and every algorithm is equal to 10 as well. Here we test the hypothesis of performance improvement of HVNS algorithm compared to the proposed VNS algorithm. For this purpose, we consider the hypothesis test below:

Null hypothesis: The average of results for the proposed VNS+SA algorithm is equal to the average of results of the proposed VNS algorithm.

Alternative hypothesis:The average of results for the proposed VNS+SA algorithm is lower than the average of results of the proposed VNS algorithm.

Since the statistical analysis in a meaningful level is 0.05 of the test, if the P-value is lower than $\alpha = 0.05$, there is no reason for accepting the null hypothesis. Table 6 represents the results of performance evaluation of the two proposed algorithms together with the results of statistical analysis of the paired-sample T test. In this table, Sig.(2-tailed) column reports the P-value and the last column reports the paired-sample T test value.

It can be observed from Table 6 that the solutions of the VNS+SA algorithm are better than VNS algorithm in all the 18 problems with 0.95 certainty level except the 6th problem; by comparing the difference percentage between the best obtained solution by VNS+SA algorithm and VNS algorithm, the same result is concluded.

Table 4 Details of the experiments for algorithm performance in examples with 25 customers and 5 candidate depots

Code example	VNS		VNS+SA		GAP (%)	Code example	VNS		VNS+SA		GAP (%)
	Objective function value	Time	Objective function value	Time			Objective function value	Time	Objective function value	Time	
MC101	1117.83	6	1077.92	7	−3.69	MR201	1160.73	6	1112.35	6	−4.35
MC102	1100.34	6	1076.85	6	−2.18	MR202	1038.93	6	1019.93	6	−1.86
MC103	1092.24	5	1069.68	7	−2.11	MR203	980.86	5	973.05	6	−0.8
MC201	1159.75	6	1136.11	7	−2.08	MRC101	1351.62	6	1306.92	6	−3.42
MC202	1152.89	5	1149.33	7	−0.31	MRC102	1211.55	6	1116.41	7	−8.52
MC203	1148.35	5	1140.49	6	−0.69	MRC103	1100.46	5	1100.46	7	0
MR101	1727.12	6	1727.12	6	0	MRC201	1124.05	6	1124.05	7	0
MR102	1573.8	4	1573.79	6	0	MRC202	1101.85	6	1101.85	6	0
MR103	1317.24	6	1317.24	6	0	MRC203	1091.17	6	1091.17	6	0

Table 5 Details of the experiments for algorithm performance in examples with 50 customers and 10 candidate depots

Code example	VNS		VNS+SA		GAP (%)	Code example	VNS		VNS+SA		GAP (%)
	Objective function value	Time	Objective function value	Time			Objective function value	Time	Objective function value	Time	
MC101	1364.28	32	1361.5	39	−0.2	MR201	1713.34	36	1612.33	40	−6.26
MC102	1365.73	33	1360.2	40	−0.41	MR202	1507.21	36	1460.5	43	−3.2
MC103	1357.67	38	1356.75	50	−0.07	MR203	1381.57	45	1320.8	45	−4.6
MC201	1288.09	53	1282.81	52	−0.41	MRC101	2221.74	28	2132.6	58	−4.18
MC202	1293.48	51	1259.63	50	−2.69	MRC102	1951.77	42	1795.27	63	−8.72
MC203	1263.36	32	1263.92	48	0.04	MRC103	1718.82	32	1651.16	60	−4.1
MR101	2584.13	35	2519.69	65	−2.56	MRC201	1581.32	35	1552.68	34	−1.84
MR102	2304.86	32	2216.65	50	−3.98	MRC202	1517.77	34	1517.03	33	−0.05
MR103	2043.92	34	1973.19	35	−3.58	MRC203	1441.66	35	1417.38	36	−1.71

Table 6 Details of experiments for algorithm performance in examples with 100 customers and 10 candidate depots.

Code example	VNS		VNS+SA		GAP (%)	P-value	Test results
	Objective function value	Time	Objective function value	Time			
MC101	2889.65	162	2709.25	148	−6.66	0.006	To accept in return for the imposition of
MC102	2727.91	131	2664.98	150	−2.36	0.004	To accept in return for the imposition of
MC103	2732.63	104	2635.32	155	−3.69	0.012	To accept in return for the imposition of
MC201	2045.2	139	1999.64	155	−2.28	0.001	To accept in return for the imposition of
MC202	2041.44	130	1927.03	144	−5.94	0.014	To accept in return for the imposition of
MC203	2053.4	137	1958.14	139	−4.86	0.000	To accept in return for the imposition of
MR101	4257.25	112	4083.84	176	−4.25	0.005	To accept in return for the imposition of
MR102	3905.81	153	3829.73	197	−1.99	0.072	Re impose zero
MR103	3391.64	142	3297.53	160	−2.85	0.042	To accept in return for the imposition of
MR201	2616.21	146	2458.96	150	−6.39	0.000	To accept in return for the imposition of
MR202	2430.31	131	2384.63	155	−1.92	0.132	Re impose zero
MR203	2210.83	127	2051.59	154	−7.76	0.000	To accept in return for the imposition of
MRC101	4033.47	124	3907.27	157	−3.23	0.031	To accept in return for the imposition of
MRC102	3800.9	126	3519.61	149	−7.99	0.013	To accept in return for the imposition of
MRC103	3313.4	124	3290.72	168	−0.69	0.124	Re impose zero
MRC201	2946.97	95	2918.49	133	−0.98	0.072	Re impose zero
MRC202	2764.36	133	2784.52	147	0.72	0.365	Re impose zero
MRC203	2435.31	102	2454.5	138	0.78	0.616	Re impose zero

4.3 *Performance Evaluation of the Proposed Algorithm in Solving the CLRP Problem*

In order to test the performance of the proposed algorithm, we used the Prodhon's problems that are available in [16]. In these examples both the routs and depots are capacitated. The algorithm is run 10 times and the best obtained solution is compared to the solutions of previous algorithms which are available in the literature. The best available algorithms in the literature for solving CLRP problem include: GRASP algorithm [2], MAPM algorithm [3], LRGTS [5] algorithm, GRASP + ELS algorithm [7] and SALRP algorithm [8].

Table 7 Results of proposed algorithm in comparison with other algorithms

Code example	BKS	GRASP	MAPM	LRGTS	GRASP+ ELS	SALRP	Proposed method	CPU (sec)	GAP (%)
20-5-1a	54,793	55,021	54,793	55,131	54,793	54,793	54,793	25	0.00
20-5-1b	39,104	39,104	39,104	39,104	39,104	39,104	39,104	22	0.00
20-5-2a	48,908	48,908	48,908	48,908	48,908	48,908	48,908	33	0.00
20-5-2b	37,542	37,542	37,542	37,542	37,542	37,542	37,542	38	0.00
50-5-1a	90,111	90,632	90,160	90,160	90,111	90,111	90,111	39	0.00
50-5-1b	63,242	64,741	63,242	63,256	63,242	63,242	63,242	40	0.00
50-5-2a	88,298	88,786	88,298	88,715	88,643	88,298	88,298	46	0.00
50-5-2b	67,308	68,042	67,893	67,698	67,308	67,308	67,308	37	0.00
50-5-2bis	84,055	84,055	84,055	84,181	84,055	84,055	84,055	33	0.00
50-5-2Bbis	51,822	52,059	51,822	51,992	51,822	51,822	51,822	48	0.00
50-5-3a	86,203	87,380	86,203	86,203	86,203	86,456	86,203	54	0.00
50-5-3b	61,830	61,890	61,830	61,830	61,830	62,700	61,830	31	0.00
100-5-1a	276,960	279,437	281,944	277,935	276,960	277,035	275,919	245	−0.38
100-5-1b	214,885	216,159	216,656	214,885	215,854	216,002	214,646	289	−0.11
100-5-2a	194,124	199,520	195,568	196,548	194,267	194,124	194,677	381	0.28
100-5-2b	157,150	159,550	157,325	157,792	157,375	157,150	157,265	245	0.07
100-5-3a	200,242	203,999	201,749	201,952	200,345	200,242	200,247	218	0.00
100-5-3b	152,467	154,596	153,322	154,709	152,528	152,467	152,503	291	0.02
100-10-1a	290,429	323,171	316,575	291,887	301,418	291,043	290,919	348	0.17
100-10-1b	234,210	271,477	270,251	235,532	269,594	234,210	233,503	281	−0.30
100-10-2a	244,265	254,087	245,123	246,708	243,778	245,813	244,253	312	0.00
100-10-2b	203,988	206,555	205,052	204,435	203,988	205,312	203,988	237	0.00
100-10-3a	250,882	270,826	253,669	258,656	253,511	250,882	251,120	388	0.09
100-10-3b	204,597	216,597	204,815	205,883	205,087	205,009	205,578	295	0.48

In Table 7, the first column shows the code for each example. The first and second characteristic numbers of the example indicate number of customers and number of candidate depots respectively. The second column shows the best obtained solution so far and the remaining columns show the obtained solution of the selected algorithms in the history of the subject and the best obtained solution of the proposed algorithm. In the end, the run time of CPU in seconds and the deviation from the best obtained solution so far (GAP) are given. Table 7 shows that the proposed algorithm in this paper, was able to achieve the best obtained solution so far for 15 out of 24 cases, it means that the deviation percentage from the best obtained solution is 0 %.

The proposed algorithm in 3 cases has obtained results better than the best obtained solution so far. The proposed algorithm has deviation between 0.02 and 0.48 % in the 6 remaining examples. Totally, the average deviation of the best obtained solution for the 24 chosen benchmark examples is equal to 0.01 %. The quality of the obtained solutions and the reasonable time to achieve the solutions by the proposed algorithm indicate that it can compete with other well-known algorithms for solving the capacitated location-routing problem without time window constraint.

5 Conclusion

The capacitated location-routing problem is an important and widely applied problem in management of supply chain and logistics. In this paper, in order to make the problem more practical and realistic, the time window constraint is added to the problem. In this paper, after setting the required assumptions, the problem was modeled with all of its constraints and assumptions. Next, by considering the exponential complexity of time for large problems and inability of exact optimization methods in solving large size problems, a meta-heuristic method was applied to solve the problem. The applied method to solve the problem is an approach based on variable neighborhood decline algorithm. Since there was no benchmark example to assess the performance of the algorithm, a number of examples were generated. Next, the performance of the proposed combinatory algorithm for capacitated-location routing problem with hard time windows and the validity of the presented model for this problem were assessed. at the end, the algorithm performance for solving capacitated location-routing problem without considering time windows was tested. The obtained results demonstrated that the algorithm was capable of achieving appropriate solutions in a reasonable time.

References

1. Nagy G, Salhi S (2007) Location-routing: issues, models and methods. Eur J Oper Res 177(2):649–672
2. Prins C, Prodhon C, Calvo RW (2006) Solving the capacitated location- routing problem by a GRASP complemented by a learning process and a path relinking. A Q J Oper Res 4(3):221–238
3. Prins C, Prodhon C, Calvo RW (2006) A memetic algorithm with population management (MA|PM) for the capacitated location-routing problem. Lecture notes in computer science, vol 3906. Springer, Berlin, pp 183–194
4. Barreto S, Ferreira C, Paixa J, Santos BS (2007) Using clustering analysis in capacitated location-routing problem. Eur J Oper Res 179:968–977
5. Prins C, Prodhon C, Ruiz A, Soriano P, Calvo RW (2007) Solving the capacitated location-routing problem by a cooperative Lagrangean relaxation granular tabu search heuristic. Transp Sci 41:470–483
6. Marinakis Y, Marinaki M (2008) A particle swarm optimization algorithm with path relinking for the location routing problem. J Math Model Algorithms 7:59–78
7. Duhamel C, Lacomme P, Prins C, Prodhon C (2010) A GRASP×ELS approach for the capacitated location-routing problem. Comput Oper Res 37:1912–1923
8. Yu VF, Lin S-W, Lee W, Ting C-J (2010) A simulated annealing heuristic for the capacitated location-routing problem. Comput Ind Eng 58:288–299
9. Nguyen V-P, Prins C, Prodhon C (2012) Solving the two-echelon location routing problem by a GRASP reinforced by a learning process and path relinking. Eur J Oper Res 216:113–126
10. Nguyen V-P, Prins C, Prodhon C (2012) A multi-start iterated local search with tabu list and path relinking for the two-echelon location-routing problem. Eng Appl Artif Intell 25:56–71
11. Jabal-Ameli MS, Ghaffari-Nasab N (2010) Location-routing problem with time windows: novel mathematical programming formulations. In: 7th International industrial engineering conference. Isfahan, Iran
12. Mladenovi'c N, Hansen P (1997) Variable neighborhood search. Comput Oper Res 24:1097–1100
13. Kirkpatrick S, Gelatti CD, Vecchi MP (1983) Optimization by simulated annealing. In: Science is currently published by American association for the advancement of science, vol 220, no. 4598, pp 671–680
14. Solomon MM (1987) Algorithms for the vehicle routing and scheduling problems with time window constraints. Oper Res 35:254–265
15. http://web.cba.neu.edu/~msolomon
16. http://prodhonc.free.fr/homepage

Printed in the United States
By Bookmasters